Student's Solutions Manual to Accompany

College Algebra
Third Edition

David Cohen
Department of Mathematics
University of California
Los Angeles, California

Prepared by
Ross Rueger
Department of Mathematics
College of the Sequoias
Visalia, California

West Publishing Company
St. Paul New York Los Angeles San Francisco

WEST'S COMMITMENT TO THE ENVIRONMENT

In 1906, West Publishing Company began recycling materials left over from the production of books. This began a tradition of efficient and responsible use of resources. Today, up to 95% of our legal books and 70% of our college texts are printed on recycled, acid-free stock. West also recycles nearly 22 million pounds of scrap paper annually—the equivalent of 181,717 trees. Since the 1960s, West has devised ways to capture and recycle waste inks, solvents, oils, and vapors created in the printing process. We also recycle plastics of all kinds, wood, glass, corrugated cardboard, and batteries, and have eliminated the use of styrofoam book packaging. We at West are proud of the longevity and the scope of our commitment to our environment.

Production, Prepress, Printing and Binding by West Publishing Company.

610 Opperman Drive
P.O. Box 64526
St. Paul, MN 55164–0526

Printed in the United States of America
99 98 97 96 95 94 93 92 8 7 6 5 4 3 2 1 0

ISBN 0–314–00575–7

Contents

Preface

This Student's Solutions Manual contains complete solutions to all odd-numbered regular and graphing calculator exercises of College Algebra by David Cohen. It also contains complete solutions to odd and even exercises for each chapter test. I have attempted to format solutions for readability and accuracy, and apologize to you for any errors that you may encounter. If you have any comments, suggestions, error corrections, or alternative solutions please feel free to drop me a note.

Please use this manual with some degree of caution. Be sure that you have attempted a solution, and re-attempted it, before you look it up in this manual. Mathematics can only be learned by **doing**, and not by observing! As you use this manual, do not just read the solution but work it along with the manual, using my solution to check your work. If you use this manual in that fashion then it should be helpful to you in your studying.

I would like to thank a number of people for their assistance in preparing this manual. Thanks go to Peter Marshall, Jane Bacon, and Mark Jacobsen at West Educational Publishing for their valuable assistance and support. Special thanks go to Chuck Heuer of Concordia College for his meticulous error-checking of my solutions. Finally, I wish to thank Joy Bishop of College of the Sequoias for her fabulous typing and editing of this entire manuscript. This project could not have been completed without her help.

I wish to express my deepest appreciation to David Cohen for continuing his tradition of excellence with this textbook. His addition of TI-81 graphing calculator exercises in this third edition will help you in conceptualizing the notion of a function. That notion is the heart of calculus and mathematics in general.

Ross Rueger
College of the Sequoias
915 South Mooney Boulevard
Visalia, CA 93277

January, 1992

Chapter One
Fundamental Concepts

1.1 Notation and Language

1. The expression is equivalent to x^5y^2.

3. The expression is equivalent to $(x + 1)^3$.

5. (a) The expression is equivalent to $4x$.

(b) The expression is equivalent to $3(x^2+1)$, or $3x^2 + 3$.

7. (a) The expression is equivalent to $3a + 2b$.
(b) The expression is equivalent to $3a^2 + 2b^2$.

9. (a) The expression is equivalent to $(2a+1)^3(2b+1)^2$.
(b) Since $1 + 2a = 2a + 1$ and $1 + 2b = 2b + 1$, the expression is equivalent to

$(2a+1)^3(2b+1)^2$.

11. Substituting $x = 4$ and $y = 6$, we have:
$$x+3y=4+3(6)=4+18=22$$

13. Substituting $a = 3$ and $b = 4$, we have:
$$a^2+b^2=(3)^2+(4)^2=9+16=25$$

15. Substituting $x = 10$ and $y = 4$, we have:
$$x^2-4y^2=(10)^2-4(4)^2=100-64=36$$

17. Substituting $x = 5$, we have:
$$x^2 - x + 1 = (5)^2 - 5 + 1 = 25 - 5 + 1 = 21$$

19. Substituting $x = 3$ and $y = 2$, we have:
$$2x^3 - 3y^2 = 2(3)^3 - 3(2)^2 = 54 - 12 = 42$$

21. Substituting $a = 1$ and $b = 1$, we have:
$$1 \div a^2 + b^2 = 1 \div 1^2 + 1^2 = 1 \div 1 + 1 = 1 + 1 = 2$$

23. Substituting $a = \frac{1}{2}$ and $b = \frac{1}{2}$, we have:
$$\frac{1}{a^2} + \frac{1}{b^2} - \frac{1}{a^2 + b^2} = \frac{1}{\left(\frac{1}{2}\right)^2} + \frac{1}{\left(\frac{1}{2}\right)^2} - \frac{1}{\left(\frac{1}{2}\right)^2 + \left(\frac{1}{2}\right)^2} = \frac{1}{\frac{1}{4}} + \frac{1}{\frac{1}{4}} - \frac{1}{\frac{1}{4} + \frac{1}{4}} = 4 + 4 - 2 = 6$$

25. (a) Substituting $A = 2$, $B = 2$ and $C = 3$, we have:
$$A^{B^C} = 2^{2^3} = 2^8 = 256$$

(b) Substituting $A = 2$, $B = 3$ and $C = 2$, we have:
$$A^{B^C} = 2^{3^2} = 2^9 = 512$$

(c) Substituting $A = 3$, $B = 2$ and $C = 2$, we have:
$$A^{B^C} = 3^{2^2} = 3^4 = 81$$

27. We compute each expression:
$$\left(2^3\right)^4 = 8^4 = 4096$$
$$2^{3^4} = 2^{81} \approx 2.4 \times 10^{24}$$

So 2^{3^4} is much larger than $\left(2^3\right)^4$.

29. (a) Substituting $A = 5$, $B = 2$ and $C = 3$, we have:
$$A^{B^C} = 5^{2^3} = 5^8 = 390{,}625$$

(b) Substituting $A = 5$, $B = 3$ and $C = 2$, we have:
$$A^{B^C} = 5^{3^2} = 5^9 = 1{,}953{,}125$$

31. (a) Since $2^5 = 32$ and $2^2 = 4$, the first quantity 2^{2^5} should be larger since it has a much larger exponent.

(b) We compute each quantity:
$$2^{2^5} = 2^{32} = 4{,}294{,}967{,}295$$
$$5^{2^2} = 5^4 = 625$$
This confirms our conjecture from part (a).

33. "Two more than x" would translate to $2 + x$, or $x + 2$.

35. "Four times the sum of a and b^2" would translate to $4(a+b^2)$.

37. "The sum of x^2 and y^2" would translate to $x^2 + y^2$.

39. "The sum of x and twice the square of x" would translate to $x + 2x^2$.

41. "The average of x, y, and z" would translate to $\frac{x+y+z}{3}$.

43. "The average of the squares of x, y, and z" would translate to $\frac{x^2+y^2+z^2}{3}$.

45. "One less than twice the product of x and y" would translate to $2xy - 1$.

47. (a) Substituting $x = 3$, we have:

$$\frac{1000}{19-\frac{12}{1-\frac{1}{x}}} = \frac{1000}{19-\frac{12}{1-\frac{1}{3}}} = \frac{1000}{19-\frac{12}{\frac{2}{3}}} = \frac{1000}{19-18} = \frac{1000}{1} = 1000$$

(b) Substituting $x = 2$, we have:

$$\frac{1}{1+\frac{1}{1+\frac{1}{1+\frac{1}{x}}}} = \frac{1}{1+\frac{1}{1+\frac{1}{1+\frac{1}{2}}}} = \frac{1}{1+\frac{1}{1+\frac{1}{\frac{3}{2}}}} = \frac{1}{1+\frac{1}{1+\frac{2}{3}}} = \frac{1}{1+\frac{1}{\frac{5}{3}}} = \frac{1}{1+\frac{3}{5}} = \frac{1}{\frac{8}{5}} = \frac{5}{8}$$

49. (a) (i) We compute each quantity:

$$6^7 = 279{,}936 \text{ while } 7^6 = 117{,}649$$

So 6^7 is larger.

(ii) We compute each quantity:

$$7^8 = 5{,}764{,}801 \text{ while } 8^7 = 2{,}097{,}152$$

So 7^8 is larger.

(iii) We compute each quantity:

$$9^8 = 43{,}046{,}721 \text{ while } 8^9 = 134{,}217{,}728$$

So 8^9 is larger.

(b) There are two such numbers. If $a = 1$, we have:

$$(a+1)^a = (1+1)^1 = 2^1 = 2$$
$$a^{a+1} = 1^{1+1} = 1^2 = 1$$

So $(a+1)^a > a^{a+1}$. If $a = 2$, we have:

$$(a+1)^a = (2+1)^2 = 3^2 = 9$$
$$a^{a+1} = 2^{2+1} = 2^3 = 8$$

So $(a+1)^a > a^{a+1}$. The two positive whole numbers are $a = 1$ and $a = 2$.

1.2 Properties of the Real Numbers

1. 8(100) + 8(4) = 800 + 32 = 832

3. 45(100) + 45(–2) = 4500 – 90 = 4410

5. (a) Using the distributive property, we have:

$$(A+B)(A+B) = A(A+B)+B(A+B) = A^2 + AB + BA + B^2 = A^2 + 2AB + B^2$$

(b) Using the distributive property, we have:

$$(A+B)(A-B) = A(A-B)+B(A-B) = A^2 - AB + BA - B^2 = A^2 - B^2$$

(c) Using the distributive property, we have:

$$(A-B)(A-B) = A(A-B)-B(A-B) = A^2 - AB - BA + B^2 = A^2 - 2AB + B^2$$

7. (a) Using the distributive property, we have:

$$\begin{aligned}(x+2y)(x+2y) &= x(x+2y)+2y(x+2y)\\ &= x^2+2xy+2yx+4y^2\\ &= x^2+4xy+4y^2\end{aligned}$$

(b) Using the distributive property, we have:

$$(x+2y)(x-2y) = x(x-2y)+2y(x-2y) = x^2-2xy+2yx-4y^2 = x^2-4y^2$$

(c) Using the distributive property, we have:

$$\begin{aligned}(x-2y)(x-2y) &= x(x-2y)-2y(x-2y)\\ &= x^2-2xy-2yx+4y^2\\ &= x^2-4xy+4y^2\end{aligned}$$

9. (a) Using the distributive property, we have:

$$A\left(x^2-x+1\right) = Ax^2 - Ax + A$$

(b) Using the distributive property, we have:

$$\begin{aligned}(x+1)\left(x^2-x+1\right) &= x\left(x^2-x+1\right)+1\left(x^2-x+1\right)\\ &= x^3-x^2+x+x^2-x+1\\ &= x^3+1\end{aligned}$$

11. commutative property of addition

13. additive inverse property

15. associative property of addition

17. associative property of multiplication

19. identity property of addition

21. closure property with respect to addition

23. distributive property

25. If $b = c = 1$, then:

$$\frac{1}{b}+\frac{1}{c}=\frac{1}{1}+\frac{1}{1}=1+1=2$$

$$\frac{1}{b+c}=\frac{1}{1+1}=\frac{1}{2}$$

So $\frac{1}{b}+\frac{1}{c}\neq\frac{1}{b+c}$.

27. If $x = 1$ and $y = 2$, then:

$$3(x+y)=3(1+2)=3(3)=9$$

$$3x+y=3(1)+2=3+2=5$$

So $3(x + y) \neq 3x + y$.

29. If $x = 2$ and $y = 1$, then:

$$x^2(x+y)=2^2(2+1)=4(3)=12$$

$$x^3+y=2^3+1=8+1=9$$

So $x^2(x + y) \neq x^3 + y$.

31. If $x = 2$ and $y = 3$, then:

$$3(xy)=3(2\bullet 3)=3(6)=18$$

$$(3x)(3y)=(3\bullet 2)(3\bullet 3)=(6)(9)=54$$

So $3(xy) \neq (3x)(3y)$.

33. If $a = 5$, $b = 3$ and $c = 1$, then:

$$a-(b-c)=5-(3-1)=5-2=3$$

$$a-b-c=5-3-1=1$$

So $a - (b - c) \neq a - b - c$.

35. $1 + 2(-3)(4) = 1 - 24 = -23$

37. $4 - (-3 + 5) = 4 - (2) = 2$

39. $-\left[(-1)^4\right]=-\left[(-1)(-1)(-1)(-1)\right]=-(1)=-1$

41. $1-\{1-[-(-1-1)]\}=1-\{1-[-(-2)]\}=1-\{1-2\}=1-(-1)=2$

43. Obtaining common denominators, we have:

$$\frac{7}{5}+\frac{2}{3}=\frac{21+10}{15}=\frac{31}{15}$$

45. Subtracting the two fractions (they already have a common denominator) yields:
$$\frac{7}{x}-\frac{12}{x}=\frac{7-12}{x}=-\frac{5}{x}$$

47. Subtracting the two fractions yields:
$$\frac{x^2+y^2}{x+y}-\frac{x^2-y^2}{x+y}=\frac{(x^2+y^2)-(x^2-y^2)}{x+y}=\frac{x^2+y^2-x^2+y^2}{x+y}=\frac{2y^2}{x+y}$$

49. Multiplying the two fractions yields:
$$\frac{7}{12}\bullet\left(\frac{-7}{3}\right)=\frac{-49}{36}=-\frac{49}{36}$$

51. Dividing the two fractions yields:
$$\frac{2}{5}\div\frac{7}{3}=\frac{2}{5}\bullet\frac{3}{7}=\frac{6}{35}$$

53. By first simplifying the parentheses, we have:
$$1\div\left(\frac{2}{3}+1\right)=1\div\left(\frac{2}{3}+\frac{3}{3}\right)=1\div\frac{5}{3}=1\bullet\frac{3}{5}=\frac{3}{5}$$

55. (a) Multiplying both numerator and denominator by 12 (the least common multiple) yields:
$$\frac{\frac{4}{3}+\frac{1}{2}}{\frac{11}{6}+\frac{5}{4}}=\frac{12\left(\frac{4}{3}+\frac{1}{2}\right)}{12\left(\frac{11}{6}+\frac{5}{4}\right)}=\frac{16+6}{22+15}=\frac{22}{37}$$

(b) Since $\frac{a-b}{b-a}=-1$, then:
$$\frac{\frac{4}{3}-\frac{1}{2}}{\frac{1}{2}-\frac{4}{3}}=-1$$

57. Since $\frac{a-b}{b-a}=-1$, then:
$$\frac{1-(x+y)}{x+y-1}=\frac{1-(x+y)}{(x+y)-1}=-1$$

59. Multiplying both numerator and denominator by 12 (the least common multiple) yields:
$$\frac{\frac{1}{2}-\frac{1}{3}+\frac{1}{4}}{\frac{1}{2}+\frac{1}{3}-\frac{1}{4}}=\frac{12\left(\frac{1}{2}-\frac{1}{3}+\frac{1}{4}\right)}{12\left(\frac{1}{2}+\frac{1}{3}-\frac{1}{4}\right)}=\frac{6-4+3}{6+4-3}=\frac{5}{7}$$

61. Substituting $x=-2$, we have:
$$(-2)^2-8(-2)-4=4+16-4=16$$

63. Substituting $x=-\frac{1}{2}$, we have:
$$8\left(-\frac{1}{2}\right)^3-4\left(-\frac{1}{2}\right)^2-4\left(-\frac{1}{2}\right)=8\left(-\frac{1}{8}\right)-4\left(\frac{1}{4}\right)+2=-1-1+2=0$$

65. Substituting $x = -1$, we have:

$$1-(-1)-(-1)^2-(-1)^3-(-1)^4 = 1+1-1+1-1 = 1$$

67. Substituting $x_2 = -1$, $x_1 = -2$, $y_2 = 6$ and $y_1 = -6$, we have:

$$\begin{aligned} d^2 &= (x_2 - x_1)^2 + (y_2 - y_1)^2 \\ &= [-1-(-2)]^2 + [6-(-6)]^2 \\ &= (-1+2)^2 + (6+6)^2 \\ &= 1+144 \\ &= 145 \end{aligned}$$

69. Using Property 1, we have:

$$\begin{aligned} \frac{a}{b} &= \frac{c}{d} \\ ad &= bc \\ \frac{ad}{cd} &= \frac{bc}{cd} \\ \frac{a}{c} &= \frac{b}{d} \end{aligned}$$

71. Using Property 1, we have:

$$\begin{aligned} \frac{a}{b} &= \frac{c}{d} \\ ad &= bc \\ ad - bd &= bc - bd \\ (a-b)d &= b(c-d) \\ \frac{(a-b)d}{bd} &= \frac{b(c-d)}{bd} \\ \frac{a-b}{b} &= \frac{c-d}{d} \end{aligned}$$

73. Substituting $x = -0.75$, we have:

$$y = x^3 - 3x^2 - 6x - 4 = (-0.75)^3 - 3(-0.75)^2 - 6(-0.75) - 4 \approx -1.6$$

So $y \approx -1.6$ (to two significant digits).

75. (a) We complete the table:

x	1	1.2500	1.5000	1.7500	2.0000	2.2500
y	0	0.2880	0.3704	0.3848	0.3750	0.3566

(b) The largest y-value is obtained when $x = 1.750$.

(c) Since $\sqrt{3} \approx 1.732...$, the x-value in the table closest to this is $x = 1.750$. The computed value of y is:

$$y = \frac{x^2 - 1}{x^3} = \frac{(\sqrt{3})^2 - 1}{(\sqrt{3})^3} = \frac{3-1}{3\sqrt{3}} = \frac{2}{3\sqrt{3}} = \frac{2}{3\sqrt{3}} \approx 0.3849$$

This largest y-value is approximately $y \approx 0.3849$. Note that it is very close to the y-value from the table corresponding to $x = 1.750$.

1.3 Sets of Real Numbers

1. (a) natural number, integer, rational number
(b) integer, rational number

3. (a) rational number
(b) irrational number

5. (a) natural number, integer, rational number
(b) rational number

7. (a) rational number
(b) rational number

9. irrational number

11. natural number, integer, rational number

13. (a) $5.4 = \frac{54}{10} = \frac{27}{5}$

(b) Let $x = 5.\overline{4}$. Then $10x = 54.\overline{4}$ and $10x - x = 49$. Thus $9x = 49$, and consequently $x = \frac{49}{9}$.

15. (a) $0.99 = \frac{99}{100}$

(b) Let $x = 0.\overline{9}$. Then $10x = 9.\overline{9}$, and $10x - x = 9$. Thus $9x = 9$, and consequently $x = 1$.

17. (a) $1.7 = \frac{17}{10}$

(b) Let $x = 1.\overline{7}$. Then $10x = 17.\overline{7}$, and $10x - x = 16$. Thus $9x = 16$, and consequently $x = \frac{16}{9}$.

19. (a) Since $\sqrt{3} \approx 1.732$, then $\sqrt{3} + 2 \approx 3.732$.

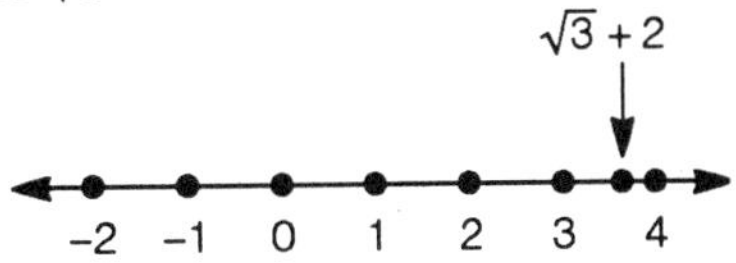

(b) Since $\sqrt{3} \approx 1.732$, then $-\frac{\sqrt{3}}{2} \approx -0.866$.

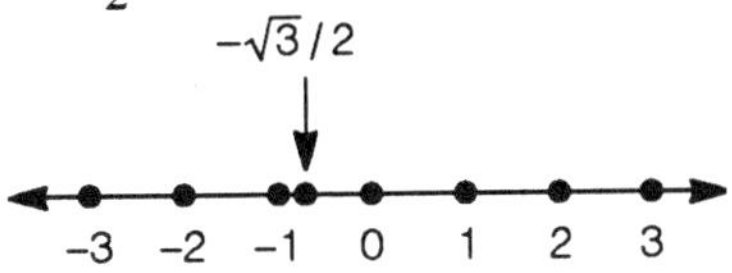

21. Since $\sqrt{3+\sqrt{7}} \approx 2.376$, we show the approximate location:

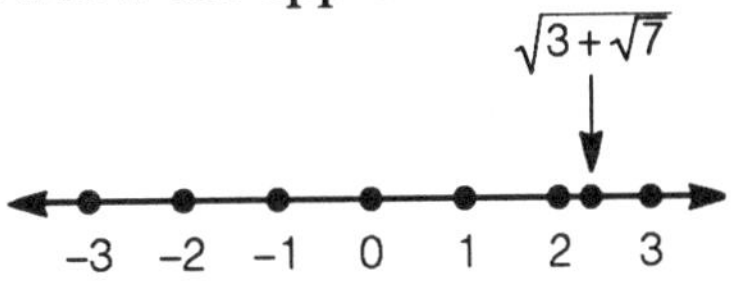

23. Since $-\pi/2 \approx -1.571$, we show the approximate location:

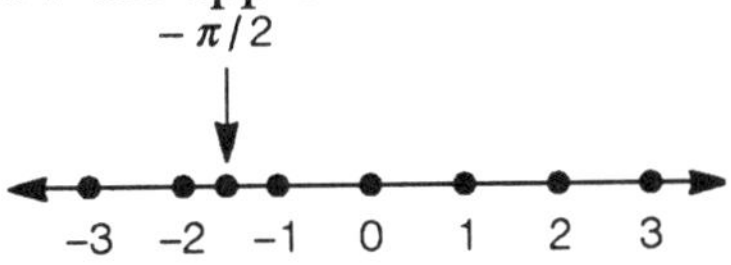

25. Since $\pi \approx 3.14$ and $2\pi \approx 6.28$, we show the approximate location:

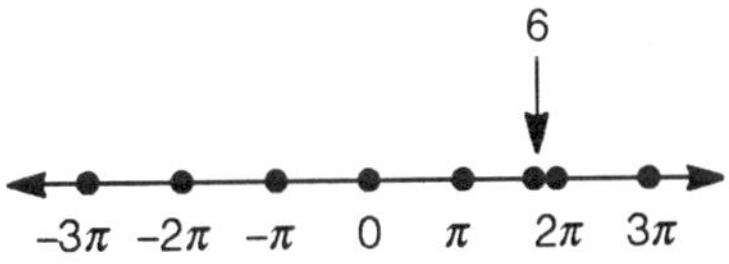

27. Since $-\pi \approx -3.14$ and $-2\pi \approx -6.28$, we show the approximate location:

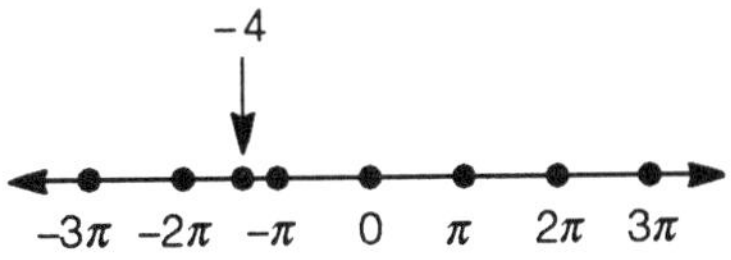

29. False, since $-5 > -100$.

31. True, since $-1 = -1$.

33. False, since $\frac{15}{16} < \frac{17}{18}$.

35. We graph the interval $2 < x < 5$:

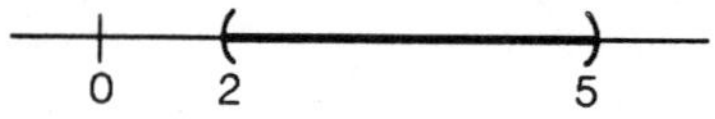

37. We graph the interval $1 \le x \le 4$:

39. We graph the interval $0 \le x < 3$:

41. We graph the interval $x > -3$:

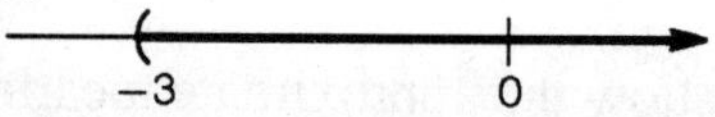

43. We graph the interval $x \ge -1$:

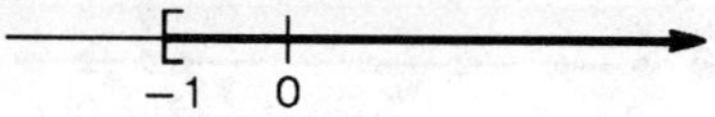

45. We graph the interval $x < 1$:

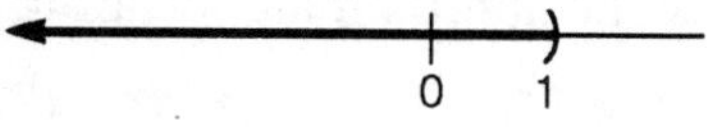

47. We graph the interval $x \le \pi$:

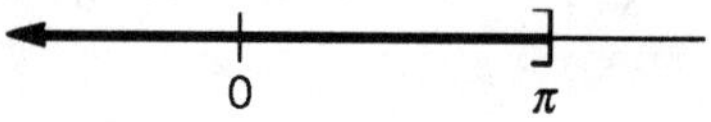

49. We graph the interval $x > -2$:

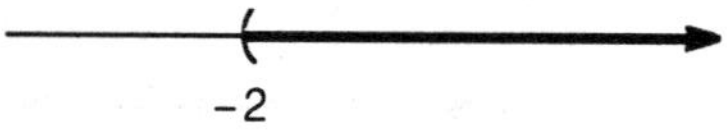

51. We graph the interval $-1 < x < 1$:

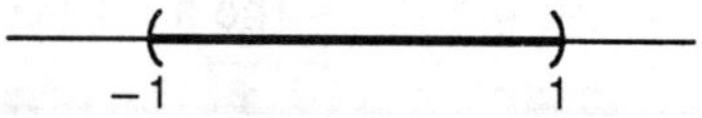

53. Let $x = 0.\overline{19}$. Then $100x - x = 19.\overline{19} - 0.\overline{19} = 19$. Thus $99x = 19$ and $x = \frac{19}{99}$.

55. Let $x = 0.3\overline{12}$. Then $1000x - 10x = 312.\overline{12} - 3.\overline{12} = 309$. Thus $990x = 309$ and $x = \frac{309}{990}$, which reduces to $\frac{103}{330}$.

57. (a) The common value to six decimal places is 3.863703.
(b) The common value to six decimal places is 3.162277.
(c) The common value to six decimal places is 1.847759.
(d) The common value to six decimal places is 2.

1.4 Absolute Value

1. $|3| = 3$

3. $|-6| = 6$

5. $|-1+3| = |2| = 2$

7. We simplify the expression:
$$\left|-\frac{4}{5}\right| - \frac{4}{5} = \frac{4}{5} - \frac{4}{5} = 0$$

9. We simplify the expression:
$$|-6+2| - |4| = |-4| - |4| = 4 - 4 = 0$$

11. We simplify the expression:
$$\big||-8| + |-9|\big| = |8+9| = |17| = 17$$

13. We simplify the expression:
$$\left|\frac{27-5}{5-27}\right| = \left|\frac{22}{-22}\right| = |-1| = 1$$

15. We simplify the expression:
$$|7(-8)| - |7||-8| = |-56| - 7(8) = 56 - 56 = 0$$

17. Substituting $a = -2$ and $b = 3$:
$$|a-b|^2 = |-2-3|^2 = |-5|^2 = (5)^2 = 25$$

19. Substituting $a = -2$, $b = 3$, and $c = -4$:
$$|c| - |b| - |a| = |-4| - |3| - |-2| = 4 - 3 - 2 = -1$$

21. Substituting $a = -2$, $b = 3$, and $c = -4$:
$$|a+b|^2 - |b+c|^2 = |-2+3|^2 - |3+(-4)|^2 = |1|^2 - |-1|^2 = (1)^2 - (1)^2 = 1 - 1 = 0$$

23. Substituting $a = -2$, $b = 3$, and $c = -4$:
$$\frac{a+b+|a-b|}{2} = \frac{-2+3+|-2-3|}{2} = \frac{1+|-5|}{2} = \frac{1+5}{2} = \frac{6}{2} = 3$$

25. Since $\sqrt{2} - 1 > 0$, we have:
$$|\sqrt{2} - 1| - 1 = (\sqrt{2} - 1) - 1 = \sqrt{2} - 2$$

27. Since $x \geq 3$, then $x - 3 \geq 0$ and thus:

$$|x-3| = x-3$$

29. Since $t^2 + 1 > 0$, then:

$$|t^2+1| = t^2+1$$

31. Since $-\sqrt{3} - 4 < 0$, then:

$$|-\sqrt{3}-4| = -(-\sqrt{3}-4) = \sqrt{3}+4$$

33. Since $x < 3$, then $x - 3 < 0$ and $x - 4 < 0$, and thus:

$$|x-3|+|x-4| = -(x-3)+[-(x-4)] = -x+3-x+4 = -2x+7$$

35. Since $3 < x < 4$, then $x - 3 > 0$ and $x - 4 < 0$, and thus:

$$|x-3|+|x-4| = (x-3)+[-(x-4)] = x-3-x+4 = 1$$

37. Since $-\frac{5}{2} < x < -\frac{3}{2}$, then $x + 1 < 0$ and $x + 3 > 0$, and thus:

$$|x+1|+4|x+3| = [-(x+1)]+4(x+3) = -x-1+4x+12 = 3x+11$$

39. The absolute value equality can be written as $|x-4| = 8$.

41. The absolute value equality can be written as $|x-1| = \frac{1}{2}$.

43. The absolute value inequality can be written as $|x-1| \geq \frac{1}{2}$.

45. The absolute value inequality can be written as $|y-(-4)| < 1$, or $|y+4| < 1$.

47. The absolute value inequality can be written as $|y-0| < 3$, or $|y| < 3$.

49. The absolute value inequality can be written as $|x^2 - a^2| < M$.

51. We graph the interval $|x| < 4$:

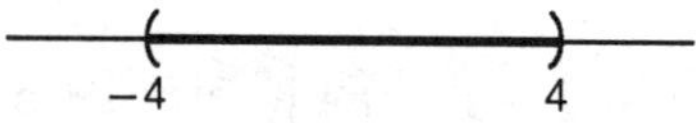

53. We graph the interval $|x| > 1$:

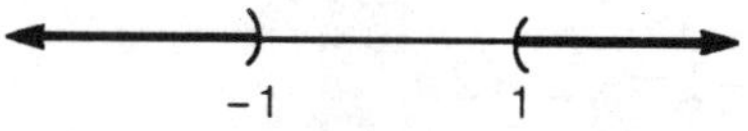

55. We graph the interval $|x-5|<3$:

57. We graph the interval $|x-3|\le 4$:

59. We graph the interval $\left|x+\frac{1}{3}\right|<\frac{3}{2}$:

61. We graph the interval $|x-5|\ge 2$:

63. (a) We graph the interval $|x-2|<1$:

(b) We graph the interval $0<|x-2|<1$, noting that $x=2$ is excluded:

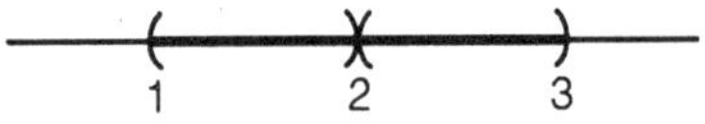

(c) The interval in (b) does not include 2.

65. (a) We know min (6, 1) = 1, now substituting $a=6$ and $b=1$:

$$\min(6,1)=\frac{6+1-|6-1|}{2}=\frac{7-5}{2}=\frac{2}{2}=1$$

Thus the equation is verified.

(b) We know min $(1,-6)=-6$, now substituting $a=1$ and $b=-6$:

$$\min(1,-6)=\frac{1+(-6)-|1-(-6)|}{2}=\frac{-5-|1+6|}{2}=\frac{-5-7}{2}=\frac{-12}{2}=-6$$

Thus the equation is verified.

(c) We know min $(-6,-6)=-6$, now substituting $a=-6$ and $b=-6$:

$$\min(-6,-6)=\frac{-6+(-6)-|-6-(-6)|}{2}=\frac{-12-|-6+6|}{2}=\frac{-12}{2}=-6$$

Thus the equation is verified.

67. Using the triangle inequality twice, we have:

$$|a+b+c| = |a+(b+c)| \le |a| + |b+c| \le |a| + |b| + |c|$$

69. We will consider three cases: $a = b$, $a > b$, and $a < b$.

case 1: If $a = b$, then $\max(a,b) = a$, now verifying:

$$\frac{a+b+|a-b|}{2} = \frac{a+a+|a-a|}{2} = \frac{2a}{2} = a$$

Thus the equation is verified.

case 2: If $a > b$, then $\max(a,b) = a$, and since $a - b > 0$, we have:

$$\frac{a+b+|a-b|}{2} = \frac{a+b+a-b}{2} = \frac{2a}{2} = a$$

Thus the equation is verified.

case 3: If $a < b$, then $\max(a,b) = b$, and since $a - b < 0$, we have:

$$\frac{a+b+|a-b|}{2} = \frac{a+b+-(a-b)}{2} = \frac{a+b-a+b}{2} = \frac{2b}{2} = b$$

Thus the equation is verified.

71. (a) Property 1(b).

(b) $a+b \le |a|+|b|$

(c) Since $(-a)+(-b) \le |a|+|b|$, then $-(a+b) \le |a|+|b|$.

(d) Since $a+b \le |a|+|b|$ and $-(a+b) \le |a|+|b|$, then $|a+b| \le |a|+|b|$, since $|a+b|$ is either $a + b$ or $-(a + b)$.

1.5 Integer Exponents, Scientific Notation

1. Substituting $x = -2$, we have:

$$2(-2)^3 - (-2) + 4 = 2(-8) + 2 + 4 = -10$$

3. Substituting $x = -\frac{1}{2}$, we have:

$$\frac{1-2\left(-\frac{1}{2}\right)^2}{1+2\left(-\frac{1}{2}\right)^3} = \frac{1-2\left(\frac{1}{4}\right)}{1+2\left(-\frac{1}{8}\right)} = \frac{1-\frac{1}{2}}{1-\frac{1}{4}} = \frac{\frac{1}{2}}{\frac{3}{4}} = \frac{2}{3}$$

5. Substituting $x = 2$, we have:

$$\frac{2^2+2^3-2^2}{2^2+3^2-3^2} = \frac{8}{4} = 2$$

7. The expression can be written as $(x + y)^2$.

9. The expression can be written as $(x+y)^2+3$.

11. The expression can be written as $\left[\frac{1}{2}(x^2-2y^3)\right]^2$.

13. The expression can be written as $|x-1|^3$.

15. (a) $a^3a^{12}=a^{3+12}=a^{15}$

(b) $(a+1)^3(a+1)^{12}=(a+1)^{15}$

(c) $(a+1)^{12}(a+1)^3=(a+1)^{15}$

17. (a) $yy^2y^8=y^{1+2+8}=y^{11}$

(b) $(y+1)(y+1)^2(y+1)^8=(y+1)^{11}$

(c) $\left[(y+1)(y+1)^8\right]^2=\left[(y+1)^9\right]^2=(y+1)^{18}$

19. (a) Using the properties of exponents, we have:

$$\frac{(x^2+3)^{10}}{(x^2+3)^9}=(x^2+3)^{10-9}=x^2+3$$

(b) Using the properties of exponents, we have:

$$\frac{(x^2+3)^9}{(x^2+3)^{10}}=(x^2+3)^{9-10}=(x^2+3)^{-1}=\frac{1}{x^2+3}$$

(c) Using the properties of exponents, we have:

$$\frac{12^{10}}{12^9}=12^{10-9}=12$$

21. (a) Using the properties of exponents, we have:

$$\frac{t^{15}}{t^9}=t^{15-9}=t^6$$

(b) Using the properties of exponents, we have:

$$\frac{t^9}{t^{15}}=t^{9-15}=t^{-6}=\frac{1}{t^6}$$

(c) Using the properties of exponents, we have:

$$\frac{(t^2+3)^{15}}{(t^2+3)^9}=(t^2+3)^{15-9}=(t^2+3)^6$$

23. (a) Using the properties of exponents, we have:

$$\frac{x^6y^{15}}{x^2y^{20}} = \frac{x^{6-2}}{y^{20-15}} = \frac{x^4}{y^5}$$

(b) Using the properties of exponents, we have:

$$\frac{x^2y^{20}}{x^6y^{15}} = \frac{y^{20-15}}{x^{6-2}} = \frac{y^5}{x^4}$$

(c) Using the properties of exponents, we have:

$$\left(\frac{x^2y^{20}}{x^6y^{15}}\right)^2 = \left(\frac{y^{20-15}}{x^{6-2}}\right)^2 = \left(\frac{y^5}{x^4}\right)^2 = \frac{y^{10}}{x^8}$$

25. (a) $4(x^3)^2 = 4x^6$

(b) $(4x^3)^2 = 4^2x^6 = 16x^6$

(c) Using the properties of exponents, we have:

$$\frac{(4x^2)^3}{(4x^3)^2} = \frac{4^3x^6}{4^2x^6} = 4^{3-2} = 4$$

27. (a) Since $x^0 = 1$ for any value of $x \neq 0$, then $64^0 = 1$.

(b) Since $x^0 = 1$ for any value of $x \neq 0$, then $(64^3)^0 = 1$.

(c) Since $x^0 = 1$ for any value of $x \neq 0$, then $(64^0)^3 = (1)^3 = 1$.

29. (a) Evaluating negative exponents, we have:

$$10^{-1} + 10^{-2} = \frac{1}{10} + \frac{1}{100} = \frac{10}{100} + \frac{1}{100} = \frac{11}{100}$$

(b) Using the result from part (a), we have:

$$(10^{-1} + 10^{-2})^{-1} = \left(\frac{11}{100}\right)^{-1} = \frac{100}{11}$$

(c) Evaluating negative exponents, we have:

$$\left[(10^{-1})(10^{-2})\right]^{-1} = \left[\frac{1}{10} \bullet \frac{1}{100}\right]^{-1} = \left(\frac{1}{1000}\right)^{-1} = 1000$$

31. Evaluating negative exponents, we have:

$$\left(\frac{1}{3}\right)^{-1}+\left(\frac{1}{4}\right)^{-1}=3+4=7$$

33. By first adding fractions, we have:

$$\left(\frac{2}{3}+\frac{3}{2}\right)^{-1}=\left(\frac{4}{6}+\frac{9}{6}\right)^{-1}=\left(\frac{13}{6}\right)^{-1}=\frac{6}{13}$$

35. (a) Evaluating negative exponents, we have:

$$5^{-2}+10^{-2}=\frac{1}{25}+\frac{1}{100}=\frac{4}{100}+\frac{1}{100}=\frac{5}{100}=\frac{1}{20}$$

(b) Evaluating negative exponents, we have:

$$(5+10)^{-2}=(15)^{-2}=\frac{1}{225}$$

37. Using properties of exponents, we have:

$$\left(a^2bc^0\right)^{-3}=\frac{1}{\left(a^2b\right)^3}=\frac{1}{a^6b^3}$$

39. Using properties of exponents, we have:

$$\left(a^{-2}b^{-1}c^3\right)^{-2}=a^4b^2c^{-6}=\frac{a^4b^2}{c^6}$$

41. Using properties of exponents, we have:

$$\left(\frac{x^3y^{-2}z}{xy^2z^{-3}}\right)^{-3}=\frac{x^{-9}y^6z^{-3}}{x^{-3}y^{-6}z^9}=\frac{y^{6-(-6)}}{x^{-3+9}z^{9+3}}=\frac{y^{12}}{x^6z^{12}}$$

43. Using properties of exponents, we have:

$$\left(\frac{x^4y^{-8}z^2}{xy^2z^{-6}}\right)^{-2}=\frac{x^{-8}y^{16}z^{-4}}{x^{-2}y^{-4}z^{12}}=\frac{y^{16+4}}{x^{-2+8}z^{12+4}}=\frac{y^{20}}{x^6z^{16}}$$

45. Using properties of exponents, we have:

$$\left(\frac{a^{-2}b^{-3}c^{-4}}{a^2b^3c^4}\right)^2=\frac{a^{-4}b^{-6}c^{-8}}{a^4b^6c^8}=\frac{1}{a^{4+4}b^{6+6}c^{8+8}}=\frac{1}{a^8b^{12}c^{16}}$$

47. Using properties of exponents, we have:

$$\frac{x^2}{y^{-3}}\div\frac{x^2}{y^3}=\frac{x^2}{y^{-3}}\bullet\frac{y^3}{x^2}=\frac{x^2y^3}{x^2y^{-3}}=y^6$$

49. Subtracting exponents, we have:

$$\frac{b^{p+1}}{b^p} = b^{p+1-p} = b$$

51. Adding exponents, we have:

$$\left(x^p\right)\left(x^p\right) = x^{p+p} = x^{2p}$$

53. Using properties of exponents, we have:

$$\left(a^{2q}b^{-p}\right)^3 = a^{6q}b^{-3p} = \frac{a^{6q}}{b^{3p}}$$

55. Factoring $9 = 3^2$ and $12 = 2^2 \bullet 3$, we have:

$$\frac{2^8 \bullet 3^{15}}{9 \bullet 3^{10} \bullet 12} = \frac{2^8 \bullet 3^{15}}{3^2 \bullet 3^{10} \bullet 2^2 \bullet 3} = \frac{2^8 \bullet 3^{15}}{2^2 \bullet 3^{13}} = 2^6 \bullet 3^2 = 576$$

57. Factoring $24 = 2^3 \bullet 3$, $32 = 2^5$, and $12 = 2^2 \bullet 3$, we have:

$$\frac{24^5}{32 \bullet 12^4} = \frac{\left(2^3 \bullet 3\right)^5}{2^5 \bullet \left(2^2 \bullet 3\right)^4} = \frac{2^{15} \bullet 3^5}{2^5 \bullet 2^8 \bullet 3^4} = \frac{2^{15} \bullet 3^5}{2^{13} \bullet 3^4} = 2^2 \bullet 3 = 12$$

59. 9.29×10^7 miles

61. 6.68×10^4 mph

63. 2.5×10^{19} miles

65. (a) 8.6688×10^1 days
(b) 1.0604772×10^4 days
(c) 8.9424×10^4 days

67. (a) 1.0×10^{-9} seconds
(b) 1.0×10^{-18} seconds
(c) 1.0×10^{-24} seconds

69. Using the properties of exponents, we have:

$$\frac{a^{3x+y}}{a^{2x}a^{x+y}} = \frac{a^{3x+y}}{a^{3x+y}} = 1$$

71. Using properties of exponents, we have:

$$\frac{\left(x^{5n+1}\right)^n}{\left(x^n\right)^{5n}} \bullet \frac{1}{x^{n-2}} = \frac{x^{5n^2+n}}{x^{5n^2}} \bullet \frac{1}{x^{n-2}} = \frac{x^{5n^2+n}}{x^{5n^2+n-2}} = x^{5n^2+n-5n^2-n+2} = x^2$$

73. Converting units, we have:

$$\frac{1\text{ sec}}{2.9979\times 10^{10}\text{ cm}}\bullet\frac{1\text{ min}}{60\text{ sec}}\bullet\frac{160930\text{ cm}}{1\text{ mi}}\bullet\left(92.9\times 10^{6}\text{ mi}\right)\approx 8\text{ min}$$

75. Raising both sides of the equation $p = b^x$ to the power of y yields $p^y = b^{xy}$. Similarly, raising both sides of $q = b^y$ to the power of x yields $q^x = b^{xy}$. Now we use these results to substitute for p^y and q^x in the equation $b^2 = \left(p^y q^x\right)^z$ to obtain:

$$b^2 = \left(b^{xy}b^{xy}\right)^z \text{ or } b^2 = b^{2xyz}$$

From this last equation we conclude (by equating exponents) that:

$$2 = 2xyz, \text{ or } xyz = 1$$

1.6 Nth Roots

1. false

3. true

5. true

7. true

9. true

11. (a) $\sqrt[3]{-64} = -4$

(b) $\sqrt[4]{-64}$ is undefined

13. (a) $\sqrt[3]{\frac{8}{125}} = \frac{2}{5}$

(b) $\sqrt[3]{-\frac{8}{125}} = -\frac{2}{5}$

15. (a) $\sqrt{-16}$ is undefined

(b) $\sqrt[4]{-16}$ is undefined

17. (a) $\sqrt[4]{\frac{256}{81}} = \frac{4}{3}$

(b) $\sqrt[3]{-\frac{27}{125}} = -\frac{3}{5}$

19. (a) $\sqrt[5]{-32} = -2$

(b) $-\sqrt[5]{-32} = -(-2) = 2$

21. (a) Factoring $18 = 9 \bullet 2$, we have:
$$\sqrt{18} = \sqrt{9}\sqrt{2} = 3\sqrt{2}$$

(b) Factoring $54 = 27 \bullet 2$, we have:
$$\sqrt[3]{54} = \sqrt[3]{27}\,\sqrt[3]{2} = 3\sqrt[3]{2}$$

23. (a) Factoring $98 = 49 \bullet 2$, we have:
$$\sqrt{98} = \sqrt{49}\sqrt{2} = 7\sqrt{2}$$

(b) Factoring $-64 = -32 \bullet 2$, we have:
$$\sqrt[5]{-64} = \sqrt[5]{-32}\,\sqrt[5]{2} = -2\sqrt[5]{2}$$

25. (a) $\sqrt{\frac{25}{4}} = \frac{5}{2}$

(b) $\sqrt[4]{\frac{16}{625}} = \frac{2}{5}$

27. (a) Factoring $8 = 4 \bullet 2$, we have:
$$\sqrt{2} + \sqrt{8} = \sqrt{2} + \sqrt{4}\sqrt{2} = \sqrt{2} + 2\sqrt{2} = 3\sqrt{2}$$

(b) Factoring $16 = 8 \bullet 2$, we have:
$$\sqrt[3]{2} + \sqrt[3]{16} = \sqrt[3]{2} + \sqrt[3]{8}\,\sqrt[3]{2} = \sqrt[3]{2} + 2\sqrt[3]{2} = 3\sqrt[3]{2}$$

29. (a) Factoring $50 = 25 \bullet 2$ and $128 = 64 \bullet 2$, we have:
$$4\sqrt{50} - 3\sqrt{128} = 4\sqrt{25}\sqrt{2} - 3\sqrt{64}\sqrt{2} = 20\sqrt{2} - 24\sqrt{2} = -4\sqrt{2}$$

(b) Factoring $32 = 16 \bullet 2$ and $162 = 81 \bullet 2$, we have:
$$\sqrt[4]{32} + \sqrt[4]{162} = \sqrt[4]{16}\,\sqrt[4]{2} + \sqrt[4]{81}\,\sqrt[4]{2} = 2\sqrt[4]{2} + 3\sqrt[4]{2} = 5\sqrt[4]{2}$$

31. (a) 0.3 [because $(0.3)^2 = 0.09$]
(b) 0.2 [because $(0.2)^3 = 0.008$]

33. Simplifying each radical yields:
$$4\sqrt{24} - 8\sqrt{54} + 2\sqrt{6} = 4\sqrt{4}\sqrt{6} - 8\sqrt{9}\sqrt{6} + 2\sqrt{6} = 8\sqrt{6} - 24\sqrt{6} + 2\sqrt{6} = -14\sqrt{6}$$

35. $\sqrt{\sqrt{64}} = \sqrt{8} = \sqrt{4}\sqrt{2} = 2\sqrt{2}$

37. (a) $\sqrt{36x^2} = 6x$, since $x > 0$

(b) $\sqrt{36y^2} = -6y$, since $y < 0$

39. (a) Multiplying radicals yields:
$$\sqrt{ab^2}\sqrt{a^2b} = \sqrt{a^3b^3} = ab\sqrt{ab}$$

(b) Multiplying radicals yields:

$$\sqrt{ab^3}\sqrt{a^3b}=\sqrt{a^4b^4}=a^2b^2$$

41. Simplifying the radical yields:

$$\sqrt{72a^3b^4c^5}=\sqrt{36a^2b^4c^4}\sqrt{2ac}=6ab^2c^2\sqrt{2ac}$$

43. Simplifying the radical yields:

$$\sqrt[4]{16a^4b^5}=\sqrt[4]{16a^4b^4}\bullet\sqrt[4]{b}=2ab\sqrt[4]{b}$$

45. Simplifying the radical yields:

$$\sqrt{18a^3b^2}=\sqrt{9a^2b^2}\sqrt{2a}=3ab\sqrt{2a}$$

47. Simplifying the radical yields:

$$\sqrt[3]{\frac{16a^{12}b^2}{c^9}}=\frac{\sqrt[3]{8a^{12}}\sqrt[3]{2b^2}}{\sqrt[3]{c^9}}=\frac{2a^4\sqrt[3]{2b^2}}{c^3}$$

49. Simplifying the radical yields:

$$\sqrt[6]{\frac{5a^7}{a^{-5}b^6}}=\sqrt[6]{\frac{5a^{12}}{b^6}}=\frac{\sqrt[6]{5}\sqrt[6]{a^{12}}}{\sqrt[6]{b^6}}=\frac{a^2\sqrt[6]{5}}{b}$$

51. Rationalizing denominators, we have:

$$\frac{4}{\sqrt{7}}\bullet\frac{\sqrt{7}}{\sqrt{7}}=\frac{4\sqrt{7}}{7}$$

53. Rationalizing denominators, we have:

$$\frac{1}{\sqrt{8}}\bullet\frac{\sqrt{2}}{\sqrt{2}}=\frac{\sqrt{2}}{\sqrt{16}}=\frac{\sqrt{2}}{4}$$

55. Rationalizing denominators, we have:

$$\frac{1}{1+\sqrt{5}}\bullet\frac{1-\sqrt{5}}{1-\sqrt{5}}=\frac{1-\sqrt{5}}{1-5}=\frac{1-\sqrt{5}}{-4}=\frac{\sqrt{5}-1}{4}$$

57. Rationalizing denominators, we have:

$$\frac{1+\sqrt{3}}{1-\sqrt{3}}\bullet\frac{1+\sqrt{3}}{1+\sqrt{3}}=\frac{1+\sqrt{3}+\sqrt{3}+3}{1-3}=\frac{4+2\sqrt{3}}{-2}=-2-\sqrt{3}$$

59. Rationalizing denominators and simplifying, we have:

$$\frac{1}{\sqrt{5}}+4\sqrt{45}=\frac{1}{\sqrt{5}}\bullet\frac{\sqrt{5}}{\sqrt{5}}+4\sqrt{9}\sqrt{5}=\frac{\sqrt{5}}{5}+12\sqrt{5}=\frac{\sqrt{5}}{5}+\frac{60\sqrt{5}}{5}=\frac{61\sqrt{5}}{5}$$

61. Rationalizing denominators, we have:

$$\frac{1}{\sqrt[3]{25}}\bullet\frac{\sqrt[3]{5}}{\sqrt[3]{5}}=\frac{\sqrt[3]{5}}{\sqrt[3]{125}}=\frac{\sqrt[3]{5}}{5}$$

63. Rationalizing denominators, we have:

$$\frac{3}{\sqrt[4]{3}} \bullet \frac{\sqrt[4]{3^3}}{\sqrt[4]{3^3}} = \frac{3\sqrt[4]{27}}{3} = \sqrt[4]{27}$$

65. Rationalizing denominators, we have:

$$\frac{1}{\sqrt[4]{2ab^5}} \bullet \frac{\sqrt[4]{8a^3b^3}}{\sqrt[4]{8a^3b^3}} = \frac{\sqrt[4]{8a^3b^3}}{\sqrt[4]{16a^4b^8}} = \frac{\sqrt[4]{8a^3b^3}}{2ab^2}$$

67. Rationalizing denominators, we have:

$$\frac{3}{\sqrt[5]{16a^4b^9}} \bullet \frac{\sqrt[5]{2ab}}{\sqrt[5]{2ab}} = \frac{3\sqrt[5]{2ab}}{\sqrt[5]{32a^5b^{10}}} = \frac{3\sqrt[5]{2ab}}{2ab^2}$$

69. Rationalizing denominators, we have:

$$\frac{x}{\sqrt{x}-2} \bullet \frac{\sqrt{x}+2}{\sqrt{x}+2} = \frac{x(\sqrt{x}+2)}{x-4}$$

71. Rationalizing denominators, we have:

$$\frac{\sqrt{x}-\sqrt{a}}{\sqrt{x}+\sqrt{a}} \bullet \frac{\sqrt{x}-\sqrt{a}}{\sqrt{x}-\sqrt{a}} = \frac{x-2\sqrt{ax}+a}{x-a}$$

73. Rationalizing numerators, we have:

$$\frac{\sqrt{x}-\sqrt{5}}{x-5} \bullet \frac{\sqrt{x}+\sqrt{5}}{\sqrt{x}+\sqrt{5}} = \frac{x-5}{(x-5)(\sqrt{x}+\sqrt{5})} = \frac{1}{\sqrt{x}+\sqrt{5}}$$

75. Rationalizing numerators, we have:

$$\frac{\sqrt{2+h}-\sqrt{2}}{h} \bullet \frac{\sqrt{2+h}+\sqrt{2}}{\sqrt{2+h}+\sqrt{2}} = \frac{2+h-2}{h(\sqrt{2+h}+\sqrt{2})} = \frac{h}{h(\sqrt{2+h}+\sqrt{2})} = \frac{1}{\sqrt{2+h}+\sqrt{2}}$$

77. Let $a = 9$ and $b = 16$, then:

$$\sqrt{a+b} = \sqrt{9+16} = \sqrt{25} = 5$$

$$\sqrt{a}+\sqrt{b} = \sqrt{9}+\sqrt{16} = 3+4 = 7$$

So the formula is not valid.

79. Let $u = 1$ and $v = 8$, then:

$$\sqrt[3]{u+v} = \sqrt[3]{1+8} = \sqrt[3]{9} \approx 2.08$$

$$\sqrt[3]{u}+\sqrt[3]{v} = \sqrt[3]{1}+\sqrt[3]{8} = 1+2 = 3$$

So the formula is not valid.

81. (a) To six decimal places, both values are 1.645751.

(b) Following the hint, we square each side:

$$(\sqrt{7}-1)^2 = 7-2\sqrt{7}+1 = 8-2\sqrt{7}$$

83. We first rationalize the denominator in each fraction:

$$\frac{\sqrt{a}}{\sqrt{a}+\sqrt{b}} = \frac{\sqrt{a}}{\sqrt{a}+\sqrt{b}} \bullet \frac{\sqrt{a}-\sqrt{b}}{\sqrt{a}-\sqrt{b}} = \frac{a-\sqrt{ab}}{a-b}$$

$$\frac{\sqrt{b}}{\sqrt{a}-\sqrt{b}} = \frac{\sqrt{b}}{\sqrt{a}-\sqrt{b}} \bullet \frac{\sqrt{a}+\sqrt{b}}{\sqrt{a}+\sqrt{b}} = \frac{\sqrt{ab}+b}{a-b}$$

Now adding the two fractions:

$$\frac{\sqrt{a}}{\sqrt{a}+\sqrt{b}} + \frac{\sqrt{b}}{\sqrt{a}-\sqrt{b}} = \frac{\sqrt{a}-\sqrt{ab}}{a-b} + \frac{\sqrt{ab}+b}{a-b} = \frac{a-\sqrt{ab}+\sqrt{ab}+b}{a-b} = \frac{a+b}{a-b}$$

85. Following the hint, we have:

$$\begin{aligned}\frac{1}{1+\sqrt{2}+\sqrt{3}} \bullet \frac{(1+\sqrt{2})-\sqrt{3}}{(1+\sqrt{2})-\sqrt{3}} &= \frac{(1+\sqrt{2})-\sqrt{3}}{(1+\sqrt{2})^2-(\sqrt{3})^2} \\ &= \frac{1+\sqrt{2}-\sqrt{3}}{1+2\sqrt{2}+2-3} \\ &= \frac{1+\sqrt{2}-\sqrt{3}}{2\sqrt{2}} \bullet \frac{\sqrt{2}}{\sqrt{2}} \\ &= \frac{\sqrt{2}+2-\sqrt{6}}{4}\end{aligned}$$

1.7 Rational Exponents

1. $a^{3/5} = \sqrt[5]{a^3} = \left(\sqrt[5]{a}\right)^3$

3. $5^{2/3} = \sqrt[3]{5^2} = \left(\sqrt[3]{5}\right)^2$

5. $\left(x^2+1\right)^{3/4} = \sqrt[4]{\left(x^2+1\right)^3} = \left(\sqrt[4]{x^2+1}\right)^3$

7. $2^{xy/3} = \sqrt[3]{2^{xy}} = \left(\sqrt[3]{2}\right)^{xy}$

9. $\sqrt[3]{p^2} = p^{2/3}$

11. $\sqrt[7]{(1+u)^4} = (1+u)^{4/7}$

13. $\sqrt[p]{\left(a^2+b^2\right)^3} = \left(a^2+b^2\right)^{3/p}$

15. $16^{1/2} = \sqrt{16} = 4$

17. $\left(\frac{1}{36}\right)^{1/2} = \sqrt{\frac{1}{36}} = \frac{1}{6}$

19. $(-16)^{1/2} = \sqrt{-16}$, which is undefined

21. $625^{1/4} = \sqrt[4]{625} = 5$

23. $8^{1/3} = \sqrt[3]{8} = 2$

25. $8^{2/3} = \left(\sqrt[3]{8}\right)^2 = (2)^2 = 4$

27. $(-32)^{1/5} = \sqrt[5]{-32} = -2$

29. $(-1000)^{1/3} = \sqrt[3]{-1000} = -10$

31. $49^{-1/2} = \left(\sqrt{49}\right)^{-1} = 7^{-1} = \frac{1}{7}$

33. $(-49)^{-1/2} = \left(\sqrt{-49}\right)^{-1}$, which is undefined

35. $(36)^{-3/2} = \left(\sqrt{36}\right)^{-3} = 6^{-3} = \frac{1}{6^3} = \frac{1}{216}$

37. $125^{2/3} = \left(\sqrt[3]{125}\right)^2 = (5)^2 = 25$

39. $(-1)^{3/5} = \left(\sqrt[5]{-1}\right)^3 = (-1)^3 = -1$

41. Evaluating the rational exponents:

$$32^{4/5} - 32^{-4/5} = \left(\sqrt[5]{32}\right)^4 - \left(\sqrt[5]{32}\right)^{-4} = 2^4 - 2^{-4} = 16 - \frac{1}{16} = \frac{255}{16}$$

43. Evaluating the rational exponents:

$$\begin{aligned}\left(\frac{9}{16}\right)^{-5/2} - \left(\frac{1000}{27}\right)^{4/3} &= \left(\frac{16}{9}\right)^{5/2} - \left(\frac{1000}{27}\right)^{4/3} \\ &= \left(\sqrt{\frac{16}{9}}\right)^5 - \left(\sqrt[3]{\frac{1000}{27}}\right)^4 \\ &= \left(\frac{4}{3}\right)^5 - \left(\frac{10}{3}\right)^4 \\ &= \frac{1024}{243} - \frac{10000}{81} \\ &= -\frac{28976}{243}\end{aligned}$$

45. $(2a^{1/3})(3a^{1/4}) = 6a^{7/12}$, since $\frac{1}{3}+\frac{1}{4}=\frac{4}{12}+\frac{3}{12}=\frac{7}{12}$

47. Writing with rational exponents, then simplifying:

$$\sqrt[4]{\frac{64a^{2/3}}{a^{1/3}}} = \sqrt[4]{64a^{1/3}} = \left(2^6 a^{1/3}\right)^{1/4} = 2^{3/2} a^{1/12}$$

49. Subtracting exponents yields:

$$\frac{\left(x^2+1\right)^{3/4}}{\left(x^2+1\right)^{-1/4}} = \left(x^2+1\right)^{4/4} = x^2+1$$

51. (a) $\sqrt{3}\,\sqrt[3]{6} = 3^{1/2}\,6^{1/3} = 3^{1/2}2^{1/3}3^{1/3} = 2^{1/3}\,3^{5/6}$

(b) $3^{1/2}6^{1/3} = 3^{3/6}6^{2/6} = \sqrt[6]{3^3 6^2} = \sqrt[6]{972}$

53. (a) $\sqrt[3]{6}\,\sqrt[4]{2} = 6^{1/3}\,2^{1/4} = 2^{1/3}3^{1/3}2^{1/4} = 2^{7/12}\,3^{1/3}$

(b) $6^{1/3}\,2^{1/4} = 6^{4/12}\,2^{3/12} = \sqrt[12]{6^4\,2^3} = \sqrt[12]{10368}$

55. (a) $\sqrt[3]{x^2}\,\sqrt[5]{y^4} = x^{2/3}\,y^{4/5}$

(b) $x^{2/3}\,y^{4/5} = x^{10/15}\,y^{12/15} = \sqrt[15]{x^{10}\,y^{12}}$

57. (a) Using rational exponents, we have:

$$\sqrt[4]{x^a}\,\sqrt[3]{x^b}\,\sqrt{x^{a/6}} = x^{a/4}\,x^{b/3}\,x^{a/12} = x^{3a/12}\,x^{4b/12}\,x^{a/12} = x^{(4a+4b)/12} = x^{(a+b)/3}$$

(b) $x^{(a+b)/3} = \sqrt[3]{x^{a+b}}$

59. $\sqrt[3]{(x+1)^2} = (x+1)^{2/3}$

61. $\left(\sqrt[5]{x+y}\right)^2 = (x+y)^{2/5}$

63. Using rational exponents, we have:

$$\sqrt[3]{\sqrt{x}} + \sqrt{\sqrt[3]{x}} = \left(x^{1/2}\right)^{1/3} + \left(x^{1/3}\right)^{1/2} = x^{1/6} + x^{1/6} = 2x^{1/6}$$

65. Using rational exponents, we have:

$$\sqrt{\sqrt[3]{x}\,\sqrt[4]{y}} = \left(x^{1/3}\,y^{1/4}\right)^{1/2} = x^{1/6}\,y^{1/8}$$

67. We compute $9^{10/9} \approx 11.5$ and $10^{9/10} \approx 7.9$. Thus $9^{10/9}$ is larger.

69. Using $a = 9$ and $b = 16$, we have:
$$(a+b)^{1/2} = (9+16)^{1/2} = 25^{1/2} = 5$$
$$a^{1/2} + b^{1/2} = 9^{1/2} + 16^{1/2} = 3+4 = 7$$
So the formula is not valid.

71. Using $u = 1$ and $v = 8$, we have:
$$(u+v)^{1/3} = (1+8)^{1/3} = 9^{1/3} \approx 2.08$$
$$u^{1/3} + v^{1/3} = 1^{1/3} + 8^{1/3} = 1+2 = 3$$
So the formula is not valid.

73. Using $x = 4$ and $m = 2$, we have:
$$x^{1/m} = 4^{1/2} = 2$$
$$\frac{1}{x^m} = \frac{1}{4^2} = \frac{1}{16}$$
So the formula is not valid.

75. We complete the table:

n	2	5	10	100	10^3	10^4	10^5	10^6
$n^{1/n}$	1.4142	1.3797	1.2589	1.0471	1.0069	1.0009	1.0001	1.0000

77. (a) Note that $2^{2/3}$ is less than 2 and $2^{3/2}$ is greater than 2, thus $2^{3/2}$ is the larger number.

(b) Note that $5^{1/2} = \sqrt{5}$ and $5^{-2} = \frac{1}{25}$, thus $5^{1/2}$ is the larger number.

(c) $2^{1/2}$ is larger. (One way to see this is to raise both numbers to the sixth power)

(d) $\left(\frac{1}{2}\right)^{1/3}$ is larger. (One way to see this is to raise both numbers to the sixth power)

(e) Note that $10^{1/10}$ is larger than 1, but $\left(\frac{1}{10}\right)^{10}$ is much less than 1, thus $10^{1/10}$ is the larger number.

79. We compute $(0.5)^{1/3} \approx 0.7937$ while $(0.5)^{1/4} \approx 0.8409$, so $(0.5)^{1/3}$ is closer to zero. Without a calculator, $(0.5)^{1/3} = (0.5)^{4/12}$ while $(0.5)^{1/4} = (0.5)^{3/12}$, and thus $(0.5)^{1/3} = (0.5)^{1/12}(0.5)^{1/4}$, and thus $(0.5)^{1/3}$ is closer to zero.

81. Using rational exponents, we have:
$$\frac{a-b}{a+b}\sqrt{\frac{a+b}{a-b}} = \frac{a-b}{a+b} \bullet \frac{(a+b)^{1/2}}{(a-b)^{1/2}} = \frac{a-b}{(a-b)^{1/2}} \bullet \frac{(a+b)^{1/2}}{a+b} = \frac{(a-b)^{1/2}}{(a+b)^{1/2}} = \left(\frac{a-b}{a+b}\right)^{1/2}$$

1.8 Polynomials

1. (a) The domain is the set of all real numbers.
(b) The domain is the set of all non-negative real numbers.

3. (a) The domain is the set of all real numbers.
(b) The domain is the set of all real numbers.

5. (a) The domain is the set of all non-negative real numbers.
(b) The domain is the set of all positive real numbers except 1.

7. (a) The degree is 3 and the coefficients are 4, –2, –6 and –1.
(b) The degree is 2 and the coefficients are a, b and c.

9. (a) The degree is 1 and the coefficient is 6.
(b) The degree is 3 and the coefficients are –1, 1 and 6.

11. $(12x^2 - 4x + 2) + (8x^2 + 6x - 1) = 20x^2 + 2x + 1$

13. $(x^2 - x - 1) - (x^2 + x + 1) = x^2 - x - 1 - x^2 - x - 1 = -2x - 2$

15. Subtracting and combining like terms, we have:

$$\begin{aligned}(2x^2 - 4x - 4) - (6x^2 + 5) - (8x - 1) &= 2x^2 - 4x - 4 - 6x^2 - 5 - 8x + 1\\ &= -4x^2 - 12x - 8\end{aligned}$$

17. Carrying out the operations and combining like terms, we have:

$$\begin{aligned}&(2x^2 - 6x - 1) + (5x^2 - 5x - 1) - (3x^2 + 8x + 12)\\ &\quad = 2x^2 - 6x - 1 + 5x^2 - 5x - 1 - 3x^2 - 8x - 12\\ &\quad = 4x^2 - 19x - 14\end{aligned}$$

19. Subtracting and combining like terms, we have:

$$(ax^2 + bx + c) - (2ax^2 - 3bx + c) = ax^2 + bx + c - 2ax^2 + 3bx - c = -ax^2 + 4bx$$

21. $2x(x^2 - 4x - 5) = 2x^3 - 8x^2 - 10x$

23. $(x - 1)(x - 2) = x^2 - 3x + 2$

25. $(2x + 4)(x + 1) = 2x^2 + 6x + 4$

27. $(x^2 + 3x)(x^2 + x) = x^4 + 4x^3 + 3x^2$

29. $(2xy - 3)(2xy - 1) = 4x^2y^2 - 8xy + 3$

31. Carrying out the multiplication, we have:

$$\left(x+\tfrac{1}{x}\right)\left(x+\tfrac{1}{x}\right)=x^2+1+1+\tfrac{1}{x^2}=x^2+2+\tfrac{1}{x^2}$$

33. Carrying out the multiplication, we have:

$$(\sqrt{a+b}+1)(\sqrt{a+b}+3)=(\sqrt{a+b})^2+4\sqrt{a+b}+3=a+b+4\sqrt{a+b}+3$$

35. Carrying out the multiplication, we have:

$$\left(x^{1/2}-1\right)\left(x^{1/2}-2\right)=\left(x^{1/2}\right)^2-3x^{1/2}+2=x-3x^{1/2}+2$$

37. Carrying out the multiplication, we have:

$$\begin{aligned}(y+2)\left(y^2-3y-5\right)&=y\left(y^2-3y-5\right)+2\left(y^2-3y-5\right)\\&=y^3-3y^2-5y+2y^2-6y-10\\&=y^3-y^2-11y-10\end{aligned}$$

39. Carrying out the multiplication, we have:

$$\begin{aligned}(x-2y+z)(x+2y-z)&=(x-(2y-z))(x+(2y-z))\\&=x^2-(2y-z)^2\\&=x^2-\left(4y^2-4yz+z^2\right)\\&=x^2-4y^2+4yz-z^2\end{aligned}$$

41. (a) Using the special product for $(A-B)(A+B)$, we have:

$$(x-y)(x+y)=x^2-y^2$$

(b) Using the special product for $(A-B)(A+B)$, we have:

$$\left(x^2-5\right)\left(x^2+5\right)=\left(x^2\right)^2-(5)^2=x^4-25$$

43. (a) Using the special product for $(A-B)(A+B)$, we have:

$$(A-4)(A+4)=A^2-16$$

(b) Using the special product for $(A-B)(A+B)$, we have:

$$[(a+b)-4][(a+b)+4]=(a+b)^2-16=a^2+2ab+b^2-16$$

45. (a) Using the special product for $(A-B)^2$, we have:

$$(x-8)^2=x^2-2(x)(8)+64=x^2-16x+64$$

(b) Using the special product for $(A-B)^2$, we have:

$$\left(2x^2-5\right)^2=\left(2x^2\right)^2-2\left(2x^2\right)(5)+25=4x^4-20x^2+25$$

47. (a) Using the special product for $(A+B)^2$, we have:

$$\left(\sqrt{x}+\sqrt{y}\right)^2=\left(\sqrt{x}\right)^2+2\left(\sqrt{x}\right)\left(\sqrt{y}\right)+\left(\sqrt{y}\right)^2=x+2\sqrt{xy}+y$$

(b) Using the special product for $(A-B)^2$, we have:

$$\left(\sqrt{x+y}-\sqrt{x}\right)^2=\left(\sqrt{x+y}\right)^2-2\left(\sqrt{x+y}\right)\left(\sqrt{x}\right)+\left(\sqrt{x}\right)^2$$
$$=x+y-2\sqrt{x^2+xy}+x$$
$$=2x+y-2\sqrt{x^2+xy}$$

49. (a) Using the special product for $(A+B)^3$, we have:

$$(a+1)^3=a^3+3\left(a^2\right)(1)+3(a)(1)^2+(1)^3=a^3+3a^2+3a+1$$

(b) Using the special product for $(A-B)^3$, we have:

$$\left(3x^2-2a^2\right)^3=\left(3x^2\right)^3-3\left(3x^2\right)^2\left(2a^2\right)+3\left(3x^2\right)\left(2a^2\right)^2-\left(2a^2\right)^3$$
$$=27x^6-54x^4a^2+36x^2a^4-8a^6$$

51. (a) Using the special product for $(A+B)\left(A^2-AB+B^2\right)$, we have:

$$(x+1)\left(x^2-x+1\right)=x^3+1$$

(b) Using the special product for $(A+B)\left(A^2-AB+B^2\right)$, we have:

$$\left(x^2+1\right)\left(x^4-x^2+1\right)=\left(x^2\right)^3+(1)^3=x^6+1$$

53. We use long division:

$$\begin{array}{r|l} & x-5 \\ \hline x-3 & x^2-8x+4 \\ & \underline{x^2-3x} \\ & \quad -5x+4 \\ & \quad \underline{-5x+15} \\ & \qquad\quad -11 \end{array}$$

The quotient is $x-5$ and the remainder is -11. We write the equation:

$$x^2-8x+4=(x-3)(x-5)-11$$

55. We use long division:

$$\begin{array}{r|l} & x-11 \\ \hline x+5 & x^2-6x-2 \\ & \underline{x^2+5x} \\ & \quad -11x-2 \\ & \quad \underline{-11x-55} \\ & \qquad\quad 53 \end{array}$$

The quotient is $x-11$ and the remainder is 53. We write the equation:

$$x^2-6x-2=(x+5)(x-11)+53$$

57. We use long division:

$$
\begin{array}{r}
3x^2-\frac{3}{2}x-\frac{1}{4} \\
2x+1\overline{\big)\,6x^3+0x^2-2x+3} \\
\underline{6x^3+3x^2} \qquad\qquad \\
-3x^2-2x \qquad \\
\underline{-3x^2-\frac{3}{2}x} \qquad \\
-\frac{1}{2}x+3 \\
\underline{-\frac{1}{2}x-\frac{1}{4}} \\
\frac{13}{4}
\end{array}
$$

The quotient is $3x^2-\frac{3}{2}x-\frac{1}{4}$ and the remainder is $\frac{13}{4}$. We write the equation:

$$6x^3-2x+3=(2x+1)\left(3x^2-\frac{3}{2}x-\frac{1}{4}\right)+\frac{13}{4}$$

59. We use long division:

$$
\begin{array}{r}
t^4-3t^3+9t^2-27t+81 \qquad \\
t+3\overline{\big)\,t^5+0t^4+0t^3+0t^2+0t+2} \\
\underline{t^5+3t^4} \qquad\qquad\qquad\qquad\qquad \\
-3t^4+0t^3 \qquad\qquad\qquad\qquad \\
\underline{-3t^4-9t^3} \qquad\qquad\qquad\qquad \\
9t^3+0t^2 \qquad\qquad\qquad \\
\underline{9t^3+27t^2} \qquad\qquad\qquad \\
-27t^2+0t \qquad\qquad \\
\underline{-27t^2-81t} \qquad\qquad \\
81t+2 \\
\underline{81t+243} \\
-241
\end{array}
$$

The quotient is $t^4-3t^3+9t^2-27t+81$ and the remainder is -241. We write the equation:

$$t^5+2=(t+3)\left(t^4-3t^3+9t^2-27t+81\right)-241$$

61. We use long division:

$$
\begin{array}{r}
u^5+2u^4+4u^3+8u^2+16u+32 \qquad \\
u-2\overline{\big)\,u^6+0u^5+0u^4+0u^3+0u^2+0u-64} \\
\underline{u^6-2u^5} \qquad\qquad\qquad\qquad\qquad\qquad \\
2u^5+0u^4 \qquad\qquad\qquad\qquad\qquad \\
\underline{2u^5-4u^4} \qquad\qquad\qquad\qquad\qquad \\
4u^4+0u^3 \qquad\qquad\qquad\qquad \\
\underline{4u^4-8u^3} \qquad\qquad\qquad\qquad \\
8u^3+0u^2 \qquad\qquad\qquad \\
\underline{8u^3-16u^2} \qquad\qquad\qquad \\
16u^2+0u \qquad\qquad \\
\underline{16u^2-32u} \qquad\qquad \\
32u-64 \\
\underline{32u-64} \\
0
\end{array}
$$

The quotient is $u^5 + 2u^4 + 4u^3 + 8u^2 + 16u + 32$ and the remainder is 0. We write the equation:

$$u^6 - 64 = (u-2)\left(u^5 + 2u^4 + 4u^3 + 8u^2 + 16u + 32\right) + 0$$

63. We use long division:

$$\begin{array}{r|l} & 5x^2 + 15x + 17 \\ \hline x^2 - 3x + 5 & 5x^4 + 0x^3 - 3x^2 + 0x + 2 \\ & \underline{5x^4 - 15x^3 + 25x^2} \\ & \quad 15x^3 - 28x^2 + 0x \\ & \quad \underline{15x^3 - 45x^2 + 75x} \\ & \qquad 17x^2 - 75x + 2 \\ & \qquad \underline{17x^2 - 51x + 85} \\ & \qquad\quad -24x - 83 \end{array}$$

The quotient is $5x^2 + 15x + 17$ and the remainder is $-24x - 83$. We write the equation:

$$5x^4 - 3x^2 + 2 = \left(x^2 - 3x + 5\right)\left(5x^2 + 15x + 17\right) + (-24x - 83)$$

65. We use long division:

$$\begin{array}{r|l} & t^2 - 2t - 4 \\ \hline t^2 - 2t + 4 & t^4 - 4t^3 + 4t^2 + 0t - 16 \\ & \underline{t^4 - 2t^3 + 4t^2} \\ & \quad -2t^3 + 0t^2 + 0t \\ & \quad \underline{-2t^3 + 4t^2 - 8t} \\ & \qquad -4t^2 + 8t - 16 \\ & \qquad \underline{-4t^2 + 8t - 16} \\ & \qquad\qquad 0 \end{array}$$

The quotient is $t^2 - 2t - 4$ and the remainder is 0. We write the equation:

$$t^4 - 4t^3 + 4t^2 - 16 = \left(t^2 - 2t + 4\right)\left(t^2 - 2t - 4\right) + 0$$

67. We use long division:

$$\begin{array}{r|l} & z^4 + z^3 + z^2 + z + 1 \\ \hline z - 1 & z^5 + 0z^4 + 0z^3 + 0z^2 + 0z - 1 \\ & \underline{z^5 - 1z^4} \\ & \quad z^4 + 0z^3 \\ & \quad \underline{z^4 - 1z^3} \\ & \qquad z^3 + 0z^2 \\ & \qquad \underline{z^3 - 1z^2} \\ & \qquad\quad z^2 + 0z \\ & \qquad\quad \underline{z^2 - 1z} \\ & \qquad\qquad z - 1 \\ & \qquad\qquad \underline{z - 1} \\ & \qquad\qquad\quad 0 \end{array}$$

The quotient is $z^4 + z^3 + z^2 + z + 1$ and the remainder is 0. We write the equation:

$$z^5 - 1 = (z-1)\left(z^4 + z^3 + z^2 + z + 1\right) + 0$$

69. Using the special product for $(A-B)(A+B)$, we have:

$$(a+b-3)(a+b+3)=(a+b)^2-(3)^2=a^2+2ab+b^2-9$$

71. Since $(a+b-x-y)=(a+b)-(x+y)$, we use the special product for $(A-B)(A+B)$ to obtain:

$$(a+b+x+y)(a+b-x-y)=(a+b)^2-(x+y)^2=a^2+2ab+b^2-x^2-2xy-y^2$$

73. Using the special product for $(A+B)^2$, we have:

$$(\sqrt{a}+\sqrt{b})^2=(\sqrt{a})^2+2(\sqrt{a})(\sqrt{b})+(\sqrt{b})^2=a+2\sqrt{ab}+b$$

75. Using the special product for $(A-B)(A^2+AB+B^2)$, we have:

$$(x^{1/3}-y^{1/3})(x^{2/3}+x^{1/3}y^{1/3}+y^{2/3})=(x^{1/3})^3-(y^{1/3})^3=x-y$$

77. (a) We complete the tables:

x	$\frac{x^2-16}{x-4}$
3.9	7.9
3.99	7.99
3.999	7.999
3.9999	7.9999
3.99999	7.99999

x	$\frac{x^2-16}{x-4}$
4.1	8.1
4.01	8.01
4.001	8.001
4.0001	8.0001
4.00001	8.00001

(b) As x approaches 4, the value of the expression $\frac{x^2-16}{x-4}$ approaches 8.

(c) Since $x^2-16=(x+4)(x-4)$, then:

$$\frac{x^2-16}{x-4}=\frac{(x+4)(x-4)}{x-4}=x+4, \text{ if } x\neq 4$$

Thus as x approaches 4, we would expect $x+4$ to approach 8.

79. We first compute the three products:

$$(b-c)(b+c-a)=b^2+bc-ab-bc-c^2+ac=b^2-c^2-ab+ac$$
$$(c-a)(c+a-b)=c^2+ac-bc-ac-a^2+ab=c^2-a^2-bc+ab$$
$$(a-b)(a+b-c)=a^2+ab-ac-ab-b^2+bc=a^2-b^2-ac+bc$$

By adding the three expressions we've obtained and combining like terms, we obtain the sum of 0, as required.

81. Using the special product for $(A-B)(A+B)$, we have:

$$(x^2+8-4x)(x^2+8+4x)=(x^2+8)^2-(4x)^2=x^4+16x^2+64-16x^2=x^4+64$$

83. (a) We use long division:

$$\begin{array}{r|l} & ax+(b+ar) \\ \hline x-r & ax^2+bx+c \\ & \underline{ax^2-arx} \\ & \quad (b+ar)x+c \\ & \quad \underline{(b+ar)x-r(b+ar)} \\ & \qquad c+r(b+ar) \end{array}$$

The remainder is $c + r(b + ar) = ar^2 + br + c$.

(b) We use long division:

$$\begin{array}{r|l} & ax^2+(b+ar)x+(c+br+ar^2) \\ \hline x-r & ax^3+bx^2+cx+d \\ & \underline{ax^3-arx^2} \\ & \quad (b+ar)x^2+cx \\ & \quad \underline{(b+ar)x^2-r(b+ar)x} \\ & \qquad (c+br+ar^2)x+d \\ & \qquad \underline{(c+br+ar^2)x-r(c+br+ar^2)} \\ & \qquad\qquad d+r(c+br+ar^2) \end{array}$$

The remainder is $d + r(c + br + ar^2) = ar^3 + br^2 + cr + d$.

(c) A reasonable guess would be $ar^4 + br^3 + cr^2 + dr + e$, which can be verified.

85. (a) We complete the table:

n	1	2	3	4	5	6
$\frac{n^2}{n+1}$	$\frac{1}{2}$	$\frac{4}{3}$	$\frac{9}{4}$	$\frac{16}{5}$	$\frac{25}{6}$	$\frac{36}{7}$

(b) We use long division:

$$\begin{array}{r|l} & n-1 \\ \hline n+1 & n^2+0n+0 \\ & \underline{n^2+1n} \\ & \quad -1n+0 \\ & \quad \underline{-1n-1} \\ & \qquad\quad 1 \end{array}$$

Since the remainder is always 1, then $\frac{n^2}{n+1}$ can never be a natural number.

1.9 Factoring

1. (a) Using a difference of squares, we have:

$$x^2 - 64 = (x+8)(x-8)$$

(b) Using a common factor of $7x^2$, we have:

$$7x^4 + 14x^2 = 7x^2\left(x^2+2\right)$$

(c) Using a common factor of z then a difference of squares, we have:
$$121z - z^3 = z(121 - z^2) = z(11 + z)(11 - z)$$

(d) Using a difference of squares, we have:
$$a^2b^2 - c^2 = (ab)^2 - (c)^2 = (ab + c)(ab - c)$$

3. (a) By trial and error, we have:
$$x^2 + 2x - 3 = (x + 3)(x - 1)$$

(b) By trial and error, we have:
$$x^2 - 2x - 3 = (x - 3)(x + 1)$$

(c) By trial and error, we find that $x^2 - 2x + 3$ is irreducible.

(d) By trial and error, we have:
$$-x^2 + 2x + 3 = (-x + 3)(x + 1)$$
Note that we could also use a common factor of –1, then trial and error:
$$-x^2 + 2x + 3 = -1(x^2 - 2x - 3) = -(x - 3)(x + 1)$$

5. (a) By the sum of cubes, we have:
$$x^3 + 1 = (x + 1)(x^2 - x + 1)$$

(b) Since $6^3 = 216$, we use the sum of cubes to obtain:
$$x^3 + 216 = x^3 + 6^3 = (x + 6)(x^2 - 6x + 36)$$

(c) Using a common factor of 8 then a difference of cubes, we have:
$$1000 - 8x^6 = 8(125 - x^6) = 8(5^3 - x^6) = 8(5 - x^2)(25 + 5x^2 + x^4)$$

(d) By the difference of cubes, we have:
$$64a^3x^3 - 125 = (4ax)^3 - 5^3 = (4ax - 5)(16a^2x^2 + 20ax + 25)$$

7. (a) Using a difference of squares, we have:
$$144 - x^2 = (12 + x)(12 - x)$$

(b) Since this is a sum of squares, the expression $144 + x^2$ is irreducible.

(c) Using a difference of squares, we have:
$$144 - (y - 3)^2 = (144 + y - 3)(144 - y + 3) = (141 + y)(147 - y)$$

9. (a) Using a common factor of h^3 then a difference of squares, we have:
$$h^3 - h^5 = h^3(1 - h^2) = h^3(1 + h)(1 - h)$$

(b) Using a common factor of h^3 then a difference of squares, we have:
$$100h^3 - h^5 = h^3(100 - h^2) = h^3(10 + h)(10 - h)$$

(c) Using a common factor of $(h + 1)^3$ then a difference of squares, we have:

$$\begin{aligned}100(h+1)^3-(h+1)^5 &= (h+1)^3[100-(h+1)^2]\\ &= (h+1)^3(10+h+1)(10-h-1)\\ &= (h+1)^3(11+h)(9-h)\end{aligned}$$

11. (a) By trial and error, we have:

$$x^2-13x+40=(x-8)(x-5)$$

(b) By trial and error, we find that $x^2 - 13x - 40$ is irreducible.

13. (a) By trial and error, we have:

$$x^2+5x-36=(x+9)(x-4)$$

(b) By trial and error, we have:

$$x^2-13x+36=(x-9)(x-4)$$

15. (a) By trial and error, we have:

$$3x^2-22x-16=(3x+2)(x-8)$$

(b) By trial and error, we find that $3x^2 - x - 16$ is irreducible.

17. (a) By trial and error, we have:

$$6x^2+13x-5=(3x-1)(2x+5)$$

(b) By trial and error, we have:

$$6x^2-x-5=(6x+5)(x-1)$$

19. (a) Using the special product for $(A + B)^2$, we have:

$$t^4+2t^2+1=\left(t^2+1\right)^2$$

(b) Using the special product for $(A - B)^2$ and the difference of squares, we have:

$$t^4-2t^2+1=\left(t^2-1\right)^2=(t+1)^2(t-1)^2$$

(c) By trial and error, we find that $t^4 - 2t^2 - 1$ is irreducible.

21. (a) Using the common factor x, we have:

$$4x^3-20x^2-25x=x\left(4x^2-20x-25\right)$$

(b) Using the common factor x and the special product for $(A - B)^2$, we have:

$$4x^3-20x^2+25x=x\left(4x^2-20x+25\right)=x(2x-5)^2$$

23. (a) Using grouping, we have:

$$ab-bc+a^2-ac=b(a-c)+a(a-c)=(a-c)(b+a)$$

(b) Using grouping, we have:

$$(u+v)x - xy + (u+v)^2 - (u+v)y = x(u+v-y) + (u+v)(u+v-y)$$
$$= (u+v-y)(x+u+v)$$

25. Using grouping, we have:

$$x^2z^2 + xzt + xyz + yt = xz(xz+t) + y(xz+t) = (xz+t)(xz+y)$$

27. Using the special product for $(A-B)^2$, we have:

$$a^4 - 4a^2b^2c^2 + 4b^4c^4 = \left(a^2 - 2b^2c^2\right)^2$$

29. Since this is a sum of squares, $A^2 + B^2$ is irreducible.

31. Since $4^3 = 64$, using the sum of cubes factoring yields:

$$x^3 + 64 = (x+4)\left(x^2 - 4x + 16\right)$$

33. Using the difference of cubes factoring, we have:

$$(x+y)^3 - y^3 = (x+y-y)\left[(x+y)^2 + (x+y)(y) + y^2\right]$$
$$= x\left(x^2 + 2xy + y^2 + xy + y^2 + y^2\right)$$
$$= x\left(x^2 + 3xy + 3y^2\right)$$

35. Using the difference of cubes factoring, then grouping, we have:

$$x^3 - y^3 + x - y = (x-y)\left(x^2 + xy + y^2\right) + (x-y)(1) = (x-y)\left(x^2 + xy + y^2 + 1\right)$$

37. (a) Using the difference of squares factoring (twice), we have:

$$p^4 - 1 = \left(p^2 + 1\right)\left(p^2 - 1\right) = \left(p^2 + 1\right)(p+1)(p-1)$$

(b) Using the difference of squares factoring (three times), we have:

$$p^8 - 1 = \left(p^4 + 1\right)\left(p^4 - 1\right)$$
$$= \left(p^4 + 1\right)\left(p^2 + 1\right)\left(p^2 - 1\right)$$
$$= \left(p^4 + 1\right)\left(p^2 + 1\right)(p+1)(p-1)$$

39. Using the special product for $(A+B)^3$, we have:

$$x^3 + 3x^2 + 3x + 1 = (x+1)^3$$

Note: If you do not recognize the special product, you can also use grouping:

$$x^3 + 3x^2 + 3x + 1 = \left(x^3 + 1\right) + \left(3x^2 + 3x\right)$$
$$= (x+1)\left(x^2 - x + 1\right) + 3x(x+1)$$
$$= (x+1)\left(x^2 - x + 1 + 3x\right)$$
$$= (x+1)\left(x^2 + 2x + 1\right)$$
$$= (x+1)(x+1)^2$$
$$= (x+1)^3$$

41. Since this is a sum of squares, $x^2 + 16y^2$ is irreducible.

43. Using the difference of squares factoring, we have:
$$\tfrac{25}{16} - c^2 = \left(\tfrac{5}{4} + c\right)\left(\tfrac{5}{4} - c\right)$$

45. Using the difference of squares factoring (twice), we have:
$$z^4 - \tfrac{81}{16} = \left(z^2 + \tfrac{9}{4}\right)\left(z^2 - \tfrac{9}{4}\right) = \left(z^2 + \tfrac{9}{4}\right)\left(z + \tfrac{3}{2}\right)\left(z - \tfrac{3}{2}\right)$$

47. Using the difference of cubes factoring, we have:
$$\frac{125}{m^3n^3} - 1 = \left(\frac{5}{mn} - 1\right)\left(\frac{25}{m^2n^2} + \frac{5}{mn} + 1\right)$$

49. Using the special product for $(A + B)^2$, we have:
$$\tfrac{1}{4}x^2 + xy + y^2 = \left(\tfrac{1}{2}x + y\right)^2$$

51. Using the common factor $x - a$, then a difference of squares, we have:
$$\begin{aligned} 64(x-a)^3 - x + a &= (x-a)\left[64(x-a)^2 - 1\right] \\ &= (x-a)[8(x-a)+1][8(x-a)-1] \\ &= (x-a)(8x-8a+1)(8x-8a-1) \end{aligned}$$

53. Using grouping, we have:
$$\begin{aligned} x^2 - a^2 + y^2 - 2xy &= \left(x^2 - 2xy + y^2\right) - a^2 \\ &= (x-y)^2 - a^2 \\ &= (x-y-a)(x-y+a) \end{aligned}$$

55. Using the common factor x, then trial and error, we have:
$$21x^3 + 82x^2 - 39x = x\left(21x^2 + 82x - 39\right) = x(7x-3)(3x+13)$$

57. Using grouping, the special product for $(A - B)^2$, and the difference of squares factoring, we have:
$$\begin{aligned} 12xy + 25 - 4x^2 - 9y^2 &= 5^2 - \left(4x^2 - 12xy + 9y^2\right) \\ &= 5^2 - (2x-3y)^2 \\ &= [5-(2x-3y)][5+(2x-3y)] \\ &= (5-2x+3y)(5+2x-3y) \end{aligned}$$

59. Factoring by trial and error, we have:
$$ax^2 + (a+b)x + b = (ax+b)(x+1)$$

61. Using the common factor $(x + 1)^{1/2}$, we have:

$$\begin{aligned}(x+1)^{1/2}-(x+1)^{3/2} &= (x+1)^{1/2}\left[1-(x+1)^{2/2}\right]\\ &= (x+1)^{1/2}(1-x-1)\\ &= (x+1)^{1/2}(-x)\\ &= -x(x+1)^{1/2}\end{aligned}$$

63. Using the common factor $(x + 1)^{-3/2}$, we have:

$$\begin{aligned}(x+1)^{-1/2}-(x+1)^{-3/2} &= (x+1)^{-3/2}\left[(x+1)^{1}-1\right]\\ &= (x+1)^{-3/2}(x)\\ &= x(x+1)^{-3/2}\end{aligned}$$

65. Using the common factor $\frac{1}{3}(2x+3)^{1/2}$, we have:

$$\begin{aligned}(2x+3)^{1/2}-\tfrac{1}{3}(2x+3)^{3/2} &= \tfrac{1}{3}(2x+3)^{1/2}\left[3-(2x+3)^{1}\right]\\ &= \tfrac{1}{3}(2x+3)^{1/2}(3-2x-3)\\ &= \tfrac{1}{3}(2x+3)^{1/2}(-2x)\\ &= -\tfrac{2}{3}x(2x+3)^{1/2}\end{aligned}$$

67. (a) Using the difference of squares factoring, we have:

$$100^2-99^2=(100+99)(100-99)=(199)(1)=199$$

(b) Using the difference of cubes factoring, we have:

$$\begin{aligned}8^3-6^3 &= (8-6)\left(8^2+8\bullet 6+6^2\right)\\ &= 2(64+48+36)\\ &= 2(148)\\ &= 296\end{aligned}$$

(c) Using the difference of squares factoring, we have:

$$1000^2-999^2=(1000+999)(1000-999)=(1999)(1)=1999$$

69. Factoring the sum of cubes and then grouping, we have:

$$\begin{aligned}A^3+B^3+3AB(A+B) &= (A+B)\left(A^2-AB+B^2\right)+(A+B)(3AB)\\ &= (A+B)\left(A^2-AB+B^2+3AB\right)\\ &= (A+B)\left(A^2+2AB+B^2\right)\\ &= (A+B)(A+B)^2\\ &= (A+B)^3\end{aligned}$$

71. Using the common factor $x\left(a^2+x^2\right)^{-3/2}$, we have:

$$\begin{aligned}2x\left(a^2+x^2\right)^{-1/2}-x^3\left(a^2+x^2\right)^{-3/2}&=x\left(a^2+x^2\right)^{-3/2}\left[2\left(a^2+x^2\right)-x^2\right]\\&=x\left(a^2+x^2\right)^{-3/2}\left(2a^2+2x^2-x^2\right)\\&=x\left(a^2+x^2\right)^{-3/2}\left(2a^2+x^2\right)\end{aligned}$$

73. Using grouping (first with fourth term, second with third term), we have:

$$\begin{aligned}y^4-p^2q^2-(p+q)y^3+pq(p+q)y&=\left(y^2+pq\right)\left(y^2-pq\right)-(p+q)(y)\left(y^2-pq\right)\\&=\left(y^2-pq\right)\left(y^2+pq-py-qy\right)\end{aligned}$$

75. We first multiply out some terms, then factor by grouping:

$$\begin{aligned}&(1+t)^2\left(1+u^2\right)-(1+u)^2\left(1+t^2\right)\\&\quad=\left[(1+t)^2-(1+u)^2\right]+\left[u^2(1+t)^2-t^2(1+u)^2\right]\\&\quad=(1+t+1+u)(1+t-1-u)+(u+ut+t+ut)(u+ut-t-ut)\\&\quad=(-2-u-t)(u-t)+(u+t+2ut)(u-t)\\&\quad=(u-t)(-2-u-t+u+t+2ut)\\&\quad=(u-t)(2ut-2)\\&\quad=2(u-t)(ut-1)\end{aligned}$$

77. Adding and subtracting $16x^2$, we have:

$$\begin{aligned}x^4+64&=x^4+16x^2+64-16x^2\\&=\left(x^2+8\right)^2-16x^2\\&=\left(x^2+8+4x\right)\left(x^2+8-4x\right)\\&=\left(x^2+4x+8\right)\left(x^2-4x+8\right)\end{aligned}$$

79. We factor as follows:

$$\begin{aligned}&(x+y)^2+(x+z)^2-(z+t)^2-(y+t)^2\\&\quad=\left[(x+y)^2-(y+t)^2\right]+\left[(x+z)^2-(z+t)^2\right]\\&\quad=\left[x+y-(y+t)\right]\left[x+y+(y+t)\right]+\left[x+z-(z+t)\right]\left[x+z+(z+t)\right]\\&\quad=(x-t)(x+2y+t)+(x-t)(x+2z+t)\\&\quad=(x-t)(x+2y+t+x+2z+t)\\&\quad=(x-t)(2x+2y+2z+2t)\\&\quad=2(x-t)(x+y+z+t)\end{aligned}$$

81. We factor as follows:

$$
\begin{aligned}
&(b-c)^3+(c-a)^3+(a-b)^3\\
&\quad = b^3-3b^2c+3bc^2-c^3+c^3-3c^2a+3ca^2-a^3+a^3-3a^2b+3ab^2-b^3\\
&\quad = -3b^2c+3bc^2-3c^2a+3ca^2-3a^2b+3ab^2\\
&\quad = \left(3bc^2-3c^2a\right)+\left(3ca^2-3b^2c\right)+\left(3ab^2-3a^2b\right)\\
&\quad = 3c^2(b-a)+3c\left(a^2-b^2\right)+3ab(b-a)\\
&\quad = 3c^2(b-a)-3c\left(b^2-a^2\right)+3ab(b-a)\\
&\quad = 3(b-a)\left[c^2-c(b+a)+ab\right]\\
&\quad = 3(b-a)\left[c^2-cb-ca+ab\right]\\
&\quad = 3(b-a)\left[c(c-b)-a(c-b)\right]\\
&\quad = 3(b-a)\left[(c-b)(c-a)\right]\\
&\quad = 3(b-a)(c-a)(c-b)
\end{aligned}
$$

1.10 Fractional Expressions

1. We factor and reduce:

$$\frac{x^2-9}{x+3}=\frac{(x-3)(x+3)}{x+3}=x-3$$

3. We factor and reduce:

$$\frac{x+2}{x^4-16}=\frac{x+2}{\left(x^2-4\right)\left(x^2+4\right)}=\frac{x+2}{(x-2)(x+2)\left(x^2+4\right)}=\frac{1}{(x-2)\left(x^2+4\right)}$$

5. We factor and reduce:

$$\frac{x^2+2x+4}{x^3-8}=\frac{x^2+2x+4}{(x-2)\left(x^2+2x+4\right)}=\frac{1}{x-2}$$

7. We factor and reduce:

$$\frac{9ab-12b^2}{6a^2-8ab}=\frac{3b(3a-4b)}{2a(3a-4b)}=\frac{3b}{2a}$$

9. We factor and reduce:

$$\frac{a^3+a^2+a+1}{a^2-1}=\frac{a^2(a+1)+(a+1)}{(a-1)(a+1)}=\frac{(a+1)\left(a^2+1\right)}{(a-1)(a+1)}=\frac{a^2+1}{a-1}$$

11. We factor and reduce:

$$\frac{x^3-y^3}{(x-y)^3}=\frac{(x-y)\left(x^2+xy+y^2\right)}{(x-y)^3}=\frac{x^2+xy+y^2}{(x-y)^2}$$

13. Multiplying, we have:

$$\frac{2}{x-2} \bullet \frac{x^2-4}{x+2} = \frac{2(x-2)(x+2)}{(x-2)(x+2)} = 2$$

15. Multiplying, we have:

$$\frac{x^2-x-2}{x^2+x-12} \bullet \frac{x^2-3x}{x^2-4x+4} = \frac{(x-2)(x+1)}{(x+4)(x-3)} \bullet \frac{x(x-3)}{(x-2)^2} = \frac{x(x+1)}{(x+4)(x-2)} = \frac{x^2+x}{(x+4)(x-2)}$$

17. Dividing, we have:

$$\frac{x^3+y^3}{x^2-4xy+3y^2} \div \frac{(x+y)^3}{x^2-2xy-3y^2} = \frac{(x+y)(x^2-xy+y^2)}{(x-3y)(x-y)} \bullet \frac{(x+y)(x-3y)}{(x+y)^3} = \frac{x^2-xy+y^2}{(x-y)(x+y)}$$

19. Multiplying, we have:

$$\frac{x^2+xy-2y^2}{x^2-5xy+4y^2} \bullet \frac{x^2-7xy+12y^2}{x^2+5xy+6y^2} = \frac{(x+2y)(x-y)}{(x-4y)(x-y)} \bullet \frac{(x-4y)(x-3y)}{(x+3y)(x+2y)} = \frac{x-3y}{x+3y}$$

21. Obtaining common denominators and subtracting, we have:

$$\frac{4}{x} - \frac{2}{x^2} = \frac{4x}{x^2} - \frac{2}{x^2} = \frac{4x-2}{x^2}$$

23. Obtaining common denominators and subtracting, we have:

$$\frac{6}{a} - \frac{a}{6} = \frac{36}{6a} - \frac{a^2}{6a} = \frac{36-a^2}{6a}$$

25. Obtaining common denominators and adding, we have:

$$\frac{1}{x+3} + \frac{3}{x+2} = \frac{x+2}{(x+3)(x+2)} + \frac{3(x+3)}{(x+3)(x+2)} = \frac{x+2+3(x+3)}{(x+3)(x+2)} = \frac{4x+11}{(x+3)(x+2)}$$

27. Obtaining common denominators and subtracting, we have:

$$\frac{3x}{x-2} - \frac{6}{x^2-4} = \frac{3x}{x-2} - \frac{6}{(x-2)(x+2)} = \frac{3x(x+2)-6}{(x-2)(x+2)} = \frac{3x^2+6x-6}{(x-2)(x+2)}$$

29. Obtaining common denominators and adding, we have:

$$\frac{a}{x-1}+\frac{2ax}{(x-1)^2}+\frac{3ax^2}{(x-1)^3}=\frac{a(x-1)^2+2ax(x-1)+3ax^2}{(x-1)^3}$$
$$=\frac{ax^2-2ax+a+2ax^2-2ax+3ax^2}{(x-1)^3}$$
$$=\frac{6ax^2-4ax+a}{(x-1)^3}$$

31. Obtaining common denominators and adding, we have:

$$\frac{x}{x^2-9}+\frac{x-1}{x^2-5x+6}=\frac{x}{(x-3)(x+3)}+\frac{x-1}{(x-3)(x-2)}$$
$$=\frac{x(x-2)+(x-1)(x+3)}{(x-3)(x+3)(x-2)}$$
$$=\frac{x^2-2x+x^2+2x-3}{(x-3)(x+3)(x-2)}$$
$$=\frac{2x^2-3}{(x-3)(x+3)(x-2)}$$

33. Factoring out –1, we have:

$$\frac{4}{x-5}-\frac{4}{5-x}=\frac{4}{x-5}+\frac{4}{x-5}=\frac{8}{x-5}$$

35. Obtaining common denominators and combining, we have:

$$\frac{a^2+b^2}{a^2-b^2}+\frac{a}{a+b}+\frac{b}{b-a}=\frac{a^2+b^2}{(a-b)(a+b)}+\frac{a}{a+b}-\frac{b}{a-b}$$
$$=\frac{a^2+b^2+a(a-b)-b(a+b)}{(a-b)(a+b)}$$
$$=\frac{a^2+b^2+a^2-ab-ab-b^2}{(a-b)(a+b)}$$
$$=\frac{2a^2-2ab}{(a-b)(a+b)}$$
$$=\frac{2a(a-b)}{(a-b)(a+b)}$$
$$=\frac{2a}{a+b}$$

37. Obtaining common denominators and subtracting, we have:

$$\frac{1}{x^2+x-20}-\frac{1}{x^2-8x+16}=\frac{1}{(x+5)(x-4)}-\frac{1}{(x-4)^2}$$
$$=\frac{x-4-(x+5)}{(x+5)(x-4)^2}$$
$$=\frac{-9}{(x+5)(x-4)^2}$$

39. Obtaining common denominators and subtracting, we have:

$$\begin{aligned}\frac{2q+p}{2p^2-9pq-5q^2}-\frac{p+q}{p^2-5pq}&=\frac{2q+p}{(2p+q)(p-5q)}-\frac{p+q}{p(p-5q)}\\&=\frac{(2q+p)p-(p+q)(2p+q)}{p(2p+q)(p-5q)}\\&=\frac{2pq+p^2-2p^2-3pq-q^2}{p(2p+q)(p-5q)}\\&=\frac{-p^2-pq-q^2}{p(2p+q)(p-5q)}\end{aligned}$$

41. Obtaining common denominators and combining, we have:

$$\begin{aligned}&\frac{x}{(x-y)(x-z)}+\frac{y}{(y-z)(y-x)}+\frac{z}{(z-x)(z-y)}\\&=\frac{x}{(x-y)(x-z)}-\frac{y}{(y-z)(x-y)}+\frac{z}{(x-z)(y-z)}\\&=\frac{x(y-z)-y(x-z)+z(x-y)}{(x-y)(x-z)(y-z)}\\&=\frac{xy-xz-xy+yz+xz-yz}{(x-y)(x-z)(y-z)}\\&=\frac{0}{(x-y)(x-z)(y-z)}\\&=0\end{aligned}$$

43. Obtaining common denominators and combining, we have:

$$\begin{aligned}&\frac{y+z}{x^2-xy-xz+yz}-\frac{x+z}{xy-xz-y^2+yz}+\frac{x+y}{xy-yz-xz+z^2}\\&=\frac{y+z}{x(x-y)-z(x-y)}-\frac{x+z}{x(y-z)-y(y-z)}+\frac{x+y}{y(x-z)-z(x-z)}\\&=\frac{y+z}{(x-y)(x-z)}-\frac{x+z}{(y-z)(x-y)}+\frac{x+y}{(x-z)(y-z)}\\&=\frac{(y+z)(y-z)-(x+z)(x-z)+(x+y)(x-y)}{(x-y)(x-z)(y-z)}\\&=\frac{y^2-z^2-x^2+z^2+x^2-y^2}{(x-y)(x-z)(y-z)}\\&=\frac{0}{(x-y)(x-z)(y-z)}\\&=0\end{aligned}$$

45. Obtaining common denominators and combining, we have:

$$\begin{aligned}\frac{x^2+x-a-1}{a^2-x^2}+\frac{x+1}{x+a}\div\frac{x-a}{x-1}&=\frac{x^2+x-a-1}{a^2-x^2}+\frac{x+1}{x+a}\bullet\frac{x-1}{x-a}\\&=\frac{x^2+x-a-1}{-(x^2-a^2)}+\frac{x^2-1}{x^2-a^2}\\&=\frac{-x^2-x+a+1+x^2-1}{x^2-a^2}\\&=\frac{-x+a}{(x+a)(x-a)}\\&=\frac{-(x-a)}{(x+a)(x-a)}\\&=\frac{-1}{x+a}\end{aligned}$$

47. Multiplying by $\frac{x}{x}$, we have:

$$\frac{\frac{1}{x}+1}{\frac{1}{x}-1}\bullet\frac{x}{x}=\frac{1+x}{1-x}$$

49. Multiplying by $\frac{xa}{xa}$, we have:

$$\frac{\frac{1}{x}-\frac{1}{a}}{x-a}\bullet\frac{xa}{xa}=\frac{a-x}{xa(x-a)}=-\frac{1}{ax}$$

51. Multiplying by $\frac{x}{x}$, we have:

$$\frac{1+\frac{4}{x}}{\frac{3}{x}-2}\bullet\frac{x}{x}=\frac{x+4}{3-2x}$$

53. Multiplying by $\frac{a}{a}$, we have:

$$\frac{a-\frac{1}{a}}{1+\frac{1}{a}}\bullet\frac{a}{a}=\frac{a^2-1}{a+1}=\frac{(a+1)(a-1)}{a+1}=a-1$$

55. Multiplying by $\frac{2(2+h)}{2(2+h)}$, we have:

$$\frac{\frac{1}{2+h}-\frac{1}{2}}{h}\bullet\frac{2(2+h)}{2(2+h)}=\frac{2-2-h}{2h(2+h)}=\frac{-h}{2h(2+h)}=-\frac{1}{4+2h}$$

57. Multiplying by $\frac{a^2x^2}{a^2x^2}$, we have:

$$\frac{\frac{a}{x^2}+\frac{x}{a^2}}{a^2-ax+x^2}\bullet\frac{a^2x^2}{a^2x^2}=\frac{a^3+x^3}{a^2x^2\left(a^2-ax+x^2\right)}=\frac{(a+x)\left(a^2-ax+x^2\right)}{a^2x^2\left(a^2-ax+x^2\right)}=\frac{a+x}{a^2x^2}$$

59. Rewriting the expression without negative exponents:

$$\left(x^{-1}+2\right)^{-1}=\frac{1}{x^{-1}+2}=\frac{1}{\frac{1}{x}+2}\bullet\frac{x}{x}=\frac{x}{1+2x}$$

61. Rewriting the expression without negative exponents:

$$\left(\frac{1}{a^{-1}}+\frac{1}{a^{-2}}\right)^{-1}=\left(a+a^2\right)^{-1}=\frac{1}{a+a^2}$$

63. Rewriting the expression without negative exponents:

$$x(x+y)^{-1}+y(x-y)^{-1}=\frac{x}{x+y}+\frac{y}{x-y}=\frac{x(x-y)+y(x+y)}{x^2-y^2}=\frac{x^2+y^2}{x^2-y^2}$$

65. Rewriting the expression without negative exponents:

$$\begin{aligned}\left(t^2+t^{-2}+2\right)\left(t+t^{-1}\right)^{-1}&=\frac{t^2+t^{-2}+2}{t+t^{-1}}\\&=\frac{t^2+\frac{1}{t^2}+2}{t+\frac{1}{t}}\bullet\frac{t^2}{t^2}\\&=\frac{t^4+1+2t^2}{t^3+t}\\&=\frac{\left(t^2+1\right)^2}{t\left(t^2+1\right)}\\&=\frac{t^2+1}{t}\end{aligned}$$

67. Combining fractions and simplifying, we have:

$$\begin{aligned}&\frac{x+y}{2\left(x^2+y^2\right)}-\frac{1}{2(x+y)}+\frac{x-y}{x^2-y^2}-\frac{x^3-y^3}{x^4-y^4}\\&=\frac{x+y}{2\left(x^2+y^2\right)}-\frac{1}{2(x+y)}+\frac{1}{x+y}-\frac{(x-y)\left(x^2+xy+y^2\right)}{(x-y)(x+y)\left(x^2+y^2\right)}\\&=\frac{x+y}{2\left(x^2+y^2\right)}-\frac{1}{2(x+y)}+\frac{1}{x+y}-\frac{x^2+xy+y^2}{(x+y)\left(x^2+y^2\right)}\\&=\frac{(x+y)(x+y)-\left(x^2+y^2\right)+2\left(x^2+y^2\right)-2\left(x^2+xy+y^2\right)}{2\left(x^2+y^2\right)(x+y)}\\&=\frac{x^2+2xy+y^2-x^2-y^2+2x^2+2y^2-2x^2-2xy-2y^2}{2\left(x^2+y^2\right)(x+y)}\\&=\frac{0}{2\left(x^2+y^2\right)(x+y)}\\&=0\end{aligned}$$

69. Multiplying by $\frac{(a-b)(a+b)}{(a-b)(a+b)}$, we have:

$$\frac{(a-b)(a+b)}{(a-b)(a+b)} \bullet \frac{\frac{a+b}{a-b}+\frac{a-b}{a+b}}{\frac{a-b}{a+b}-\frac{a+b}{a-b}} \bullet \frac{ab\left(b^2-a^2\right)}{a^2+b^2} = \frac{(a+b)^2+(a-b)^2}{(a-b)^2-(a+b)^2} \bullet \frac{ab(b-a)(b+a)}{a^2+b^2}$$

$$= \frac{2a^2+2b^2}{-4ab} \bullet \frac{ab(b-a)(b+a)}{a^2+b^2}$$

$$= \frac{2\left(a^2+b^2\right)}{-4} \bullet \frac{(b-a)(b+a)}{a^2+b^2}$$

$$= \frac{(b-a)(b+a)}{-2}$$

$$= \frac{b^2-a^2}{-2}$$

$$= \frac{a^2-b^2}{2}$$

71. Rewriting the expression without negative exponents:

$$\left[x-\left(x+x^{-1}\right)^{-1}\right]^{-1}-\left[x+\left(x-x^{-1}\right)^{-1}\right]^{-1}$$

$$= \frac{1}{x-\left(x+x^{-1}\right)^{-1}} - \frac{1}{x+\left(x-x^{-1}\right)^{-1}}$$

$$= \frac{1}{x-\frac{1}{x+x^{-1}}} - \frac{1}{x+\frac{1}{x-x^{-1}}}$$

$$= \frac{1}{x-\frac{1}{x+x^{-1}}} \bullet \frac{x+x^{-1}}{x+x^{-1}} - \frac{1}{x+\frac{1}{x-x^{-1}}} \bullet \frac{x-x^{-1}}{x-x^{-1}}$$

$$= \frac{x+x^{-1}}{x^2+1-1} - \frac{x-x^{-1}}{x^2-1+1}$$

$$= \frac{x+\frac{1}{x}}{x^2} - \frac{x-\frac{1}{x}}{x^2}$$

$$= \frac{x^2+1}{x^3} - \frac{x^2-1}{x^3}$$

$$= \frac{x^2+1-x^2+1}{x^3}$$

$$= \frac{2}{x^3}$$

73. Combining fractions and simplifying, we have:

$$\begin{aligned}
&\frac{ap+q}{ax-bx-a^2+ab}+\frac{bp+q}{bx-ax-b^2+ab}\\
&=\frac{ap+q}{x(a-b)-a(a-b)}+\frac{bp+q}{x(b-a)-b(b-a)}\\
&=\frac{ap+q}{(a-b)(x-a)}+\frac{bp+q}{(b-a)(x-b)}\\
&=\frac{ap+q}{(a-b)(x-a)}-\frac{bp+q}{(a-b)(x-b)}\\
&=\frac{(ap+q)(x-b)-(bp+q)(x-a)}{(a-b)(x-a)(x-b)}\\
&=\frac{apx+qx-abp-bq-bpx-qx+apb+aq}{(a-b)(x-a)(x-b)}\\
&=\frac{apx-bq-bpx+aq}{(a-b)(x-a)(x-b)}\\
&=\frac{a(px+q)-b(px+q)}{(a-b)(x-a)(x-b)}\\
&=\frac{(px+q)(a-b)}{(a-b)(x-a)(x-b)}\\
&=\frac{px+q}{(x-a)(x-b)}
\end{aligned}$$

75. Simplifying the complex fractions:

$$\begin{aligned}
&\frac{\frac{1}{a}-\frac{a-x}{a^2+x^2}}{\frac{1}{x}-\frac{x-a}{x^2+a^2}}+\frac{\frac{1}{a}-\frac{a+x}{a^2+x^2}}{\frac{1}{x}-\frac{x+a}{x^2+a^2}}=\frac{ax\left(a^2+x^2\right)}{ax\left(a^2+x^2\right)}\bullet\frac{\frac{1}{a}-\frac{a-x}{a^2+x^2}}{\frac{1}{x}-\frac{x-a}{x^2+a^2}}+\frac{ax\left(a^2+x^2\right)}{ax\left(a^2+x^2\right)}\bullet\frac{\frac{1}{a}-\frac{a+x}{a^2+x^2}}{\frac{1}{x}-\frac{x+a}{x^2+a^2}}\\
&=\frac{x\left(a^2+x^2\right)-(a-x)ax}{a\left(a^2+x^2\right)-(x-a)ax}+\frac{x\left(a^2+x^2\right)-(a+x)ax}{a\left(a^2+x^2\right)-(x+a)ax}\\
&=\frac{a^2x+x^3-a^2x+ax^2}{a^3+ax^2-ax^2+a^2x}+\frac{a^2x+x^3-a^2x-ax^2}{a^3+ax^2-ax^2-a^2x}\\
&=\frac{x^3+ax^2}{a^3+a^2x}+\frac{x^3-ax^2}{a^3-a^2x}\\
&=\frac{x^2(x+a)}{a^2(a+x)}+\frac{x^2(x-a)}{a^2(a-x)}\\
&=\frac{x^2}{a^2}+\frac{x^2}{a^2}(-1)\\
&=0
\end{aligned}$$

77. Combining fractions, we have:

$$\frac{x^2-qr}{(p-q)(p-r)}+\frac{x^2-rp}{(q-r)(q-p)}+\frac{x^2-pq}{(r-p)(r-q)}$$

$$=\frac{x^2-qr}{(p-q)(p-r)}-\frac{x^2-rp}{(q-r)(p-q)}+\frac{x^2-pq}{(p-r)(q-r)}$$

$$=\frac{(x^2-qr)(q-r)-(x^2-rp)(p-r)+(x^2-pq)(p-q)}{(p-q)(p-r)(q-r)}$$

$$=\frac{qx^2-q^2r-rx^2+qr^2-px^2+p^2r+rx^2-pr^2+px^2-p^2q-qx^2+pq^2}{(p-q)(p-r)(q-r)}$$

$$=\frac{-q^2r+qr^2+p^2r-pr^2-p^2q+pq^2}{(p-q)(p-r)(q-r)}$$

$$=\frac{(p^2r-q^2r)-(pr^2-qr^2)-(p^2q-pq^2)}{(p-q)(p-r)(q-r)}$$

$$=\frac{r(p-q)(p+q)-r^2(p-q)-pq(p-q)}{(p-q)(p-r)(q-r)}$$

$$=\frac{(p-q)[r(p+q)-r^2-pq]}{(p-q)(p-r)(q-r)}$$

$$=\frac{rp+rq-r^2-pq}{(p-r)(q-r)}$$

$$=\frac{p(r-q)+r(q-r)}{(p-r)(q-r)}$$

$$=\frac{-p(q-r)+r(q-r)}{(p-r)(q-r)}$$

$$=\frac{(q-r)(-p+r)}{(p-r)(q-r)}$$

$$=-1$$

79. (a) The sum and the product both equal $\frac{1}{42}$.

(b) After adding the three fractions and combining the many like terms appearing in the resulting numerator, we obtain:

$$\begin{aligned}
\frac{b-c}{1+bc}+\frac{c-a}{1+ca}+\frac{a-b}{1+ab} &= \frac{ab^2-ac^2+bc^2-a^2b+a^2c-b^2c}{(1+bc)(1+ca)(1+ab)} \\
&= \frac{\left(ab^2-b^2c\right)+\left(a^2c-ac^2\right)+\left(bc^2-a^2b\right)}{(1+bc)(1+ca)(1+ab)} \\
&= \frac{b^2(a-c)+ac(a-c)+b\left(c^2-a^2\right)}{(1+bc)(1+ca)(1+ab)} \\
&= \frac{b^2(a-c)+ac(a-c)-b(a-c)(a+c)}{(1+bc)(1+ca)(1+ab)} \\
&= \frac{(a-c)\left(b^2+ac-b(a+c)\right)}{(1+bc)(1+ca)(1+ab)} \\
&= \frac{(a-c)\left(b^2+ac-ab-bc\right)}{(1+bc)(1+ca)(1+ab)} \\
&= \frac{(a-c)\left(c(a-b)-b(a-b)\right)}{(1+bc)(1+ca)(1+ab)} \\
&= \frac{(a-c)\left((a-b)(c-b)\right)}{(1+bc)(1+ca)(1+ab)} \\
&= \frac{(c-a)(a-b)(b-c)}{(1+bc)(1+ca)(1+ab)} \\
&= \frac{b-c}{1+bc}\bullet\frac{c-a}{1+ca}\bullet\frac{a-b}{1+ab}
\end{aligned}$$

81. For convenience, let $z=x-y$. Then $-z=y-x$, and we have:

$$\begin{aligned}
\left(1+a^z\right)^{-1}+\left(1+a^{-z}\right)^{-1} &= \frac{1}{1+a^z}+\frac{1}{1+\frac{1}{a^z}} \\
&= \frac{1}{1+a^z}+\frac{a^z}{a^z}\bullet\frac{1}{1+\frac{1}{a^z}} \\
&= \frac{1}{1+a^z}+\frac{a^z}{a^z+1} \\
&= \frac{1+a^z}{1+a^z} \\
&= 1
\end{aligned}$$

1.11 The Complex Number System

1. We complete the table:

i^2	i^3	i^4	i^5	i^6	i^7	i^8
-1	$-i$	1	i	-1	$-i$	1

3. (a) The real part is 4 and the imaginary part is 5.
(b) The real part is 4 and the imaginary part is –5.

(c) The real part is $\frac{1}{2}$ and the imaginary part is –1.
(d) The real part is 0 and the imaginary part is 16.

5. Equating the real parts gives $2c = 8$, and therefore $c = 4$. Similarly, equating the imaginary parts yields $d = -3$.

7. (a) $(5-6i)+(9+2i) = (5+9)+(-6+2)i = 14-4i$
(b) $(5-6i)-(9+2i) = (5-9)+(-6-2)i = -4-8i$

9. (a) $(3-4i)(5+i) = 15-17i-4i^2 = 19-17i$

(b) $(5+i)(3-4i) = 19-17i$, from part (a)

(c) We compute the quotient:
$$\frac{3-4i}{5+i}\bullet\frac{5-i}{5-i}=\frac{15-23i+4i^2}{25-i^2}=\frac{11-23i}{26}=\frac{11}{26}-\frac{23}{26}i$$

(d) We compute the quotient:
$$\frac{5+i}{3-4i}\bullet\frac{3+4i}{3+4i}=\frac{15+23i+4i^2}{9-16i^2}=\frac{11+23i}{25}=\frac{11}{25}+\frac{23}{25}i$$

11. (a) $z+w=(2+3i)+(9-4i)=11-i$

(b) $\bar{z}+w=(2-3i)+(9-4i)=11-7i$

(c) $z+\bar{z}=(2+3i)+(2-3i)=4$

13. We evaluate the expression:
$$(z+w)+w_1=[(2+3i)+(9-4i)]+(-7-i)=(11-i)+(-7-i)=4-2i$$

15. We compute the product:
$$zw=(2+3i)(9-4i)=18+19i-12i^2=30+19i$$

17. $z\bar{z}=(2+3i)(2-3i)=4-9i^2=13$

19. We first compute the product:

$$ww_1 = (9-4i)(-7-i) = -63+19i+4i^2 = -67+19i$$

Therefore:

$$z(ww_1) = (2+3i)(-67+19i) = -134-163i+57i^2 = -191-163i$$

21. We first compute the sum:

$$w+w_1 = (9-4i)+(-7-i) = 2-5i$$

Therefore:

$$z(w+w_1) = (2+3i)(2-5i) = 4-4i-15i^2 = 19-4i$$

23. We first compute the powers:

$$z^2 = (2+3i)(2+3i) = 4+12i+9i^2 = -5+12i$$

$$w^2 = (9-4i)(9-4i) = 81-72i+16i^2 = 65-72i$$

Therefore:

$$z^2-w^2 = (-5+12i)-(65-72i) = -70+84i$$

25. Since $zw = 30 + 19i$ (from Exercise 15), then:

$$(zw)^2 = (30+19i)(30+19i) = 900+1140i+361i^2 = 539+1140i$$

27. Since $z^2 = -5 + 12i$ (from Exercise 23), then:

$$z^3 = z \bullet z^2 = (2+3i)(-5+12i) = -10+9i+36i^2 = -46+9i$$

29. We compute the quotient:

$$\frac{z}{w} = \frac{2+3i}{9-4i} \bullet \frac{9+4i}{9+4i} = \frac{18+35i+12i^2}{81-16i^2} = \frac{6+35i}{97} = \frac{6}{97}+\frac{35}{97}i$$

31. Using the values of $\bar{z}$ and $\bar{w}$, we have:

$$\frac{\bar{z}}{\bar{w}} = \frac{2-3i}{9+4i} \bullet \frac{9-4i}{9-4i} = \frac{18-35i+12i^2}{81-16i^2} = \frac{6-35i}{97} = \frac{6}{97}-\frac{35}{97}i$$

33. Using the value of $\bar{z}$, we have:

$$\frac{z}{\bar{z}} = \frac{2+3i}{2-3i} \bullet \frac{2+3i}{2+3i} = \frac{4+12i+9i^2}{4-9i^2} = \frac{-5+12i}{13} = -\frac{5}{13}+\frac{12}{13}i$$

35. Since $w-\bar{w} = (9-4i)-(9+4i) = -8i$, then:

$$\frac{w-\bar{w}}{2i} = \frac{-8i}{2i} = -4$$

37. We compute the quotient:

$$\frac{i}{5+i} \bullet \frac{5-i}{5-i} = \frac{5i-i^2}{25-i^2} = \frac{1+5i}{26} = \frac{1}{26}+\frac{5}{26}i$$

39. We compute the quotient:

$$\frac{1}{i}\bullet\frac{i}{i}=\frac{i}{i^2}=\frac{i}{-1}=-i$$

41. By writing the numbers in complex form, we have:

$$\sqrt{-49}+\sqrt{-9}+\sqrt{-4}=7i+3i+2i=12i$$

43. By writing the numbers in complex form, we have:

$$\begin{aligned}\sqrt{-20}-3\sqrt{-45}+\sqrt{-80}&=\sqrt{4}\sqrt{-5}-3\sqrt{9}\sqrt{-5}+\sqrt{16}\sqrt{-5}\\&=2\sqrt{5}i-9\sqrt{5}i+4\sqrt{5}i\\&=-3\sqrt{5}i\end{aligned}$$

45. By writing the numbers in complex form, we have:

$$1+\sqrt{-36}\sqrt{-36}=1+(6i)(6i)=1+36i^2=-35$$

47. By writing the numbers in complex form, we have:

$$\begin{aligned}3\sqrt{-128}-4\sqrt{-18}&=3\sqrt{-64}\sqrt{2}-4\sqrt{-9}\sqrt{2}\\&=3(8i)\sqrt{2}-4(3i)\sqrt{2}\\&=24\sqrt{2}i-12\sqrt{2}i\\&=12\sqrt{2}i\end{aligned}$$

49. (a) Notice that the result does agree with the definition:

$$z+w=(a+bi)+(c+di)=(a+c)+(b+d)i$$

(b) Notice that the result does agree with the definition:

$$z-w=(a+bi)-(c+di)=(a-c)+(b-d)i$$

(c) Notice that the result does agree with the definition:

$$zw=(a+bi)(c+di)=ac+bci+adi+bdi^2=(ac-bd)+(bc+ad)i$$

(d) Notice that the result does agree with the definition:

$$\frac{z}{w}=\frac{a+bi}{c+di}\bullet\frac{c-di}{c-di}=\frac{ac+bci-adi-bdi^2}{c^2-d^2i^2}=\frac{ac+bd}{c^2+d^2}+\frac{bc-ad}{c^2+d^2}i$$

51. (a) We compute each power:

$$\begin{aligned}z^3&=\left(\frac{-1+\sqrt{3}i}{2}\right)^3\\&=\frac{(-1+\sqrt{3}i)^3}{8}\\&=\frac{(-1+\sqrt{3}i)(-1+\sqrt{3}i)^2}{8}\\&=\frac{(-1+\sqrt{3}i)(1-2\sqrt{3}i+3i^2)}{8}\\&=\frac{(-1+\sqrt{3}i)(-2-2\sqrt{3}i)}{8}\\&=\frac{2-2\sqrt{3}i+2\sqrt{3}i-6i^2}{8}\\&=\frac{8}{8}\\&=1\end{aligned}$$

$$\begin{aligned}w^3&=\left(\frac{-1-\sqrt{3}i}{2}\right)^3\\&=\frac{(-1-\sqrt{3}i)^3}{8}\\&=\frac{(-1-\sqrt{3}i)(-1-\sqrt{3}i)^2}{8}\\&=\frac{(-1-\sqrt{3}i)(1+2\sqrt{3}i+3i^2)}{8}\\&=\frac{(-1-\sqrt{3}i)(-2+2\sqrt{3}i)}{8}\\&=\frac{2+2\sqrt{3}i-2\sqrt{3}i-6i^2}{8}\\&=\frac{8}{8}\\&=1\end{aligned}$$

(b) We compute the product:

$$zw = \left(\frac{-1+\sqrt{3}i}{2}\right)\left(\frac{-1-\sqrt{3}i}{2}\right) = \frac{1-\sqrt{3}i+\sqrt{3}i-3i^2}{4} = \frac{4}{4} = 1$$

(c) We compute each power:

$$w^2 = \left(\frac{-1-\sqrt{3}i}{2}\right)^2 = \frac{1+2\sqrt{3}i+3i^2}{4} = \frac{-2+2\sqrt{3}i}{4} = \frac{-1+\sqrt{3}i}{2} = z$$

$$z^2 = \left(\frac{-1+\sqrt{3}i}{2}\right)^2 = \frac{1-2\sqrt{3}i+3i^2}{4} = \frac{-2-2\sqrt{3}i}{4} = \frac{-1-\sqrt{3}i}{2} = w$$

Note: Another approach (using parts (a) and (b)) is to recognize:

$$z^3 = zw, \text{ so } z^2 = \frac{z^3}{z} = \frac{zw}{z} = w$$

$$w^3 = zw, \text{ so } w^2 = \frac{w^3}{w} = \frac{zw}{w} = z$$

(d) We compute the product:

$$\begin{aligned}(1-z+z^2)(1+z-z^2) &= (1-z+w)(1+z-w)\\ &= \left(\frac{2}{2}-\frac{-1+\sqrt{3}i}{2}+\frac{-1-\sqrt{3}i}{2}\right)\left(\frac{2}{2}+\frac{-1+\sqrt{3}i}{2}-\frac{-1-\sqrt{3}i}{2}\right)\\ &= \left(\frac{2-2\sqrt{3}i}{2}\right)\left(\frac{2+2\sqrt{3}i}{2}\right)\\ &= (1-\sqrt{3}i)(1+\sqrt{3}i)\\ &= 1-3i^2\\ &= 4\end{aligned}$$

53. (a) Let $z = a + bi$, then:

$$0+z = (0+0i)+(a+bi) = a+bi = z$$

$$z+0 = (a+bi)+(0+0i) = a+bi = z$$

(b) Let $z = a + bi$, then:

$$0 \bullet z = (0+0i)(a+bi) = 0+0i = 0$$

$$z \bullet 0 = (a+bi)(0+0i) = 0$$

55. (a) We compute each sum:

$$z+w = (a+bi)+(c+di) = (a+c)+(b+d)i$$

$$w+z = (c+di)+(a+bi) = (c+a)+(d+b)i = z+w$$

(b) We compute each product:

$$zw = (a+bi)(c+di) = ac+bci+adi+bdi^2 = (ac-bd)+(bc+ad)i$$

$$wz = (c+di)(a+bi) = ac+adi+bci+bdi^2 = (ac-bd)+(bc+ad)i = zw$$

57. We compute each quotient then add the complex numbers:

$$\frac{a+bi}{a-bi}+\frac{a-bi}{a+bi}=\frac{a+bi}{a-bi}\bullet\frac{a+bi}{a+bi}+\frac{a-bi}{a+bi}\bullet\frac{a-bi}{a-bi}$$
$$=\frac{(a+bi)^2}{a^2+b^2}+\frac{(a-bi)^2}{a^2+b^2}$$
$$=\frac{a^2+2abi-b^2+a^2-2abi-b^2}{a^2+b^2}$$
$$=\frac{2a^2-2b^2}{a^2+b^2}$$

Thus the real part is $\frac{2a^2-2b^2}{a^2+b^2}$, and the imaginary part is 0.

59. We compute each quotient then add the complex numbers:

$$\frac{(a+bi)^2}{a-bi}-\frac{(a-bi)^2}{a+bi}=\frac{(a+bi)^3-(a-bi)^3}{(a-bi)(a+bi)}$$
$$=\frac{\left(a^3+3a^2bi+3ab^2i^2+b^3i^3\right)-\left(a^3-3a^2bi+3ab^2i^2-b^3i^3\right)}{a^2-b^2i^2}$$
$$=\frac{\left(a^3-3ab^2\right)+\left(3a^2b-b^3\right)i-\left(a^3-3ab^2\right)+\left(3a^2b-b^3\right)}{a^2+b^2}$$
$$=\frac{6a^2b-2b^3}{a^2+b^2}i$$

Thus the real part is 0.

61. We will first compute zw:

$$zw=(a+bi)(c+di)=ac+bci+adi+bdi^2=(ac-bd)+(bc+ad)i$$

Since $zw=0$, then $ac-bd=0$ and $bc+ad=0$. Now assuming $a\neq 0$, we can solve the first equation for c to obtain $c=\frac{bd}{a}$. Substituting into the second equation yields:

$$b\left(\tfrac{bd}{a}\right)+ad=0$$
$$b^2d+a^2d=0$$
$$\left(b^2+a^2\right)d=0$$

But if $a\neq 0$, then $b^2+a^2\neq 0$ and thus $d=0$. Since $c=\frac{bd}{a}$, then $c=0$. Thus $w=0$.

Chapter One Review Exercises

1. Factoring as a difference of squares:

$$a^2-16b^2=a^2-(4b)^2=(a-4b)(a+4b)$$

3. Factoring as a difference of cubes:

$$\begin{aligned} 8-(a+1)^3 &= [2-(a+1)]\left[4+2(a+1)+(a+1)^2\right] \\ &= (1-a)\left(4+2a+2+a^2+2a+1\right) \\ &= (1-a)\left(7+4a+a^2\right) \end{aligned}$$

5. Factoring the common factor x then using the special product for $(A+B)^2$:

$$a^2x^3+2ax^2b+b^2x = x\left(a^2x^2+2abx+b^2\right) = x(ax+b)^2$$

7. Factoring by trial and error:

$$8x^2+6x+1 = (4x+1)(2x+1)$$

9. Factoring the common factor a^4x^4 then as a difference of squares (twice):

$$\begin{aligned} a^4x^4-x^8a^8 &= a^4x^4\left(1-x^4a^4\right) \\ &= a^4x^4\left(1-x^2a^2\right)\left(1+x^2a^2\right) \\ &= a^4x^4(1-xa)(1+xa)\left(1+x^2a^2\right) \end{aligned}$$

11. Factoring by grouping:

$$\begin{aligned} 8+12a+6a^2+a^3 &= \left(8+a^3\right)+\left(12a+6a^2\right) \\ &= (2+a)\left(4-2a+a^2\right)+6a(2+a) \\ &= (2+a)\left(4-2a+a^2+6a\right) \\ &= (2+a)\left(4+4a+a^2\right) \\ &= (2+a)(2+a)^2 \\ &= (2+a)^3 \end{aligned}$$

13. Factoring the common factor z^3 then by trial and error:

$$4x^2y^2z^3-3xyz^3-z^3 = z^3\left(4x^2y^2-3xy-1\right) = z^3(4xy+1)(xy-1)$$

15. Factoring as a difference of squares then as a sum and difference of cubes:

$$\begin{aligned} 1-x^6 &= \left(1^3\right)^2-\left(x^3\right)^2 \\ &= \left(1-x^3\right)\left(1+x^3\right) \\ &= (1-x)\left(1+x+x^2\right)(1+x)\left(1-x+x^2\right) \end{aligned}$$

17. Factoring by grouping the first three terms:

$$a^2x^2+2abx+b^2-4a^2b^2x^2 = (ax+b)^2-(2abx)^2 = (ax+b-2abx)(ax+b+2abx)$$

19. Factoring by grouping pairs of terms:

$$\begin{aligned} a^2-b^2+ac-bc+a^2b-b^2a &= (a-b)(a+b)+c(a-b)+ab(a-b) \\ &= (a-b)(a+b+c+ab) \end{aligned}$$

21. $4^{3/2} = (\sqrt{4})^3 = 2^3 = 8$

23. $\left((3025)^{1/2}\right)^0 = 1$ since $x^0 = 1$ for any real number x

25. Evaluating the rational exponent:

$$8^{-4/3} = \left(\sqrt[3]{8}\right)^{-4} = 2^{-4} = \frac{1}{2^4} = \frac{1}{16}$$

27. Evaluating the rational exponent:

$$(-243)^{-2/5} = \left(\sqrt[5]{-243}\right)^{-2} = (-3)^{-2} = \frac{1}{(-3)^2} = \frac{1}{9}$$

29. Using properties of exponents:

$$\left(a^2b^6c^8\right)^{1/2} = a^{2/2}b^{6/2}c^{8/2} = ab^3c^4$$

31. Multiplying radicals, we have:

$$\sqrt{a^3b^5}\sqrt{4ab^3} = \sqrt{4a^4b^8} = 2a^2b^4$$

33. Factoring $16 = 8 \bullet 2$ and $-54 = -27 \bullet 2$, we have:

$$\sqrt[3]{16} - \sqrt[3]{-54} = \sqrt[3]{8}\,\sqrt[3]{2} - \sqrt[3]{-27}\,\sqrt[3]{2} = 2\sqrt[3]{2} - (-3)\sqrt[3]{2} = 5\sqrt[3]{2}$$

35. Simplifying radicals, we have:

$$\sqrt{24a^2b^3} + ba\sqrt{54b} = \sqrt{4a^2b^2}\sqrt{6b} + ba\sqrt{9}\sqrt{6b} = 2ab\sqrt{6b} + 3ab\sqrt{6b} = 5ab\sqrt{6b}$$

37. $\sqrt{t^2} = |t|$

39. $\sqrt[4]{16x^4} = |2x| = 2|x|$

41. (a) Using rational exponents, we have:

$$\sqrt[3]{x}\,\sqrt[4]{x^3} = x^{1/3}x^{3/4} = x^{4/12}x^{9/12} = x^{13/12}$$

(b) Writing as a radical:

$$x^{13/12} = \sqrt[12]{x^{13}} = x\sqrt[12]{x}$$

43. (a) Using rational exponents, we have:

$$\sqrt{\sqrt[3]{t}\,\sqrt[5]{t^4}} = \left(t^{1/3}t^{4/5}\right)^{1/2} = \left(t^{5/15}t^{12/15}\right)^{1/2} = \left(t^{17/15}\right)^{1/2} = t^{17/30}$$

(b) Writing as a radical:

$$t^{17/30} = \sqrt[30]{t^{17}}$$

45. Rationalizing denominators, we have:

$$\frac{6}{\sqrt{3}} \bullet \frac{\sqrt{3}}{\sqrt{3}} = \frac{6\sqrt{3}}{3} = 2\sqrt{3}$$

47. Rationalizing denominators, we have:

$$\frac{1}{\sqrt{6}-\sqrt{3}} \bullet \frac{\sqrt{6}+\sqrt{3}}{\sqrt{6}+\sqrt{3}} = \frac{\sqrt{6}+\sqrt{3}}{6-3} = \frac{\sqrt{6}+\sqrt{3}}{3}$$

49. Rationalizing denominators, we have:

$$\begin{aligned}\frac{\sqrt{a^2+x^2}+\sqrt{a^2-x^2}}{\sqrt{a^2+x^2}-\sqrt{a^2-x^2}} \bullet \frac{\sqrt{a^2+x^2}+\sqrt{a^2-x^2}}{\sqrt{a^2+x^2}+\sqrt{a^2-x^2}} &= \frac{\left(a^2+x^2\right)+2\sqrt{a^4-x^4}+\left(a^2-x^2\right)}{\left(a^2+x^2\right)-\left(a^2-x^2\right)} \\ &= \frac{2a^2+2\sqrt{a^4-x^4}}{2x^2} \\ &= \frac{a^2+\sqrt{a^4-x^4}}{x^2}\end{aligned}$$

51. Rationalizing denominators, we have:

$$\begin{aligned}\frac{1}{\sqrt{2}+\sqrt{3}+\sqrt{5}} \bullet \frac{\sqrt{2}+\sqrt{3}-\sqrt{5}}{\sqrt{2}+\sqrt{3}-\sqrt{5}} &= \frac{\sqrt{2}+\sqrt{3}-\sqrt{5}}{\left(\sqrt{2}+\sqrt{3}\right)^2-5} \\ &= \frac{\sqrt{2}+\sqrt{3}-\sqrt{5}}{2+2\sqrt{6}+3-5} \\ &= \frac{\sqrt{2}+\sqrt{3}-\sqrt{5}}{2\sqrt{6}} \bullet \frac{\sqrt{6}}{\sqrt{6}} \\ &= \frac{\sqrt{12}+\sqrt{18}-\sqrt{30}}{2(6)} \\ &= \frac{2\sqrt{3}+3\sqrt{2}-\sqrt{30}}{12}\end{aligned}$$

53. Rationalizing numerators, we have:

$$\frac{\sqrt{t}-a}{t-a^2} \bullet \frac{\sqrt{t}+a}{\sqrt{t}+a} = \frac{t-a^2}{\left(t-a^2\right)\left(\sqrt{t}+a\right)} = \frac{1}{\sqrt{t}+a}$$

55. Factoring the denominator, we have:

$$\frac{\sqrt[3]{x}-2}{x-8} = \frac{x^{1/3}-2}{\left(x^{1/3}\right)^3-2^3} = \frac{x^{1/3}-2}{\left(x^{1/3}-2\right)\left(x^{2/3}+2x^{1/3}+4\right)} = \frac{1}{\sqrt[3]{x^2}+2\sqrt[3]{x}+4}$$

57. Using the special product for $(A+B)^2$, we have:

$$(3x+1)^2 = (3x)^2 + 2(3x)(1) + (1)^2 = 9x^2 + 6x + 1$$

59. Using the special product for $(A+B)^2$, we have:

$$\left(3x^2+y^2\right)^2 = \left(3x^2\right)^2 + 2\left(3x^2\right)\left(y^2\right) + \left(y^2\right)^2 = 9x^4 + 6x^2y^2 + y^4$$

61. Using the special product for $(A+B)^3$, we have:

$$(2a+3)^3=(2a)^3+3(2a)^2(3)+3(2a)(3)^2+3^3=8a^3+36a^2+54a+27$$

63. Using the special product for $(A-B)(A^2+AB+B^2)$, we have:

$$(1-3a)(1+3a+9a^2)=1^3-(3a)^3=1-27a^3$$

65. Using the special product for $(A-B)(A+B)$, we have:

$$(x^{1/2}-y^{1/2})(x^{1/2}+y^{1/2})=(x^{1/2})^2-(y^{1/2})^2=x-y$$

67. Using the special product for $(A+B)^3$, we have:

$$\begin{aligned}(x^{1/3}+y^{1/3})^3&=(x^{1/3})^3+3(x^{1/3})^2(y^{1/3})+3(x^{1/3})(y^{1/3})^2+(y^{1/3})^3\\&=x+3x^{2/3}y^{1/3}+3x^{1/3}y^{2/3}+y\end{aligned}$$

69. Using the special product for $(A-B)^2$, we have:

$$\begin{aligned}(x^2-3x+1)^2&=[x^2-(3x-1)]^2\\&=x^4-2x^2(3x-1)+(3x-1)^2\\&=x^4-6x^3+2x^2+9x^2-6x+1\\&=x^4-6x^3+11x^2-6x+1\end{aligned}$$

71. Performing the indicated operations yields:

$$(3-2i)(3+2i)+(1+3i)^2=9-4i^2+1+6i+9i^2=9+4+1+6i-9=5+6i$$

73. Performing the indicated operations yields:

$$(1+i\sqrt{2})(1-i\sqrt{2})+(\sqrt{2}+i)(\sqrt{2}-i)=1-2i^2+2-i^2=1+2+2+1=6$$

75. Performing the indicated operations yields:

$$\frac{3-i\sqrt{3}}{3+i\sqrt{3}}\bullet\frac{3-i\sqrt{3}}{3-i\sqrt{3}}=\frac{9-6i\sqrt{3}-3}{9+3}=\frac{6-6i\sqrt{3}}{12}=\frac{1-i\sqrt{3}}{2}=\frac{1}{2}-\frac{\sqrt{3}}{2}i$$

77. Writing in complex form, we have:

$$-\sqrt{-2}\sqrt{-9}+\sqrt{-8}-\sqrt{-72}=-(i\sqrt{2})(3i)+2i\sqrt{2}-6i\sqrt{2}=3\sqrt{2}-4i\sqrt{2}=3\sqrt{2}-4\sqrt{2}i$$

79. Let $z=a+bi$, so $\overline{z}=a-bi$. Therefore:

$$\frac{z+\overline{z}}{2}=\frac{a+bi+a-bi}{2}=\frac{2a}{2}=a=\text{Re}(z)$$

81. (a) We compute the absolute values:

$$|6+2i|=\sqrt{6^2+2^2}=\sqrt{36+4}=\sqrt{40}=2\sqrt{10}$$
$$|6-2i|=\sqrt{6^2+(-2)^2}=\sqrt{36+4}=\sqrt{40}=2\sqrt{10}$$

(b) Using the new definition:

$$|-3| = |-3+0i| = \sqrt{(-3)^2+0^2} = \sqrt{9} = 3$$

(c) Let $z = a+bi$, so $\bar{z} = a-bi$. We now compute:

$$z\bar{z} = (a+bi)(a-bi) = a^2 - b^2i^2 = a^2 + b^2$$

$$|z|^2 = \left(\sqrt{a^2+b^2}\right)^2 = a^2 + b^2$$

Thus $z\bar{z} = |z|^2$.

83. distributive property (used twice), associative property of addition

85. distributive property

87. Using absolute value, the equality is $|x-6| = 2$.

89. Using absolute value, the equality is $|a-b| = 3$.

91. Using absolute value, the inequality is $|x-0| > 10$, or $|x| > 10$.

93. Since $\sqrt{6} - 2 > 0$, then:

$$|\sqrt{6}-2| = \sqrt{6}-2$$

95. Since $x^4 + x^2 + 1 > 0$, then:

$$|x^4+x^2+1| = x^4+x^2+1$$

97. (a) If $x < 2$, then $x - 2 < 0$ and $x - 3 < 0$, so:

$$|x-2|+|x-3| = -(x-2)-(x-3) = -2x+5$$

(b) If $2 < x < 3$, then $x - 2 > 0$ and $x - 3 < 0$, so:

$$|x-2|+|x-3| = x-2-(x-3) = x-2-x+3 = 1$$

(c) If $x > 3$, then $x - 2 > 0$ and $x - 3 > 0$, so:

$$|x-2|+|x-3| = x-2+x-3 = 2x-5$$

99. (a) true
(b) true
(c) true
(d) false (since $0.9 < 1$)

101. We sketch the interval $(3,5]$:

103. We sketch the interval $(-\infty, 4)$:

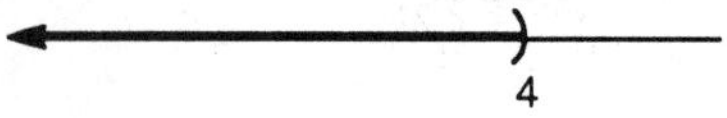

105. 1.4×10^{-3}

107. 1.2001×10^{1}

109. Simplifying parentheses, we obtain:

$$1-2[3-4(1-5)]=1-2[3-4(-4)]=1-2(19)=-37$$

111. Multiplying out parentheses, we have:

$$\begin{aligned}1+3\left(x^2-5x-4\right)-\left[1-\left(15x-3x^2\right)\right] &= 1+3x^2-15x-12-\left(1-15x+3x^2\right)\\ &= 3x^2-15x-11-1+15x-3x^2\\ &= -12\end{aligned}$$

113. Multiplying out parentheses, we have:

$$(x-1)(x+1)(x+2)=\left(x^2-1\right)(x+2)=x^3-x+2x^2-2=x^3+2x^2-x-2$$

115. Multiplying out parentheses, we have:

$$x^2+4-(x-1)(x-2)=x^2+4-\left(x^2-3x+2\right)=x^2+4-x^2+3x-2=3x+2$$

117. We use long division:

$$\begin{array}{r|l} & x^2-x \\ \hline x-1 & x^3-2x^2+x+4 \\ & \underline{x^3-1x^2} \\ & \quad -1x^2+1x \\ & \quad \underline{-1x^2+1x} \\ & \qquad\qquad 4 \end{array}$$

The quotient is x^2-x and the remainder is 4.

119. We use long division:

$$\begin{array}{r|l} & x^2-2 \\ \hline x^2+1 & x^4-\ x^2+1 \\ & \underline{x^4+1x^2} \\ & \quad -2x^2+1 \\ & \quad \underline{-2x^2-2} \\ & \qquad\qquad 3 \end{array}$$

The quotient is x^2-2 and the remainder is 3.

121. We sketch the interval $(3, 9)$:

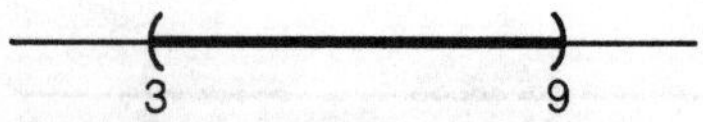

123. We sketch the intervals $(-\infty,-2]$ and $[0,\infty)$:

125. (a) We sketch the interval $(-1,9)$, excluding the point $x=4$:

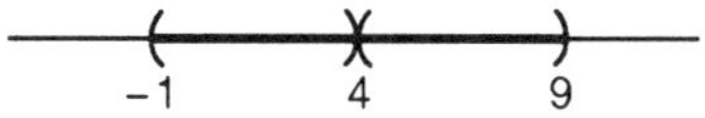

(b) We sketch the interval $(-1,9)$:

127. Let $x = 12.\overline{71}$, so $100x = 1271.\overline{71}$. Then $100x - x = 1259$, so $99x = 1259$, and thus $x = \frac{1259}{99}$.

129. We complete the table:

x	$1+\frac{x}{2}$	$\sqrt{1+x}$
0.1	1.05	1.048808
0.01	1.005	1.004987
0.001	1.0005	1.000499

131. (a) To six decimal places, its value is 0.447214.

(b) To six decimal places, its value is 0.447214.

(c) We rationalize the denominator:

$$\frac{\sqrt{3-\sqrt{5}}}{\sqrt{2}+\sqrt{7-3\sqrt{5}}}\bullet\frac{\sqrt{2}-\sqrt{7-3\sqrt{5}}}{\sqrt{2}-\sqrt{7-3\sqrt{5}}}=\frac{\sqrt{6-2\sqrt{5}}-\sqrt{(3-\sqrt{5})(7-3\sqrt{5})}}{2-(7-3\sqrt{5})}$$
$$=\frac{\sqrt{6-2\sqrt{5}}-\sqrt{21-16\sqrt{5}+15}}{-5+3\sqrt{5}}$$
$$=\frac{\sqrt{6-2\sqrt{5}}-\sqrt{36-16\sqrt{5}}}{-5+3\sqrt{5}}$$
$$=\frac{\sqrt{6-2\sqrt{5}}-2\sqrt{9-4\sqrt{5}}}{-5+3\sqrt{5}}$$

(d) We square each side:

$(\sqrt{5}-1)^2 = 5-2\sqrt{5}+1 = 6-2\sqrt{5}$, so $\sqrt{5}-1=\sqrt{6-2\sqrt{5}}$

$(\sqrt{5}-2)^2 = 5-4\sqrt{5}+4 = 9-4\sqrt{5}$, so $\sqrt{5}-2=\sqrt{9-4\sqrt{5}}$

(e) The expression from (c) becomes:

$$\frac{\sqrt{5}-1-2(\sqrt{5}-2)}{-5+3\sqrt{5}}=\frac{\sqrt{5}-1-2\sqrt{5}+4}{-5+3\sqrt{5}}$$
$$=\frac{3-\sqrt{5}}{-5+3\sqrt{5}}\bullet\frac{-5-3\sqrt{5}}{-5-3\sqrt{5}}$$
$$=\frac{-15-4\sqrt{5}+15}{25-45}$$
$$=\frac{-4\sqrt{5}}{-20}$$
$$=\frac{\sqrt{5}}{5}$$

133. Combining the fractions, we have:

$$\frac{1}{a-bi}-\frac{1}{a+bi}=\frac{(a+bi)-(a-bi)}{(a-bi)(a+bi)}=\frac{2bi}{a^2-b^2i^2}=\frac{2bi}{a^2+b^2}$$

135. Combining the fractions, we have:

$$\frac{a+bi}{a-bi}-\frac{a-bi}{a+bi}=\frac{(a+bi)^2-(a-bi)^2}{(a-bi)(a+bi)}$$
$$=\frac{a^2+2abi-b^2-a^2+2abi+b^2}{a^2+b^2}$$
$$=\frac{4abi}{a^2+b^2}$$

137. The LCM is $(a-b)(a-c)(b-c)$. We first build each fraction:

$$\frac{bc(x-a)^2}{(a-b)(a-c)}\bullet\frac{b-c}{b-c}=\frac{(b^2c-bc^2)(x-a)^2}{(a-b)(a-c)(b-c)}$$
$$\frac{ca(x-b)^2}{(b-c)(b-a)}\bullet\frac{-1(a-c)}{-1(a-c)}=\frac{(ac^2-a^2c)(x-b)^2}{(a-b)(a-c)(b-c)}$$
$$\frac{ab(x-c)^2}{(c-a)(c-b)}\bullet\frac{(-1)(-1)(a-b)}{(-1)(-1)(a-b)}=\frac{(a^2b-ab^2)(x-c)^2}{(a-b)(a-c)(b-c)}$$

Adding the fractions, multiplying out all numerators, and grouping terms results in the fraction:

$$\frac{\left[a^2(b-c)+b^2(c-a)+c^2(a-b)\right]x^2}{(a-b)(a-c)(b-c)}=\frac{(b-c)(a-b)(a-c)x^2}{(a-b)(a-c)(b-c)}=x^2$$

Chapter One Test

1. (a) Factoring by difference of cubes, we have:

$$64-(x-2)^3=[4-(x-2)]\left[16+4(x-2)+(x-2)^2\right]$$
$$=(4-x+2)(16+4x-8+x^2-4x+4)$$
$$=(6-x)(12+x^2)$$

(b) Since this is a sum of squares, $w^2y^2 + 16$ is irreducible.

2. (a) Using trial and error, $2x^2 + 3x - 8$ is irreducible.

(b) Factoring the common factor z and using trial and error, we have:

$$6x^2y^2z+xyz-z=z(6x^2y^2+xy-1)=z(3xy-1)(2xy+1)$$

3. Factoring the common factor $3x(1-x^2)^{-3/2}$, we have:

$$6x(1-x^2)^{-1/2}-3x^2(1-x^2)^{-3/2}=3x(1-x^2)^{-3/2}\left[2(1-x^2)-x\right]$$
$$=3x(1-x^2)^{-3/2}(2-2x^2-x)$$
$$=-3x(1-x^2)^{-3/2}(2x^2+x-2)$$

4. Factoring using grouping, we have:

$$9ax+6bx-6ay-4by=3x(3a+2b)-2y(3a+2b)=(3a+2b)(3x-2y)$$

5. Factoring $54=27\bullet 2$ and $-2000=-1000\bullet 2$, we have:

$$\sqrt[3]{54}-\sqrt[3]{-2000}=\sqrt[3]{27}\bullet\sqrt[3]{2}-\sqrt[3]{-1000}\bullet\sqrt[3]{2}$$
$$=3\bullet\sqrt[3]{2}+10\bullet\sqrt[3]{2}$$
$$=13\sqrt[3]{2}$$

6. (a) Using rational exponents, we have:

$$\sqrt[5]{x^2}\,\sqrt[7]{x^3}=x^{2/5}x^{3/7}=x^{14/35+15/35}=x^{29/35}$$

(b) Writing in radical form, we have:

$$x^{29/35}=\sqrt[35]{x^{29}}$$

7. Rationalizing the denominator, we have:

$$\frac{\sqrt{x^2+1}+\sqrt{x^2-1}}{\sqrt{x^2+1}-\sqrt{x^2-1}}\bullet\frac{\sqrt{x^2+1}+\sqrt{x^2-1}}{\sqrt{x^2+1}+\sqrt{x^2-1}}=\frac{(x^2+1)+2\sqrt{x^4-1}+(x^2-1)}{(x^2+1)-(x^2-1)}$$
$$=\frac{2x^2+2\sqrt{x^4-1}}{2}$$
$$=x^2+\sqrt{x^4-1}$$

8. Using the special product for $(A+B)^3$, we have:

$$\begin{aligned}(2a+3b^2)^3 &= (2a)^3+3(2a)^2(3b^2)+3(2a)(3b^2)^2+(3b^2)^3\\ &= 8a^3+36a^2b^2+54ab^4+27b^6\end{aligned}$$

9. We simplify the expression:

$$\begin{aligned}(4+2i)(4-2i)+\frac{3+i}{1+2i} &= 16-4i^2+\frac{3+i}{1+2i}\bullet\frac{1-2i}{1-2i}\\ &= 16+4+\frac{3-5i-2i^2}{1-4i^2}\\ &= 20+\frac{5-5i}{5}\\ &= 20+1-i\\ &= 21-i\end{aligned}$$

10. (a) distributive property
(b) commutative property of multiplication, associative property of multiplication

11. If $6<x<7$, then $x-6>0$ and $x-7<0$, so:

$$|x-6|+|x-7|=(x-6)-(x-7)=x-6-x+7=1$$

12. (a) 3.96×10^9 light-years
(b) 6.750×10^{-7} meters

13. We use long division:

$$\begin{array}{r|l} & x^2-\tfrac{3}{2}x+\tfrac{1}{4}\\ \hline 2x+3 & 2x^3+0x^2-4x-9\\ & \underline{2x^3+3x^2}\\ & \quad -3x^2-4x\\ & \quad \underline{-3x^2-\tfrac{9}{2}x}\\ & \qquad \tfrac{1}{2}x-9\\ & \qquad \underline{\tfrac{1}{2}x+\tfrac{3}{4}}\\ & \qquad\quad -\tfrac{39}{4}\end{array}$$

The quotient is $x^2-\frac{3}{2}x+\frac{1}{4}$ and the remainder is $-\frac{39}{4}$.

14. Let $x=31.\overline{2}$, so $10x=312.\overline{2}$. Subtracting, we have:

$$\begin{aligned}10x-x&=281\\ 9x&=281\\ x&=\tfrac{281}{9}\end{aligned}$$

15. (a) Since the distance from x to 4 is less than $\frac{1}{10}$, then:

$$4-\frac{1}{10}<x<4+\frac{1}{10}$$
$$\frac{39}{10}<x<\frac{41}{10}$$

The interval is $\left(\frac{39}{10},\frac{41}{10}\right)$.

(b) Since $x \geq -2$, the interval is $[-2,\infty)$.

Chapter Two
Equations and Inequalities

2.1 Setting Up Equations (A Warm Up)

1. (a) The problem asks us to find two numbers. Let x and y denote the two numbers.

(b) We complete the table:

in English	in algebraic notation
Find two numbers.	x and y
Their sum is -9.	$x+y=-9$
The sum of their squares is 153.	$x^2+y^2=153$

(c) From part (b) we have the equations:

$$x+y=-9$$
$$x^2+y^2=153$$

From the first equation, we have $y=-9-x$, and therefore:

$$x^2+(-9-x)^2=153$$
$$x^2+81+18x+x^2=153$$
$$2x^2+18x-72=0$$
$$x^2+9x-36=0$$

3. (a) The problem asks us to find two numbers. Let x and y denote the two numbers.

(b) We complete the table:

in English	in algebraic notation
Find two numbers.	x and y
Their sum is 17.	$x+y=17$
Their product is 52.	$xy=52$

(c) From part (b) we have the equations:

$$x+y=17$$
$$xy=52$$

From the first equation, we have $y = 17 - x$, and therefore:

$$x(17-x)=52$$
$$17x-x^2=52$$
$$-x^2+17x-52=0$$
$$x^2-17x+52=0$$

5. (a) The problem asks us to find two consecutive integers. Let x and $x + 1$ denote these consecutive integers.

(b) We complete the table:

in English	in algebraic notation
Find two consecutive integers.	x and $x+1$
The sum of their cubes is 341.	$x^3+(x+1)^3=341$

(c) Working from the equation $x^3 + (x + 1)^3 = 341$ obtained in part (b), we have:

$$x^3+x^3+3x^2+3x+1=341$$
$$2x^3+3x^2+3x-340=0$$

7. (a) The problem asks us to find two numbers. Let x and y denote the two numbers.

(b) We complete the table:

in English	in algebraic notation
Find two numbers.	x and y
Their average is 4.	$\frac{1}{2}(x+y)=4$
The average of their squares is 80.	$\frac{1}{2}(x^2+y^2)=80$

(c) From part (b) we have the equations

$$\tfrac{1}{2}(x+y)=4$$
$$\tfrac{1}{2}(x^2+y^2)=80$$

From the first equation we have $x + y = 8$, so $y = 8 - x$, and therefore:

$$\frac{1}{2}\left[x^2 + (8-x)^2\right] = 80$$
$$x^2 + 64 - 16x + x^2 = 160$$
$$2x^2 - 16x - 96 = 0$$
$$x^2 - 8x - 48 = 0$$

9. (a) The problem is asking for the dimensions of the rectangle. We label the length l and the width w, and draw the sketch:

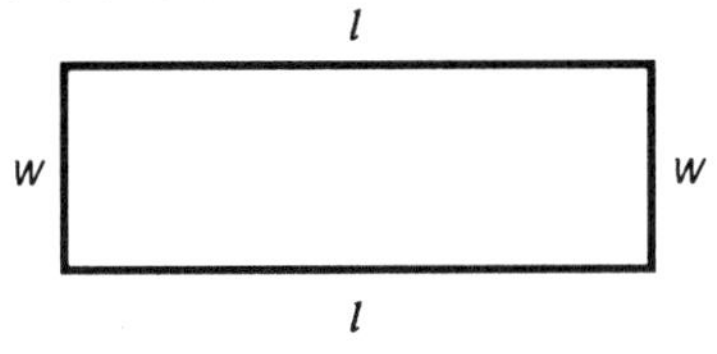

(b) We complete the table:

in English	in algebraic notation
Find the width and length of a rectangle.	w and l
The perimeter is 40 cm.	$2w + 2l = 40$
The area is 64 cm^2.	$lw = 64$

(c) From part (b) we have the equations:

$$2w + 2l = 40$$
$$lw = 64$$

From the first equation we have $w + l = 20$, so $l = 20 - w$, and therefore:

$$(20 - w)w = 64$$
$$20w - w^2 = 64$$
$$-w^2 + 20w - 64 = 0$$
$$w^2 - 20w + 64 = 0$$

11. (a) The problem asks for the number of years required until the total interest earned on all three accounts is \$8,760. Let t denote the required number of years.

(b) We apply the formula $I = Prt$ to each account. After t years, the interest on the first account is $(3000)(0.07)t$, the interest on the second account is $(5000)(0.08)t$, and the interest on the third account is $(10000)(0.085)t$. Since we want the sum of these three amounts to be \$8,760, the required equation is:

$$(3000)(0.07)t + (5000)(0.08)t + (10000)(0.085)t = 8760$$

13. Let r represent the interest rate for the smaller account (as a percent), so $r + 1$ will be the rate for the larger account. Applying the formula $I = Prt$, the first account will generate $(7400)\left(\frac{r}{100}\right)(10)$ in interest, so the total amount in the account (including the initial principal) is $7400 + (7400)\left(\frac{r}{100}\right)(10) = 7400\left(1 + \frac{r}{10}\right)$. Similarly, the second account will generate $(12600)\left(\frac{1+r}{100}\right)(10)$ in interest, so the total amount in the account is $12600 + (12600)\left(\frac{1+r}{100}\right)(10) = 12600\left(1 + \frac{1+r}{10}\right)$. Since we want the sum of these amounts to be \$31,260, the required equation is:

$$7400\left(1 + \frac{r}{10}\right) + 12600\left(1 + \frac{1+r}{10}\right) = 31260$$

15. (a) Since the height is four times the radius, we can write $h = 4r$ where h represents the height and r represents the radius.

(b) The general formula for the volume is $V = \pi r^2 h$. Replacing h with $4r$, we obtain:

$$V = \pi r^2(4r) = 4\pi r^3$$

(c) Since $h = 4r$, then $r = \frac{h}{4}$. Replacing r with $\frac{h}{4}$, we obtain:

$$V = \pi\left(\frac{h}{4}\right)^2 \bullet h = \frac{1}{16}\pi h^3$$

17. (a) Since the sum of the radius r and the height h is 4 cm, then $r + h = 4$.

(b) Since $r + h = 4$, then $h = 4 - r$. Using the general formula for the volume, $V = \pi r^2 h$, and replacing h with $4 - r$, we obtain:

$$V = \pi r^2(4 - r)$$

(c) We complete the table:

r	1.00	1.50	2.00	2.50	3.00	3.50
V	9.42	17.67	25.13	29.45	28.27	19.24

(d) From the table, $r = 2.5$ cm yields the largest volume.

(e) When $r = 2.6$ cm, $V \approx 29.73$ cm^3. When $r = 2.7$ cm, $V \approx 29.77$ cm^3.

(f) No. The largest value for V is approximately 29.77 cm^3, occurring when $r = 2.7$ cm.

19. (a) Since the volume is given by $V = \frac{1}{3}\pi r^2 h$, we have $\pi = \frac{1}{3}\pi r^2 h$, so $r^2 h = 3$.

(b) Solving the equation from part (a) for h, we have $h = \frac{3}{r^2}$. Using the formula for the lateral surface area S, we have:

$$S = \pi r\sqrt{r^2 + h^2} = \pi r\sqrt{r^2 + \left(\frac{3}{r^2}\right)^2} = \pi r\sqrt{r^2 + \frac{9}{r^4}} = \frac{\pi r\sqrt{r^6 + 9}}{r^2} = \frac{\pi}{r}\sqrt{r^6 + 9}$$

(c) We complete the table:

r	0.250	0.500	0.750	1.000	1.250	1.500	1.750	2.000
S	37.700	18.866	12.690	9.935	8.997	9.457	11.026	13.421

(d) The least surface area occurs when $r = 1.25$ cm.

(e) When $r = 1.30$ cm, we have $S \approx 8.986 \text{ cm}^2$.

(f) Yes. When $r = 1.29$ cm, we have $S \approx 8.984 \text{ cm}^2$.

21. Since the first piece is bent into a square, calling the side of the square s we have $4s = x$, so $s = \frac{x}{4}$. Thus the area of the square is given by:

$$A_{\text{square}} = s^2 = \left(\tfrac{x}{4}\right)^2 = \tfrac{1}{16}x^2$$

Since the second piece is bent into a rectangle, calling the length l and the width w we have $2w + 2l = 1 - x$. But since the length is twice the width, then $l = 2w$ and thus:

$$\begin{aligned} 2w + 2(2w) &= 1 - x \\ 6w &= 1 - x \\ w &= \tfrac{1}{6} - \tfrac{1}{6}x \\ l &= 2\left(\tfrac{1}{6} - \tfrac{1}{6}x\right) = \tfrac{1}{3} - \tfrac{1}{3}x \end{aligned}$$

Thus the area of the rectangle is given by:

$$A_{\text{rectangle}} = w \bullet l = \left(\tfrac{1}{6} - \tfrac{1}{6}x\right)\left(\tfrac{1}{3} - \tfrac{1}{3}x\right) = \tfrac{1}{18} - \tfrac{1}{9}x + \tfrac{1}{18}x^2$$

So the total combined area A is given by:

$$A = A_{\text{square}} + A_{\text{rectangle}} = \tfrac{1}{16}x^2 + \tfrac{1}{18} - \tfrac{1}{9}x + \tfrac{1}{18}x^2 = \tfrac{17}{144}x^2 - \tfrac{1}{9}x + \tfrac{1}{18}$$

2.2 Linear Equations

1. Solving for x, we have:

$$\begin{aligned} 2x - 3 &= -5 \\ 2x &= -2 \\ x &= -1 \end{aligned}$$

3. Solving for x, we have:

$$\begin{aligned} 8x - 4 &= 20 \\ 8x &= 24 \\ x &= 3 \end{aligned}$$

5. Solving for y, we have:

$$\begin{aligned} 1-y&=12 \\ -y&=11 \\ y&=-11 \end{aligned}$$

7. Solving for m, we have:

$$\begin{aligned} 2m-1+3m+5&=6m-8 \\ 5m+4&=6m-8 \\ -m&=-12 \\ m&=12 \end{aligned}$$

9. Solving for x, we have:

$$\begin{aligned} (x+2)(x+1)&=x^2+11 \\ x^2+3x+2&=x^2+11 \\ 3x+2&=11 \\ 3x&=9 \\ x&=3 \end{aligned}$$

11. Solving for x, we have:

$$\begin{aligned} 1-(x-2)^2&=-x^2+5x+3 \\ 1-\left(x^2-4x+4\right)&=-x^2+5x+3 \\ 1-x^2+4x-4&=-x^2+5x+3 \\ 4x-3&=5x+3 \\ -x&=6 \\ x&=-6 \end{aligned}$$

13. Solving for x, we have:

$$\begin{aligned} (x-1)^2-\left(x^2-1\right)&=16 \\ x^2-2x+1-x^2+1&=16 \\ -2x+2&=16 \\ -2x&=14 \\ x&=-7 \end{aligned}$$

15. Multiplying by 15 and then solving for x, we have:

$$\begin{aligned} \frac{x}{3}+\frac{2x}{5}&=-\frac{11}{5} \\ 15\left(\frac{x}{3}\right)+15\left(\frac{2x}{5}\right)&=15\left(-\frac{11}{5}\right) \\ 5x+6x&=-33 \\ 11x&=-33 \\ x&=-3 \end{aligned}$$

17. Multiplying by 3 and then solving for *x*, we have:

$$1-\frac{y}{3}=6$$
$$3(1)-3\left(\frac{y}{3}\right)=3(6)$$
$$3-y=18$$
$$-y=15$$
$$y=-15$$

19. Multiplying by 36 and then solving for *x*, we have:

$$3-\frac{x-1}{4}=\frac{x}{9}$$
$$36\left(3-\frac{x-1}{4}\right)=36\left(\frac{x}{9}\right)$$
$$108-9(x-1)=4x$$
$$108-9x+9=4x$$
$$-13x=-117$$
$$x=9$$

21. Multiplying by 4 and then solving for *x*, we have:

$$\frac{x-1}{4}+\frac{2x+3}{-1}=0$$
$$4\left(\frac{x-1}{4}+\frac{2x+3}{-1}\right)=4(0)$$
$$x-1-4(2x+3)=0$$
$$x-1-8x-12=0$$
$$-7x=13$$
$$x=-\tfrac{13}{7}$$

23. Multiplying by $x^2-9=(x+3)(x-3)$ and then solving for *x*, we have:

$$\frac{1}{x-3}-\frac{2}{x+3}=\frac{1}{x^2-9}$$
$$\frac{(x-3)(x+3)}{x-3}-\frac{2(x-3)(x+3)}{x+3}=\frac{(x-3)(x+3)}{(x-3)(x+3)}$$
$$x+3-2(x-3)=1$$
$$x+3-2x+6=1$$
$$-x=-8$$
$$x=8$$

Check: Replacing *x* by 8 in the original equation yields:

$\frac{1}{5}-\frac{2}{11}=\frac{1}{55}$

$\frac{11}{55}-\frac{10}{55}=\frac{1}{55}$, which is true

25. Multiplying by $x^2 - 25 = (x + 5)(x - 5)$ and then solving for x, we have:

$$\frac{1}{x-5} + \frac{1}{x+5} = \frac{2x}{x^2-25}$$

$$\frac{(x-5)(x+5)}{x-5} + \frac{(x-5)(x+5)}{x+5} = \frac{2x(x-5)(x+5)}{(x-5)(x+5)}$$

$$x+5+x-5 = 2x$$

$$2x = 2x$$

This last equation is an identity; it is satisfied by all real numbers. However, in the original equation, note that x cannot take on the values ± 5, since that leads to denominators that are zero. Thus the original equation is satisfied by all real numbers except 5 and –5.

27. Multiplying by $x^2 - 4 = (x + 2)(x - 2)$ and then solving for x, we have:

$$\frac{4}{x+2} + \frac{1}{x-2} = \frac{4}{x^2-4}$$

$$\frac{4(x-2)(x+2)}{x+2} + \frac{(x-2)(x+2)}{x-2} = \frac{4(x-2)(x+2)}{(x-2)(x+2)}$$

$$4(x-2)+(x+2) = 4$$

$$4x-8+x+2 = 4$$

$$5x = 10$$

$$x = 2$$

This shows that if the original equation has a solution, it must be $x = 2$. However, the value $x = 2$ does not check in the original equation (it produces zeros in two of the denominators). Consequently the original equation has no solutions.

29. Multiplying by $x - 1$ and then solving for x, we have:

$$\frac{2(x+1)}{x-1} - 3 = \frac{5x-1}{x-1}$$

$$2(x+1)-3(x-1) = 5x-1$$

$$2x+2-3x+3 = 5x-1$$

$$-x+5 = 5x-1$$

$$-6x = -6$$

$$x = 1$$

This shows that if the original equation has a solution, it must be $x = 1$. However, the value $x = 1$ does not satisfy the original equation, since it yields zeros in the denominators. Consequently the original equation has no solution.

31. Multiplying by x and then solving for x, we have:

$$\frac{1}{x} = \frac{4}{x} - 1$$

$$1 = 4 - x$$

$$x = 3$$

Check: Replacing x by 3 in the original equation yields:

$\frac{1}{3} = \frac{4}{3} - 1$

$\frac{1}{3} = \frac{4}{3} - \frac{3}{3}$, which is true

33. (a) Multiplying by $3x$ and then solving for x, we have:

$$3x \bullet \frac{2}{3x} = 3x \bullet \frac{3}{x}$$
$$2 = 9$$

Since there are no values of x which can make this last statement true, there is no solution.

(b) Multiplying by $3x(x + 1)$ and then solving for x, we have:

$$3x(x+1) \bullet \frac{2}{3x} = 3x(x+1) \bullet \frac{3}{x+1}$$
$$2x + 2 = 9x$$
$$2 = 7x$$
$$x = \tfrac{2}{7}$$

Check: Replacing x by $\frac{2}{7}$ in the original equation yields:

$$\frac{2}{\frac{6}{7}} = \frac{3}{\frac{2}{7}+1}$$
$$\frac{7}{3} = \frac{7}{3}, \text{ which is true}$$

(c) Multiplying by $3x$ and then solving for x, we have:

$$3x \bullet \frac{2}{3x} = 3x\left(\frac{3}{x} + 1\right)$$
$$2 = 9 + 3x$$
$$-7 = 3x$$
$$x = -\frac{7}{3}$$

Check: Replacing x by $-\frac{7}{3}$ in the original equation yields:

$$\frac{2}{-7} = \frac{3}{-\frac{7}{3}} + 1$$
$$-\frac{2}{7} = -\frac{9}{7} + 1, \text{ which is true}$$

35. (a) Multiplying by $3x - 2$ and then solving for x, we have:

$$(3x-2) \bullet \frac{1}{3x-2} = (3x-2) \bullet 1$$
$$1 = 3x - 2$$
$$3 = 3x$$
$$x = 1$$

Check: Replacing x by 1 in the original equation yields:

$$\tfrac{1}{3-2} = 3 - 2$$
$$1 = 1, \text{ which is true}$$

(b) Multiplying by $x(3x-2)$ and then solving for x, we have:

$$x(3x-2)\bullet\frac{1}{3x-2}=x(3x-2)\bullet\frac{1}{x}$$
$$x=3x-2$$
$$-2x=-2$$
$$x=1$$

Check: Replacing x by 1 in the original equation yields:

$$\tfrac{1}{3-2}=\tfrac{1}{1}$$
$1=1$, which is true

(c) Multiplying by $(3x-2)(3x+2)$ and then solving for x, we have:

$$(3x-2)(3x+2)\bullet\frac{1}{3x-2}=(3x-2)(3x+2)\bullet\frac{1}{3x+2}$$
$$3x+2=3x-2$$
$$2=-2$$

Since there are no values of x which can make this last statement true, there is no solution.

37. If $x-5\geq 0$, the equation becomes:

$$x-5=1$$
$$x=6$$

If $x-5<0$, the equation becomes:

$$-(x-5)=1$$
$$-x+5=1$$
$$-x=-4$$
$$x=4$$

The solutions of the original equation are 6 and 4.

39. If $x+6\geq 0$, the equation becomes:

$$x+6=\tfrac{1}{2}$$
$$x=-\tfrac{11}{2}$$

If $x+6<0$, the equation becomes:

$$-(x+6)=\tfrac{1}{2}$$
$$-x-6=\tfrac{1}{2}$$
$$-x=\tfrac{13}{2}$$
$$x=-\tfrac{13}{2}$$

The solutions of the original equation are $-\tfrac{11}{2}$ and $-\tfrac{13}{2}$.

41. If $6x-5\geq 0$, the equation becomes:

$$6x-5=25$$
$$6x=30$$
$$x=5$$

If $6x-5<0$, the equation becomes:

$$\begin{aligned}-(6x-5)&=25\\-6x+5&=25\\-6x&=20\\x&=-\tfrac{10}{3}\end{aligned}$$

The solutions of the original equation are 5 and $-\tfrac{10}{3}$.

43. Solving for x, we have:

$$\begin{aligned}4.50x-9.11&=x\\3.50x&=9.11\\x&=\tfrac{9.11}{3.50}\approx 2.60\end{aligned}$$

45. Multiplying by $(x+0.23)(x-2.24)$ and solving for x, we have:

$$\begin{aligned}\frac{x-0.18}{x+0.23}&=\frac{x+1.16}{x-2.24}\\(x-0.18)(x-2.24)&=(x+0.23)(x+1.16)\\x^2-2.42x+0.4032&=x^2+1.39x+0.2668\\-2.42x+0.4032&=1.39x+0.2668\\-3.81x&=-0.1364\\x&\approx 0.04\end{aligned}$$

47. Solving for x, we have:

$$\begin{aligned}3ax-2b&=b+3\\3ax&=3b+3\\ax&=b+1\\x&=\frac{b+1}{a}\end{aligned}$$

49. Solving for x, we have:

$$\begin{aligned}ax+b&=bx+a\\ax-bx&=a-b\\x(a-b)&=a-b\\x&=\frac{a-b}{a-b}\\x&=1\end{aligned}$$

51. Multiplying by x and solving for x, we have:

$$\begin{aligned}\frac{1}{x}&=a+b\\1&=x(a+b)\\x&=\frac{1}{a+b}\end{aligned}$$

53. Solving for x, we have:

$$\begin{aligned}(x+b)(x+c)-a(a+b)&=(x-a)(x+a)+bc\\x^2+bx+cx+bc-a^2-ab&=x^2-a^2+bc\\bx+cx-ab&=0\\x(b+c)&=ab\\x&=\frac{ab}{b+c}\end{aligned}$$

55. Solving for x, we have:

$$\begin{aligned}(a-x)^2&=x^2+b^2\\a^2-2ax+x^2&=x^2+b^2\\a^2-2ax&=b^2\\-2ax&=b^2-a^2\\x&=\frac{b^2-a^2}{-2a}\\x&=\frac{a^2-b^2}{2a}\end{aligned}$$

57. (a) Multiplying by bx and solving for x, we have:

$$\begin{aligned}bx\bullet\frac{a}{bx}&=bx\bullet d\\a&=bdx\\x&=\frac{a}{bd}\end{aligned}$$

(b) Multiplying by $bx+c$ and solving for x, we have:

$$\begin{aligned}(bx+c)\bullet\frac{a}{bx+c}&=(bx+c)\bullet d\\a&=bdx+cd\\a-cd&=bdx\\x&=\frac{a-cd}{bd}\end{aligned}$$

59. (a) Obtaining a common denominator of $15x$, we have:

$$\frac{1}{3x}\bullet\frac{5}{5}-\frac{2}{5x}\bullet\frac{3}{3}=\frac{5}{15x}-\frac{6}{15x}=-\frac{1}{15x}$$

(b) Using the result from part (a), and multiplying by $15x$, we have:

$$\begin{aligned}-\frac{1}{15x}&=-2\\-1&=-30x\\x&=\tfrac{1}{30}\end{aligned}$$

61. (a) Obtaining a common denominator of $5x^2$, we have:

$$\frac{8}{5x} \bullet \frac{x}{x} + \frac{1}{x^2} \bullet \frac{5}{5} = \frac{8x}{5x^2} + \frac{5}{5x^2} = \frac{8x+5}{5x^2}$$

(b) Using the result from part (a), and multiplying by $10x^2$, we have:

$$\begin{aligned} \frac{8x+5}{5x^2} &= -\frac{3}{2x} \\ 16x+10 &= -15x \\ 31x &= -10 \\ x &= -\tfrac{10}{31} \end{aligned}$$

63. Multiplying by 6, we have:

$$\begin{aligned} 3|x-3| + 2|x-3| &= -10x \\ 5|x-3| &= -10x \end{aligned}$$

There are two cases to consider.

case 1: If $x-3 \geq 0$, then $|x-3| = x-3$, so:

$$\begin{aligned} 5(x-3) &= -10x \\ 5x-15 &= -10x \\ 15x &= 15 \\ x &= 1, \text{ which is not in the interval } x \geq 3 \end{aligned}$$

case 2: If $x-3 < 0$, then $|x-3| = -(x-3)$, so:

$$\begin{aligned} -5(x-3) &= -10x \\ -5x+15 &= -10x \\ 5x &= -15 \\ x &= -3 \end{aligned}$$

So the solution is –3.

65. Solving for x, we have:

$$\begin{aligned} a^2(a-x) &= b^2(b+x) - 2abx \\ a^3 - a^2x &= b^3 + b^2x - 2abx \\ a^3 - b^3 &= a^2x + b^2x - 2abx \\ a^3 - b^3 &= x\left(a^2 + b^2 - 2ab\right) \\ a^3 - b^3 &= x(a-b)^2 \\ x &= \frac{(a-b)\left(a^2+ab+b^2\right)}{(a-b)^2} \\ x &= \frac{a^2+ab+b^2}{a-b} \end{aligned}$$

67. Multiplying by $(a-b)(b-c)$ and solving for x, we have:

$$\begin{aligned}
\frac{a-x}{a-b}-2&=\frac{c-x}{b-c}\\
(a-x)(b-c)-2(a-b)(b-c)&=(c-x)(a-b)\\
ab-bx-ac+cx-2ab+2b^2+2ac-2bc&=ac-ax-bc+bx\\
-ab-bx+ac+cx+2b^2-2bc&=ac-ax-bc+bx\\
ax-2bx+cx&=ab-2b^2+bc\\
x(a-2b+c)&=b(a-2b+c)\\
x&=b \qquad \text{(assuming } a-2b+c\neq 0\text{)}
\end{aligned}$$

69. Multiplying by $(x-b)(a-x)$ and solving for x, we have:

$$\begin{aligned}
\frac{x-a}{x-b}&=\frac{b-x}{a-x}\\
(x-a)(a-x)&=(b-x)(x-b)\\
ax-a^2-x^2+ax&=bx-x^2-b^2+bx\\
2ax-a^2&=2bx-b^2\\
2ax-2bx&=a^2-b^2\\
2x(a-b)&=a^2-b^2\\
x&=\frac{a^2-b^2}{2(a-b)}\\
x&=\frac{(a-b)(a+b)}{2(a-b)}\\
x&=\frac{a+b}{2} \qquad \text{(assuming } a\neq b\text{)}
\end{aligned}$$

71. Multiplying by $1+rt$ and solving for r, we have:

$$\begin{aligned}
d&=\frac{r}{1+rt}\\
d(1+rt)&=r\\
d+drt&=r\\
drt-r&=-d\\
r(dt-1)&=-d\\
r&=\frac{-d}{dt-1}\\
r&=\frac{d}{1-dt}
\end{aligned}$$

73. Solving for h, we have:

$$\begin{aligned}
S&=2\pi r^2+2\pi rh\\
S-2\pi r^2&=2\pi rh\\
h&=\frac{S-2\pi r^2}{2\pi r}
\end{aligned}$$

75. Multiplying by $r-l$ and solving for r, we have:

$$\begin{aligned} S &= \frac{rl-a}{r-l} \\ S(r-l) &= rl-a \\ Sr-Sl &= rl-a \\ Sr-rl &= Sl-a \\ r(S-l) &= Sl-a \\ r &= \frac{Sl-a}{S-l} \end{aligned}$$

77. Replacing x by $\frac{ab}{a+b}$ yields:

$$\begin{aligned} \frac{1}{\frac{ab}{a+b}} &= \frac{1}{a}+\frac{1}{b} \\ \frac{a+b}{ab} &= \frac{1}{a}+\frac{1}{b} \\ \frac{a}{ab}+\frac{b}{ab} &= \frac{1}{a}+\frac{1}{b} \\ \frac{1}{b}+\frac{1}{a} &= \frac{1}{a}+\frac{1}{b}, \text{ which is true} \end{aligned}$$

79. Rather than clearing the equation of fractions directly, we show an alternate approach. First, add the two fractions on the left-hand side of the equation:

$$\begin{aligned} \frac{(x-q)-(x-p)}{(x-p)(x-q)} &= \frac{p-q}{x^2-pq} \\ \frac{-q+p}{(x-p)(x-q)} &= \frac{p-q}{x^2-pq} \end{aligned}$$

Now divide both sides by $p-q$ (assuming $p \neq q$) to obtain:

$$\begin{aligned} \frac{1}{(x-p)(x-q)} &= \frac{1}{x^2-pq} \\ x^2-px-qx+pq &= x^2-pq \\ -x(p+q) &= -2pq \\ x &= \frac{2pq}{p+q} \end{aligned}$$

2.3 Quadratic Equations

1. We substitute -3 for x:

$$\begin{aligned} 2(-3)^2-6(-3)-36 &= 0 \\ 18+18-36 &= 0, \text{ which is true} \end{aligned}$$

So $x=-3$ is a solution.

3. We substitute $-\frac{1}{4}$ for x:

$$4\left(-\tfrac{1}{4}\right)^2 - 1 = 0$$
$$4\left(\tfrac{1}{16}\right) - 1 = 0$$
$$\tfrac{1}{4} - 1 = 0, \text{ which is false}$$

5. We substitute $-1-\sqrt{7}$ for x:

$$(-1-\sqrt{7})^2 - 2(-1-\sqrt{7}) - 6 = 0$$
$$1 + 2\sqrt{7} + 7 + 2 + 2\sqrt{7} - 6 = 0$$
$$4\sqrt{7} + 4 = 0, \text{ which is false}$$

7. Factoring, we have:

$$x^2 - 5x - 6 = 0$$
$$(x-6)(x+1) = 0$$
$$x = 6 \text{ or } x = -1$$

9. Factoring, we have:

$$x^2 - 100 = 0$$
$$(x-10)(x+10) = 0$$
$$x = \pm 10$$

11. Factoring, we have:

$$25x^2 - 60x + 36 = 0$$
$$(5x-6)^2 = 0$$
$$5x - 6 = 0$$
$$x = \tfrac{6}{5} \text{ (double root)}$$

13. Factoring, we have:

$$10z^2 - 13z - 3 = 0$$
$$(5z+1)(2z-3) = 0$$
$$z = -\tfrac{1}{5} \text{ or } z = \tfrac{3}{2}$$

15. Factoring, we have:

$$(x+1)^2 - 4 = 0$$
$$(x+1)^2 - 2^2 = 0$$
$$(x+1-2)(x+1+2) = 0$$
$$(x-1)(x+3) = 0$$
$$x = 1 \text{ or } x = -3$$

17. Factoring, we have:

$$x(2x-13) = -6$$
$$2x^2 - 13x + 6 = 0$$
$$(2x-1)(x-6) = 0$$
$$x = \tfrac{1}{2} \text{ or } x = 6$$

19. Factoring, we have:

$$\begin{aligned} x(x+1) &= 156 \\ x^2+x-156 &= 0 \\ (x+13)(x-12) &= 0 \\ x &= -13 \text{ or } x = 12 \end{aligned}$$

21. Factoring, we have:

$$\begin{aligned} x^2-49 &= 0 \\ (x+7)(x-7) &= 0 \\ x &= \pm 7 \end{aligned}$$

23. Factoring, we have:

$$\begin{aligned} \tfrac{1}{4}-x^2 &= 0 \\ \left(\tfrac{1}{2}+x\right)\left(\tfrac{1}{2}-x\right) &= 0 \\ x &= \pm\tfrac{1}{2} \end{aligned}$$

25. (a) Completing the square, we have:

$$\begin{aligned} x^2+10x+9 &= 0 \\ x^2+10x &= -9 \\ x^2+10x+25 &= 16 \\ (x+5)^2 &= 16 \\ x+5 &= \pm 4 \\ x &= -5+4 = -1 \text{ or } x = -5-4 = -9 \end{aligned}$$

(b) Using the quadratic formula with $a = 1, b = 10, c = 9$, we have:

$$x = \frac{-10 \pm \sqrt{100-36}}{2} = \frac{-10 \pm \sqrt{64}}{2} = \frac{-10 \pm 8}{2} = -1 \text{ or } -9$$

27. Completing the square, we have:

$$\begin{aligned} x^2-x-5 &= 0 \\ x^2-x &= 5 \\ x^2-x+\frac{1}{4} &= \frac{21}{4} \\ \left(x-\frac{1}{2}\right)^2 &= \frac{21}{4} \\ x-\frac{1}{2} &= \pm\sqrt{\frac{21}{4}} \\ x-\frac{1}{2} &= \pm\frac{\sqrt{21}}{2} \\ x &= \frac{1 \pm \sqrt{21}}{2} \end{aligned}$$

(b) Using the quadratic formula with $a = 1$, $b = -1$, $c = -5$, we have:

$$x = \frac{1 \pm \sqrt{1 - 4(1)(-5)}}{2(1)} = \frac{1 \pm \sqrt{21}}{2}$$

29. (a) Completing the square, we have:

$$\begin{aligned} 2x^2 + 3x - 4 &= 0 \\ x^2 + \frac{3}{2}x &= 2 \\ x^2 + \frac{3}{2}x + \frac{9}{16} &= \frac{41}{16} \\ \left(x + \frac{3}{4}\right)^2 &= \frac{41}{16} \\ x + \frac{3}{4} &= \pm\frac{\sqrt{41}}{4} \\ x &= \frac{-3 \pm \sqrt{41}}{4} \end{aligned}$$

(b) Using the quadratic formula with $a = 2$, $b = 3$, $c = -4$, we have:

$$x = \frac{-3 \pm \sqrt{9 - 4(2)(-4)}}{2(2)} = \frac{-3 \pm \sqrt{41}}{4}$$

31. (a) Completing the square, we have:

$$\begin{aligned} 12x^2 + 32x + 5 &= 0 \\ x^2 + \frac{8}{3}x &= -\frac{5}{12} \\ x^2 + \frac{8}{3}x + \frac{16}{9} &= \frac{49}{36} \\ \left(x + \frac{4}{3}\right)^2 &= \frac{49}{36} \\ x + \frac{4}{3} &= \pm\frac{7}{6} \\ x &= -\frac{4}{3} + \frac{7}{6} = -\frac{1}{6} \text{ or } x = -\frac{4}{3} - \frac{7}{6} = -\frac{5}{2} \end{aligned}$$

(b) Using the quadratic formula with $a = 12$, $b = 32$, $c = 5$, we have:

$$x = \frac{-32 \pm \sqrt{(32)^2 - 4(12)(5)}}{2(12)} = \frac{-32 \pm \sqrt{784}}{24} = \frac{-32 \pm 28}{24}$$

$$x = \frac{-32 + 28}{24} = \frac{-4}{24} = -\frac{1}{6} \text{ or } x = \frac{-32 - 28}{24} = \frac{-60}{24} = -\frac{5}{2}$$

33. (a) Completing the square, we have:

$$2x^2 = x + 5$$
$$2x^2 - x - 5 = 0$$
$$x^2 - \tfrac{1}{2}x = \tfrac{5}{2}$$
$$x^2 - \tfrac{1}{2}x + \tfrac{1}{16} = \tfrac{41}{16}$$
$$\left(x - \tfrac{1}{4}\right)^2 = \tfrac{41}{16}$$
$$x - \tfrac{1}{4} = \pm\tfrac{\sqrt{41}}{4}$$
$$x = \tfrac{1 \pm \sqrt{41}}{4}$$

(b) Using the quadratic formula with $a = 2$, $b = -1$, $c = -5$, we have:

$$x = \frac{1 \pm \sqrt{1 - 4(2)(-5)}}{2(2)} = \frac{1 \pm \sqrt{41}}{4}$$

35. (a) Completing the square, we have:

$$-6x^2 + 12x = -1$$
$$-6x^2 + 12x + 1 = 0$$
$$x^2 - 2x = \frac{1}{6}$$
$$x^2 - 2x + 1 = \frac{7}{6}$$
$$(x - 1)^2 = \frac{7}{6}$$
$$x - 1 = \pm\frac{\sqrt{7}}{\sqrt{6}}$$
$$x - 1 = \pm\frac{\sqrt{42}}{6}$$
$$x = 1 \pm \frac{\sqrt{42}}{6} \text{ or } x = \frac{6 \pm \sqrt{42}}{6}$$

(b) Using the quadratic formula with $a = -6$, $b = 12$, $c = 1$, we have:

$$x = \frac{-12 \pm \sqrt{144 - 4(-6)(1)}}{2(-6)} = \frac{-12 \pm \sqrt{168}}{-12} = \frac{-12 \pm 2\sqrt{42}}{-12} = \frac{6 \pm \sqrt{42}}{6}$$

37. Solving for x, we have:

$$x(x + 18) = 81$$
$$x^2 + 18x - 81 = 0$$

Using the quadratic formula with $a = 1$, $b = 18$, $c = -81$, we have:

$$x = \frac{-18 \pm \sqrt{(18)^2 - 4(1)(-81)}}{2(1)} = \frac{-18 \pm \sqrt{648}}{2} = \frac{-18 \pm 18\sqrt{2}}{2} = -9 \pm 9\sqrt{2}$$

39. Solving for t, we have:

$$t^2+6=2t$$
$$t^2-2t+6=0$$

Using the quadratic formula with $a=1$, $b=-2$, $c=6$, we have:

$$t=\frac{-(-2)\pm\sqrt{(-2)^2-4(1)(6)}}{2(1)}=\frac{2\pm\sqrt{-20}}{2}=\frac{2\pm 2i\sqrt{5}}{2}=1\pm i\sqrt{5}$$

41. Solving for u, we have:

$$10=11u+8u^2$$
$$8u^2+11u-10=0$$
$$(8u-5)(u+2)=0$$
$$u=\tfrac{5}{8}\text{ or }u=-2$$

43. Solving for y, we have:

$$12+12y^2=10y$$
$$12y^2-10y+12=0$$
$$6y^2-5y+6=0$$

Using the quadratic formula with $a=6$, $b=-5$, $c=6$, we have:

$$y=\frac{-(-5)\pm\sqrt{(-5)^2-4(6)(6)}}{2(6)}=\frac{5\pm\sqrt{-119}}{12}=\frac{5\pm i\sqrt{119}}{12}$$

45. Using the quadratic formula, we have:

$$x^2=24$$
$$x^2-24=0$$
$$x=\frac{0\pm\sqrt{0-4(-24)}}{2}=\frac{\pm\sqrt{96}}{2}=\frac{\pm 4\sqrt{6}}{2}=\pm 2\sqrt{6}$$

47. Using the quadratic formula, we have:

$$x(x-1)=1$$
$$x^2-x=1$$
$$x^2-x-1=0$$
$$x=\frac{1\pm\sqrt{1-4(-1)}}{2}=\frac{1\pm\sqrt{5}}{2}$$

49. Multiply by 6 to clear fractions, then use the quadratic formula:

$$6\left(\tfrac{1}{2}x^2-x-\tfrac{1}{3}\right)=6(0)$$
$$3x^2-6x-2=0$$
$$x=\frac{6\pm\sqrt{36-4(3)(-2)}}{2(3)}=\frac{6\pm\sqrt{60}}{6}=\frac{6\pm 2\sqrt{15}}{6}=\frac{3\pm\sqrt{15}}{3}$$

51. Multiply by $\sqrt{5}$ to simplify the equation:

$$\sqrt{5}\left(2\sqrt{5}x^2-x-2\sqrt{5}\right)=\sqrt{5}(0)$$
$$10x^2-\sqrt{5}x-10=0$$

Using the quadratic formula, we have:

$$x=\frac{\sqrt{5}\pm\sqrt{5-4(10)(-10)}}{2(10)}=\frac{\sqrt{5}\pm\sqrt{405}}{20}=\frac{\sqrt{5}\pm 9\sqrt{5}}{20}$$

$$x=\frac{\sqrt{5}+9\sqrt{5}}{20}=\frac{10\sqrt{5}}{20}=\frac{\sqrt{5}}{2} \text{ or } x=\frac{\sqrt{5}-9\sqrt{5}}{20}=\frac{-8\sqrt{5}}{20}=-\frac{2\sqrt{5}}{5}$$

53. Solving for x by factoring, we have:

$$2y^2x^2-3yx+1=0$$
$$(2yx-1)(yx-1)=0$$
$$x=\frac{1}{2y} \text{ or } x=\frac{1}{y}$$

55. Solving for x by factoring, we have:

$$(x-p)^2+(x-q)^2=p^2+q^2$$
$$x^2-2px+p^2+x^2-2qx+q^2=p^2+q^2$$
$$2x^2-2px-2qx=0$$
$$x^2-px-qx=0$$
$$x(x-p-q)=0$$
$$x=0 \text{ or } x=p+q$$

57. Solving for x by factoring, we have:

$$12x^2=ax+20a^2$$
$$12x^2-ax-20a^2=0$$
$$(4x+5a)(3x-4a)=0$$
$$x=-\frac{5a}{4} \text{ or } x=\frac{4a}{3}$$

59. Solving for r using the quadratic formula, we have:

$$2\pi r^2+2\pi rh=20\pi$$
$$2\pi r^2+2\pi rh-20\pi=0$$
$$2\pi\left(r^2+rh-10\right)=0$$
$$r^2+rh-10=0$$
$$r=\frac{-h\pm\sqrt{h^2-4(-10)}}{2(1)}=\frac{-h\pm\sqrt{h^2+40}}{2}$$

61. Solving for t by factoring, we have:

$$-16t^2+v_0t=0$$
$$t\left(-16t+v_0\right)=0$$
$$t=0 \text{ or } t=\frac{v_0}{16}$$

63. Solving for x by factoring:

$$x^3 + bx^2 - 2b^2x = 0$$
$$x(x^2 + bx - 2b^2) = 0$$
$$x(x + 2b)(x - b) = 0$$
$$x = 0, -2b, b$$

65. Given $a = 1$, $b = -12$, $c = 16$, we compute the discriminant:

$$b^2 - 4ac = 144 - 4(1)(16) = 80 > 0$$

The equation has two real roots.

67. Given $a = 4$, $b = -5$, $c = -\frac{1}{2}$, we compute the discriminant:

$$b^2 - 4ac = 25 - 4(4)\left(-\tfrac{1}{2}\right) = 33 > 0$$

The equation has two real roots.

69. Given $a = 1$, $b = \sqrt{3}$, $c = \frac{3}{4}$, we compute the discriminant:

$$b^2 - 4ac = 3 - 4(1)\left(\tfrac{3}{4}\right) = 0$$

The equation has one real root.

71. Given $a = 1$, $b = -\sqrt{5}$, $c = 1$, we compute the discriminant:

$$b^2 - 4ac = 5 - 4(1)(1) = 1 > 0$$

The equation has two real roots.

73. We set the discriminant equal to 0:

$$b^2 - 4ac = 0$$
$$144 - 4(1)(k) = 0$$
$$4k = 144$$
$$k = 36$$

75. We set the discriminant equal to 0:

$$b^2 - 4ac = 0$$
$$k^2 - 4(1)(5) = 0$$
$$k = \pm\sqrt{20} = \pm 2\sqrt{5}$$

77. Using the quadratic formula, we have:

$$x = \frac{-3 \pm \sqrt{9 - 4(1)(2.249)}}{2(1)} \approx -1.47, -1.53$$

79. Multiplying through by the least common denominator $x(x + 5)$ gives us:

$$\begin{aligned} 3x+4(x+5) &= 2x(x+5) \\ 7x+20 &= 2x^2+10x \\ -2x^2-3x+20 &= 0 \\ 2x^2+3x-20 &= 0 \\ (2x-5)(x+4) &= 0 \\ x = \tfrac{5}{2} &\text{ or } x = -4 \end{aligned}$$

Both of these values check in the original equation.

81. Multiplying through by the least common denominator $6x + 1$ gives us:

$$\begin{aligned} -6x^2+5x-1 &= 0 \\ 6x^2-5x+1 &= 0 \\ (3x-1)(2x-1) &= 0 \\ x = \tfrac{1}{3} &\text{ or } x = \tfrac{1}{2} \end{aligned}$$

Both values check in the original equation.

83. Multiplying through by the least common denominator $(x + 1)(x - 2)(x - 3)$ gives us:

$$\begin{aligned} 3x^2-6x-3+(5-2x)(x+1) &= 0 \\ 3x^2-6x-3-2x^2+3x+5 &= 0 \\ x^2-3x+2 &= 0 \\ (x-2)(x-1) &= 0 \\ x = 2 &\text{ or } x = 1 \end{aligned}$$

Of these two values, only $x = 1$ checks in the original equation.

85. Multiplying through by the least common denominator $(x - 1)(x + 1)(x - 3)$ gives us:

$$\begin{aligned} 2x(x+3)-(x-1)(x+1) &= 0 \\ 2x^2+6x-x^2+1 &= 0 \\ x^2+6x+1 &= 0 \\ x = \frac{-6\pm\sqrt{36-4}}{2} &= \frac{-6\pm4\sqrt{2}}{2} = -3\pm2\sqrt{2} \end{aligned}$$

Both values check in the original equation.

87. (a) Making the desired substitution $x = y + k$, we have:

$$\begin{aligned} (y+k)^2+(y+k)-1 &= 0 \\ y^2+2ky+k^2+y+k-1 &= 0 \\ y^2+(2k+1)y &= 1-k-k^2 \end{aligned}$$

(b) When $k = -\frac{1}{2}$, we have:

$$\begin{aligned} y^2+(-1+1)y &= 1+\tfrac{1}{2}-\tfrac{1}{4} \\ y^2 &= \tfrac{5}{4} \end{aligned}$$

(c) Solving for y, we have $y = \pm\frac{\sqrt{5}}{2}$. Since $x = y + k$, $x = -\frac{1}{2}\pm\frac{\sqrt{5}}{2}$.

89. If the roots are A and B, then:

$$x^2 + Ax + B = (x - A)(x - B)$$
$$x^2 + Ax + B = x^2 - (A + B)x + AB$$

Setting the coefficients equal, we have:

$$-(A + B) = A \quad \text{and} \quad AB = B$$
$$-A - B = A \qquad B(A - 1) = 0$$
$$2A + B = 0 \qquad B = 0 \text{ or } A = 1$$

If $B = 0$, then $A = 0$, and if $A = 1$, then $B = -2$. So the values of A and B are:

$$A = 0 \text{ and } B = 0$$
$$A = 1 \text{ and } B = -2$$

91. Let $\alpha = \dfrac{-b + \sqrt{b^2 - 4ac}}{2a}$ and $\beta = \dfrac{-b - \sqrt{b^2 - 4ac}}{2a}$.

(a) Working from the left-hand side, we have:

$$\alpha + \beta = \frac{-2b}{2a} = \frac{-b}{a}$$

(b) Working from the left-hand side, we have:

$$\alpha\beta = \frac{b^2 - (b^2 - 4ac)}{4a^2} = \frac{b^2 - b^2 + 4ac}{4a^2} = \frac{4ac}{4a^2} = \frac{c}{a}$$

(c) Using the hint, we have:

$$\alpha^2 + \beta^2 = (\alpha + \beta)^2 - 2\alpha\beta = \left(-\frac{b}{a}\right)^2 - 2\left(\frac{c}{a}\right) = \frac{b^2}{a^2} - \frac{2ac}{a^2} = \frac{b^2 - 2ac}{a^2}$$

(d) Using the hint, we have:

$$\frac{1}{\alpha^2} + \frac{1}{\beta^2} = \frac{\alpha^2 + \beta^2}{\alpha^2\beta^2} = \frac{\alpha^2 + \beta^2}{(\alpha\beta)^2} = \frac{\frac{b^2 - 2ac}{a^2}}{\left(\frac{c}{a}\right)^2} = \frac{b^2 - 2ac}{c^2}$$

93. (a) Let r be one root, thus $r^2 + pr + q = 0$. Since $3r$ will be another root, then $(3r)^2 + p(3r) + q = 0$, thus $9r^2 + 3pr + q = 0$. Therefore:

$$r^2 + pr = 9r^2 + 3pr$$
$$8r^2 = -2pr$$
$$8r^2 + 2pr = 0$$
$$2r(4r + p) = 0$$
$$r = 0 \text{ or } r = -\frac{p}{4}$$

Now $r = 0$ implies $q = 0$, which is impossible. So $r = \frac{-p}{4}$, and:

$$\left(-\frac{p}{4}\right)^2 + p\left(-\frac{p}{4}\right) + q = 0$$

$$\frac{p^2}{16} - \frac{p^2}{4} + q = 0$$

$$\frac{-3p^2}{16} + q = 0$$

$$-3p^2 + 16q = 0$$

$$16q = 3p^2$$

$$\frac{p^2}{q} = \frac{16}{3}$$

(b) Let r be one root, so $r^2 + pr + q = 0$. Since nr will be another root, then $(nr)^2 + p(nr) + q = 0$, so $n^2r^2 + npr + q = 0$. Therefore:

$$r^2 + pr = n^2r^2 + npr$$

$$r + p = n^2r + np \quad (\text{since } r \neq 0)$$

$$p - np = n^2r - r$$

$$p(1-n) = \left(n^2 - 1\right)r$$

$$r = \frac{p(1-n)}{n^2 - 1} = \frac{-p}{n+1}$$

Therefore:

$$\left(\frac{-p}{n+1}\right)^2 + p\left(\frac{-p}{n+1}\right) + q = 0$$

$$\frac{p^2}{(n+1)^2} - \frac{p^2}{n+1} + q = 0$$

$$p^2 - (n+1)p^2 + (n+1)^2 q = 0$$

$$(n+1)^2 q = np^2$$

$$\frac{p^2}{q} = \frac{(n+1)^2}{n}$$

95. The equation can be solved by means of the quadratic formula. Alternately, we have the following solution by factoring:

$$[abx - (a+b)][(a+b)x + 2] = 0$$

$$abx = a + b \quad \text{or} \quad (a+b)x + 2 = 0$$

$$x = \frac{a+b}{ab} \quad \text{or} \quad x = \frac{-2}{a+b}$$

The two solutions are $\frac{a+b}{ab}$ and $\frac{-2}{a+b}$.

2.4 Applications

1. Denote the two numbers by x and $7-x$. Since the difference is 17, we have:

$$x-(7-x)=17$$
$$2x=24$$
$$x=12$$

Then $7-x=7-12=-5$. The two numbers are 12 and -5.

3. Denoting the two numbers by x and $17-x$, we have:

$$x(17-x)=52$$
$$x^2-17x+52=0$$
$$(x-13)(x-4)=0$$
$$x=13 \text{ or } x=4$$

If $x=13$, then $17-x=4$. Similarly, if $x=4$, then $17-x=13$. So in either case, we find the two numbers are 13 and 4.

5. Let x denote the number. Then we have:

$$x+2x=63$$
$$3x=63$$
$$x=21$$

The number is 21.

7. Let x denote the positive number. Then:

$$x(x+4)=96$$
$$x^2+4x-96=0$$
$$(x+12)(x-8)=0$$
$$x=-12 \text{ or } x=8$$

Since x is supposed to be positive, we discard the first solution. The required positive number is 8.

9. Letting x denote the required number, we have:

$$\frac{3+x}{7+x}=\frac{2}{3}$$
$$9+3x=14+2x$$
$$x=14-9$$
$$x=5$$

The required number is 5.

11. Call the numbers x and y. Then we have $x-y=\frac{1}{3}(x+y)$. Solving this for x in terms of y yields $x=2y$. Since the sum of the squares is 180, we have:

$$(2y)^2+y^2=180$$
$$5y^2=180$$
$$y^2=36$$
$$y=\pm 6$$

Since y is supposed to be positive, we have $y=6$ and $x=2(6)=12$. The two numbers are 12 and 6.

13. Letting x denote the required score, we have:

$$\frac{70+77+75+x}{4}=75$$
$$222+x=300$$
$$x=300-222$$
$$x=78$$

The required score is 78.

15. Letting x denote the fourth number, we have:

$$\tfrac{1}{4}(80+22+62+x)=52$$
$$164+x=208$$
$$x=208-164$$
$$x=44$$

The fourth number is 44.

17. Let x denote the length of the hypotenuse. Then the lengths of the other two sides are $x-8$ and $x-1$, and we have:

$$(x-1)^2+(x-8)^2=x^2$$
$$x^2-18x+65=0$$
$$(x-13)(x-5)=0$$
$$x=13 \text{ or } x=5$$

If $x = 13$, the other two sides are $13 - 8 = 5$, and $13 - 1 = 12$. The second solution $x = 5$ must be discarded, for in that case $x - 8$ is negative. The required lengths are 5 cm, 12 cm, and 13 cm.

19. Letting x denote the length of a side in the original square, we have:

$$(x+2)^2=x^2+14$$
$$4x=10$$
$$x=\tfrac{5}{2}$$

The length of a side in the original square is 2.5 cm.

21. If x and y are the dimensions of the rectangle, we have:

$$2x+2y=104$$
$$x+y=52$$
$$y=52-x$$

Since the area is 640 cm^2, we have:

$$x(52-x)=640$$
$$-x^2+52x=640$$
$$x^2-52x+640=0$$
$$(x-20)(x-32)=0$$
$$x=20 \text{ or } x=32$$

Now if $x = 20$, then $y = 52 - 20 = 32$. Similarly, if $x = 32$, then $y = 52 - 32 = 20$. So in either case the shorter side is 20 cm.

23. If the shortest side is 2 in., there are 10 in. left for the other two sides. Call those sides x and $10 - x$. Then:

$$2^2 + x^2 = (10 - x)^2$$
$$4 + x^2 = 100 - 20x + x^2$$
$$20x = 96$$
$$x = \tfrac{96}{20}$$
$$x = 4.8$$

The other two sides of the triangle are 4.8 in. and 10 – 4.8 = 5.2 in.

25. Let x denote the length of the base. Then the legs each have length $3x$ and we have:

$$x + 3x + 3x = 70$$
$$7x = 70$$
$$x = 10$$

The three sides are 10 cm, 30 cm, and 30 cm.

27. Letting t represent the number of years until the total interest is \$5840, we have:

$$(4000)(0.07)t + (5000)(0.09)t = 5840$$
$$280t + 450t = 5840$$
$$730t = 5840$$
$$t = \tfrac{5840}{730}$$
$$t = 8$$

Thus it will take 8 years until the interest is \$5840.

29. Let t be the number of years until the balance in the account reaches \$21,000. Then we have:

$$10{,}000 + (10{,}000)(0.11)t = 21{,}000$$
$$10{,}000 + 1100t = 21{,}000$$
$$100 + 11t = 210$$
$$11t = 110$$
$$t = 10$$

It will take 10 years for the balance to reach \$21,000.

31. Let i be the overall interest rate on the entire investment of \$11,500. Then we have:

$$(11{,}500)(i)(1) = (7500)(0.075)(1) + (4000)(0.05)(1)$$
$$11500i = 762.5$$
$$i = \tfrac{762.5}{11500}$$
$$i \approx 0.06630$$

To convert this value to a percent, we multiply by 100. Rounding off to the nearest tenth of one percent, we obtain $i \approx 6.6\%$.

33. Let the interest rate for the \$8000 account be $r\%$. Then the rate for the \$9000 account is $(r + 1)\%$. Since the \$9000 account earns \$165 more than the \$8000 account each year, we have:

$$\begin{aligned}(8000)\left(\tfrac{r}{100}\right)+165&=(9000)\left(\tfrac{r+1}{100}\right)\\ 80r+165&=90r+90\\ -10r&=-75\\ r&=7.5\end{aligned}$$

The interest rate on the \$8000 account is 7.5%, and the rate on the \$9000 account is 8.5%.

35. Let x be the number of cc of the 10% solution, and let $200-x$ be the number of cc of the 35% solution. Note the following table:

type of solution	cc	% acid	total acid (cc)
10% solution	x	10%	$0.10(x)$
35% solution	$200-x$	35%	$0.35(200-x)$
mixture	200	25%	$0.25(200)$

Since the total acid in the mixture comes from that in the two given solutions, we have:

$$\begin{aligned}0.10(x)+0.35(200-x)&=0.25(200)\\ 10x+35(200-x)&=25(200)\\ 10x+7000-35x&=5000\\ -25x&=-2000\\ x&=80\end{aligned}$$

Then $200-x=200-80=120$. The student should use 80 cm^3 of the 10% solution and 120 cm^3 of the 35% solution.

37. Let x denote the required number of pounds of the first type of coffee. Note the following table:

type of coffee	pounds	cost / pound	total cost
first type	x	\$5.20	$5.20(x)$
second type	5	\$5.80	$5.80(5)$
blend	$x+5$	\$5.35	$5.35(x+5)$

Looking at the total cost column in our table, we obtain the equation:

$$\begin{aligned}5.20(x)+5.80(5)&=5.35(x+5)\\ 520x+580(5)&=535(x+5)\\ 520x+2900&=535x+2675\\ -15x&=-225\\ x&=15\end{aligned}$$

Thus 15 pounds of the first type of coffee are required.

39. Let x denote the required number of tons of the second ore. Note the following table:

type of ore	tons	% iron	total iron (tons)
first type	16	20%	0.20(16)
second type	x	35%	0.35(x)
blend	$16+x$	25%	$0.25(16+x)$

By considering the total iron, we have:

$$0.20(16)+0.35(x)=0.25(16+x)$$
$$20(16)+35x=25(16+x)$$
$$320+35x=400+25x$$
$$10x=80$$
$$x=8$$

Thus 8 tons of the second ore are required.

41. Denoting the lengths of the two shorter sides by x and $x-7$, we have:

$$\tfrac{1}{2}x(x-7)=60$$
$$x^2-7x-120=0$$
$$(x-15)(x+8)=0$$
$$x=15 \text{ or } x=-8$$

We discard the negative root since x represents a length. Thus $x=15$ and $x-7=8$. Therefore the hypotenuse is:

$$\sqrt{15^2+8^2}=\sqrt{289}=17 \text{ cm}$$

43. The area of the square and the rectangle are $(1-x)^2$ and $(1)(x)$, respectively. Therefore:

$$(1-x)^2=x$$
$$x^2-3x+1=0$$
$$x=\frac{3\pm\sqrt{9-4}}{2}=\frac{3\pm\sqrt{5}}{2}$$

We choose the negative root here, since the quantity $1-x$ is supposed to be positive.

Therefore $x=\frac{3-\sqrt{5}}{2}$.

45. We set up the proportion:

$$\frac{AC}{AB}=\frac{AB}{BC}$$
$$AC \bullet BC=(AB)^2=1^2=1$$
$$AC \bullet (AC-1)=1$$
$$(AC)^2-AC-1=0$$
$$AC=\frac{1\pm\sqrt{1-4(-1)}}{2}=\frac{1\pm\sqrt{5}}{2}$$

We choose the positive root here since $AC>0$. The required ratio is, therefore:

$$\frac{AC}{AB}=\frac{AC}{1}=\frac{1+\sqrt{5}}{2}$$

47. (a) When the ball lands $h = 0$. Therefore:

$$-16t^2 + 96t = 0$$
$$-16t(t-6) = 0$$
$$t = 0 \text{ or } t = 6$$

The value $t = 0$ gives the time when the ball is first thrown. Consequently the ball lands after 6 seconds.

(b) With $h = 80$, the equation becomes:

$$80 = -16t^2 + 96t$$
$$16t^2 - 96t + 80 = 0$$
$$t^2 - 6t + 5 = 0$$
$$(t-5)(t-1) = 0$$
$$t = 5 \text{ or } t = 1$$

At $t = 1$ second the ball is 80 ft. high and rising. At $t = 5$ seconds the ball is 80 ft high and falling.

49. After t minutes, the distance covered by Emily and Ernie are $\frac{t}{9}$ and $\frac{t}{11}$, respectively.

Emily will lap Ernie when her distance exceeds Ernie's by $\frac{1}{4}$ mile. In that case we have:

$$\frac{t}{9} = \frac{t}{11} + \frac{1}{4}$$
$$44t = 36t + 99$$
$$8t = 99$$
$$t = \frac{99}{8}$$

It will take $12\frac{3}{8}$ minutes for Emily to lap Ernie on the track. This is about 12 minutes and 22 seconds.

51. Let r and $r - 5$ denote the speeds of the eastbound and westbound trains, respectively. When $t = 6$ hours the total distance traveled by the two trains is 630 mi, so we have:

$$r(6) + (r-5)6 = 630$$
$$6r + 6r - 30 = 630$$
$$12r = 660$$
$$r = 55$$

The speed of the eastbound train is 55 mph.

53. Let r and $r + 50$ represent the speeds of the slower and faster planes, respectively. When $t = 2$ the total distance covered is 1500, so we have:

$$r(2) + (r+50)2 = 1500$$
$$4r = 1400$$
$$r = 350$$

The speed of the slower plane is 350 mph.

55. Let D be the distance that the boat can travel out to sea, and let t be the time for that part of the trip. Then the time for the return trip is $2 - t$. For the trip out to sea, we have $D = 20t$, and for the trip back $D = 12(2 - t)$. Solving the first equation for t yields $t = \frac{D}{20}$, and then substituting in the second equation:

$$\begin{aligned} D &= 12\left(2 - \tfrac{D}{20}\right) \\ D &= 24 - \tfrac{3D}{5} \\ 5D &= 120 - 3D \\ 8D &= 120 \\ D &= 15 \end{aligned}$$

The boat can travel 15 miles out to sea before heading back.

57. Let x represent the additional amount that must be invested. Then the total of the two deposits and one investment (in dollars) is $2650 + 3300 + x$, or $5950 + x$. Since the overall return on this investment is to be 10%, we have:

$$\begin{aligned} 0.1(5950 + x) &= 0.081(2650) + 0.092(3300) + 0.115x \\ 595 + 0.1x &= 214.65 + 303.60 + 0.115x \\ -0.015x &= -76.75 \\ x &\approx 5116.67 \end{aligned}$$

The additional investment at 11.5% should be \$5117.

59. Let Linda's initial rate be r. When Larry has covered 2 miles, the elapsed time is $\frac{2}{1/15} = 30$ minutes. In that 30 minutes Linda has walked a distance of $r(30)$. At 1:15 (75 minutes past noon), the distances covered by Larry and Linda are equal, and consequently:

Larry's total distance	=	Linda's distance in first 30 min.	+	Linda's distance in next 45 min.
$\left(\frac{1}{15}\right)(75)$	=	$30r$	+	$(2r)(45)$

Therefore:

$$\begin{aligned} 5 &= 30r + 90r \\ 5 &= 120r \\ r &= \tfrac{5}{120} \\ r &= \tfrac{1}{24} \end{aligned}$$

Linda's initial rate was $\frac{1}{24}$ mile/minute.

61. Let D denote the distance from town A to town B. Then the time to go from town A to town B is $\frac{D}{v_1}$, and the time for the return trip is $\frac{D}{v_2}$. Thus we have:

$$\text{average speed} = \frac{\text{total distance}}{\text{total time}} = \frac{2D}{\frac{D}{v_1} + \frac{D}{v_2}} = \frac{2}{\frac{1}{v_1} + \frac{1}{v_2}} = \frac{2v_1v_2}{v_1 + v_2} \text{ mph}$$

63. Let t denote the required number of years. Then:

$$\begin{aligned} d\left(\tfrac{r}{100}\right)t + D\left(\tfrac{R}{100}\right)t &= d + D \\ drt + DRt &= 100(d + D) \\ t(dr + DR) &= 100(d + D) \\ t &= \frac{100(d + D)}{dr + DR} \text{ years} \end{aligned}$$

65. Let x denote the length of the hypotenuse. Since the perimeter is a and the length of the shortest side is b, the length of the third side is $a - b - x$. By the Pythagorean theorem we have:

$$\begin{aligned} x^2 &= (a - b - x)^2 + b^2 \\ x^2 &= (a - b)^2 - 2(a - b)x + x^2 + b^2 \\ 2(a - b)x &= a^2 - 2ab + 2b^2 \\ x &= \frac{a^2 - 2ab + 2b^2}{2(a - b)} \text{ inches} \end{aligned}$$

67. Denoting the length of a side of the original square by x, we have:

$$\begin{aligned} (ax)^2 &= x^2 + b^2 \\ a^2x^2 - x^2 &= b^2 \\ x^2\left(a^2 - 1\right) &= b^2 \\ x^2 &= \frac{b^2}{a^2 - 1} \\ x &= \frac{b}{\sqrt{a^2 - 1}} \end{aligned}$$

The length of each side in the original square is $\frac{b}{\sqrt{a^2-1}}$ units. This quantity can also be written as $\frac{b\sqrt{a^2-1}}{a^2-1}$ units.

69. Let x and $16 - x$ denote the lengths of the two pieces. So the two circumferences are x and $16 - x$, respectively. We can find the corresponding radii by writing the formula $C = 2\pi r$ as $r = \frac{C}{2\pi}$. Thus the corresponding radii are $\frac{x}{2\pi}$ and $\frac{16-x}{2\pi}$, respectively. Now if the sum of the two areas is 12 cm^2, we have:

$$\begin{aligned} \pi\left(\frac{x}{2\pi}\right)^2 + \pi\left(\frac{16 - x}{2\pi}\right)^2 &= 12 \\ \frac{x^2}{4\pi} + \frac{256 - 32x + x^2}{4\pi} &= 12 \\ 2x^2 - 32x + 256 &= 48\pi \\ x^2 - 16x + 128 &= 24\pi \\ x^2 - 16x + (128 - 24\pi) &= 0 \end{aligned}$$

Applying the quadratic formula with $a = 1$, $b = -16$, and $c = 128 - 24\pi$ yields:

$$\begin{aligned} x &= \frac{16 \pm \sqrt{256 - 4(128 - 24\pi)}}{2} \\ &= \frac{16 \pm \sqrt{96\pi - 256}}{2} \\ &= \frac{16 \pm 4\sqrt{6\pi - 16}}{2} \\ &= 8 \pm 2\sqrt{6\pi - 16} \end{aligned}$$

Both roots here yield positive values for x. First consider the positive sign. Then $x = 8 + 2\sqrt{6\pi - 16}$ and consequently $16 - x = 8 - 2\sqrt{6\pi - 16}$ (which is the same as the value obtained for x using the negative root). Similarly, if we begin with $x = 8 - 2\sqrt{6\pi - 16}$, we find that $16 - x$ is $8 + 2\sqrt{6\pi - 16}$. So in either case the lengths are:

$$8 + 2\sqrt{6\pi - 16} \quad \text{and} \quad 8 - 2\sqrt{6\pi - 16}$$

The smaller of these two numbers is $8 - 2\sqrt{6\pi - 16}$. Using a calculator and rounding to two decimal places, we find the required length to be 4.62 cm.

71. Call the side x, thus we have the following figure:

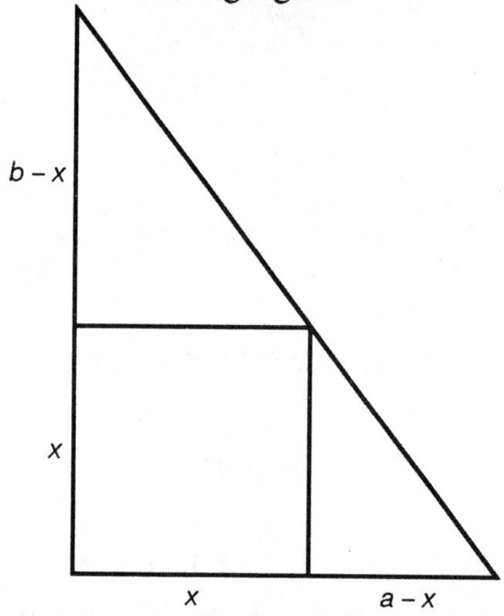

By similar triangles, we have:

$$\begin{aligned} \frac{x}{a - x} &= \frac{b - x}{x} \\ x^2 &= ab - bx - ax + x^2 \\ (a + b)x &= ab \\ x &= \frac{ab}{a + b} \end{aligned}$$

2.5 Other Types of Equations

1. Factoring, we have:

$$\begin{aligned} 3x^2 - 48x &= 0 \\ 3x(x-16) &= 0 \\ x &= 0, 16 \end{aligned}$$

The solutions are 0 and 16.

3. Factoring, we have:

$$\begin{aligned} t^3 - 125 &= 0 \\ (t-5)\left(t^2 + 5t + 25\right) &= 0 \end{aligned}$$

Setting the first factor equal to zero yields $t = 5$. Setting the second factor equal to zero yields a quadratic equation with no real roots. The only solution is 5.

5. Factoring, we have:

$$\begin{aligned} 7x^4 - 28x^2 &= 0 \\ 7x^2\left(x^2 - 4\right) &= 0 \\ 7x^2(x-2)(x+2) &= 0 \\ x &= 0, 2, -2 \end{aligned}$$

The solutions are 0, 2, and –2.

7. Factoring, we have:

$$\begin{aligned} 225(x-1) - x^2(x-1) &= 0 \\ (x-1)\left(225 - x^2\right) &= 0 \\ (x-1)(15-x)(15+x) &= 0 \\ x &= 1, 15, -15 \end{aligned}$$

The solutions are 1, 15, and –15.

9. Factoring, we have:

$$\begin{aligned} 4y^3 - 20y^2 + 25y &= 0 \\ y\left(4y^2 - 20y + 25\right) &= 0 \\ y(2y-5)^2 &= 0 \\ y &= 0, \tfrac{5}{2} \end{aligned}$$

The solutions are 0 and $\frac{5}{2}$.

11. Factoring, we have:

$$\begin{aligned} t^4 + 2t^3 - 3t^2 &= 0 \\ t^2\left(t^2 + 2t - 3\right) &= 0 \\ t^2(t+3)(t-1) &= 0 \\ t &= 0, -3, 1 \end{aligned}$$

The solutions are 0, –3, and 1.

13. Factoring, we have:

$$6x = 23x^2 + 4x^3$$
$$4x^3 + 23x^2 - 6x = 0$$
$$x(4x^2 + 23x - 6) = 0$$
$$x(4x - 1)(x + 6) = 0$$
$$x = 0, \tfrac{1}{4}, -6$$

The solutions are 0, $\frac{1}{4}$, and -6.

15. Factoring, we have:

$$x^4 - x^2 = 6$$
$$x^4 - x^2 - 6 = 0$$
$$(x^2 - 3)(x^2 + 2) = 0$$
$$x = \pm\sqrt{3}$$

Note that $x^2 + 2 \neq 0$. The solutions are $-\sqrt{3}$ and $\sqrt{3}$.

17. Factoring, we have:

$$4y^2 = 5 - y^4$$
$$y^4 + 4y^2 - 5 = 0$$
$$(y^2 + 5)(y^2 - 1) = 0$$
$$(y^2 + 5)(y + 1)(y - 1) = 0$$
$$y = -1, 1$$

Note that $y^2 + 5 \neq 0$. The solutions are -1 and 1.

19. Factoring, we have:

$$3t^2 + 2 = 9t^4$$
$$9t^4 - 3t^2 - 2 = 0$$
$$(3t^2 + 1)(3t^2 - 2) = 0$$
$$3t^2 = 2$$
$$t = \pm\sqrt{\tfrac{2}{3}} = \pm\tfrac{\sqrt{6}}{3}$$

Note that $3t^2 + 1 \neq 0$. The solutions are $\pm\frac{\sqrt{6}}{3}$.

21. Taking roots, we have:

$$(x - 2)^3 - 5 = 0$$
$$(x - 2)^3 = 5$$
$$x - 2 = \sqrt[3]{5}$$
$$x = 2 + \sqrt[3]{5} \approx 3.71$$

The only solution is $2 + \sqrt[3]{5} \approx 3.71$.

23. Taking roots, we have:

$$\begin{aligned}(x+4)^5+16&=0\\(x+4)^5&=-16\\x+4&=-\sqrt[5]{16}\\x&=-4-\sqrt[5]{16}\approx-5.74\end{aligned}$$

The only solution is $-4-\sqrt[5]{16}\approx-5.74$.

25. (a) Taking roots, we have:

$$\begin{aligned}(x-3)^4-30&=0\\(x-3)^4&=30\\x-3&=\pm\sqrt[4]{30}\\x&=3\pm\sqrt[4]{30}\approx5.34,0.66\end{aligned}$$

The solutions are $3+\sqrt[4]{30}\approx5.34$ and $3-\sqrt[4]{30}\approx0.66$.

(b) Solving for x, we have:

$$\begin{aligned}(x-3)^4+30&=0\\(x-3)^4&=-30\end{aligned}$$

Since $(x-3)^4\geq0$ for real values of x, there are no solutions to this equation.

27. Factoring, we have:

$$\begin{aligned}y^4+4y^2-5&=0\\(y^2-1)(y^2+5)&=0\\(y-1)(y+1)(y^2+5)&=0\\y&=1,-1\end{aligned}$$

Note that $y^2+5\neq0$. The solutions are ±1.

29. Factoring, we have:

$$\begin{aligned}9t^4-3t^2-2&=0\\(3t^2-2)(3t^2+1)&=0\\3t^2-2&=0\\t^2&=\tfrac{2}{3}\\t&=\pm\sqrt{\tfrac{2}{3}}=\pm\tfrac{\sqrt{6}}{3}\end{aligned}$$

Note that $3t^2+1\neq0$. The solutions are $\pm\frac{\sqrt{6}}{3}$.

31. Factoring, we have:

$$\begin{aligned}x^6-10x^4+24x^2&=0\\x^2(x^4-10x^2+24)&=0\\x^2(x^2-6)(x^2-4)&=0\\x^2(x^2-6)(x-2)(x+2)&=0\\x&=0,\pm\sqrt{6},\pm2\end{aligned}$$

The solutions are 0, $\pm\sqrt{6}$, and ±2.

33. Letting $x^2 = t$ and $x^4 = t^2$ yields:

$$t^2 + t - 1 = 0$$

$$t = \frac{-1 \pm \sqrt{1 - 4(-1)}}{2} = \frac{-1 \pm \sqrt{5}}{2}$$

Since t must be non-negative (because $t = x^2$) we choose the value:

$$t = \frac{-1 + \sqrt{5}}{2}$$

Therefore:

$$x^2 = \frac{-1 + \sqrt{5}}{2}$$

$$x = \pm\sqrt{\frac{-1 + \sqrt{5}}{2}}$$

The solutions are $\pm\sqrt{\frac{-1+\sqrt{5}}{2}}$.

35. Letting $x^2 = t$ and $x^4 = t^2$ yields:

$$t^2 + 3t - 2 = 0$$

$$t = \frac{-3 \pm \sqrt{9 - 4(-2)}}{2} = \frac{-3 \pm \sqrt{17}}{2}$$

Since t must be non-negative (because $t = x^2$) we choose the value:

$$t = \frac{-3 + \sqrt{17}}{2}$$

Therefore:

$$x^2 = \frac{-3 + \sqrt{17}}{2}$$

$$x = \pm\sqrt{\frac{-3 + \sqrt{17}}{2}}$$

The solutions are $\pm\sqrt{\frac{-3+\sqrt{17}}{2}}$.

37. Letting $x^3 = t$ and $x^6 = t^2$, we find that the equation becomes:

$$t^2 + 7t - 8 = 0$$

$$(t + 8)(t - 1) = 0$$

$$t = -8, 1$$

So $x^3 = -8$ or $x^3 = 1$, thus $x = -2$ or $x = 1$. The solutions are -2 and 1.

39. Let $t^{-1} = x$ and $t^{-2} = x^2$. Then the equation becomes:

$$x^2 - 7x + 12 = 0$$

$$(x - 3)(x - 4) = 0$$

$$x = 3, 4$$

So $t^{-1} = 3$ or $t^{-1} = 4$, thus $t = \frac{1}{3}$ or $t = \frac{1}{4}$. The solutions are $\frac{1}{3}$ and $\frac{1}{4}$.

41. Letting $y^{-1} = t$ and $y^{-2} = t^2$, we have the equation:

$$12t^2 - 23t + 5 = 0$$
$$(3t - 5)(4t - 1) = 0$$
$$t = \frac{5}{3}, \frac{1}{4}$$

So $y^{-1} = \frac{5}{3}$ or $y^{-1} = \frac{1}{4}$, thus $y = \frac{3}{5}$ or $y = 4$. The solutions are $\frac{3}{5}$ and 4.

43. Letting $x^{-2} = t$ and $x^{-4} = t^2$, we have the equation:

$$4t^2 - 33t - 27 = 0$$
$$(4t + 3)(t - 9) = 0$$
$$t = -\frac{3}{4}, 9$$

So $x^{-2} = -\frac{3}{4}$ or $x^{-2} = 9$. If $x^{-2} = -\frac{3}{4}$ then $x^2 = -\frac{4}{3}$, which is impossible for real values of x since $x^2 \geq 0$. If $x^{-2} = 9$ then $x^2 = \frac{1}{9}$ and thus $x = \pm\frac{1}{3}$. The solutions are $\pm\frac{1}{3}$.

45. Raising both sides of the given equation to the power $\frac{3}{2}$ yields:

$$t = \pm 9^{3/2} = \pm\left(\sqrt{9}\right)^3 = \pm 27$$

Since $t = \pm 27$ both satisfy the original equation, the solutions are ± 27.

47. Taking fourth roots, we have $x = \pm\sqrt[4]{81} = \pm 3$. Since $x = \pm 3$ both satisfy the original equation, the solutions are ± 3.

49. Taking cube roots, we have:

$$(y - 1)^3 = 7$$
$$y - 1 = \sqrt[3]{7}$$
$$y = 1 + \sqrt[3]{7}$$

The solution is $1 + \sqrt[3]{7}$.

51. Taking fifth roots, we have:

$$(t + 1)^5 = -243$$
$$t + 1 = \sqrt[5]{-243}$$
$$t + 1 = -3$$
$$t = -4$$

The solution is -4.

53. Letting $x^{2/3} = t$ and $x^{4/3} = t^2$, the equation becomes:

$$9t^2 - 10t + 1 = 0$$
$$(t - 1)(9t - 1) = 0$$
$$t = 1, \frac{1}{9}$$

So $x^{2/3} = 1$ or $x^{2/3} = \frac{1}{9}$. Thus we have either of the two equations:

$$\begin{aligned} x^{2/3} &= 1 \\ \left(\sqrt[3]{x}\right)^2 &= 1 \\ \sqrt[3]{x} &= \pm 1 \\ x &= \pm 1 \end{aligned} \qquad\qquad \begin{aligned} x^{2/3} &= \tfrac{1}{9} \\ \left(\sqrt[3]{x}\right)^2 &= \tfrac{1}{9} \\ \sqrt[3]{x} &= \pm\tfrac{1}{3} \\ x &= \pm\tfrac{1}{27} \end{aligned}$$

The solutions are ± 1 and $\pm\frac{1}{27}$.

55. Squaring both sides yields:

$$\begin{aligned} \sqrt{1-3x} &= 2 \\ 1-3x &= 4 \\ -3x &= 3 \\ x &= -1 \end{aligned}$$

Upon checking, we find that –1 is the solution.

57. Isolating the radical and squaring yields:

$$\begin{aligned} \sqrt{x}+6 &= x \\ \sqrt{x} &= x-6 \\ x &= x^2-12x+36 \\ 0 &= x^2-13x+36 \\ 0 &= (x-9)(x-4) \\ x &= 4, 9 \end{aligned}$$

Upon checking, we find that 9 is the solution ($x = 4$ does not check).

59. Isolating the radical and squaring yields:

$$\begin{aligned} x-\sqrt{3-x} &= -3 \\ x+3 &= \sqrt{3-x} \\ x^2+6x+9 &= 3-x \\ x^2+7x+6 &= 0 \\ (x+6)(x+1) &= 0 \\ x &= -1, -6 \end{aligned}$$

Upon checking, we find that –1 is the solution ($x = -6$ does not check).

61. Isolating the radical and squaring yields:

$$\begin{aligned} 4x+\sqrt{2x+5} &= 0 \\ \sqrt{2x+5} &= -4x \\ 2x+5 &= 16x^2 \\ 0 &= 16x^2-2x-5 \\ 0 &= (8x-5)(2x+1) \\ x &= -\tfrac{1}{2}, \tfrac{5}{8} \end{aligned}$$

Upon checking, we find that $-\frac{1}{2}$ is the solution ($x = \frac{5}{8}$ does not check).

63. Letting $x^2 = t$ and $x^4 = t^2$, we have:

$$\sqrt{t^2-13t+37} = 1$$
$$t^2-13t+37 = 1$$
$$t^2-13t+36 = 0$$
$$(t-9)(t-4) = 0$$
$$t = 4, 9$$

Since $x^2 = t$, then $x^2 = 4$ or $x^2 = 9$, thus $x = \pm 2$ or $x = \pm 3$. Upon checking, we find that ± 2 and ± 3 are solutions.

65. Isolating $\sqrt{1-2x}$ and squaring yields:

$$\sqrt{1-2x} = 4-\sqrt{x+5}$$
$$1-2x = 16-8\sqrt{x+5}+x+5$$
$$8\sqrt{x+5} = 3x+20$$
$$64(x+5) = 9x^2+120x+400$$
$$0 = 9x^2+56x+80$$
$$0 = (x+4)(9x+20)$$
$$x = -4, -\tfrac{20}{9}$$

Upon checking, we find that -4 and $-\frac{20}{9}$ are solutions.

67. Isolating $\sqrt{3+2t}$ and squaring yields:

$$\sqrt{3+2t} = 1-\sqrt{-1+4t}$$
$$3+2t = 1-2\sqrt{-1+4t}-1+4t$$
$$2\sqrt{-1+4t} = 2t-3$$
$$4(-1+4t) = 4t^2-12t+9$$
$$0 = 4t^2-28t+13$$
$$0 = (2t-1)(2t-13)$$
$$t = \tfrac{1}{2}, \tfrac{13}{2}$$

Upon checking, we find that neither of these values are solutions.

69. Isolating $\sqrt{3y+3}$ and squaring yields:

$$\sqrt{2y-3}+\sqrt{3y-2} = \sqrt{3y+3}$$
$$2y-3+2\sqrt{2y-3}\sqrt{3y-2}+3y-2 = 3y+3$$
$$2\sqrt{6y^2-13y+6} = -2y+8$$
$$\sqrt{6y^2-13y+6} = -y+4$$
$$6y^2-13y+6 = y^2-8y+16$$
$$5y^2-5y-10 = 0$$
$$y^2-y-2 = 0$$
$$(y-2)(y+1) = 0$$
$$y = -1, 2$$

Upon checking, we find that 2 is the solution ($y = -1$ does not check).

71. Letting $x^2 = t$ and $x^4 = t^2$, we have:

$$t^2 - 2t - 4 = 0$$
$$t = \frac{2 \pm \sqrt{20}}{2} = \frac{2 \pm 2\sqrt{5}}{2} = 1 \pm \sqrt{5}$$

We want the positive value for t here, because $t = x^2$. Thus:

$$x^2 = 1 + \sqrt{5}$$
$$x = \pm\sqrt{1+\sqrt{5}}$$

Using a calculator and rounding to two decimal places, we find these values of x to be ± 1.80.

73. Isolating $2x\sqrt{x^2+4}$ and squaring yields:

$$2x\sqrt{x^2+4} + 2x\sqrt{x^2+1} = 3$$
$$2x\sqrt{x^2+4} = 3 - 2x\sqrt{x^2+1}$$
$$4x^2(x^2+4) = 9 - 12x\sqrt{x^2+1} + 4x^2(x^2+1)$$
$$12x^2 - 9 = -12x\sqrt{x^2+1}$$
$$4x^2 - 3 = -4x\sqrt{x^2+1}$$
$$16x^4 - 24x^2 + 9 = 16x^4 + 16x^2$$
$$-40x^2 = -9$$
$$x = \pm\sqrt{\frac{9}{40}} = \pm\frac{3}{2\sqrt{10}} = \pm\frac{3\sqrt{10}}{20} \approx \pm 0.47$$

The solution is 0.47 ($x = -0.47$ does not check).

75. Squaring both sides yields:

$$\sqrt{\sqrt{x}+\sqrt{a}} + \sqrt{\sqrt{x}-\sqrt{a}} = \sqrt{2\sqrt{x}+2\sqrt{b}}$$
$$\left(\sqrt{\sqrt{x}+\sqrt{a}} + \sqrt{\sqrt{x}-\sqrt{a}}\right)^2 = \left(\sqrt{2\sqrt{x}+2\sqrt{b}}\right)^2$$
$$\sqrt{x}+\sqrt{a} + 2\sqrt{\sqrt{x}+\sqrt{a}}\sqrt{\sqrt{x}-\sqrt{a}} + \sqrt{x}-\sqrt{a} = 2\sqrt{x}+2\sqrt{b}$$
$$\sqrt{\sqrt{x}+\sqrt{a}}\sqrt{\sqrt{x}-\sqrt{a}} = \sqrt{b}$$
$$(\sqrt{x}+\sqrt{a})(\sqrt{x}-\sqrt{a}) = b$$
$$x - a = b$$
$$x = a + b$$

Upon checking, we find that this value of x satisfies the original equation. Note: the following fact is useful in carrying out the check. Two non-negative quantities are equal if and only if their squares are equal.

77. Let $t = \sqrt{x}$. Then the given equation can be written as:

$$\frac{t-a}{t} = \frac{t+a}{t-b}$$
$$t^2 + at = t^2 - at - bt + ab$$
$$2at + bt = ab$$
$$t(2a+b) = ab$$
$$t = \frac{ab}{2a+b}$$

Thus $x = t^2 = \frac{a^2b^2}{(2a+b)^2}$. This value of x checks in the original equation.

79. Let $t = x^2 - x - 1$. Then the given equation becomes:

$$\sqrt{t} - \frac{2}{\sqrt{t}} = 1$$
$$t - 2 = \sqrt{t}$$
$$t^2 - 4t + 4 = t$$
$$t^2 - 5t + 4 = 0$$
$$(t-4)(t-1) = 0$$
$$t = 4 \text{ or } t = 1$$

Upon checking, we find that $t = 4$ satisfies the original equation involving t, but $t = 1$ does not. With $t = 4$ we have:

$$x^2 - x - 1 = 4$$
$$x^2 - x - 5 = 0$$

Now applying the quadratic formula:

$$x = \frac{1 \pm \sqrt{1 - 4(-5)}}{2} = \frac{1 \pm \sqrt{21}}{2}$$

81. With $t = \frac{x-a}{x}$ the given equation becomes:

$$\sqrt{t} + \frac{4}{\sqrt{t}} = 5$$
$$t + 4 = 5\sqrt{t}$$
$$t^2 + 8t + 16 = 25t$$
$$t^2 - 17t + 16 = 0$$
$$(t-1)(t-16) = 0$$
$$t = 1, 16$$

Note that both $t = 1$ and $t = 16$ satisfy the original equation involving t.
If $t = 1$, we have:

$$1 = \frac{x-a}{x}$$
$$x = x - a$$
$$0 = -a$$

Since $a \neq 0$, we discard this case. If $t = 16$, we have:

$$16 = \frac{x-a}{x}$$
$$16x = x - a$$
$$15x = -a$$
$$x = -\frac{a}{15}$$

83. (a) We'll show the verfication for $x=\sqrt{3+\sqrt{5}}$. (The other three cases are similar.) With $x=\sqrt{3+\sqrt{5}}$ the equation becomes:

$$\begin{aligned}
\left(\sqrt{3+\sqrt{5}}\right)^4-6\left(\sqrt{3+\sqrt{5}}\right)^2+4&=0\\
(3+\sqrt{5})^2-6(3+\sqrt{5})+4&=0\\
9+6\sqrt{5}+5-18-6\sqrt{5}+4&=0\\
(9+5-18+4)+6\sqrt{5}-6\sqrt{5}&=0\\
0&=0
\end{aligned}$$

(b) With $x=\frac{1}{8}$ we have:

$$\begin{aligned}
4\left(\tfrac{1}{8}\right)^{4/3}+15\left(\tfrac{1}{8}\right)^{2/3}-4&=0\\
4\left(\sqrt[3]{\tfrac{1}{8}}\right)^4+15\left(\sqrt[3]{\tfrac{1}{8}}\right)^2-4&=0\\
4\left(\tfrac{1}{2}\right)^4+15\left(\tfrac{1}{2}\right)^2-4&=0\\
\tfrac{4}{16}+\tfrac{15}{4}-4&=0\\
4-4&=0
\end{aligned}$$

The verification for $x=-\frac{1}{8}$ is similar.

85. We could begin by squaring both sides of the given equation. However, the following preliminary simplification saves some steps (and avoids the introduction of an extraneous root later):

$$\begin{aligned}
\sqrt{8+2t}+\sqrt{5+t}&=\sqrt{3(5+t)}\\
\sqrt{8+2t}&=\sqrt{3}\sqrt{5+t}-\sqrt{5+t}\\
\sqrt{8+2t}&=(\sqrt{3}-1)(\sqrt{5+t})
\end{aligned}$$

Now we square both sides to obtain:

$$\begin{aligned}
8+2t&=(\sqrt{3}-1)^2(5+t)\\
8+2t&=(4-2\sqrt{3})(5+t)\\
8+2t&=20-10\sqrt{3}+(4-2\sqrt{3})t\\
(2-4+2\sqrt{3})t&=12-10\sqrt{3}\\
t&=\frac{12-10\sqrt{3}}{-2+2\sqrt{3}}=\frac{6-5\sqrt{3}}{-1+\sqrt{3}}=\frac{\sqrt{3}-9}{2}
\end{aligned}$$

After a lengthy check, we find that $\frac{\sqrt{3}-9}{2}$ satisfies the original equation.

87. Letting $x^2 + x = t$, the given equation becomes:

$$\begin{aligned}\sqrt{t} + \frac{1}{\sqrt{t}} &= \frac{5}{2}\\ 2t + 2 &= 5\sqrt{t}\\ 4t^2 + 8t + 4 &= 25t\\ 4t^2 - 17t + 4 &= 0\\ (4t - 1)(t - 4) &= 0\\ t &= \tfrac{1}{4} \text{ or } t = 4\end{aligned}$$

Both of these values for t check in the equation $\sqrt{t} + \frac{1}{\sqrt{t}} = \frac{5}{2}$.

If $t = \frac{1}{4}$ we have:

$$\begin{aligned}x^2 + x &= \tfrac{1}{4}\\ 4x^2 + 4x - 1 &= 0\end{aligned}$$

Using the quadratic formula yields:

$$x = \frac{-4 \pm \sqrt{16 - 4(4)(-1)}}{2(4)} = \frac{-4 \pm \sqrt{32}}{8} = \frac{-4 \pm 4\sqrt{2}}{8} = \frac{-1 \pm \sqrt{2}}{2} \approx 0.21, -1.21$$

If $t = 4$ we have:

$$\begin{aligned}x^2 + x &= 4\\ x^2 + x - 4 &= 0\end{aligned}$$

Using the quadratic formula yields:

$$x = \frac{-1 \pm \sqrt{1 - 4(1)(-4)}}{2(1)} = \frac{-1 \pm \sqrt{17}}{2} \approx 1.56, -2.56$$

The solutions are 0.21, –1.21, 1.56, and –2.56.

89. If we rewrite using rational exponents, we obtain:

$$\begin{aligned}(a + x)^{2/m} + 2(a - x)^{2/m} &= 3(a + x)^{1/m}(a - x)^{1/m}\\ (a + x)^{2/m} - 3(a + x)^{1/m}(a - x)^{1/m} + 2(a - x)^{2/m} &= 0\\ \left[(a + x)^{1/m} - 2(a - x)^{1/m}\right]\left[(a + x)^{1/m} - (a - x)^{1/m}\right] &= 0\end{aligned}$$

Setting the first factor equal to 0 yields:

$$\begin{aligned}(a + x)^{1/m} &= 2(a - x)^{1/m}\\ a + x &= 2^m(a - x)\\ a + x &= 2^m a - 2^m x\\ (2^m + 1)x &= a(2^m - 1)\\ x &= \frac{a(2^m - 1)}{2^m + 1}\end{aligned}$$

Setting the second factor equal to 0 yields:

$$\begin{aligned}(a + x)^{1/m} &= (a - x)^{1/m}\\ a + x &= a - x\\ 2x &= 0\\ x &= 0\end{aligned}$$

The solutions are $x = 0$ and $x = \frac{a(2^m - 1)}{2^m + 1}$.

2.6 Inequalities

1. We solve the inequality:

$$x+5<4$$
$$x<-1$$

The solution set is $(-\infty,-1)$.

3. We solve the inequality:

$$1-3x\le 0$$
$$-3x\le -1$$
$$x\ge \tfrac{1}{3}$$

The solution set is $\left[\tfrac{1}{3},\infty\right)$.

5. We solve the inequality:

$$4x+6<3(x-1)-x$$
$$4x+6<2x-3$$
$$2x<-9$$
$$x<-\tfrac{9}{2}$$

The solution set is $\left(-\infty,-\tfrac{9}{2}\right)$.

7. We solve the inequality:

$$1-2(t+3)-t\le 1-2t$$
$$1-2t-6-t\le 1-2t$$
$$-6-t\le 0$$
$$-t\le 6$$
$$t\ge -6$$

The solution set is $[-6,\infty)$.

9. Multiplying by 15, we solve the inequality:

$$\frac{3x}{5}-\frac{x-1}{3}<1$$
$$9x-5(x-1)<15$$
$$4x+5<15$$
$$4x<10$$
$$x<\tfrac{5}{2}$$

The solution set is $\left(-\infty,\tfrac{5}{2}\right)$.

11. Multiplying by 20, we solve the inequality:

$$\begin{aligned} \frac{x-1}{4}-\frac{2x+3}{5} &\le x \\ 5(x-1)-4(2x+3) &\le 20x \\ 5x-5-8x-12 &\le 20x \\ -23x &\le 17 \\ x &\ge -\tfrac{17}{23} \end{aligned}$$

The solution set is $\left[-\frac{17}{23},\infty\right)$.

13. We solve the double inequality:

$$\begin{aligned} -2 \le x-6 \le 0 \\ 4 \le x \le 6 \end{aligned}$$

The solution set is $[4,6]$.

15. Multiplying by 3, we solve the double inequality:

$$\begin{aligned} -1 \le \frac{1-4t}{3} \le 1 \\ -3 \le 1-4t \le 3 \\ -4 \le -4t \le 2 \\ 1 \ge t \ge -\tfrac{1}{2} \\ -\tfrac{1}{2} \le t \le 1 \end{aligned}$$

The solution set is $\left[-\frac{1}{2},1\right]$.

17. Multiplying by 2, we solve the double inequality:

$$\begin{aligned} 0.99 < \frac{x}{2}-1 < 0.999 \\ 1.98 < x-2 < 1.998 \\ 3.98 < x < 3.998 \end{aligned}$$

The solution set is $(3.98, 3.998)$.

19. (a) The inequality is equivalent to $-\frac{1}{2} \le x \le \frac{1}{2}$, so the solution set is $\left[-\frac{1}{2},\frac{1}{2}\right]$.

(b) The inequality is equivalent to $x > \frac{1}{2}$ or $x < -\frac{1}{2}$, so the solution set is $\left(-\infty,-\frac{1}{2}\right)\cup\left(\frac{1}{2},\infty\right)$.

21. (a) The inequality is equivalent to $x > 0$ or $x < 0$, so the solution set is $(-\infty,0)\cup(0,\infty)$.

(b) Since $|x| \ge 0$ for all x, there is no solution to this inequality.

23. (a) We solve the inequality:

$$\begin{aligned} x-2 &< 1 \\ x &< 3 \end{aligned}$$

The solution set is $(-\infty, 3)$.

(b) We solve the inequality:

$$|x-2|<1$$
$$-1<x-2<1$$
$$1<x<3$$

The solution set is $(1,3)$.

(c) We solve the inequality:

$$|x-2|>1$$
$$x-2>1 \quad \text{or} \quad x-2<-1$$
$$x>3 \quad \text{or} \quad x<1$$

The solution set is $(-\infty,1)\cup(3,\infty)$.

25. (a) We solve the inequality:

$$1-x\le 5$$
$$-x\le 4$$
$$x\ge -4$$

The solution set is $[-4,\infty)$.

(b) We solve the inequality:

$$|1-x|\le 5$$
$$-5\le 1-x\le 5$$
$$-6\le -x\le 4$$
$$6\ge x\ge -4$$

The solution set is $[-4,6]$.

(c) We solve the inequality:

$$|1-x|>5$$
$$1-x>5 \quad \text{or} \quad 1-x<-5$$
$$-x>4 \quad \text{or} \quad -x<-6$$
$$x<-4 \quad \text{or} \quad x>6$$

The solution set is $(-\infty,-4)\cup(6,\infty)$.

27. (a) We solve the inequality:

$$a-x<c$$
$$-x<c-a$$
$$x>a-c$$

The solution set is $(a-c,\infty)$.

(b) We solve the inequality:

$$|a-x|<c$$
$$-c<a-x<c$$
$$-c-a<-x<c-a$$
$$a+c>x>a-c$$

The solution set is $(a-c,a+c)$.

(c) We solve the inequality:

$$|a-x| \geq c$$

$$a-x \geq c \quad \text{or} \quad a-x \leq -c$$

$$-x \geq c-a \quad \text{or} \quad -x \leq -c-a$$

$$x \leq a-c \quad \text{or} \quad x \geq a+c$$

The solution set is $(-\infty, a-c] \cup [a+c, \infty)$.

29. We solve the inequality:

$$\left|\frac{x-2}{3}\right| < 4$$

$$-4 < \frac{x-2}{3} < 4$$

$$-12 < x-2 < 12$$

$$-10 < x < 14$$

The solution set is $(-10, 14)$.

31. We solve the inequality:

$$\left|\frac{x+1}{2} - \frac{x-1}{3}\right| < 1$$

$$-1 < \frac{x+1}{2} - \frac{x-1}{3} < 1$$

$$-6 < 3(x+1) - 2(x-1) < 6$$

$$-6 < x+5 < 6$$

$$-11 < x < 1$$

The solution set is $(-11, 1)$.

33. (a) We solve the inequality:

$$\left|(x+h)^2 - x^2\right| < 3h^2$$

$$\left|x^2 + 2xh + h^2 - x^2\right| < 3h^2$$

$$\left|2xh + h^2\right| < 3h^2$$

$$-3h^2 < 2xh + h^2 < 3h^2$$

$$-4h^2 < 2xh < 2h^2$$

Since $h > 0$, we can divide by $2h$ to obtain $-2h < x < h$. The solution set is $(-2h, h)$.

(b) Since $h < 0$, when we divide the inequality in (a) by $2h$, we reverse the inequalities to obtain $-2h > x > h$. The solution set is $(h, -2h)$.

35. We solve the inequality:

$$1.86 < 4.11x - 0.15 < 1.95$$

$$2.01 < 4.11x < 2.1$$

$$0.49 < x < 0.51$$

The solution set is $(0.49, 0.51)$.

37. Let x be the amount to be invested at 7%. Then $6000-x$ will be invested at 9%. Since the total dividends are to be at least \$500, we have:

$$(x)(7\%)+(6000-x)(9\%)\geq 500$$
$$.07x+.09(6000-x)\geq 500$$
$$7x+9(6000-x)\geq 50000$$
$$7x+54000-9x\geq 50000$$
$$-2x\geq -4000$$
$$x\leq 2000$$

At most \$2000 can be invested at 7%.

39. First solve the equation $F=\frac{9}{5}C+32$ for C. This yields:

$$C=\frac{5F-160}{9}$$

The following inequalities are then equivalent:

$$-25\leq C\leq 475$$
$$-25\leq \frac{5F-160}{9}\leq 475$$
$$-225\leq 5F-160\leq 4275$$
$$-65\leq 5F\leq 4435$$
$$-13^\circ\leq F\leq 887^\circ$$

The corresponding range on the Fahrenheit scale is $[-13^\circ,887^\circ]$.

41. Using the expression for C found in Exercise 39, we have:

$$5100\leq C\leq 6500$$
$$5100\leq \frac{5F-160}{9}\leq 6500$$
$$45900\leq 5F-160\leq 58500$$
$$46060\leq 5F\leq 58660$$
$$9212\leq F\leq 11732$$

Rounding to the nearest 100°, the corresponding range is $[9200^\circ,11700^\circ]$.

43. (a) We complete the table:

a	b	$\sqrt{ab}$ (G.M.)	$\frac{a+b}{2}$ (A.M.)	Which is larger, G.M. or A.M.?
1	2	1.4142	1.5	A.M.
1	3	1.7320	2.0	A.M.
1	4	2.0000	2.5	A.M.
2	3	2.4495	2.5	A.M.
3	4	3.4641	3.5	A.M.
5	10	7.0711	7.5	A.M.
9	10	9.4868	9.5	A.M.
99	100	99.4987	99.5	A.M.
999	1000	999.4999	999.5	A.M.

(b) Using the hint and squaring each side, proving the inequality is equivalent to proving the following inequality:

$$ab \le \left(\frac{a+b}{2}\right)^2$$

We note that the inequality $(a-b)^2 \ge 0$ is true for all a and b, thus:

$$\begin{aligned}(a-b)^2 &\ge 0\\ a^2 - 2ab + b^2 &\ge 0\\ a^2 + b^2 &\ge 2ab\end{aligned}$$

Working from the right-hand side of the original inequality:

$$\left(\frac{a+b}{2}\right)^2 = \frac{a^2+2ab+b^2}{4} = \frac{a^2+b^2}{4} + \frac{ab}{2} \ge \frac{2ab}{4} + \frac{ab}{2} = \frac{ab}{2} + \frac{ab}{2} = ab$$

This proves the required inequality.

45. (a) We complete the table:

x	y	$\frac{x}{y}+\frac{y}{x}$	True or False $\frac{x}{y}+\frac{y}{x} \ge 2$
1	1	2.0000	true
2	3	2.1667	true
3	5	2.2667	true
4	7	2.3214	true
5	9	2.3556	true
9	10	2.0111	true
49	50	2.0004	true
99	100	2.0001	true

(b) Using the inequality in Exercise 43(b) and taking $a = \frac{x}{y}$ and $b = \frac{y}{x}$, we have:

$$\begin{aligned}\sqrt{ab} &\le \frac{a+b}{2}\\ \sqrt{\frac{x}{y} \bullet \frac{y}{x}} &\le \frac{\frac{x}{y}+\frac{y}{x}}{2}\\ 1 &\le \frac{\frac{x}{y}+\frac{y}{x}}{2}\\ 2 &\le \frac{x}{y}+\frac{y}{x}\end{aligned}$$

This proves the required inequality.

47. Since $x(y-3)+3(x-2)=xy-3x+3x-6=xy-6$, then:

$$\begin{aligned}|xy-6| &= |x(y-3)+3(x-2)| \\ &\le |x||y-3|+3|x-2| \\ &\le |x|\bullet(0.01)+3(0.1) \\ &= 0.01|x|+0.3\end{aligned}$$

We still need to eliminate the $|x|$ term. Note that:

$$|x-2|<0.1$$
$$-0.1<x-2<0.1$$
$$1.9<x<2.1$$

Thus $|x|<2.1$. Now we complete the proof:

$$|xy-6|\le 0.01|x|+0.3\le(0.01)(2.1)+0.3=0.321$$

49. (a) Using the result of Exercise 43(b), we have:

$$\sqrt{abcd}=\sqrt{(ab)(cd)}\le\frac{ab+cd}{2}$$

(b) Working with the squares of each side and using part (a), we have:

$$\begin{aligned}\left(\sqrt{ab}+\sqrt{cd}\right)^2 &= ab+2\sqrt{(ad)(bc)}+cd \\ &\le ab+2\left(\frac{ad+bc}{2}\right)+cd \\ &= ab+ad+bc+cd \\ &= (a+c)(b+d)\end{aligned}$$

Therefore $\sqrt{ab}+\sqrt{cd}\le\sqrt{(a+c)(b+d)}$.

51. (a) Let l be the length, so $xl=25$, and $l=\frac{25}{x}$. So the perimeter is given by:

$$P=2x+2\left(\tfrac{25}{x}\right)=2x+\tfrac{50}{x}$$

(b) Using the result of Exercise 43(b), we have:

$$\frac{2x+\frac{50}{x}}{2}\ge\sqrt{2x\bullet\tfrac{50}{x}}$$
$$2x+\tfrac{50}{x}\ge 2\sqrt{100}=20$$

(c) We set $P=20$:

$$\begin{aligned}2x+\tfrac{50}{x}&=20\\ 2x^2+50&=20x\\ x^2-10x+25&=0\\ (x-5)^2&=0\\ x&=5\end{aligned}$$

Since the length is also 5, the rectangle is a square of dimensions 5 ft by 5 ft.

2.7 More on Inequalities

1. We first factor $x^2 + x - 6 = (x + 3)(x - 2)$, so the key numbers are -3 and 2. Now draw the sign chart:

Interval	Test Number	$x+3$	$x-2$	$(x+3)(x-2)$
$(-\infty,-3)$	-4	neg.	neg.	pos.
$(-3,2)$	0	pos.	neg.	neg.
$(2,\infty)$	3	pos.	pos.	pos.

So $x^2 + x - 6 < 0$ on the interval $(-3, 2)$.

3. We first factor $x^2 - 11x + 18 = (x - 2)(x - 9)$, so the key numbers are 2 and 9. Now draw the sign chart:

Interval	Test Number	$x-2$	$x-9$	$(x-2)(x-9)$
$(-\infty,2)$	0	neg.	neg.	pos.
$(2,9)$	3	pos.	neg.	neg.
$(9,\infty)$	10	pos.	pos.	pos.

So $x^2 - 11x + 18 > 0$ on the intervals $(-\infty,2)\cup(9,\infty)$.

5. We first simplify and factor the inequality:

$$9x - x^2 \le 20$$
$$-x^2 + 9x - 20 \le 0$$
$$x^2 - 9x + 20 \ge 0$$
$$(x-4)(x-5) \ge 0$$

So the key numbers are 4 and 5. Now draw the sign chart:

Interval	Test Number	$x-4$	$x-5$	$(x-4)(x-5)$
$(-\infty,4)$	0	neg.	neg.	pos.
$(4,5)$	$\frac{9}{2}$	pos.	neg.	neg.
$(5,\infty)$	6	pos.	pos.	pos.

So $x^2 - 9x + 20 \ge 0$ (and consequently the original inequality is satisfied) on the intervals $(-\infty,4]\cup[5,\infty)$.

7. We first factor $x^2 - 16 = (x + 4)(x - 4)$, so the key numbers are -4 and 4. Now draw the sign chart:

Interval	Test Number	$x-4$	$x+4$	$(x-4)(x+4)$
$(-\infty,-4)$	-5	neg.	neg.	pos.
$(-4,4)$	0	neg.	pos.	neg.
$(4,\infty)$	5	pos.	pos.	pos.

So $x^2 - 16 \ge 0$ on the intervals $(-\infty,-4]\cup[4,\infty)$.

9. We first simplify and factor the inequality:

$$16x^2 + 24x < -9$$
$$16x^2 + 24x + 9 < 0$$
$$(4x+3)^2 < 0$$

Since $(4x + 3)^2 \geq 0$ for all real values of x, this inequality has no solution.

11. We first factor $x^3 + 13x^2 + 42x = x\left(x^2 + 13x + 42\right) = x(x+7)(x+6)$, so the key numbers are 0, –7 and –6. Now draw the sign chart:

Interval	Test Number	x	$x+7$	$x+6$	$x(x+7)(x+6)$
$(-\infty,-7)$	–8	neg.	neg.	neg.	neg.
$(-7,-6)$	$-\frac{13}{2}$	neg.	pos.	neg.	pos.
$(-6,0)$	–1	neg.	pos.	pos.	neg.
$(0,\infty)$	1	pos.	pos.	pos.	pos.

So $x^3 + 13x^2 + 42x > 0$ on the intervals $(-7,-6) \cup (0,\infty)$.

13. We first simplify and factor the inequality:

$$225x \leq x^3$$
$$225x - x^3 \leq 0$$
$$x\left(225 - x^2\right) \leq 0$$
$$x(15+x)(15-x) \leq 0$$

So the key numbers are 0, –15, and 15. Now draw the sign chart:

Interval	Test Number	x	$15-x$	$15+x$	$x(15-x)(15+x)$
$(-\infty,-15)$	–16	neg.	pos.	neg.	pos.
$(-15,0)$	–1	neg.	pos.	pos.	neg.
$(0,15)$	1	pos.	pos.	pos.	pos.
$(15,\infty)$	16	pos.	neg.	pos.	neg.

So $225 - x^3 \leq 0$ (and consequently the original inequality is satisfied) on the intervals $[-15,0] \cup [15,\infty)$.

15. Since $2x^2 + 1 \geq 1$ for all real numbers x, the inequality is satisfied for the interval $(-\infty,\infty)$.

17. We first factor $12x^3+17x^2+6x = x(12x^2+17x+6) = x(4x+3)(3x+2)$, so the key numbers are $0, -\frac{3}{4}$ and $-\frac{2}{3}$. Now draw the sign chart:

Interval	Test Number	x	$4x+3$	$3x+2$	$x(4x+3)(3x+2)$
$\left(-\infty,-\frac{3}{4}\right)$	-1	neg.	neg.	neg.	neg.
$\left(-\frac{3}{4},-\frac{2}{3}\right)$	$-\frac{17}{24}$	neg.	pos.	neg.	pos.
$\left(-\frac{2}{3},0\right)$	$-\frac{1}{3}$	neg.	pos.	pos.	neg.
$(0,\infty)$	1	pos.	pos.	pos.	pos.

So $12x^3 + 17x^2 + 6x < 0$ on the intervals $\left(-\infty,-\frac{3}{4}\right)\cup\left(-\frac{2}{3},0\right)$.

19. The key numbers are found by solving the equation $x^2+x-1=0$. Using the quadratic formula we have:

$$x=\frac{-1\pm\sqrt{1-4(-1)}}{2}=\frac{-1\pm\sqrt{5}}{2}$$

For purposes of picking appropriate test numbers, note that:

$$\frac{-1+\sqrt{5}}{2}\approx 0.6 \quad\text{and}\quad \frac{-1-\sqrt{5}}{2}\approx -1.6$$

We draw the sign chart:

Interval	Test Number	$x+1.6$	$x-0.6$	x^2+x-1
$\left(-\infty,\frac{-1-\sqrt{5}}{2}\right)$	-2	neg.	neg.	pos.
$\left(\frac{-1-\sqrt{5}}{2},\frac{-1+\sqrt{5}}{2}\right)$	0	pos.	neg.	neg.
$\left(\frac{-1+\sqrt{5}}{2},\infty\right)$	1	pos.	pos.	pos.

So $x^2+x-1>0$ on the intervals $\left(-\infty,\frac{-1-\sqrt{5}}{2}\right)\cup\left(\frac{-1+\sqrt{5}}{2},\infty\right)$.

21. The key numbers are found by using the quadratic formula to solve $x^2-8x+2=0$. We have:

$$x=\frac{8\pm\sqrt{64-4(2)}}{2}=\frac{8\pm\sqrt{56}}{2}=\frac{8\pm 2\sqrt{14}}{2}=4\pm\sqrt{14}$$

For purposes of picking appropriate test numbers, note that $4+\sqrt{14}\approx 7.7$ and $4-\sqrt{14}\approx 0.3$. We draw the sign chart:

Interval	Test Number	$x-7.7$	$x-0.3$	x^2-8x+2
$(-\infty,4-\sqrt{14})$	0	neg.	neg.	pos.
$(4-\sqrt{14},4+\sqrt{14})$	4	neg.	pos.	neg.
$(4+\sqrt{14},\infty)$	8	pos.	pos.	pos.

So $x^2-8x+2\le 0$ on the intervals $\left[4-\sqrt{14},4+\sqrt{14}\right]$.

23. The key numbers are -4, -3, and 1. We draw the sign chart:

Interval	Test Number	$x-1$	$x+3$	$x+4$	$(x-1)(x+3)(x+4)$
$(-\infty,-4)$	-5	neg.	neg.	neg.	neg.
$(-4,-3)$	$-\frac{7}{2}$	neg.	neg.	pos.	pos.
$(-3,1)$	0	neg.	pos.	pos.	neg.
$(1,\infty)$	2	pos.	pos.	pos.	pos.

So $(x-1)(x+3)(x+4) \geq 0$ on the intervals $[-4,-3] \cup [1,\infty)$.

25. The key numbers are -4, -5, and -6. We draw the sign chart:

Interval	Test Number	$x+4$	$x+5$	$x+6$	$(x+4)(x+5)(x+6)$
$(-\infty,-6)$	-7	neg.	neg.	neg.	neg.
$(-6,-5)$	$-\frac{11}{2}$	neg.	neg.	pos.	pos.
$(-5,-4)$	$-\frac{9}{2}$	neg.	pos.	pos.	neg.
$(-4,\infty)$	0	pos.	pos.	pos.	pos.

So $(x+4)(x+5)(x+6) < 0$ on the intervals $(-\infty,-6) \cup (-5,-4)$.

27. The key numbers are $-\frac{1}{3}, \frac{1}{3}$ and 2. We draw the sign chart:

Interval	Test Number	$x-2$	$3x+1$	$3x-1$	$(x-2)^2(3x+1)^3(3x-1)$
$\left(-\infty,-\frac{1}{3}\right)$	-1	neg.	neg.	neg.	pos.
$\left(-\frac{1}{3},\frac{1}{3}\right)$	0	neg.	pos.	neg.	neg.
$\left(\frac{1}{3},2\right)$	1	neg.	pos.	pos.	pos.
$(2,\infty)$	3	pos.	pos.	pos.	pos.

So $(x-2)^2(3x+1)^3(3x-1) > 0$ on the intervals $\left(-\infty,-\frac{1}{3}\right) \cup \left(\frac{1}{3},2\right) \cup (2,\infty)$.

29. The key numbers are $3, -1, -\frac{1}{2}$ and $-\frac{2}{3}$. We draw the sign chart:

Interval	Test Number	$x-3$	$x+1$	$2x+1$	$3x+2$	product
$(-\infty,-1)$	-2	neg.	neg.	neg.	neg.	neg.
$\left(-1,-\frac{2}{3}\right)$	$-\frac{3}{4}$	neg.	pos.	neg.	neg.	neg.
$\left(-\frac{2}{3},-\frac{1}{2}\right)$	$-\frac{7}{12}$	neg.	pos.	neg.	pos.	pos.
$\left(-\frac{1}{2},3\right)$	0	neg.	pos.	pos.	pos.	pos.
$(3,\infty)$	4	pos.	pos.	pos.	pos.	pos.

So $(x-3)^2(x+1)^4(2x+1)^4(3x+2) \leq 0$ on the intervals $(-\infty,-1] \cup \left[-1,-\frac{2}{3}\right]$, which is equivalent to the interval $\left(-\infty,-\frac{2}{3}\right]$. It is also zero at 3 and $-1/2$, so the inequality is satisfied for the values $\left(-\infty,-\frac{2}{3}\right] \cup \left\{3,-\frac{1}{2}\right\}$.

31. We first simplify and factor the inequality:

$$20 \geq x^2(9 - x^2)$$
$$20 \geq 9x^2 - x^4$$
$$x^4 - 9x^2 + 20 \geq 0$$
$$(x^2 - 4)(x^2 - 5) \geq 0$$
$$(x - 2)(x + 2)(x - \sqrt{5})(x + \sqrt{5}) \geq 0$$

The key numbers are -2, 2, $-\sqrt{5}$, and $\sqrt{5}$. We draw the sign chart:

Interval	Test Number	$x - \sqrt{5}$	$x - 2$	$x + 2$	$x + \sqrt{5}$	$x^4 - 9x^2 + 20$
$(-\infty, -\sqrt{5})$	-3	neg.	neg.	neg.	neg.	pos.
$(-\sqrt{5}, -2)$	-2.1	neg.	neg.	neg.	pos.	neg.
$(-2, 2)$	0	neg.	neg.	pos.	pos.	pos.
$(2, \sqrt{5})$	2.1	neg.	pos.	pos.	pos.	neg.
$(\sqrt{5}, \infty)$	3	pos.	pos.	pos.	pos.	pos.

So $x^4 - 9x^2 + 20 \geq 0$ (and consequently the original inequality is satisfied) on the intervals $(-\infty, -\sqrt{5}] \cup [-2, 2] \cup [\sqrt{5}, \infty)$.

33. We first simplify and factor the inequality:

$$9(x - 4) - x^2(x - 4) < 0$$
$$(x - 4)(9 - x^2) < 0$$
$$(x - 4)(3 + x)(3 - x) < 0$$

The key numbers are 4, -3 and 3. We draw the sign chart:

Interval	Test Number	$x - 4$	$3 + x$	$3 - x$	$(x - 4)(3 + x)(3 - x)$
$(-\infty, -3)$	-4	neg.	neg.	pos.	pos.
$(-3, 3)$	0	neg.	pos.	pos.	neg.
$(3, 4)$	3.5	neg.	pos.	neg.	pos.
$(4, \infty)$	6	pos.	pos.	neg.	neg.

So $(x + 4)(3 + x)(3 - x) < 0$ (and consequently the original inequality is satisfied) on the intervals $(-3, 3) \cup (4, \infty)$.

35. We first simplify and factor the inequality:

$$4(x^2 - 9) - (x^2 - 9)^2 > -5$$
$$4x^2 - 36 - x^4 + 18x^2 - 81 > -5$$
$$-x^4 + 22x^2 - 112 > 0$$
$$x^4 - 22x^2 + 112 < 0$$
$$(x^2 - 8)(x^2 - 14) < 0$$
$$(x - 2\sqrt{2})(x + 2\sqrt{2})(x - \sqrt{14})(x + \sqrt{14}) < 0$$

The key numbers are $2\sqrt{2}, -2\sqrt{2}, \sqrt{14}$, and $-\sqrt{14}$. We draw the sign chart:

Interval	Test Number	x^2-8	x^2-14	x^4-22x^2+112
$(-\infty,-\sqrt{14})$	−5	pos.	pos.	pos.
$(-\sqrt{14},-2\sqrt{2})$	−3	pos.	neg.	neg.
$(-2\sqrt{2},2\sqrt{2})$	0	neg.	neg.	pos.
$(2\sqrt{2},\sqrt{14})$	3	pos.	neg.	neg.
$(\sqrt{14},\infty)$	5	pos.	pos.	pos.

So $x^4 - 22x^2 + 112 < 0$ (and consequently the original inequality is satisfied) on the intervals $(-\sqrt{14},-2\sqrt{2})\cup(2\sqrt{2},\sqrt{14})$.

37. One of the key numbers is 4. The others are found by solving the equation $2x^2 - 6x - 1 = 0$. With the quadratic formula, we have:

$$x=\frac{6\pm\sqrt{36-4(2)(-1)}}{2(2)}=\frac{6\pm\sqrt{44}}{4}=\frac{6\pm2\sqrt{11}}{4}=\frac{3\pm\sqrt{11}}{2}$$

Thus, the key numbers are 4 and $\frac{3\pm\sqrt{11}}{2}$. For purposes of picking appropriate test numbers, note that $\frac{3+\sqrt{11}}{2}\approx 3.2$ and that $\frac{3-\sqrt{11}}{2}\approx -0.2$. We draw the sign chart:

Interval	Test Number	$x-4$	$x-3.2$	$x+0.2$	$(x-4)(2x^2-6x-1)$
$(-\infty,-0.2)$	-1	neg.	neg.	neg.	neg.
$(-0.2,3.2)$	0	neg.	neg.	pos.	pos.
$(3.2,4)$	3.5	neg.	pos.	pos.	neg.
$(4,\infty)$	5	pos.	pos.	pos.	pos.

So $(x-4)(2x^2-6x-1)<0$ on the intervals $\left(-\infty,\frac{3-\sqrt{11}}{2}\right)\cup\left(\frac{3+\sqrt{11}}{2},4\right)$.

39. We first factor the inequality:

$$\begin{aligned} x^3+2x^2-x-2&>0\\ x^2(x+2)-(x+2)&>0\\ (x+2)(x^2-1)&>0\\ (x+2)(x+1)(x-1)&>0 \end{aligned}$$

The key numbers are −2, −1 and 1. We draw the sign chart:

Interval	Test Number	$x-1$	$x+1$	$x+2$	$(x-1)(x+1)(x+2)$
$(-\infty,-2)$	-3	neg.	neg.	neg.	neg.
$(-2,-1)$	-1.5	neg.	neg.	pos.	pos.
$(-1,1)$	0	neg.	pos.	pos.	neg.
$(1,\infty)$	2	pos.	pos.	pos.	pos.

So $x^3 + 2x^2 - x - 2 > 0$ on the intervals $(-2,-1)\cup(1,\infty)$.

41. The key numbers are –1 and 1. We draw the sign chart:

Interval	Test Number	$x-1$	$x+1$	$\frac{x-1}{x+1}$
$(-\infty,-1)$	–2	neg.	neg.	pos.
$(-1,1)$	0	neg.	pos.	neg.
$(1,\infty)$	2	pos.	pos.	pos.

Thus the quotient is negative on (–1, 1) and it is zero when $x = 1$. So $\frac{x-1}{x+1} \le 0$ on the interval (–1, 1].

43. The key numbers are 2 and $\frac{3}{2}$. We draw the sign chart:

Interval	Test Number	$2-x$	$3-2x$	$\frac{2-x}{3-2x}$
$(-\infty,\frac{3}{2})$	0	pos.	pos.	pos.
$(\frac{3}{2},2)$	$\frac{7}{4}$	pos.	neg.	neg.
$(2,\infty)$	3	neg.	neg.	pos.

Thus the quotient is positive on $\left(-\infty,\frac{3}{2}\right)\cup(2,\infty)$ and it is zero when $x = 2$. So $\frac{2-x}{3-2x} \ge 0$ on the intervals $\left(-\infty,\frac{3}{2}\right)\cup[2,\infty)$.

45. Factoring, we have $\frac{(x+1)(x-9)}{x} < 0$. So the key numbers are –1, 0, and 9. We draw the sign chart:

Interval	Test Number	$x-9$	x	$x+1$	$\frac{(x-9)(x+1)}{x}$
$(-\infty,-1)$	-2	neg.	neg.	neg.	neg.
$(-1,0)$	$-\frac{1}{2}$	neg.	neg.	pos.	pos.
$(0,9)$	1	neg.	pos.	pos.	neg.
$(9,\infty)$	10	pos.	pos.	pos.	pos.

Thus the quotient is negative on the intervals $(-\infty,-1)\cup(0,9)$.

47. Factoring the inequality yields:

$$\frac{2x^3+5x^2-7x}{3x^2+7x+4} > 0$$

$$\frac{x(2x^2+5x-7)}{(3x+4)(x+1)} > 0$$

$$\frac{x(2x+7)(x-1)}{(3x+4)(x+1)} > 0$$

So the key numbers are $0, -\frac{7}{2}, 1, -\frac{4}{3}$ and -1. We draw the sign chart:

Interval	Test Number	$x-1$	x	$x+1$	$3x+4$	$2x+7$	$\frac{x(2x+7)(x-1)}{(3x+4)(x+1)}$
$\left(-\infty,-\frac{7}{2}\right)$	-4	neg.	neg.	neg.	neg.	neg.	neg.
$\left(-\frac{7}{2},-\frac{4}{3}\right)$	-2	neg.	neg.	neg.	neg.	pos.	pos.
$\left(-\frac{4}{3},-1\right)$	$-\frac{7}{6}$	neg.	neg.	neg.	pos.	pos.	neg.
$(-1,0)$	$-\frac{1}{2}$	neg.	neg.	pos.	pos.	pos.	pos.
$(0,1)$	$\frac{1}{2}$	neg.	pos.	pos.	pos.	pos.	neg.
$(1,\infty)$	2	pos.	pos.	pos.	pos.	pos.	pos.

Thus the quotient is positive on the intervals $\left(-\frac{7}{2},-\frac{4}{3}\right)\cup(-1,0)\cup(1,\infty)$.

49. We first simplify the inequality:

$$\frac{x}{x+1}-1>0$$
$$\frac{x-x-1}{x+1}>0$$
$$\frac{-1}{x+1}>0$$

This is true whenever $x + 1 < 0$, so $x < -1$. So the solution is $(-\infty,-1)$.

51. We first simplify the inequality:

$$\frac{1}{x}-\frac{1}{x+1}\le 0$$
$$\frac{x+1-x}{x(x+1)}\le 0$$
$$\frac{1}{x(x+1)}\le 0$$

So the key numbers are 0 and -1. We draw the sign chart:

Interval	Test Number	x	$x+1$	$\frac{1}{x(x+1)}$
$(-\infty,-1)$	-2	neg.	neg.	pos.
$(-1,0)$	-0.5	neg.	pos.	neg.
$(0,\infty)$	1	pos.	pos.	pos.

So $\frac{1}{x(x+1)}\le 0$ (and consequently the original inequality is satisfied) on the interval $(-1,0)$.

53. We first simplify the inequality:

$$\frac{x+2}{x+5}-1\le 0$$
$$\frac{x+2-x-5}{x+5}\le 0$$
$$\frac{-3}{x+5}\le 0$$

This is true whenever $x + 5 > 0$, so $x > -5$. So the solution is $(-5,\infty)$.

55. We first simplify the inequality:

$$\frac{1}{x-2}-\frac{1}{x-1}-\frac{1}{6}\ge 0$$
$$\frac{6(x-1)-6(x-2)-(x-2)(x-1)}{6(x-2)(x-1)}\ge 0$$
$$\frac{-x^2+3x+4}{6(x-2)(x-1)}\ge 0$$
$$\frac{x^2-3x-4}{6(x-2)(x-1)}\le 0$$
$$\frac{(x-4)(x+1)}{6(x-2)(x-1)}\le 0$$

So the key numbers are –1, 1, 2, and 4. We draw the sign chart:

Interval	Test Number	$x-4$	$x-2$	$x-1$	$x+1$	$\frac{(x-4)(x+1)}{6(x-2)(x-1)}$
$(-\infty,-1)$	-2	neg.	neg.	neg.	neg.	pos.
$(-1,1)$	0	neg.	neg.	neg.	pos.	neg.
$(1,2)$	$\frac{3}{2}$	neg.	neg.	pos.	pos.	pos.
$(2,4)$	3	neg.	pos.	pos.	pos.	neg.
$(4,\infty)$	5	pos.	pos.	pos.	pos.	pos.

So the quotient is negative on the intervals $(-1,1)\cup(2,4)$ and zero at $x = 4$ and $x = -1$, so our original inequality is satisfied on the intervals $[-1,1)\cup(2,4]$.

57. We first simplify the inequality:

$$\frac{1+x}{1-x}-\frac{1-x}{1+x}+1<0$$
$$\frac{(1+x)^2-(1-x)^2+(1-x^2)}{(1-x)(1+x)}<0$$
$$\frac{-x^2+4x+1}{(1-x)(1+x)}<0$$
$$\frac{x^2-4x-1}{(x-1)(x+1)}<0$$

The denominator is zero when $x = \pm 1$. Using the quadratic formula, we find the numerator is zero when $x = 2 \pm \sqrt{5}$. Thus, the key numbers are ± 1 and $2 \pm \sqrt{5}$. For purposes of picking appropriate test numbers, note that $2 + \sqrt{5} \approx 4.2$ and that $2 - \sqrt{5} \approx -0.2$. We draw the sign chart:

Interval	Test Number	$x-4.2$	$x-1$	$x+0.2$	$x+1$	$\frac{x^2-4x-1}{(x-1)(x+1)}$
$(-\infty,-1)$	-2	neg.	neg.	neg.	neg.	pos.
$(-1,-0.2)$	-0.5	neg.	neg.	neg.	pos.	neg.
$(-0.2,1)$	0	neg.	neg.	pos.	pos.	pos.
$(1,4.2)$	2	neg.	pos.	pos.	pos.	neg.
$(4.2,\infty)$	5	pos.	pos.	pos.	pos.	pos.

So $\frac{x^2-4x-1}{(x-1)(x+1)} < 0$ (and consequently the original inequality is satisfied) on the intervals

$(-1,2-\sqrt{5}) \cup (1,2+\sqrt{5})$.

59. We first simplify the inequality:

$$\frac{3-2x}{3+2x} - \frac{1}{x} > 0$$

$$\frac{x(3-2x)-(3+2x)}{x(3+2x)} > 0$$

$$\frac{x-2x^2-3}{x(3+2x)} > 0$$

$$\frac{2x^2-x+3}{x(3+2x)} < 0$$

The denominator is zero when $x = 0$ and $x = -\frac{3}{2}$. By using the quadratic formula, we find that there are no real numbers for which the numerator is zero. Also note that

$2x^2 - x + 3$ is positive for all x. Thus 0 and $-\frac{3}{2}$ are the only key numbers. We draw the sign chart:

Interval	Test Number	$2x+3$	x	$\frac{2x^2-x+3}{x(2x+3)}$
$\left(-\infty,-\frac{3}{2}\right)$	-2	neg.	neg.	pos.
$\left(-\frac{3}{2},0\right)$	-1	pos.	neg.	neg.
$(0,\infty)$	1	pos.	pos.	pos.

So $\frac{2x^2-x+3}{x(2x+3)} < 0$ (and consquently the original inequality is satisfied) on the interval

$\left(-\frac{3}{2},0\right)$.

61. We first simplify the inequality:

$$1+\frac{1}{x}-\frac{1}{1+x}\geq 0$$

$$\frac{x(1+x)+(1+x)-x}{x(1+x)}\geq 0$$

$$\frac{x^2+x+1}{x(1+x)}\geq 0$$

The only key numbers are 0 and –1. By using the quadratic formula, we find that there are no real numbers for which the numerator x^2+x+1 is zero. Note that x^2+x+1 is positive for all values of x. We draw the sign chart:

Interval	Test Number	$x+1$	x	$\frac{x^2+x+1}{x(x+1)}$
$(-\infty,-1)$	-2	neg.	neg.	pos.
$(-1,0)$	$-\frac{1}{2}$	pos.	neg.	neg.
$(0,\infty)$	1	pos.	pos.	pos.

So $\frac{x^2+x+1}{x(1+x)}\geq 0$ (and consequently the original inequality is satisfied) on the intervals $(-\infty,-1)\cup(0,\infty)$.

63. (a) We solve the inequality:

$$x^2-4x-5\geq 0$$

$$(x-5)(x+1)\geq 0$$

The key numbers are 5 and –1. We draw the sign chart:

Interval	Test Number	$x-5$	$x+1$	$(x-5)(x+1)$
$(-\infty,-1)$	-2	neg.	neg.	pos.
$(-1,5)$	0	neg.	pos.	neg.
$(5,\infty)$	6	pos.	pos.	pos.

The solution set for $(x-5)(x+1)\geq 0$ is $(-\infty,-1]\cup[5,\infty)$.

(b) We solve the inequality $\frac{1}{x^2-4x-5}\geq 0$. This will have the same solutions as in (a), except the endpoints $x=-1$ and $x=5$ are not included. The solution set for this is $(-\infty,-1)\cup(5,\infty)$.

65. The solutions will be real provided the discriminant b^2-4ac is non-negative. Thus, we have:

$$b^2-4\geq 0$$

$$(b-2)(b+2)\geq 0$$

The key numbers are ±2. We draw the sign chart:

Interval	Test Number	$b-2$	$b+2$	$(b-2)(b+2)$
$(-\infty,-2)$	-3	neg.	neg.	pos.
$(-2,2)$	0	neg.	pos.	neg.
$(2,\infty)$	3	pos.	pos.	pos.

The solution set consists of the two intervals $(-\infty,-2]\cup[2,\infty)$. For values of b in either of these two intervals, the equation $x^2+bx+1=0$ will have real solutions.

67. If $x=1$ is a solution of $\frac{2a+x}{x-2a}<1$, then we have:

$$\frac{2a+1}{1-2a}<1$$
$$\frac{2a+1}{1-2a}-1<0$$
$$\frac{2a+1-(1-2a)}{1-2a}<0$$
$$\frac{4a}{1-2a}<0$$

The key numbers are 0 and $\frac{1}{2}$. We draw the sign chart:

Interval	Test Number	$4a$	$1-2a$	$\frac{4a}{1-2a}$
$(-\infty,0)$	-1	neg.	pos.	neg.
$\left(0,\frac{1}{2}\right)$	$\frac{1}{4}$	pos.	pos.	pos.
$\left(\frac{1}{2},\infty\right)$	1	pos.	neg.	neg.

The allowable values of a are those numbers in either of the intervals $(-\infty,0)\cup\left(\frac{1}{2},\infty\right)$.

69. Using the Pythagorean theorem, we find the hypotenuse is $\sqrt{x^2+(1-x)^2}$, or $\sqrt{2x^2-2x+1}$. If this is less than $\frac{\sqrt{17}}{5}$, then we have:

$$\sqrt{2x^2-2x+1}<\frac{\sqrt{17}}{5}$$
$$\left(5\sqrt{2x^2-2x+1}\right)^2<\left(\sqrt{17}\right)^2$$
$$50x^2-50x+25<17$$
$$50x^2-50x+8<0$$
$$25x^2-25x+4<0$$
$$(5x-1)(5x-4)<0$$

The key numbers here are $\frac{1}{5}$ and $\frac{4}{5}$. We draw the sign chart:

Interval	Test Number	$5x-1$	$5x-4$	$(5x-1)(5x-4)$
$(-\infty, \frac{1}{5})$	0	neg.	neg.	pos.
$\left(\frac{1}{5}, \frac{4}{5}\right)$	$\frac{2}{5}$	pos.	neg.	neg.
$\left(\frac{4}{5}, \infty\right)$	1	pos.	pos.	pos.

The solution set is the interval $\left(\frac{1}{5}, \frac{4}{5}\right)$.

71. For the cylinder, we have:

$$\frac{V}{S} = \frac{\pi r^2 h}{2\pi r^2 + 2\pi rh} = \frac{\pi r^2}{2\pi r^2 + 2\pi r} = \frac{r}{2r+2}$$

The condition $\frac{V}{S} < \frac{1}{3}$ then becomes:

$$\frac{r}{2r+2} < \frac{1}{3}$$

Since r is positive here, we can multiply through by the positive quantity $3(2r+2)$. This yields:

$$3r < 2r + 2$$
$$r < 2$$

Since $r > 0$, the possible values of r are in the interval $(0, 2)$.

73. (a) We complete the table:

x	x^2+1000	$2x^2+x$	$\frac{x^2+1000}{2x^2+x}$
1	1,001	3	333.67
2	1,004	10	100.40
5	1,025	55	18.64
10	1,100	210	5.24
100	11,000	20,100	0.55
200	41,000	80,200	0.51
10,000	100,001,000	200,010,000	0.50

(b) We simplify the inequality:

$$\frac{x^2+1000}{2x^2+x} < \frac{1}{2}$$
$$\frac{x^2+1000}{x(2x+1)} - \frac{1}{2} < 0$$
$$\frac{2(x^2+1000) - x(2x+1)}{2x(2x+1)} < 0$$
$$\frac{-x+2000}{2x(2x+1)} < 0$$

Thus the key numbers are 2000, 0, and $-\frac{1}{2}$. We draw the sign chart:

Interval	Test Number	$2x+1$	x	$-x+2000$	$\frac{-x+2000}{2x(2x+1)}$
$\left(-\infty,-\frac{1}{2}\right)$	-1	neg.	neg.	pos.	pos.
$\left(-\frac{1}{2},0\right)$	$-\frac{1}{4}$	pos.	neg.	pos.	neg.
$(0,2000)$	1	pos.	pos.	pos.	pos.
$(2000,\infty)$	2001	pos.	pos.	neg.	neg.

The solution set for $\frac{x^2+1000}{2x^2+x}<\frac{1}{2}$ is $\left(-\frac{1}{2},0\right)\cup(2000,\infty)$.

(c) The inequality holds when $n>2000$, and the smallest natural number fulfilling this condition is $n=2001$.

75. We multiply this out to obtain:

$$x^2-2ax+a^2-x^2+2bx-b^2>\frac{(a-b)^2}{4}$$

$$(2b-2a)x+(a^2-b^2)>\frac{(a-b)^2}{4}$$

$$8(b-a)x+4(a+b)(a-b)>(a-b)^2$$

Since $a>b$, then $a-b>0$, so we can divide by $a-b$ and preserve the inequality. Thus:

$$-8x+4(a+b)>a-b$$

$$-8x+4a+4b>a-b$$

$$-8x>-3a-5b$$

$$x<\frac{3a+5b}{8}$$

So the solution is $\left(-\infty,\frac{3a+5b}{8}\right)$.

Chapter Two Review Exercises

1. T

3. F (For instance: $x=5$)

5. F (For instance: $x=1$)

7. T

9. F (For instance: $x=0$, $y=-1$)

11. F (For instance: $a=-5$, $b=1$)

13. We solve the equation:

$$\begin{aligned} 5-9x &= 2 \\ -9x &= -3 \\ x &= \tfrac{1}{3} \end{aligned}$$

15. We solve the equation:

$$\begin{aligned} (t-4)(t+3) &= (t+5)^2 \\ t^2-t-12 &= t^2+10t+25 \\ -11t &= 37 \\ t &= -\tfrac{37}{11} \end{aligned}$$

17. We solve the equation:

$$\begin{aligned} \frac{2t-1}{t+2} &= 5 \\ 2t-1 &= 5t+10 \\ -3t &= 11 \\ t &= -\tfrac{11}{3} \end{aligned}$$

19. We solve the equation:

$$\begin{aligned} \frac{2y-5}{4y+1} &= \frac{y-1}{2y+5} \\ 4y^2-25 &= 4y^2-3y-1 \\ 3y &= 24 \\ y &= 8 \end{aligned}$$

21. Since $|y+4|=2$, then $y+4=2$ or $y+4=-2$, and thus $y=-2$ or $y=-6$.

23. We solve the equation:

$$\begin{aligned} 12x^2+2x-2 &= 0 \\ 6x^2+x-1 &= 0 \\ (3x-1)(2x+1) &= 0 \\ x &= \tfrac{1}{3}, -\tfrac{1}{2} \end{aligned}$$

25. Multiplying by 2, we solve the equation:

$$\begin{aligned} \tfrac{1}{2}x^2+x-12 &= 0 \\ x^2+2x-24 &= 0 \\ (x+6)(x-4) &= 0 \\ x &= -6, 4 \end{aligned}$$

27. We solve the equation:

$$\frac{x}{5-x}=\frac{-2}{11-x}$$
$$11x-x^2=-10+2x$$
$$-x^2+9x+10=0$$
$$x^2-9x-10=0$$
$$(x-10)(x+1)=0$$
$$x=10,-1$$

29. We solve the equation:

$$\frac{1}{3x-7}-\frac{2}{5x-5}-\frac{3}{3x+1}=0$$
$$(5x-5)(3x+1)-2(3x-7)(3x+1)-3(3x-7)(5x-5)=0$$
$$15x^2-10x-5-18x^2+36x+14-45x^2+150x-105=0$$
$$-48x^2+176x-96=0$$
$$3x^2-11x+6=0$$
$$(3x-2)(x-3)=0$$
$$x=\tfrac{2}{3},3$$

31. We multiply by 2 to obtain $2t^2+2t-1=0$. Now using the quadratic formula:

$$t=\frac{-2\pm\sqrt{4-4(2)(-1)}}{2(2)}=\frac{-2\pm\sqrt{12}}{4}=\frac{-2\pm2\sqrt{3}}{4}=\frac{-1\pm\sqrt{3}}{2}$$

33. Since $y^4=9$, then $y^2=3$ or $y^2=-3$ (which has no real solutions). For $y^2=3$, we have $y=\pm\sqrt{3}$.

35. Factoring yields $x(x^4-2x^2-2)=0$, so $x=0$ is one solution. Using the quadratic formula, we have:

$$x^2=\frac{2\pm\sqrt{4-4(-2)}}{2}=\frac{2\pm\sqrt{12}}{2}=\frac{2\pm2\sqrt{3}}{2}=1\pm\sqrt{3}$$

Note that $1-\sqrt{3}$ is negative, so no real solutions are obtained from $x^2=1-\sqrt{3}$. Consequently, $x^2=1+\sqrt{3}$ and $x=\pm\sqrt{1+\sqrt{3}}$. So the solutions are $\pm\sqrt{1+\sqrt{3}}$ and 0.

37. Multiplying by x^2, we have:

$$x^2+14x+48=0$$
$$(x+6)(x+8)=0$$
$$x=-6,-8$$

39. Let $t=x^{1/4}$ and $t^2=x^{1/2}$. Then:

$$t^2-13t+36=0$$
$$(t-9)(t-4)=0$$
$$t=9 \text{ or } t=4$$

If $t=9$, then $x^{1/4}=9$ and $x=9^4=6561$. If $t=4$, then $x^{1/4}=4$ and $x=4^4=256$. So the solutions are 6561and 256.

41. Squaring, we have:

$$\begin{aligned}\sqrt{4-3x}&=5\\ 4-3x&=25\\ -3x&=21\\ x&=-7\end{aligned}$$

This value checks in the original equation.

43. Squaring, we have:

$$\begin{aligned}\sqrt{4x+3}&=\sqrt{11-8x}-1\\ 4x+3&=11-8x-2\sqrt{11-8x}+1\\ 12x-9&=-2\sqrt{11-8x}\\ 144x^2-216x+81&=44-32x\\ 144x^2-184x+37&=0\\ (4x-1)(36x-37)&=0\\ x&=\tfrac{1}{4}\text{ or }x=\tfrac{37}{36}\end{aligned}$$

Upon checking, we find only $x=\frac{1}{4}$ checks in the original equation.

45. Isolating the radical and squaring, we have:

$$\begin{aligned}\sqrt{x+48}-\sqrt{x}&=4\\ \sqrt{x+48}&=\sqrt{x}+4\\ x+48&=x+8\sqrt{x}+16\\ 32&=8\sqrt{x}\\ 4&=\sqrt{x}\\ 16&=x\end{aligned}$$

This value checks in the original equation.

47. Multiplying by $\sqrt{x+6}\sqrt{x-6}$, we obtain:

$$\begin{aligned}\frac{2}{\sqrt{x^2-36}}+\frac{1}{\sqrt{x+6}}-\frac{1}{\sqrt{x-6}}&=0\\ 2+\sqrt{x-6}-\sqrt{x+6}&=0\\ 2+\sqrt{x-6}&=\sqrt{x+6}\\ 4+4\sqrt{x-6}+x-6&=x+6\\ 4\sqrt{x-6}&=8\\ \sqrt{x-6}&=2\\ x-6&=4\\ x&=10\end{aligned}$$

This value checks in the original equation.

49. Squaring, we have:

$$
\begin{aligned}
\sqrt{x+7}-\sqrt{x+2} &= \sqrt{x-1}-\sqrt{x-2} \\
x+7-2\sqrt{x+7}\sqrt{x+2}+x+2 &= x-1-2\sqrt{x-1}\sqrt{x-2}+x-2 \\
-2\sqrt{x+7}\sqrt{x+2}+9 &= -2\sqrt{x-1}\sqrt{x-2}-3 \\
-2\sqrt{x+7}\sqrt{x+2} &= -2\sqrt{x-1}\sqrt{x-2}-12 \\
\sqrt{x+7}\sqrt{x+2} &= \sqrt{x-1}\sqrt{x-2}+6 \\
(x+7)(x+2) &= (x-1)(x-2)+12\sqrt{x-1}\sqrt{x-2}+36 \\
x^2+9x+14 &= x^2-3x+2+12\sqrt{x-1}\sqrt{x-2}+36 \\
12x-24 &= 12\sqrt{x-1}\sqrt{x-2} \\
x-2 &= \sqrt{x-1}\sqrt{x-2} \\
x^2-4x+4 &= x^2-3x+2 \\
-x &= -2 \\
x &= 2
\end{aligned}
$$

This value checks in the original equation.

51. We solve the equation:

$$
\begin{aligned}
(a^2+b^2)x &= a^3+b^3+abx \\
(a^2+b^2)x-abx &= a^3+b^3 \\
x(a^2-ab+b^2) &= a^3+b^3 \\
x &= \frac{a^3+b^3}{a^2-ab+b^2} = \frac{(a+b)(a^2-ab+b^2)}{a^2-ab+b^2} = a+b
\end{aligned}
$$

53. Multiplying by ax, we have:

$$
\begin{aligned}
\frac{x}{a}+\frac{a}{x} &= 2 \\
x^2+a^2 &= 2ax \\
x^2-2ax+a^2 &= 0 \\
(x-a)^2 &= 0 \\
x &= a
\end{aligned}
$$

55. We solve the equation for x:

$$
\begin{aligned}
4x^2y^2-4xy &= -1 \\
4x^2y^2-4xy+1 &= 0 \\
(2xy-1)^2 &= 0 \\
2xy-1 &= 0 \\
x &= \frac{1}{2y}
\end{aligned}
$$

57. Multiplying through by a^2bx yields:

$$
\begin{aligned}
a^2bx^2+abx-a^2x &= 2b+2a \\
a^2bx^2+(ab-a^2)x-2(b+a) &= 0
\end{aligned}
$$

Applying the quadratic formula:

$$x=\frac{-(ab-a^2)\pm\sqrt{(ab-a^2)^2-4(a^2b)[-2(b+a)]}}{2(a^2b)}$$
$$=\frac{-ab+a^2\pm\sqrt{a^4+6a^3b+9a^2b^2}}{2a^2b}$$
$$=\frac{-ab+a^2\pm\sqrt{(a^2+3ab)^2}}{2a^2b}$$
$$=\frac{-ab+a^2\pm(a^2+3ab)}{2a^2b}$$

So the two solutions are:

$$x=\frac{-ab+a^2+a^2+3ab}{2a^2b}=\frac{2a^2+2ab}{2a^2b}=\frac{2a(a+b)}{2a^2b}=\frac{a+b}{ab}$$
$$x=\frac{-ab+a^2-a^2-3ab}{2a^2b}=\frac{-4ab}{2a^2b}=-\frac{2}{a}$$

Both of these values satisfy the original equation.

59. Multiplying by $abx(x + a + b)$ yields:

$$\frac{1}{x+a+b}=\frac{1}{x}+\frac{1}{a}+\frac{1}{b}$$
$$abx=ab(x+a+b)+bx(x+a+b)+ax(x+a+b)$$
$$abx=abx+a^2b+ab^2+bx^2+abx+b^2x+ax^2+a^2x+abx$$
$$0=(a+b)x^2+(a^2+2ab+b^2)x+(a^2b+ab^2)$$
$$0=(a+b)x^2+(a+b)^2x+ab(a+b)$$
$$0=x^2+(a+b)x+ab \qquad \text{(since } a+b\neq 0)$$
$$0=(x+a)(x+b)$$
$$x=-a \text{ or } x=-b$$

Both of these values satisfy the original equation.

61. Letting $t=\frac{1}{\sqrt{x}}$ and $t^2=\frac{1}{x}$, we have:

$$(a^2-b^2)t^2-2(a^2+b^2)t+(a^2-b^2)=0$$
$$[(a-b)t-(a+b)][(a+b)t-(a-b)]=0$$

Setting each factor equal to zero, we have:

$$(a-b)t-(a+b)=0 \qquad\qquad (a+b)t-(a-b)=0$$
$$t=\frac{a+b}{a-b} \qquad\qquad t=\frac{a-b}{a+b}$$
$$\frac{1}{\sqrt{x}}=\frac{a+b}{a-b} \qquad\qquad \frac{1}{\sqrt{x}}=\frac{a-b}{a+b}$$
$$\sqrt{x}=\frac{a-b}{a+b} \qquad\qquad \sqrt{x}=\frac{a+b}{a-b}$$
$$x=\left(\frac{a-b}{a+b}\right)^2 \qquad\qquad x=\left(\frac{a+b}{a-b}\right)^2$$

Both of these values check in the original equation.

63. We solve the inequality:

$$-1<\frac{1-2(1+x)}{3}<1$$
$$-3<1-2-2x<3$$
$$-2<-2x<4$$
$$1>x>-2$$

The solution set is (–2, 1).

65. The inequality is equivalent to $-\frac{1}{2}\leq x\leq\frac{1}{2}$, so the solution set is $\left[-\frac{1}{2},\frac{1}{2}\right]$.

67. We solve the inequality:

$$|x+4|<\tfrac{1}{10}$$
$$-\tfrac{1}{10}<x+4<\tfrac{1}{10}$$
$$-\tfrac{41}{10}<x<-\tfrac{39}{10}$$

The solution set is $\left(-\frac{41}{10},-\frac{39}{10}\right)$.

69. We solve the inequality $|2x-1|\geq 5$:

$$\begin{aligned} 2x-1&\geq 5 \\ 2x&\geq 6 \\ x&\geq 3 \end{aligned} \qquad \text{or} \qquad \begin{aligned} -(2x-1)&\geq 5 \\ 2x-1&\leq -5 \\ 2x&\leq -4 \\ x&\leq -2 \end{aligned}$$

The solution set is $(-\infty,-2]\cup[3,\infty)$.

71. We factor $x^2 + 3x - 40 = (x - 5)(x + 8)$, so the key numbers are 5 and –8. Now draw the sign chart:

Interval	Test Number	$x+8$	$x-5$	$(x+8)(x-5)$
$(-\infty,-8)$	–9	neg.	neg.	pos.
$(-8,5)$	0	pos.	neg.	neg.
$(5,\infty)$	6	pos.	pos.	pos.

So the product is negative on the interval $(-8,5)$.

73. The key numbers are found by using the quadratic formula to solve $x^2 - 6x - 1 = 0$. The result is $x=3\pm\sqrt{10}$. For purposes of picking appropriate test numbers, note that $3+\sqrt{10}\approx 6.2$ and that $3-\sqrt{10}\approx -0.2$. Now draw the sign chart:

Interval	Test Number	$x-6.2$	$x+0.2$	x^2-6x-1
$(-\infty,3-\sqrt{10})$	–1	neg.	neg.	pos.
$(3-\sqrt{10},3+\sqrt{10})$	0	neg.	pos.	neg.
$(3+\sqrt{10},\infty)$	7	pos.	pos.	pos.

So the product is negative on the interval $(3-\sqrt{10},3+\sqrt{10})$.

75. We first factor the inequality:

$$x^4 - 34x^2 + 225 < 0$$
$$\left(x^2 - 9\right)\left(x^2 - 25\right) < 0$$
$$(x-3)(x+3)(x-5)(x+5) < 0$$

The key numbers are –3, 3, –5, and 5. Now draw the sign chart:

Interval	Test Number	$x-5$	$x-3$	$x+3$	$x+5$	$\left(x^2-9\right)\left(x^2-25\right)$
$(-\infty,-5)$	–6	neg.	neg.	neg.	neg.	pos.
$(-5,-3)$	–4	neg.	neg.	neg.	pos.	neg.
$(-3,3)$	0	neg.	neg.	pos.	pos.	pos.
$(3,5)$	4	neg.	pos.	pos.	pos.	neg.
$(5,\infty)$	6	pos.	pos.	pos.	pos.	pos.

So the product is negative on the intervals $(-5,-3)\cup(3,5)$.

77. The key numbers are 7 and –2. Now draw the sign chart:

Interval	Test Number	$x+2$	$x-7$	$\frac{(x-7)^2}{(x+2)^2}$
$(-\infty,-2)$	–3	neg.	neg.	neg.
$(-2,7)$	0	pos.	neg.	pos.
$(7,\infty)$	8	pos.	pos.	pos.

The quotient is positive on the intervals $(-2,7)\cup(7,\infty)$, and zero at $x=7$. The single interval $(-2,\infty)$ satisfies the inequality.

79. We factor the inequality as $\frac{(x-1)(x-9)}{(x+1)(x^2-x+1)} \le 0$. The key numbers are 1, 9, and –1. We draw the sign chart:

Interval	Test Number	$x-9$	$x-1$	$x+1$	$\frac{(x-1)(x-9)}{x^3+1}$
$(-\infty,-1)$	–2	neg.	neg.	neg.	neg.
$(-1,1)$	0	neg.	neg.	pos.	pos.
$(1,9)$	2	neg.	pos.	pos.	neg.
$(9,\infty)$	10	pos.	pos.	pos.	pos.

The quotient is negative on the intervals $(-\infty,-1)\cup(1,9)$, and is zero at $x=1$ and $x=9$. The solution set is $(-\infty,-1)\cup[1,9]$.

81. We first simplify and factor the inequality:

$$\frac{1-2x}{1+2x}-\frac{1}{2}\le 0$$
$$\frac{2(1-2x)-(1+2x)}{2(1+2x)}\le 0$$
$$\frac{-6x+1}{2(1+2x)}\le 0$$

The key numbers are $\frac{1}{6}$ and $-\frac{1}{2}$. Now draw the sign chart:

Interval	Test Number	$2x+1$	$-6x+1$	$\frac{-6x+1}{2(2x+1)}$
$\left(-\infty,-\frac{1}{2}\right)$	-1	neg.	pos.	neg.
$\left(-\frac{1}{2},\frac{1}{6}\right)$	0	pos.	pos.	pos.
$\left(\frac{1}{6},\infty\right)$	1	pos.	neg.	neg.

The quotient is negative on the intervals $\left(-\infty,-\frac{1}{2}\right)\cup\left(\frac{1}{6},\infty\right)$, and zero at $\frac{1}{6}$. The solution set is $\left(-\infty,-\frac{1}{2}\right)\cup\left[\frac{1}{6},\infty\right)$.

83. We first simplify the inequality:

$$\sqrt{x}-\frac{5}{\sqrt{x}}-4\le 0$$
$$\frac{x-5-4\sqrt{x}}{\sqrt{x}}\le 0$$

Looking at the denominator, we see that one key number is $x=0$. The others, if any, are found by solving the equation:

$$x-5-4\sqrt{x}=0$$

Writing this as $x-5=4\sqrt{x}$, and then squaring both sides, we have:

$$x^2-10x+25=16x$$
$$x^2-26x+25=0$$
$$(x-25)(x-1)=0$$
$$x=25 \text{ or } x=1$$

The value $x=25$ satisfies the equation $x-5-4\sqrt{x}=0$, but $x=1$ is an extraneous root. Thus the key numbers are 0 and 25. Now draw the sign chart:

Interval	Test Number	$\sqrt{x}$	$x-5-4\sqrt{x}$	$\frac{x-5-4\sqrt{x}}{\sqrt{x}}$
$(-\infty,0)$	-1		undefined	
$(0,25)$	1	pos.	neg.	neg.
$(25,\infty)$	36	pos.	pos.	pos.

The quotient is negative on the interval $(0,25)$ and zero at $x=25$. So the solution set is $(0,25]$.

85. The discriminant is $b^2-4ac=1-4k^2$. We solve for when this discriminant is non-negative:

$$1-4k^2\geq 0$$
$$(1-2k)(1+2k)\geq 0$$

The key numbers here are $k=\pm\frac{1}{2}$. We draw the sign chart:

Interval	Test Number	$1+2k$	$1-2k$	$1-4k^2$
$\left(-\infty,-\frac{1}{2}\right)$	-1	neg.	pos.	neg.
$\left(-\frac{1}{2},\frac{1}{2}\right)$	0	pos.	pos.	pos.
$\left(\frac{1}{2},\infty\right)$	1	pos.	neg.	neg.

The product is positive on $\left(-\frac{1}{2},\frac{1}{2}\right)$, and zero at $x=-\frac{1}{2}$ and $x=\frac{1}{2}$. So the values of k are chosen from the interval $\left[-\frac{1}{2},\frac{1}{2}\right]$.

87. The discriminant is:

$$b^2-4ac=(k+1)^2-4(2k)=k^2+2k+1-8k=k^2-6k+1$$

To find the key numbers for $k^2-6k+1\geq 0$, we use the quadratic formula:

$$k=\frac{6\pm\sqrt{36-4}}{2}=\frac{6\pm\sqrt{32}}{2}=\frac{6\pm 4\sqrt{2}}{2}=3\pm 2\sqrt{2}$$

For purposes of picking appropriate test numbers, note that $3+2\sqrt{2}\approx 5.8$ and that $3-2\sqrt{2}\approx 0.2$. Now draw the sign chart:

Interval	Test Number	$k-5.8$	$k-0.2$	k^2-6k+1
$(-\infty,3-2\sqrt{2})$	0	neg.	neg.	pos.
$(3-2\sqrt{2},3+2\sqrt{2})$	3	neg.	pos.	neg.
$(3+2\sqrt{2},\infty)$	7	pos.	pos.	pos.

The expression is positive on the interval $(-\infty,3-2\sqrt{2})\cup(3+2\sqrt{2},\infty)$, and zero at $k=3-2\sqrt{2}$ and $k=3+2\sqrt{2}$. So the values of k are chosen from the interval $(-\infty,3-2\sqrt{2}]\cup[3+2\sqrt{2},\infty)$.

89. Let x and $11-x$ denote the units digit and tens digit, respectively. Then we have:

$$\text{original number }+27=\text{ new number}$$
$$10(11-x)+x+27=10x+(11-x)$$
$$110-10x+x+27=10x+11-x$$
$$-18x=-126$$
$$x=7$$
$$11-x=4$$

The original number is 47.

91. We first re-draw the figure:

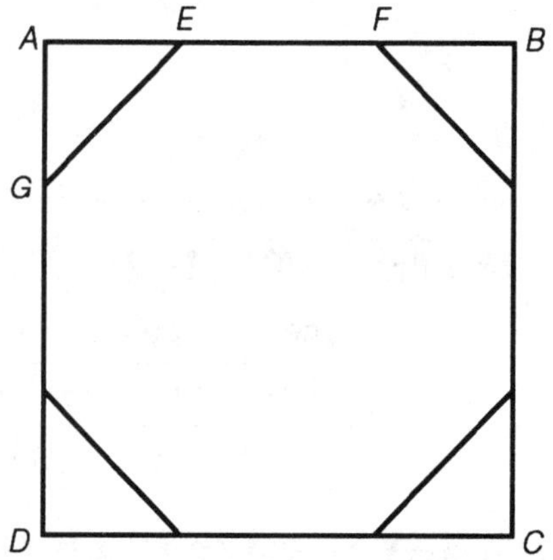

Since $AE = FB$, we have:

$$AE + EF + FB = 1$$
$$EF + 2AE = 1$$
$$AE = \frac{1 - EF}{2}$$

By the Pythagorean theorem and the fact that $AG = AE$ and $EF = GE$, we have:

$$(AG)^2 + (AE)^2 = (GE)^2$$
$$(AE)^2 + (AE)^2 = (EF)^2$$
$$2(AE)^2 = (EF)^2$$
$$2\left(\frac{1 - EF}{2}\right)^2 = EF^2$$

Now, for convenience, let $EF = x$. Then we have:

$$2\left(\frac{1 - 2x + x^2}{4}\right) = x^2$$
$$1 - 2x + x^2 = 2x^2$$
$$0 = x^2 + 2x - 1$$

The quadratic formula then gives us:

$$x = \frac{-2 \pm \sqrt{4+4}}{2} = \frac{-2 \pm \sqrt{8}}{2} = \frac{-2 \pm 2\sqrt{2}}{2} = -1 \pm \sqrt{2}$$

We choose the positive root, since $x > 0$. Thus $x = EF = -1 + \sqrt{2}$ cm.

93. Given that the length of a side is $\frac{x}{4}$, we compute the area and the perimeter:

$$\text{area} = \left(\frac{x}{4}\right)^2 = \frac{x^2}{16}$$

$$\text{perimeter} = 4\left(\frac{x}{4}\right) = x$$

If the area is numerically greater than the perimeter, then we have:

$$\frac{x^2}{16} > x$$
$$x^2 > 16x$$
$$x^2 - 16x > 0$$
$$x(x-16) > 0$$

Since x must be positive, this is satisfied only if $x - 16 > 0$, and hence $x > 16$. We conclude that the area will be numerically greater than the perimeter when $x > 16$ cm, which is the interval (16 cm, ∞).

95. Denoting the integers by $x - 1$, x and $x + 1$, we have:

$$(x-1)^2 + x^2 + (x+1)^2 = 1454$$
$$x^2 - 2x + 1 + x^2 + x^2 + 2x + 1 = 1454$$
$$3x^2 = 1452$$
$$x^2 = 484$$
$$x = \sqrt{484}$$
$$x = 22$$

Thus $x - 1 = 21$ and $x + 1 = 23$, so the three integers are 21, 22, and 23.

97. Let Joan's speed for the first four miles be r mph. Then the time for those four miles is $\frac{4}{r}$. Since her rate for the next six miles is $r + \frac{1}{2}$, her time on that portion is $\frac{6}{r+1/2}$ or $\frac{12}{2r+1}$, after simplifying. Thus her time for the race is $\frac{4}{r} + \frac{12}{2r+1}$ hours. On the other hand, had she run the entire ten miles at the faster pace, her time would have been $\frac{10}{r+1/2}$ or $\frac{20}{2r+1}$, after simplifying. Since this time is $\frac{2}{60} = \frac{1}{30}$ of an hour faster than her actual time, we can write:

$$\frac{4}{r} + \frac{12}{2r+1} - \frac{1}{30} = \frac{20}{2r+1}$$
$$\frac{4}{r} - \frac{8}{2r+1} - \frac{1}{30} = 0$$
$$4(30)(2r+1) - 8(30)r - r(2r+1) = 0$$
$$120 - 2r^2 - r = 0$$
$$2r^2 + r - 120 = 0$$
$$(2r-15)(r+8) = 0$$

We choose the positive value here, since $r > 0$. Thus Joan's speed for the first four miles is 7.5 mph, and for the last six miles it is 8 mph. Her time for the entire race is $\left(\frac{4}{7.5} + \frac{6}{8}\right)$ hours. This simplifies to $\frac{77}{60}$ hours, or 1 hour and 17 minutes.

99. We first draw a figure:

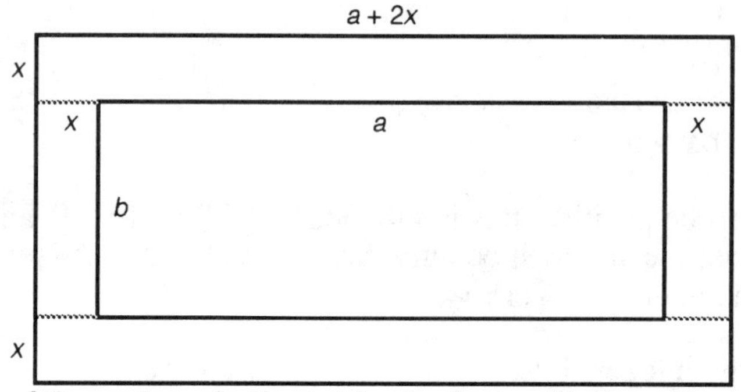

Let x denote the width of the path, as indicated in the figure. To compute the area of the path, we divide it into four rectangular portions, as indicated by the dotted lines in the figure. Then the area of the path is:

$$2[x(a+2x)]+2(bx)=4x^2+2ax+2bx$$

Since the area of the path and the flower garden are equal, we have:

$$4x^2+2ax+2bx=ab$$
$$4x^2+2(a+b)x-ab=0$$
$$x=\frac{-2(a+b)\pm\sqrt{4(a+b)^2+16ab}}{8}$$

Choosing the positive root here (since $x>0$), we have:

$$x=\frac{-2(a+b)+2\sqrt{(a+b)^2+4ab}}{8}=\frac{-(a+b)+\sqrt{a^2+6ab+b^2}}{4}$$

101. For each ball we will find at what time the height is 0. For the first ball we have $h=-16t^2+40t+50$, so setting $h=0$ yields $-16t^2+40t+50=0$, or $8t^2-20t-25=0$. Using the quadratic formula and choosing the positive root yields:

$$t=\frac{20+\sqrt{400+32(25)}}{2(8)}=\frac{20+20\sqrt{3}}{16}=\frac{5+5\sqrt{3}}{4}\approx 3.4\text{ sec}$$

For the second ball we have $h=-16t^2+5t+100$, so setting $h=0$ yields $-16t^2+5t+100=0$, or $16t^2-5t-100=0$. Using the quadratic formula and choosing the postive root yields:

$$t=\frac{5+\sqrt{25+6400}}{2(16)}=\frac{5+\sqrt{6425}}{32}=\frac{5+5\sqrt{257}}{32}\approx 2.7\text{ sec}$$

Therefore the second ball (the one thrown from 100 ft) hits the ground first.

103. We re-draw the figure:

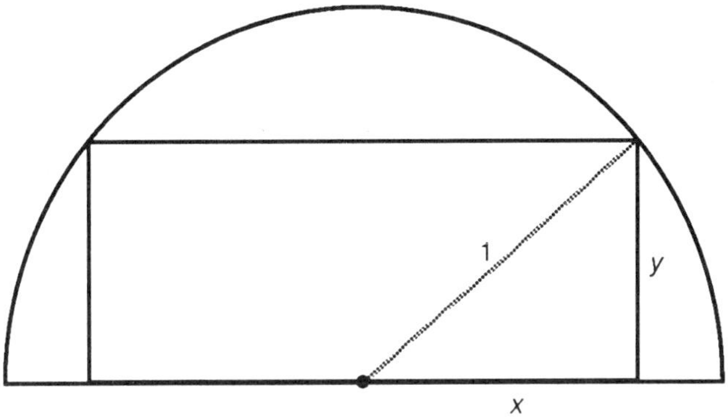

Let y be the height of the rectangle, so the area of the rectangle (using the Pythagorean theorem) is:

$$\text{area} = 2xy = 2x\sqrt{1-x^2}$$

If the area is 1 cm^2, then we have:

$$\begin{aligned} 2x\sqrt{1-x^2} &= 1 \\ 4x^2(1-x^2) &= 1 \\ 4x^2 - 4x^4 - 1 &= 0 \\ 4x^4 - 4x^2 + 1 &= 0 \\ (2x^2-1)^2 &= 0 \\ 2x^2 - 1 &= 0 \\ x^2 &= \tfrac{1}{2} \\ x &= \sqrt{\tfrac{1}{2}} = \tfrac{\sqrt{2}}{2} \end{aligned}$$

This value checks in the original equation.

105. Let x be the integer, so $x + 1$ is the next highest integer. Given some integer N, we must find where:

$$\begin{aligned} x^2 + x^3 &= N(x+1)^2 \\ x^2(1+x) &= N(x+1)^2 \\ \frac{x^2}{x+1} &= N \end{aligned}$$

But, by Exercise 85 of Section 1.8, $\frac{x^2}{x+1}$ cannot be a natural number (positive integer), thus there is no solution to the equation. This proves the desired result.

Chapter Two Test

1. Multiplying by $(x+4)(x-4)$, we have:

$$\frac{2}{x+4}-\frac{1}{x-4}=\frac{-7}{x^2-16}$$
$$2(x-4)-(x+4)=-7$$
$$2x-8-x-4=-7$$
$$x=5$$

2. Multiplying by $6(1-x)(2-x)$, we have:

$$\frac{1}{1-x}+\frac{4}{2-x}=\frac{11}{6}$$
$$6(2-x)+4(6)(1-x)=11(1-x)(2-x)$$
$$12-6x+24-24x=22-33x+11x^2$$
$$-11x^2+3x+14=0$$
$$11x^2-3x-14=0$$
$$(11x-14)(x+1)=0$$
$$x=\tfrac{14}{11},-1$$

3. (a) We solve by factoring:

$$x^2+4x=5$$
$$x^2+4x-5=0$$
$$(x+5)(x-1)=0$$
$$x=-5,1$$

(b) We solve by writing the equation as $x^2+4x-1=0$, then using the quadratic formula:

$$x=\frac{-4\pm\sqrt{(4)^2-4(1)(-1)}}{2(1)}=\frac{-4\pm\sqrt{20}}{2}=\frac{-4\pm2\sqrt{5}}{2}=-2\pm\sqrt{5}$$

4. We solve by factoring:

$$x^2\left(x^2-7\right)+12=0$$
$$x^4-7x^2+12=0$$
$$\left(x^2-4\right)\left(x^2-3\right)=0$$

So $x^2=4$ or $x^2=3$, and thus $x=\pm2$ or $x=\pm\sqrt{3}$.

5. Let $t=x^{2/3}$ and $t^2=x^{4/3}$ so that the equation becomes $4t^2-13t+9=0$. Factoring yields:

$$(4t-9)(t-1)=0$$
$$t=\tfrac{9}{4},1$$

With $t = \frac{9}{4}$ we have:

$$x^{2/3} = \tfrac{9}{4}$$
$$x^{1/3} = \pm\tfrac{3}{2}$$
$$x = \pm\tfrac{27}{8}$$

With $t = 1$ we have:

$$x^{2/3} = 1$$
$$x^{1/3} = \pm 1$$
$$x = \pm 1$$

So the solutions are ± 1 and $\pm\frac{27}{8}$.

6. Isolating $\sqrt{5-2x}$ and squaring, we have:

$$\sqrt{5-2x} - \sqrt{2-x} - \sqrt{3-x} = 0$$
$$\sqrt{5-2x} = \sqrt{2-x} + \sqrt{3-x}$$
$$5-2x = 2-x+2\sqrt{2-x}\sqrt{3-x}+3-x$$
$$0 = 2\sqrt{2-x}\sqrt{3-x}$$

Thus $\sqrt{2-x} = 0$ and consequently $x = 2$ or $\sqrt{3-x} = 0$ and consequently $x = 3$.The value $x = 2$ checks in the original equation, but (considering only the real number system) the value $x = 3$ does not check.

7. Given $|3x-1| = 2$, either $3x - 1 = 2$ or $3x - 1 = -2$. Therefore:

$$3x-1=2 \quad \text{or} \quad 3x-1=-2$$
$$3x=3 \qquad\qquad 3x=-1$$
$$x=1 \qquad\qquad x=-\tfrac{1}{3}$$

The solutions are 1 and $-\frac{1}{3}$.

8. Multiplying by $cx + d$, we have:

$$\frac{ax+b}{cx+d} = e$$
$$ax+b = cex+de$$
$$ax-cex = de-b$$
$$x(a-ce) = de-b$$
$$x = \frac{de-b}{a-ce}$$

9. Solving for x, we obtain:

$$ax = bx+a^2-b^2$$
$$ax-bx = a^2-b^2$$
$$x(a-b) = (a+b)(a-b)$$
$$x = \frac{(a+b)(a-b)}{a-b} = a+b$$

10. We solve the inequality:

$$4(1+x)-3(2x-1)\ge 1$$
$$4+4x-6x+3\ge 1$$
$$-2x\ge -6$$
$$x\le 3$$

The solution set is $(-\infty, 3]$.

11. Solving the double inequality, we have:

$$\frac{3}{5}<\frac{3-2x}{-4}<\frac{4}{5}$$
$$-12>15-10x>-16$$
$$-27>-10x>-31$$
$$\tfrac{27}{10}<x<\tfrac{31}{10}$$

The solution set is $\left(\frac{27}{10},\frac{31}{10}\right)$.

12. Solving $|3x-8|\le 1$, we have:

$$-1\le 3x-8\le 1$$
$$7\le 3x\le 9$$
$$\tfrac{7}{3}\le x\le 3$$

The solution set is $\left[\frac{7}{3},3\right]$.

13. The key numbers are 4 and -8. We draw the sign chart:

Interval	Test Number	$x+8$	$x-4$	$(x-4)^2(x+8)^3$
$(-\infty,-8)$	-9	neg.	neg.	neg.
$(-8,4)$	0	pos.	neg.	pos.
$(4,\infty)$	5	pos.	pos.	pos.

The product is positive on $(-8,4)\cup(4,\infty)$ and zero at $x=-8$ and $x=4$. So the inequality is satisfied on $[-8,4]\cup[4,\infty)$, which combine as $[-8,\infty)$.

14. Simplifying the inequality and factoring, we obtain:

$$\frac{(x+1)(x+2)+x(x+2)+x(x+1)}{x(x+1)(x+2)}\ge 0$$
$$\frac{3x^2+6x+2}{x(x+1)(x+2)}\ge 0$$

Three of the key numbers are 0, -1 and -2. The other two are found by using the quadratic formula to solve $3x^2+6x+2=0$. The roots are found to be $\frac{-3\pm\sqrt{3}}{3}$. For purposes of picking appropriate test numbers, note that $\frac{-3+\sqrt{3}}{3}\approx -0.4$ and that $\frac{-3-\sqrt{3}}{3}\approx -1.6$.

We draw the sign chart:

Interval	Test Number	x	$x+0.4$	$x+1$	$x+1.6$	$x+2$	$\frac{3x^2+6x+2}{x(x+1)(x+2)}$
$(-\infty,-2)$	-3	neg.	neg.	neg.	neg.	neg.	neg.
$(-2,-1.6)$	-1.8	neg.	neg.	neg.	neg.	pos.	pos.
$(-1.6,-1)$	-1.5	neg.	neg.	neg.	pos.	pos.	neg.
$(-1,-0.4)$	-0.5	neg.	neg.	pos.	pos.	pos.	pos.
$(-0.4,0)$	-0.1	neg.	pos.	pos.	pos.	pos.	neg.
$(0,\infty)$	2	pos.	pos.	pos.	pos.	pos.	pos.

So the solution set is $\left(-2,\frac{-3-\sqrt{3}}{3}\right]\cup\left(-1,\frac{-3+\sqrt{3}}{3}\right]\cup(0,\infty)$.

15. Let x and $280-x$ denote the required number of pounds of Kona and Columbian beans, respectively. We organize the data in a chart:

type of coffee	pounds	cost / pound	total cost
Kona	x	\$6.00	$6(x)$
Columbian	$280-x$	\$5.00	$5(280-x)$
Blend	280	\$5.60	$5.6(280)$

We therefore have the equation:

$$6x+5(280-x)=5.6(280)$$
$$6x+1400-5x=1568$$
$$x=168$$
$$280-x=112$$

Thus 168 pounds of Kona beans and 112 pounds of Columbian beans are required.

16. In the three hours from 6 a.m. to 9 a.m., the train from Los Angeles covers a distance of $(50)(3)=150$ miles. Now let $t=0$ correspond to 9 a.m. Then in t hours the train from Los Angeles has covered a total distance of $150+50t$, while the train from San Francisco has traveled $40t$ miles. When the trains pass one another the total distance must add up to 450 miles, so we have:

$$(150+50t)+40t=450$$
$$90t=300$$
$$t=3\tfrac{1}{3}\text{ hours}$$

So the trains meet 3 hours and 20 minutes after 9 a.m., which is 12:20 p.m.

17. The discriminant is $b^2-4ac=(3)^2-4(1)(k^2)=9-4k^2$. For the equation to have real solutions, this discriminant must be non-negative:

$$9-4k^2\geq 0$$
$$(3+2k)(3-2k)\geq 0$$

The key numbers are $-\frac{3}{2}$ and $\frac{3}{2}$. We draw the sign chart:

Interval	Test Number	$3+2k$	$3-2k$	$9-4k^2$
$\left(-\infty,-\frac{3}{2}\right)$	−2	neg.	pos.	neg.
$\left(-\frac{3}{2},\frac{3}{2}\right)$	0	pos.	pos.	pos.
$\left(\frac{3}{2},\infty\right)$	2	pos.	neg.	neg.

So the product is positive on $\left(-\frac{3}{2},\frac{3}{2}\right)$, and zero at $k=-\frac{3}{2}$ and $k=\frac{3}{2}$. Thus the equation will have real solutions if k is in the interval $\left[-\frac{3}{2},\frac{3}{2}\right]$.

18. (a) Since $h=2r$, then $r=\frac{h}{2}$. Substituting for r, we obtain:

$$S=2\pi\left(\frac{h}{2}\right)^2+2\pi\left(\frac{h}{2}\right)\bullet h=\frac{\pi h^2}{2}+\pi h^2=\frac{3}{2}\pi h^2$$

(b) Solving our result from (a) for h, we obtain:

$$\frac{3}{2}\pi h^2=S$$

$$h^2=\frac{2S}{3\pi}$$

$$h=\sqrt{\frac{2S}{3\pi}}$$

We were able to choose the positive root since $h\geq 0$.

Chapter Three
Coordinates and Graphs

3.1 Rectangular Coordinates

1. We plot the points:

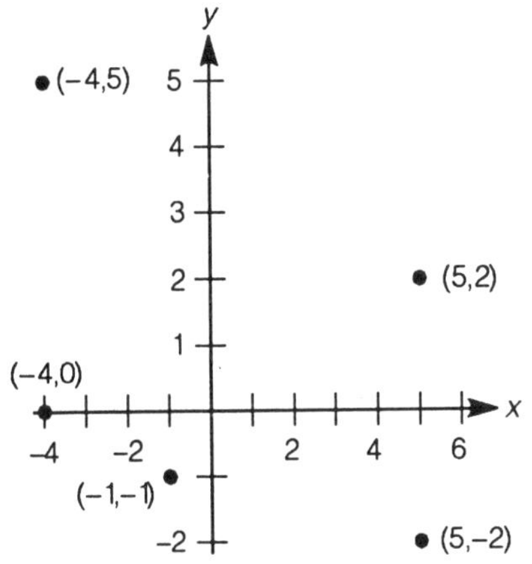

3. (a) We draw the right triangle PQR:

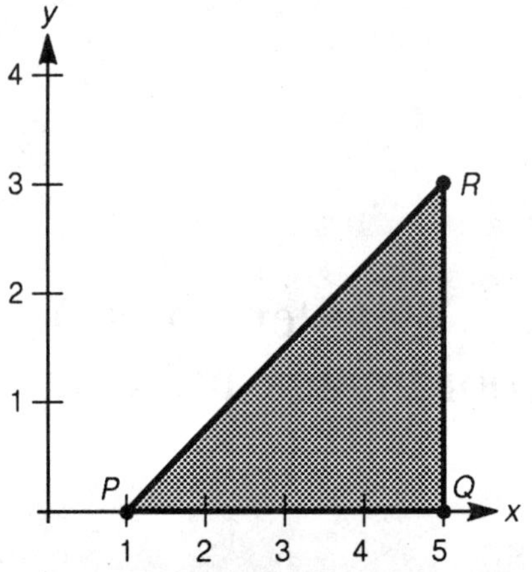

(b) Since the base is $b = 5 - 1 = 4$ and the height is $h = 3 - 0 = 3$, then the area is given by:

$$A = \frac{1}{2}bh = \frac{1}{2}(4)(3) = 6 \text{ square units}$$

5. (a) Here $(x_1, y_1) = (0, 0)$ and $(x_2, y_2) = (-3, 4)$, so by the distance formula:

$$d = \sqrt{(-3-0)^2 + (4-0)^2} = \sqrt{9+16} = \sqrt{25} = 5$$

(b) Here $(x_1, y_1) = (2, 1)$ and $(x_2, y_2) = (7, 13)$, so:

$$d = \sqrt{(7-2)^2 + (13-1)^2} = \sqrt{25+144} = \sqrt{169} = 13$$

7. (a) Here $(x_1, y_1) = (-5, 0)$ and $(x_2, y_2) = (5, 0)$, so:

$$d = \sqrt{[5-(-5)]^2 + (0-0)^2} = \sqrt{100+0} = \sqrt{100} = 10$$

(b) Here $(x_1, y_1) = (0, -8)$ and $(x_2, y_2) = (0, 1)$, so:

$$d = \sqrt{(0-0)^2 + [1-(-8)]^2} = \sqrt{0+81} = \sqrt{81} = 9$$

Note that we really don't need to use the distance formula for either (a) or (b), since in each case one of the coordinates (either x or y) is the same. Draw quick graphs and you can find the distance by inspection:

(a) This graph indicates the distance is 10:

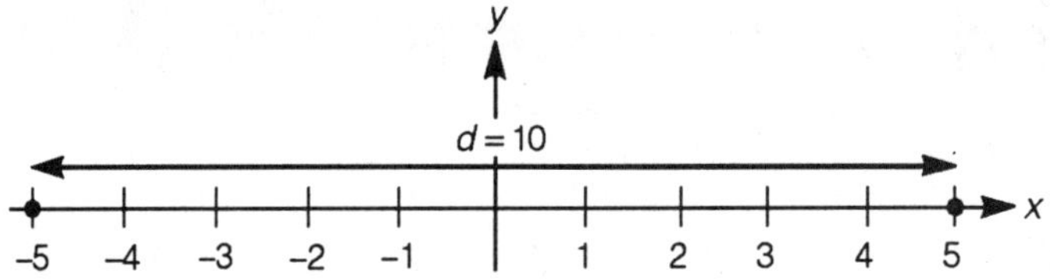

(b) This graph indicates the distance is 9:

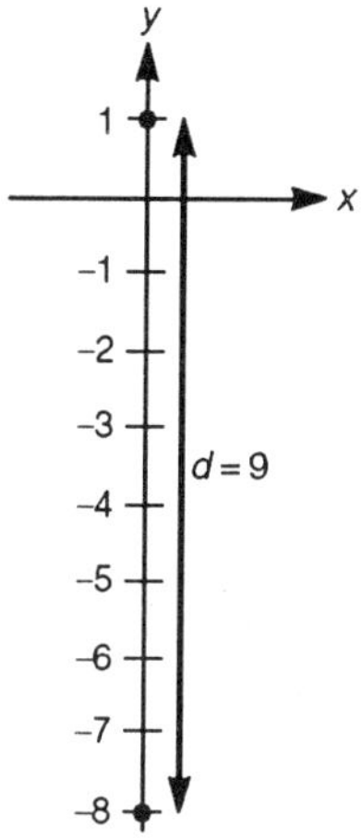

9. Here $(x_1,y_1)=(1,\sqrt{3})$ and $(x_2,y_2)=(-1,-\sqrt{3})$, so:

$$d=\sqrt{(-1-1)^2+(-\sqrt{3}-\sqrt{3})^2}=\sqrt{(-2)^2+(-2\sqrt{3})^2}=\sqrt{4+12}=4$$

11. (a) We calculate the distance of each point from the origin:

$(3,-2)$: $d=\sqrt{(3-0)^2+(-2-0)^2}=\sqrt{9+4}=\sqrt{13}$

$(4,\frac{1}{2})$: $d=\sqrt{(4-0)^2+(\frac{1}{2}-0)^2}=\sqrt{16+\frac{1}{4}}=\sqrt{16.25}$

So $(4,\frac{1}{2})$ is farther from the origin.

(b) We calculate the distance of each point from the origin:

$(-6,7)$: $d=\sqrt{(-6-0)^2+(7-0)^2}=\sqrt{36+49}=\sqrt{85}$

$(9,0)$: $d=\sqrt{(9-0)^2+(0-0)^2}=\sqrt{81+0}=\sqrt{81}$

So $(-6,7)$ is farther from the origin.

13. We will graph each triangle and then determine (using the converse of the Pythagorean theorem) whether $a^2+b^2=c^2$.

(a) We graph the points:

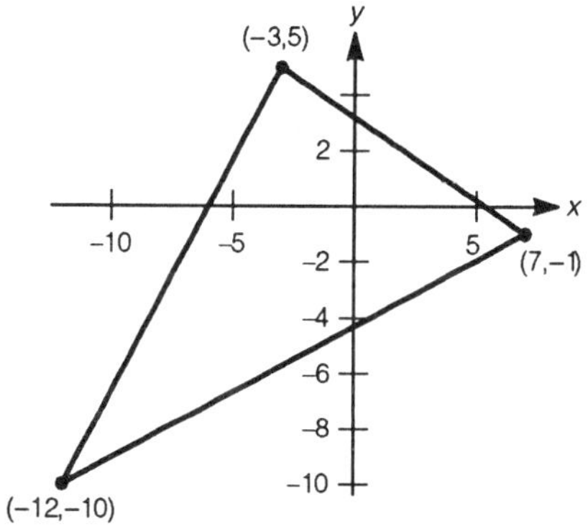

Now calculate the distances:

$$a = \sqrt{(-3-7)^2 + [5-(-1)]^2} = \sqrt{100+36} = \sqrt{136}$$
$$b = \sqrt{[-12-(-3)]^2 + (-10-5)^2} = \sqrt{81+225} = \sqrt{306}$$
$$c = \sqrt{(-12-7)^2 + [-10-(-1)]^2} = \sqrt{361+81} = \sqrt{442}$$

Now check the converse of the Pythagorean theorem:

$$a^2 + b^2 = 136 + 306 = 442 = c^2$$

So the triangle is a right triangle.

(b) We graph the points:

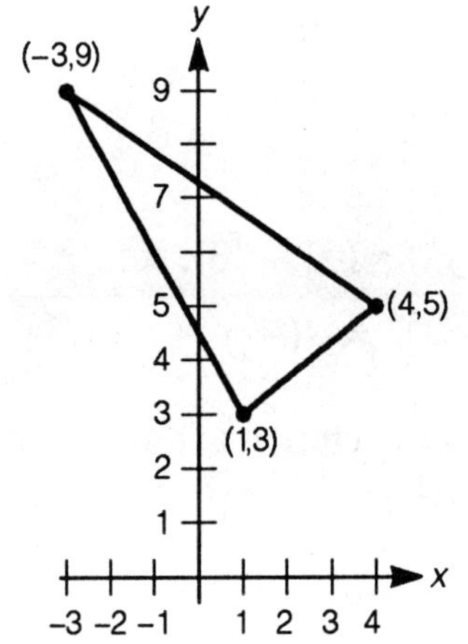

Now calculate the distances:

$$a = \sqrt{(-3-1)^2 + (9-3)^2} = \sqrt{16+36} = \sqrt{52}$$
$$b = \sqrt{(4-1)^2 + (5-3)^2} = \sqrt{9+4} = \sqrt{13}$$
$$c = \sqrt{(-3-4)^2 + (9-5)^2} = \sqrt{49+16} = \sqrt{65}$$

Now check the converse of the Pythagorean theorem:

$$a^2 + b^2 = 52 + 13 = 65 = c^2$$

So the triangle is a right triangle.

(c) We graph the points:

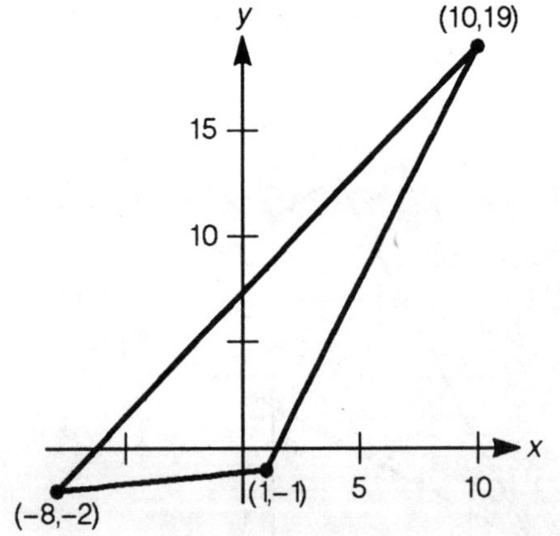

Now calculate the distances:

$$a=\sqrt{(-8-1)^2+[-2-(-1)]^2}=\sqrt{81+1}=\sqrt{82}$$
$$b=\sqrt{(10-1)^2+[19-(-1)]^2}=\sqrt{81+400}=\sqrt{481}$$
$$c=\sqrt{[10-(-8)]^2+[19-(-2)]^2}=\sqrt{324+441}=\sqrt{765}$$

Now check the converse of the Pythagorean theorem:

$$a^2+b^2=82+481=563\neq 765=c^2$$

So the triangle is not a right triangle.

15. Let $(x_1,y_1)=(1,-4)$, $(x_2,y_2)=(5,3)$, and $(x_3,y_3)=(13,17)$, so using the formula from Exercise 14(b) we have:

$$\begin{aligned}\text{Area}&=\left(\tfrac{1}{2}\right)\left|1(3)-5(-4)+5(17)-13(3)+13(-4)-1(17)\right|\\&=\left(\tfrac{1}{2}\right)\left|3+20+85-39-52-17\right|\\&=\left(\tfrac{1}{2}\right)\left|0\right|\\&=0\end{aligned}$$

For the area of the triangle to be 0 it must be that these three points do not form a triangle. The only way that could occur is if the three points are collinear; that is, they all lie on the same line.

17. (a) Here $(x_1,y_1)=(3,2)$ and $(x_2,y_2)=(9,8)$, so by the midpoint formula:

$$M=\left(\frac{3+9}{2},\frac{2+8}{2}\right)=\left(\frac{12}{2},\frac{10}{2}\right)=(6,5)$$

(b) Here $(x_1,y_1)=(-4,0)$ and $(x_2,y_2)=(5,-3)$, so by the midpoint formula:

$$M=\left(\frac{-4+5}{2},\frac{0-3}{2}\right)=\left(\frac{1}{2},-\frac{3}{2}\right)$$

(c) Here $(x_1,y_1)=(3,-6)$ and $(x_2,y_2)=(-1,-2)$, so by the midpoint formula:

$$M=\left(\frac{3-1}{2},\frac{-6-2}{2}\right)=\left(\frac{2}{2},\frac{-8}{2}\right)=(1,-4)$$

19. Label $C(x,y)$. Since B is the midpoint of $\overline{AC}$, then:

$$\frac{-1+x}{2}=5 \quad\text{and}\quad \frac{2+y}{2}=-3$$

We solve each of these for x and y:

$$\begin{aligned}\frac{-1+x}{2}&=5 &\quad\text{and}\quad \frac{2+y}{2}&=-3\\-1+x&=10 & 2+y&=-6\\x&=11 & y&=-8\end{aligned}$$

So the coordinates of C must be $(11,-8)$.

21. First find the distances from $P(x,y)$ to $(-2,4)$ and $(3,0)$:

$$d_1 = \sqrt{(x+2)^2+(y-4)^2} = \sqrt{x^2+4x+4+y^2-8y+16} = \sqrt{x^2+4x+y^2-8y+20}$$
$$d_2 = \sqrt{(x-3)^2+(y-0)^2} = \sqrt{x^2-6x+9+y^2} = \sqrt{x^2-6x+y^2+9}$$

Since $d_1 = d_2$, we have:

$$\sqrt{x^2+4x+y^2-8y+20} = \sqrt{x^2-6x+y^2+9}$$
$$x^2+4x+y^2-8y+20 = x^2-6x+y^2+9$$
$$10x-8y+11=0$$

23. First find the distance from $P(x,y)$ to $O(0,0)$:

$$d = \sqrt{(x-0)^2+(y-0)^2} = \sqrt{x^2+y^2}$$

Since $d = 5$, we have:

$$\sqrt{x^2+y^2} = 5$$
$$x^2+y^2 = 25$$

25. First find the distances from $P(x,y)$ to $(1,0)$ and $(-1,0)$:

$$d_1 = \sqrt{(x-1)^2+(y-0)^2} = \sqrt{x^2-2x+1+y^2} = \sqrt{x^2-2x+y^2+1}$$
$$d_2 = \sqrt{(x+1)^2+(y-0)^2} = \sqrt{x^2+2x+1+y^2} = \sqrt{x^2+2x+y^2+1}$$

Since $d_1 = 2d_2$, we have:

$$\sqrt{x^2-2x+y^2+1} = 2\sqrt{x^2+2x+y^2+1}$$
$$x^2-2x+y^2+1 = 4x^2+8x+4y^2+4$$
$$0 = 3x^2+10x+3y^2+3$$

27. First find the distance from $P(x,y)$ to $(-6,5)$:

$$d = \sqrt{(x+6)^2+(y-5)^2}$$

Since $d = 1$, we have:

$$\sqrt{(x+6)^2+(y-5)^2} = 1$$
$$(x+6)^2+(y-5)^2 = 1$$

29. (a) We sketch the parallelogram $ABCD$:

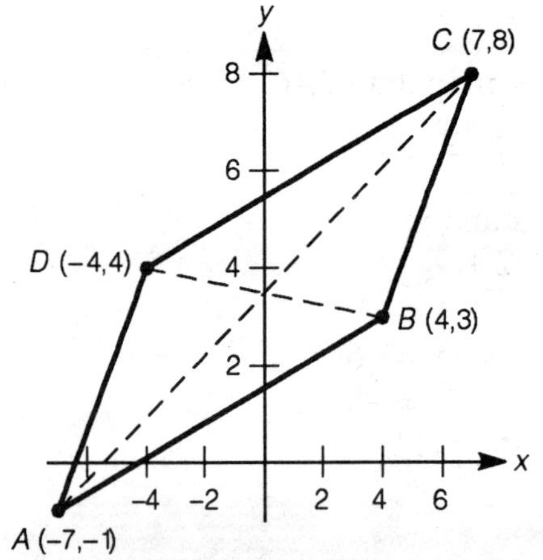

(b) For AC, let $(x_1, y_1) = (-7, -1)$ and $(x_2, y_2) = (7, 8)$, so the midpoint is:

$$M_1 = \left(\frac{-7+7}{2}, \frac{-1+8}{2}\right) = \left(0, \frac{7}{2}\right)$$

For BD, let $(x_1, y_1) = (4, 3)$ and $(x_2, y_2) = (-4, 4)$, so the midpoint is:

$$M_2 = \left(\frac{4-4}{2}, \frac{3+4}{2}\right) = \left(0, \frac{7}{2}\right)$$

(c) It appears that the midpoints of the two diagonals of a parallelogram are the same. Stated more concisely, the diagonals of a parallelogram bisect each other.

31. (a) Since $(AB)^2 = 68$, $(BC)^2 = 40$, and $(AC)^2 = 20$, then:

$$(AB)^2 + (BC)^2 + (AC)^2 = 128$$

(b) Let the midpoints be X, Y, and Z, so:

$$X = M_{AB} = (5,2)$$
$$Y = M_{BC} = (6,4)$$
$$Z = M_{AC} = (2,3)$$

Then the medians are:

$$AY = \sqrt{(6-1)^2 + (4-1)^2} = \sqrt{25+9} = \sqrt{34}$$
$$BZ = \sqrt{(2-9)^2 + (3-3)^2} = \sqrt{49+0} = \sqrt{49} = 7$$
$$CX = \sqrt{(5-3)^2 + (2-5)^2} = \sqrt{4+9} = \sqrt{13}$$

So the required value is:

$$(AY)^2 + (BZ)^2 + (CX)^2 = 34 + 49 + 13 = 96$$

(c) This ratio is $\frac{128}{96} = \frac{4}{3}$.

33. Using the Pythagorean theorem, we have:

$$a^2 = 1^2 + 1^2 = 2, \text{ so } a = \sqrt{2}$$
$$b^2 = 1^2 + \left(\sqrt{2}\right)^2 = 1 + 2 = 3, \text{ so } b = \sqrt{3}$$
$$c^2 = 1^2 + \left(\sqrt{3}\right)^2 = 1 + 3 = 4, \text{ so } c = 2$$
$$d^2 = 1^2 + 2^2 = 1 + 4 = 5, \text{ so } d = \sqrt{5}$$
$$e^2 = 1^2 + \left(\sqrt{5}\right)^2 = 1 + 5 = 6, \text{ so } e = \sqrt{6}$$
$$f^2 = 1^2 + \left(\sqrt{6}\right)^2 = 1 + 6 = 7, \text{ so } f = \sqrt{7}$$
$$g^2 = 1^2 + \left(\sqrt{7}\right)^2 = 1 + 7 = 8, \text{ so } g = \sqrt{8} = 2\sqrt{2}$$

35. Note that point B has coordinates $(a + b, c)$. We find the lengths of the diagonals:

$$OB = \sqrt{(a+b-0)^2 + (c-0)^2} = \sqrt{(a+b)^2 + c^2}$$
$$AC = \sqrt{(b-a)^2 + (c-0)^2} = \sqrt{(b-a)^2 + c^2}$$

Therefore:

$$\begin{aligned} S_1 &= OB^2 + AC^2 \\ &= (a+b)^2 + c^2 + (b-a)^2 + c^2 \\ &= a^2 + 2ab + b^2 + b^2 - 2ab + a^2 + 2c^2 \\ &= 2a^2 + 2b^2 + 2c^2 \end{aligned}$$

Now find the lengths of each side:

$$\begin{aligned} OA &= \sqrt{(a-0)^2 + (0-0)^2} = \sqrt{a^2} \\ AB &= \sqrt{(a+b-a)^2 + (c-0)^2} = \sqrt{b^2 + c^2} \\ BC &= \sqrt{(a+b-b)^2 + (c-c)^2} = \sqrt{a^2} \\ OC &= \sqrt{(b-0)^2 + (c-0)^2} = \sqrt{b^2 + c^2} \end{aligned}$$

Therefore:

$$\begin{aligned} S_2 &= OA^2 + AB^2 + BC^2 + OC^2 \\ &= a^2 + b^2 + c^2 + a^2 + b^2 + c^2 \\ &= 2a^2 + 2b^2 + 2c^2 \end{aligned}$$

This proves the result.

37. We first draw a figure:

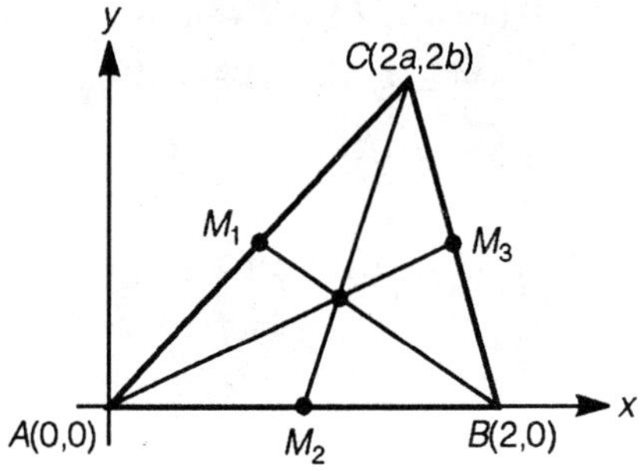

We find the sides of the triangle:

$$\begin{aligned} AB &= 2 \\ BC &= \sqrt{(2a-2)^2 + (2b-0)^2} = \sqrt{(2a-2)^2 + 4b^2} \\ AC &= \sqrt{(2a-0)^2 + (2b-0)^2} = \sqrt{4a^2 + 4b^2} \end{aligned}$$

Therefore:

$$\begin{aligned} S_1 &= AB^2 + BC^2 + AC^2 \\ &= 4 + (2a-2)^2 + 4b^2 + 4a^2 + 4b^2 \\ &= 4 + 4a^2 - 8a + 4 + 4b^2 + 4a^2 + 4b^2 \\ &= 8a^2 + 8b^2 + 8 - 8a \end{aligned}$$

Now we must do the same for the medians. First find the midpoints:

$$M_1 = \left(\frac{0+2a}{2}, \frac{0+2b}{2}\right) = (a,b)$$

$$M_2 = \left(\frac{0+2}{2}, \frac{0+0}{2}\right) = (1,0)$$

$$M_3 = \left(\frac{2a+2}{2}, \frac{2b+0}{2}\right) = (a+1,b)$$

Now compute the distances:

$$AM_3 = \sqrt{(a+1-0)^2 + (b-0)^2} = \sqrt{(a+1)^2 + b^2}$$
$$BM_1 = \sqrt{(a-2)^2 + (b-0)^2} = \sqrt{(a-2)^2 + b^2}$$
$$CM_2 = \sqrt{(2a-1)^2 + (2b-0)^2} = \sqrt{(2a-1)^2 + 4b^2}$$

Therefore:

$$\begin{aligned} S_2 &= \left(AM_3\right)^2 + \left(BM_1\right)^2 + \left(CM_2\right)^2 \\ &= (a+1)^2 + b^2 + (a-2)^2 + b^2 + (2a-1)^2 + 4b^2 \\ &= a^2 + 2a + 1 + a^2 - 4a + 4 + 4a^2 - 4a + 1 + 6b^2 \\ &= 6a^2 + 6b^2 + 6 - 6a \end{aligned}$$

We must show $S_1 = \frac{4}{3} S_2$:

$$\tfrac{4}{3} S_2 = \tfrac{4}{3}\left(6a^2 + 6b^2 + 6 - 6a\right) = 8a^2 + 8b^2 + 8 - 8a = S_1$$

This proves the desired result. Note that although it may appear that we have only proved the result for these three specific vertices, they are arbitrary enough that we have proved the result in general.

39. (a) First find the distances from $P(x,y)$ to $(-2,0)$ and $(2,0)$:

$$d_1 = \sqrt{(x+2)^2 + (y-0)^2} = \sqrt{x^2 + 4x + 4 + y^2}$$
$$d_2 = \sqrt{(x-2)^2 + (y-0)^2} = \sqrt{x^2 - 4x + 4 + y^2}$$

Since $d_1 + d_2 = 6$, we have:

$$\begin{aligned} \sqrt{x^2 + 4x + 4 + y^2} + \sqrt{x^2 - 4x + 4 + y^2} &= 6 \\ \sqrt{x^2 - 4x + 4 + y^2} &= 6 - \sqrt{x^2 + 4x + 4 + y^2} \end{aligned}$$

Squaring, we have:

$$\begin{aligned} x^2 - 4x + 4 + y^2 &= 36 - 12\sqrt{x^2 + 4x + 4 + y^2} + x^2 + 4x + 4 + y^2 \\ -8x - 36 &= -12\sqrt{x^2 + 4x + 4 + y^2} \\ 2x + 9 &= 3\sqrt{x^2 + 4x + 4 + y^2} \\ 4x^2 + 36x + 81 &= 9x^2 + 36x + 36 + 9y^2 \\ 45 &= 5x^2 + 9y^2 \end{aligned}$$

(b) First find the distances from $P(x,y)$ to $(-2,1)$ and $(2,-1)$:

$$d_1 = \sqrt{(x+2)^2 + (y-1)^2} = \sqrt{x^2 + 4x + 4 + y^2 - 2y + 1} = \sqrt{x^2 + 4x + y^2 - 2y + 5}$$
$$d_2 = \sqrt{(x-2)^2 + (y+1)^2} = \sqrt{x^2 - 4x + 4 + y^2 + 2y + 1} = \sqrt{x^2 - 4x + y^2 + 2y + 5}$$

Since $d_1 + d_2 = 6$, we have the equation:

$$\sqrt{x^2 + 4x + y^2 - 2y + 5} + \sqrt{x^2 - 4x + y^2 + 2y + 5} = 6$$

Isolating the radical and squaring, we have:

$$\begin{aligned} \sqrt{x^2 - 4x + y^2 + 2y + 5} &= 6 - \sqrt{x^2 + 4x + y^2 - 2y + 5} \\ -8x + 4y - 36 &= -12\sqrt{x^2 + 4x + y^2 - 2y + 5} \\ 2x - y + 9 &= 3\sqrt{x^2 + 4x + y^2 - 2y + 5} \\ (2x - y + 9)^2 &= 9x^2 + 36x + 9y^2 - 18y + 45 \\ 4x^2 - 4xy + 36x + y^2 - 18y + 81 &= 9x^2 + 36x + 9y^2 - 18y + 45 \\ 36 &= 5x^2 + 4xy + 8y^2 \end{aligned}$$

41. Using the hint and squaring, we have:

$$13 = \sqrt{(12-0)^2 + (t-2)^2}$$
$$169 = 144 + (t-2)^2$$
$$25 = (t-2)^2$$

So either $t - 2 = 5$ and thus $t = 7$, or $t - 2 = -5$ and thus $t = -3$. The values of t are 7 and -3.

43. Using the distance formula and squaring, we have:

$$\sqrt{(t+2)^2 + (1-3)^2} = 1$$
$$(t+2)^2 + 4 = 1$$
$$(t+2)^2 = -3$$

This equation has no solutions, since $(t + 2)^2 \geq 0$.

45. (a) Note that $\angle CAD = \angle DCB$ since they are both complementary to the same angle ($\angle ABC$). Thus ΔBAC and ΔBCD are similar, as all angles have the same measure.

(b) Comparing hypotenuse to short side, we have:

$$\frac{a}{y} = \frac{c}{a}, \text{ so } a^2 = cy$$

(c) Note that ΔACD is similar to ΔABC, since their angles have the same measure. Comparing hypotenuse to long side, we have:

$$\frac{b}{c-y} = \frac{c}{b}, \text{ so } b^2 = c^2 - cy$$

(d) Since $b^2 = c^2 - cy$ and $cy = a^2$, we have $b^2 = c^2 - a^2$ and thus $a^2 + b^2 = c^2$.

3.2 Graphs and Symmetry

1. Substituting $x = 8$, we have:

$$y = \tfrac{1}{2}(8) + 3 = 4 + 3 = 7 \neq 6$$

The point $(8, 6)$ does not lie on the graph.

3. Substituting $(x, y) = (4, 3)$, we have:

$$3(4)^2 + (3)^2 = 3(16) + 9 = 48 + 9 = 57 \neq 52$$

The point $(4, 3)$ does not lie on the graph.

5. Substituting $x = a$, we have:

$$y = 4a$$

The point $(a, 4a)$ lies on the graph.

7. (a) We graph the equation:

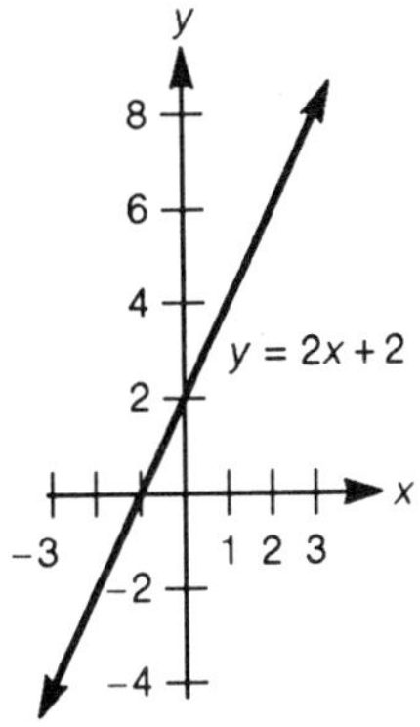

(b) We find the distance from $P(x, 2x + 2)$ to $O(0, 0)$:

$$d = \sqrt{(x-0)^2 + (2x+2-0)^2} = \sqrt{x^2 + 4x^2 + 8x + 4} = \sqrt{5x^2 + 8x + 4}$$

Setting $d = 5$, we have:

$$\begin{aligned}\sqrt{5x^2 + 8x + 4} &= 5\\ 5x^2 + 8x + 4 &= 25\\ 5x^2 + 8x - 21 &= 0\\ (5x-7)(x+3) &= 0\\ x &= \tfrac{7}{5}, -3\\ y &= \tfrac{24}{5}, -4\end{aligned}$$

The points are $\left(\frac{7}{5}, \frac{24}{5}\right)$ and $(-3, -4)$.

9. (a) We graph the equation:

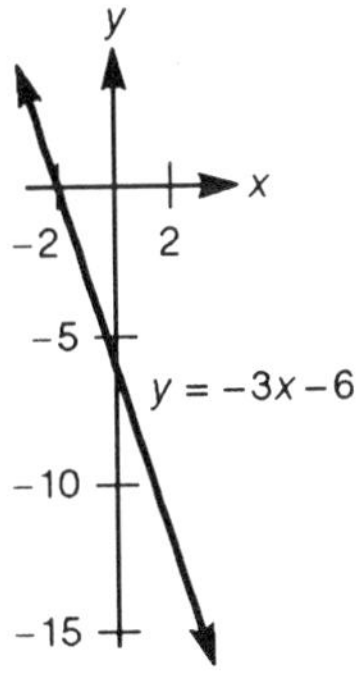

(b) We find where $x = \frac{1}{2}y$:

$$\begin{aligned}x &= \tfrac{1}{2}(-3x-6)\\ 2x &= -3x - 6\\ 5x &= -6\\ x &= -\tfrac{6}{5}\\ y &= -\tfrac{12}{5}\end{aligned}$$

The point is $\left(-\frac{6}{5}, -\frac{12}{5}\right)$.

11. To find the x-intercept, we substitute $y = 0$ into the equation $3x + 4y = 12$:

$$3x + 4(0) = 12$$
$$3x = 12$$
$$x = 4$$

To find the y-intercept, we substitute $x = 0$ into the equation $3x + 4y = 12$:

$$3(0) + 4y = 12$$
$$4y = 12$$
$$y = 3$$

Now graph the line:

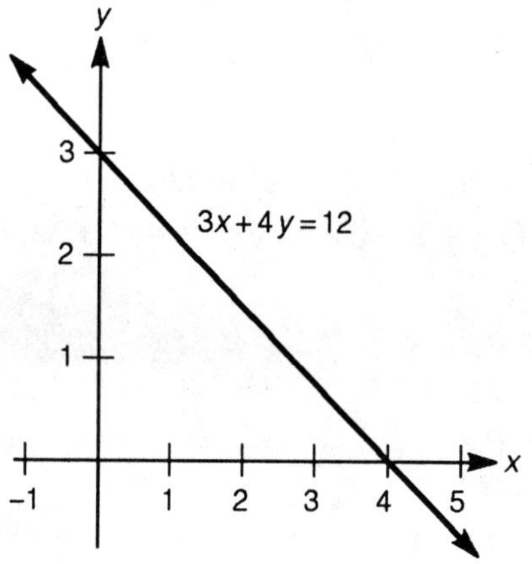

13. To find the x-intercept, we substitute $y = 0$ into the equation $y = 2x - 4$:

$$0 = 2x - 4$$
$$-2x = -4$$
$$x = 2$$

To find the y-intercept, we substitute $x = 0$ into the equation $y = 2x - 4$:

$$y = 2(0) - 4$$
$$y = -4$$

Now graph the line:

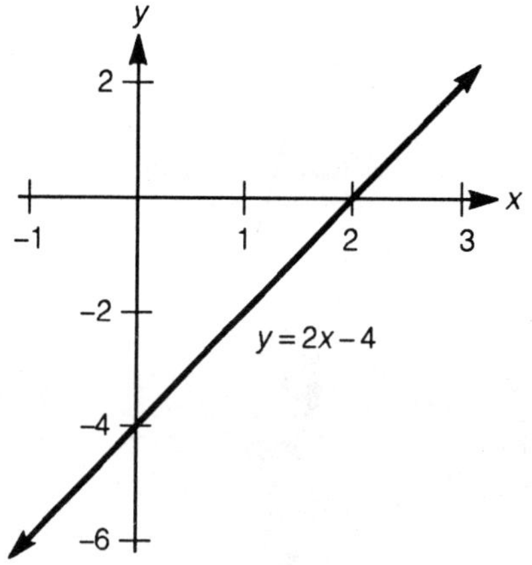

15. To find the x-intercept, we substitute $y = 0$ into the equation $x + y = 1$:

$$x + 0 = 1$$
$$x = 1$$

To find the y-intercept, we substitute $x = 0$ into the equation $x + y = 1$:

$$0 + y = 1$$
$$y = 1$$

Now graph the line:

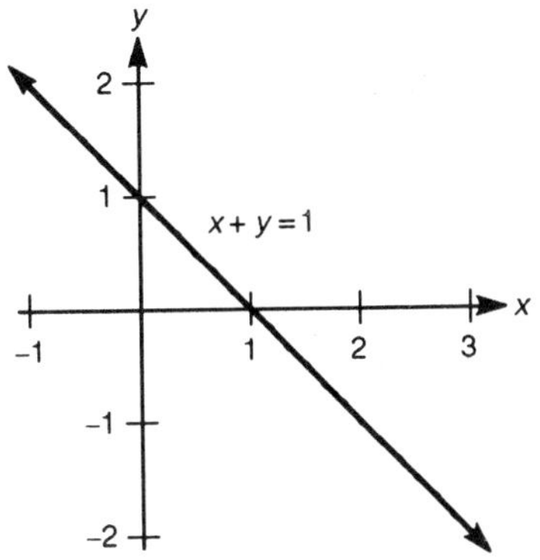

17. (a) To find the y-intercepts, let $x = 0$:

$$y = (0)^2 + 3(0) + 2 = 0 + 0 + 2 = 2$$

To find the x-intercepts, let $y = 0$:

$$x^2 + 3x + 2 = 0$$
$$(x + 2)(x + 1) = 0$$
$$x = -1, -2$$

(b) To find the y-intercepts, let $x = 0$:

$$y = (0)^2 + 2(0) + 3 = 0 + 0 + 3 = 3$$

To find the x-intercepts, we must solve the equation $x^2 + 2x + 3 = 0$. Using the quadratic formula, we obtain:

$$x = \frac{-2 \pm \sqrt{(2)^2 - 4(1)(3)}}{2(1)} = \frac{-2 \pm \sqrt{4 - 12}}{2} = \frac{-2 \pm \sqrt{-8}}{2}$$

Since this equation has no real solutions, there are no x-intercepts.

19. (a) To find the y-intercepts, let $x = 0$:

$$y = (0)^2 + 0 - 1 = 0 + 0 - 1 = -1$$

To find the x-intercepts, we must solve the equation $x^2 + x - 1 = 0$. Using the quadratic formula, we obtain:

$$x = \frac{-1 \pm \sqrt{(1)^2 - 4(1)(-1)}}{2(1)} = \frac{-1 \pm \sqrt{1 + 4}}{2} = \frac{-1 \pm \sqrt{5}}{2}$$

(b) To find the y-intercepts, let $x = 0$:

$$y = (0)^2 + 0 + 1 = 0 + 0 + 1 = 1$$

To find the x-intercepts, we must solve the equation $x^2 + x + 1 = 0$. Using the quadratic formula, we obtain:

$$x = \frac{-1 \pm \sqrt{(1)^2 - 4(1)(1)}}{2(1)} = \frac{-1 \pm \sqrt{1 - 4}}{2} = \frac{-1 \pm \sqrt{-3}}{2}$$

Since this equation has no real solutions, there are no x-intercepts.

21. To find the y-intercepts, let $x = 0$:

$$\tfrac{0}{2} + \tfrac{y}{3} = 1$$
$$\tfrac{y}{3} = 1$$
$$y = 3$$

To find the x-intercepts, let $y = 0$:

$$\tfrac{x}{2} + \tfrac{0}{3} = 1$$
$$\tfrac{x}{2} = 1$$
$$x = 2$$

23. To find the y-intercepts, let $x = 0$:

$$3(0) - 5y = 10$$
$$-5y = 10$$
$$y = -2$$

To find the x-intercepts, let $y = 0$:

$$3x - 5(0) = 10$$
$$3x = 10$$
$$x = \tfrac{10}{3}$$

25. To find the y-intercepts, let $x = 0$:

$$y = (0)^3 - 8 = 0 - 8 = -8$$

To find the x-intercepts, let $y = 0$:

$$x^3 - 8 = 0$$
$$x^3 = 8$$
$$x = 2$$

27. To find the y-intercepts, let $x = 0$:

$$y = \sqrt{0-1} - 0 + 3 = \sqrt{-1} + 3$$

Since this is not a real number, there are no y-intercepts.
To find the x-intercepts, let $y = 0$:

$$0 = \sqrt{x-1} - x + 3$$
$$x - 3 = \sqrt{x-1}$$
$$(x-3)^2 = x - 1$$
$$x^2 - 6x + 9 = x - 1$$
$$x^2 - 7x + 10 = 0$$
$$(x-2)(x-5) = 0$$
$$x = 2, 5$$

Note that $x = 5$ satisfies the required equation, but $x = 2$ does not. So $x = 5$ is the only x-intercept.

29. To find the y-intercepts, let $x = 0$:

$$y = 11(0) - 2(0)^2 - (0)^3 = 0 - 0 - 0 = 0$$

To find the x-intercepts, let $y = 0$:

$$0 = 11x - 2x^2 - x^3$$
$$0 = x(11 - 2x - x^2)$$

So $x = 0$ is clearly one x-intercept. We find the other two using the quadratic formula:

$$x = \frac{2 \pm \sqrt{(-2)^2 - 4(11)(-1)}}{2(-1)} = \frac{2 \pm \sqrt{48}}{-2} = -1 \pm 2\sqrt{3} \approx 2.46, -4.46$$

31. To find the y-intercepts, let $x = 0$:

$$y = (0)^4 - 2(0)^2 - 3 = 0 - 0 - 3 = -3$$

To find the x-intercepts, let $y = 0$:

$$0 = x^4 - 2x^2 - 3$$
$$0 = (x^2 - 3)(x^2 + 1)$$

Setting the first factor equal to zero yields $x^2 - 3 = 0$, so $x^2 = 3$ and thus $x = \pm\sqrt{3} \approx \pm 1.73$. Setting the second factor equal to zero yields $x^2 + 1 = 0$, so $x^2 = -1$, which has no real solutions.

33. (a) The reflection of $\overline{AB}$ about the x-axis is denoted by $\overline{A'B'}$:

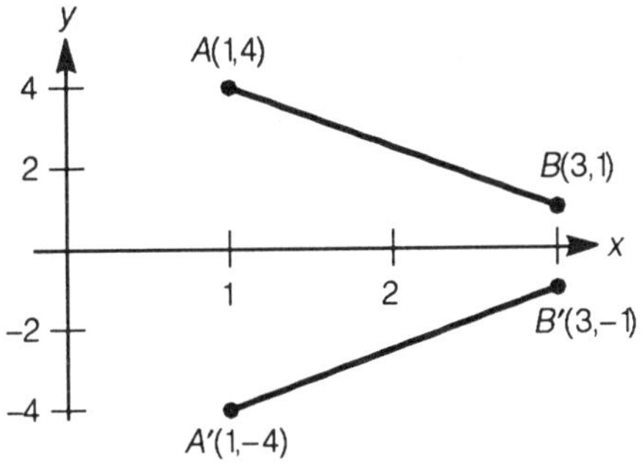

(b) The reflection of $\overline{AB}$ about the y-axis is denoted by $\overline{A'B'}$:

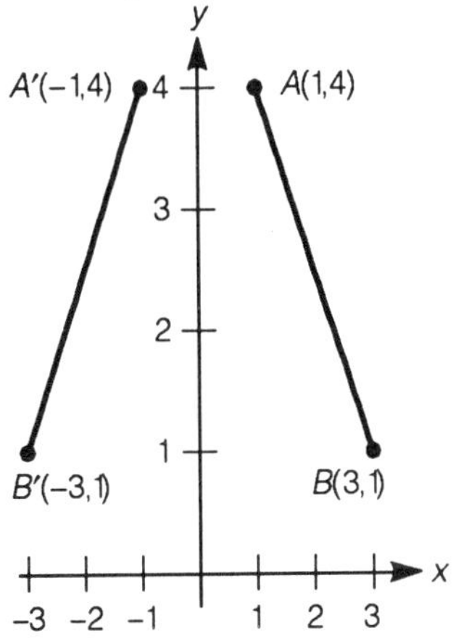

(c) The reflection of $\overline{AB}$ about the origin is denoted by $\overline{A'B'}$:

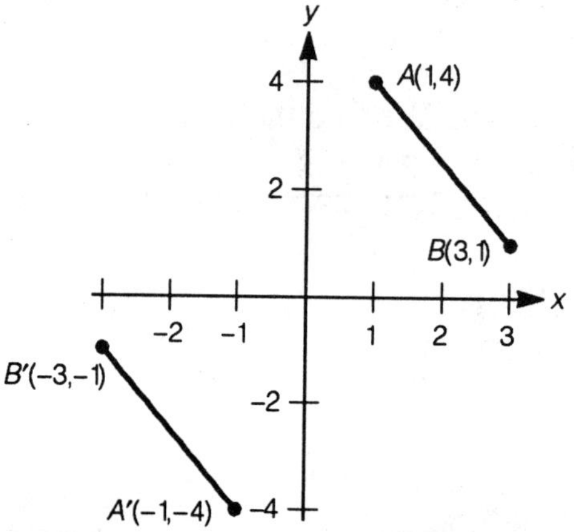

35. (a) The reflection of $\overline{AB}$ about the x-axis is denoted by $\overline{A'B'}$:

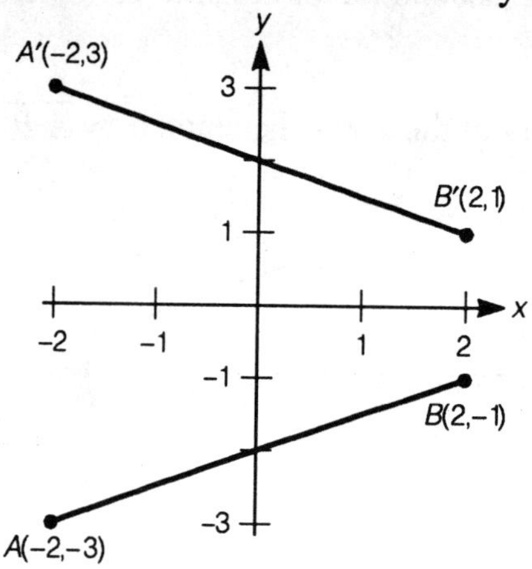

(b) The reflection of $\overline{AB}$ about the y-axis is denoted by $\overline{A'B'}$:

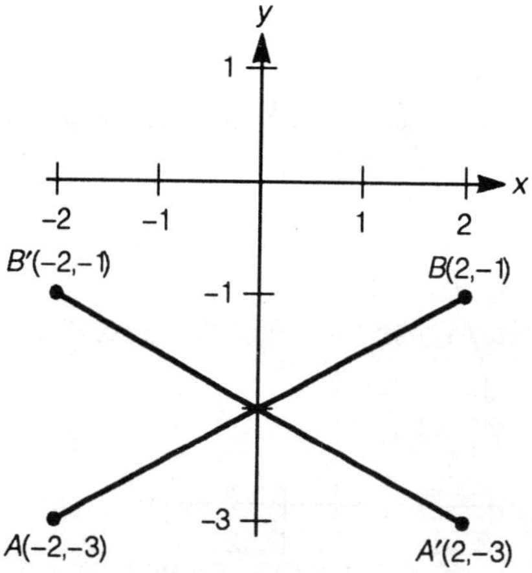

(c) The reflection of $\overline{AB}$ about the origin is denoted by $\overline{A'B'}$:

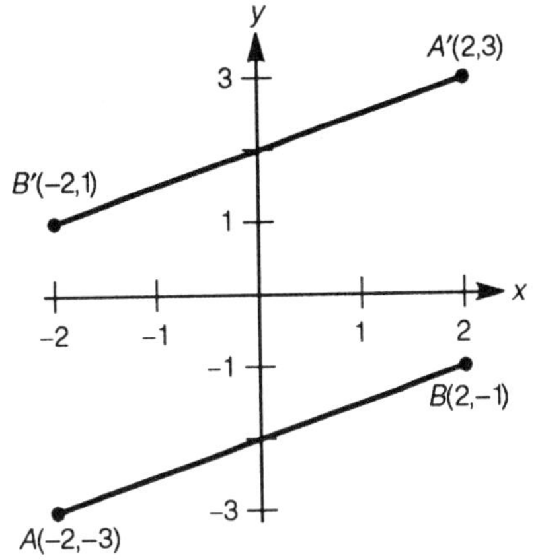

37. (a) To test the equation for symmetry about the x-axis, we replace the point (x, y) with the point $(x, -y)$:

$$3x^2 + (-y) = 16$$
$$3x^2 - y = 16$$

Since the equation is changed, it is not symmetric about the x-axis. To test the equation for symmetry about the y-axis, we replace the point (x, y) with the point $(-x, y)$:

$$3(-x)^2 + y = 16$$
$$3x^2 + y = 16$$

Since the equation remains unchanged, it is symmetric about the y-axis. To test the equation for symmetry about the origin, we replace the point (x, y) with the point $(-x, -y)$:

$$3(-x)^2 + (-y) = 16$$
$$3x^2 - y = 16$$

Since the equation is changed, it is not symmetric about the origin.

(b) To test the equation for symmetry about the x-axis, we replace the point (x, y) with the point $(x, -y)$:

$$3x + (-y)^2 = 16$$
$$3x + y^2 = 16$$

Since the equation remains unchanged, it is symmetric about the x-axis. To test the equation for symmetry about the y-axis, we replace the point (x, y) with the point $(-x, y)$:

$$3(-x) + y^2 = 16$$
$$-3x + y^2 = 16$$

Since the equation is changed, it is not symmetric about the y-axis. To test the equation for symmetry about the origin, we replace the point (x, y) with the point $(-x, -y)$:

$$3(-x) + (-y)^2 = 16$$
$$-3x + y^2 = 16$$

Since the equation is changed, it is not symmetric about the origin.

39. (a) To test for x-axis symmetry, we replace the point (x, y) with the point $(x, -y)$:

$$(-y) = x^4 - 3$$
$$y = -x^4 + 3$$

Since the equation is changed, there is no x-axis symmetry. To test for y-axis symmetry, we replace the point (x, y) with the point $(-x, y)$:

$$y = (-x)^4 - 3$$
$$y = x^4 - 3$$

Since the equation remains unchanged, it is symmetric about the y-axis. To test for origin symmetry, we replace the point (x, y) with the point $(-x, -y)$:

$$(-y) = (-x)^4 - 3$$
$$-y = x^4 - 3$$
$$y = -x^4 + 3$$

Since the equation is changed, there is no origin symmetry.

(b) To test for x-axis symmetry, we replace the point (x, y) with the point $(x, -y)$:

$$(-y) = x^3 - 3$$
$$y = -x^3 + 3$$

Since the equation is changed, there is no x-axis symmetry. To test for y-axis symmetry, we replace the point (x, y) with the point $(-x, y)$:

$$y = (-x)^3 - 3$$
$$y = -x^3 - 3$$

Since the equation is changed, there is no y-axis symmetry. To test for origin symmetry, we replace the point (x, y) with the point $(-x, -y)$:

$$(-y) = (-x)^3 - 3$$
$$-y = -x^3 - 3$$
$$y = x^3 + 3$$

Since the equation is changed, there is no origin symmetry.

(c) To test for x-axis symmetry, we replace the point (x, y) with the point $(x, -y)$:

$$(-y)^2 = x^3 - 3$$
$$y^2 = x^3 - 3$$

Since the equation remains unchanged, it is symmetric about the x-axis. To test for y-axis symmetry, we replace the point (x, y) with the point $(-x, y)$:

$$y^2 = (-x)^3 - 3$$
$$y^2 = -x^3 - 3$$

Since the equation is changed, there is no y-axis symmetry. To test for origin symmetry, we replace the point (x, y) with the point $(-x, -y)$:

$$(-y)^2 = (-x)^3 - 3$$
$$y^2 = -x^3 - 3$$

Since the equation is changed, there is no origin symmetry.

(d) To test for x-axis symmetry, we replace the point (x, y) with the point $(x, -y)$:

$$(-y)^3 = x$$
$$-y^3 = x$$
$$y^3 = -x$$

Since the equation is changed, there is no x-axis symmetry. To test for y-axis symmetry, we replace the point (x, y) with the point $(-x, y)$:

$$y^3 = -x$$

Since the equation is changed, there is no y-axis symmetry. To test for origin symmetry, we replace the point (x, y) with the point $(-x, -y)$:

$$(-y)^3 = -x$$
$$-y^3 = -x$$
$$y^3 = x$$

Since the equation remains unchanged, it is symmetric about the origin.

41. To test for x-axis symmetry, we replace the point (x, y) with the point $(x, -y)$:

$$x^2 + (-y)^2 = 16$$
$$x^2 + y^2 = 16$$

Since the equation remains unchanged, it is symmetric about the x-axis. To test for y-axis symmetry, we replace the point (x, y) with the point $(-x, y)$:

$$(-x)^2 + y^2 = 16$$
$$x^2 + y^2 = 16$$

Since the equation remains unchanged, it is symmetric about the y-axis. To test for origin symmetry, we replace the point (x, y) with the point $(-x, -y)$:

$$(-x)^2 + (-y)^2 = 16$$
$$x^2 + y^2 = 16$$

Since the equation remains unchanged, it is symmetric about the origin.

43. To test for x-axis symmetry, we replace the point (x, y) with the point $(x, -y)$:

$$(-y) = x^2 + x^3$$
$$y = -x^2 - x^3$$

Since the equation is changed, there is no x-axis symmetry. To test for y-axis symmetry, we replace the point (x, y) with the point $(-x, y)$:

$$y = (-x)^2 + (-x)^3$$
$$y = x^2 - x^3$$

Since the equation is changed, there is no y-axis symmetry. To test for origin symmetry, we replace the point (x, y) with the point $(-x, -y)$:

$$(-y) = (-x)^2 + (-x)^3$$
$$-y = x^2 - x^3$$
$$y = -x^2 + x^3$$

Since the equation is changed, there is no origin symmetry.

45. To test for x-axis symmetry, we replace the point (x, y) with the point $(x, -y)$:

$$x + (-y) = 1$$
$$x - y = 1$$

Since the equation is changed, there is no x-axis symmetry. To test for y-axis symmetry, we replace the point (x, y) with the point $(-x, y)$:

$$(-x) + y = 1$$
$$-x + y = 1$$

Since the equation is changed, there is no y-axis symmetry. To test for origin symmetry, we replace the point (x, y) with the point $(-x, -y)$:

$$(-x) + (-y) = 1$$
$$-x - y = 1$$
$$x + y = -1$$

Since the equation is changed, there is no origin symmetry.

47. The x- and y-intercepts are both 0. The graph is symmetric about the y-axis.

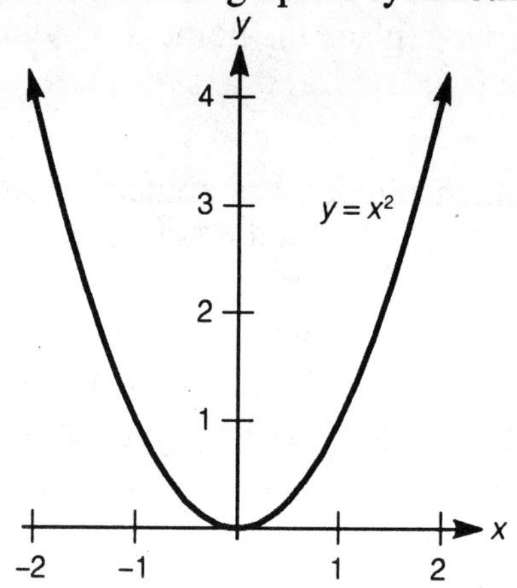

49. There are no x- or y-intercepts. The graph is symmetric about the origin.

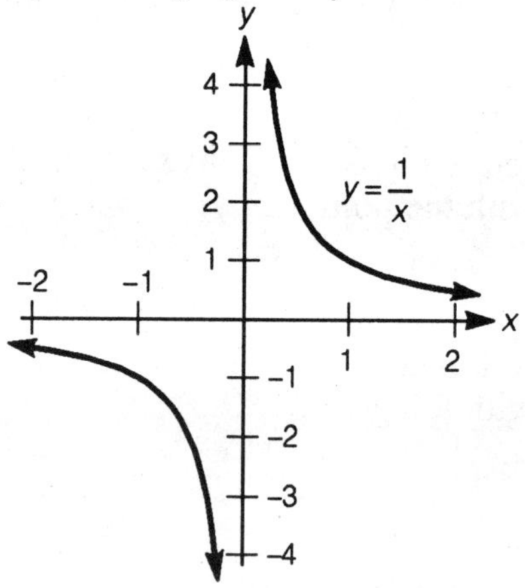

51. The x- and y-intercepts are both 0. The graph is symmetric about the y-axis.

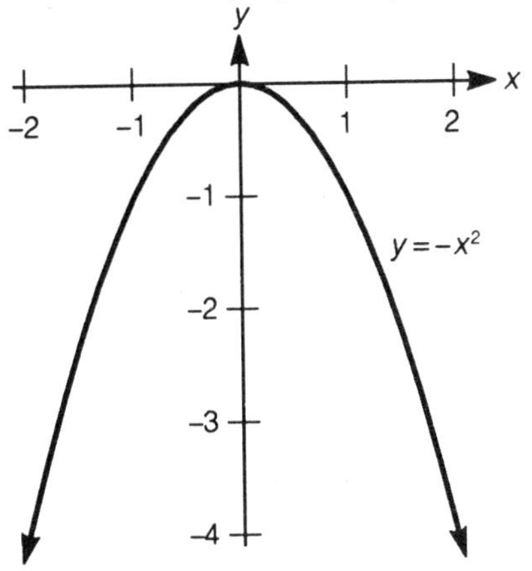

53. There are no x- or y-intercepts. The graph is symmetric about the origin.

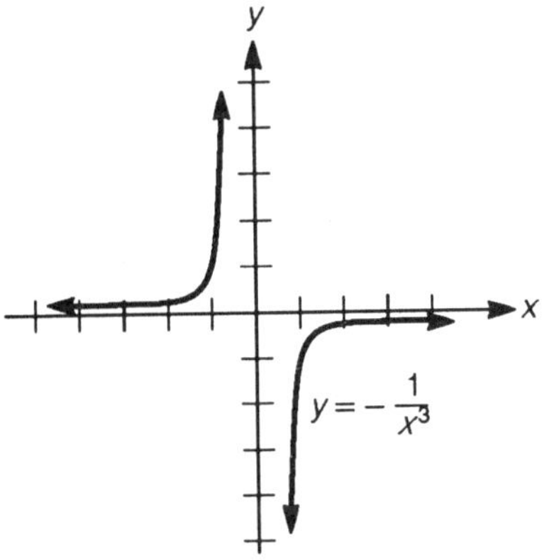

55. The x- and y-intercepts are both 0. The graph is symmetric about the y-axis.

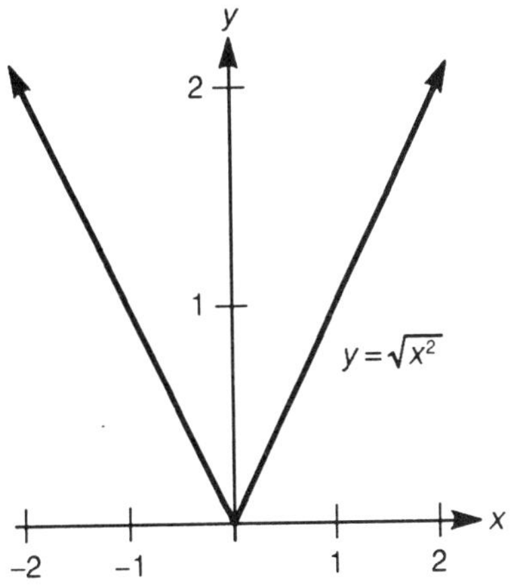

57. The x- and y-intercepts are both 1. The graph does not possess any of the three types of symmetry.

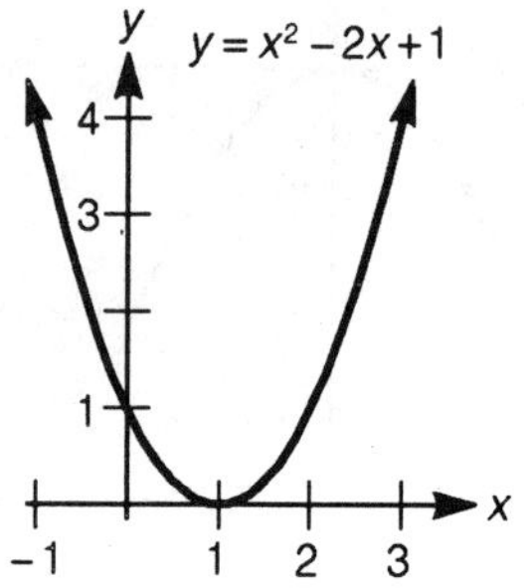

59. The x-intercept is 2 and there is no y-intercept. The graph is symmetric about the x-axis.

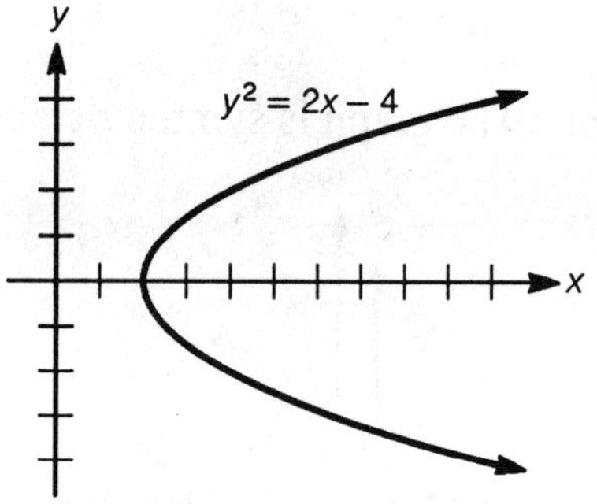

61. To find the x-intercepts, we use the quadratic formula:

$$x = \frac{-1 \pm \sqrt{(1)^2 - 4(2)(-4)}}{2(2)} = \frac{-1 \pm \sqrt{33}}{4}$$

The y-intercept is -4. The graph does not possess any of the three types of symmetry.

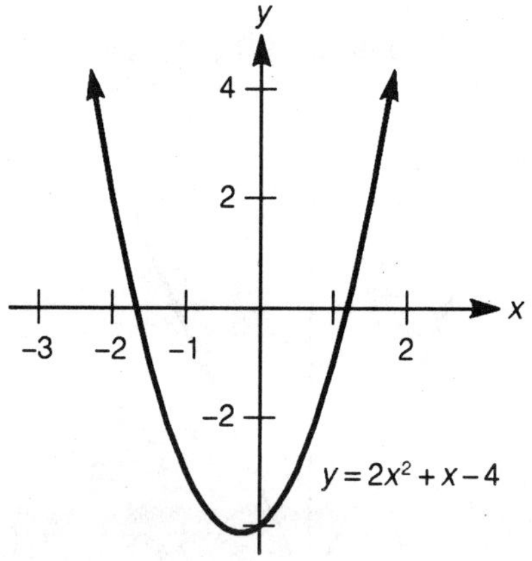

63. (a) The x-intercept is 2 and the y-intercept is -6. The graph does not possess any of the three types of symmetry.

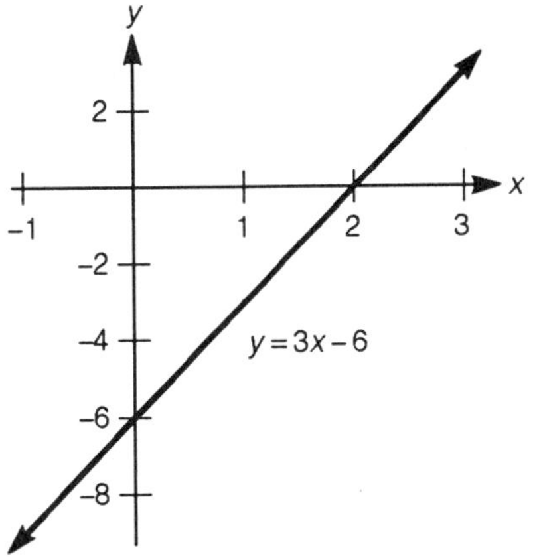

(b) The x-intercept is 2 and the y-intercept is 6. The graph does not possess any of the three types of symmetry.

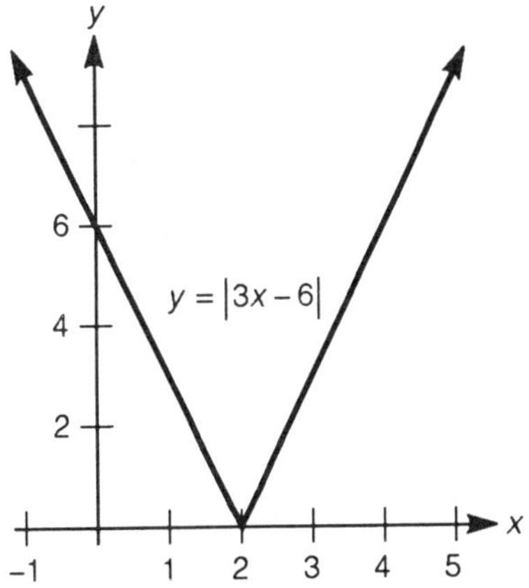

65. The circle is in standard form, so its center is (3, 1) and its radius is $\sqrt{25} = 5$. The x-intercepts are found by setting $y = 0$:

$$(x-3)^2 + 1 = 25$$
$$(x-3)^2 = 24$$
$$x - 3 = \pm\sqrt{24} = \pm 2\sqrt{6}$$
$$x = 3 \pm 2\sqrt{6}$$

The y-intercepts are found by setting $x = 0$:

$$9 + (y-1)^2 = 25$$
$$(y-1)^2 = 16$$
$$y - 1 = \pm 4$$
$$y = 1 \pm 4 = 5 \text{ and } -3$$

67. The circle is in standard form, so its center is (0, 0) and its radius is $\sqrt{\sqrt{2}} = \sqrt[4]{2}$. The x-intercepts are found by setting $y = 0$:

$$x^2 = \sqrt{2}$$
$$x = \pm\sqrt{\sqrt{2}} = \pm\sqrt[4]{2}$$

The y-intercepts are found by setting $x = 0$:

$$y^2 = \sqrt{2}$$
$$y = \pm\sqrt{\sqrt{2}} = \pm\sqrt[4]{2}$$

69. We must complete the square:

$$x^2 + 8x + y^2 - 6y = -24$$
$$\left(x^2 + 8x + 16\right) + \left(y^2 - 6y + 9\right) = -24 + 16 + 9$$
$$(x+4)^2 + (y-3)^2 = 1$$

The center is (–4, 3) and the radius is $\sqrt{1} = 1$. The x-intercepts are found by setting $y = 0$:

$$(x+4)^2 + 9 = 1$$
$$(x+4)^2 = -8$$

Since this equation has no real solution, there are no x-intercepts. The y-intercepts are found by setting $x = 0$:

$$16 + (y-3)^2 = 1$$
$$(y-3)^2 = -15$$

Since this equation has no real solution, there are no y-intercepts.

71. We divide by 9 and complete the square:

$$x^2 + 6x + y^2 - \tfrac{2}{3}y = -\tfrac{64}{9}$$
$$\left(x^2 + 6x + 9\right) + \left(y^2 - \tfrac{2}{3}y + \tfrac{1}{9}\right) = -\tfrac{64}{9} + 9 + \tfrac{1}{9}$$
$$(x+3)^2 + \left(y - \tfrac{1}{3}\right)^2 = 2$$

The center is $\left(-3, \tfrac{1}{3}\right)$ and the radius is $\sqrt{2}$. The x-intercepts are found by setting $y = 0$:

$$(x+3)^2 + \tfrac{1}{9} = 2$$
$$(x+3)^2 = \tfrac{17}{9}$$
$$|x+3| = \sqrt{\tfrac{17}{9}} = \tfrac{\sqrt{17}}{3}$$
$$x + 3 = \pm\tfrac{\sqrt{17}}{3}$$
$$x = -3 \pm \tfrac{\sqrt{17}}{3}$$

The y-intercepts are found by setting $x = 0$:

$$9 + \left(y - \tfrac{1}{3}\right)^2 = 2$$
$$\left(y - \tfrac{1}{3}\right)^2 = -7$$

Since this equation has no real solution, there are no y-intercepts.

73. We must find the radius by completing the square:

$$x^2 - 2ax + y^2 - 2by = c^2 - a^2 - b^2$$
$$(x^2 - 2ax + a^2) + (y^2 - 2by + b^2) = c^2 - a^2 - b^2 + a^2 + b^2$$
$$(x - a)^2 + (y - b)^2 = c^2$$

Since $c > 0$, then the radius is c. Then $A = \pi(c)^2 = \pi c^2$ and $C = 2\pi(c) = 2\pi c$.

75. If $\overline{PQ}$ is the diameter, then its midpoint must be the center of the circle:

$$M = \left(\frac{-3+2}{2}, \frac{-4+8}{2}\right) = \left(-\frac{1}{2}, \frac{4}{2}\right) = \left(-\frac{1}{2}, 2\right)$$

Furthermore, the diameter is twice the length of the radius, so we can use the distance formula to find:

$$d = \sqrt{(-3-2)^2 + (-4-8)^2} = \sqrt{25+144} = \sqrt{169} = 13$$

Thus the radius is $\frac{1}{2}d = \frac{13}{2}$. Thus the equation of the circle in standard form is:

$$\left(x + \tfrac{1}{2}\right)^2 + (y - 2)^2 = \tfrac{169}{4}$$

77. The circle will have an equation of $(x + 3)^2 + (y - 2)^2 = r^2$. We now find the y-intercepts of the other circle by setting $x = 0$:

$$y^2 - 4 = 0$$
$$(y + 2)(y - 2) = 0$$
$$y = \pm 2$$

So the radius is $2(4) = 8$, and thus the equation is $(x + 3)^2 + (y - 2)^2 = 64$.

79. (a) We set up a table of values:

x	0	1	2	3
y	0	16	64	144

Now graph the curve:

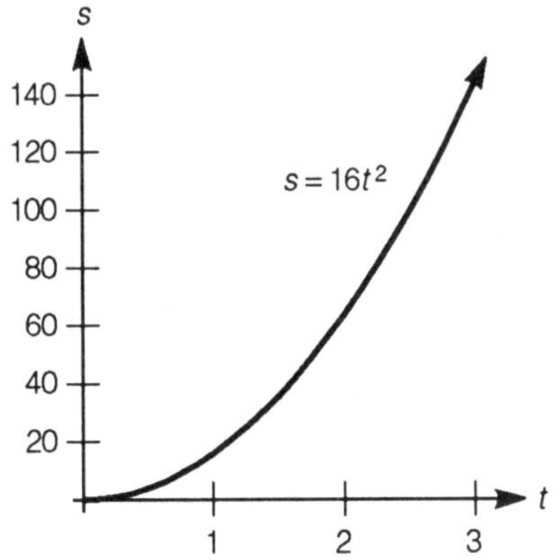

(b) When $0 \le t \le 1$, we have $0 \le s \le 16$, from the graph:

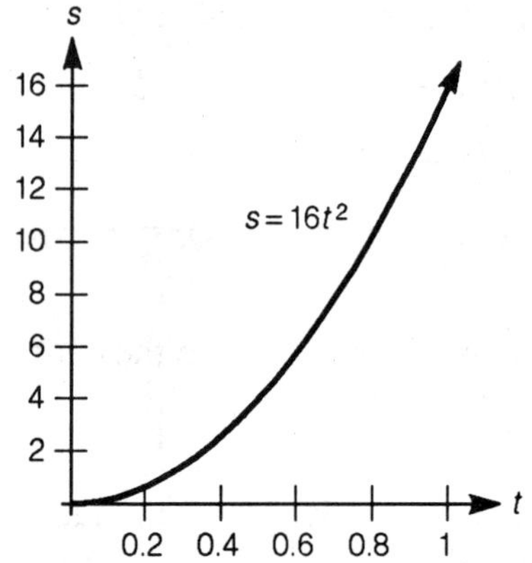

(c) When $1 \le t \le 2$, we have $16 \le s \le 64$, from the graph:

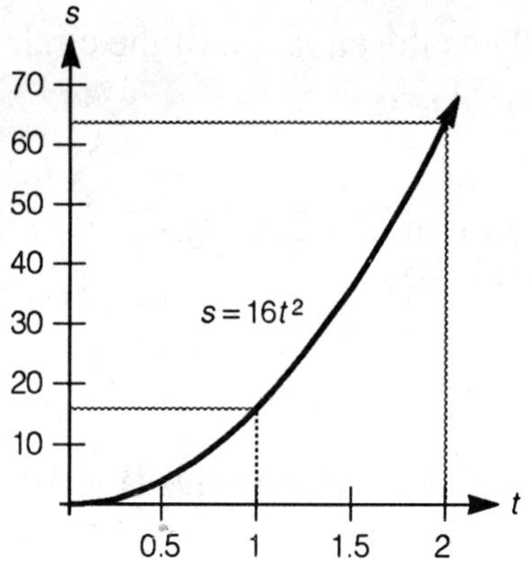

81. (a) For $y = x$, we set up a table of values:

x	0	$\frac{1}{4}$	$\frac{1}{3}$	$\frac{1}{2}$	$\frac{3}{4}$	1
y	0	$\frac{1}{4}$	$\frac{1}{3}$	$\frac{1}{2}$	$\frac{3}{4}$	1

(b) For $y = x^2$, we set up a table of values:

x	0	$\frac{1}{4}$	$\frac{1}{3}$	$\frac{1}{2}$	$\frac{3}{4}$	1
y	0	$\frac{1}{16}$	$\frac{1}{9}$	$\frac{1}{4}$	$\frac{9}{16}$	1

(c) For $y = x^3$, we set up a table of values:

x	0	$\frac{1}{4}$	$\frac{1}{3}$	$\frac{1}{2}$	$\frac{3}{4}$	1
y	0	$\frac{1}{64}$	$\frac{1}{27}$	$\frac{1}{8}$	$\frac{27}{64}$	1

(d) For $y = x^4$, we set up a table of values:

x	0	$\frac{1}{4}$	$\frac{1}{3}$	$\frac{1}{2}$	$\frac{3}{4}$	1
y	0	$\frac{1}{256}$	$\frac{1}{81}$	$\frac{1}{16}$	$\frac{81}{256}$	1

Now graph the four curves:

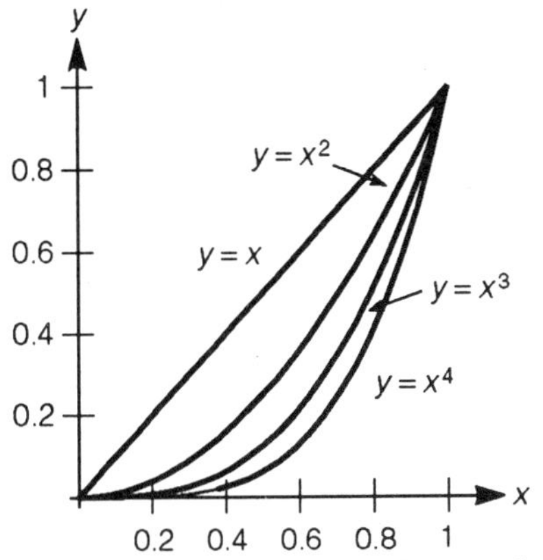

The pattern seems to be as n gets larger on $y = x^n$, the graph flattens out more and more as in the figure. We could guess $y = x^{100}$ would look something like:

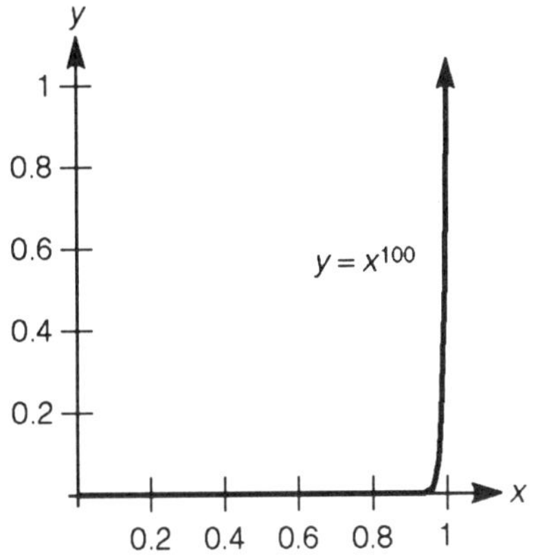

83. (a) Tracing up to the curve from $t = 0$, we get $N = 500$ bacteria.

(b) Since $(0, 500)$ lies on the curve, we find where $(t, 1000)$ would be on the curve, as the population would now be double. Tracing down from $N = 1000$, we get $t = 1.5$ hours. So the population will double in 1.5 hours.

(c) As in (b), we find where $(t, 2500)$ would be on the curve. Tracing down from $N = 2500$, we get $t = 3.5$ hours.

(d) Between $t = 0$ and $t = 1$, the population has grown from $N = 500$ to $N = 800$, so it has increased at an average rate of 300 bacteria per hour. Between $t = 3$ and $t = 4$, the population has grown from $N = 2000$ to $N = 3000$, so it has increased at an average rate of 1000 bacteria per hour. So the population has increased more rapidly between $t = 3$ and $t = 4$.

85. The x-intercepts are found by setting $y = 0$:

$$x^3 - 27x = 0$$
$$x(x - 3\sqrt{3})(x + 3\sqrt{3}) = 0$$
$$x = 0, \pm 3\sqrt{3} \approx 0, \pm 5.2$$

When $x = 0$, $y = (0)^3 - 27(0) = 0$, so the y-intercept is 0. Using the point $(-x, -y)$, we find the graph is symmetric with respect to the origin:

$$-y = (-x)^3 - 27(-x)$$
$$-y = -x^3 + 27x$$
$$y = x^3 - 27x$$

We set up a table of values:

x	0	0.5	1	1.5	2	2.5	3	3.5	4	4.5	5	5.5	6
y	0	−13.4	−26	−37.1	−46	−51.9	−54	−51.6	−44	−30.4	−10	17.9	54

Now graph the curve:

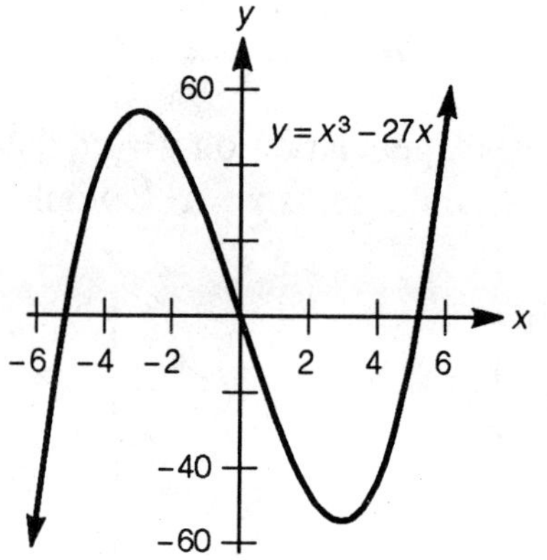

3.3 Equations of Lines

1. (a) Here $(x_1, y_1) = (-3, 2)$ and $(x_2, y_2) = (1, -6)$, so:

$$\text{slope} = \frac{-6-2}{1-(-3)} = \frac{-8}{4} = -2$$

(b) Here $(x_1, y_1) = (2, -5)$ and $(x_2, y_2) = (4, 1)$, so:

$$\text{slope} = \frac{1-(-5)}{4-2} = \frac{6}{2} = 3$$

(c) Here $(x_1, y_1) = (-2, 7)$ and $(x_2, y_2) = (1, 0)$, so:

$$\text{slope} = \frac{0-7}{1-(-2)} = -\frac{7}{3}$$

(d) Here $(x_1, y_1) = (4, 5)$ and $(x_2, y_2) = (5, 8)$, so:

$$\text{slope} = \frac{8-5}{5-4} = \frac{3}{1} = 3$$

3. (a) Here $(x_1, y_1) = (1, 1)$ and $(x_2, y_2) = (-1, -1)$, so:

$$\text{slope} = \frac{-1-1}{-1-1} = \frac{-2}{-2} = 1$$

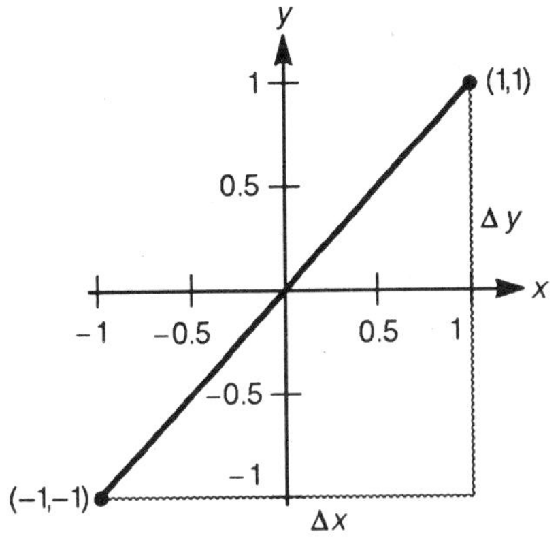

(b) Here $(x_1, y_1) = (0, 5)$ and $(x_2, y_2) = (-8, 5)$, so:

$$\text{slope} = \frac{5-5}{-8-0} = \frac{0}{-8} = 0$$

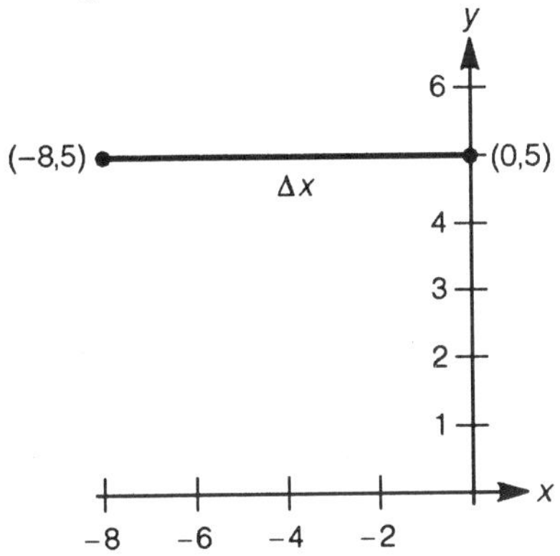

(c) Here $(x_1, y_1) = (-1, 1)$ and $(x_2, y_2) = (1, -1)$, so:

$$\text{slope} = \frac{-1-1}{1-(-1)} = \frac{-2}{2} = -1$$

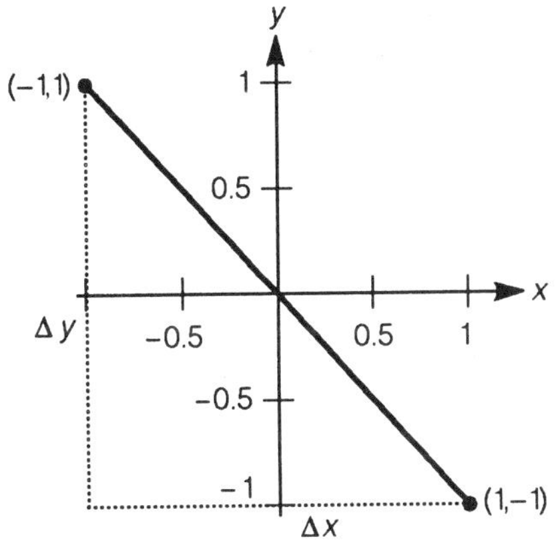

(d) Here $(x_1, y_1) = (a, b)$ and $(x_2, y_2) = (b, a)$, so:

$$\text{slope} = \frac{a-b}{b-a} = -1$$

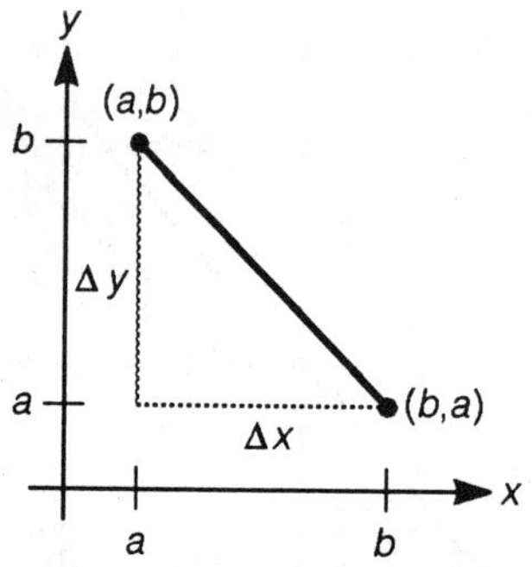

5. m_3 is smallest (since it is negative)
m_2 is next (it appears to be near zero)
m_4 is next (it is positive but not as steep as m_1)
m_1 is largest (it is steeper than m_4)

7. We compute the slopes:

$$\text{slope}_{AB} = \frac{\frac{1}{2}-(-2)}{2-(-8)} = \frac{\frac{5}{2}}{10} = \frac{1}{4}$$

$$\text{slope}_{BC} = \frac{-1-\frac{1}{2}}{11-2} = \frac{-\frac{3}{2}}{9} = -\frac{1}{6}$$

Since these slopes are different the three points cannot be collinear.

9. We compute the slopes:

$$\text{slope}_{AB} = \frac{4-(-5)}{3-0} = \frac{9}{3} = 3$$

$$\text{slope}_{BC} = \frac{-8-4}{-1-3} = \frac{-12}{-4} = 3$$

Since these slopes are equal the three points are collinear.

11. (a) Here $(x_1, y_1) = (-2, 1)$ and $m = -5$, so by the point-slope formula:

$$\begin{aligned} y-1 &= -5[x-(-2)] \\ y-1 &= -5(x+2) \\ y-1 &= -5x-10 \\ y &= -5x-9 \end{aligned}$$

(b) Here $(x_1, y_1) = (4, -4)$ and $m = 4$, so by the point-slope formula:

$$\begin{aligned} y-(-4) &= 4(x-4) \\ y+4 &= 4x-16 \\ y &= 4x-20 \end{aligned}$$

(c) Here $(x_1, y_1) = \left(-6, -\frac{2}{3}\right)$ and $m = \frac{1}{3}$, so by the point-slope formula:

$$y - \left(-\frac{2}{3}\right) = \left(\frac{1}{3}\right)[x - (-6)]$$
$$y + \frac{2}{3} = \left(\frac{1}{3}\right)(x + 6)$$
$$y + \frac{2}{3} = \frac{1}{3}x + 2$$
$$y = \frac{1}{3}x + \frac{4}{3}$$

(d) Here $(x_1, y_1) = (0, 1)$ and $m = -1$, so by the point-slope formula:

$$y - 1 = -1(x - 0)$$
$$y - 1 = -x$$
$$y = -x + 1$$

Note in (d) that, since the given point is the y-intercept, we could have immediately written $y = -x + 1$ using the slope-intercept formula.

13. (a) First we find the slope:

$$m = \frac{-6 - 8}{-3 - 4} = \frac{-14}{-7} = 2$$

Using $(x_1, y_1) = (4, 8)$ in the point-slope formula, we have:

$$y - 8 = 2(x - 4)$$
$$y - 8 = 2x - 8$$
$$y = 2x$$

(b) First we find the slope:

$$m = \frac{-10 - 0}{3 - (-2)} = \frac{-10}{5} = -2$$

Using $(x_1, y_1) = (-2, 0)$ in the point-slope formula, we have:

$$y - 0 = -2[x - (-2)]$$
$$y = -2(x + 2)$$
$$y = -2x - 4$$

(c) First we find the slope:

$$m = \frac{-1 - (-2)}{4 - (-3)} = \frac{1}{7}$$

Using $(x_1, y_1) = (4, -1)$ in the point-slope formula, we have:

$$y - (-1) = \frac{1}{7}(x - 4)$$
$$y + 1 = \frac{1}{7}x - \frac{4}{7}$$
$$y = \frac{1}{7}x - \frac{11}{7}$$

15. Since vertical lines have the form $x =$ constant, and $(-3, 4)$ is on the line, then the equation is $x = -3$.

17. Since horizontal lines have the form $y =$ constant, and $(-3, 4)$ is on the line, then the equation is $y = 4$.

19. The y-axis is vertical, so its equation must have the form x = constant. Since $(0,0)$ is on the y-axis, the equation is $x = 0$.

21. (a) Using the slope-intercept formula with $m = -4$ and $b = 7$, we have $y = -4x + 7$.

(b) Using the slope-intercept formula with $m = 2$ and $b = \frac{3}{2}$, we have $y = 2x + \frac{3}{2}$.

(c) Using the slope-intercept formula with $m = -\frac{4}{3}$ and $b = 14$, we have $y = -\frac{4}{3}x + 14$.

23. (a) We use the point-slope formula:

$$y - (-1) = 4[x - (-3)]$$
$$y + 1 = 4(x + 3)$$
$$y + 1 = 4x + 12$$
$$y = 4x + 11$$

Now draw the graph:

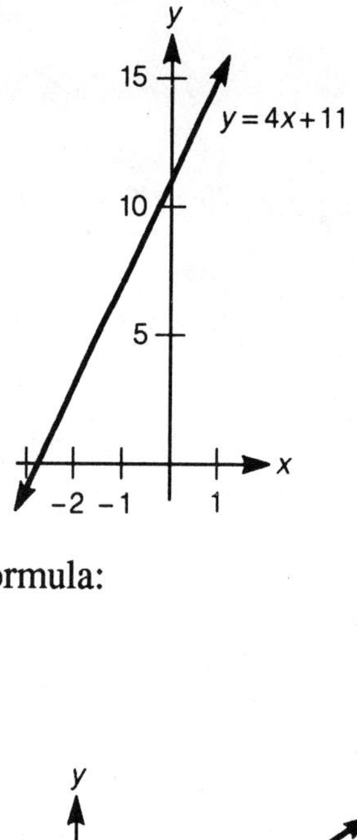

(b) We use the point-slope formula:

$$y - 0 = \tfrac{1}{2}\left(x - \tfrac{5}{2}\right)$$
$$y = \tfrac{1}{2}x - \tfrac{5}{4}$$

Now draw the graph:

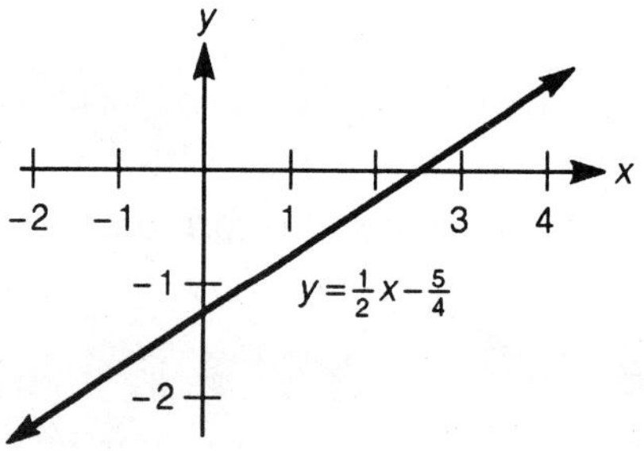

(c) We find the slope between the points $(6,0)$ and $(0,5)$:

$$m = \frac{5-0}{0-6} = -\frac{5}{6}$$

Since 5 is the y-intercept, then by the slope-intercept formula:

$$y = -\tfrac{5}{6}x + 5$$

Now draw the graph:

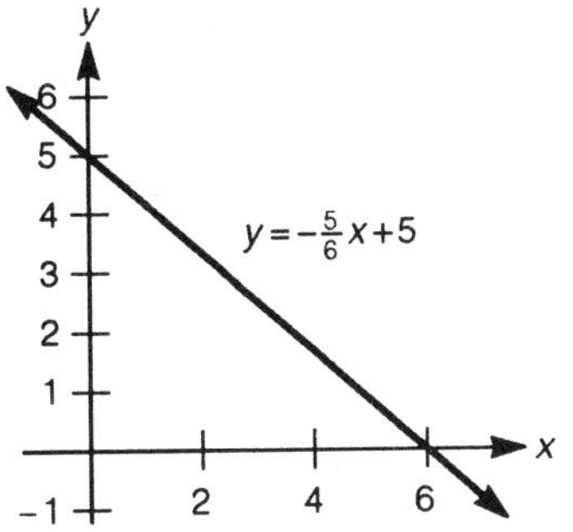

(d) We use the point-slope formula with the point (–2, 0):

$$y-0=\tfrac{3}{4}[x-(-2)]$$
$$y=\tfrac{3}{4}(x+2)$$
$$y=\tfrac{3}{4}x+\tfrac{3}{2}$$

Now draw the graph:

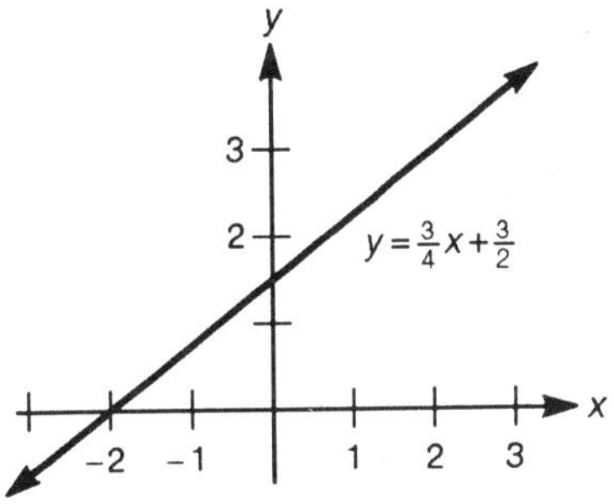

(e) We first find the slope:

$$m=\frac{6-2}{2-1}=\frac{4}{1}=4$$

Using the point (1, 2) in the point-slope formula, we have:

$$y-2=4(x-1)$$
$$y-2=4x-4$$
$$y=4x-2$$

Now draw the graph:

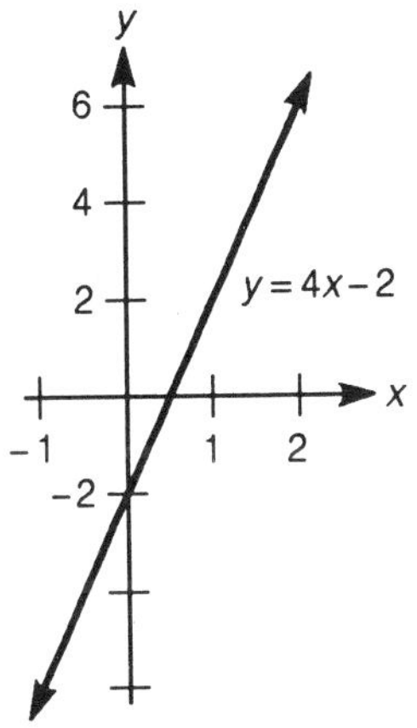

25. We first find the x- and y-intercepts of the circle by letting $x = 0$ and $y = 0$:

$$\begin{array}{ll} y = 0 & x = 0 \\ x^2 + 4x + 4 = 0 & y^2 - 4y + 4 = 0 \\ (x+2)^2 = 0 & (y-2)^2 = 0 \\ x = -2 & y = 2 \end{array}$$

So the line passes through the points $(-2, 0)$ and $(0, 2)$. We find its slope:

$$m = \frac{2-0}{0-(-2)} = \frac{2}{2} = 1$$

So $y = x + 2$ is the equation (in slope-intercept form) of the line. We draw the graph:

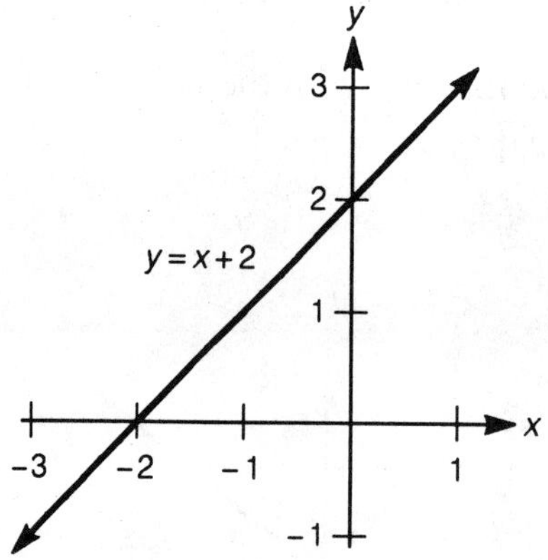

27. If the line is parallel to the x-axis, then its slope is 0. Since this line is of the form $y =$ constant, then for $(-3, 4)$ to lie on the line the equation is $y = 4$, or $y - 4 = 0$ in the desired form. We draw the graph:

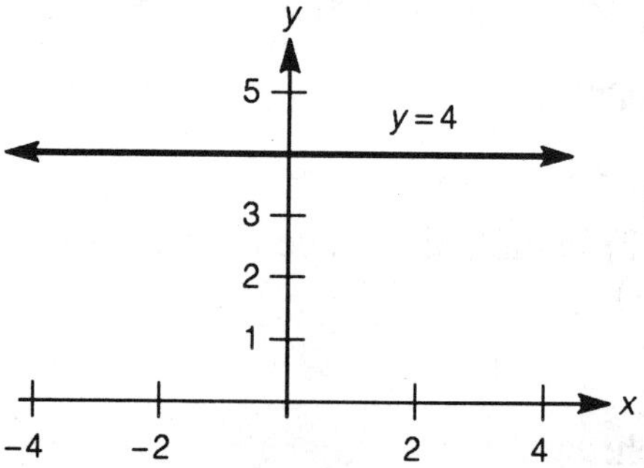

29. (a) If $x = 0$, then $y = 3$ and if $y = 0$, then $x = 5$. So the x-intercept is 5 and the y-intercept is 3. Thus we have:

$$\text{area} = \tfrac{1}{2}(5)(3) = \tfrac{15}{2}$$
$$\text{perimeter} = 5 + 3 + \sqrt{5^2 + 3^2} = 8 + \sqrt{34}$$

(b) If $x = 0$, then $y = -3$ and if $y = 0$, then $x = 5$. So the x-intercept is 5 and the y-intercept is -3. Thus we have:

$$\text{area} = \tfrac{1}{2}(5)(3) = \tfrac{15}{2}$$
$$\text{perimeter} = 5 + 3 + \sqrt{5^2 + 3^2} = 8 + \sqrt{34}$$

31. (a) We find the slopes of each:

$$\begin{aligned} 3x - 4y &= 12 \\ -4y &= -3x + 12 \\ y &= \tfrac{3}{4}x - 3 \\ m &= \tfrac{3}{4} \end{aligned} \qquad \begin{aligned} 4x - 3y &= 12 \\ -3y &= -4x + 12 \\ y &= \tfrac{4}{3}x - 4 \\ m &= \tfrac{4}{3} \end{aligned}$$

The lines are not parallel (slopes aren't the same), the lines are not perpendicular $\left(\frac{3}{4} \bullet \frac{4}{3} = 1, \text{ not } -1\right)$, so they are neither.

(b) We find the slopes of each:

$$\begin{aligned} y &= 5x - 16 \\ m &= 5 \end{aligned} \qquad \begin{aligned} y &= 5x + 2 \\ m &= 5 \end{aligned}$$

Since these slopes are the same, the lines are parallel.

(c) We find the slopes of each:

$$\begin{aligned} 5x - 6y &= 25 \\ -6y &= -5x + 25 \\ y &= \tfrac{5}{6}x - \tfrac{25}{6} \\ m &= \tfrac{5}{6} \end{aligned} \qquad \begin{aligned} 6x + 5y &= 0 \\ 5y &= -6x \\ y &= -\tfrac{6}{5}x \\ m &= -\tfrac{6}{5} \end{aligned}$$

Since $\left(\frac{5}{6}\right)\left(-\frac{6}{5}\right) = -1$, the lines are perpendicular.

(d) We find the slopes of each:

$$\begin{aligned} y &= -\tfrac{2}{3}x - 1 \\ m &= -\tfrac{2}{3} \end{aligned} \qquad \begin{aligned} y &= \tfrac{3}{2}x - 1 \\ m &= \tfrac{3}{2} \end{aligned}$$

Since $\left(-\frac{2}{3}\right)\left(\frac{3}{2}\right) = -1$, the lines are perpendicular.

(e) We find the slopes of each:

$$\begin{aligned} -2x - 5y &= 1 \\ -5y &= 2x + 1 \\ y &= -\tfrac{2}{5}x - \tfrac{1}{5} \\ m &= -\tfrac{2}{5} \end{aligned} \qquad \begin{aligned} y - \tfrac{2}{5}x - 4 &= 0 \\ y &= \tfrac{2}{5}x + 4 \\ m &= \tfrac{2}{5} \end{aligned}$$

The lines are not parallel (slopes aren't the same), the lines are not perpendicular $\left(-\frac{2}{5} \bullet \frac{2}{5} = -\frac{4}{25}, \text{ not } -1\right)$, so they are neither.

(f) We find the slopes of each:

$$\begin{aligned} x &= 8y + 3 \\ 8y &= x - 3 \\ y &= \tfrac{1}{8}x - \tfrac{3}{8} \\ m &= \tfrac{1}{8} \end{aligned} \qquad \begin{aligned} 4y - \tfrac{1}{2}x &= 32 \\ 4y &= \tfrac{1}{2}x + 32 \\ y &= \tfrac{1}{8}x + 8 \\ m &= \tfrac{1}{8} \end{aligned}$$

Since these slopes are the same, the lines are parallel.

33. We first find the slope:

$$2x - 5y = 10$$
$$-5y = -2x + 10$$
$$y = \tfrac{2}{5}x - 2$$

So the slope is $\tfrac{2}{5}$. We use the point (–1, 2) in the point-slope formula:

$$y - 2 = \tfrac{2}{5}[x - (-1)]$$
$$y - 2 = \tfrac{2}{5}(x + 1)$$
$$y - 2 = \tfrac{2}{5}x + \tfrac{2}{5}$$
$$y = \tfrac{2}{5}x + \tfrac{12}{5} \quad \text{[form } y = mx + b\text{]}$$
$$5y = 2x + 12$$
$$2x - 5y + 12 = 0 \quad \text{[form } Ax + By + C = 0\text{]}$$

35. We first find the slope of $4y - 3x = 1$:

$$4y = 3x + 1$$
$$y = \tfrac{3}{4}x + \tfrac{1}{4}$$

Since this slope is $\tfrac{3}{4}$, the perpendicular line slope is $-\tfrac{4}{3}$. We now use the point (4, 0) in the point-slope formula:

$$y - 0 = -\tfrac{4}{3}(x - 4)$$
$$y = -\tfrac{4}{3}x + \tfrac{16}{3} \quad \text{[form } y = mx + b\text{]}$$
$$3y = -4x + 16$$
$$4x + 3y - 16 = 0 \quad \text{[form } Ax + By + C = 0\text{]}$$

37. We first find the slope of $3x - 5y = 25$:

$$-5y = -3x + 25$$
$$y = \tfrac{3}{5}x - 5$$

So the slope is $\tfrac{3}{5}$. We now find the y-intercept of $6x - y + 11 = 0$:

$$y = 6x + 11$$

So the y-intercept is 11. We write the equation:

$$y = \tfrac{3}{5}x + 11 \quad \text{[form } y = mx + b\text{]}$$
$$5y = 3x + 55$$
$$3x - 5y + 55 = 0 \quad \text{[form } Ax + By + C = 0\text{]}$$

39. (a) The center is (0, 0) and the radius is 5. We draw the graph:

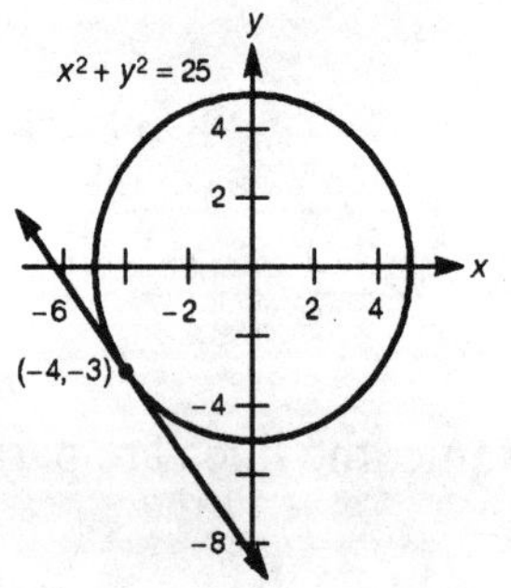

(b) The tangent line will be perpendicular to the radius drawn from the center $(0,0)$ and the point $(-4,-3)$. We find the slope:

$$m=\frac{-3-0}{-4-0}=\frac{-3}{-4}=\frac{3}{4}$$

So the perpendicular line will have slope $=-\frac{4}{3}$. We use the point $(-4,-3)$ in the point-slope formula:

$$\begin{aligned} y-(-3)&=-\tfrac{4}{3}[x-(-4)] \\ y+3&=-\tfrac{4}{3}(x+4) \\ y+3&=-\tfrac{4}{3}x-\tfrac{16}{3} \\ y&=-\tfrac{4}{3}x-\tfrac{25}{3} \end{aligned}$$

41. Let $(x_1,y_1)=(3,9)$ and $(x_2,y_2)=[3+h,(3+h)^2]$, so:

$$m=\frac{(3+h)^2-9}{3+h-3}=\frac{9+6h+h^2-9}{h}=\frac{6h+h^2}{h}=6+h$$

43. Let $(x_1,y_1)=(x,x^3)$ and $(x_2,y_2)=[x+h,(x+h)^3]$. Therefore the slope is given by:

$$\begin{aligned} \frac{(x+h)^3-x^3}{x+h-x}&=\frac{x^3+3x^3h+3xh^2+h^3-x^3}{h} \\ &=\frac{3x^2h+3xh^2+h^3}{h} \\ &=\frac{h(3x^2+3xh+h^2)}{h} \\ &=3x^2+3xh+h^2 \end{aligned}$$

45. Let $(x_1,y_1)=\left(x,\frac{1}{x}\right)$ and $(x_2,y_2)=\left(x+h,\frac{1}{x+h}\right)$, so:

$$\text{slope}=\frac{\frac{1}{x+h}-\frac{1}{x}}{x+h-x}=\frac{\frac{1}{x+h}-\frac{1}{x}}{h}$$

We multiply the numerator and denominator by $x(x+h)$ to obtain:

$$\text{slope}=\frac{x-x-h}{hx(x+h)}=\frac{-h}{hx(x+h)}=\frac{-1}{x(x+h)}$$

47. We draw the graph:

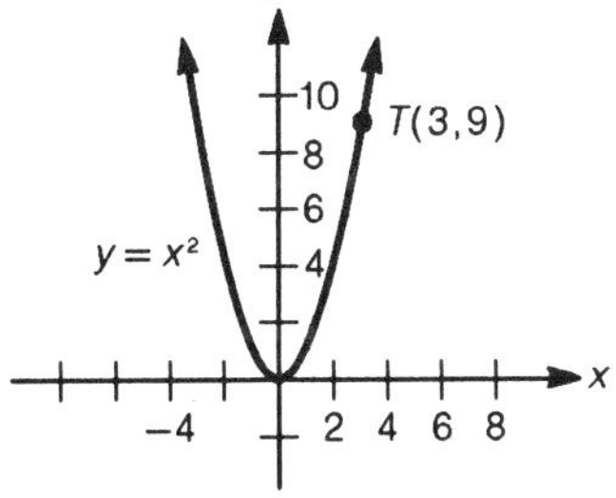

The completed table is:

x	2.5	2.9	2.99	2.999	2.9999
y	6.25	8.41	8.9401	8.994001	8.99940001
Δx	0.5	0.1	0.01	0.001	0.0001
Δy	2.75	0.59	0.0599	0.005999	0.00059999
m	5.5	5.9	5.99	5.999	5.9999

It would appear that as x approaches 3, the slope of these lines (called secant lines) approaches 6. So we would estimate that the slope of the tangent line to the curve $y = x^2$ at $T(3,9)$ is 6.

49. Let the coordinates of P be (x, x^3). Since the slope of the line passing through P and $(1,1)$ is $\frac{3}{4}$, we have:

$$\frac{x^3-1}{x-1} = \frac{3}{4}$$
$$\frac{(x-1)(x^2+x+1)}{x-1} = \frac{3}{4}$$
$$x^2+x+1 = \frac{3}{4}$$
$$4x^2+4x+4 = 3$$
$$4x^2+4x+1 = 0$$
$$(2x+1)^2 = 0$$
$$2x+1 = 0$$
$$x = -\frac{1}{2}$$
$$y = \left(-\frac{1}{2}\right)^3 = -\frac{1}{8}$$

Thus the point P is $\left(-\frac{1}{2}, -\frac{1}{8}\right)$.

51. Let the coordinates of P be $\left(x, \frac{1}{x}\right)$. Since the slope of the line through P and $\left(2, \frac{1}{2}\right)$ is $-\frac{1}{16}$, we have:

$$\frac{\frac{1}{x}-\frac{1}{2}}{x-2} = -\frac{1}{16}$$
$$\frac{2-x}{2x(x-2)} = -\frac{1}{16}$$
$$\frac{-1}{2x} = -\frac{1}{16}$$
$$2x = 16$$
$$x = 8$$

So the point P is $\left(8, \frac{1}{8}\right)$.

53. We use the point-slope formula to find the equation of the line:

$$y-6=-5(x-3)$$
$$y-6=-5x+15$$
$$y=-5x+21$$

To find the x-intercept, let $y=0$:

$$-5x+21=0$$
$$5x=21$$
$$x=\frac{21}{5}$$

To find the y-intercept, let $x=0$, so $y=21$. The area of the triangle is given by:

$$\text{Area}=\frac{1}{2}(\text{base})(\text{height})=\frac{1}{2}\bullet\frac{21}{5}\bullet 21=44.1 \text{ square units}$$

55. (a) We draw the graph:

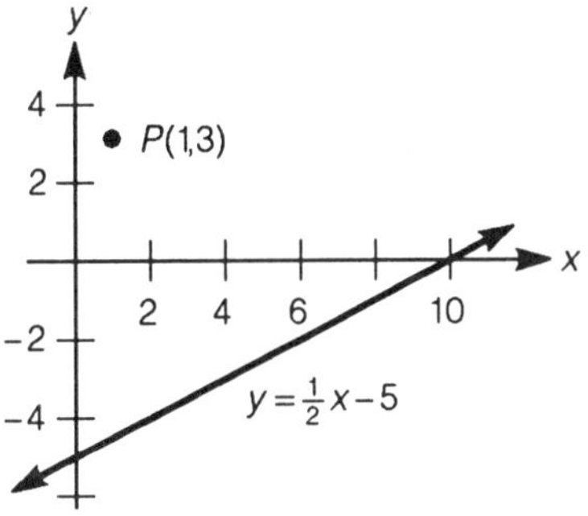

(b) Since the slope must be –2, we use (1, 3) in the point-slope formula:

$$y-3=-2(x-1)$$
$$y-3=-2x+2$$
$$y=-2x+5$$

(c) Set the two equations equal:

$$-2x+5=\frac{1}{2}x-5$$
$$-4x+10=x-10$$
$$-5x=-20$$
$$x=4$$
$$y=-2(4)+5=-3$$

The intersection point is (4, –3).

(d) We find the distance from (1, 3) to (4, –3):

$$d=\sqrt{(4-1)^2+(-3-3)^2}=\sqrt{9+36}=\sqrt{45}=3\sqrt{5}$$

57. (a) Since the slope is m and the point (2, 1) lies on the line, we use the point-slope formula to obtain $y-1=m(x-2)$.

(b) For the x-intercept, let $y = 0$:

$$-1 = m(x-2)$$
$$-1 = mx - 2m$$
$$mx = 2m - 1$$
$$x = \frac{2m-1}{m}$$

For the y-intercept, let $x = 0$:

$$y - 1 = m(-2)$$
$$y - 1 = -2m$$
$$y = -2m + 1$$

(c) The area is given by:

$$\text{Area} = \tfrac{1}{2}(\text{base})(\text{height}) = \frac{1}{2}\left(\frac{2m-1}{m}\right)(-2m+1)$$

(d) Since the area is 4, we have:

$$\frac{1}{2}\left(\frac{2m-1}{m}\right)(-2m+1) = 4$$
$$(2m-1)(-2m+1) = 8m$$
$$-4m^2 + 4m - 1 = 8m$$
$$4m^2 + 4m + 1 = 0$$

(e) We solve for m:

$$(2m+1)^2 = 0$$
$$2m + 1 = 0$$
$$2m = -1$$
$$m = -\tfrac{1}{2}$$

59. (a) Since $\frac{\Delta x}{\Delta P} = -\frac{2}{3}$, we have the equation:

$$x = 280 - \tfrac{2}{3}(P - 195) = 410 - \tfrac{2}{3}P$$

(b) If $P = 270$, we have:

$$x = 410 - \tfrac{2}{3}(270) = 410 - 180 = 230$$

So 230 units can be sold in a month.

(c) If $x = 205$, we have:

$$205 = 410 - \tfrac{2}{3}P$$
$$\tfrac{2}{3}P = 205$$
$$P = \tfrac{3}{2}(205) = 307.5$$

The price would be \$307.50 per unit.

61. (a) Since A and C lie on the line $y = m_1x + b_1$, and the x-coordinates of A and C are 0 and 1, respectively, then the y-coordinates will be b_1 and $m_1 + b_1$, so $A = (0, b_1)$ and $C = (1, m_1 + b_1)$. Similarly, the points B and D are $B = (0, b_2)$ and $D = (1, m_2 + b_2)$.

(b) We verify the results:

$$AB = \sqrt{(0-0)^2 + (b_1 - b_2)^2} = \sqrt{(b_1 - b_2)^2} = |b_1 - b_2|,$$

which is $b_1 - b_2$ since $b_1 > b_2$

$$\begin{aligned} CD &= \sqrt{(1-1)^2 + [(m_1 + b_1) - (m_2 + b_2)]^2} \\ &= \sqrt{[(m_1 + b_1) - (m_2 + b_2)]^2} \\ &= |(m_1 + b_1) - (m_2 + b_2)| \\ &= m_1 + b_1 - (m_2 + b_2) \text{ since } m_1 + b_1 > m_2 + b_2 \end{aligned}$$

(c) Since $AB = CD$, we have:

$$\begin{aligned} b_1 - b_2 &= m_1 + b_1 - m_2 - b_2 \\ 0 &= m_1 - m_2 \\ m_1 &= m_2 \end{aligned}$$

63. (a) When $y = mx + b$ is reflected about the x-axis, its slope becomes $-m$ and its y-intercept becomes $-b$, so the new equation is $y = -mx - b$. Alternate approach: Use $(x, -y)$ to get $-y = mx + b$, or $y = -mx - b$.

(b) When $y = mx + b$ is reflected about the y-axis, its slope becomes $-m$ and its y-intercept remains b, so the new equation is $y = -mx + b$. Alternate approach: Use $(-x, y)$ to get $y = m(-x) + b$, or $y = -mx + b$.

(c) When $y = mx + b$ is reflected about the origin, its slope remains m and its y-intercept becomes $-b$, so the new equation is $y = mx - b$. Alternate approach: Use $(-x, -y)$ to get $-y = m(-x) + b$, or $y = mx - b$.

65. (a) See the figure:

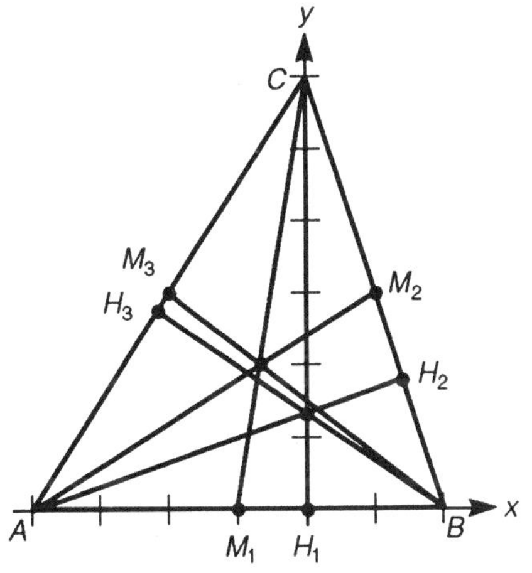

We find the midpoints:

$M_1 = (-1,0)$, $M_2 = (1,3)$, $M_3 = (-2,3)$

We substitute in the equation $(x-h)^2 + (y-k)^2 = r^2$, and simplify to obtain the equations:

$$1 + 2h + h^2 + k^2 = r^2$$
$$10 - 2h + h^2 - 6k + k^2 = r^2$$
$$13 + 4h + h^2 - 6k + k^2 = r^2$$

Subtracting the first equation from the other two yields:

$$4h + 6k = 9$$
$$2h - 6k = -12$$

Adding these yields $h = -\frac{1}{2}$, and substituting into the first equation yields $k = \frac{11}{6}$, and finally we find that $r^2 = \frac{65}{18}$. So the circle is $\left(x+\frac{1}{2}\right)^2 + \left(y-\frac{11}{6}\right)^2 = \frac{65}{18}$. This can be written in the form $3x^2 + 3y^2 + 3x - 11y = 0$.

(b) We now find H_1, H_2 and H_3. $H_1 = (0,0)$ by inspection. Since BC has a slope $= -3$, then AH_2 has a slope$=\frac{1}{3}$, thus $y = \frac{1}{3}x + \frac{4}{3}$. BC has an equation of $y = -3x + 6$, so we find the intersection point to be $\left(\frac{7}{5}, \frac{9}{5}\right)$. So $H_2 = \left(\frac{7}{5}, \frac{9}{5}\right)$. Similarly AC has a slope $= \frac{3}{2}$, so BH_3 has a slope$= -\frac{2}{3}$, thus $y = -\frac{2}{3}x + \frac{4}{3}$. AC has an equation of $y = \frac{3}{2}x + 6$, so we find the intersection point to be $\left(-\frac{28}{13}, \frac{36}{13}\right)$. So $H_3 = \left(-\frac{28}{13}, \frac{36}{13}\right)$. We find the equation of the circle passing through these three points and get $\left(x+\frac{1}{2}\right)^2 + \left(y-\frac{11}{6}\right)^2 = \frac{130}{36}$, which simplifies to $3x^2 + 3y^2 + 3x - 11y = 0$.

(c) The equations of the three altitudes are $x = 0$, $y = \frac{1}{3}x + \frac{4}{3}$, and $y = -\frac{2}{3}x + \frac{4}{3}$. Since $x = 0$, we get $y = \frac{4}{3}$, so they all intersect at the point $P\left(0, \frac{4}{3}\right)$.

(d) We compute $N_1 = \left(-2, \frac{2}{3}\right)$, $N_2 = \left(1, \frac{2}{3}\right)$, and $N_3 = \left(0, \frac{11}{3}\right)$. We check that N_1, N_2, and N_3 lie on $3x^2 + 3y^2 + 3x - 11y = 0$, which they all do.

(e) Rather than using the hint, we substitute in $(x-h)^2 + (y-k)^2 = r^2$ directly and simplify, yielding:

$$36 + h^2 - 12k + k^2 = r^2$$
$$16 + h^2 + 8h + k^2 = r^2$$
$$4 - 4h + h^2 + k^2 = r^2$$

Subtracting the first equation from the other two yields:

$$2h + 3k = 5$$
$$-h + 3k = 8$$

This has a solution of $h = -1$ and $k = \frac{7}{3}$. So the center of the circle is $\left(-1, \frac{7}{3}\right)$.

(f) We find the radius of the circumcircle using the point $(0,6)$ and the center $\left(-1,\frac{7}{3}\right)$:

$$r^2 = (0+1)^2 + \left(6-\frac{7}{3}\right)^2 = \frac{130}{9}, \text{ so } r = \frac{\sqrt{130}}{3}$$

From (a) we have the radius of the nine-point circle is $\frac{\sqrt{130}}{6}$, which is one-half of the radius of the circumcircle.

(g) The midpoint of QP is $\left(-\frac{1}{2},\frac{11}{6}\right)$. Note that this is the center of the nine-point circle from (a).

(h) We first find the medians:

CM_1: $m = 6$, so $y = 6x+6$

BM_3: $m = -\frac{3}{4}$, so $y = -\frac{3}{4}x + \frac{3}{2}$

AM_2: $m = \frac{3}{5}$, so $y = \frac{3}{5}x + \frac{12}{5}$

Now find the intersections:

$6x + 6 = -\frac{3}{4}x + \frac{3}{2}$, so $x = -\frac{2}{3}$ and $y = 2$

$6x + 6 = \frac{3}{5}x + \frac{12}{5}$, so $x = -\frac{2}{3}$ and $y = 2$

$-\frac{3}{4}x + \frac{3}{2} = \frac{3}{5}x + \frac{12}{5}$, so $x = -\frac{2}{3}$ and $y = 2$

So the intersection point is $\left(-\frac{2}{3},2\right)$.

(i) We find the equation of QP using $m = -1$, so $y = -x + \frac{4}{3}$. By the distance formula:

$$PG = \sqrt{\left(-\frac{2}{3}-0\right)^2 + \left(2-\frac{4}{3}\right)^2} = \frac{2\sqrt{2}}{3}$$

$$GQ = \sqrt{\left(-1+\frac{2}{3}\right)^2 + \left(\frac{7}{3}-2\right)^2} = \frac{\sqrt{2}}{3}$$

So $PG = 2 \bullet GQ$.

Chapter Three Review Exercises

1. We find the slope:

$$m = \frac{6-2}{-6-(-4)} = \frac{4}{-2} = -2$$

Using the point $(-4,2)$ in the point-slope formula:

$$\begin{aligned} y-2 &= -2[x-(-4)] \\ y-2 &= -2(x+4) \\ y-2 &= -2x-8 \\ y &= -2x-6 \end{aligned}$$

3. Using the point-slope formula:

$$y-(-3)=\tfrac{1}{4}[x-(-2)]$$
$$y+3=\tfrac{1}{4}(x+2)$$
$$y+3=\tfrac{1}{4}x+\tfrac{1}{2}$$
$$y=\tfrac{1}{4}x-\tfrac{5}{2}$$

5. We find the slope between the points $(-4,0)$ and $(0,8)$:

$$m=\frac{8-0}{0-(-4)}=\frac{8}{4}=2$$

Using the slope-intercept formula, we have $y = 2x + 8$.

7. Since the line is parallel to the x-axis, its equation will be of the form y = constant (horizontal line). Since $(0,-2)$ is on the line, its equation is $y = -2$.

9. We find the slope of $x + y + 1 = 0$ by writing it in slope-intercept form, and obtain $y = -x - 1$. So its slope is -1, and thus the perpendicular slope would be 1. We use the point $(1,2)$ in the point-slope formula:

$$y-2=1(x-1)$$
$$y-2=x-1$$
$$y=x+1$$

11. We find the center of each circle by completing the square:

$$x^2+4x+y^2+2y=0$$
$$(x^2+4x+4)+(y^2+2y+1)=0+4+1$$
$$(x+2)^2+(y+1)^2=5$$
center: $(-2,-1)$

$$x^2-4x+y^2-16y=0$$
$$(x^2-4x+4)+(y^2-16y+64)=0+4+64$$
$$(x-2)^2+(y-8)^2=68$$
center: $(2,8)$

We find the slope between $(-2,-1)$ and $(2,8)$:

$$m=\frac{8-(-1)}{2-(-2)}=\frac{9}{4}$$

We use the point $(2,8)$ in the point-slope formula:

$$y-8=\tfrac{9}{4}(x-2)$$
$$4y-32=9x-18$$
$$-9x+4y-14=0$$
$$9x-4y+14=0$$

13. We find the midpoint of the line segment joining $(-2,-3)$ and $(6,-5)$:

$$M=\left(\frac{-2+6}{2},\frac{-3-5}{2}\right)=\left(\frac{4}{2},-\frac{8}{2}\right)=(2,-4)$$

Now find the slope between $(0,0)$ and $(2,-4)$:

$$m=\frac{-4-0}{2-0}=\frac{-4}{2}=-2$$

Using the slope-intercept formula, we obtain $y=-2x$, or $2x+y=0$.

15. First find the center of the circle by completing the square:

$$\begin{aligned}x^2-6x+y^2+8y&=0\\(x^2-6x+9)+(y^2+8y+16)&=0+9+16\\(x-3)^2+(y+4)^2&=25\end{aligned}$$

So the center is $(3,-4)$. Now find the slope of the radius drawn from $(3,-4)$ to $(0,0)$:

$$m=\frac{0-(-4)}{0-3}=\frac{4}{-3}=-\frac{4}{3}$$

So the slope of the perpendicular tangent line is $\frac{3}{4}$. Using the slope-intercept formula, we obtain $y=\frac{3}{4}x$, or $3x-4y=0$.

17. Call the x-intercept a. Then the y-intercept is $2-a$, and we find the slope from each to $(2,-1)$:

$$m=\frac{-1-0}{2-a}=\frac{-1}{2-a}$$

$$m=\frac{-1-(2-a)}{2-0}=\frac{-1-2+a}{2}=\frac{a-3}{2}$$

Since these two slopes must be equal, we have:

$$\begin{aligned}\frac{-1}{2-a}&=\frac{a-3}{2}\\-2&=(2-a)(a-3)\\-2&=-a^2+5a-6\\0&=a^2-5a+4\\0&=(a-4)(a-1)\\a&=1\text{ or }a=4\end{aligned}$$

When $a=1$, we have $m=\frac{1-3}{2}=\frac{-2}{2}=-1$. When $a=4$, we have $m=\frac{4-3}{2}=\frac{1}{2}$. Using the point $(2,-1)$ and each of these slopes in the point-slope formula:

$$\begin{aligned}y-(-1)&=-1(x-2)\\y+1&=-x+2\\x+y-1&=0\end{aligned}\qquad\qquad\begin{aligned}y-(-1)&=\tfrac{1}{2}(x-2)\\2y+2&=x-2\\x-2y-4&=0\end{aligned}$$

Both of these lines satisfy the given conditions.

19. Using the distance formula with $(x_1,y_1)=(-1,2)$ and $(x_2,y_2)=(4,-10)$:

$$d=\sqrt{[4-(-1)]^2+(-10-2)^2}=\sqrt{25+144}=\sqrt{169}=13$$

21. We find the distance of each from the origin:

$$d_1 = \sqrt{(15-0)^2 + (6-0)^2} = \sqrt{225+36} = \sqrt{261}$$
$$d_2 = \sqrt{(16-0)^2 + (2-0)^2} = \sqrt{256+4} = \sqrt{260}$$

The point $(15, 6)$ is farther from the origin.

23. For x-axis symmetry, we replace (x, y) with $(x, -y)$:

$$-y = x^4 - 2x^2$$
$$y = -x^4 + 2x^2$$

The graph is not symmetric about the x-axis.
For y-axis symmetry, we replace (x, y) with $(-x, y)$:

$$y = (-x)^4 - 2(-x)^2 = x^4 - 2x^2$$

The graph is symmetric about the y-axis.
For origin symmetry, we replace (x, y) with $(-x, -y)$:

$$-y = (-x)^4 - 2(-x)^2$$
$$-y = x^4 - 2x^2$$
$$y = -x^4 + 2x^2$$

The graph is not symmetric about the origin.

25. For x-axis symmetry, we replace (x, y) with $(x, -y)$:

$$-y = x^3 + 5x$$
$$y = -x^3 - 5x$$

The graph is not symmetric about the x-axis.
For y-axis symmetry, we replace (x, y) with $(-x, y)$:

$$y = (-x)^3 + 5(-x) = -x^3 - 5x$$

The graph is not symmetric about the y-axis.
For origin symmetry, we replace (x, y) with $(-x, -y)$:

$$-y = (-x)^3 + 5(-x)$$
$$-y = -x^3 - 5x$$
$$y = x^3 + 5x$$

The graph is symmetric about the origin.

27. For x-axis symmetry, we replace (x, y) with $(x, -y)$:

$$(-y)^2 = [x + (-y)]^4$$
$$y^2 = (x - y)^4$$

The graph is not symmetric about the x-axis.
For y-axis symmetry, we replace (x, y) with $(-x, y)$:

$$y^2 = (-x + y)^4$$

The graph is not symmetric about the y-axis.
For origin symmetry, we replace (x, y) with $(-x, -y)$:

$$(-y)^2 = [-x + (-y)]^4$$
$$y^2 = (-x - y)^4$$
$$y^2 = (x + y)^4$$

The graph is symmetric about the origin.

29. For x-axis symmetry, we replace (x,y) with $(x,-y)$:

$$-y = 3x - \tfrac{1}{x}$$
$$y = -3x + \tfrac{1}{x}$$

The graph is not symmetric about the x-axis.
For y-axis symmetry, we replace (x,y) with $(-x,y)$:

$$y = 3(-x) - \tfrac{1}{-x} = -3x + \tfrac{1}{x}$$

The graph is not symmetric about the y-axis.
For origin symmetry, we replace (x,y) with $(-x,-y)$:

$$-y = 3(-x) - \tfrac{1}{-x}$$
$$-y = -3x + \tfrac{1}{x}$$
$$y = 3x - \tfrac{1}{x}$$

The graph is symmetric about the origin.

31. For x-axis symmetry, we replace (x,y) with $(x,-y)$:

$$-y = 2^x - 2^{-x}$$
$$y = -2^x + 2^{-x}$$

The graph is not symmetric about the x-axis.
For y-axis symmetry, we replace (x,y) with $(-x,y)$:

$$y = 2^{-x} - 2^{-(-x)} = 2^{-x} - 2^x$$

The graph is not symmetric about the y-axis.
For origin symmetry, we replace (x,y) with $(-x,-y)$:

$$-y = 2^{-x} - 2^{-(-x)}$$
$$-y = 2^{-x} - 2^x$$
$$y = 2^x - 2^{-x}$$

The graph is symmetric about the origin.

33. The graph is symmetric about the x-axis only.

35. The graph is symmetric about the x-axis, the y-axis, and the origin.

37. The graph is symmetric about the x-axis only.

39. The graph is symmetric about the x-axis only.

41. Using the distance formula and substituting $y = -\tfrac{2}{3}x + 1$, we have:

$$\sqrt{x^2 + \left(-\tfrac{2}{3}x + 1\right)^2} = \sqrt{10}$$
$$x^2 + \tfrac{4}{9}x^2 - \tfrac{4}{3}x + 1 = 10$$
$$\tfrac{13}{9}x^2 - \tfrac{4}{3}x - 9 = 0$$
$$13x^2 - 12x - 81 = 0$$
$$(13x + 27)(x - 3) = 0$$
$$x = -\tfrac{27}{13}, 3$$
$$y = \tfrac{31}{13}, -1$$

The two points are $\left(-\tfrac{27}{13}, \tfrac{31}{13}\right)$ and $(3,-1)$.

43. Since the sum of the x- and y-coordinates is 2 and $y = x^2$, we have:

$$\begin{aligned} x + x^2 &= 2 \\ x^2 + x - 2 &= 0 \\ (x+2)(x-1) &= 0 \\ x &= -2, 1 \\ y &= 4, 1 \end{aligned}$$

The two points are $(-2,4)$ and $(1,1)$.

45. We draw the graph:

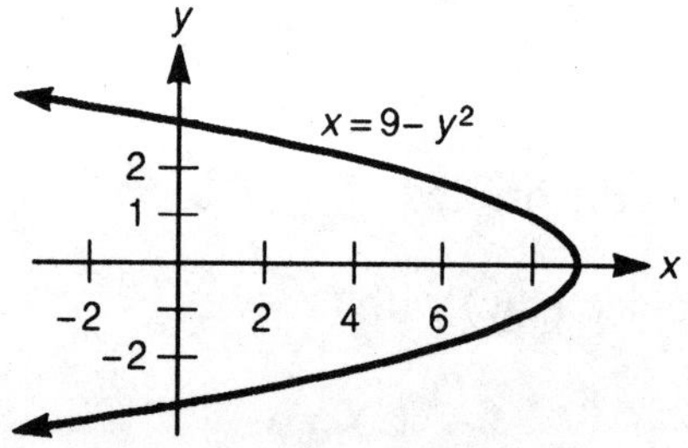

47. This is a circle with center = $(0,0)$ and radius = 1. We draw the graph:

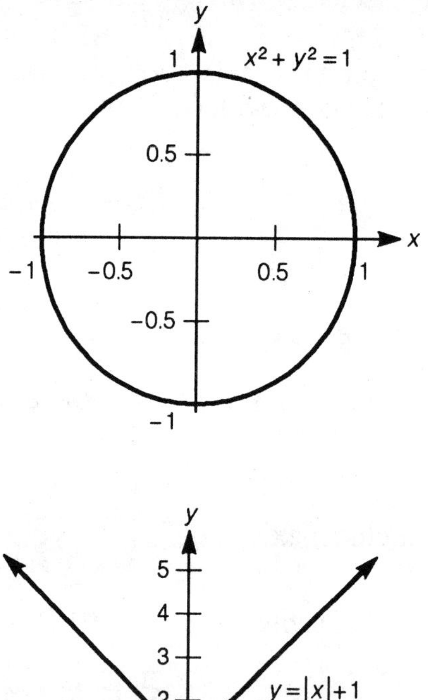

49. We draw the graph:

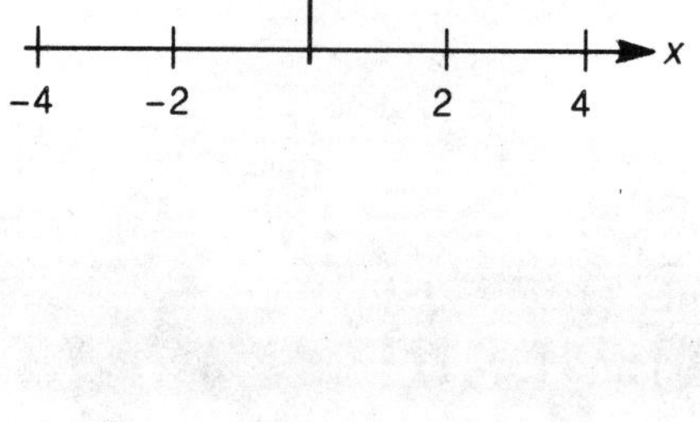

51. We draw the graph:

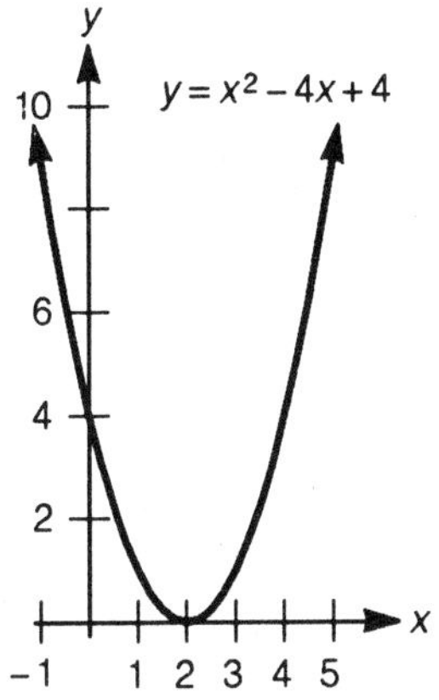

53. We draw the graph:

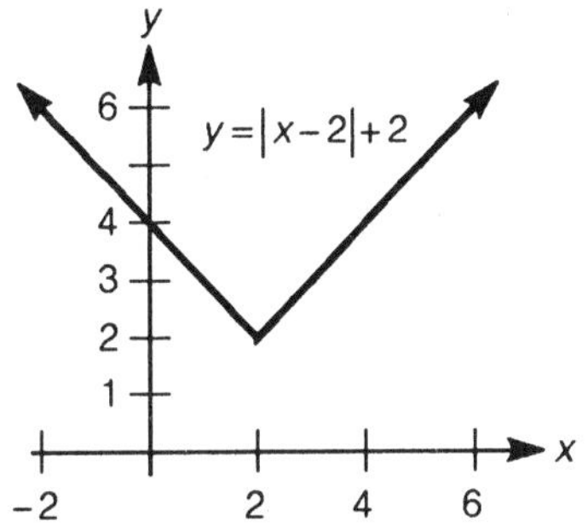

55. We write the line in slope-intercept form:

$$3x + y = 0$$
$$y = -3x$$

Now draw the graph:

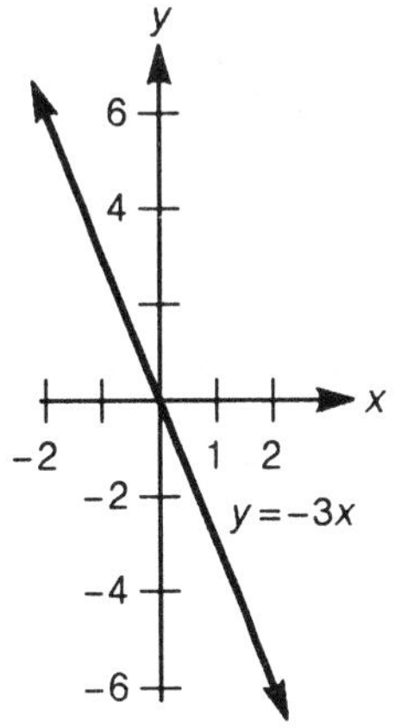

57. Complete the square:

$$x^2 - 4x + y^2 + 6y = 0$$
$$\left(x^2 - 4x + 4\right) + \left(y^2 + 6y + 9\right) = 4 + 9$$
$$(x-2)^2 + (y+3)^2 = 13$$

This is a circle with center = (2,–3) and radius = $\sqrt{13}$.

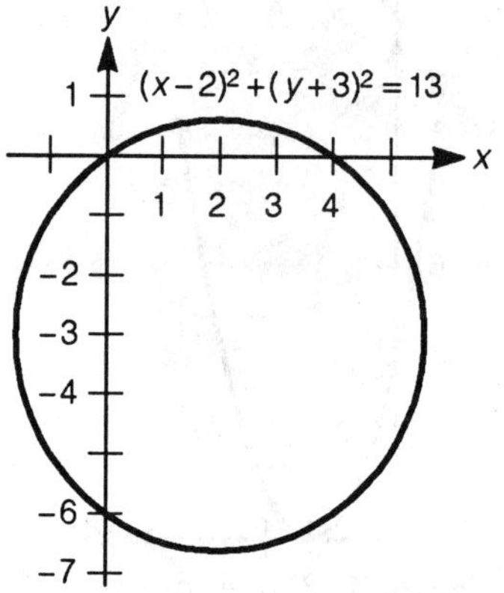

59. As long as $x \neq 4$, we can simplify:

$$y = \frac{x^2 - 16}{x - 4} = \frac{(x+4)(x-4)}{(x-4)} = x + 4$$

This is a line in slope-intercept form. Now draw the graph:

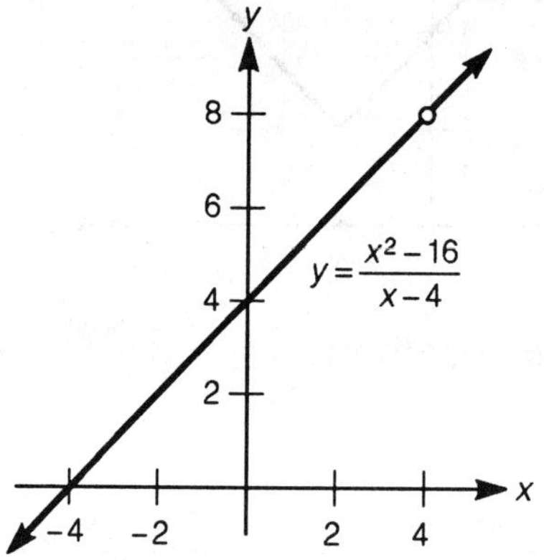

61. Setting each factor equal to 0, we have $4x - y + 4 = 0$ and thus $y = 4x + 4$, or $4x + y - 4 = 0$ and thus $y = -4x + 4$. So the graph consists of two lines:

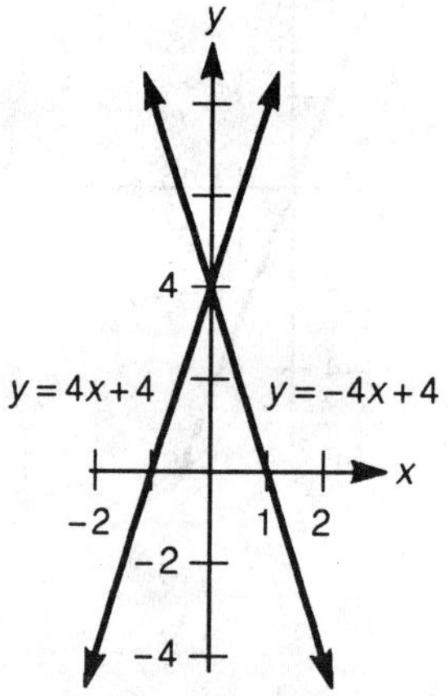

63. We set up the equation involving slope:

$$\frac{t-1}{5-2}=6$$
$$\frac{t-1}{3}=6$$
$$t-1=18$$
$$t=19$$

65. The slope between the points $(-2,-8)$ and (x,x^3) is given by:

$$m=\frac{x^3-(-8)}{x-(-2)}=\frac{x^3+8}{x+2}=\frac{(x+2)(x^2-2x+4)}{x+2}=x^2-2x+4$$

67. The midpoint of the hypotenuse BC would be:

$$M=\left(\frac{0+2c}{2},\frac{2b+0}{2}\right)=\left(\frac{2c}{2},\frac{2b}{2}\right)=(c,b)$$

We compute the distances:

$$MA=\sqrt{(c-0)^2+(b-0)^2}=\sqrt{c^2+b^2}$$
$$MB=\sqrt{(c-0)^2+(b-2b)^2}=\sqrt{c^2+b^2}$$
$$MC=\sqrt{(c-2c)^2+(b-0)^2}=\sqrt{c^2+b^2}$$

These distances are all the same.

69. (a) We compute the distance:

$$d=\sqrt{(5-2)^2+(-6-5)^2}=\sqrt{9+121}=\sqrt{130}$$

(b) We compute the slope:

$$m=\frac{-6-5}{5-2}=-\frac{11}{3}$$

(c) We compute the midpoint:

$$M=\left(\frac{2+5}{2},\frac{5-6}{2}\right)=\left(\frac{7}{2},-\frac{1}{2}\right)$$

71. (a) We compute the distance:

$$d=\sqrt{\left(-\tfrac{\sqrt{3}}{2}-\tfrac{\sqrt{3}}{2}\right)^2+\left(-\tfrac{1}{2}-\tfrac{1}{2}\right)^2}$$
$$=\sqrt{(-\sqrt{3})^2+(-1)^2}$$
$$=\sqrt{3+1}$$
$$=\sqrt{4}$$
$$=2$$

(b) We compute the slope:

$$m=\frac{-\frac{1}{2}-\frac{1}{2}}{-\frac{\sqrt{3}}{2}-\frac{\sqrt{3}}{2}}=\frac{-1}{-\sqrt{3}}=\frac{1}{\sqrt{3}}=\frac{\sqrt{3}}{3}$$

(c) We compute the midpoint:

$$M = \left(\frac{\frac{\sqrt{3}}{2} - \frac{\sqrt{3}}{2}}{2}, \frac{\frac{1}{2} - \frac{1}{2}}{2} \right) = (0,0)$$

73. We find the slope:

$$m = \frac{1-2}{4-1} = -\frac{1}{3}$$

Using $(1,2)$ in the point-slope formula:

$$y - 2 = -\tfrac{1}{3}(x-1)$$
$$y - 2 = -\tfrac{1}{3}x + \tfrac{1}{3}$$
$$y = -\tfrac{1}{3}x + \tfrac{7}{3}$$

So the x- and y-intercepts are $\frac{7}{3}$ (if $x = 0$) and 7 (if $y = 0$). Then the area is given by:

$$\text{Area} = \tfrac{1}{2}(\text{base})(\text{height}) = \tfrac{1}{2} \bullet 7 \bullet \tfrac{7}{3} = \tfrac{49}{6}$$

75. (a) We find each midpoint:

$$M_1 = \left(\frac{-5+7}{2}, \frac{3+7}{2} \right) = (1,5)$$
$$M_2 = \left(\frac{3+7}{2}, \frac{1+7}{2} \right) = (5,4)$$
$$M_3 = \left(\frac{-5+3}{2}, \frac{3+1}{2} \right) = (-1,2)$$

We now use the result from Exercise 74 to find the desired points:

$$CM_1 = \left(\frac{1}{3}(3) + \frac{2}{3}(1), \frac{1}{3}(1) + \frac{2}{3}(5) \right) = \left(1 + \frac{2}{3}, \frac{1}{3} + \frac{10}{3} \right) = \left(\frac{5}{3}, \frac{11}{3} \right)$$
$$AM_2 = \left(\frac{1}{3}(-5) + \frac{2}{3}(5), \frac{1}{3}(3) + \frac{2}{3}(4) \right) = \left(-\frac{5}{3} + \frac{10}{3}, 1 + \frac{8}{3} \right) = \left(\frac{5}{3}, \frac{11}{3} \right)$$
$$BM_3 = \left(\frac{1}{3}(7) + \frac{2}{3}(-1), \frac{1}{3}(7) + \frac{2}{3}(2) \right) = \left(\frac{7}{3} - \frac{2}{3}, \frac{7}{3} + \frac{4}{3} \right) = \left(\frac{5}{3}, \frac{11}{3} \right)$$

They all intersect at the same point.

(b) We find each midpoint:

$$M_1 = \left(\frac{0+2b}{2}, \frac{0+2c}{2} \right) = (b,c)$$
$$M_1 = \left(\frac{2b+2a}{2}, \frac{2c+0}{2} \right) = (a+b,c)$$
$$M_3 = \left(\frac{0+2a}{2}, \frac{0+0}{2} \right) = (a,0)$$

We now use the result from Exercise 74 to find the desired point:

$$CM_3 = \left(\frac{1}{3}(2b) + \frac{2}{3}(a), \frac{1}{3}(2c) + \frac{2}{3}(0)\right) = \left(\frac{2b}{3} + \frac{2a}{3}, \frac{2c}{3}\right)$$

$$BM_1 = \left(\frac{1}{3}(2a) + \frac{2}{3}(b), \frac{1}{3}(0) + \frac{2}{3}(c)\right) = \left(\frac{2a}{3} + \frac{2b}{3}, \frac{2c}{3}\right)$$

$$AM_2 = \left(\frac{1}{3}(0) + \frac{2}{3}(a+b), \frac{1}{3}(0) + \frac{2}{3}(c)\right) = \left(\frac{2a}{3} + \frac{2b}{3}, \frac{2c}{3}\right)$$

Again all these intersect at the same point. All medians of a triangle intersect at a point which is $\frac{2}{3}$ of the distance from each vertex to the midpoint on the opposite side.

77. We first complete the square to find the center and radius:

$$x^2 - 4x + y^2 + 2y = 15$$
$$(x^2 - 4x + 4) + (y^2 + 2y + 1) = 15 + 4 + 1$$
$$(x-2)^2 + (y+1)^2 = 20$$

So the center is $(2,-1)$ and the radius is $\sqrt{20} = 2\sqrt{5}$. Since $(1,-2)$ is the midpoint of the chord, the chord must be perpendicular to the radius line drawn to that point. We find the distance a of the line segment joining $(2,-1)$ and $(1,-2)$:

$$a = \sqrt{(1-2)^2 + [-2-(-1)]^2} = \sqrt{(-1)^2 + (-1)^2} = \sqrt{1+1} = \sqrt{2}$$

We can find d (half of the length of the chord) by the Pythagorean theorem:

$$d^2 + (\sqrt{2})^2 = (\sqrt{20})^2$$
$$d^2 + 2 = 20$$
$$d^2 = 18$$
$$d = \sqrt{18} = 3\sqrt{2}$$

The length of the chord is $2d = 2(3\sqrt{2}) = 6\sqrt{2}$.

79. Using the hint, we have:

$$m^2 = (a+c-0)^2 + (b-0)^2$$
$$m^2 = a^2 + 2ac + c^2 + b^2$$
$$s^2 = (2a-0)^2 + (2b-0)^2 = 4a^2 + 4b^2$$
$$t^2 = (2c)^2 = 4c^2$$
$$u^2 = (2a-2c)^2 + (2b-0)^2 = 4a^2 - 8ac + 4c^2 + 4b^2$$

Working from the right-hand side, we have:

$$\begin{aligned}\tfrac{1}{2}(s^2 + t^2) - \tfrac{1}{4}u^2 &= \tfrac{1}{2}(4a^2 + 4b^2 + 4c^2) - \tfrac{1}{4}(4a^2 - 8ac + 4c^2 + 4b^2) \\ &= 2a^2 + 2b^2 + 2c^2 - a^2 + 2ac - c^2 - b^2 \\ &= a^2 + b^2 + c^2 + 2ac \\ &= m^2\end{aligned}$$

81. Here $(x_0, y_0) = (1,2), m = \frac{1}{2}$, and $b = -5$:

$$d = \frac{2 - \frac{1}{2}(1) - (-5)}{\sqrt{1 + \left(\frac{1}{2}\right)^2}} = \frac{2 - \frac{1}{2} + 5}{\sqrt{1 + \frac{1}{4}}} = \frac{\frac{13}{2}}{\sqrt{\frac{5}{4}}} = \frac{13\sqrt{5}}{5}$$

83. Here $(x_0, y_0) = (-1,-3), A = 2, B = 3$ and $C = -6$:

$$d = \frac{|2(-1) + 3(-3) + (-6)|}{\sqrt{2^2 + 3^2}} = \frac{|-2 - 9 - 6|}{\sqrt{13}} = \frac{17}{\sqrt{13}} = \frac{17\sqrt{13}}{13}$$

Chapter Three Graphing Calculator Exercises

1. We draw the graph:

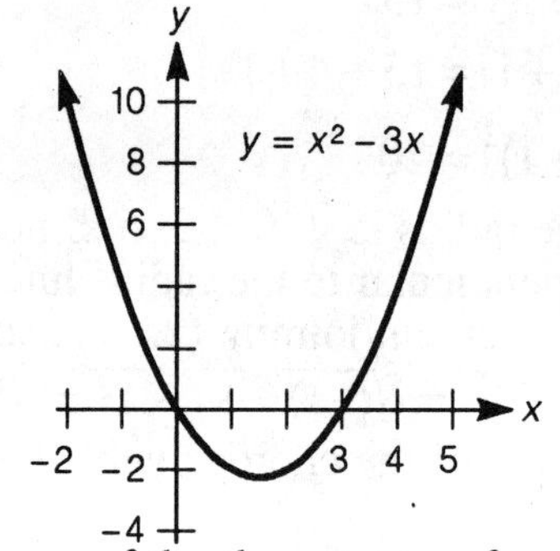

The graph does not possess any of the three types of symmetry.

3. We draw the graph:

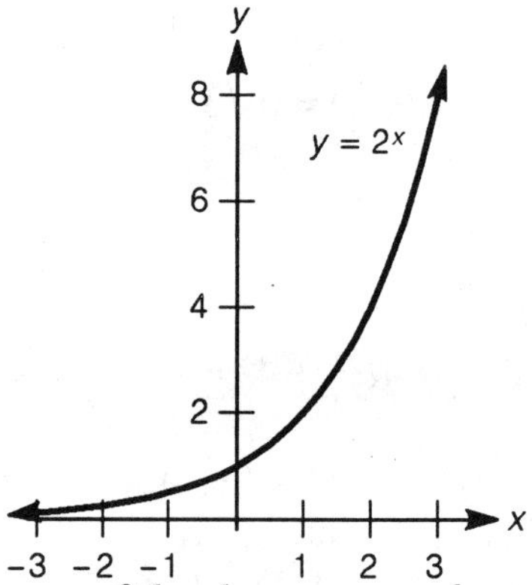

The graph does not possess any of the three types of symmetry.

5. We draw the graph:

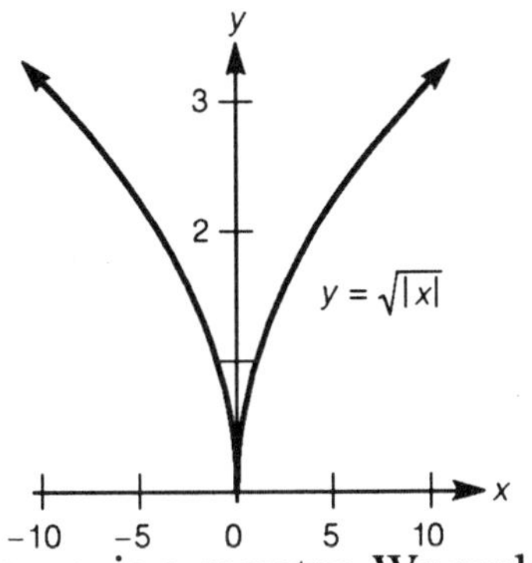

The graph appears to possess y-axis symmetry. We replace (x, y) with $(-x, y)$:

$$y = \sqrt{|-x|} = \sqrt{|x|}$$

So the graph does possess y-axis symmetry.

7. We draw the graph:

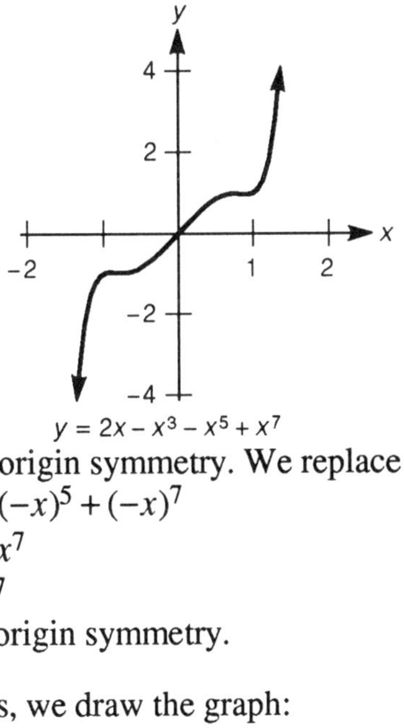

The graph appears to have origin symmetry. We replace (x, y) with $(-x, -y)$:

$$-y = 2(-x) - (-x)^3 - (-x)^5 + (-x)^7$$
$$-y = -2x + x^3 + x^5 - x^7$$
$$y = 2x - x^3 - x^5 + x^7$$

So the graph does possess origin symmetry.

9. Using the indicated settings, we draw the graph:

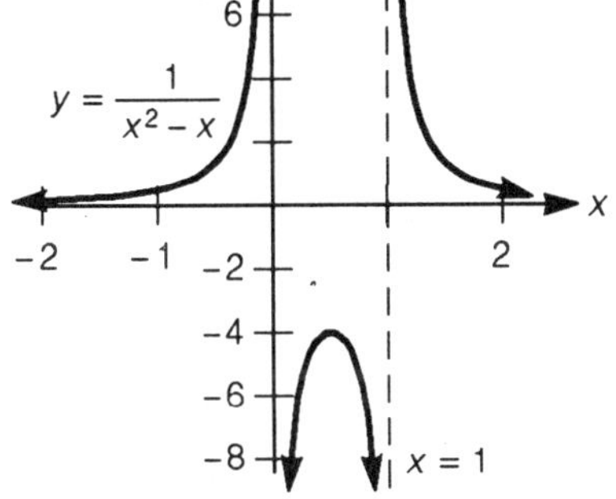

11. (a) We draw the graph:

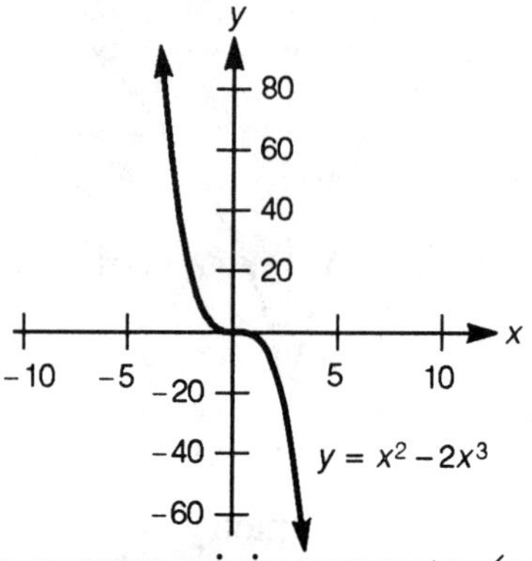

The graph "appears" to possess origin symmetry (see part (b)).

(b) Using the indicated settings, we draw the graph:

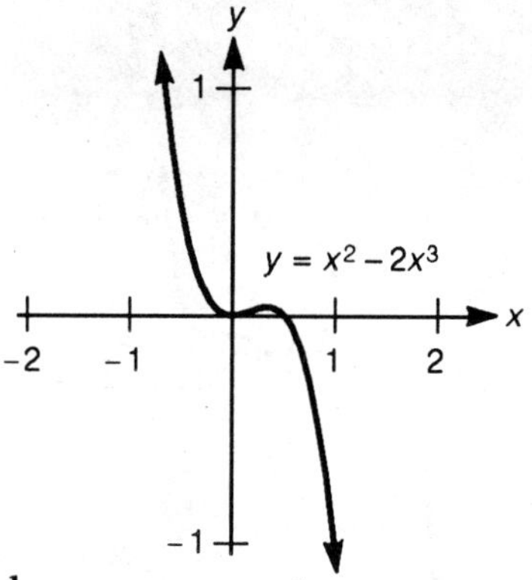

Note that the graph no longer appears to possess origin symmetry.

(c) We replace (x, y) with $(-x, -y)$:

$$-y = (-x)^2 - 2(-x)^3$$
$$-y = x^2 + 2x^3$$
$$y = -x^2 - 2x^3$$

The graph does not possess origin symmetry.

13. Using $x_{min} = -3$, $x_{max} = 3$, $y_{min} = -3$, and $y_{max} = 3$, we draw the graph:

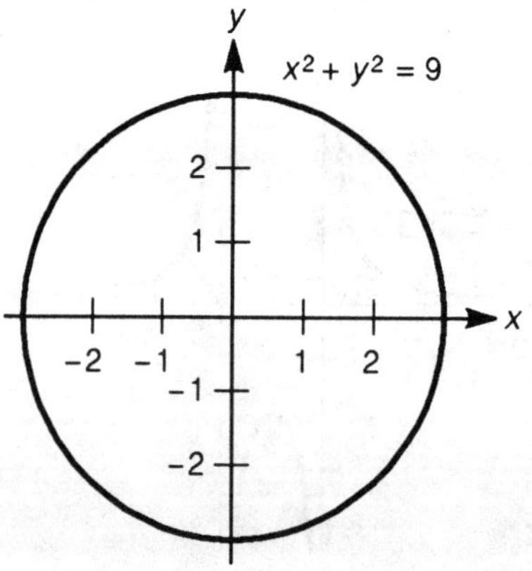

15. Solving for y, we obtain:

$$(x-2)^2+(y-1)^2=16$$
$$(y-1)^2=16-(x-2)^2$$
$$y-1=\pm\sqrt{16-(x-2)^2}$$
$$y=1\pm\sqrt{16-(x-2)^2}$$

Using $x_{min}=-2$, $x_{max}=6$, $y_{min}=-3$, and $y_{max}=5$, we draw the graph:

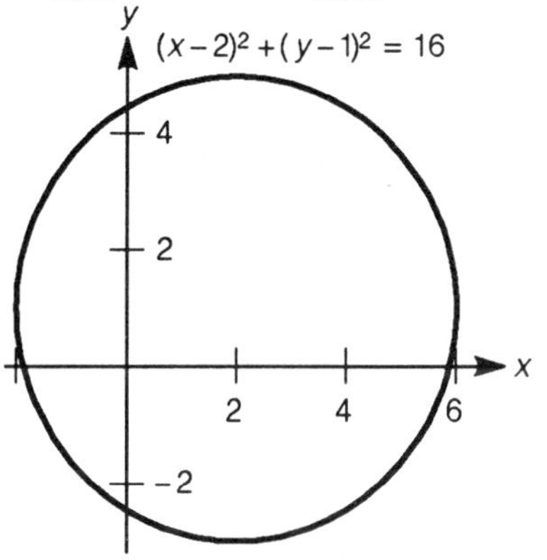

17. (a) Solving for y and graphing each curve, and using the settings $x_{min}=-3$, $x_{max}=3$, $y_{min}=-3$, and $y_{max}=3$, we draw the graph:

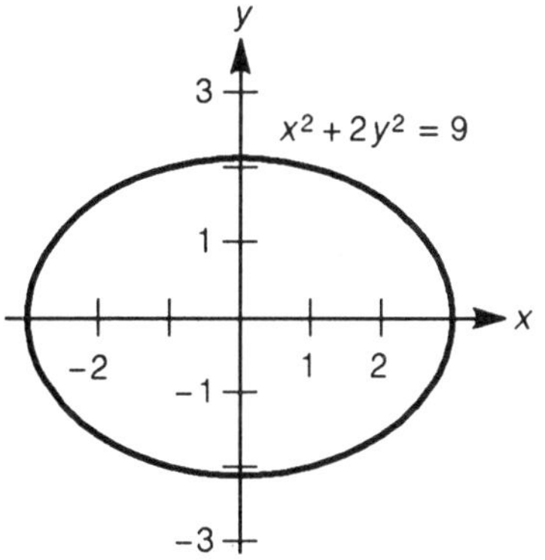

(b) The curves appear to intersect at $(-3,0)$ and $(3,0)$:

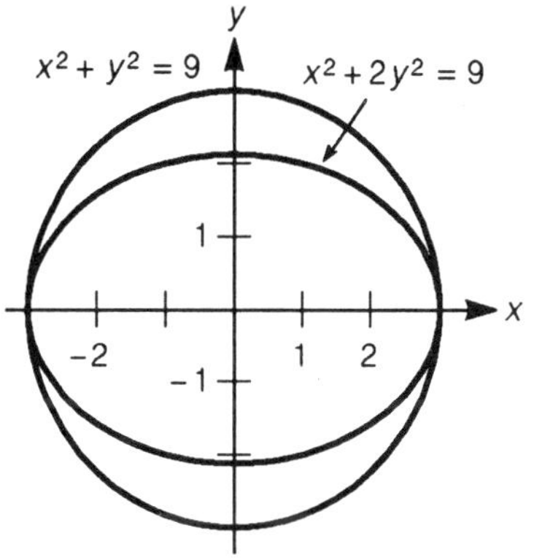

(c) Substituting the points $(\pm 3, 0)$ into the two equations:

$$(\pm 3)^2 + (0)^2 = 9$$
$$(\pm 3)^2 + 2(0)^2 = 9$$

Thus the intersection points are indeed $(-3, 0)$ and $(3, 0)$.

19. (a) Using the settings $x_{min} = -6$, $x_{max} = 6$, $y_{min} = -6$, and $y_{max} = 6$, we draw the graph:

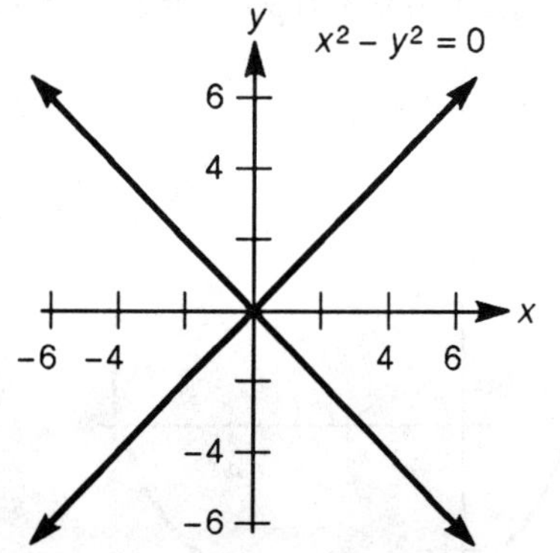

(b) We graph both curves on the same axes:

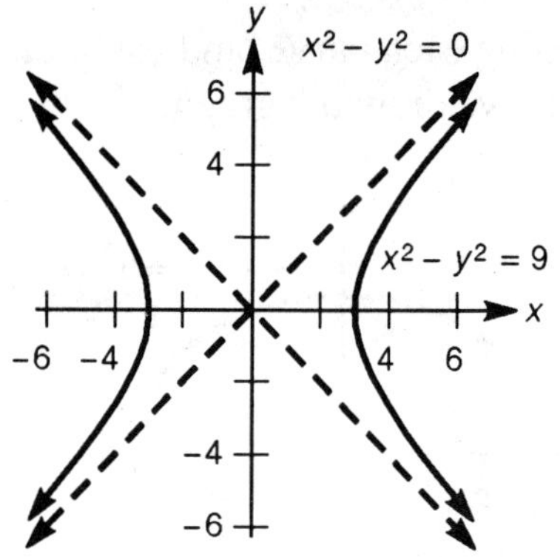

21. (c) Using the settings $x_{min} = -5$, $x_{max} = 5$, $y_{min} = -5$, and $y_{max} = 5$, line segment $\overline{DE}$ appears as:

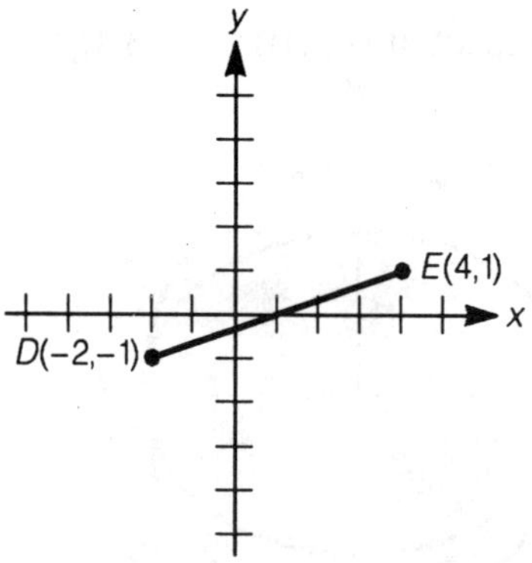

(d) Graphing $\overline{DE}$ and $\overline{EF}$ together appears as:

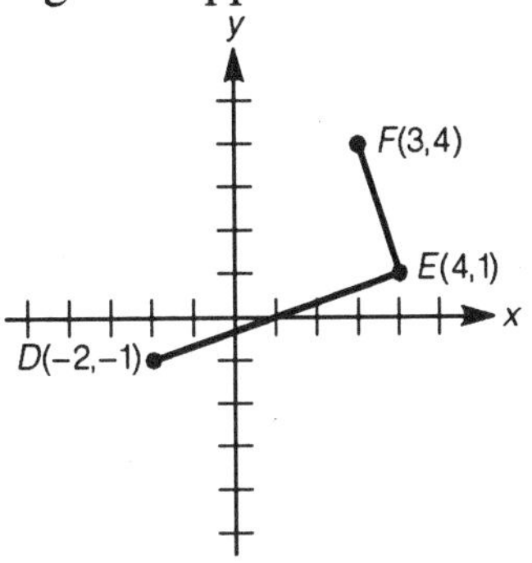

(e) Now graphing all three line segments together:

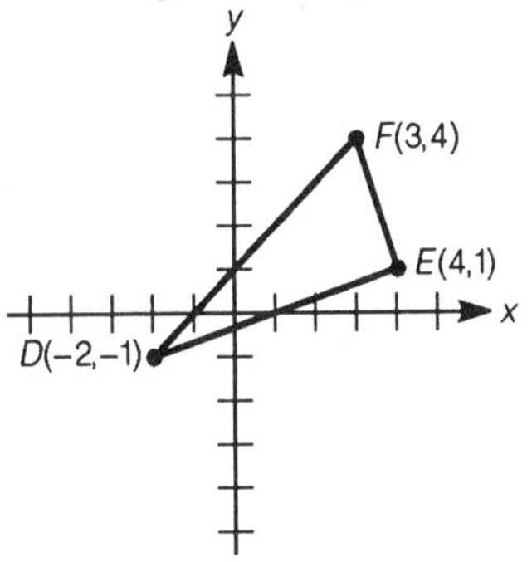

23. The equations are $y = -x$, $y = -4x$, and $y = -8x$. We graph these three lines:

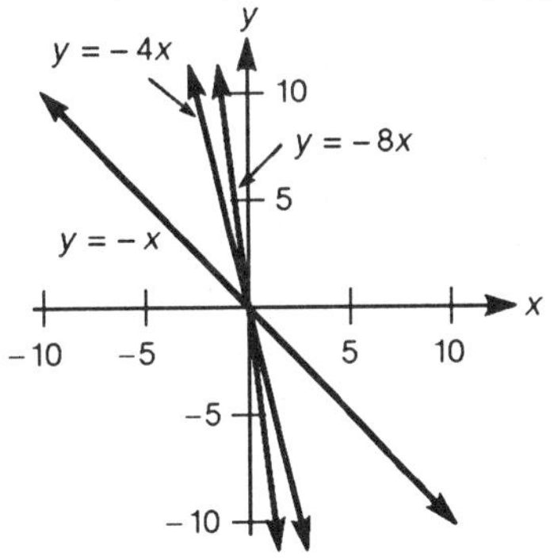

25. We graph the three parallel lines:

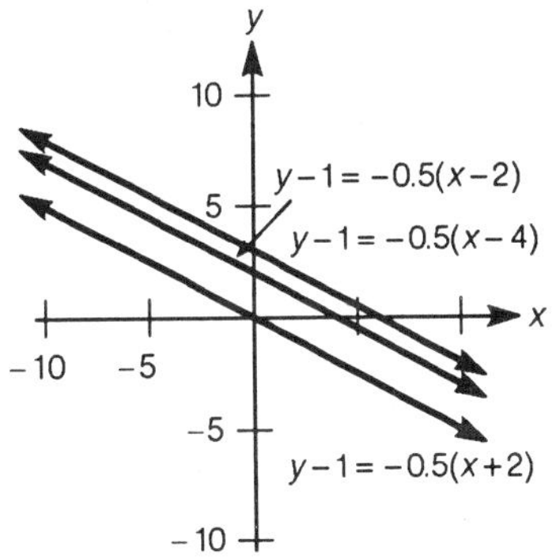

27. (a) Solving for y, we have:

$$\frac{x}{2}+\frac{y}{3}=1$$
$$3x+2y=6$$
$$2y=-3x+6$$
$$y=-\tfrac{3}{2}x+3$$

Now graph the line:

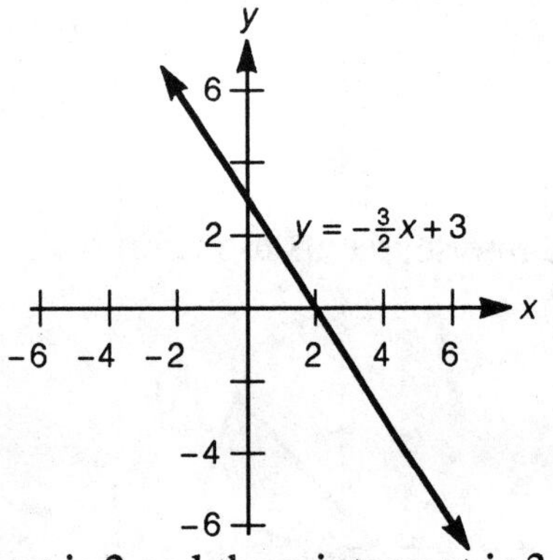

Note that the x-intercept is 2 and the y-intercept is 3.

(b) Solving for y, we have:

$$\frac{x}{-2}+\frac{y}{-3}=1$$
$$3x+2y=-6$$
$$2y=-3x-6$$
$$y=-\tfrac{3}{2}x-3$$

Now graph the line:

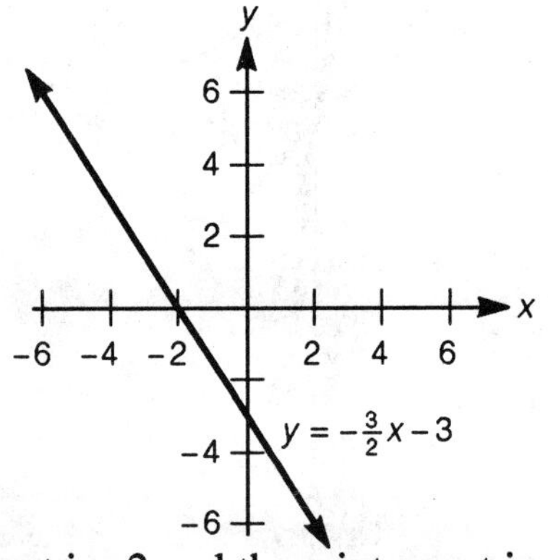

Note that the x-intercept is –2 and the y-intercept is –3.

(c) Solving for y, we have:

$$\frac{x}{6}+\frac{y}{5}=1$$
$$5x+6y=30$$
$$6y=-5x+30$$
$$y=-\tfrac{5}{6}x+5$$

Now graph the line:

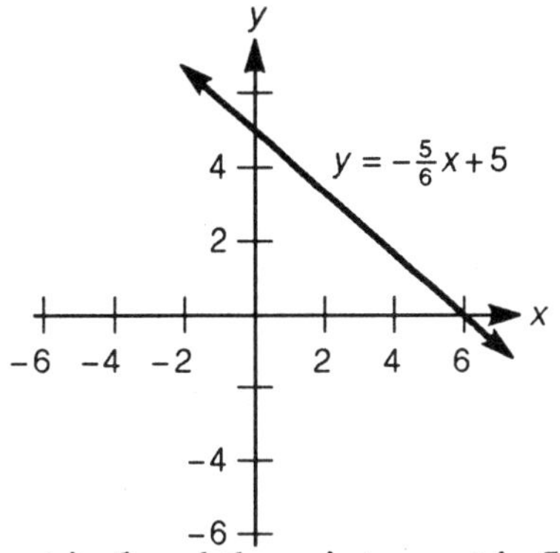

Note that the x-intercept is 6 and the y-intercept is 5.

(d) Solving for y, we have:

$$\frac{x}{-6} + \frac{y}{-5} = 1$$
$$5x + 6y = -30$$
$$6y = -5x - 30$$
$$y = -\tfrac{5}{6}x - 5$$

Now graph the line:

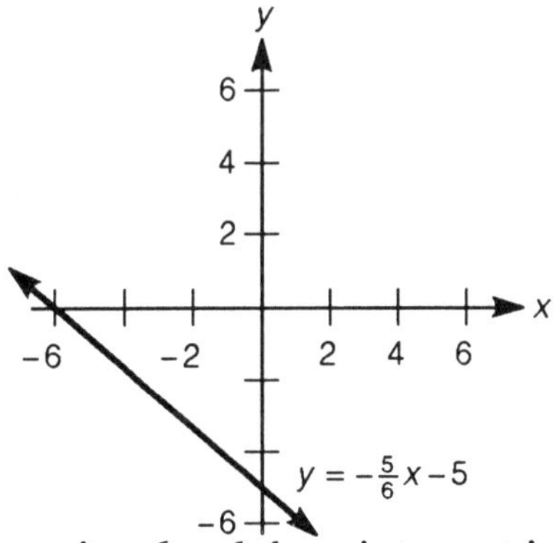

Note that the x-intercept is -6 and the y-intercept is -5.

(e) It appears that $\frac{x}{a} + \frac{y}{b} = 1$ is a line with an x-intercept of a and y-intercept of b.

Chapter Three Test

1. We first find the slope:

$$m = \frac{8-(-2)}{3-1} = \frac{10}{2} = 5$$

Now using the point (3, 8) in the point-slope formula, we have:

$$y - 8 = 5(x - 3)$$
$$y - 8 = 5x - 15$$
$$y = 5x - 7$$

2. We first write $5x + 6y = 30$ in slope-intercept form:

$$\begin{aligned} 5x+6y&=30 \\ 6y&=-5x+30 \\ y&=-\tfrac{5}{6}x+5 \end{aligned}$$

So the perpendicular slope is $\frac{6}{5}$. Now using the point $(2,-1)$ in the point-slope formula, we have:

$$\begin{aligned} y-(-1)&=\tfrac{6}{5}(x-2) \\ y+1&=\tfrac{6}{5}x-\tfrac{12}{5} \\ y&=\tfrac{6}{5}x-\tfrac{17}{5} \end{aligned}$$

3. We first find the midpoint of the line segment joining the points $(-5,-2)$ and $(3,8)$:

$$M=\left(\frac{-5+3}{2},\frac{-2+8}{2}\right)=\left(\frac{-2}{2},\frac{6}{2}\right)=(-1,3)$$

Since the line passes through the origin, its slope is given by:

$$m=\frac{3-0}{-1-0}=\frac{3}{-1}=-3$$

So the equation is $y=-3x$.

4. We first complete the square to find the center of the circle:

$$\begin{aligned} x^2-8x+y^2+10y&=0 \\ x^2-8x+16+y^2+10y+25&=16+25 \\ (x-4)^2+(y+5)^2&=41 \end{aligned}$$

So the center of the circle is $(4,-5)$. The slope from this center to the origin is therefore:

$$m=\frac{-5-0}{4-0}=\frac{-5}{4}$$

So the tangent line slope, which is perpendicular to this radius slope, is $\frac{4}{5}$. So the equation of the tangent line is $y=\frac{4}{5}x$.

5. We compute the distance from each point to the origin:

$$\begin{aligned} d_1&=\sqrt{(3-0)^2+(9-0)^2}=\sqrt{9+81}=\sqrt{90} \\ d_2&=\sqrt{(5-0)^2+(8-0)^2}=\sqrt{25+64}=\sqrt{89} \end{aligned}$$

So $(3,9)$ is farther from the origin.

6. (a) To test for symmetry about the x-axis, we replace (x,y) with $(x,-y)$:

$$\begin{aligned} -y&=x^3+5x \\ y&=-x^3-5x \end{aligned}$$

Since the equation is changed, there is no x-axis symmetry. To test for symmetry about the y-axis, we replace (x,y) with $(-x,y)$:

$$y=(-x)^3+5(-x)=-x^3-5x$$

Since the equation is changed, there is no y-axis symmetry.

To test for symmetry about the origin, we replace (x, y) with $(-x, -y)$:

$$-y = (-x)^3 + 5(-x)$$
$$-y = -x^3 - 5x$$
$$y = x^3 + 5x$$

Since the equation is unchanged, there is origin symmetry.

(b) To test for symmetry about the x-axis, we replace (x, y) with $(x, -y)$:

$$-y = 3^x + 3^{-x}$$
$$y = -3^x - 3^{-x}$$

Since the equation is changed, there is no x-axis symmetry. To test for symmetry about the y-axis, we replace (x, y) with $(-x, y)$:

$$y = 3^{-x} + 3^{-(-x)} = 3^{-x} + 3^x$$

Since the equation is unchanged, there is y-axis symmetry. To test for symmetry about the origin, we replace (x, y) with $(-x, -y)$:

$$-y = 3^{-x} + 3^{-(-x)}$$
$$-y = 3^{-x} + 3^x$$
$$y = -3^{-x} - 3^x$$

Since the equation is changed, there is no origin symmetry.

(c) To test for symmetry about the x-axis, we replace (x, y) with $(x, -y)$:

$$(-y)^2 = 5x^2 + x$$
$$y^2 = 5x^2 + x$$

Since the equation is unchanged, there is x-axis symmetry. To test for symmetry about the y-axis, we replace (x, y) with $(-x, y)$:

$$y^2 = 5(-x)^2 + (-x) = 5x^2 - x$$

Since the equation is changed, there is no y-axis symmetry. To test for symmetry about the origin, we replace (x, y) with $(-x, -y)$:

$$(-y)^2 = 5(-x)^2 + (-x)$$
$$y^2 = 5x^2 - x$$

Since the equation is changed, there is no origin symmetry.

7. To find the x-intercept, let $y = 0$:

$$3x = 15$$
$$x = 5$$

To find the y-intercept, let $x = 0$:

$$-5y = 15$$
$$y = -3$$

Now draw the graph:

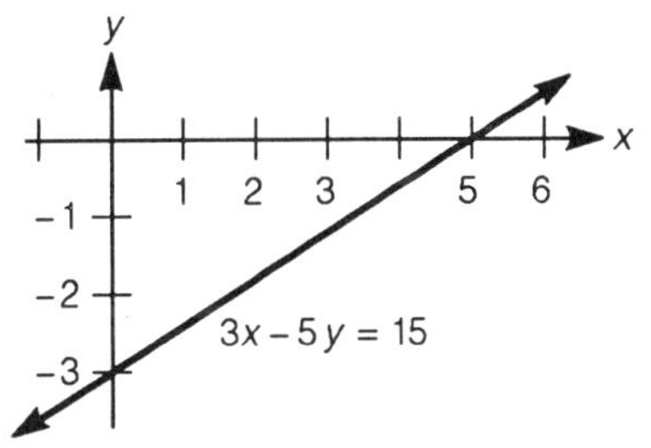

8. Dividing by 2, we complete the square:

$$x^2 - 5x + y^2 + 3y = \tfrac{17}{2}$$
$$x^2 - 5x + \tfrac{25}{4} + y^2 + 3y + \tfrac{9}{4} = \tfrac{17}{2} + \tfrac{25}{4} + \tfrac{9}{4}$$
$$\left(x - \tfrac{5}{2}\right)^2 + \left(y + \tfrac{3}{2}\right)^2 = 17$$

This is the graph of a circle centered at $\left(\frac{5}{2}, -\frac{3}{2}\right)$ with radius $\sqrt{17}$.
To find the x-intercepts, let $y = 0$:

$$\left(x - \tfrac{5}{2}\right)^2 + \tfrac{9}{4} = 17$$
$$\left(x - \tfrac{5}{2}\right)^2 = \tfrac{59}{4}$$
$$x - \tfrac{5}{2} = \tfrac{\pm\sqrt{59}}{2}$$
$$x = \tfrac{5 \pm \sqrt{59}}{2}$$

To find the y-intercepts, let $x = 0$:

$$\tfrac{25}{4} + \left(y + \tfrac{3}{2}\right)^2 = 17$$
$$\left(y + \tfrac{3}{2}\right)^2 = \tfrac{43}{4}$$
$$y + \tfrac{3}{2} = \tfrac{\pm\sqrt{43}}{2}$$
$$x = \tfrac{-3 \pm \sqrt{43}}{2}$$

Now draw the graph:

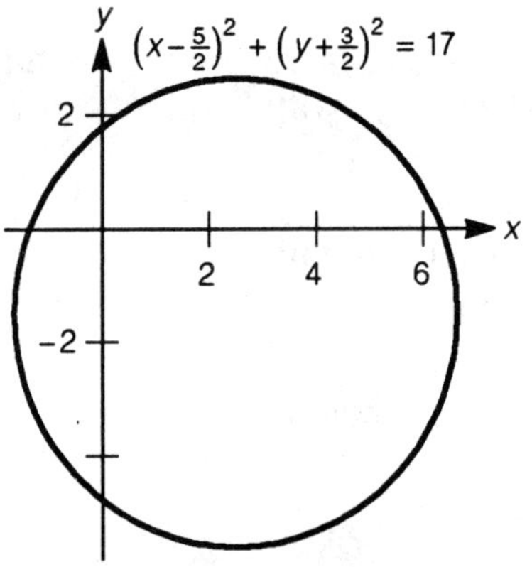

9. To find the x-intercept, let $y = 0$:

$$0 = |x + 1| + 2$$
$$-2 = |x + 1|$$

Since the equation has no solution, there is no x-intercept.
To find the y-intercept, let $x = 0$:

$$y = |0 + 1| + 2 = 1 + 2 = 3$$

Now draw the graph:

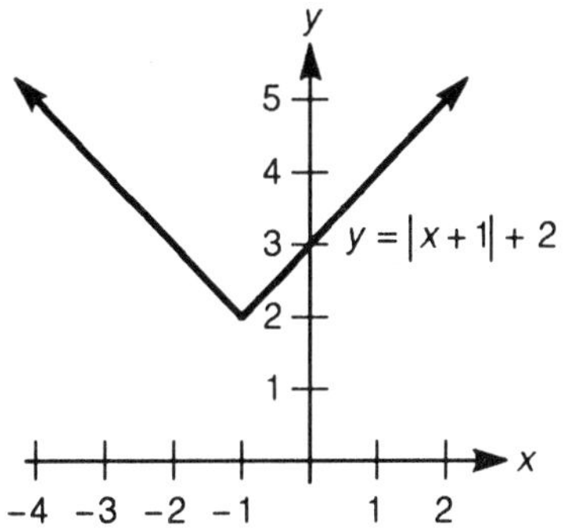

10. The slope of the line passing through $(-4,3)$ and $(t,13)$ is given by:

$$m=\frac{13-3}{t-(-4)}=\frac{10}{t+4}$$

Setting the slope $m=2$, we have:

$$\begin{aligned}\frac{10}{t+4}&=2\\ 10&=2t+8\\ 2&=2t\\ 1&=t\end{aligned}$$

11. Substituting $y=x^2$ into $x+y=12$, we have:

$$\begin{aligned}x+x^2&=12\\ x^2+x-12&=0\\ (x+4)(x-3)&=0\\ x&=-4,3\\ y&=16,9\end{aligned}$$

The points are $(-4,16)$ and $(3,9)$.

12. Substituting $x=-\frac{1}{2}$ into $y=4x^2-8x$:

$$y=4\left(-\tfrac{1}{2}\right)^2-8\left(-\tfrac{1}{2}\right)=4\left(\tfrac{1}{4}\right)+4=1+4=5$$

So $\left(-\frac{1}{2},5\right)$ lies on the graph of $y=4x^2-8x$.

13. We find the distances from $P(x,y)$ to $(-3,-4)$ and $(7,10)$:

$$\begin{aligned}d_1&=\sqrt{(x+3)^2+(y+4)^2}\\ &=\sqrt{x^2+6x+9+y^2+8y+16}\\ &=\sqrt{x^2+6x+y^2+8y+25}\\ d_2&=\sqrt{(x-7)^2+(y-10)^2}\\ &=\sqrt{x^2-14x+49+y^2-20y+100}\\ &=\sqrt{x^2-14x+y^2-20y+149}\end{aligned}$$

Now setting $d_1 = d_2$, we have:

$$x^2 + 6x + y^2 + 8y + 25 = x^2 - 14x + y^2 - 20y + 149$$
$$20x + 28y = 124$$
$$5x + 7y = 31$$

14. We first find the slope of the line:

$$m = \frac{4-2}{1-6} = -\frac{2}{5}$$

Using the point-slope formula with the point $(1,4)$, we have:

$$y - 4 = -\tfrac{2}{5}(x-1)$$
$$5y - 20 = -2x + 2$$
$$2x + 5y = 22$$

Now we find the x- and y-intercepts:

let $x = 0$	let $y = 0$
$5y = 22$	$2x = 22$
$y = \frac{22}{5}$	$x = 11$

So the area is given by:

$$\text{Area} = \tfrac{1}{2}(\text{base})(\text{height}) = \tfrac{1}{2} \bullet 11 \bullet \tfrac{22}{5} = \tfrac{121}{5}$$

Chapter Four
Functions

4.1 The Definition of a Function

1. (a) Since $-5x + 1$ is defined for all real numbers, the domain is all real numbers, or $(-\infty, \infty)$.

(b) For $\sqrt{-5x+1}$ to be defined, we must solve $-5x + 1 \geq 0$:

$$-5x \geq -1$$
$$x \leq \tfrac{1}{5}$$

So the domain is $\left(-\infty, \tfrac{1}{5}\right]$.

(c) For the expression to be defined, we must be sure $|-5x+1| > 0$. Since $|-5x+1| \geq 0$ because it is an absolute value, we must solve $-5x + 1 = 0$:

$$-5x = -1$$
$$x = \tfrac{1}{5}$$

So the domain is all real numbers except $\tfrac{1}{5}$, or $\left(-\infty, \tfrac{1}{5}\right) \cup \left(\tfrac{1}{5}, \infty\right)$.

3. (a) The expression is undefined when its denominator is equal to 0, which occurs when $x = 4$. So the domain is all real numbers except 4, or $(-\infty, 4) \cup (4, \infty)$.

(b) The expression is undefined when its denominator is equal to 0, so we must solve $x^2 - 4 = 0$:

$$x^2 - 4 = 0$$
$$(x+2)(x-2) = 0$$
$$x = -2, 2$$

So the domain is all real numbers except -2 and 2, or $(-\infty, -2) \cup (-2, 2) \cup (2, \infty)$.

(c) We must be sure the quantity inside the radical is non-negative:

$$\frac{x^2+4}{x^2-4} \geq 0$$

$$\frac{x^2+4}{(x+2)(x-2)} \geq 0$$

The key numbers are –2 and 2. We draw the sign chart:

Interval	Test Number	$x+2$	$x-2$	x^2+4	$\frac{x^2+4}{(x+2)(x-2)}$
$(-\infty,-2)$	–3	neg.	neg.	pos.	pos.
$(-2,2)$	0	pos.	neg.	pos.	neg.
$(2,\infty)$	3	pos.	pos.	pos.	pos.

So the expression is positive on the intervals $(-\infty,-2)\cup(2,\infty)$, and it is never equal to 0. Thus the domain is $(-\infty,-2)\cup(2,\infty)$.

5. (a) We must be sure the quantity inside the radical is non-negative:

$$x^2-4x-5 \geq 0$$

$$(x-5)(x+1) \geq 0$$

The key numbers are –1 and 5. We draw the sign chart:

Interval	Test Number	$x-5$	$x+1$	$(x-5)(x+1)$
$(-\infty,-1)$	–2	neg.	neg.	pos.
$(-1,5)$	0	neg.	pos.	neg.
$(5,\infty)$	6	pos.	pos.	pos.

So the expression is positive on the intervals $(-\infty,-1)\cup(5,\infty)$, and it is 0 at $x=-1$ and $x=5$. Thus the domain is $(-\infty,-1]\cup[5,\infty)$.

(b) We must solve $x^2-4x-5>0$, which has the same solutions as in part (a) but excluding the endpoints. So the domain is $(-\infty,-1)\cup(5,\infty)$.

(c) There are no restrictions on $\sqrt[3]{x^2-4x-5}$, so the domain is all real numbers, or $(-\infty,\infty)$.

7. (a) To find the range, we first solve for x:

$$y=\frac{x+3}{x-5}$$

$$y(x-5)=x+3$$

$$yx-5y=x+3$$

$$yx-x=5y+3$$

$$x(y-1)=5y+3$$

$$x=\frac{5y+3}{y-1}$$

So the range is all real numbers except 1.

(b) To find the range, we first solve for x:

$$y = \frac{x-5}{x+3}$$
$$y(x+3) = x-5$$
$$yx + 3y = x - 5$$
$$yx - x = -3y - 5$$
$$x(y-1) = -3y - 5$$
$$x = \frac{-3y-5}{y-1}$$
$$x = \frac{3y+5}{1-y}$$

So the range is all real numbers except 1.

9. (a) To find the range, we first solve for x:

$$y = x^2 + 4$$
$$y - 4 = x^2$$
$$x = \pm\sqrt{y-4}$$

In order for the radical to be defined, we must have $y - 4 \geq 0$, so $y \geq 4$. So the range is $[4, \infty)$. Another approach is to note that since x^2 will assume values from 0 to ∞, then $x^2 + 4$ will assume values from 4 to ∞. So the range is $[4, \infty)$.

(b) To find the range, we first solve for x:

$$y = x^3 + 4$$
$$y - 4 = x^3$$
$$x = \sqrt[3]{y-4}$$

Since the cube root is defined for all real numbers, the range is all real numbers. Another approach is to note that since x^3 will assume values from $-\infty$ to ∞, then so will $x^3 + 4$. So the range is all real numbers.

11. Rules f, g, F and H are functions. Rule h is not a function since $h(x) = 1$ and $h(x) = 2$, which violates the definition of a function. Rule G is not a function, since $G(y)$ has not been assigned a value.

13. (a) The range of f is $\{1, 2, 3\}$, the range of g is $\{2, 3\}$, the range of F is $\{1\}$, and the range of H is $\{1, 2\}$.

(b) The range of g is $\{i, j\}$, the range of F is $\{i, j\}$, and the range of G is $\{k\}$.

15. (a) $y = (x-3)^2$
(b) $y = x^2 - 3$
(c) $y = (3x)^2$
(d) $y = 3x^2$

17. (a) We compute $f(1)$:

$$f(1) = (1)^2 - 3(1) + 1 = 1 - 3 + 1 = -1$$

(b) We compute $f(0)$:

$$f(0) = (0)^2 - 3(0) + 1 = 0 - 0 + 1 = 1$$

(c) We compute $f(-1)$:

$$f(-1) = (-1)^2 - 3(-1) + 1 = 1 + 3 + 1 = 5$$

(d) We compute $f\left(\frac{3}{2}\right)$:

$$f\left(\tfrac{3}{2}\right) = \left(\tfrac{3}{2}\right)^2 - 3\left(\tfrac{3}{2}\right) + 1 = \tfrac{9}{4} - \tfrac{9}{2} + 1 = -\tfrac{5}{4}$$

(e) We compute $f(z)$:

$$f(z) = (z)^2 - 3(z) + 1 = z^2 - 3z + 1$$

(f) We compute $f(x + 1)$:

$$\begin{aligned} f(x+1) &= (x+1)^2 - 3(x+1) + 1 \\ &= x^2 + 2x + 1 - 3x - 3 + 1 \\ &= x^2 - x - 1 \end{aligned}$$

(g) We compute $f(a + 1)$:

$$\begin{aligned} f(a+1) &= (a+1)^2 - 3(a+1) + 1 \\ &= a^2 + 2a + 1 - 3a - 3 + 1 \\ &= a^2 - a - 1 \end{aligned}$$

(h) We compute $f(-x)$:

$$f(-x) = (-x)^2 - 3(-x) + 1 = x^2 + 3x + 1$$

(i) Using our result from (a):

$$|f(1)| = |-1| = 1$$

(j) We compute $f(\sqrt{3})$:

$$f(\sqrt{3}) = (\sqrt{3})^2 - 3(\sqrt{3}) + 1 = 3 - 3\sqrt{3} + 1 = 4 - 3\sqrt{3}$$

(k) We compute $f(1+\sqrt{2})$:

$$f(1+\sqrt{2}) = (1+\sqrt{2})^2 - 3(1+\sqrt{2}) + 1 = 1 + 2\sqrt{2} + 2 - 3 - 3\sqrt{2} + 1 = 1 - \sqrt{2}$$

(l) We compute $|1 - f(2)|$:

$$|1 - f(2)| = \left|1 - [(2)^2 - 3(2) + 1]\right| = |1 - [4 - 6 + 1]| = |1 - (-1)| = |1 + 1| = 2$$

19. (a) We compute $f(2x)$:

$$f(2x) = 3(2x)^2 = 3(4x^2) = 12x^2$$

(b) We compute $2f(x)$:

$$2f(x) = 2(3x^2) = 6x^2$$

(c) We compute $f(x^2)$:

$$f(x^2) = 3(x^2)^2 = 3x^4$$

(d) We compute $[f(x)]^2$:

$$[f(x)]^2 = (3x^2)^2 = 9x^4$$

(e) We compute $f\left(\frac{x}{2}\right)$:

$$f\left(\frac{x}{2}\right) = 3\left(\frac{x}{2}\right)^2 = 3 \bullet \frac{x^2}{4} = \frac{3}{4}x^2$$

(f) We compute $\frac{f(x)}{2}$:

$$\frac{f(x)}{2} = \frac{3x^2}{2} = \tfrac{3}{2}x^2$$

21. (a) We compute $H(0)$:

$$H(0) = 1 - 2(0)^2 = 1 - 0 = 1$$

(b) We compute $H(2)$:

$$H(2) = 1 - 2(2)^2 = 1 - 2(4) = 1 - 8 = -7$$

(c) We compute $H(\sqrt{2})$:

$$H(\sqrt{2}) = 1 - 2(\sqrt{2})^2 = 1 - 2(2) = 1 - 4 = -3$$

(d) We compute $H\left(\frac{5}{6}\right)$:

$$H\left(\tfrac{5}{6}\right) = 1 - 2\left(\tfrac{5}{6}\right)^2 = 1 - 2\left(\tfrac{25}{36}\right) = 1 - \tfrac{25}{18} = -\tfrac{7}{18}$$

(e) We compute $H(x+1)$:

$$\begin{aligned} H(x+1) &= 1-2(x+1)^2 \\ &= 1-2(x^2+2x+1) \\ &= 1-2x^2-4x-2 \\ &= -2x^2-4x-1 \end{aligned}$$

(f) We compute $H(x+h)$:

$$H(x+h) = 1-2(x+h)^2 = 1-2(x^2+2xh+h^2) = 1-2x^2-4xh-2h^2$$

(g) Using our result from part (f), we have:

$$\begin{aligned} H(x+h)-H(x) &= (1-2x^2-4xh-2h^2)-(1-2x^2) \\ &= 1-2x^2-4xh-2h^2-1+2x^2 \\ &= -4xh-2h^2 \end{aligned}$$

(h) Using our result from part (g), we have:

$$\frac{H(x+h)-H(x)}{h} = \frac{-4xh-2h^2}{h} = \frac{h(-4x-2h)}{h} = -4x-2h$$

23. (a) For the domain, we must exclude those values which make $x-2=0$, or $x=2$. So the domain is all real numbers except 2. For the range, we solve for x:

$$\begin{aligned} y &= \frac{2x-1}{x-2} \\ y(x-2) &= 2x-1 \\ yx-2y &= 2x-1 \\ yx-2x &= 2y-1 \\ x(y-2) &= 2y-1 \\ x &= \frac{2y-1}{y-2} \end{aligned}$$

Since the denominator cannot be zero, $y=2$ is excluded. So the range is all real numbers except 2.

(b) We compute $R(0)$:

$$R(0) = \frac{2(0)-1}{0-2} = \frac{-1}{-2} = \frac{1}{2}$$

(c) We compute $R\left(\frac{1}{2}\right)$:

$$R\left(\tfrac{1}{2}\right) = \frac{2\left(\frac{1}{2}\right)-1}{\frac{1}{2}-2} = \frac{1-1}{-\frac{3}{2}} = 0$$

(d) We compute $R(-1)$:

$$R(-1)=\frac{2(-1)-1}{-1-2}=\frac{-2-1}{-3}=\frac{-3}{-3}=1$$

(e) We compute $R(x^2)$:

$$R(x^2)=\frac{2(x^2)-1}{x^2-2}=\frac{2x^2-1}{x^2-2}$$

(f) We compute $R\left(\frac{1}{x}\right)$:

$$R\left(\frac{1}{x}\right)=\frac{2\left(\frac{1}{x}\right)-1}{\frac{1}{x}-2}=\frac{\frac{2}{x}-1}{\frac{1}{x}-2}=\frac{2-x}{1-2x}$$

(g) We compute $R(a)$:

$$R(a)=\frac{2(a)-1}{a-2}=\frac{2a-1}{a-2}$$

(h) We compute $R(x-1)$:

$$R(x-1)=\frac{2(x-1)-1}{(x-1)-2}=\frac{2x-2-1}{x-3}=\frac{2x-3}{x-3}$$

25. (a) We compute the indicated values:

$$d(1)=-16(1)^2+96(1)=-16+96=80$$
$$d\left(\tfrac{3}{2}\right)=-16\left(\tfrac{3}{2}\right)^2+96\left(\tfrac{3}{2}\right)=-16\left(\tfrac{9}{4}\right)+144=108$$
$$d(2)=-16(2)^2+96(2)=-64+192=128$$
$$d(t_0)=-16t_0^{\,2}+96t_0$$

(b) We set $d(t)=0$:

$$-16t^2+96t=0$$
$$-16t(t-6)=0$$
$$t=0,6$$

(c) We set $d(t)=1$:

$$-16t^2+96t=1$$
$$16t^2-96t+1=0$$

Using the quadratic formula, we have:

$$t=\frac{96\pm\sqrt{(-96)^2-4(16)(1)}}{2(16)}$$
$$=\frac{96\pm\sqrt{9216-64}}{32}$$
$$=\frac{96\pm\sqrt{9152}}{32}$$
$$=\frac{96\pm8\sqrt{143}}{32}$$
$$=\frac{12\pm\sqrt{143}}{4}$$

27. We compute the given values:

$$g(3) = |3-4| = |-1| = 1$$
$$g(x+4) = |x+4-4| = |x|$$

29. (a) We first compute:

$$f(x+h) = (x+h)^2 = x^2 + 2xh + h^2$$

Therefore:

$$\frac{f(x+h)-f(x)}{h} = \frac{(x^2+2xh+h^2)-(x^2)}{h} = \frac{2xh+h^2}{h} = \frac{h(2x+h)}{h} = 2x+h$$

(b) We first compute:

$$\begin{aligned} f(x+h) &= 2(x+h)^2 - 3(x+h) + 1 \\ &= 2(x^2+2xh+h^2) - 3x - 3h + 1 \\ &= 2x^2 + 4xh + 2h^2 - 3x - 3h + 1 \end{aligned}$$

Therefore:

$$\begin{aligned} \frac{f(x+h)-f(x)}{h} &= \frac{(2x^2+4xh+2h^2-3x-3h+1)-(2x^2-3x+1)}{h} \\ &= \frac{2x^2+4xh+2h^2-3x-3h+1-2x^2+3x-1}{h} \\ &= \frac{4xh+2h^2-3h}{h} \\ &= \frac{h(4x+2h-3)}{h} \\ &= 4x+2h-3 \end{aligned}$$

(c) We first compute:

$$f(x+h) = (x+h)^3 = x^3 + 3x^2h + 3xh^2 + h^3$$

Therefore:

$$\begin{aligned} \frac{f(x+h)-f(x)}{h} &= \frac{(x^3+3x^2h+3xh^2+h^3)-(x^3)}{h} \\ &= \frac{3x^2h+3xh^2+h^3}{h} \\ &= \frac{h(3x^2+3xh+h^2)}{h} \\ &= 3x^2+3xh+h^2 \end{aligned}$$

31. (a) We compute the difference quotient:

$$\begin{aligned} \frac{f(x)-f(a)}{x-a} &= \frac{\frac{x}{x-1}-\frac{a}{a-1}}{x-a} \bullet \frac{(x-1)(a-1)}{(x-1)(a-1)} \\ &= \frac{x(a-1)-a(x-1)}{(x-a)(x-1)(a-1)} \\ &= \frac{ax-x-ax+a}{(x-a)(x-1)(a-1)} \\ &= \frac{-(x-a)}{(x-a)(x-1)(a-1)} \\ &= \frac{-1}{(x-1)(a-1)} \end{aligned}$$

(b) Using our result from (a) with $a = 3$, we have:
$$\frac{f(x)-f(3)}{x-3} = \frac{-1}{(x-1)(3-1)} = \frac{-1}{2(x-1)}$$

(c) We compute the difference quotient:
$$\begin{aligned}\frac{f(x+h)-f(x)}{h} &= \frac{\frac{x+h}{x+h-1}-\frac{x}{x-1}}{h} \bullet \frac{(x-1)(x+h-1)}{(x-1)(x+h-1)}\\ &= \frac{(x+h)(x-1)-x(x+h-1)}{h(x-1)(x+h-1)}\\ &= \frac{x^2+hx-x-h-x^2-xh+x}{h(x-1)(x+h-1)}\\ &= \frac{-h}{h(x-1)(x+h-1)}\\ &= \frac{-1}{(x-1)(x+h-1)}\end{aligned}$$

(d) Using our result from (c) with $x = 3$, we have:
$$\frac{f(3+h)-f(3)}{h} = \frac{-1}{(3-1)(3+h-1)} = \frac{-1}{2(2+h)}$$

33. (a) We set $f(x_0) = g(x_0)$:
$$\begin{aligned}4x_0 - 3 &= 8 - x_0\\ 5x_0 &= 11\\ x_0 &= \tfrac{11}{5}\end{aligned}$$

(b) We set $f(x_0) = g(x_0)$:
$$\begin{aligned}x_0^2 - 4 &= 4 - x_0^2\\ 2x_0^2 &= 8\\ x_0^2 &= 4\\ x_0 &= \pm 2\end{aligned}$$

(c) We set $f(x_0) = g(x_0)$:
$$\begin{aligned}x_0^2 &= x_0^3\\ x_0^3 - x_0^2 &= 0\\ x_0^2(x_0 - 1) &= 0\\ x_0 &= 0, 1\end{aligned}$$

(d) We set $f(x_0) = g(x_0)$:

$$2x_0^2 - x_0 = 3$$
$$2x_0^2 - x_0 - 3 = 0$$
$$(2x_0 - 3)(x_0 + 1) = 0$$
$$x_0 = \tfrac{3}{2}, -1$$

35. (a) We compute $A(1)$ and $A(0)$:

$$A(1) = 1000\left(1 + \tfrac{0.12}{4}\right)^{4(1)} \approx \$1125.51$$
$$A(0) = 1000\left(1 + \tfrac{0.12}{4}\right)^{4(0)} = \$1000.00$$

So $A(1) - A(0) \approx 1125.51 - 1000 = \$125.51$.

(b) We compute $A(10)$ and $A(9)$:

$$A(10) = 1000\left(1 + \tfrac{0.12}{4}\right)^{4(10)} \approx \$3262.04$$
$$A(9) = 1000\left(1 + \tfrac{0.12}{4}\right)^{4(9)} \approx \$2898.28$$

So $A(10) - A(9) \approx 3262.04 - 2898.28 = \$363.76$.

37. (a) We complete the table:

n	2	3	4	5	6	7	8
$g(n)$	1.4142	1.4422	1.4142	1.3797	1.3480	1.3205	1.2968

(b) Computing $g(15) = 1.19786$ and $g(14) = 1.20744$, so 15 is the smallest natural number n such that $g(n) < 1.2$.

39. (a) We compute $f(a)$, $f(2a)$ and $f(3a)$:

$$f(a) = \frac{a-a}{a+a} = \frac{0}{2a} = 0$$
$$f(2a) = \frac{2a-a}{2a+a} = \frac{a}{3a} = \frac{1}{3}$$
$$f(3a) = \frac{3a-a}{3a+a} = \frac{2a}{4a} = \frac{1}{2}$$

Since $\frac{1}{2} \neq 0 + \frac{1}{3}$, then $f(3a) \neq f(a) + f(2a)$.

(b) We compute $f(5a)$:

$$f(5a) = \frac{5a-a}{5a+a} = \frac{4a}{6a} = \frac{2}{3}$$

Since $f(2a) = \frac{1}{3}$, then $f(5a) = 2f(2a)$.

41. We simplify $\phi(y^2)$ and $[\phi(y)]^2$:

$$\phi(y^2) = 2(y^2) - 3 = 2y^2 - 3$$
$$[\phi(y)]^2 = (2y-3)^2 = 4y^2 - 12y + 9$$

So $\phi(y^2) \neq [\phi(y)]^2$.

43. We simplify $f(ax+b) = 2(ax+b) + 3 = 2ax + 2b + 3$. Since $f(ax+b) = x$, we have $2ax + 2b + 3 = x$. Since a and b are constants, we must have:

$$2ax = x \qquad\qquad 2b + 3 = 0$$
$$2a = 1 \qquad\qquad 2b = -3$$
$$a = \tfrac{1}{2} \qquad\qquad b = -\tfrac{3}{2}$$

So $a = \frac{1}{2}$ and $b = -\frac{3}{2}$.

45. We first compute:

$$f(x+y) = \frac{(x+y)-x}{(x+y)+y} = \frac{y}{x+2y} \text{ and } f(x-y) = \frac{(x-y)-x}{(x-y)+y} = \frac{-y}{x}$$

Therefore:

$$\begin{aligned} f(x+y) + f(x-y) &= \frac{y}{x+2y} - \frac{y}{x} \\ &= \frac{yx - y(x+2y)}{x(x+2y)} \\ &= \frac{yx - yx - 2y^2}{x^2 + 2xy} \\ &= \frac{-2y^2}{x^2+2xy} \end{aligned}$$

47. We compute $F\left(\frac{ax+b}{cx-a}\right)$:

$$\begin{aligned} F\left(\tfrac{ax+b}{cx-a}\right) &= \frac{a\left(\frac{ax+b}{cx-a}\right) + b}{c\left(\frac{ax+b}{cx-a}\right) - a} \\ &= \frac{a(ax+b) + b(cx-a)}{c(ax+b) - a(cx-a)} \\ &= \frac{a^2x + ab + bcx - ab}{acx + bc - acx + a^2} \\ &= \frac{a^2x + bcx}{a^2 + bc} \\ &= \frac{x(a^2+bc)}{a^2+bc} \\ &= x \end{aligned}$$

49. Since $g(1)=(1)^2-3(1)k-4=1-3k-4=-3k-3$, then if $g(1)=-2$, we have:

$$\begin{aligned} -3k-3&=-2\\ -3k&=1\\ k&=-\tfrac{1}{3} \end{aligned}$$

51. (a) We have $L(1)=0$, since $2^0=1$.

(b) We have $L(2)=1$, since $2^1=2$.

(c) We have $L(4)=2$, since $2^2=4$.

(d) We have $L(64)=6$, since $2^6=64$.

(e) We have $L\left(\tfrac{1}{2}\right)=-1$, since $2^{-1}=\tfrac{1}{2}$.

(f) We have $L\left(\tfrac{1}{4}\right)=-2$, since $2^{-2}=\tfrac{1}{4}$.

(g) We have $L\left(\tfrac{1}{64}\right)=-6$, since $2^{-6}=\tfrac{1}{64}$.

(h) We have $L(\sqrt{2})=\tfrac{1}{2}$, since $2^{1/2}=\sqrt{2}$.

53. Actually, we already know the answer to this question. Since $\frac{-b+\sqrt{b^2-4ac}}{2a}$ is one of the roots to the quadratic equation $q(x)=0$, we know $q\left(\frac{-b+\sqrt{b^2-4ac}}{2a}\right)=0$. Let's check our answer manually:

$$\begin{aligned} q\left(\frac{-b+\sqrt{b^2-4ac}}{2a}\right)&=a\left(\frac{-b+\sqrt{b^2-4ac}}{2a}\right)^2+b\left(\frac{-b+\sqrt{b^2-4ac}}{2a}\right)+c\\ &=a\left(\frac{b^2-2b\sqrt{b^2-4ac}+b^2-4ac}{4a^2}\right)+\left(\frac{-b^2+b\sqrt{b^2-4ac}}{2a}\right)+c\\ &=\frac{2b^2-4ac-2b\sqrt{b^2-4ac}}{4a}+\frac{-b^2+b\sqrt{b^2-4ac}}{2a}+c\\ &=\frac{b^2-2ac-b\sqrt{b^2-4ac}-b^2+b\sqrt{b^2-4ac}}{2a}+c\\ &=\frac{-2ac}{2a}+c\\ &=-c+c\\ &=0 \end{aligned}$$

55. (a) We compute $f[f(x)]$:

$$\begin{aligned} f[f(x)] &= f\left(\frac{3x-4}{x-3}\right) \\ &= \frac{3\left(\frac{3x-4}{x-3}\right)-4}{\left(\frac{3x-4}{x-3}\right)-3} \\ &= \frac{3(3x-4)-4(x-3)}{(3x-4)-3(x-3)} \\ &= \frac{9x-12-4x+12}{3x-4-3x+9} \\ &= \frac{5x}{5} \\ &= x \end{aligned}$$

(b) Since $f[f(x)] = x$, then $f\left[f\left(\frac{22}{7}\right)\right] = \frac{22}{7}$.

57. The problem here is the word "nearest." $G(4) = 3$, but also $G(4) = 5$, since both 3 and 5 are equally "near" 4. So G is not a function, since it assigns more than one value to $x = 4$. To alter the definition of G, one could define G to assign the closest prime number less than or equal to x (or, for that matter, greater than or equal to x). This would provide G with a way of "deciding" between 3 and 5, in the previous example. Why do we need the "or equal to" portion of the definition? [Hint: How would you define $G(5)$ otherwise?]

59. We compute the values:

$P(1) = (1)^2 - 1 + 17 = 17$
$P(2) = (2)^2 - 2 + 17 = 19$
$P(3) = (3)^2 - 3 + 17 = 23$
$P(4) = (4)^2 - 4 + 17 = 29$

The first natural number x such that $P(x)$ is not prime is $x = 17$. We compute:

$P(17) = (17)^2 - 17 + 17 = 17^2$

4.2 The Graph of a Function

1. The domain is $[-4, 2]$ and the range is $[-3, 3]$.

3. The domain is $[-4, -1) \cup (-1, 4]$ and the range is $[-2, 3)$.

5. The domain is $[-4, 4]$ and the range is $[-2, 2)$.

7. The domain is $[-4, 3]$ and the range is $\{2\}$.

9. (a) Since $(-5, 1)$ lies on the graph of F, then $F(-5) = 1$.
(b) Since $(2, -3)$ lies on the graph of F, then $F(2) = -3$.
(c) Since $F(1) \approx -2$, then $F(1)$ is negative, not positive.

(d) Since $(2,-3)$ lies on the graph of F, then $F(x) = -3$ when $x = 2$.
(e) Since $F(2) = -3$ and $F(-2) = -2$, then $F(2) - F(-2) = -3 - (-2) = -1$.

11. (a) positive
(b) $f(-2) = 4; f(1) = 1; f(2) = 2; f(3) = 0$
(c) $f(2)$, since $f(2) > 0$ and $f(4) < 0$
(d) $f(4) - f(1) = -2 - 1 = -3$
(e) $|f(4) - f(1)| = |-3| = 3$
(f) domain $= [-2, 4]$; range $= [-2, 4]$

13. (a) $f(-2) = 0$ and $g(-2) = 1$, so $g(-2)$ is larger.

(b) $f(0) - g(0) = 2 - (-3) = 2 + 3 = 5$

(c) We compute the three values:
$$f(1) - g(1) = 1 - (-1) = 2$$
$$f(2) - g(2) = 1 - 0 = 1$$
$$f(3) - g(3) = 4 - 1 = 3$$
So $f(2) - g(2)$ is the smallest.

(d) Since $f(1) = 1$, we look for where $g(x) = 1$. This occurs at $x = -2$ or $x = 3$

(e) Since $(3, 4)$ is a point on the graph of f, then 4 is in the range of f.

15. The range of f is $[0, 4]$ and the range of g is $[-3, 3]$.

17. The completed table is given below:

Function	$\lvert x\rvert$	x^2	x^3
Turning Point	(0,0)	(0,0)	none
Maximum Value	none	none	none
Minimum Value	0	0	none
Interval(s) where Increasing	$[0,\infty)$	$[0,\infty)$	$(-\infty,\infty)$
Interval(s) where Decreasing	$(-\infty,0]$	$(-\infty,0]$	none

19. (a) The range is $[-1, 1]$.
(b) The maximum value is 1 (occurring at $x = 1$).
(c) The minimum value is -1 (occurring at $x = 3$).
(d) The function is increasing on the intervals $[0, 1]$ and $[3, 4]$.
(e) The function is decreasing on the interval $[1, 3]$.

21. (a) The range is $[-3, 0]$.
(b) The maximum value is 0 (occurring at $x = 0$ and $x = 4$).
(c) The minimum value is -3 (occurring at $x = 2$).
(d) The function is increasing on the interval $[2, 4]$.
(e) The function is decreasing on the interval $[0, 2]$.

23. (a) Since any vertical line drawn intersects the graph in at most one point, this is the graph of a function.

(b) Since a vertical line can be drawn which intersects the graph in two points, this cannot be the graph of a function.

(c) Since a vertical line can be drawn which intersects the graph in two points, this cannot be the graph of a function.

(d) Since any vertical line drawn intersects the graph in only one point, this is the graph of a function.

25. The slope is given by:

$$m = \frac{f(4) - f(3)}{4 - 3} = \frac{4^2 - 3^2}{1} = \frac{16 - 9}{1} = 7$$

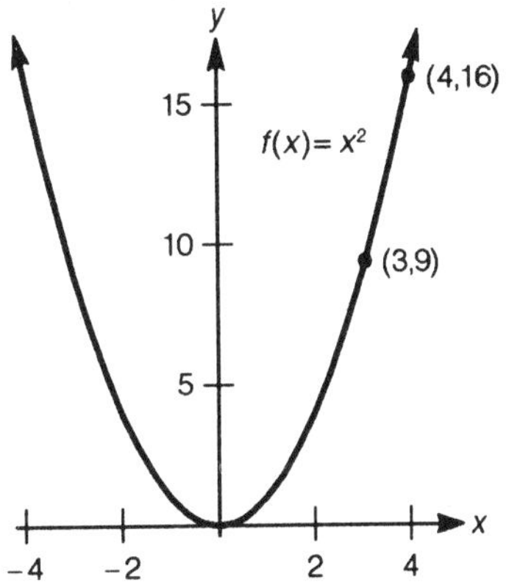

27. We find the two slopes:

$$m_1 = \frac{T(4) - T(1)}{4 - 1} = \frac{\sqrt{4} - \sqrt{1}}{3} = \frac{2 - 1}{3} = \frac{1}{3}$$

$$m_2 = \frac{T(9) - T(4)}{9 - 4} = \frac{\sqrt{9} - \sqrt{4}}{5} = \frac{3 - 2}{5} = \frac{1}{5}$$

Since $m_1 > m_2$, the line between $(1, T(1))$ and $(4, T(4))$ has the larger slope.

29. We graph the function:

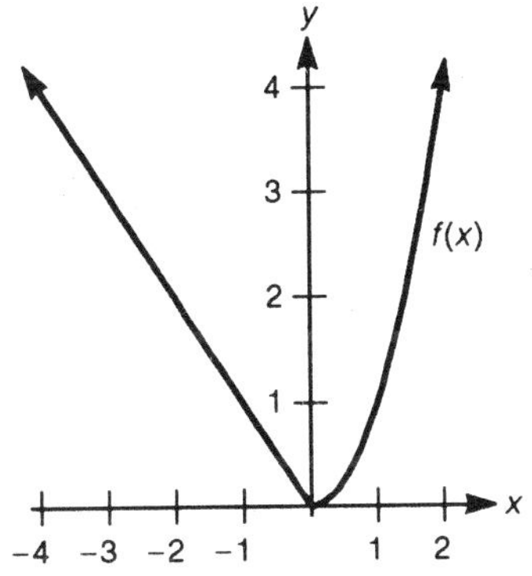

31. We graph the function:

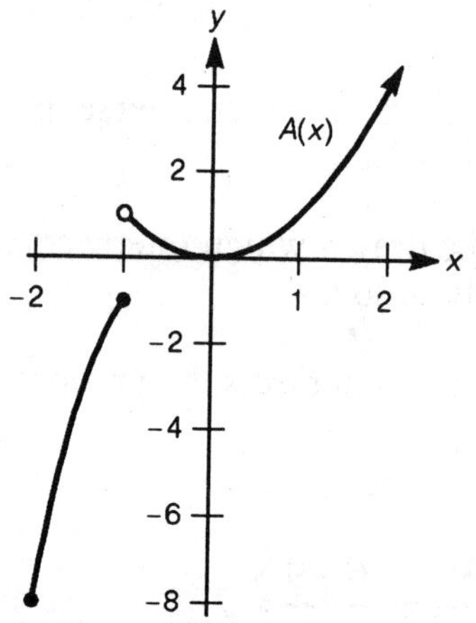

33. We graph the function:

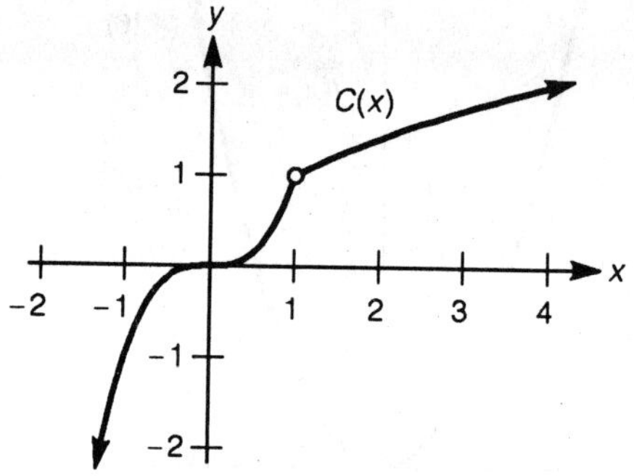

35. (a) We graph the function:

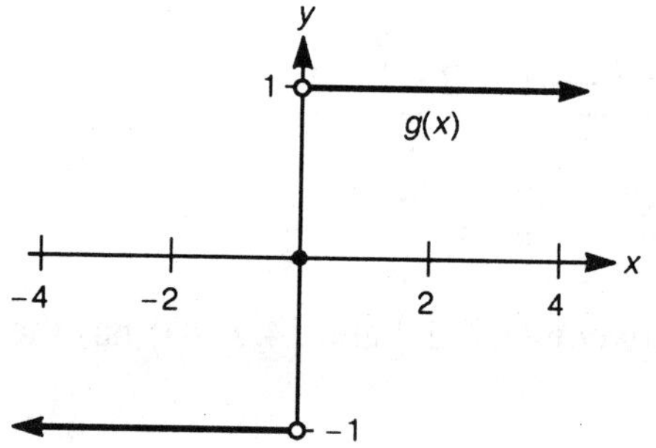

(b) We graph the function:

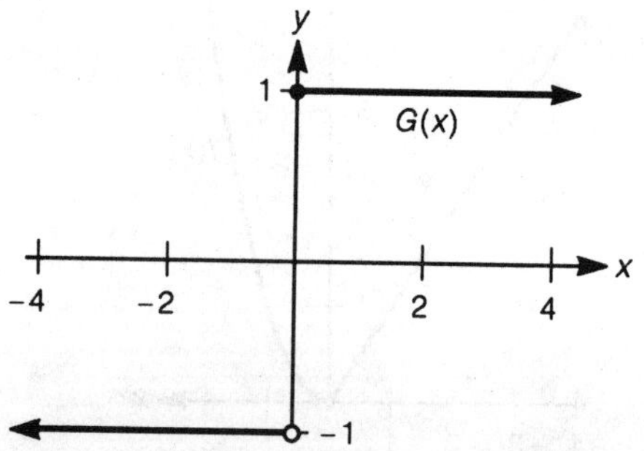

37. We graph the function:

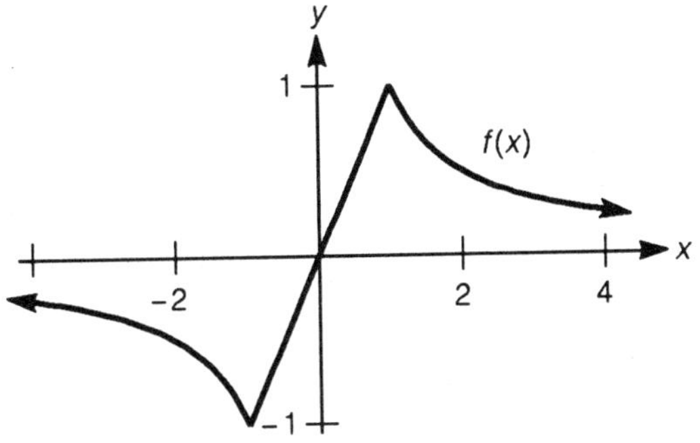

39. Let the coordinates of T be $\left(t, \frac{1}{t}\right)$. Then:

$$m = \frac{\frac{1}{t} - 1}{t - 1} = \frac{1 - t}{t(t-1)} = \frac{-1}{t}$$

Since $m = -\frac{1}{5}$, we have:

$$\frac{-1}{t} = -\frac{1}{5}$$
$$-t = -5$$
$$t = 5$$

So the point T is $\left(5, \frac{1}{5}\right)$. We draw the sketch:

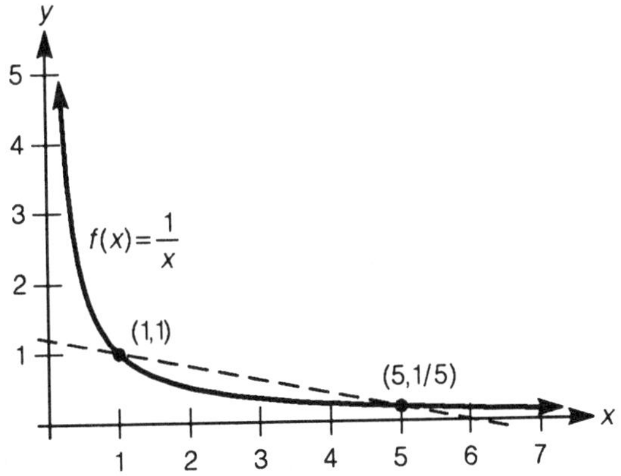

41. The coordinates of B are $(x+h, f(x+h))$. Since A is $(x, f(x))$, then the slope is:

$$m = \frac{f(x+h) - f(x)}{h}$$

43. (a) We can see from the figure that the slope of the line segment joining $(a, f(a))$ and $(b, f(b))$ is $m = \frac{f(b)-f(a)}{b-a}$:

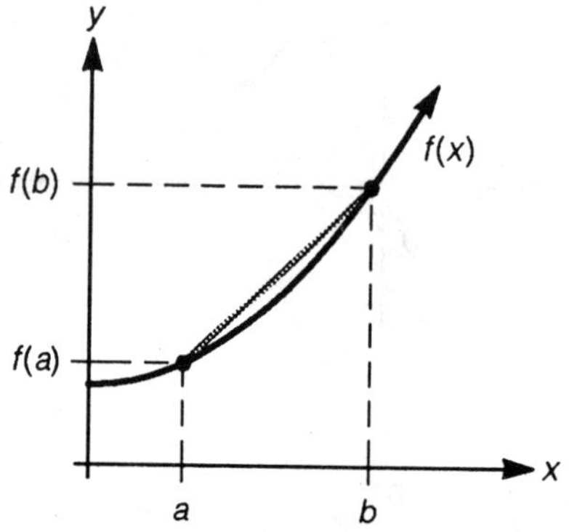

(b) We compute the slope between (0, 0) and (1, 1):

$$m = \frac{1-0}{1-0} = \frac{1}{1} = 1$$

We compute the slope between (10, 100) and (11, 121):

$$m = \frac{121-100}{11-10} = \frac{21}{1} = 21$$

The average rate of change between 10 and 11 is larger.

(c) We compute the average rate of change for each function:

$f(x) = x^2$: the points are (3, 9) and (4, 16):

$$m = \frac{16-9}{4-3} = \frac{7}{1} = 7$$

$g(x) = x^3$: the points are (3, 27) and (4, 64):

$$m = \frac{64-27}{4-3} = \frac{37}{1} = 37$$

$h(x) = \frac{1}{x}$: the points are $\left(3, \frac{1}{3}\right)$ and $\left(4, \frac{1}{4}\right)$:

$$m = \frac{\frac{1}{4}-\frac{1}{3}}{4-3} = \frac{-\frac{1}{12}}{1} = -\frac{1}{12}$$

So $g(x) = x^3$ has the largest average rate of change between 3 and 4.

(d) (i) Points are (1, 1) and (1.1, 1.21):

$$m = \frac{1.21-1}{1.1-1} = \frac{0.21}{0.1} = 2.1$$

(ii) Points are (1, 1) and (1.01, 1.0201):

$$m = \frac{1.0201-1}{1.01-1} = \frac{0.0201}{0.01} = 2.01$$

(iii) Points are (1, 1) and (1.001, 1.002001):

$$m = \frac{1.002001-1}{1.001-1} = \frac{0.002001}{0.001} = 2.001$$

It appears that, as b approaches 1, the average rate of change is approaching 2.

45. (a) We compute $f(2)-f(1)$:

$$\begin{aligned} f(2)-f(1) &= 16(2)^2-16(1)^2 \\ &= 16(4)-16(1) \\ &= 64-16 \\ &= 48 \text{ ft} \end{aligned}$$

Now draw the sketch:

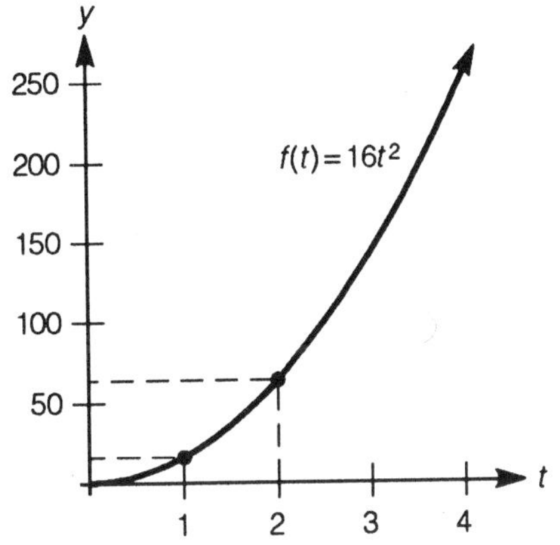

(b) We compute $f(3)-f(2)$:

$$\begin{aligned} f(3)-f(2) &= 16(3)^2-16(2)^2 \\ &= 16(9)-16(4) \\ &= 144-64 \\ &= 80 \text{ ft} \end{aligned}$$

Now draw the sketch:

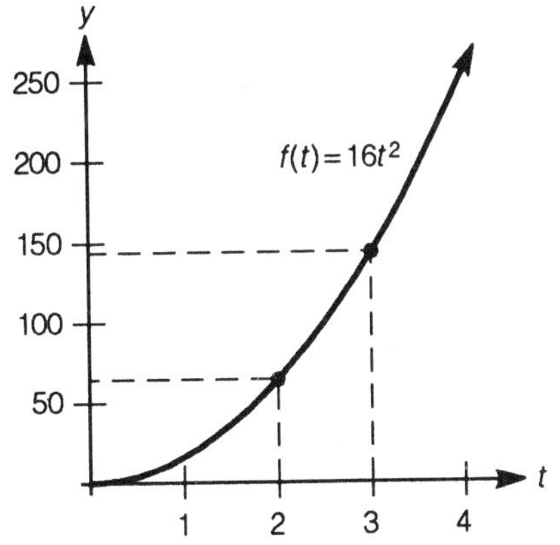

(c) We compute the average velocity:

$$\frac{f(3)-f(1)}{3-1}=\frac{144-16}{3-1}=\frac{128}{2}=64 \text{ ft / sec}$$

This is the same as the slope of the line joining the points $(1, f(1))$ and $(3, f(3))$. We draw the sketch:

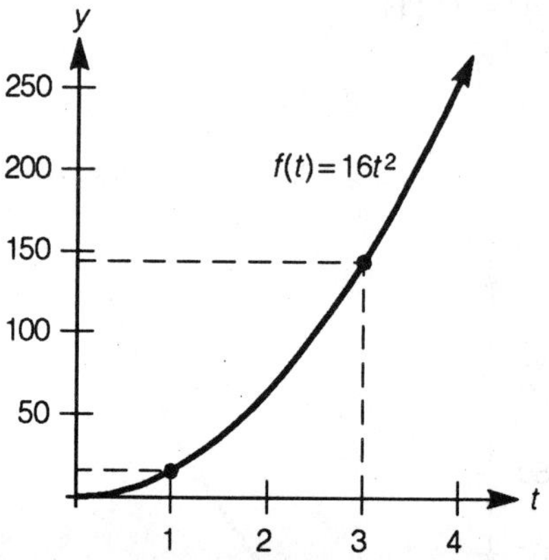

(d) (i) We compute the average velocity:

$$v_{avg} = \frac{f(1.1) - f(1)}{1.1 - 1} = \frac{19.36 - 16}{0.1} = \frac{3.36}{0.1} = 33.6$$

(ii) We compute the average velocity:

$$v_{avg} = \frac{f(1.001) - f(1)}{1.001 - 1} = \frac{16.032016 - 16}{0.001} = 32.016$$

(iii) We compute the average velocity:

$$v_{avg} = \frac{f(1.00001) - f(1)}{1.00001 - 1} = \frac{16.0003200016 - 16}{0.00001} = 32.00016$$

As b approaches 1, the average velocity seems to be approaching 32 ft/sec. Case (iii) gives the clearest indication of what these values are approaching.

4.3 Techniques in Graphing

1. (a) C (b) F (c) I (d) A (e) J (f) K
(g) D (h) B (i) E (j) H (k) G

3. This graph will be that of $y = x^3$ translated down 3 units:

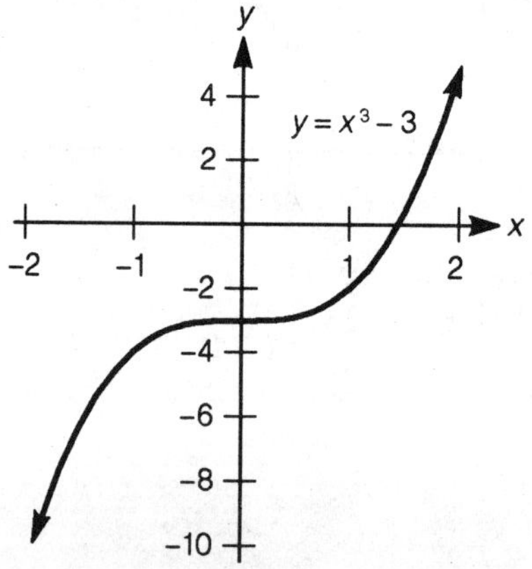

5. This graph will be that of $y = x^2$ translated to the left 4 units:

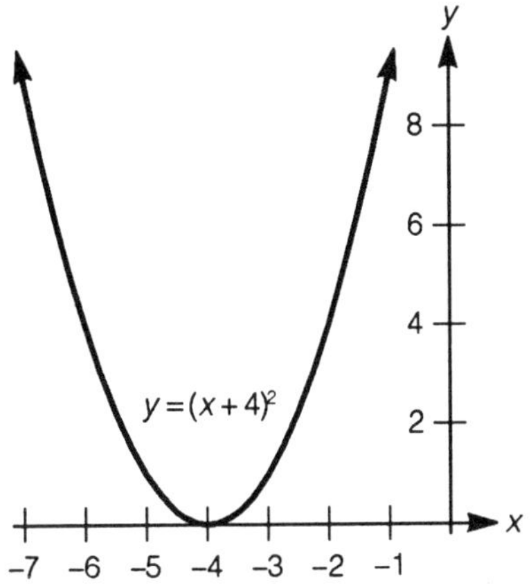

7. This graph will be that of $y = x^2$ translated to the right 4 units:

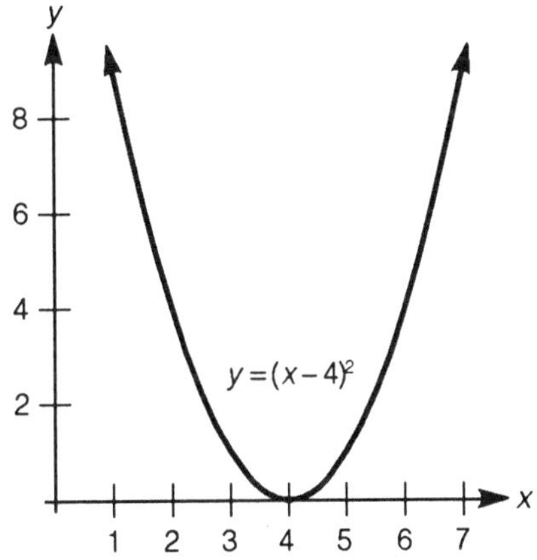

9. This graph will be that of $y = x^2$ reflected across the x-axis:

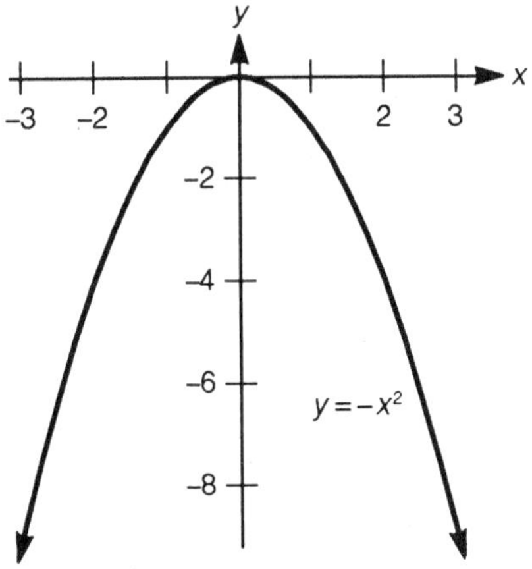

11. This graph will be that of $y = x^2$ translated to the right 3 units, then reflected across the x-axis:

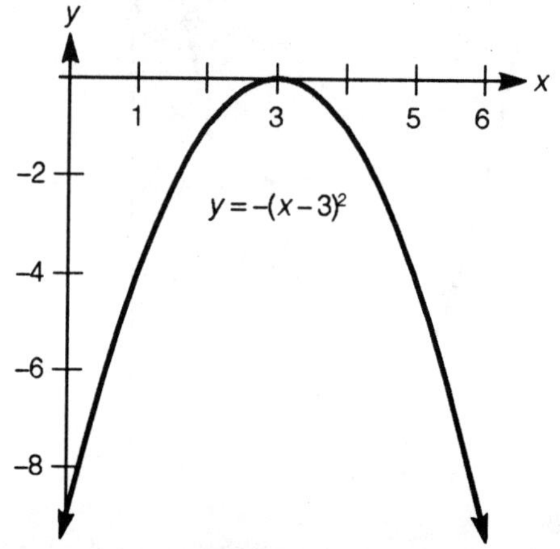

13. This graph will be that of $y = \sqrt{x}$ translated to the right 3 units:

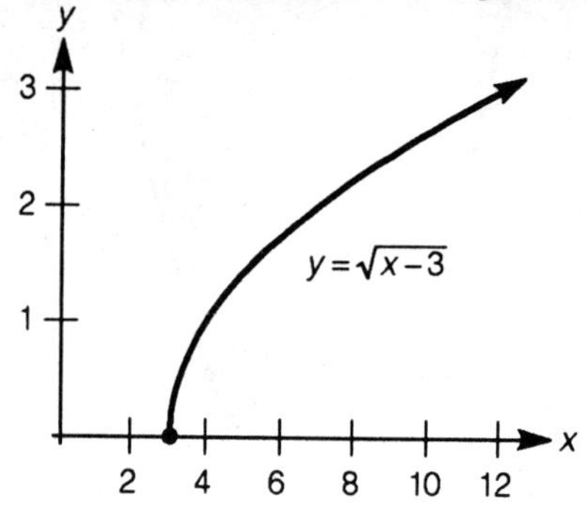

15. This graph will be that of $y = \sqrt{x}$ translated to the left 1 unit, then reflected across the x-axis:

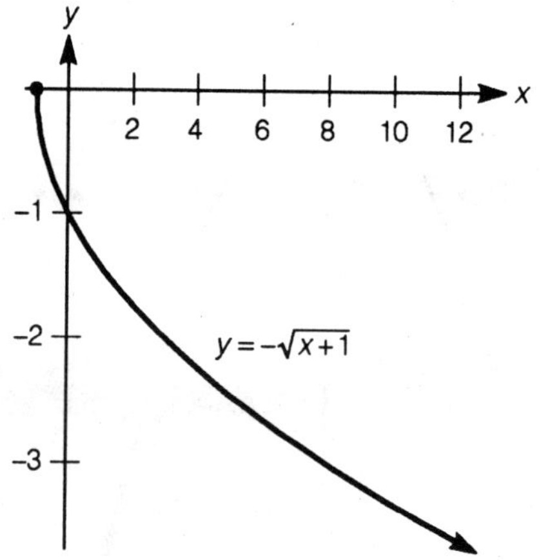

17. This graph will be that of $y = \frac{1}{x}$ translated to the left 2 units, then translated up 2 units:

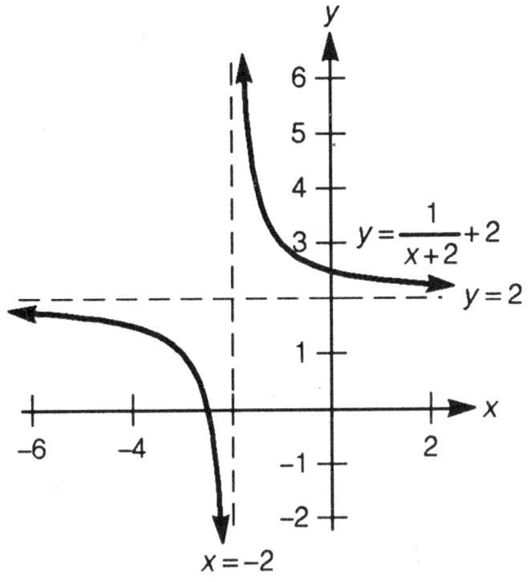

19. This graph will be that of $y = x^3$ translated to the right 2 units:

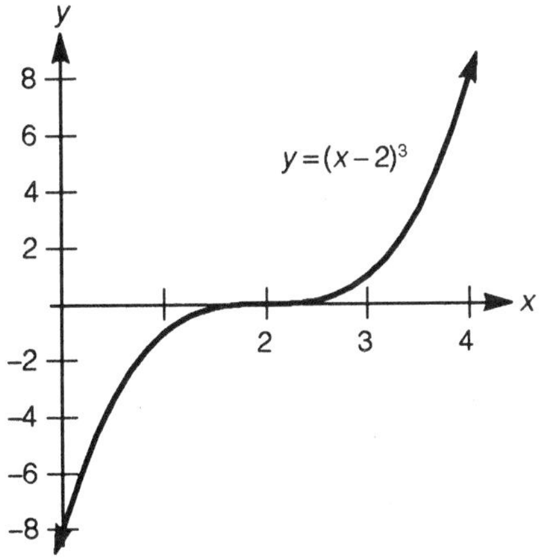

21. This graph will be that of $y = x^3$ reflected across the x-axis, then translated up 4 units:

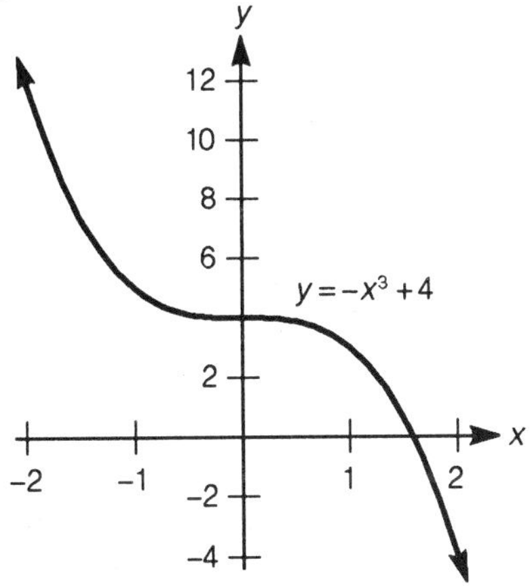

23. (a) This graph will be that of $y = |x|$ translated to the left 4 units:

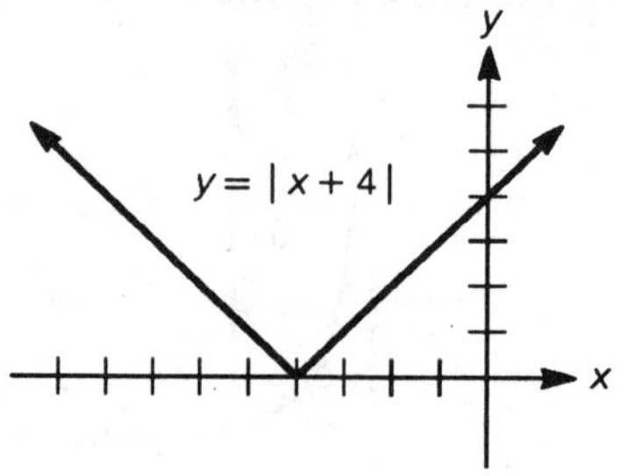

(b) This graph will be that of $y = |x|$ reflected across the y-axis, then translated to the right 4 units. Note that the reflection has no effect on the graph, since $y = |x|$ is symmetric about the y-axis:

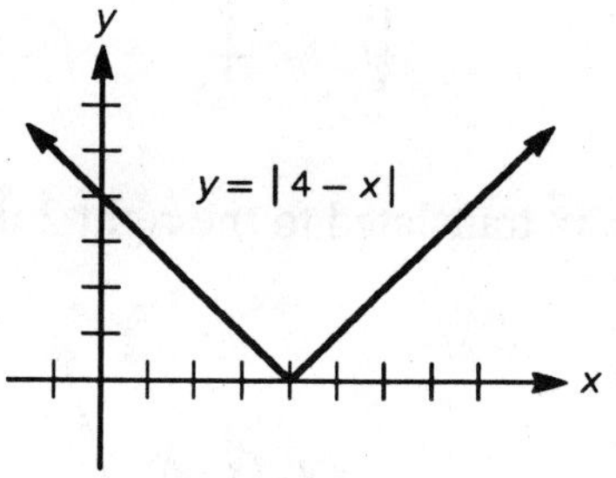

(c) This graph will be that of $y = |x|$ reflected across the y-axis, translated to the right 4 units, reflected across the x-axis, then translated up 1 unit:

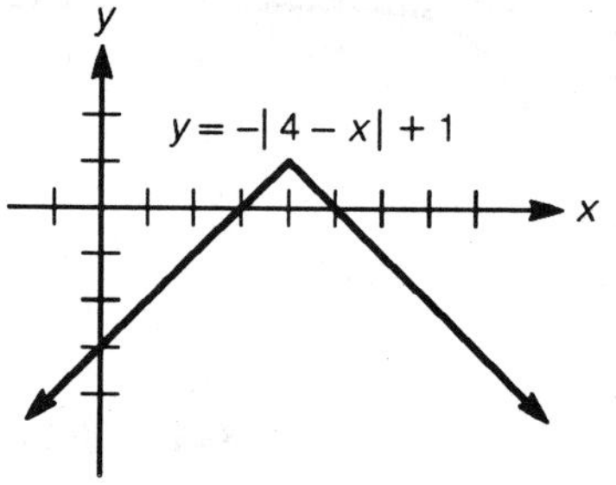

25. This is $f(x) = |x|$ translated to the right 5 units:

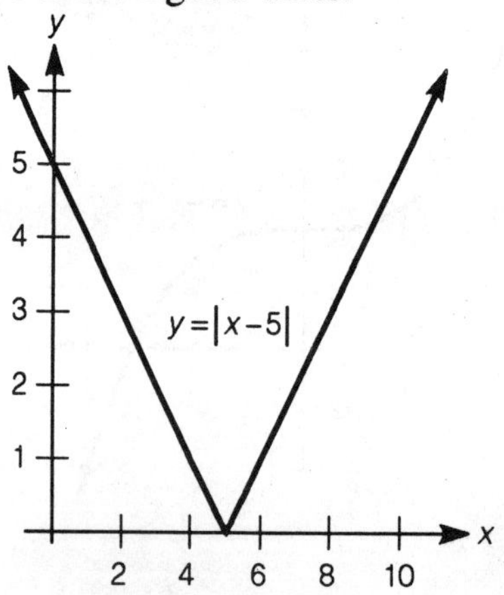

27. This is $f(x) = |x|$ reflected across the y-axis, then translated to the right 5 units. Note that the reflection has no effect on the graph, since $y = |x|$ is symmetric about the y-axis:

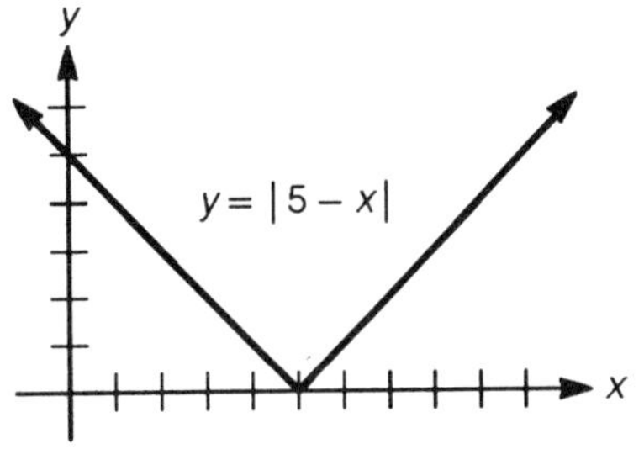

29. This is $f(x) = |x|$ translated to the right 5 units, reflected across the x-axis, then translated up 1 unit:

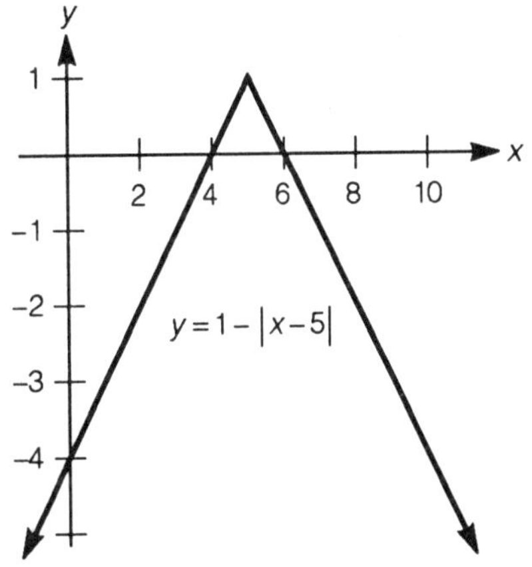

31. This is $F(x) = \frac{1}{x}$ translated to the left 3 units:

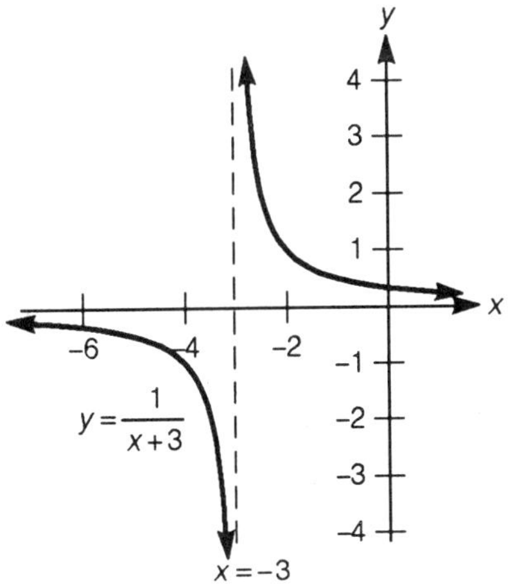

33. This is $F(x) = \dfrac{1}{x}$ translated to the left 3 units, then reflected across the x-axis:

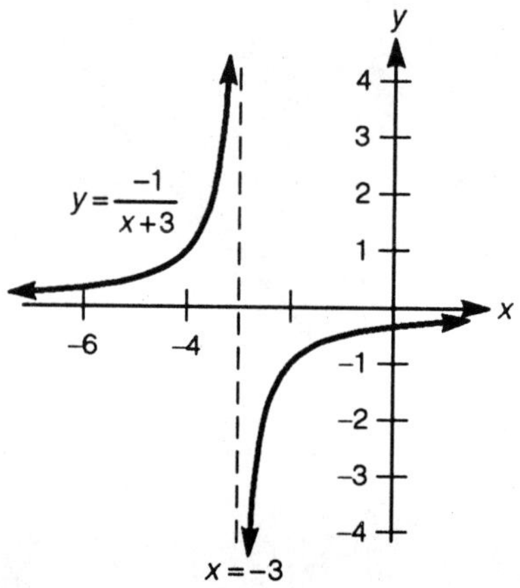

35. This is $g(x) = \sqrt{1 - x^2}$ translated to the right 2 units:

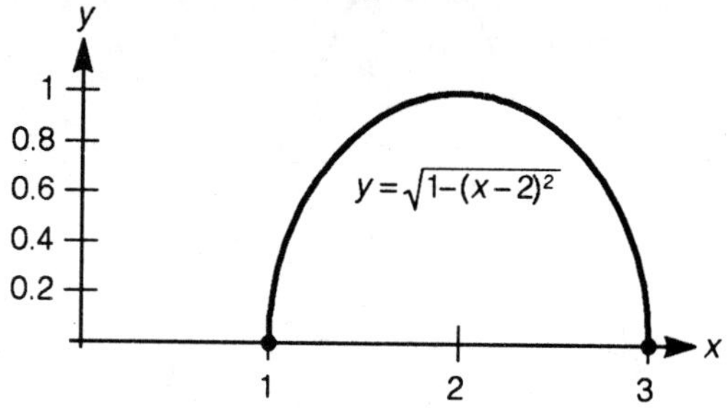

37. This is $g(x) = \sqrt{1 - x^2}$ translated to the right 2 units, reflected across the x-axis, then translated up 1 unit:

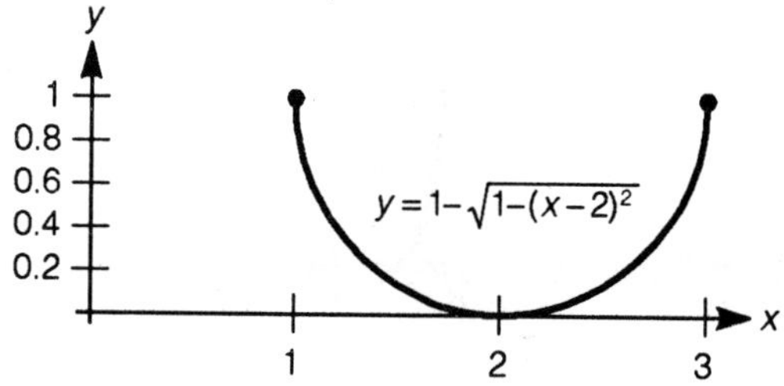

39. This is $g(x) = \sqrt{1 - x^2}$ reflected across the y-axis, then translated to the right 2 units:

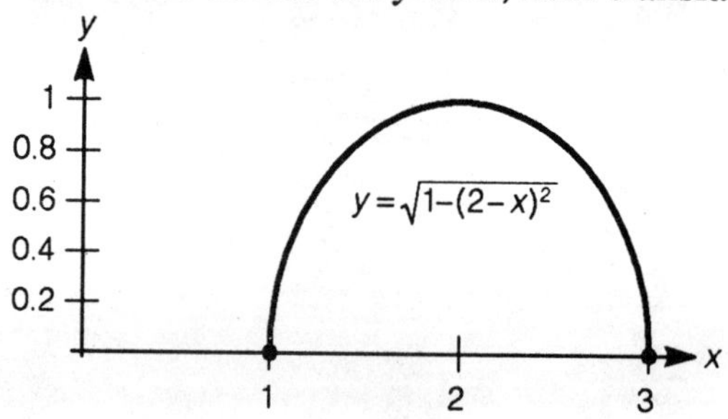

41. (a) This is $y = f(x)$ reflected across the x-axis:

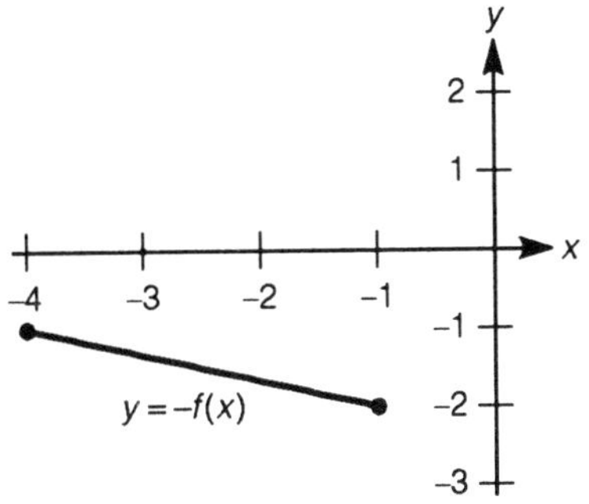

(b) This is $y = f(x)$ reflected across the y-axis:

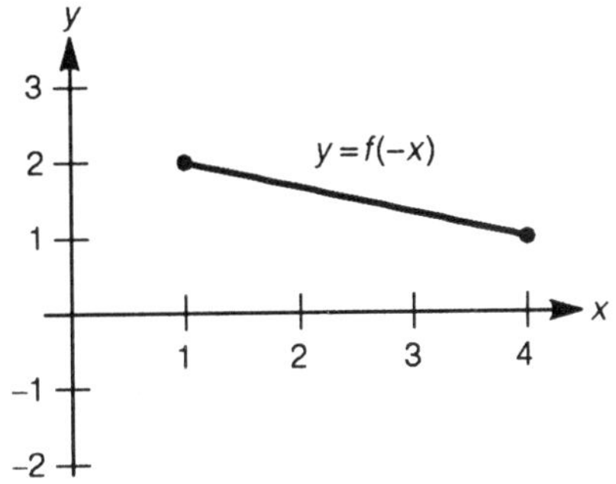

(c) This is $y = f(x)$ reflected across the y-axis, then reflected across the x-axis:

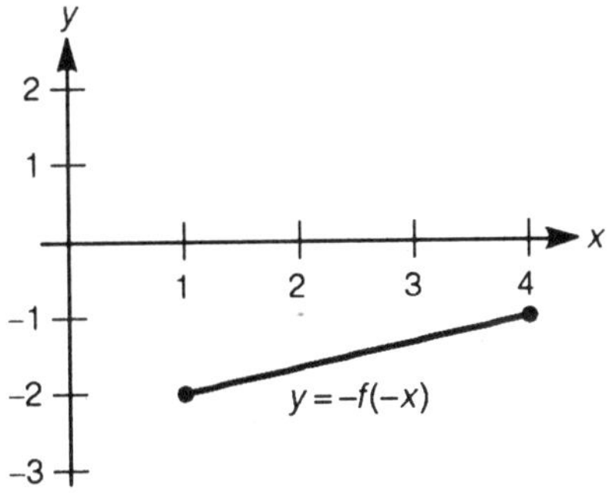

43. (a) This is $y = g(x)$ reflected across the y-axis:

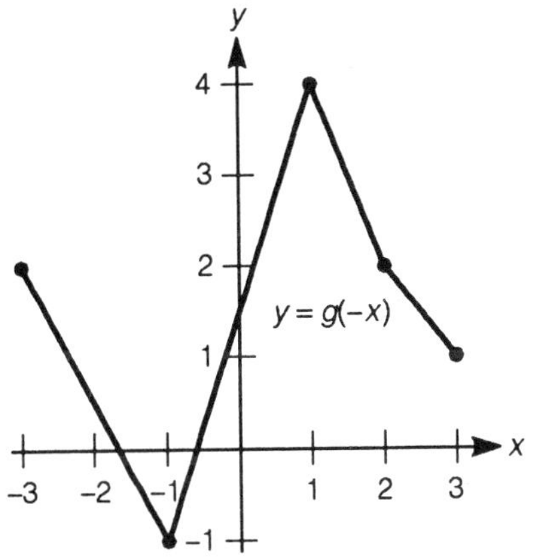

(b) This is $y = g(x)$ reflected across the x-axis:

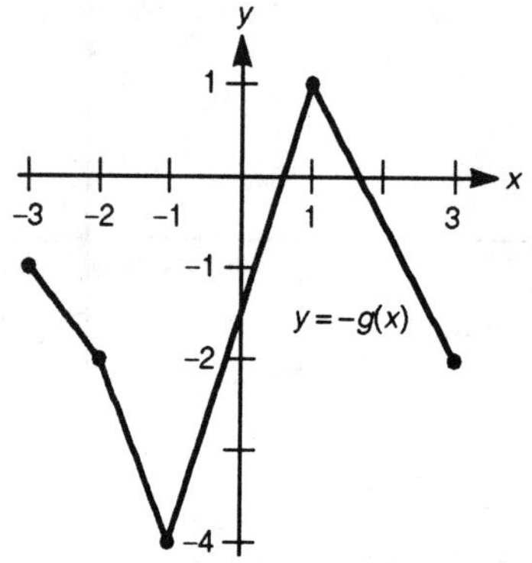

(c) This is $y = g(x)$ reflected across the y-axis, then reflected across the x-axis:

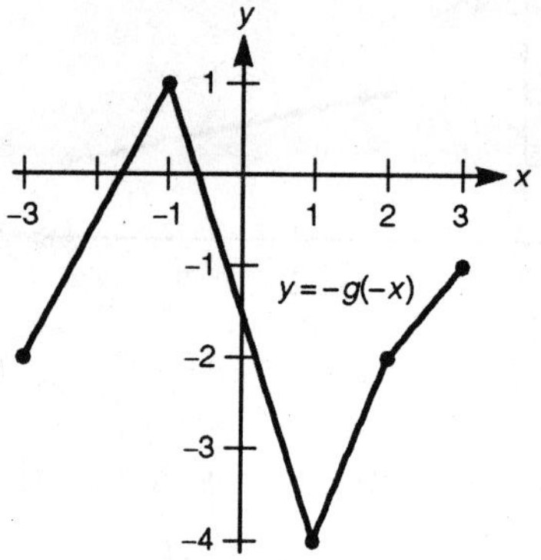

45. (a) We complete the table:

x	x^2	$x^2 - 1$	$x^2 + 1$
0	0	−1	1
±1	1	0	2
±2	4	3	5
±3	9	8	10

(b) Notice that the graph of $y = x^2 - 1$ is a vertical displacement down one unit (from $y = x^2$), while the graph of $y = x^2 + 1$ is a vertical displacement up one unit (from $y = x^2$):

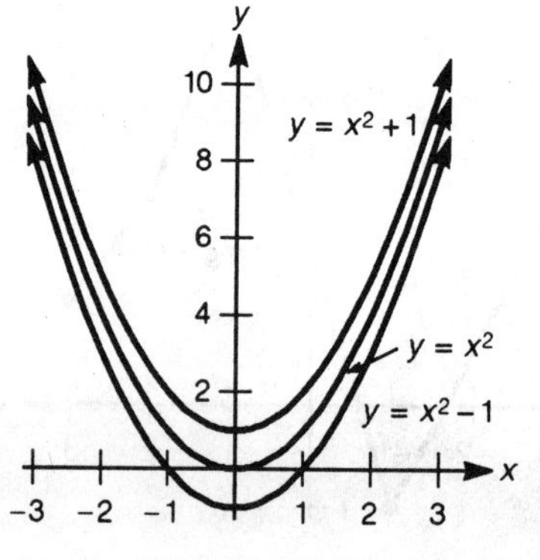

47. (a) We complete the table:

x	$\sqrt{x}$	$-\sqrt{x}$
0	0.0	0.0
1	1.0	−1.0
2	1.4	−1.4
3	1.7	−1.7
4	2.0	−2.0
5	2.2	−2.2

(b) Notice that the graph of $y = -\sqrt{x}$ is a reflection of $y = \sqrt{x}$ across the x-axis:

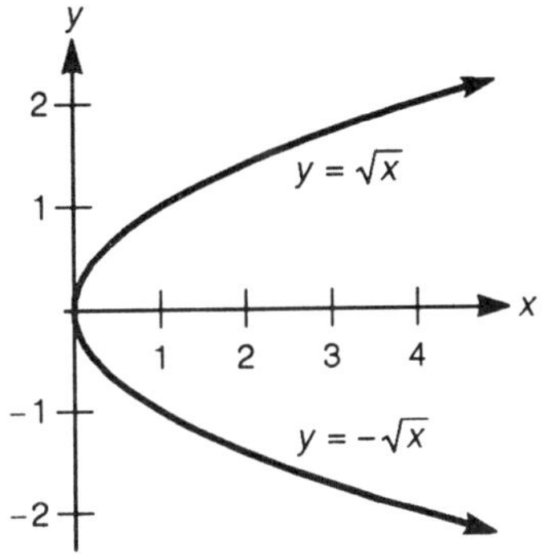

49. (a) The point is $(a+3, b)$, since $f(a+3-3) = f(a) = b$.
(b) The point is $(a, b-3)$, since $f(a) - 3 = b - 3$.
(c) The point is $(a+3, b-3)$, since $f(a+3-3) - 3 = f(a) - 3 = b - 3$.
(d) The point is $(a, -b)$, since $-f(a) = -b$.
(e) The point is $(-a, b)$, since $f[-(-a)] = f(a) = b$.
(f) The point is $(-a, -b)$, since $-f[-(-a)] = -f(a) = -b$.
(g) The point is $(-a+3, b)$, since $f[3-(-a+3)] = f(3+a-3) = f(a) = b$.
(h) The point is $(-a+3, -b+1)$, since $-f[3-(-a+3)] + 1 = -f(a) + 1 = -b + 1$.

51. We work from the right-hand side of the equality:

$$\frac{1}{x-1} + 1 = \frac{1+x-1}{x-1} = \frac{x}{x-1}$$

Since $g(x) = \frac{1}{x-1} + 1$, then we graph $y = \frac{1}{x}$ displaced 1 unit to the right and 1 unit up. See graph:

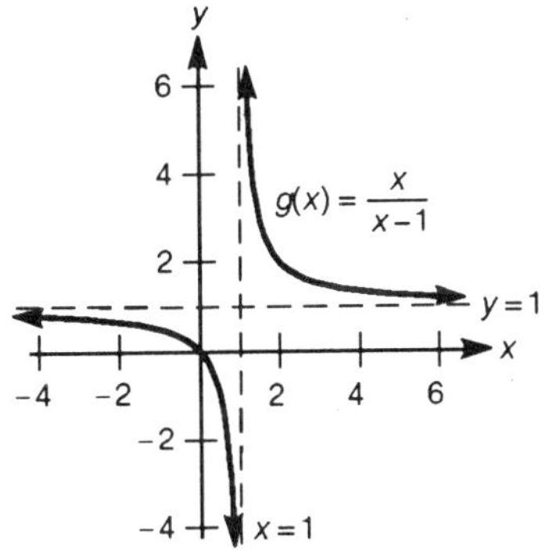

53. We rationalize the denominator (which is valid for $0 < |x| \le 1$):

$$\frac{x^2}{1-\sqrt{1-x^2}} \bullet \frac{1+\sqrt{1-x^2}}{1+\sqrt{1-x^2}} = \frac{x^2\left(1+\sqrt{1-x^2}\right)}{1-1+x^2} = \frac{x^2\left(1+\sqrt{1-x^2}\right)}{x^2} = 1+\sqrt{1-x^2}$$

Now, since $0 \le \sqrt{1-x^2} < 1$, then $1 \le 1+\sqrt{1-x^2} < 2$. So the range is $[1, 2)$.

55. (a) Call $y = f(x)$. We replace x by $-x$ and y by $-y$:

$$\begin{aligned} -y &= f(-x) \\ -y &= -f(x) \qquad [\text{since } f(-x) = -f(x)] \\ y &= f(x) \end{aligned}$$

So the resulting equation is identical to the original equation, and thus the graph of $y = f(x)$ is symmetric about the origin.

(b) (i) We compute $f(-x)$:

$$f(-x) = (-x)^3 = -x^3 = -f(x)$$

(ii) We compute $f(-x)$:

$$\begin{aligned} f(-x) &= -2(-x)^5 + 4(-x)^3 - (-x) \\ &= 2x^5 - 4x^3 + x \\ &= -\left(-2x^5 + 4x^3 - x\right) \\ &= -f(x) \end{aligned}$$

(iii) We compute $f(-x)$:

$$f(-x) = \frac{|-x|}{(-x)+(-x)^7} = \frac{|x|}{-x-x^7} = -\frac{|x|}{x+x^7} = -f(x)$$

57. (a) Since $y = f(-x)$ is a reflection across the y-axis, then the new function will have these intervals reversed with a change in sign:

increasing: $(-\infty, -4) \cup (-2, \infty)$
decreasing: $(-4, -2)$

(b) Since $y = -f(x)$ is a reflection across the x-axis, then the new function will have these intervals reversed:

increasing: $(-\infty, 2) \cup (4, \infty)$
decreasing: $(2, 4)$

4.4 Methods of Combining Functions

1. (a) We compute $(f+g)(x)$:

$$\begin{aligned} (f+g)(x) &= f(x) + g(x) \\ &= (2x-1) + \left(x^2 - 3x - 6\right) \\ &= x^2 - x - 7 \end{aligned}$$

(b) We compute $(f-g)(x)$:

$$\begin{aligned}(f-g)(x) &= f(x)-g(x)\\ &= (2x-1)-\left(x^2-3x-6\right)\\ &= -x^2+5x+5\end{aligned}$$

(c) Using our answer from part (b), we have:

$$\begin{aligned}(f-g)(0) &= -(0)^2+5(0)+5\\ &= 0+0+5\\ &= 5\end{aligned}$$

3. (a) We compute $(m-f)(x)$:

$$\begin{aligned}(m-f)(x) &= m(x)-f(x)\\ &= \left(x^2-9\right)-(2x-1)\\ &= x^2-9-2x+1\\ &= x^2-2x-8\end{aligned}$$

(b) We compute $(f-m)(x)$:

$$\begin{aligned}(f-m)(x) &= f(x)-m(x)\\ &= (2x-1)-\left(x^2-9\right)\\ &= 2x-1-x^2+9\\ &= -x^2+2x+8\end{aligned}$$

5. (a) We compute $(fk)(x)$:

$$(fk)(x) = f(x)k(x) = (2x-1)(2) = 4x-2$$

(b) We compute $(kf)(x)$:

$$(kf)(x) = k(x)f(x) = 2(2x-1) = 4x-2$$

(c) Using our results from parts (a) and (b):

$$(fk)(1)-(kf)(2) = [4(1)-2]-[4(2)-2] = 2-6 = -4$$

7. (a) We compute $(f/m)(x)-(m/f)(x)$:

$$\begin{aligned}\frac{f}{m}(x)-\frac{m}{f}(x) &= \frac{f(x)}{m(x)}-\frac{m(x)}{f(x)}\\ &= \frac{[f(x)]^2-[m(x)]^2}{f(x)m(x)}\\ &= \frac{(2x-1)^2-\left(x^2-9\right)^2}{\left(x^2-9\right)(2x-1)}\\ &= \frac{\left(4x^2-4x+1\right)-\left(x^4-18x^2+81\right)}{2x^3-x^2-18x+9}\\ &= \frac{-x^4+22x^2-4x-80}{2x^3-x^2-18x+9}\end{aligned}$$

(b) Using our result from part (a):

$$\frac{f}{m}(0)-\frac{m}{f}(0)=\frac{-0^4+22(0)^2-4(0)-80}{2(0)^3-0^2-18(0)+9}=-\frac{80}{9}$$

9. (a) We compute $[m \bullet (k-h)](x)$:

$$\begin{aligned}[m \bullet (k-h)](x) &= m(x)[k(x)-h(x)] \\ &= (x^2-9)(2-x^3) \\ &= 2x^2-18-x^5+9x^3 \\ &= -x^5+9x^3+2x^2-18\end{aligned}$$

(b) We compute $(mk)(x)-(mh)(x)$:

$$\begin{aligned}(mk)(x)-(mh)(x) &= m(x)k(x)-m(x)h(x) \\ &= (x^2-9)(2)-(x^2-9)(x^3) \\ &= 2x^2-18-x^5+9x^3 \\ &= -x^5+9x^3+2x^2-18\end{aligned}$$

(c) Using our result from part (b):

$$\begin{aligned}(mk)(-1)-(mh)(-1) &= -(-1)^5+9(-1)^3+2(-1)^2-18 \\ &= -(-1)+9(-1)+2(1)-18 \\ &= 1-9+2-18 \\ &= -24\end{aligned}$$

11. (a) We compute $(f \circ g)(x)$:

$$\begin{aligned}(f \circ g)(x) &= f[g(x)] \\ &= f(-2x-5) \\ &= 3(-2x-5)+1 \\ &= -6x-15+1 \\ &= -6x-14\end{aligned}$$

(b) Using our result from part (a):

$$(f \circ g)(10) = -6(10)-14 = -60-14 = -74$$

(c) We compute $(g \circ f)(x)$:

$$\begin{aligned}(g \circ f)(x) &= g[f(x)] \\ &= g(3x+1) \\ &= -2(3x+1)-5 \\ &= -6x-2-5 \\ &= -6x-7\end{aligned}$$

(d) Using our result from part (a):

$$(g \circ f)(10) = -6(10)-7 = -60-7 = -67$$

13. (a) We find the indicated compositions:

$$(f\circ g)(x)=f[g(x)]=f(1+x)=1-(1+x)=1-1-x=-x$$
$$(f\circ g)(-2)=-(-2)=2$$
$$(g\circ f)(x)=g[f(x)]=g(1-x)=1+(1-x)=2-x$$
$$(g\circ f)(-2)=2-(-2)=2+2=4$$

(b) We find the indicated compositions:

$$\begin{aligned}(f\circ g)(x)&=f[g(x)]\\&=f(2-3x)\\&=(2-3x)^2-3(2-3x)-4\\&=4-12x+9x^2-6+9x-4\\&=9x^2-3x-6\end{aligned}$$
$$(f\circ g)(-2)=9(-2)^2-3(-2)-6=36+6-6=36$$
$$\begin{aligned}(g\circ f)(x)&=g[f(x)]\\&=g\left(x^2-3x-4\right)\\&=2-3\left(x^2-3x-4\right)\\&=2-3x^2+9x+12\\&=-3x^2+9x+14\end{aligned}$$
$$(g\circ f)(-2)=-2(-2)^2+9(-2)+14=-12-18+14=-16$$

(c) We find the indicated compositions:

$$(f\circ g)(x)=f[g(x)]=f\left(1-x^4\right)=\frac{1-x^4}{3}$$
$$(f\circ g)(-2)=\frac{1-(-2)^4}{3}=\frac{1-16}{3}=\frac{-15}{3}=-5$$
$$(g\circ f)(x)=g[f(x)]=g\left(\frac{x}{3}\right)=1-\left(\frac{x}{3}\right)^4=1-\frac{x^4}{81}$$
$$(g\circ f)(-2)=1-\frac{(-2)^4}{81}=1-\frac{16}{81}=\frac{65}{81}$$

(d) We find the indicated compositions:

$$(f\circ g)(x)=f[g(x)]=f\left(x^2+1\right)=2^{x^2+1}$$
$$(f\circ g)(-2)=2^{(-2)^2+1}=2^{4+1}=2^5=32$$
$$(g\circ f)(x)=g[f(x)]=g\left(2^x\right)=\left(2^x\right)^2+1=2^{2x}+1$$
$$(g\circ f)(-2)=2^{2(-2)}+1=2^{-4}+1=\tfrac{1}{16}+1=\tfrac{17}{16}$$

(e) We find the indicated compositions:

$$(f\circ g)(x)=f[g(x)]=f\left(3x^5-4x^2\right)=3x^5-4x^2$$
$$(f\circ g)(-2)=3(-2)^5-4(-2)^2=-96-16=-112$$
$$(g\circ f)(x)=g[f(x)]=g(x)=3x^5-4x^2$$
$$(g\circ f)(-2)=3(-2)^5-4(-2)^2=-96-16=-112$$

(f) We find the indicated compositions:

$$(f\circ g)(x)=f[g(x)]=f\left(\frac{x+4}{3}\right)=3\left(\frac{x+4}{3}\right)-4=x+4-4=x$$

$$(f\circ g)(-2)=-2$$

$$(g\circ f)(x)=g[f(x)]=g(3x-4)=\frac{3x-4+4}{3}=\frac{3x}{3}=x$$

$$(g\circ f)(-2)=-2$$

15. (a) We compute $(F\circ G)(x)$:

$$\begin{aligned}(F\circ G)(x)&=F[G(x)]\\&=F\left(\tfrac{x+1}{x-1}\right)\\&=\frac{3\left(\frac{x+1}{x-1}\right)-4}{3\left(\frac{x+1}{x-1}\right)+3}\\&=\frac{3(x+1)-4(x-1)}{3(x+1)+3(x-1)}\\&=\frac{3x+3-4x+4}{3x+3+3x-3}\\&=\frac{-x+7}{6x}\end{aligned}$$

(b) Using our result from part (a):

$$F[G(t)]=\frac{-t+7}{6t}$$

(c) Using our result from part (a);

$$(F\circ G)(2)=F[G(2)]=\frac{-2+7}{6(2)}=\frac{5}{12}$$

(d) We compute $(G\circ F)(x)$:

$$\begin{aligned}(G\circ F)(x)&=G[F(x)]\\&=G\left(\tfrac{3x-4}{3x+3}\right)\\&=\frac{\frac{3x-4}{3x+3}+1}{\frac{3x-4}{3x+3}-1}\\&=\frac{(3x-4)+1(3x+3)}{(3x-4)-1(3x+3)}\\&=\frac{3x-4+3x+3}{3x-4-3x-3}\\&=\frac{6x-1}{-7}\\&=\frac{1-6x}{7}\end{aligned}$$

(e) Using our result from part (d):

$$G[F(y)] = \frac{1-6y}{7}$$

(f) Using our result from part (d):

$$(G \circ F)(2) = G[F(2)] = \frac{1-6(2)}{7} = -\frac{11}{7}$$

17. (a) We compute $M(7)$ and $M[M(7)]$:

$$M(7) = \frac{2(7)-1}{7-2} = \frac{14-1}{5} = \frac{13}{5}$$

$$M[M(7)] = M\left(\tfrac{13}{5}\right) = \frac{2\left(\frac{13}{5}\right)-1}{\frac{13}{5}-2} = \frac{\frac{26}{5}-1}{\frac{3}{5}} = \frac{\frac{21}{5}}{\frac{3}{5}} = 7$$

(b) We compute $(M \circ M)(x)$:

$$\begin{aligned}(M \circ M)(x) &= M[M(x)] \\ &= M\left(\tfrac{2x-1}{x-2}\right) \\ &= \frac{2\left(\frac{2x-1}{x-2}\right)-1}{\left(\frac{2x-1}{x-2}\right)-2} \\ &= \frac{2(2x-1)-1(x-2)}{(2x-1)-2(x-2)} \\ &= \frac{4x-2-x+2}{2x-1-2x+4} \\ &= \frac{3x}{3} \\ &= x\end{aligned}$$

(c) Using our result from part (b), we have $(M \circ M)(7) = M[M(7)] = 7$. This agrees with our answer from part (a).

19. (a) $f[g(3)] = f(0) = 1$
(b) $g[f(3)] = g(4) = -3$
(c) $f[h(3)] = f(2) = -1$
(d) $(h \circ g)(2) = h[g(2)] = h(1) = 2$
(e) $h\{f[g(3)]\} = h[f(0)] = h(1) = 2$
(f) $(g \circ f \circ h \circ f)(2) = (g \circ f \circ h)(-1) = (g \circ f)(3) = g(4) = -3$

21. (a) We compute each composition:

$$(T \circ I)(x) = T[I(x)] = T(x) = 4x^3 - 3x^2 + 6x - 1$$

$$(I \circ T)(x) = I[T(x)] = I(4x^3 - 3x^2 + 6x - 1) = 4x^3 - 3x^2 + 6x - 1$$

(b) We compute each composition:

$$(G \circ I)(x) = G[I(x)] = G(x) = ax^2 + bx + c$$
$$(I \circ G)(x) = I[G(x)] = I(ax^2 + bx + c) = ax^2 + bx + c$$

(c) In general, given any function $f(x)$ and $I(x) = x$, then $(f \circ I)(x) = f(x)$ and $(I \circ f)(x) = f(x)$. The function $I(x) = x$ is called the identity function.

23. We compute the compositions:

$$(f \circ g)(0) = f[g(0)] = f(3) = 1$$
$$(f \circ g)(1) = f[g(1)] = f(2) = 3$$
$$(f \circ g)(2) = f[g(2)] = f(0) = 2$$
$$(f \circ g)(3) = f[g(3)] = f(4) = \text{undefined}$$
$$(f \circ g)(4) = f[g(4)] = f(-1) = 2$$

Thus we have the table:

x	0	1	2	3	4
$(f \circ g)(x)$	1	3	2	undef.	2

We compute the compositions:

$$(g \circ f)(-1) = g[f(-1)] = g(2) = 0$$
$$(g \circ f)(0) = g[f(0)] = g(2) = 0$$
$$(g \circ f)(1) = g[f(1)] = g(0) = 3$$
$$(g \circ f)(2) = g[f(2)] = g(3) = 4$$
$$(g \circ f)(3) = g[f(3)] = g(1) = 2$$
$$(g \circ f)(4) = g[f(4)] = \text{undefined}$$

Thus we have the table:

x	−1	0	1	2	3	4
$(g \circ f)(x)$	0	0	3	4	2	undef.

25. (a) We compute $(f \circ g)(x)$:

$$(f \circ g)(x) = f[g(x)] = f(3x - 4) = 2(3x - 4) + 1 = 6x - 8 + 1 = 6x - 7$$

We draw the graph:

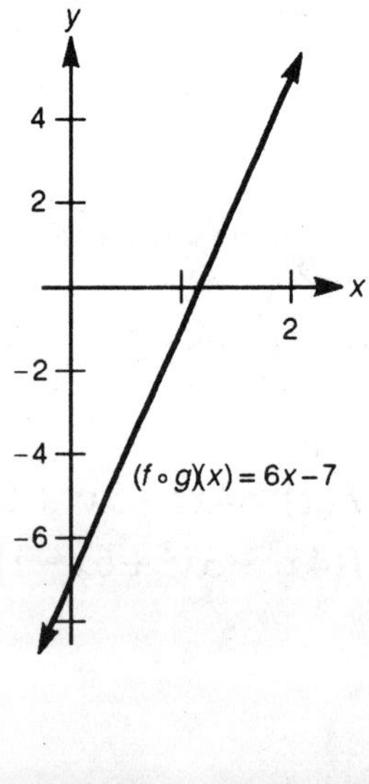

(b) We compute $(g \circ f)(x)$:

$$(g \circ f)(x) = g[f(x)] = g(2x+1) = 3(2x+1) - 4 = 6x + 3 - 4 = 6x - 1$$

We draw the graph:

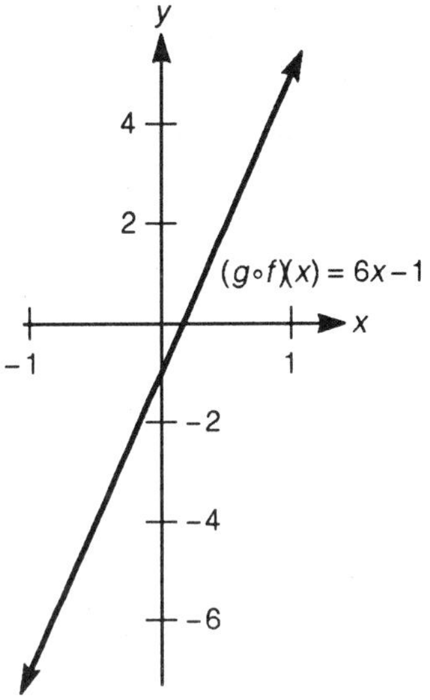

27. (a) The domain is $[0, \infty)$ and the range is $[-3, \infty)$:

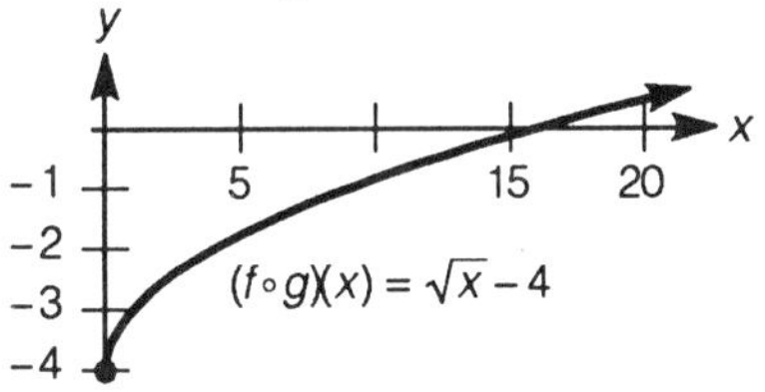

(b) The domain is $(-\infty, \infty)$ and the range is $(-\infty, \infty)$:

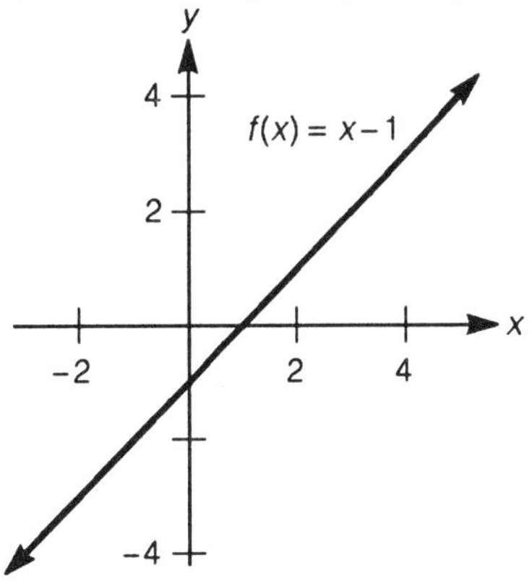

(c) We first compute $(f \circ g)(x)$:

$$(f \circ g)(x) = f[g(x)] = f\left(\sqrt{x} - 3\right) = \left(\sqrt{x} - 3\right) - 1 = \sqrt{x} - 4$$

The domain is $[0,\infty)$ and the range is $[-4,\infty)$:

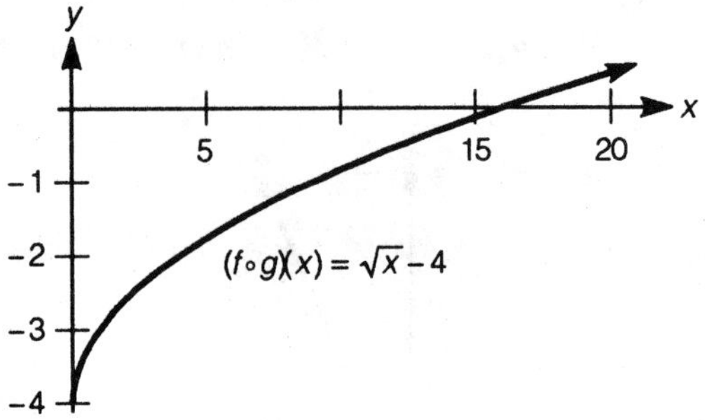

(d) We first compute $g[f(x)]$:

$$g[f(x)] = g(x-1) = \sqrt{x-1} - 3$$

The domain is $[1,\infty)$.

(e) We draw the graph:

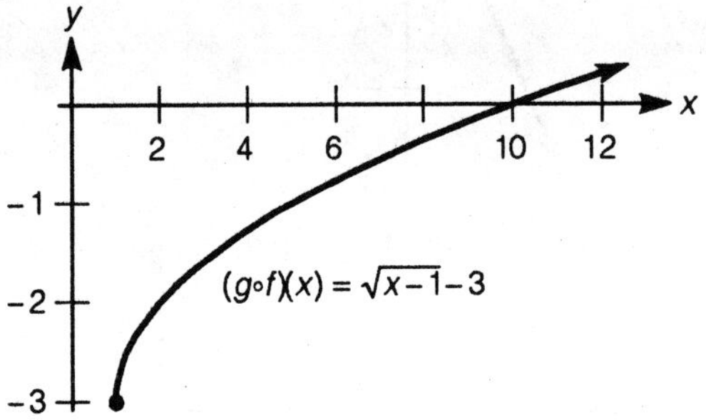

29. Let $f(x) = x^4$ and $g(x) = 3x - 1$. Then $C(x) = (f \circ g)(x)$, since:

$$(f \circ g)(x) = f[g(x)] = f(3x-1) = (3x-1)^4$$

31. (a) Let $f(x) = \sqrt[3]{x}$ and $g(x) = 3x + 4$. Then $F(x) = (f \circ g)(x)$, since:

$$(f \circ g)(x) = f[g(x)] = f(3x+4) = \sqrt[3]{3x+4}$$

(b) Let $f(x) = |x|$ and $g(x) = 2x - 3$. Then $G(x) = (f \circ g)(x)$, since:

$$(f \circ g)(x) = f[g(x)] = f(2x-3) = |2x-3|$$

(c) Let $f(x) = x^5$ and $g(x) = ax + b$. Then $H(x) = (f \circ g)(x)$, since:

$$(f \circ g)(x) = f[g(x)] = f(ax+b) = (ax+b)^5$$

(d) Let $f(x) = \frac{1}{x}$ and $g(x) = \sqrt{x}$. Then $T(x) = (f \circ g)(x)$, since:

$$(f \circ g)(x) = f[g(x)] = f(\sqrt{x}) = \frac{1}{\sqrt{x}}$$

33. (a) We have $f(x) = (b \circ c)(x)$, since:

$$(b \circ c)(x) = b[c(x)] = b(2x+1) = \sqrt[3]{2x+1}$$

(b) We have $g(x) = (a \circ d)(x)$, since:

$$(a \circ d)(x) = a[d(x)] = a\left(x^2\right) = \frac{1}{x^2}$$

(c) We have $h(x) = (c \circ d)(x)$, since:

$$(c \circ d)(x) = c[d(x)] = c\left(x^2\right) = 2x^2 + 1$$

(d) We have $K(x) = (c \circ b)(x)$, since:

$$(c \circ b)(x) = c[b(x)] = c\left(\sqrt[3]{x}\right) = 2\sqrt[3]{x} + 1$$

(e) We have $l(x) = (c \circ a)(x)$, since:

$$(c \circ a)(x) = c[a(x)] = c\left(\frac{1}{x}\right) = 2\left(\frac{1}{x}\right) + 1 = \frac{2}{x} + 1$$

(f) We have $m(x) = (a \circ c)(x)$, since:

$$(a \circ c)(x) = a[c(x)] = a(2x+1) = \frac{1}{2x+1}$$

(g) We have $n(x) = (b \circ d)(x)$, since:

$$(b \circ d)(x) = b[d(x)] = b\left(x^2\right) = \sqrt[3]{x^2} = x^{2/3}$$

Note that we could also use $n(x) = (d \circ b)(x)$, since:

$$(d \circ b)(x) = d[b(x)] = d\left(\sqrt[3]{x}\right) = \left(\sqrt[3]{x}\right)^2 = x^{2/3}$$

35. We first compute $(C \circ f)(t)$:

$$(C \circ f)(t) = C[f(t)] = C\left(\frac{1}{t^2+1}\right) = 2\pi\left(\frac{1}{t^2+1}\right) = \frac{2\pi}{t^2+1}$$

When $t = 3$, we have:

$$(C \circ f)(3) = \frac{2\pi}{3^2+1} = \frac{2\pi}{10} = \frac{\pi}{5} \text{ ft}$$

37. (a) We compute $(C \circ f)(t)$:

$$(C \circ f)(t) = C[f(t)] = C(5t) = 100 + 90(5t) - (5t)^2 = 100 + 450t - 25t^2$$

(b) When $t = 3$ hr, we have:

$$C[f(3)] = 100 + 450(3) - 25(3)^2 = \$1225$$

(c) When $t = 6$ hr, we have:

$$C[f(6)] = 100 + 450(6) - 25(6)^2 = \$1900$$

No, the cost is not twice as much for 6 hours.

39. Call $y = f(x)$. Since $(g \circ f)(x) = g[f(x)] = g(y)$, then:

$$\begin{aligned} g(y) &= x + 5 \\ 4y - 1 &= x + 5 \\ 4y &= x + 6 \\ y &= \frac{x+6}{4} \end{aligned}$$

So $f(x) = \frac{x+6}{4}$.

41. We set $f[g(x)] = x$:

$$\begin{aligned} f(ax+b) &= x \\ -2(ax+b)+1 &= x \\ -2ax - 2b + 1 &= x \end{aligned}$$

Since a and b are constants, we can equate components:

$$\begin{aligned} -2a &= 1 \\ a &= -\tfrac{1}{2} \end{aligned} \qquad \text{and} \qquad \begin{aligned} -2b + 1 &= 0 \\ -2b &= -1 \\ b &= \tfrac{1}{2} \end{aligned}$$

So $a = -\frac{1}{2}$ and $b = \frac{1}{2}$.

43. (a) We compute the difference quotient:

$$\begin{aligned} \frac{f[g(x)] - f[g(a)]}{g(x) - g(a)} &= \frac{f(2x-1) - f(2a-1)}{(2x-1) - (2a-1)} \\ &= \frac{(2x-1)^2 - (2a-1)^2}{2x - 1 - 2a + 1} \\ &= \frac{[(2x-1) + (2a-1)][(2x-1) - (2a-1)]}{2x - 2a} \\ &= \frac{(2x + 2a - 2)(2x - 2a)}{2x - 2a} \\ &= 2x + 2a - 2 \end{aligned}$$

(b) We compute the difference quotient:

$$\frac{f[g(x)] - f[g(a)]}{x - a} = \frac{(2x + 2a - 2)(2x - 2a)}{x - a} = \frac{4(x + a - 1)(x - a)}{x - a} = 4x + 4a - 4$$

45. (a) We compute $(g \circ h \circ f)(x)$:

$$(g \circ h \circ f)(x) = g\{h[f(x)]\} = g[h(x^2)] = g\left(\frac{x^2}{2}\right) = \frac{x^2}{2} + 1$$

(b) We compute $(h \circ f \circ g)(x)$:

$$(h \circ f \circ g)(x) = h\{f[g(x)]\} = h[f(x+1)] = h[(x+1)^2] = \frac{(x+1)^2}{2}$$

(c) We compute $(g \circ f \circ h)(x)$:

$$(g \circ f \circ h)(x) = g\{f[h(x)]\} = g\left[f\left(\frac{x}{2}\right)\right] = g\left[\left(\frac{x}{2}\right)^2\right] = g\left(\frac{x^2}{4}\right) = \frac{x^2}{4} + 1$$

(d) We compute $(f \circ h \circ g)(x)$:

$$(f \circ h \circ g)(x) = f\{h[g(x)]\} = f[h(x+1)] = f\left(\frac{x+1}{2}\right) = \left(\frac{x+1}{2}\right)^2 = \frac{(x+1)^2}{4}$$

(e) We compute $(h \circ g \circ f)(x)$:

$$(h \circ g \circ f)(x) = h\{g[f(x)]\} = h[g(x^2)] = h(x^2+1) = \frac{x^2+1}{2}$$

47. (a) We have $p(x) = (g \circ f \circ h)(x)$, since:

$$(g \circ f \circ h)(x) = g\{f[h(x)]\} = g[f(3x)] = g[(3x)^2] = g(9x^2) = 1 - 9x^2$$

(b) We have $q(x) = (h \circ g \circ f)(x)$, since:

$$(h \circ g \circ f)(x) = h\{g[f(x)]\} = h[g(x^2)] = h(1-x^2) = 3(1-x^2) = 3 - 3x^2$$

(c) We have $r(x) = (f \circ g \circ h)(x)$, since:

$$(f \circ g \circ h)(x) = f\{g[h(x)]\} = f[g(3x)] = f(1-3x) = (1-3x)^2 = 1 - 6x + 9x^2$$

(d) We have $s(x) = (h \circ f \circ g)(x)$, since:

$$\begin{aligned}(h \circ f \circ g)(x) &= h\{f[g(x)]\} \\ &= h[f(1-x)] \\ &= h[(1-x)^2] \\ &= h(1-2x+x^2) \\ &= 3(1-2x+x^2) \\ &= 3 - 6x + 3x^2\end{aligned}$$

49. (a) We compute $F'[G(x)] \bullet G'(x)$:

$$F'[G(x)] \bullet G'(x) = F'(x^2+2x+2) \bullet (2x+2)$$
$$= \frac{1}{2\sqrt{x^2+2x+2}} \bullet (2x+2)$$
$$= \frac{x+1}{\sqrt{x^2+2x+2}}$$

(b) Using our result from part (a), we have:

$$F'[G(9)] \bullet G'(9) = \frac{9+1}{\sqrt{9^2+2(9)+2}} = \frac{10}{\sqrt{81+18+2}} = \frac{10}{\sqrt{101}} = \frac{10\sqrt{101}}{101}$$

51. (a) We have the following compositions:

$$(i \circ i)(x) = i(x)$$
$$(i \circ a)(x) = i(-x) = -x = a(x)$$
$$(i \circ b)(x) = i\left(\frac{1}{x}\right) = \frac{1}{x} = b(x)$$
$$(i \circ c)(x) = i\left(-\frac{1}{x}\right) = -\frac{1}{x} = c(x)$$
$$(b \circ i)(x) = b(x)$$
$$(b \circ a)(x) = b(-x) = -\frac{1}{x} = c(x)$$
$$(b \circ b)(x) = b\left(\frac{1}{x}\right) = x = i(x)$$
$$(b \circ c)(x) = b\left(-\frac{1}{x}\right) = -x = a(x)$$
$$(a \circ i)(x) = a(x)$$
$$(a \circ a)(x) = a(-x) = x = i(x)$$
$$(a \circ b)(x) = a\left(\frac{1}{x}\right) = -\frac{1}{x} = c(x)$$
$$(a \circ c)(x) = a\left(-\frac{1}{x}\right) = \frac{1}{x} = b(x)$$
$$(c \circ i)(x) = c(x)$$
$$(c \circ a)(x) = c(-x) = \frac{1}{x} = b(x)$$
$$(c \circ b)(x) = c\left(\frac{1}{x}\right) = -x = a(x)$$
$$(c \circ c)(x) = c\left(-\frac{1}{x}\right) = x = i(x)$$

So we have the composition table:

$\circ$	i	a	b	c
i	i	a	b	c
a	a	i	c	b
b	b	c	i	a
c	c	b	a	i

(b) yes

(c) $a^2 = i$; $b^2 = i$; $c^2 = i$; $c^3 = c \circ c^2 = c \circ i = c$

(d) Since $(a \circ b) \circ c = (c) \circ c = i$ and $a \circ (b \circ c) = a \circ (a) = i$, then $(a \circ b) \circ c = a \circ (b \circ c)$.

(e) Since $i^2 = a^2 = b^2 = c^2 = i$, then the range of f is $\{i\}$.

(f) We compute:

$$f(a \circ b) = f(c) = c^2 = i$$
$$[f(a)] \circ [f(b)] = a^2 \circ b^2 = i \circ i = i,$$

So $f(a \circ b) = [f(a)] \circ [f(b)]$.

(g) We compute:

$$f(a \circ b \circ c) = f(a \circ a) = f(i) = i^2 = i$$
$$[f(a)] \circ [f(b)] \circ [f(c)] = a^2 \circ b^2 \circ c^2 = i \circ (i \circ i) = i,$$

So $f(a \circ b \circ c) = [f(a)] \circ [f(b)] \circ [f(c)]$.

4.5 Inverse Functions

1. (a) We must show that $f[g(x)] = x$ and $g[f(x)] = x$:

$$f[g(x)] = f\left(\tfrac{x}{3}\right) = 3\left(\tfrac{x}{3}\right) = x$$
$$g[f(x)] = g(3x) = \tfrac{3x}{3} = x$$

So $f(x)$ and $g(x)$ are inverse functions.

(b) We must show that $f[g(x)] = x$ and $g[f(x)] = x$:

$$f[g(x)] = f\left(\tfrac{x+1}{4}\right) = 4\left(\tfrac{x+1}{4}\right) - 1 = x + 1 - 1 = x$$
$$g[f(x)] = g(4x-1) = \tfrac{(4x-1)+1}{4} = \tfrac{4x}{4} = x$$

So $f(x)$ and $g(x)$ are inverse functions.

(c) We must show that $g[h(x)] = x$ and $h[g(x)] = x$:

$$g[h(x)] = g(x^2) = \sqrt{x^2} = x, \quad \text{since } x \geq 0$$
$$h[g(x)] = h(\sqrt{x}) = (\sqrt{x})^2 = x$$

So $g(x)$ and $h(x)$ are inverse functions.

3. (a) Since $f[f^{-1}(x)] = x$ for all x in the domain of f^{-1}, then $f[f^{-1}(4)] = 4$ since the domain of f^{-1} is $(-\infty, \infty)$.

(b) Since $f^{-1}[f(x)] = x$ as long as $f(x)$ is in the domain of f^{-1}, then $f^{-1}[f(-1)] = -1$ since the domain of f^{-1} is $(-\infty, \infty)$.

(c) Since $f[f^{-1}(x)] = x$ for all x in the domain of f^{-1}, then $f[f^{-1}(\sqrt{2})] = \sqrt{2}$ since the domain of f^{-1} is $(-\infty, \infty)$.

(d) Since $f[f^{-1}(x)] = x$ for all x in the domain of f^{-1}, then $f[f^{-1}(t+1)] = t + 1$ since the domain of f^{-1} is $(-\infty, \infty)$.

(e) We first compute $f(0)$:

$$f(0) = (0)^3 + 2(0) + 1 = 0 + 0 + 1 = 1$$

Since the inverse function interchanges inputs and outputs, and $f(0) = 1$, then $f^{-1}(1) = 0$.

(f) We first compute $f(-1)$:

$$f(-1) = (-1)^3 + 2(-1) + 1 = -1 - 2 + 1 = -2$$

Since the inverse function interchanges inputs and outputs, and $f(-1) = -2$, then $f^{-1}(-2) = -1$.

5. (a) Let $y = 3x - 1$. We switch the roles of x and y and solve the resulting equation for y:

$$x = 3y - 1$$
$$3y = x + 1$$
$$y = \tfrac{x+1}{3}$$

So the inverse is $f^{-1}(x) = \tfrac{x+1}{3}$.

(b) We verify that $f[f^{-1}(x)] = x$ and $f^{-1}[f(x)] = x$:

$$f\left[f^{-1}(x)\right] = f\left(\tfrac{x+1}{3}\right) = 3\left(\tfrac{x+1}{3}\right) - 1 = x + 1 - 1 = x$$
$$f^{-1}\left[f(x)\right] = f^{-1}(3x-1) = \frac{(3x-1)+1}{3} = \frac{3x}{3} = x$$

(c) The graphs of each line are given below. Note the symmetry of the two lines about the line $y = x$:

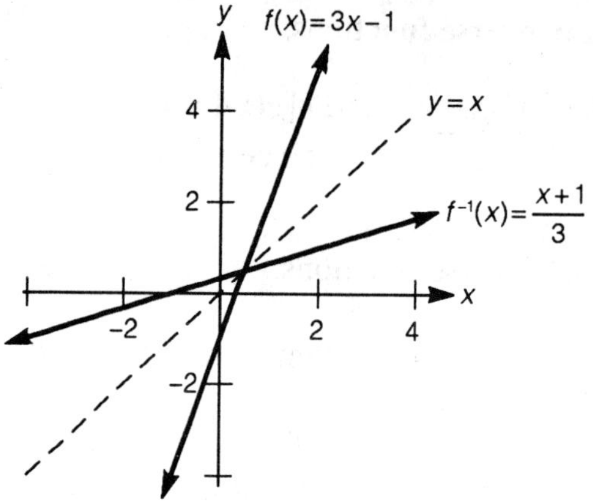

7. (a) Let $y = \sqrt{x-1}$. We switch the roles of x and y and solve the resulting equation for y:

$$x = \sqrt{y-1}$$
$$x^2 = y - 1$$
$$y = x^2 + 1$$

So the inverse is $f^{-1}(x) = x^2 + 1$ for $x \geq 0$.

(b) We verify that $f[f^{-1}(x)] = x$ and $f^{-1}[f(x)] = x$:

$$f\left[f^{-1}(x)\right] = f\left(x^2+1\right) = \sqrt{\left(x^2+1\right)-1} = \sqrt{x^2} = x \quad \text{(since } x \geq 0\text{)}$$
$$f^{-1}\left[f(x)\right] = f^{-1}\left(\sqrt{x-1}\right) = \left(\sqrt{x-1}\right)^2 + 1 = x - 1 + 1 = x$$

(c) The graphs of each curve are given below. Note the symmetry of the curves about the line $y = x$:

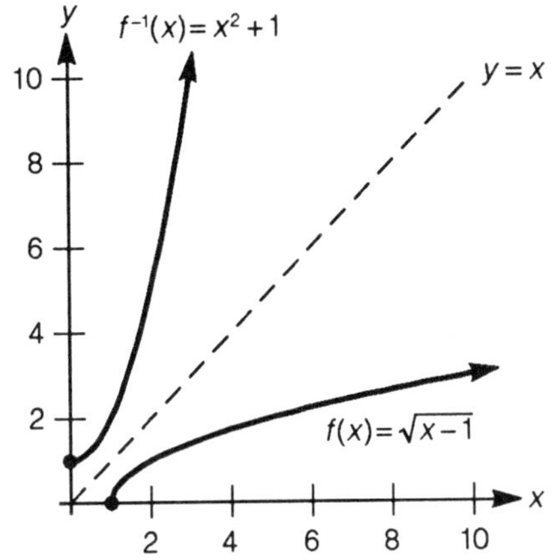

9. (a) Since the denominator is 0 when $x = 3$, the domain of f is the set of all real numbers except 3. To find the range of f, we first solve for x:

$$\begin{aligned} y &= \frac{x+2}{x-3} \\ y(x-3) &= x+2 \\ yx - 3y &= x+2 \\ yx - x &= 3y+2 \\ x(y-1) &= 3y+2 \\ x &= \frac{3y+2}{y-1} \end{aligned}$$

Since the denominator is 0 when $y = 1$, the range of f is the set of all real numbers except 1.

(b) Let $y = \dfrac{x+2}{x-3}$. We switch the roles of x and y and solve the resulting equation for y:

$$\begin{aligned} x &= \frac{y+2}{y-3} \\ x(y-3) &= y+2 \\ xy - 3x &= y+2 \\ xy - y &= 3x+2 \\ y(x-1) &= 3x+2 \\ y &= \frac{3x+2}{x-1} \end{aligned}$$

So the inverse is $f^{-1}(x) = \dfrac{3x+2}{x-1}$.

(c) Since the denominator is 0 when $x = 1$, the domain of f^{-1} is the set of all real numbers except 1. To find the range of f^{-1}, we first solve for x:

$$y = \frac{3x+2}{x-1}$$
$$y(x-1) = 3x+2$$
$$yx - y = 3x+2$$
$$yx - 3x = y+2$$
$$x(y-3) = y+2$$
$$x = \frac{y+2}{y-3}$$

Since the denominator is 0 when $y = 3$, the range of f^{-1} is the set of all real numbers except 3. We observe that the domain of f is equal to the range of f^{-1}, and that the range of f is equal to the domain of f^{-1}.

11. Let $y = 2x^3 + 1$. We switch the roles of x and y and solve the resulting equation for y:

$$x = 2y^3 + 1$$
$$2y^3 = x - 1$$
$$y^3 = \tfrac{x-1}{2}$$
$$y = \sqrt[3]{\tfrac{x-1}{2}}$$

So the inverse is $f^{-1}(x) = \sqrt[3]{\frac{x-1}{2}}$.

13. We compute $f[f^{-1}(x)]$ and $f^{-1}[f(x)]$:

$$f[f^{-1}(x)] = f\left(\frac{x+4}{3}\right) = 3\left(\frac{x+4}{3}\right) - 4 = x + 4 - 4 = x$$
$$f^{-1}[f(x)] = f^{-1}(3x-4) = \frac{(3x-4)+4}{3} = \frac{3x}{3} = x$$

This verfies that f and f^{-1} are inverse functions. We sketch the graph:

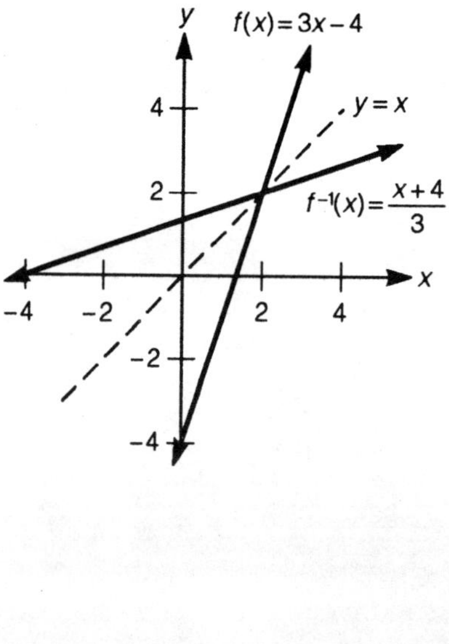

15. (a) Let $y = (x-3)^3 - 1$. We switch the roles of x and y and solve the resulting equation for y:

$$x = (y-3)^3 - 1$$
$$(y-3)^3 = x+1$$
$$y-3 = \sqrt[3]{x+1}$$
$$y = \sqrt[3]{x+1}+3$$

So the inverse is $f^{-1}(x) = \sqrt[3]{x+1}+3$.

(b) Note that $f(x)$ and $f^{-1}(x)$ are symmetric about the line $y = x$:

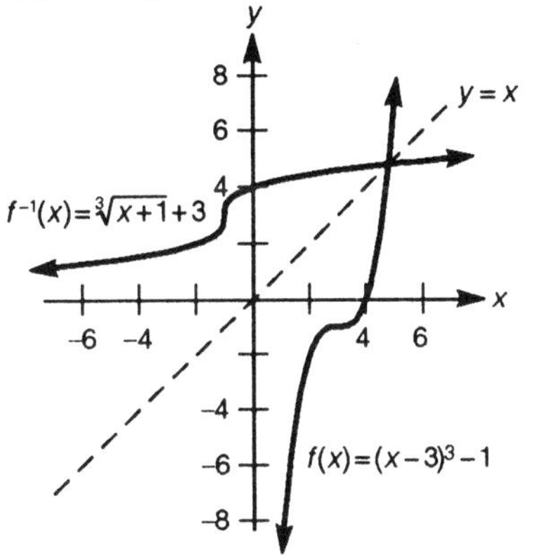

17. (a) The graph of $y = g^{-1}(x)$ is a reflection of $g(x)$ across the line $y = x$:

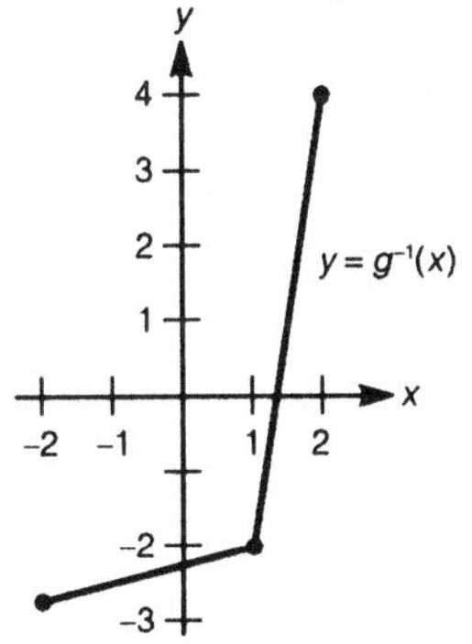

(b) The graph of $y = g^{-1}(x) - 1$ is a displacement of $g^{-1}(x)$ down 1 unit:

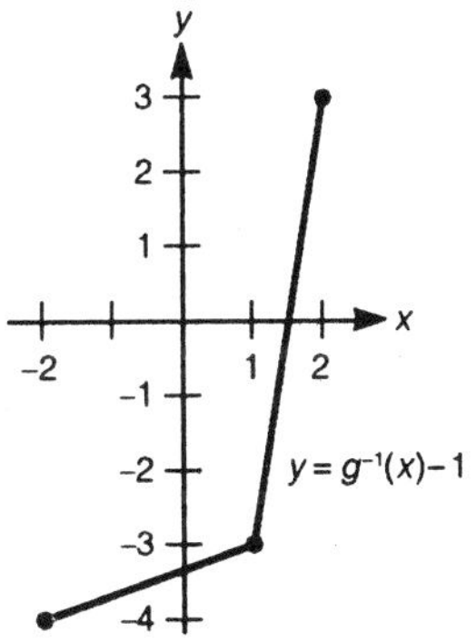

(c) The graph of $y = g^{-1}(x-1)$ is a displacement of $g^{-1}(x)$ to the right 1 unit:

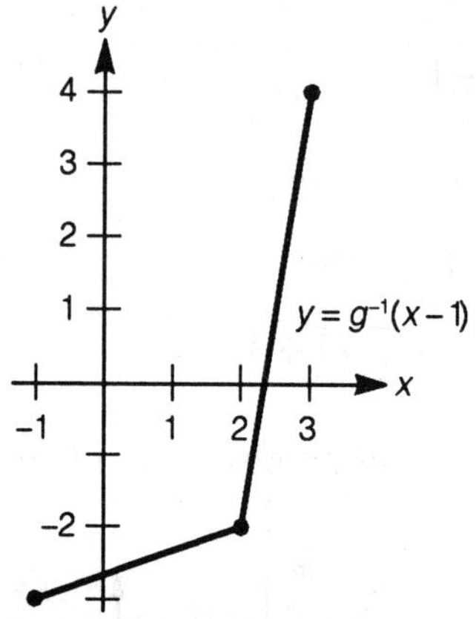

(d) The graph of $y = g^{-1}(-x)$ is a reflection of $g^{-1}(x)$ across the y-axis:

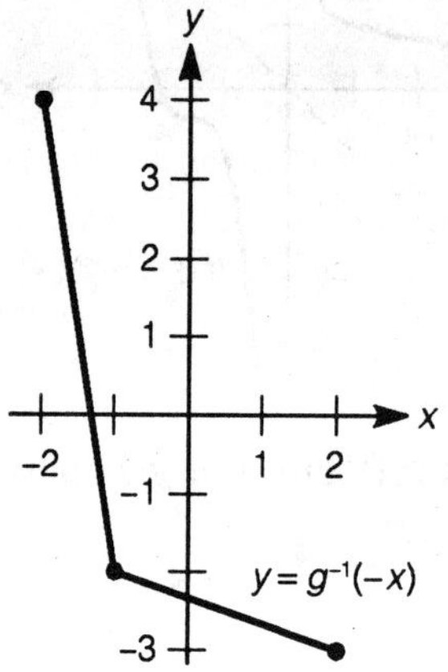

(e) The graph of $y = -g^{-1}(x)$ is a reflection of $g^{-1}(x)$ across the x-axis:

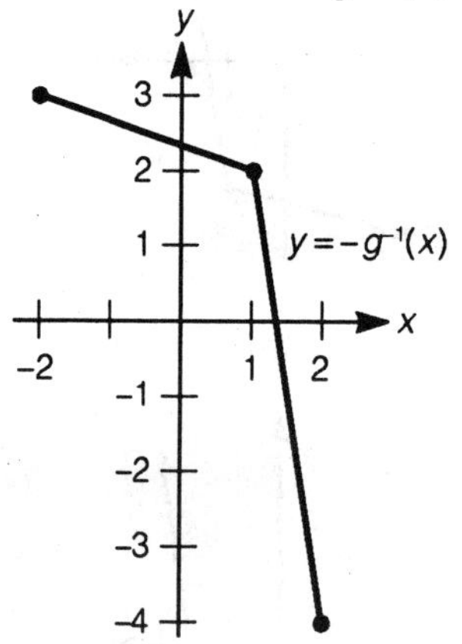

(f) The graph of $y = -g^{-1}(-x)$ is a reflection of $g^{-1}(x)$ across the y-axis and across the x-axis:

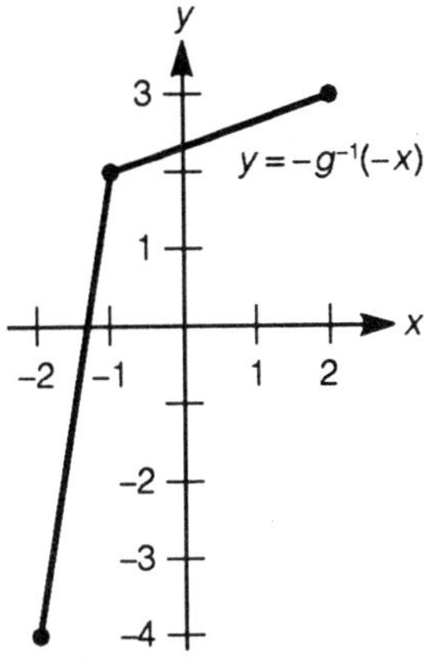

19. The graph of $y = x^2 + 1$ fails the horizontal line test, so it is not one-to-one:

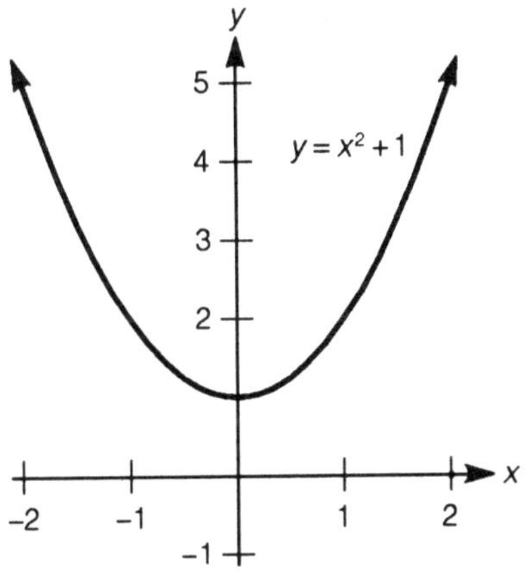

21. The graph of $y = \frac{1}{x}$ passes the horizontal line test, so it is one-to-one:

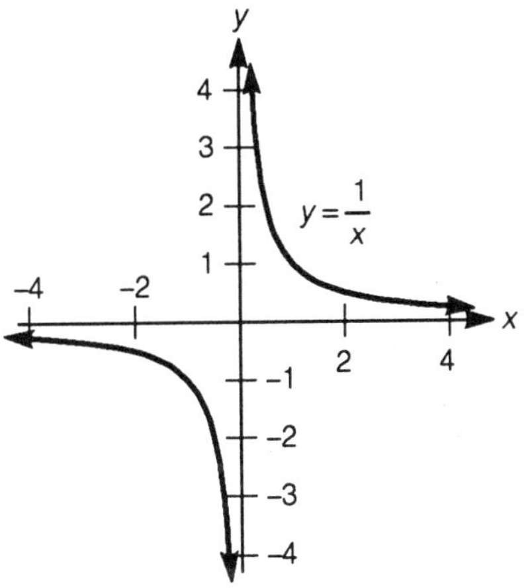

23. The graph of $y = x^3$ passes the horizontal line test, so it is one-to-one:

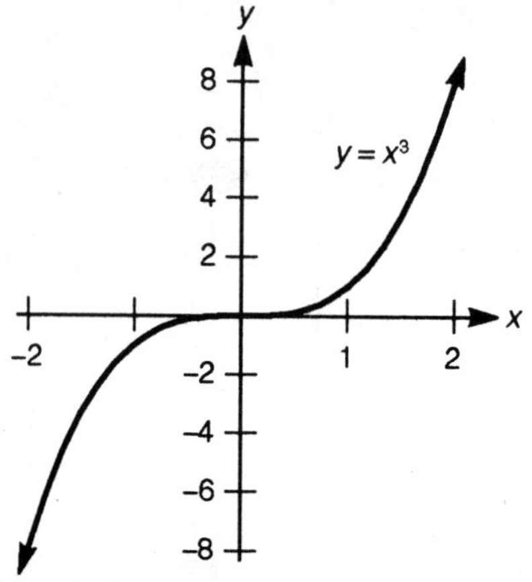

25. The graph of $y = \sqrt{1 - x^2}$ fails the horizontal line test, so it is not one-to-one:

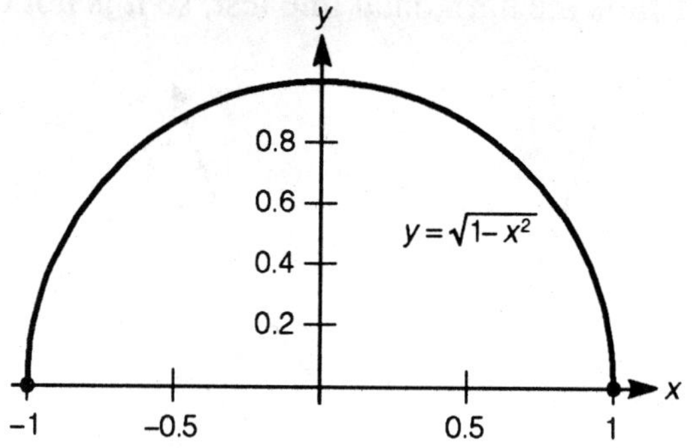

27. The graph of $g(x) = 5$ fails the horizontal line test (it <u>is</u> a horizontal line), so it is not one-to-one:

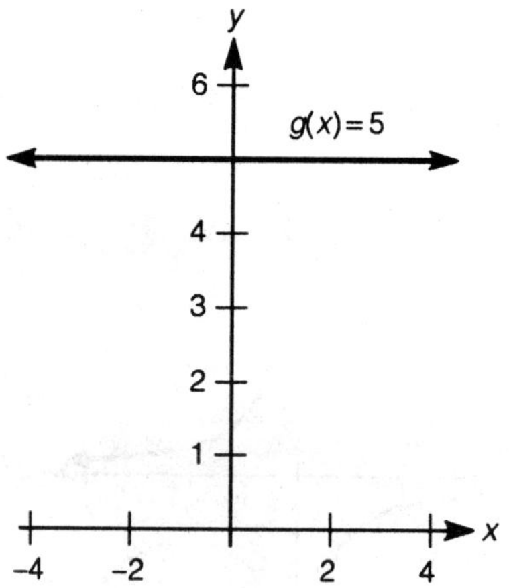

29. The graph of $f(x)$ passes the horizontal line test, so it is one-to-one:

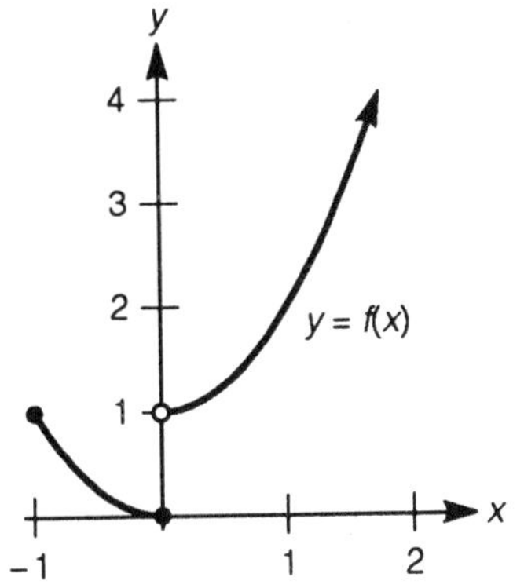

31. We find $f[f(x)]$:

$$\begin{aligned} f[f(x)] &= f\left(\tfrac{3x-2}{5x-3}\right) \\ &= \frac{3\left(\frac{3x-2}{5x-3}\right)-2}{5\left(\frac{3x-2}{5x-3}\right)-3} \\ &= \frac{3(3x-2)-2(5x-3)}{5(3x-2)-3(5x-3)} \\ &= \frac{9x-6-10x+6}{15x-10-15x+9} \\ &= \frac{-x}{-1} \\ &= x \end{aligned}$$

Thus $f^{-1}(x) = f(x)$.

33. (a) We solve the equation:

$$\begin{aligned} 7+f^{-1}(x-1) &= 9 \\ f^{-1}(x-1) &= 2 \\ x-1 &= f(2) \\ x-1 &= 6 \\ x &= 7 \end{aligned}$$

(b) We solve the equation:

$$\begin{aligned} 4+f(x+3) &= -3 \\ f(x+3) &= -7 \\ x+3 &= f^{-1}(-7) \\ x+3 &= 0 \\ x &= -3 \end{aligned}$$

35. We solve the equation:

$$f^{-1}\left(\frac{t+1}{t-2}\right)=12$$
$$\frac{t+1}{t-2}=f(12)$$
$$\frac{t+1}{t-2}=13$$
$$t+1=13t-26$$
$$27=12t$$
$$t=\tfrac{27}{12}=\tfrac{9}{4}$$

37. (a) Let $y=\sqrt{x}$. We switch the roles of x and y and solve the resulting equation for y:

$$x=\sqrt{y}$$
$$x^2=y$$

So the inverse function is $f^{-1}(x)=x^2$. Since $x=\sqrt{y}$, then $x\geq 0$. So the domain of $f^{-1}(x)$ is $[0,\infty)$.

(b) (i) Since $2=\sqrt{4}$, then $(4,2)$ lies on the graph of f.
(ii) Since $(4,2)$ lies on the graph of f, then $(2,4)$ lies on the graph of f^{-1}.
(iii) Since $\sqrt{5}=\sqrt{5}$, then $(5,\sqrt{5})$ lies on the graph of f.
(iv) Since $(5,\sqrt{5})$ lies on the graph of f, then $(\sqrt{5},5)$ lies on the graph of f^{-1}.
(v) Since $f(a)=f(a)$, then $(a,f(a))$ lies on the graph of f.
(vi) Since $(a,f(a))$ lies on the graph of f, then $(f(a),a)$ lies on the graph of f^{-1}.
(vii) Since the graph of f^{-1} is all points $(x,f^{-1}(x))$, then $(b,f^{-1}(b))$ lies on the graph of f^{-1}.
(viii) Since $b=f[f^{-1}(b)]$, then $(f^{-1}(b),b)$ lies on the graph of f.

39. Let the points $P=(7,-1)$ and $Q=(-1,7)$. We must first show that the line segment $\overline{PQ}$ is perpendicular to the line $y=x$. Line segment $\overline{PQ}$ has a slope of:

$$m=\frac{7-(-1)}{-1-7}=\frac{8}{-8}=-1$$

Since $y=x$ has a slope of 1, and $1(-1)=-1$, then the two lines are perpendicular. Next, we must show that P and Q are equidistant from $y=x$. We find the midpoint of line segment $\overline{PQ}$:

$$M=\left(\tfrac{7-1}{2},\tfrac{-1+7}{2}\right)=(3,3)$$

Since $(3,3)$ lies on the line $y=x$, and since $\overline{PQ}$ is perpendicular to $y=x$, and $PM=QM$, then by the definition of symmetry P and Q are symmetric about the line $y=x$.

41. Let the points $P = (a, b)$ and $Q = (b, a)$. We must first show that the line segment $\overline{PQ}$ is perpendicular to the line $y = x$. Line segment $\overline{PQ}$ has a slope of:

$$m = \frac{a-b}{b-a} = -1$$

Since $y = x$ has a slope of 1, and $1(-1) = -1$, then the two lines are perpendicular. Next, we must show that P and Q are equidistant from $y = x$. We find the midpoint of line segment $\overline{PQ}$:

$$M = \left(\tfrac{a+b}{2}, \tfrac{b+a}{2}\right) = \left(\tfrac{a+b}{2}, \tfrac{a+b}{2}\right)$$

Since $\left(\tfrac{a+b}{2}, \tfrac{a+b}{2}\right)$ lies on the line $y = x$, and since $\overline{PQ}$ is perpendicular to $y = x$, and $PM = QM$, then by the definition of symmetry P and Q are symmetric about the line $y = x$.

43. Let's pick the points $P(0,3)$, $Q(2,4)$, $R(4,5)$. Then the reflected points will be $P'(3,0)$, $Q'(4,2)$, and $R'(5,4)$. We will show that P', Q', and R' are collinear by computing the slopes:

$$m_{P'Q'} = \frac{2-0}{4-3} = \frac{2}{1} = 2$$

$$m_{Q'R'} = \frac{4-2}{5-4} = \frac{2}{1} = 2$$

So P', Q', and R' all are collinear (lie on the same line). There are two ways to find the equation of the line.

1st way: We use the point $(3,0)$ and $m = 2$ in the point-slope formula:

$$y - 0 = 2(x-3)$$
$$y = 2x - 6$$

2nd way: We realize that this line must be the inverse function of $y = \frac{1}{2}x + 3$. Exchange x and y and solve for y:

$$x = \tfrac{1}{2}y + 3$$
$$x - 3 = \tfrac{1}{2}y$$
$$y = 2x - 6$$

In either case, we obtain the same equation.

4.6 Variation

1. (a) $y = kx$

(b) $A = \dfrac{k}{B}$

3. (a) $x = kuv^2$
(b) $z = kA^2B^3$

5. (a) $F = \dfrac{k}{r^2}$

(b) $V^2 = k(U^2 + T^2)$

7. (a) Since A varies inversely as B, then $A = \frac{k}{B}$. We find k by substituting $A = -1$ and $B = 2$:

$$-1 = \frac{k}{2}$$
$$k = -2$$

(b) $A = \dfrac{-2}{B}$

(c) When $B = \frac{5}{4}$, we have:

$$A = \frac{-2}{\frac{5}{4}} = -\frac{8}{5}$$

9. Since x varies jointly as y and z, then $x = kyz$. We find k by substituting $x = 9$, $y = 2$, and $z = -3$:

$$9 = k(2)(-3)$$
$$9 = -6k$$
$$k = -\tfrac{3}{2}$$

So $x = -\frac{3}{2}yz$. Thus, when $y = 4$ and $z = 4$:

$$x = -\tfrac{3}{2}(4)(4) = -24$$

11. Since A varies jointly as B and C, then $A = kBC$. Now we substitute $3B$ and $2C$ in for B and C, respectively:

$$A = k(3B)(2C)$$
$$A = 6(kBC)$$

So A is six times its original value.

13. Since x varies jointly as B and C and inversely as $\sqrt{A}$, then $x = \frac{kBC}{\sqrt{A}}$. Now we substitute $4B$, $4C$ and $4A$ in for B, C and A, respectively:

$$x = \frac{k(4B)(4C)}{\sqrt{4A}} = \frac{16kBC}{2\sqrt{A}} = 8\left(\frac{kBC}{\sqrt{A}}\right)$$

So x is eight times its original value.

15. (a) Let S = surface area and r = radius. Then $S = kr^2$. When $r = 2$ and $S = 16\pi$, we have:

$$16\pi = k(2)^2$$
$$16\pi = 4k$$
$$k = 4\pi$$

So $S = 4\pi r^2$.

(b) When $r = \sqrt{3}$, we have:

$$S = 4\pi(\sqrt{3})^2 = 12\pi \text{ cm}^2$$

17. Let m_1 and m_2 be the two masses, F = force, and d = distance. Then $F = \frac{km_1m_2}{d^2}$. Now we substitute $3m_1$, $4m_2$ and $\frac{1}{2}d$ in for m_1, m_2 and d, respectively:

$$F = \frac{k(3m_1)(4m_2)}{\left(\frac{d}{2}\right)^2} = \frac{12m_1m_2}{\frac{d^2}{4}} = 48\left(\frac{km_1m_2}{d^2}\right)$$

So the force is 48 times its original value.

19. Let d = distance and t = time. Then $d = kt^2$. We find k by substituting $d = 490$ and $t = 10$:

$$490 = k(10)^2$$
$$490 = 100k$$
$$k = \frac{490}{100} = 4.9$$

So $d = 4.9t^2$. When $t = 5$, we have:

$$d = (4.9)(5)^2 = (4.9)(25) = 122.5 \text{ m}$$

21. (a) Let V = volume and P = pressure. Then $V = \frac{k}{P}$. We find k by substituting $V = 2$ and $P = 1.025$:

$$2 = \frac{k}{1.025}$$
$$k = 2.05$$

So $V = \frac{2.05}{P}$.

(b) When $P = 1$, we have $V = \frac{2.05}{1} = 2.05$ liters.

23. (a) Let E = kinetic energy, m = mass and r = radius. Then $E = \frac{km}{r}$. Now we substitute $\frac{1}{2}r$ and $3m$ for r and m, respectively:

$$E = \frac{k(3m)}{\frac{1}{2}r} = \frac{6km}{r} = 6\left(\frac{km}{r}\right)$$

So the kinetic energy is six times its original value.

(b) Let V = velocity and r = radius. Then $V = \frac{k}{\sqrt{r}}$. Now we substitute $\frac{1}{2}r$ for r:

$$V = \frac{k}{\sqrt{\frac{1}{2}r}} = \frac{k\sqrt{2}}{\sqrt{r}} = \sqrt{2}\left(\frac{k}{\sqrt{r}}\right)$$

So the velocity is $\sqrt{2}$ times its original value.

25. Let W = weight and d = distance. Then $W = \frac{k}{d^2}$. We find k by substituting $W = 140$ and $d = 4000$:

$$140 = \frac{k}{(4000)^2}$$
$$k = 140(4000)^2$$
$$k = 2.24 \times 10^9$$

When $d = 4500$, we have:

$$W = \frac{2.24 \times 10^9}{(4500)^2} = 110.6 \text{ lbs.}$$

27. (a) We need a graph which is linear with a negative slope. Such a graph is F.

(b) We need a graph for $y = \frac{k}{x}$. Such a graph is C.
(c) We need a graph which is a parabola pointed upward. Such a graph is D.
(d) We need a graph which is linear with a positive slope. Such a graph is A.
(e) We need a graph which is a parabola pointed downward. Such a graph is B.

29. Let P = period and L = length. Then $P = k\sqrt{L}$. We denote the new length and period by L' and P', respectively, where $P' = \frac{P}{2}$. Since $P' = k\sqrt{L'}$, we have:

$$\frac{P}{2} = k\sqrt{L'}$$
$$\frac{k\sqrt{L}}{2} = k\sqrt{L'}$$
$$\frac{k^2L}{4} = k^2L'$$
$$L' = \frac{L}{4}$$

So the new length L' should be $\frac{1}{4}$ of the original length. Thus it must be shortened by $\frac{3}{4}L$.

31. Let I = intensity and d = distance. Then $I = \frac{k}{d^2}$. When $d = 6$, we have:

$$I = \frac{k}{6^2} = \frac{k}{36}$$

We are asked to find the value of d when $I = 2\left(\frac{k}{36}\right) = \frac{k}{18}$:

$$\frac{k}{18} = \frac{k}{d^2}$$
$$d^2 = 18$$
$$d = 3\sqrt{2} \approx 4.24$$

At a distance of 4.24 ft, the illumination will be twice as great.

33. The area of the triangle is:

$$A = \tfrac{1}{2}(\text{base})(\text{height}) = \tfrac{1}{2}(x)(y) = \tfrac{1}{2}(x)(mx) = \tfrac{1}{2}mx^2$$

Thus A varies jointly as m and x^2.

35. Let t = time and d = distance. Then $t^2 = kd^3$. We first find k by using $d = 1.000000$ and $t = 365.2564$ (information for earth):

$$t^2 = kd^3$$
$$k = \frac{t^2}{d^3} = \frac{(365.2564)^2}{(1.000000)^3} = 133412.2377$$

So $t = \sqrt{kd^3} = \sqrt{133412.2377d^3}$. We now compute t for each planet:

Mercury:	87.9693 days
Venus:	224.7007 days
Earth:	365.2564 days
Mars:	686.9786 days
Jupiter:	4336.6159 days
Saturn:	10826.9994 days
Uranus:	30873.7244 days
Neptune:	60300.6863 days
Pluto:	91814.3739 days

37. (a) Let $x = k_1z$ and $y = k_2z$. Then:

$$x + y = k_1z + k_2z = (k_1 + k_2)z$$

Call $K = k_1 + k_2$. Then $x + y = Kz$. This shows that $x + y$ varies directly as z.

(b) Let $x = k_1z$ and $y = k_2z$. Then:

$$xy = (k_1z)(k_2z) = k_1k_2z^2$$

Call $K = k_1k_2$. Then $xy = Kz^2$. So xy does <u>not</u> vary directly as z.

(c) Let $x = k_1z$ and $y = k_2z$. Then:

$$\sqrt{xy} = \sqrt{k_1k_2z^2} = \sqrt{k_1k_2}\, z \text{ since } z > 0$$

Call $K = \sqrt{k_1k_2}$. Then $\sqrt{xy} = Kz$. This shows that $\sqrt{xy}$ varies directly as z.

39. (a) We have:

$$\begin{aligned} x+y &= k(x-y) \\ x+y &= kx-ky \\ y+ky &= kx-x \\ y(1+k) &= x(k-1) \\ y &= \tfrac{k-1}{k+1}x \end{aligned}$$

Setting $K = \frac{k-1}{k+1}$, we have $y = Kx$. Thus y varies directly as x.

(b) We have:

$$x^2+y^2 = x^2+(Kx)^2 = x^2+K^2x^2 = \left(K^2+1\right)\left(x^2\right)$$

Since $y = Kx$, then $x = \frac{y}{K}$, so:

$$x^2+y^2 = \left(K^2+1\right)\left(x^2\right) = \left(K^2+1\right)\left(\frac{y}{k}\right)(x) = \frac{K^2+1}{K}(xy)$$

Setting $A = \frac{K^2+1}{K}$, we have $x^2+y^2 = Axy$. Thus x^2+y^2 varies jointly as x and y.

(c) We have:

$$\begin{aligned} x^3+y^3 &= (x+y)\left(x^2-xy+y^2\right) \\ &= (x+y)\left(x^2+y^2-xy\right) \\ &= (x+y)(Axy-xy), \text{ from part (b)} \\ &= (x+y)(xy)(A-1) \\ &= (A-1)(x+y)(xy) \end{aligned}$$

Let $B = A-1$. Then $x^3+y^3 = Bxy(x+y)$. So x^3+y^3 varies jointly as x, y, and $x+y$.

41. The area of the triangle is:

$$A = \tfrac{1}{2}(\text{base})(\text{height}) = \tfrac{1}{2}(x)(y) = \tfrac{1}{2}(x)\left(\frac{a^2}{x}\right) = \tfrac{1}{2}a^2$$

Thus A varies directly as a^2.

Chapter Four Review Exercises

1. (a) We must guarantee that the quantity within the radical is non-negative:

$$\begin{aligned} 15-5x &\geq 0 \\ -5x &\geq -15 \\ x &\leq 3 \end{aligned}$$

So the domain is $(-\infty,3]$.

(b) We solve $y = \frac{3+x}{2x-5}$ for x:

$$y(2x-5) = 3+x$$
$$2xy - 5y = 3+x$$
$$2xy - x = 5y+3$$
$$x(2y-1) = 5y+3$$
$$x = \frac{5y+3}{2y-1}$$

So the range is all real numbers except $\frac{1}{2}$, or $\left(-\infty, \frac{1}{2}\right) \cup \left(\frac{1}{2}, \infty\right)$.

3. Yes. Let $f(x) = ax + b$ and $g(x) = cx + d$, then:

$$(f \circ g)(x) = f(cx+d) = a(cx+d)+b = acx + (ad+b)$$

So $(f \circ g)(x)$ is a linear function.

5. (a) We compute the difference quotient:

$$\frac{F(x)-F(a)}{x-a} = \frac{\frac{1}{x}-\frac{1}{a}}{x-a} \bullet \frac{ax}{ax} = \frac{a-x}{ax(x-a)} = -\frac{1}{ax}$$

(b) We first compute:

$$g(x+h) = (x+h) - 2(x+h)^2 = x + h - 2x^2 - 4xh - 2h^2$$

So we have:

$$\frac{g(x+h)-g(x)}{h} = \frac{\left(x+h-2x^2-4xh-2h^2\right)-\left(x-2x^2\right)}{h}$$
$$= \frac{x+h-2x^2-4xh-2h^2-x+2x^2}{h}$$
$$= \frac{h-4xh-2h^2}{h}$$
$$= 1-4x-2h$$

7. (a) We switch the roles of x and y and solve the resulting equation for y:

$$x = \frac{1-5y}{3y}$$
$$3xy = 1-5y$$
$$3xy + 5y = 1$$
$$y(3x+5) = 1$$
$$y = \frac{1}{3x+5}$$

So $g^{-1}(x) = \frac{1}{3x+5}$.

(b) We graph f^{-1}:

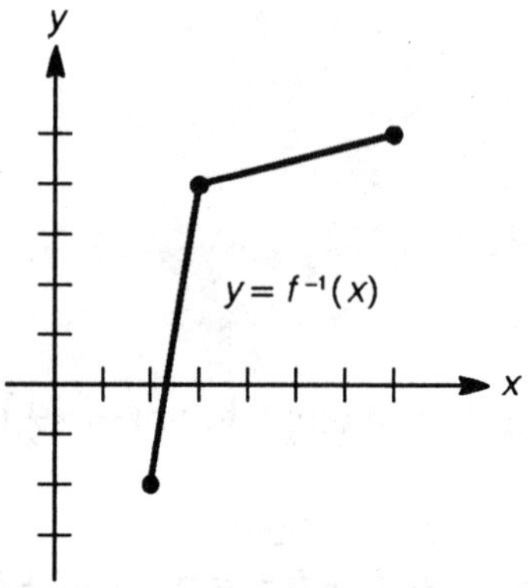

9. (a) We displace $y = |x|$ two units to the left and 3 units down. The x-intercepts are -5 and 1, and the y-intercept is -1:

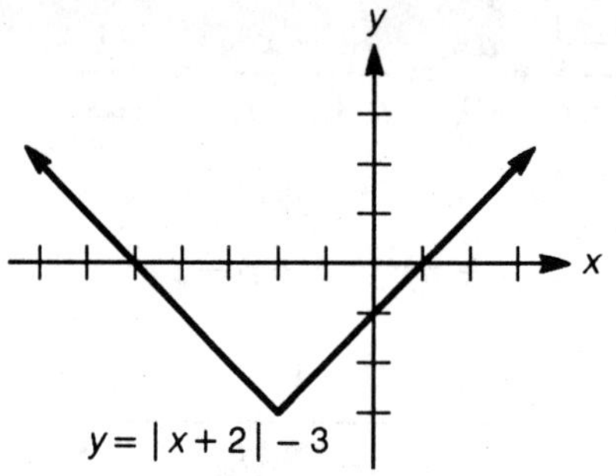

(b) We displace $y = \frac{1}{x}$ two units to the left and 1 unit down. The x-intercept is -1 and the y-intercept is $-\frac{1}{2}$:

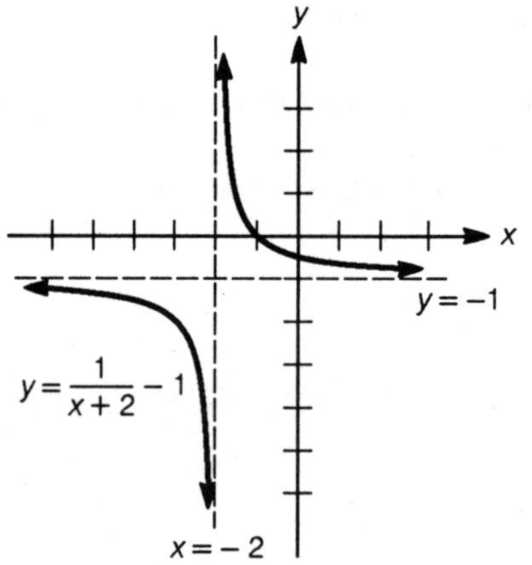

11. (a) We compute $f(-1)$:

$$f(-1) = 3(-1)^2 - 2(-1) = 3 + 2 = 5$$

(b) We compute $f(1-\sqrt{2})$:

$$\begin{aligned} f(1-\sqrt{2}) &= 3(1-\sqrt{2})^2 - 2(1-\sqrt{2}) \\ &= 3(1-2\sqrt{2}+2) - 2 + 2\sqrt{2} \\ &= 9 - 6\sqrt{2} - 2 + 2\sqrt{2} \\ &= 7 - 4\sqrt{2} \end{aligned}$$

13. The slope is given by:

$$m = \frac{(5+h)^2 - 25}{5+h-5} = \frac{25 + 10h + h^2 - 25}{h} = \frac{10h + h^2}{h} = 10 + h$$

Using functional notation, we have $m(h) = 10 + h$.

15. The graph of $y = f(-x)$ will result in a reflection across the y-axis:

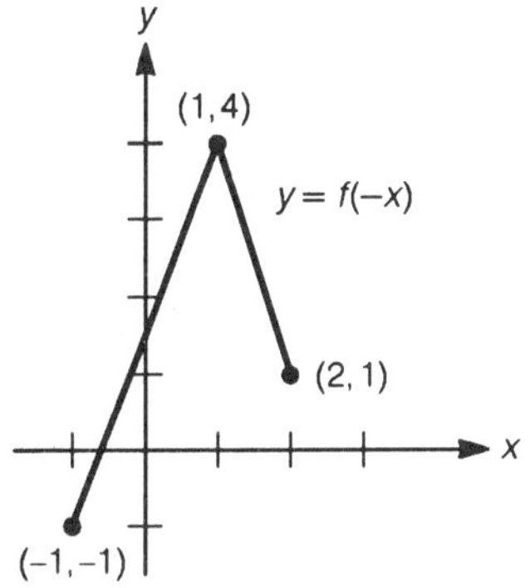

17. (a) Since the graph of $y = f(x) + 1$ is obtained by translating the graph of $y = f(x)$ up one unit, then the point (a, b) is translated to the point $(a, b + 1)$. Thus E is the correct answer.

(b) Since the graph of $y = f(x + 1)$ is obtained by translating the graph of $y = f(x)$ to the left one unit, then the point (a, b) is translated to the point $(a - 1, b)$. Thus C is the correct answer.

(c) Since the graph of $y = f(x - 1) + 1$ is obtained by translating the graph of $y = f(x)$ to the right one unit and up one unit, then the point (a, b) is translated to the point $(a + 1, b + 1)$. Thus L is the correct answer.

(d) Since the graph of $y = f(-x)$ is obtained by reflecting the graph of $y = f(x)$ across the y-axis, then the point (a, b) is reflected to the point $(-a, b)$. Thus A is the correct answer.

(e) Since the graph of $y = -f(x)$ is obtained by reflecting the graph of $y = f(x)$ across the x-axis, then the point (a, b) is reflected to the point $(a, -b)$. Thus J is the correct answer.

(f) Since the graph of $y = -f(-x)$ is obtained by reflecting the graph of $y = f(x)$ across the x-axis and across the y-axis, then the point (a, b) is reflected to the point $(-a, -b)$. Thus G is the correct answer.

(g) Since the graph of $y = f^{-1}(x)$ is obtained by reflecting the graph of $y = f(x)$ across the line $y = x$, then the point (a, b) is reflected to the point (b, a). Thus B is the correct answer.

(h) Since the graph of $y = f^{-1}(x) + 1$ is obtained by reflecting the graph of $y = f(x)$ across the line $y = x$, then translating up one unit, then the point (a, b) is reflected to the point (b, a) then translated to the point $(b, a + 1)$. Thus M is the correct answer.

(i) Since the graph of $y = f^{-1}(x - 1)$ is obtained by reflecting the graph of $y = f(x)$ across the line $y = x$, then translating to the right one unit, then the point (a, b) is reflected to the point (b, a) then translated to the point $(b + 1, a)$. Thus K is the correct answer.

(j) Since the graph of $y = f^{-1}(-x) + 1$ is obtained by reflecting the graph of $y = f(x)$ across the line $y = x$, then across the y-axis, then translating up one unit, then the point (a, b) is reflected to the point (b, a), then reflected to the point $(-b, a)$, then translated to the point $(-b, a + 1)$. Thus D is the correct answer.

(k) Since the graph of $y = -f^{-1}(x)$ is obtained by reflecting the graph of $y = f(x)$ across the line $y = x$ then across the x-axis, then the point (a, b) is reflected to the point (b, a) then reflected to the point $(b, -a)$. Thus I is the correct answer.

(l) Since the graph of $y = -f^{-1}(-x) + 1$ is obtained by reflecting the graph of $y = f(x)$ across the line $y = x$, the x-axis, and the y-axis, then translating up one unit, then the point (a, b) is reflected to the point (b, a), then reflected to the point $(-b, a)$ and then $(-b, -a)$, and finally translated to the point $(-b, -a + 1)$. Thus H is the correct answer.

(m) Since the graph of $y = 1 - f^{-1}(x)$ is obtained by reflecting the graph of $y = f(x)$ across the line $y = x$ and across the x-axis, then translated up one unit, then the point (a, b) is reflected to the point (b, a), then reflected to the point $(b, -a)$, then translated to the point $(b, -a + 1)$. Thus N is the correct answer.

(n) Since the graph of $y = f(1 - x)$ is obtained by reflecting the graph of $y = f(x)$ across the y-axis, then translating to the right one unit, then the point (a, b) is reflected to the point $(-a, b)$ then translated to the point $(-a + 1, b)$. Thus F is the correct answer.

19. To find the x-intercepts, we let $y = 0$:

$$
\begin{aligned}
-(x-1)^2 + 2 &= 0 \\
-(x-1)^2 &= -2 \\
x - 1 &= \pm\sqrt{2} \\
x &= 1 \pm \sqrt{2}
\end{aligned}
$$

To find the y-intercept, we let $x = 0$:

$$y = -(0-1)^2 + 2 = -1 + 2 = 1$$

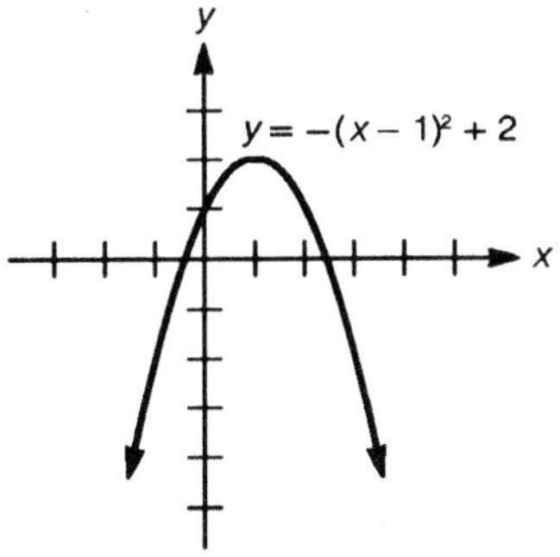

21. To find the x-intercepts, we let $y = 0$:

$$\frac{1}{x+1} = 0$$
$$1 = 0$$

Clearly this is impossible, so there are no x-intercepts. To find the y-intercept, we let $x = 0$:

$$y = \frac{1}{0+1} = 1$$

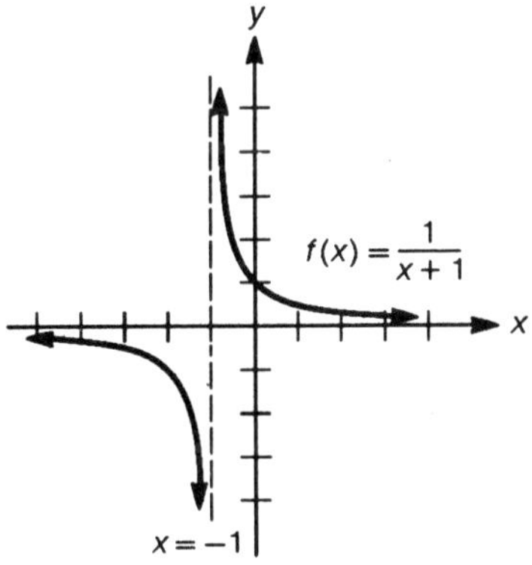

23. To find the x-intercepts, we let $y = 0$:

$$|x+3| = 0$$
$$x = -3$$

To find the y-intercept, we let $x = 0$:

$$y = |0+3| = 3$$

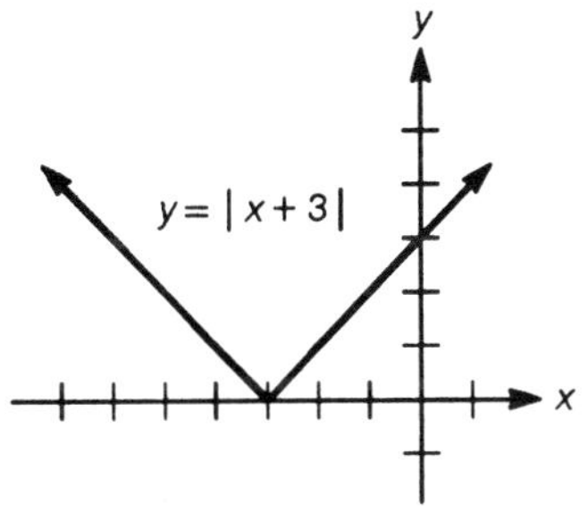

25. To find the x-intercepts, we let $y = 0$:

$$\sqrt{1-x^2} = 0$$
$$1 - x^2 = 0$$
$$-x^2 = -1$$
$$x = \pm 1$$

To find the y-intercept, we let $x = 0$:

$$y = \sqrt{1-0} = 1$$

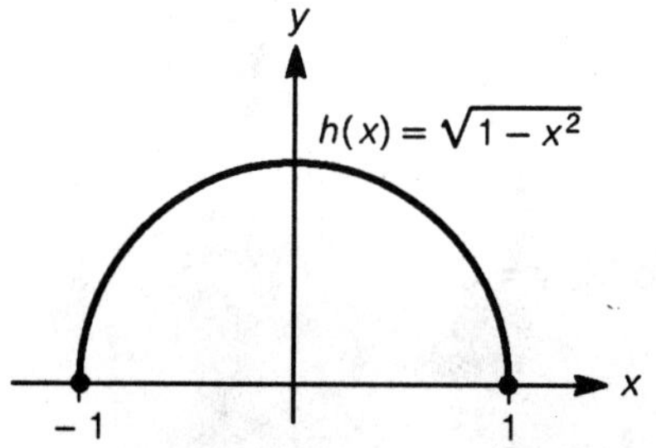

27. To find the x-intercepts, we let $y = 0$:

$$1 - (x+1)^3 = 0$$
$$-(x+1)^3 = -1$$
$$(x+1)^3 = 1$$
$$x + 1 = 1$$
$$x = 0$$

To find the y-intercept, we let $x = 0$:

$$y = 1 - (0+1)^3 = 1 - 1 = 0$$

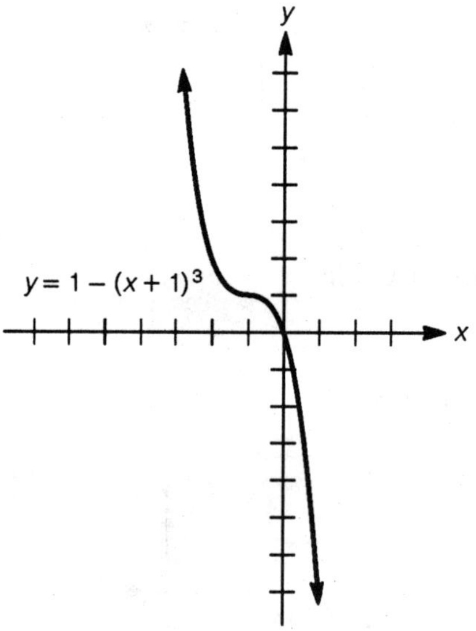

29. The x-intercept is 0 and the y-intercept is 0:

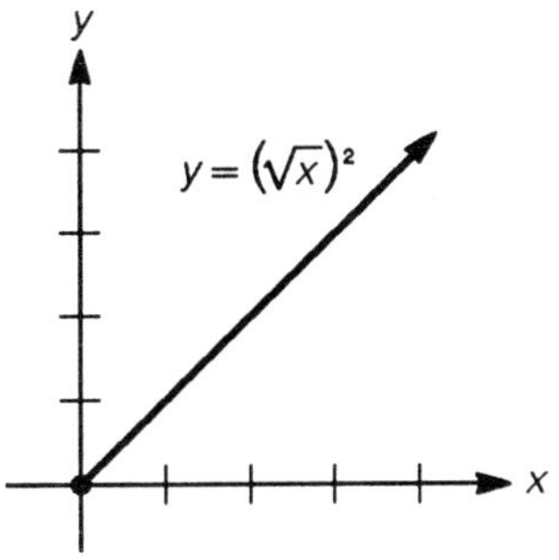

31. We first compute $(f \circ g)(x)$:

$$(f \circ g)(x) = f\left(\sqrt{x-1}\right) = -\left(\sqrt{x-1}\right)^2 = 1 - x \text{ for } x \geq 1$$

The x-intercept is 1 and there is no y-intercept:

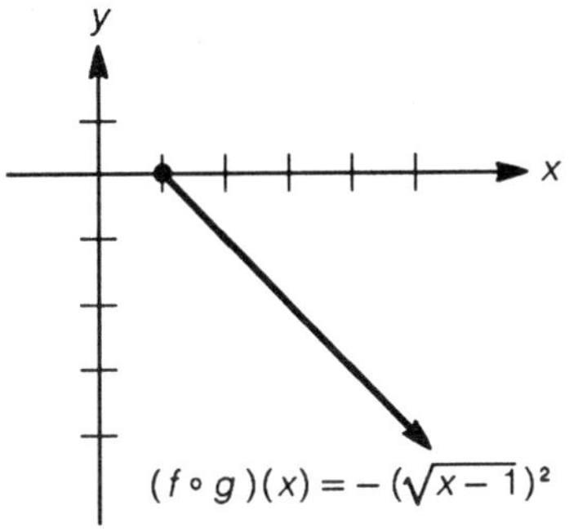

33. The x-intercepts are –1 and 0, and the y-intercept is 0:

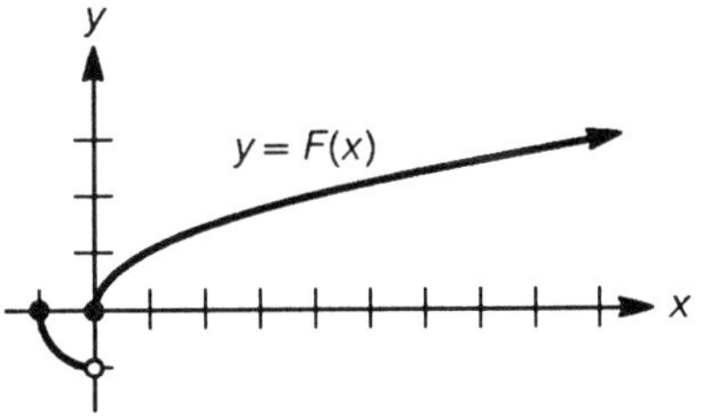

35. There are no x- or y-intercepts:

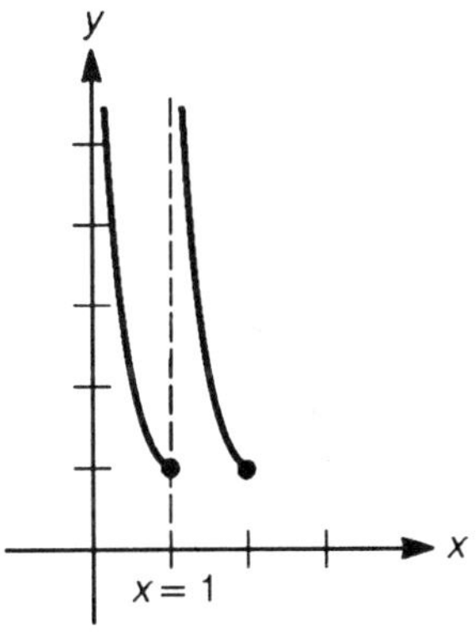

37. To find the inverse function, we switch the roles of x and y and solve the resulting equation for y:

$$x = \frac{y+1}{2}$$
$$2x = y+1$$
$$2x-1 = y$$

So $f^{-1}(x) = 2x - 1$. To find the x-intercepts, we let $y = 0$:

$$2x-1=0$$
$$2x=1$$
$$x=\tfrac{1}{2}$$

To find the y-intercept, we let $x = 0$:

$$y = 2(0)-1=-1$$

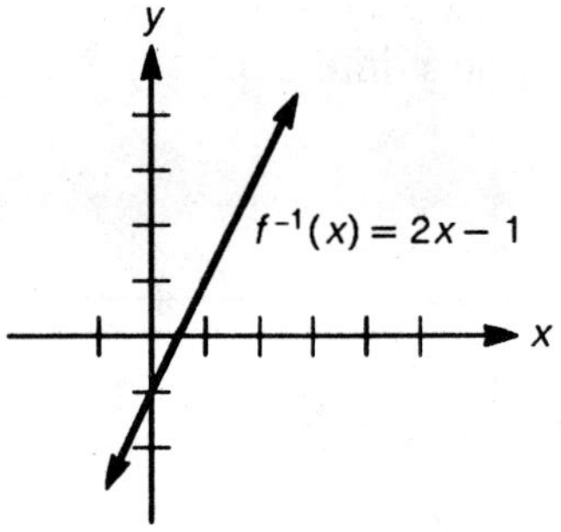

39. We compute $\left(f \circ f^{-1}\right)(x) = x$ for $x \geq 0$. The x-intercept is 0 and the y-intercept is 0:

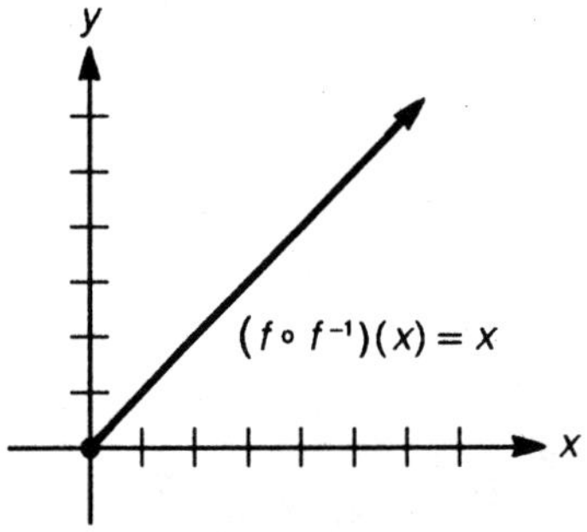

41. We must make sure that $x^2 - 9 \neq 0$. We find the points to exclude:

$$x^2-9=0$$
$$x^2=9$$
$$x=\pm 3$$

So the domain is all real numbers except 3 and –3.

43. We must be sure that the quantity inside the radical is non-negative:

$$8-2x \geq 0$$
$$-2x \geq -8$$
$$x \leq 4$$

So the domain is $(-\infty, 4]$.

45. We must be sure that the quantity inside the radical is non-negative, so $|2-5x| \ge 0$. But this is true for all real numbers, so the domain is all real numbers, or $(-\infty,\infty)$.

47. We must be sure that the quantity inside the radical is non-negative:

$$x^2-2x-3 \ge 0$$
$$(x-3)(x+1) \ge 0$$

The key numbers are −1 and 3. We draw the sign chart:

Interval	Test Number	$x-3$	$x+1$	$(x-3)(x+1)$
$(-\infty,-1)$	−2	neg.	neg.	pos.
$(-1,3)$	0	neg.	pos.	neg.
$(3,\infty)$	4	pos.	pos.	pos.

So the product is positive on the intervals $(-\infty,-1)\cup(3,\infty)$, and it is zero when $x=-1$ and $x=3$. So the domain is $(-\infty,-1]\cup[3,\infty)$.

49. We must be sure that $x \ne 0$. So the domain is all real numbers except 0, or $(-\infty,0)\cup(0,\infty)$.

51. We solve for x:

$$y=\frac{x+4}{3x-1}$$
$$y(3x-1)=x+4$$
$$3xy-y=x+4$$
$$3xy-x=y+4$$
$$x(3y-1)=y+4$$
$$x=\frac{y+4}{3y-1}$$

Now $3y-1=0$ when $y=\frac{1}{3}$, so the range is all real numbers except $\frac{1}{3}$.

53. We first compute $(f\circ g)(x)$:

$$(f\circ g)(x)=f(3x+4)=\frac{1}{3x+4}$$

We solve $y=\frac{1}{3x+4}$ for x:

$$y=\frac{1}{3x+4}$$
$$y(3x+4)=1$$
$$3xy+4y=1$$
$$3xy=1-4y$$
$$x=\frac{1-4y}{3y}$$

Now $3y=0$ when $y=0$, so the range is all real numbers except 0, or $(-\infty,0)\cup(0,\infty)$.

55. Since the range of f^{-1} is the domain of f, we must exclude the values of x where $3x - 6 = 0$, or $x = 2$. So the range is all real numbers except 2.

57. We have $a(x) = (f \circ g)(x)$, since:
$$(f \circ g)(x) = f[g(x)] = f(x-1) = \frac{1}{x-1}$$

59. We have $c(x) = (G \circ g)(x)$, since:
$$(G \circ g)(x) = G[g(x)] = G(x-1) = \sqrt{x-1}$$

61. We have $A(x) = (g \circ f \circ G)(x)$, since:
$$(g \circ f \circ G)(x) = (g \circ f)(\sqrt{x}) = g\left(\frac{1}{\sqrt{x}}\right) = \frac{1}{\sqrt{x}} - 1$$

63. We have $C(x) = (g \circ G \circ G)(x)$, since:
$$(g \circ G \circ G)(x) = (g \circ G)(\sqrt{x}) = g\left(\sqrt{\sqrt{x}}\right) = g\left(\sqrt[4]{x}\right) = \sqrt[4]{x} - 1$$

65. Using $f(x)$, we compute $f(-3)$:
$$f(-3) = (-3)^2 - (-3) = 9 + 3 = 12$$

67. Using $F(x)$, we compute $F\left(\frac{3}{4}\right)$:
$$F\left(\tfrac{3}{4}\right) = \frac{\frac{3}{4} - 3}{\frac{3}{4} + 4} = \frac{3 - 3(4)}{3 + 4(4)} = \frac{3 - 12}{3 + 16} = -\frac{9}{19}$$

69. Using $f(x)$, we compute $f(-t)$:
$$f(-t) = (-t)^2 - (-t) = t^2 + t$$

71. Using $f(x)$, we compute $f(x-2)$:
$$f(x-2) = (x-2)^2 - (x-2) = x^2 - 4x + 4 - x + 2 = x^2 - 5x + 6$$

73. Using $g(x)$, we first compute $g(2)$ and $g(0)$:
$$g(2) = 1 - 2(2) = 1 - 4 = -3$$
$$g(0) = 1 - 2(0) = 1 - 0 = 1$$
So $g(2) - g(0) = -3 - 1 = -4$

75. Using $f(x)$, we first compute $f(1)$ and $f(3)$:
$$f(1) = (1)^2 - 1 = 1 - 1 = 0$$
$$f(3) = (3)^2 - 3 = 9 - 3 = 6$$
So $|f(1) - f(3)| = |0 - 6| = |-6| = 6$.

77. Using $f(x)$, we compute $f\left(x^2\right)$:

$$f\left(x^2\right)=\left(x^2\right)^2-x^2=x^4-x^2$$

79. We compute $[f(x)][g(x)]$:

$$[f(x)][g(x)]=\left(x^2-x\right)(1-2x)=x^2-2x^3-x+2x^2=-2x^3+3x^2-x$$

81. We compute $f[g(x)]$:

$$f[g(x)]=f(1-2x)=(1-2x)^2-(1-2x)=1-4x+4x^2-1+2x=4x^2-2x$$

83. We compute $(g\circ f)(x)$:

$$\begin{aligned}(g\circ f)(x)&=g[f(x)]\\&=g\left(x^2-x\right)\\&=1-2\left(x^2-x\right)\\&=2-2x^2+2x\\&=-2x^2+2x+1\end{aligned}$$

85. We compute $(F\circ g)(x)$:

$$\begin{aligned}(F\circ g)(x)&=F[g(x)]\\&=F(1-2x)\\&=\frac{(1-2x)-3}{(1-2x)+4}\\&=\frac{-2x-2}{-2x+5}\text{ or }\frac{2x+2}{2x-5}\end{aligned}$$

87. We find the difference quotient:

$$\frac{f(x+h)-f(x)}{h}=\frac{2xh+h^2-h}{h}=\frac{h(2x+h-1)}{h}=2x+h-1$$

89. Let $y=\dfrac{x-3}{x+4}$. To find $F^{-1}(x)$, we switch the roles of x and y and solve the resulting equation for y:

$$\begin{aligned}x&=\frac{y-3}{y+4}\\x(y+4)&=y-3\\xy+4x&=y-3\\xy-y&=-4x-3\\y(x-1)&=-4x-3\\y&=\frac{-4x-3}{x-1}\end{aligned}$$

So $F^{-1}(x)=\dfrac{-4x-3}{x-1}=\dfrac{4x+3}{1-x}$.

91. $F^{-1}[F(x)] = x$ for $x \neq 4$, by definition of $F^{-1}(x)$.

93. The value is x, since $\left(g \circ g^{-1}\right)(x) = x$ for all x.

95. Since $g^{-1}(x) = \dfrac{1-x}{2}$, we have:

$$g^{-1}(-x) = \frac{1-(-x)}{2} = \frac{1+x}{2}$$

97. Since $\frac{22}{7}$ is in the domain of $F(x)$, then $F^{-1}\left[F\left(\frac{22}{7}\right)\right] = \frac{22}{7}$.

99. Since $f(0) = -2$, it is negative.

101. Since the point $(-3, -1)$ lies on the graph, then $f(-3) = -1$.

103. Since $f(0) = -2$ and $f(8) = -1$, then $f(0) - f(8) = -2 - (-1) = -2 + 1 = -1$.

105. The coordinates of the turning points are $(0, -2)$ and $(5, 1)$.

107. $f(x)$ is decreasing on the intervals $[-6, 0]$ and $[5, 8]$.

109. Since $|x| \leq 2$ corresponds to the interval $-2 \leq x \leq 2$, the largest value of $f(x)$ is 0, occurring at $x = 2$.

111. Since $f(x)$ is not a one-to-one function (it does not pass the horizontal line test) then it does not possess an inverse function.

113. From the graph, we note that $f(x) = g(x)$ when $x = 4$.

115. (a) From the graph, we note that $f(x) = 0$ when $x = 10$.
(b) From the graph, we note that $g(x) = 0$ when $x = 0$.

117. (a) $(f + g)(8) = f(8) + g(8) = 1 + 4 = 5$
(b) $(f - g)(8) = f(8) - g(8) = 1 - 4 = -3$
(c) $fg(8) = f(8) \bullet g(8) = 1 \bullet 4 = 4$
(d) $(f / g)(x) = \dfrac{f(8)}{g(8)} = \dfrac{1}{4}$

119. We compute each value:

$$(f \circ f)(10) = f[f(10)] = f(0) = 5$$
$$(g \circ g)(10) = g[g(10)] = g(3) = 1$$

So $(f \circ f)(10)$ is larger.

121. We note that $f(x) \geq 3$ for $0 \leq x \leq 4$, which is the interval $[0,4]$.

123. The maximum point of $g(x)$ is $(6,5)$, so the largest number in the range of g is 5.

125. We note that $g(x)$ is decreasing when $1 < x < 3$ and $6 < x < 10$, which are the intervals $(1,3) \cup (6,10)$.

127. We consider two cases:
If $4 < x < 5$, then $f(x) > f(5)$ and $x < 5$, thus $f(x) - f(5) > 0$ while $x - 5 < 0$.

Thus $\dfrac{f(x) - f(5)}{x - 5} < 0$.
If $5 < x < 7$, then $f(x) < f(5)$ and $x > 5$, thus $f(x) - f(5) < 0$ while $x - 5 > 0$.

Thus $\dfrac{f(x) - f(5)}{x - 5} < 0$.
Therefore, the quantity is negative for all x-values in the interval $(4,7)$.

129. If A varies jointly as x and y^2, then $A = kxy^2$. We replace x by $3x$ and y by $2y$:
$$A = k(3x)(2y)^2 = k(3x)(4y^2) = 12(kxy^2)$$
So the value of A is 12 times its original value.

131. Let I = intensity and d = distance, so $I = \frac{k}{d^2}$. If $d = 10$ ft, then:
$$I = \frac{k}{10^2} = \frac{k}{100}$$
We are asked to find the value of d when $I = 5\left(\frac{k}{100}\right) = \frac{k}{20}$, so:
$$\frac{k}{20} = \frac{k}{d^2}$$
$$\frac{1}{20} = \frac{1}{d^2}$$
$$d^2 = 20$$
$$d = \sqrt{20} = 2\sqrt{5}$$
So the intensity will be five times as great when $d = 2\sqrt{5} = 4.47$ feet.

133. Let L = length and A = area, so $L = k\sqrt{A}$. We find k by substituting $A = 16\sqrt{3}$ and $L = 8$:
$$8 = k\sqrt{16\sqrt{3}}$$
$$8 = k \bullet 4\sqrt[4]{3}$$
$$k = \frac{8}{4\sqrt[4]{3}}$$
$$k = \frac{2}{\sqrt[4]{3}}$$

So $L = \frac{2}{\sqrt[4]{3}}\sqrt{A}$. When $A = 20$ we have:

$$L = \frac{2}{\sqrt[4]{3}}\sqrt{20} = \frac{4\sqrt{5}}{\sqrt[4]{3}} = \frac{4}{3}\sqrt{5}\,\sqrt[4]{27} \approx 6.80 \text{ cm}$$

Chapter Four Graphing Calculator Exercises

1. (a) Using Zoom-6 settings, we obtain the graph:

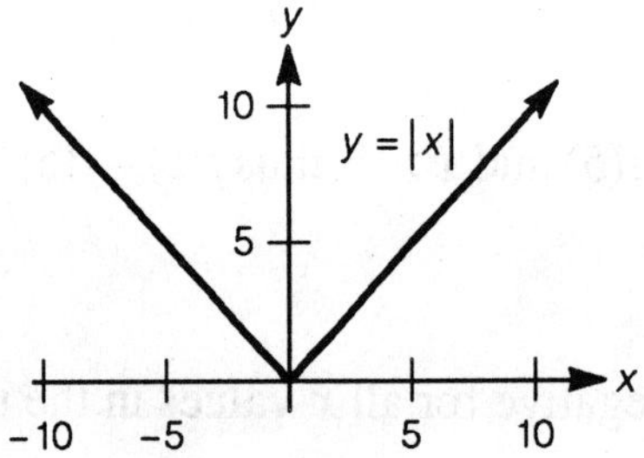

(b) Using the indicated settings, we obtain the graph:

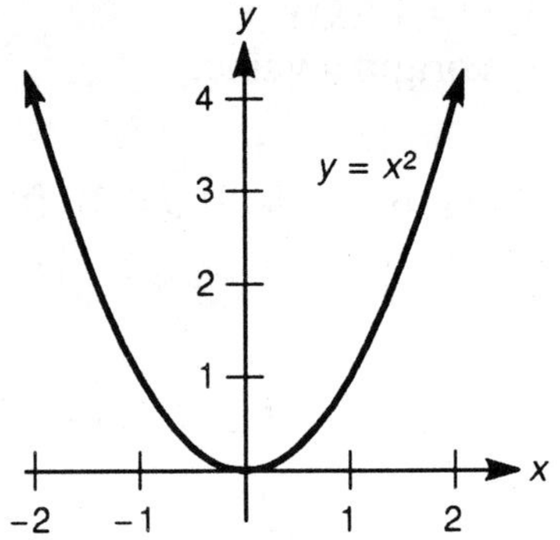

(c) Using the indicated settings, we obtain the graph:

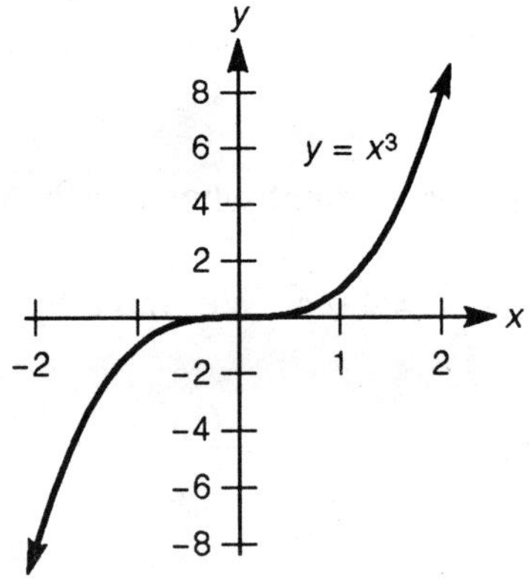

(d) Using the indicated settings, we obtain the graphs:

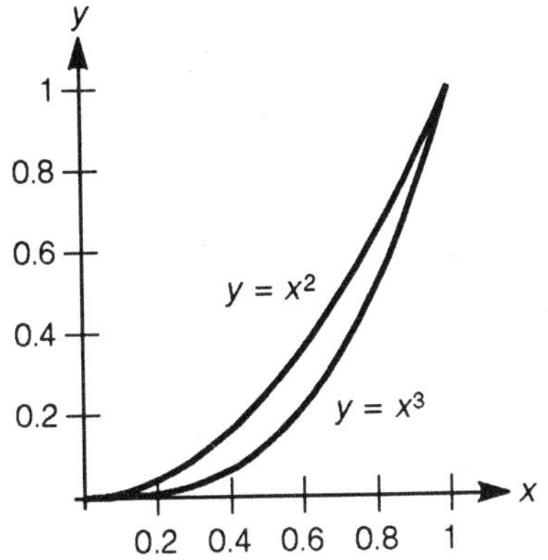

The graphs intersect at $(0, 0)$ and $(1, 1)$. In the interval $(0, 1)$, it appears that x^2 is larger than x^3.

(e) Using the settings $x_{min} = 1$, $x_{max} = 3$, $y_{min} = 0$, and $y_{max} = 30$, we obtain the graphs:

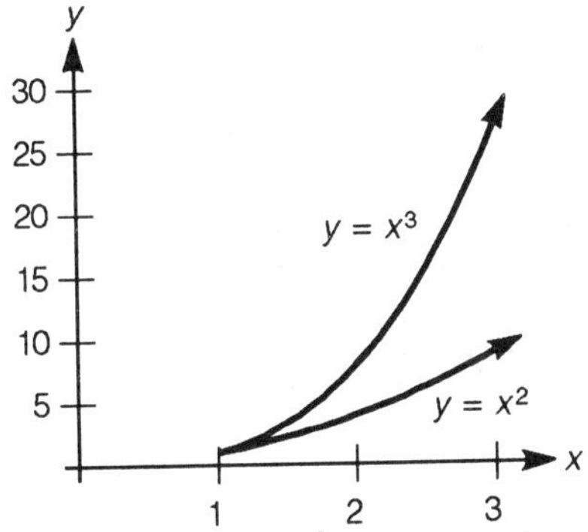

In the interval $(1, \infty)$, it appears that x^3 is larger than x^2.

3. (a) Following the indicated keystrokes, we obtain the graph:

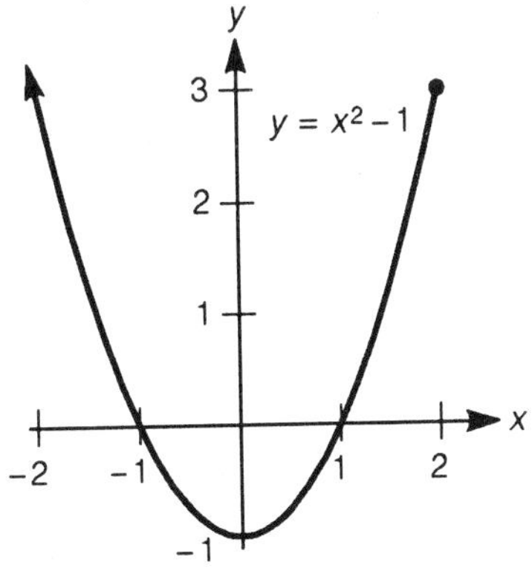

(b) After entering the second function, we obtain the graph:

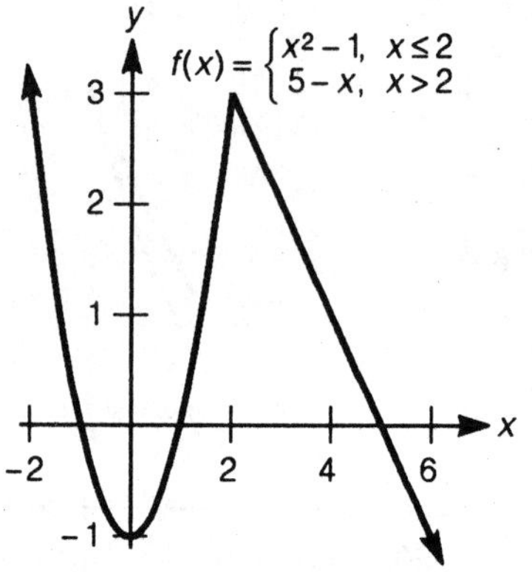

5. Using the illustrated technique, we obtain the graph:

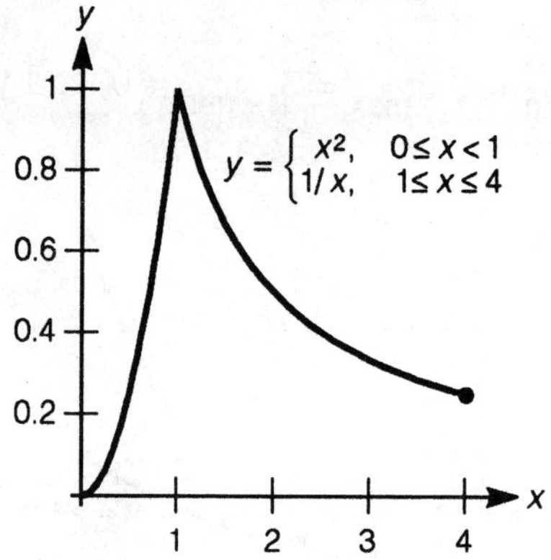

7. Using the illustrated technique, we obtain the graph:

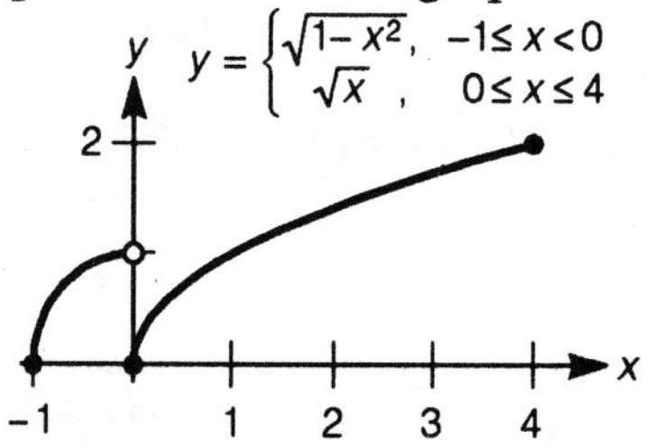

9. Using the illustrated technique, we obtain the graph:

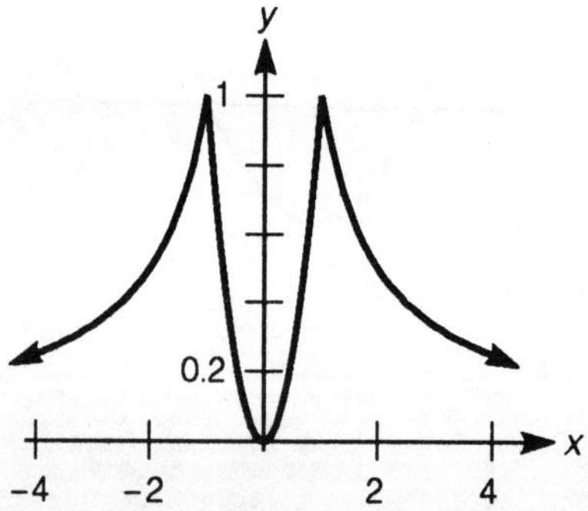

11. (a) We draw the graph:

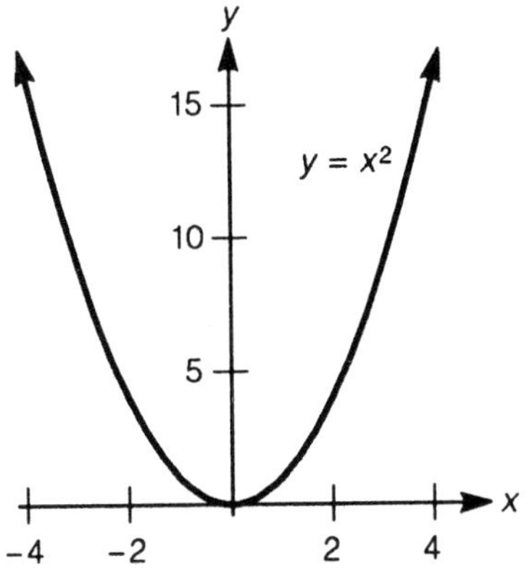

(b) This will be a translation of $y = x^2$ to the left 3 units:

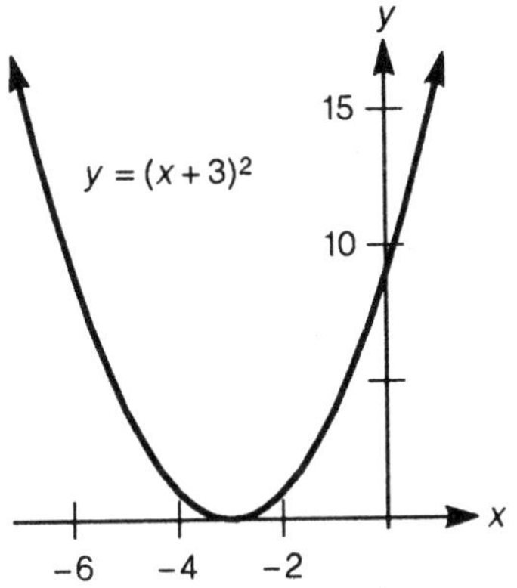

(c) This will be a translation of $y = x^2$ to the left 3 units, then a reflection across the x-axis:

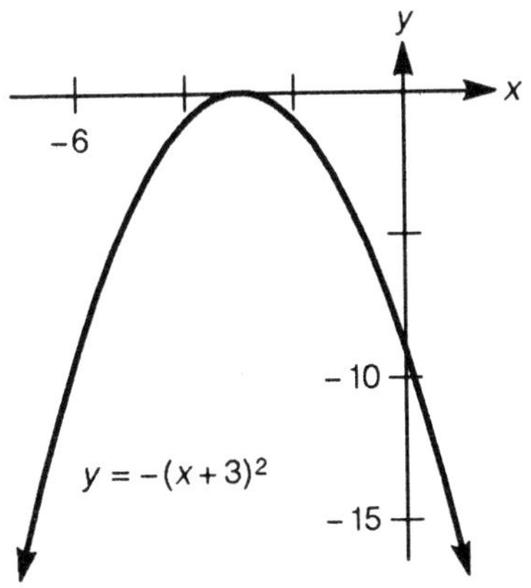

(d) This will be a translation of $y = x^2$ to the left 3 units, then a reflection across the y-axis:

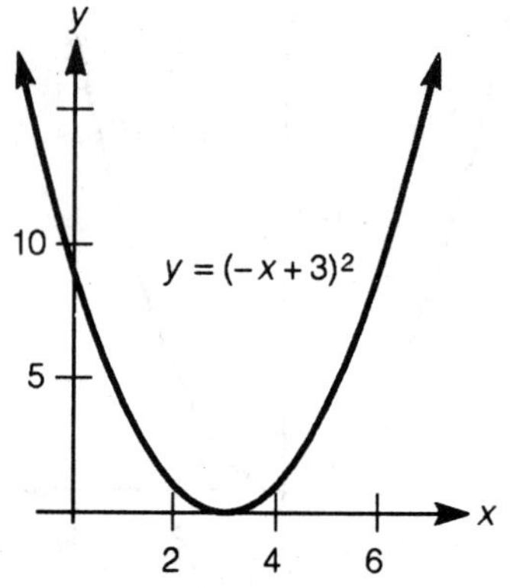

13. (a) We draw the graph:

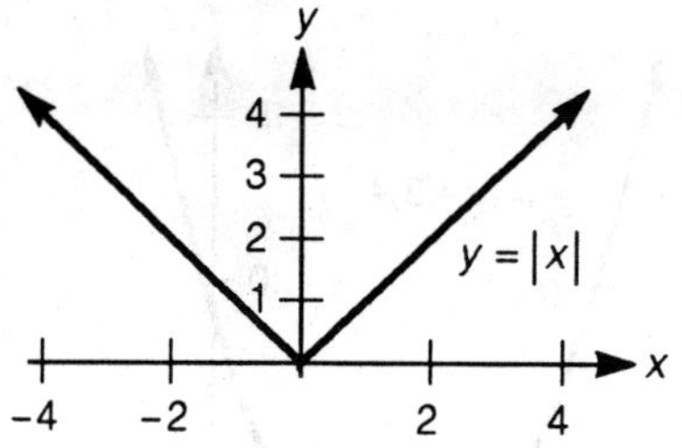

(b) This will be a translation of $y = |x|$ to the left 4 units:

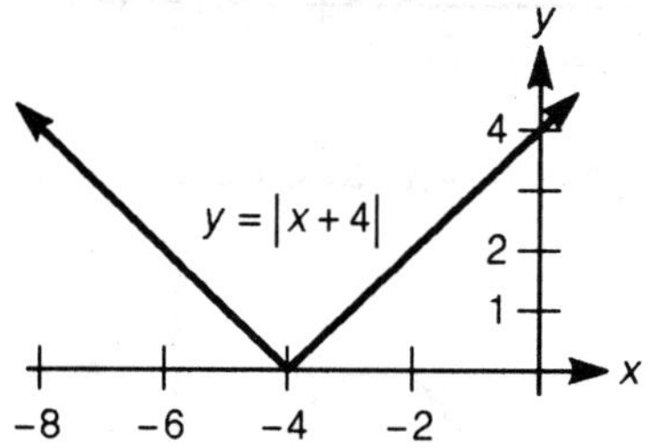

(c) This will be a translation of $y = |x|$ to the left 4 units, then a reflection across the y-axis:

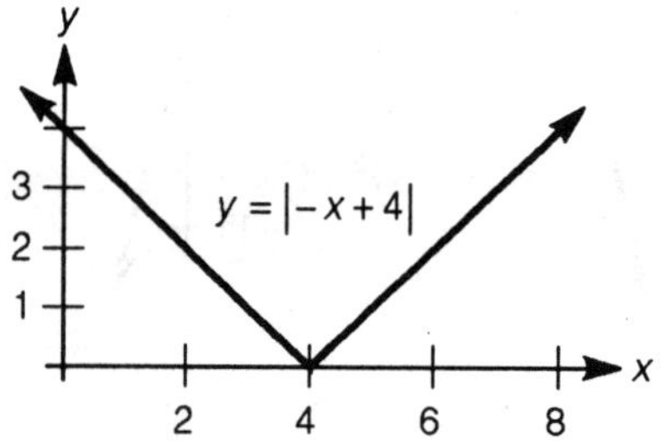

(d) This will be a translation of $y = |x|$ to the left 4 units, then a reflection across the y-axis, then a translation down 2 units:

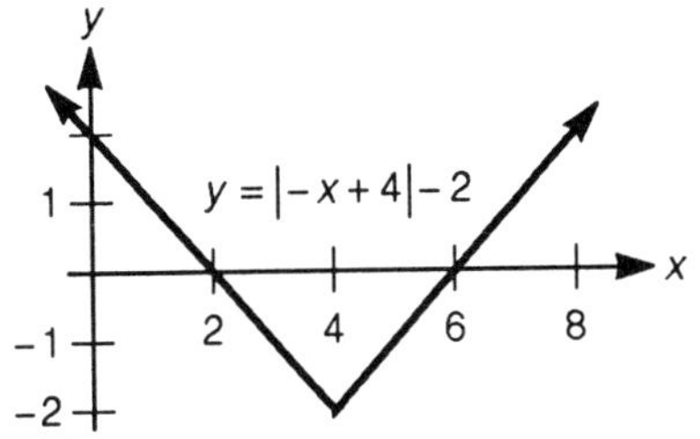

15. (a) We draw the graph:

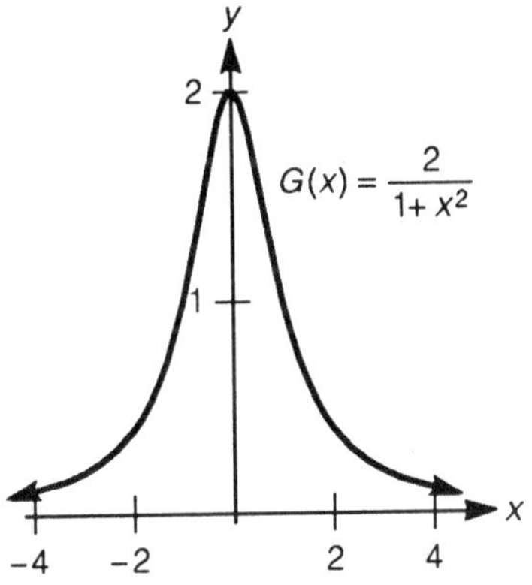

(b) This will be a translation of $G(x)$ to the left 3 units. We compute $G(x + 3)$:

$$G(x+3) = \frac{2}{1+(x+3)^2}$$

Now draw the graph:

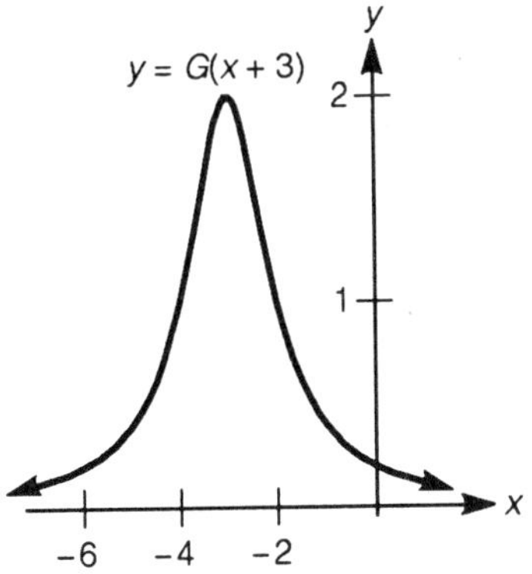

(c) This will be a translation of $G(x)$ to the right 3 units. We compute $G(x - 3)$:

$$G(x-3) = \frac{2}{1+(x-+3)^2}$$

Now draw the graph:

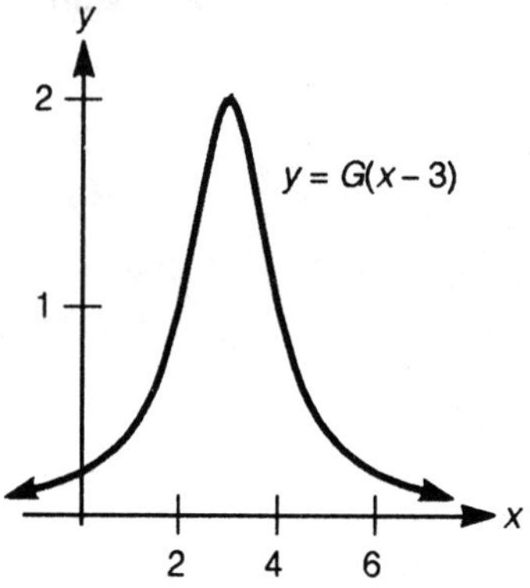

(d) This will be a reflection of $G(x)$ across the x-axis, then a translation up 2 units. We compute $-G(x) + 2$:

$$-G(x)+2=\frac{-2}{1+x^2}+2$$

Now draw the graph:

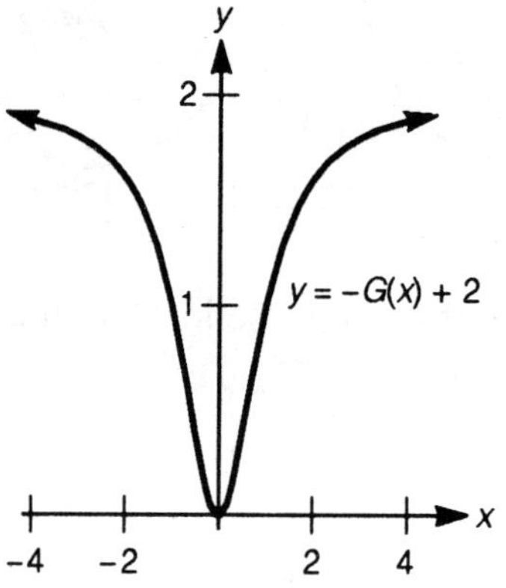

17. We draw the graphs, noting the symmetry about $y = x$:

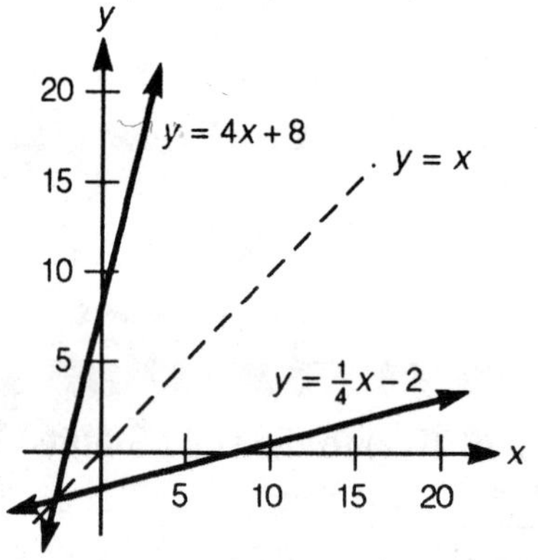

19. We draw the graphs, noting the symmetry about $y = x$:

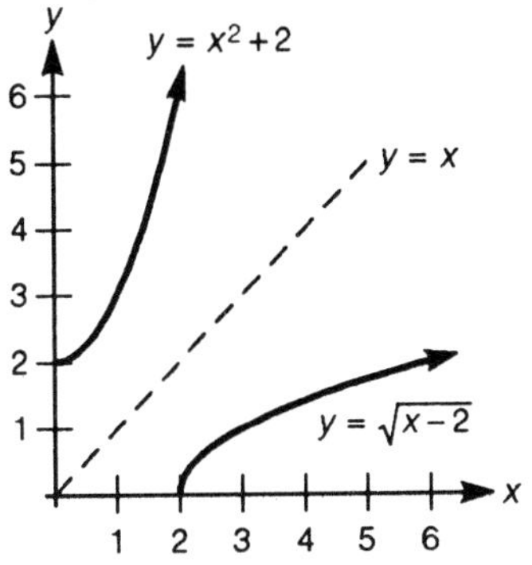

21. We draw the graphs, noting the symmetry about $y = x$:

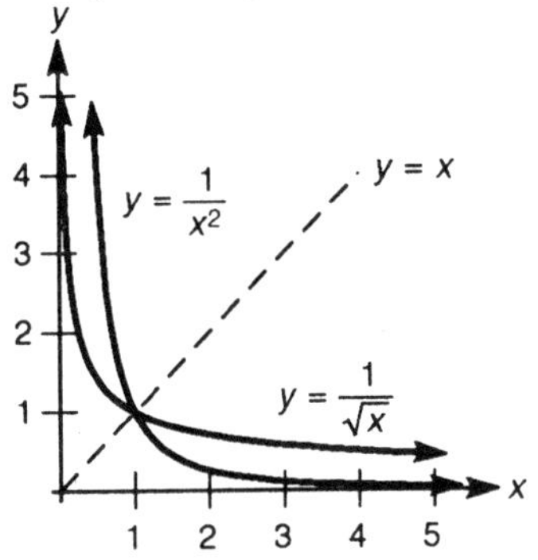

23. The function is not one-to-one. We draw the graph:

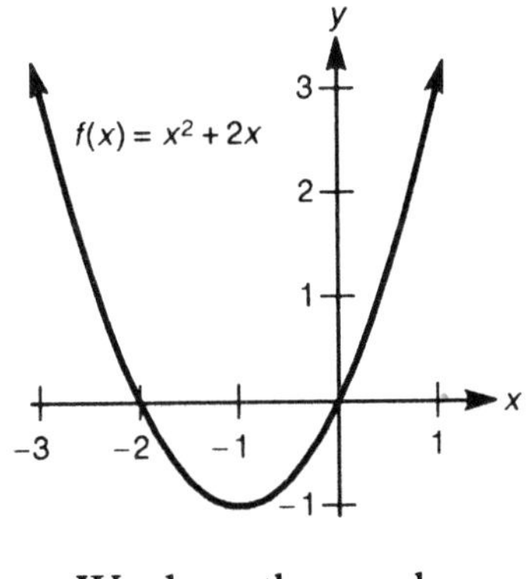

25. The function is not one-to-one. We draw the graph:

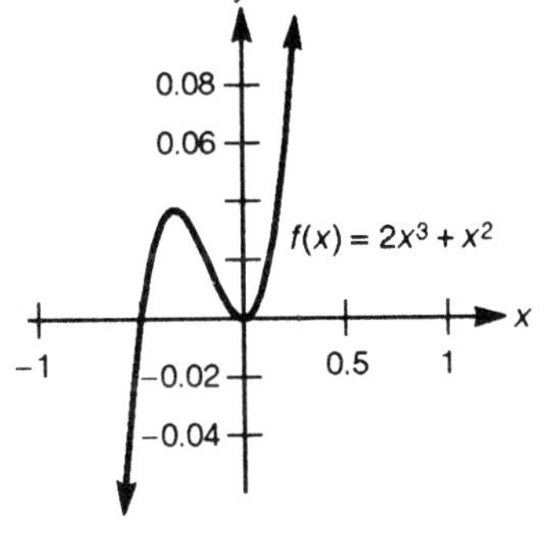

27. (a) The function is one-to-one. We draw the graph:

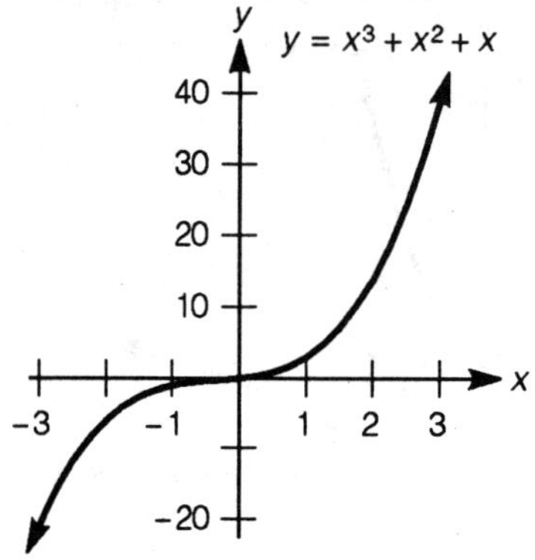

(b) The function is one-to-one. We draw the graph:

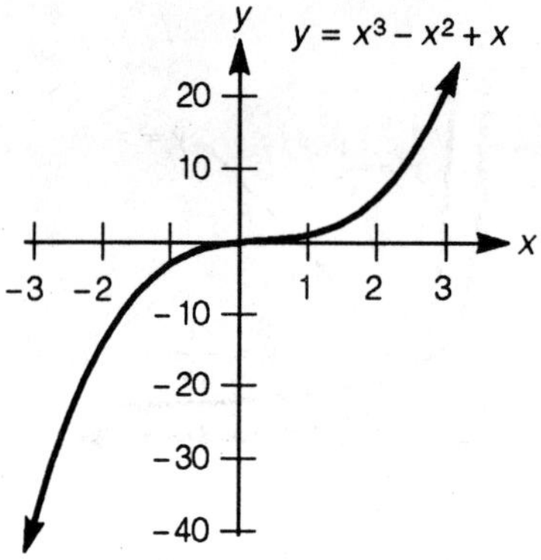

(c) The function is not one-to-one. We draw the graph:

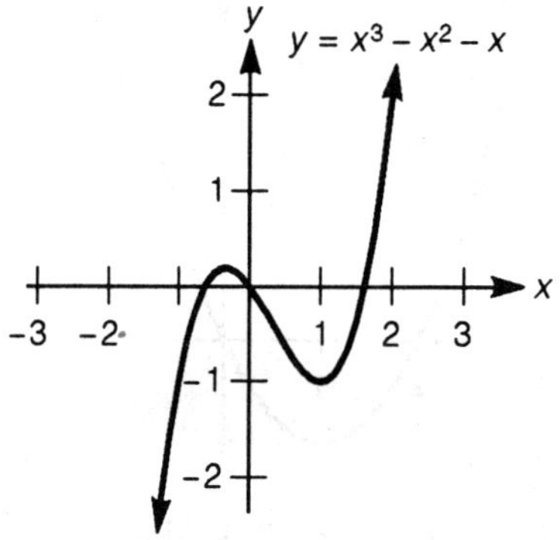

29. We first find $F^{-1}(x)$. Let $y = F(x)$, then switch x and y and solve the resulting equation for y:

$$x = \sqrt{1-y^2}$$
$$x^2 = 1 - y^2$$
$$y^2 = 1 - x^2$$
$$y = -\sqrt{1-x^2}$$

Note that $y \leq 0$ since $x \leq 0$ in the original function $F(x)$. So $F^{-1}(x) = -\sqrt{1-x^2}$ with domain $0 \leq x \leq 1$. We draw the graph using Zoom-5 to observe the symmetry:

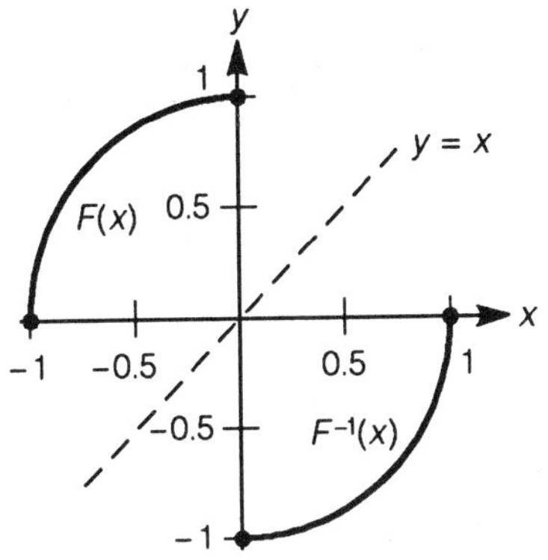

31. (a) We graph the function, noting that it passes the horizontal line test:

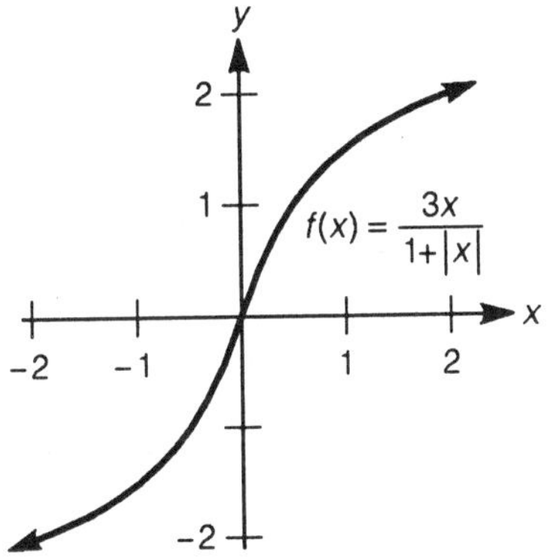

(b) If $x \geq 0$, then $y = \frac{3x}{1+x}$ since $|x| = x$. Switching the roles of x and y, and solving the resulting equation for y:

$$\frac{3y}{1+y} = x$$
$$3y = x + xy$$
$$y(3-x) = x$$
$$y = \frac{x}{3-x} \quad \text{(for } x < 3\text{)}$$

If $x < 0$, then $y = \frac{3x}{1-x}$ since $|x| = -x$. Switching the roles of x and y, and solving the resulting equation for y:

$$\frac{3y}{1-y} = x$$
$$3y = x - xy$$
$$y(3+x) = x$$
$$y = \frac{x}{3+x} \quad \text{(for } x > -3\text{)}$$

Thus the inverse function is:

$$f^{-1}(x)=\begin{cases} x/(3-x) & \text{if } 0 \le x < 3 \\ x/(3+x) & \text{if } -3 < x < 0 \end{cases}$$

Now graph both functions, noting the symmetry about $y = x$:

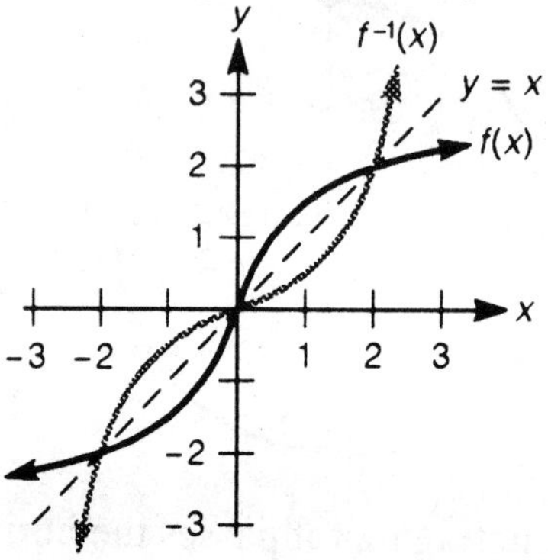

Chapter Four Test

1. We must be sure the quantity inside the radical is non-negative:

$$x^2-5x-6 \ge 0$$
$$(x-6)(x+1) \ge 0$$

The key numbers are –1 and 6. We draw the sign chart:

Interval	Test Number	$x-6$	$x+1$	$(x-6)(x+1)$
$(-\infty,-1)$	–2	neg.	neg.	pos.
$(-1,6)$	0	neg.	pos.	neg.
$(6,\infty)$	8	pos.	pos.	pos.

The product is positive on the intervals $(-\infty,-1)\cup(6,\infty)$ and 0 at $x=-1$ and $x=6$. So the domain is $(-\infty,-1]\cup[6,\infty)$.

2. Let $y=\frac{2x-8}{3x+5}$. We solve for x:

$$\frac{2x-8}{3x+5}=y$$
$$2x-8=3yx+5y$$
$$2x-3yx=5y+8$$
$$x(2-3y)=5y+8$$
$$x=\frac{5y+8}{2-3y}$$

So $y \ne \frac{2}{3}$.The range is $\left(-\infty,\frac{2}{3}\right)\cup\left(\frac{2}{3},\infty\right)$.

3. (a) We compute $(f-g)(x)$:
$$\begin{aligned}(f-g)(x) &= f(x)-g(x)\\ &= (2x^2-3x)-(2-x)\\ &= 2x^2-3x-2+x\\ &= 2x^2-2x-2\end{aligned}$$

(b) We compute $(f \circ g)(x)$:
$$\begin{aligned}(f\circ g)(x) &= f[g(x)]\\ &= f(2-x)\\ &= 2(2-x)^2-3(2-x)\\ &= 8-8x+2x^2-6+3x\\ &= 2x^2-5x+2\end{aligned}$$

(c) Using our result from part (b):
$$f[g(-4)] = 2(-4)^2-5(-4)+2 = 32+20+2 = 54$$

4. We compute the difference quotient:
$$\frac{f(t)-f(a)}{t-a} = \frac{\frac{2}{t}-\frac{2}{a}}{t-a} \bullet \frac{at}{at} = \frac{2a-2t}{at(t-a)} = \frac{-2(t-a)}{at(t-a)} = -\frac{2}{at}$$

5. We first compute $g(x+h)$:
$$g(x+h) = 2(x+h)^2-5(x+h) = 2x^2+4xh+2h^2-5x-5h$$
Now compute the difference quotient:
$$\begin{aligned}\frac{g(x+h)-g(x)}{h} &= \frac{2x^2+4xh+2h^2-5x-5h-2x^2+5x}{h}\\ &= \frac{4xh+2h^2-5h}{h}\\ &= 4x+2h-5\end{aligned}$$

6. Let $y = \frac{-4x}{6x+1}$. We switch the roles of x and y, then solve the resulting equation for y:
$$\begin{aligned}\frac{-4y}{6y+1} &= x\\ -4y &= 6xy+x\\ -x &= 6xy+4y\\ -x &= y(6x+4)\\ \frac{-x}{6x+4} &= y\end{aligned}$$
So $g^{-1}(x) = -\frac{x}{6x+4}$.

7. Since $f^{-1}(x)$ will be the line segment joining the points $(1,-3)$ and $(6,5)$, then $y = -f^{-1}(x)$ will be the line segment joining the points $(1,3)$ and $(6,-5)$. We compute the slope:
$$m = \frac{-5-3}{6-1} = \frac{-8}{5}$$

8. (a) The x-intercepts are 4 and 2, and the y-intercept is –2:

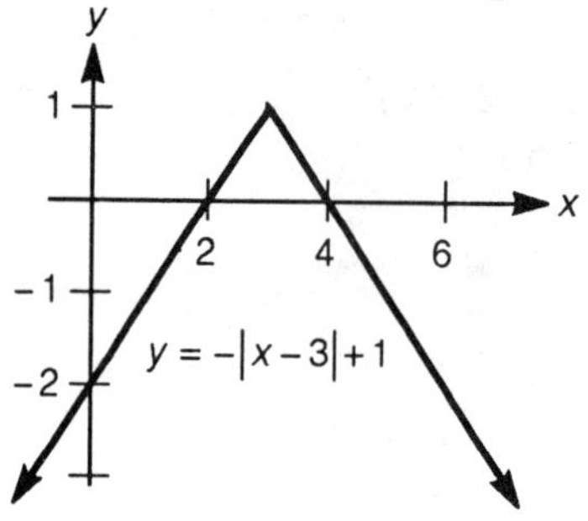

(b) The x-intercept is $-\frac{5}{2}$ and the y-intercept is $-\frac{5}{3}$:

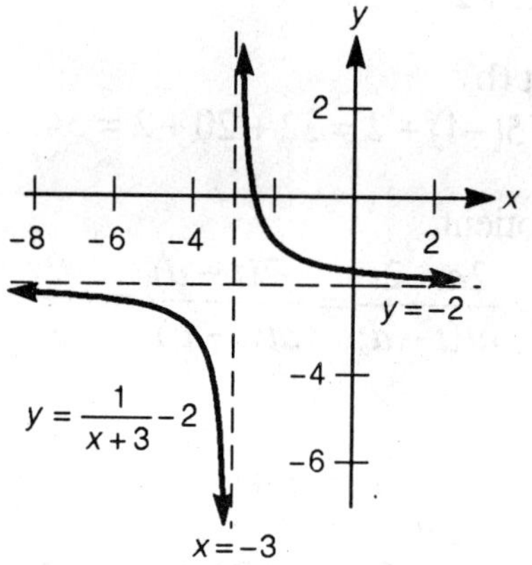

9. (a) The range of g is $[-3, 1]$.
(b) The turning point has coordinates $(-2, 1)$.
(c) The minimum value of g is -3, which occurs where $x = 2$.
(d) The maximum value of g is 1, occurring at $x = -2$.
(e) For $-2 < x < 2$, g is decreasing. So g is decreasing on $(-2, 2)$.

10. (a) We compute $f\left(-\frac{3}{2}\right)$:

$$f\left(-\tfrac{3}{2}\right) = \left(-\tfrac{3}{2}\right)^2 - 3\left(-\tfrac{3}{2}\right) - 1 = \tfrac{9}{4} + \tfrac{9}{2} - 1 = \tfrac{23}{4}$$

(b) We compute $f(\sqrt{3} - 2)$:

$$\begin{aligned} f(\sqrt{3} - 2) &= (\sqrt{3} - 2)^2 - 3(\sqrt{3} - 2) - 1 \\ &= 3 - 4\sqrt{3} + 4 - 3\sqrt{3} + 6 - 1 \\ &= 12 - 7\sqrt{3} \end{aligned}$$

11. The domain is all real numbers except -1, or $(-\infty,-1)\cup(-1,\infty)$. We draw the graph:

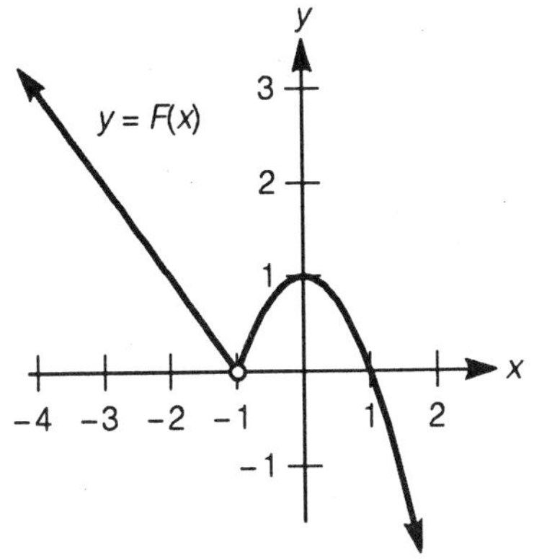

12. Since $y=f(-x)$ is a reflection of $y=f(x)$ across the y-axis, it will be the line segment joining the points $(-1,3)$ and $(-5,-2)$:

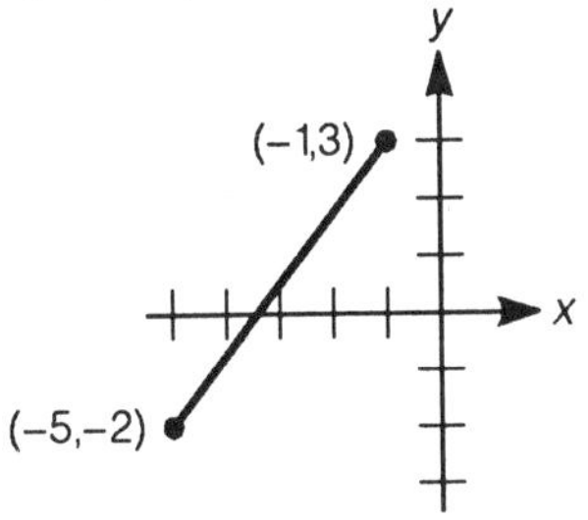

13. We solve the equation:

$$\begin{aligned} 5+f(4t-3)&=2\\ f(4t-3)&=-3\\ 4t-3&=f^{-1}(-3)\\ 4t-3&=1\\ 4t&=4\\ t&=1 \end{aligned}$$

14. The variation law is $z=\frac{kx^2}{\sqrt{t}}$. Given $x=6$, $t=9$, $z=12$, we find k:

$$\begin{aligned} 12&=\frac{k\bullet(6)^2}{\sqrt{9}}\\ 12&=\frac{36k}{3}\\ 1&=k \end{aligned}$$

So the variation law is $z=\frac{x^2}{\sqrt{t}}$. Now substitute $x=3$ and $t=16$:

$$z=\frac{(3)^2}{\sqrt{16}}=\frac{9}{4}$$

Chapter Five
Polynomial and Rational Functions: Applications to Optimization

5.1 Linear Functions

1. We first find the slope between the points $(-1,0)$ and $(5,4)$:

$$m=\frac{4-0}{5-(-1)}=\frac{4}{6}=\frac{2}{3}$$

Now use the point $(-1,0)$ in the point-slope formula:

$$y-0=\tfrac{2}{3}[x-(-1)]$$
$$y=\tfrac{2}{3}(x+1)$$
$$y=\tfrac{2}{3}x+\tfrac{2}{3}$$

Using functional notation we have $f(x)=\tfrac{2}{3}x+\tfrac{2}{3}$.

3. We first find the slope between the points $(0,0)$ and $(1,\sqrt{2})$:

$$m=\frac{\sqrt{2}-0}{1-0}=\sqrt{2}$$

Now, since $(0,0)$ is the y-intercept, we use the slope-intercept formula to write $y=\sqrt{2}x$. Using functional notation we have $g(x)=\sqrt{2}x$.

5. We find the slope of $x-y=1$:

$$-y=-x+1$$
$$y=x-1$$

So the parallel slope is 1. Use the point $\left(\frac{1}{2},-3\right)$ in the point-slope formula:

$$y-(-3)=1\left(x-\tfrac{1}{2}\right)$$
$$y+3=x-\tfrac{1}{2}$$
$$y=x-\tfrac{7}{2}$$

Using functional notation we have $f(x)=x-\frac{7}{2}$.

7. We set $x=0$ to find the y-intercepts of the circle:

$$0-0+y^2-3=0$$
$$y^2=3$$
$$y=\pm\sqrt{3}$$

So a horizontal line passing through $(0,\sqrt{3})$ is $y=\sqrt{3}$. Using functional notation we have $f(x)=\sqrt{3}$.

9. Since the points $(-1,2)$ and $(0,4)$ lie on the graph of the inverse function, then $(2,-1)$ and $(4,0)$ must lie on the graph of the function. We find the slope:

$$m=\frac{0-(-1)}{4-2}=\frac{1}{2}$$

We use the point $(4,0)$ in the point-slope formula:

$$y-0=\tfrac{1}{2}(x-4)$$
$$y=\tfrac{1}{2}x-2$$

Using functional notation we have $f(x)=\frac{1}{2}x-2$.

11. We compute $(f\circ g)(x)$:

$$(f\circ g)(x)=f[g(x)]=f(1-2x)=3(1-2x)-4=3-6x-4=-6x-1$$

So $f\circ g$ is a linear function (it is in standard form).

13. Call $V(t)$ the value of the machine after t years. When $t=0$ we have $V=20{,}000$ and when $t=8$ we have $V=1{,}000$. We find the slope of the line between the points $(0,20000)$ and $(8,1000)$:

$$m=\frac{1000-20000}{8-0}=\frac{-19000}{8}=-2375$$

Since $(0,20000)$ is the y-intercept, we use the slope-intercept formula to write $V=-2375t+20000$. Using functional notation we have $V(t)=-2375t+20000$.

15. (a) Call $V(t)$ the value of the machine after t years. Now $V=60{,}000$ when $t=0$ and $V=0$ when $t=5$. We find the slope of the line between $(0,60000)$ and $(5,0)$:

$$m=\frac{0-60000}{5-0}=\frac{-60000}{5}=-12000$$

Since $(0,60000)$ is the y-intercept, we use the slope-intercept formula to write $V=-12000t+60000$. Using functional notation we have $V(t)=-12000t+60000$.

(b) The completed schedule is:

End of Year	Yearly Depreciation	Accumulated Depreciation	Value V
0	0	0	60,000
1	12,000	12,000	48,000
2	12,000	24,000	36,000
3	12,000	36,000	24,000
4	12,000	48,000	12,000
5	12,000	60,000	0

17. (a) We let $x = 10$, so $C(10) = 450 + 8x = 450 + 8(10) = \$530$.

(b) We let $x = 11$, so $C(11) = 450 + 8(11) = \$538$.

(c) There are two ways to find the marginal cost. One way is to recognize that the marginal cost will be the slope of the line, which is \$8/fan. Another way would be to use the definition of the marginal cost: it is cost of producing the next unit. Since $C(10)$ is the cost of producing 10 fans and $C(11)$ is the cost of producing 11 fans, then the marginal cost would be $C(11) - C(10) = 538 - 530 = \$8$/fan. We get the same answer using either approach.

19. (a) We first compute $C(n + 1)$:

$$C(n+1) = 400 + 50(n+1) = 400 + 50n + 50 = 450 + 50n$$

Therefore:

$$C(n+1) - C(n) = (450 + 50n) - (400 + 50n) = 450 + 50n - 400 - 50n = 50$$

(b) The marginal cost, which represents the cost to produce the next unit, is the slope of the line. So the marginal cost is \$50/player.

(c) The answers are the same. Notice that (a) directly computed marginal cost from its definition.

21. Since the velocity will be the slope of the distance line, we compute the slope of each line.

(a) Let $(x_1, y_1) = (1, 4)$ and $(x_2, y_2) = (6, 8)$. Then:

$$m = \frac{8-4}{6-1} = \frac{4}{5}$$

So the velocity is $\frac{4}{5}$ ft/sec.

(b) Let $(x_1, y_1) = (2, 4)$ and $(x_2, y_2) = (5, 4)$. Then:

$$m = \frac{4-4}{5-2} = \frac{0}{3} = 0$$

So the velocity is 0 cm/sec.

(c) Let $(x_1, y_1) = (0, 0)$ and $(x_2, y_2) = (2, 16)$. Then:

$$m = \frac{16-0}{2-0} = \frac{16}{2} = 8$$

So the velocity is 8 mph.

23. (a) Since the velocity of A is 3 units/sec and the velocity of B is 20 units/sec, then B is traveling faster.

(b) When $t = 0$, A is at $x = 100$ and B is at $x = -36$. So A is farther to the right.

(c) We set the two x-coordinates equal:

$$\begin{aligned} 3t + 100 &= 20t - 36 \\ 136 &= 17t \\ 8 &= t \end{aligned}$$

When $t = 8$ sec, A and B have the same x-coordinate.

25. (a) We plot the points:

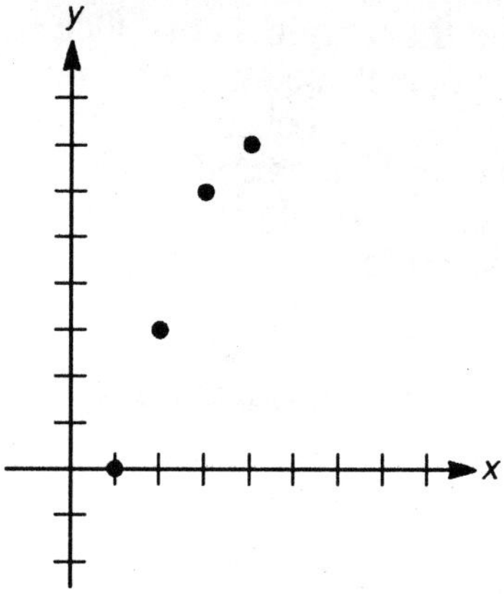

(b) The slope appears to be approximately 2.5 and the y-intercept is –2:

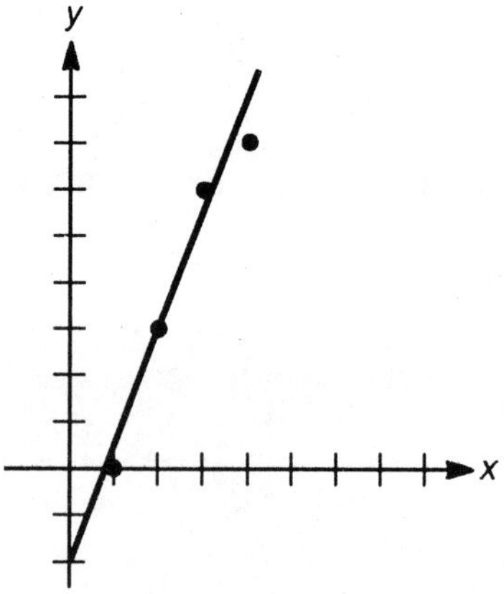

(c) We graph the regression line $y = 2.4x - 2$:

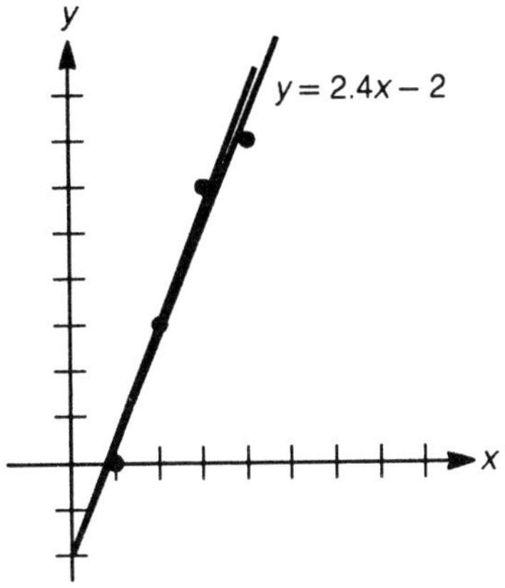

27. (a) We let $x = 2000$:

$$y = 37025(2000) - 70226566 = 3823434$$

It would appear that Los Angeles will not reach a population of 4 million by the year 2000.

(b) We let $x = 2010$:

$$y = 37025(2010) - 70226566 = 4195000$$

The population in Los Angeles will be approximately 4195000 in the year 2010.

29. (a) We let $x = 1996$:

$$y = (0.3428)(1996) - 680.6114 = 3.6174$$

The value of the toy imports from China in 1996 will be approximately 3.6174 billion dollars.

(b) We let $x = 2000$:

$$y = (0.3428)(2000) - 680.6114 = 4.9886$$

The value of the toy imports will not exceed 5 billion dollars by the year 2000.

31. We first simplify $(f \circ f)(x)$:

$$(f \circ f)(x) = f(mx + b) = m(mx + b) + b = m^2x + (mb + b)$$

Since $(f \circ f)(x) = 9x + 4$, and since m and b are constants, we must have the two equations (equating x-coefficients and constants):

$$m^2 = 9 \quad \text{and} \quad mb + b = 4$$

Since $m > 0$, then $m = 3$ and:

$$3b + b = 4$$
$$4b = 4$$
$$b = 1$$

So $f(x) = 3x + 1$.

33. (a) Since f is a linear function, then $f(x) = mx + b$. We compute each side of the equality:

$$f\left(\frac{x_1+x_2}{2}\right)=m\left(\frac{x_1+x_2}{2}\right)+b=\tfrac{m}{2}x_1+\tfrac{m}{2}x_2+b$$

$$\frac{f(x_1)+f(x_2)}{2}=\frac{(mx_1+b)+(mx_2+b)}{2}=\frac{mx_1+mx_2+2b}{2}=\tfrac{m}{2}x_1+\tfrac{m}{2}x_2+b$$

Since each quantity is equal, we have shown the desired result.

(b) We compute each side of the equality:

$$f\left(\frac{x_1+x_2}{2}\right)=\left(\frac{x_1+x_2}{2}\right)^2=\frac{x_1^2+2x_1x_2+x_2^2}{4}=\tfrac{1}{4}x_1^2+\tfrac{1}{2}x_1x_2+\tfrac{1}{4}x_2^2$$

$$\frac{f(x_1)+f(x_2)}{2}=\frac{x_1^2+x_2^2}{2}=\tfrac{1}{2}x_1^2+\tfrac{1}{2}x_2^2$$

Clearly the two quantities are not equal.

35. (a) We compute the sums:

$$\sum x=1+2+3+4=10$$
$$\sum y=0+3+6+7=16$$

(b) We compute the sums:

$$\sum x^2=1+4+9+16=30$$
$$\sum xy=0+6+18+28=52$$

(c) We multiply the first equation by -3:

$$-12b-30m=-48$$
$$10b+30m=52$$

Adding, we obtain:

$$-2b=4$$
$$b=-2$$

Substituting into the first equation:

$$4b+10m=16$$
$$-8+10m=16$$
$$10m=24$$
$$m=2.4$$

So the regression line is $y = 2.4x - 2$.

37. We first do the computations:

$$\sum x=1+2+3+4+5=15$$
$$\sum y=2+3+9+9+11=34$$
$$\sum x^2=1+4+9+16+25=55$$
$$\sum xy=2+6+27+36+55=126$$

So the system of equations becomes:

$$5b + 15m = 34$$
$$15b + 55m = 126$$

We multiply the first equation by –3:

$$-15b - 45m = -102$$
$$15b + 55m = 126$$

Adding, we obtain:

$$10m = 24$$
$$m = 2.4$$

Substituting into the first equation:

$$5b + 36 = 34$$
$$5b = -2$$
$$b = -0.4$$

So the regression line is $y = 2.4x - 0.4$.

39. We first do the computations:

$$\sum x = 520 + 740 + 560 + 610 + 650 = 3080$$
$$\sum y = 81 + 98 + 83 + 88 + 95 = 445$$
$$\sum x^2 = 270400 + 547600 + 313600 + 372100 + 422500 = 1926200$$
$$\sum xy = 42120 + 72520 + 46480 + 53680 + 61750 = 276550$$

So the system of equations becomes:

$$5b + 3080m = 445$$
$$3080b + 1926200m = 276550$$

Divide the first equation by 5 and the second equation by 10:

$$b + 616m = 89$$
$$308b + 192620m = 27655$$

We multiply the first equation by –308:

$$-308b - 189728m = -27412$$
$$308b + 192620m = 27655$$

Adding, we obtain:

$$2892m = 243$$
$$m \approx 0.084$$

Substituting into the first equation:

$$5b + 258.8 = 445$$
$$5b = 186.2$$
$$b \approx 37.241$$

So the regression line is $y = 0.084x + 37.241$.

5.2 Quadratic Functions

1. The vertex is $(-2, 0)$, the axis of symmetry is $x = -2$, the minimum value is 0, the x-intercept is -2, and the y-intercept is 4. We draw the graph:

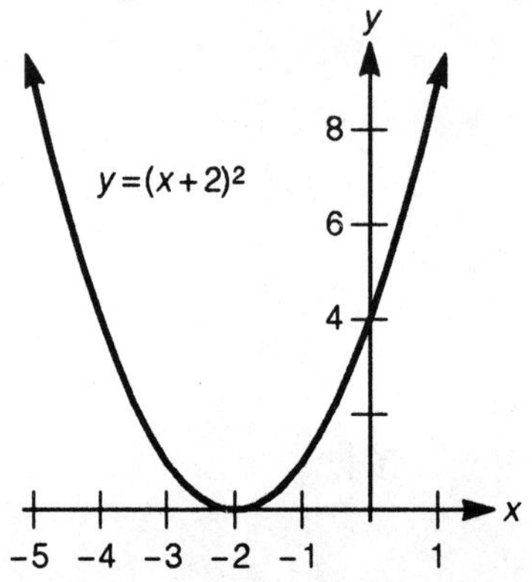

3. The vertex is $(-2, 0)$, the axis of symmetry is $x = -2$, the minimum value is 0, the x-intercept is -2, and the y-intercept is 8. We draw the graph:

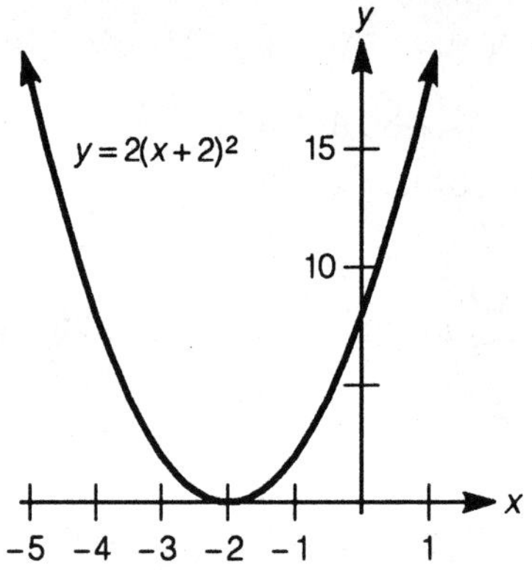

5. The vertex is $(-2, 4)$, the axis of symmetry is $x = -2$, the maximum value is 4, the x-intercepts are $-2 \pm \sqrt{2}$, and the y-intercept is -4. We draw the graph:

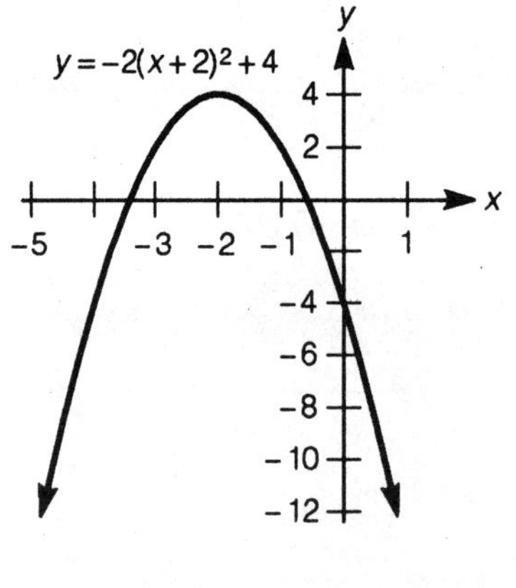

7. We first complete the square:

$$f(x) = x^2 - 4x = \left(x^2 - 4x + 4\right) - 4 = (x-2)^2 - 4$$

The vertex is $(2,-4)$, the axis of symmetry is $x = 2$, the minimum value is -4, the x-intercepts are 0 and 4, and the y-intercept is 0. We draw the graph:

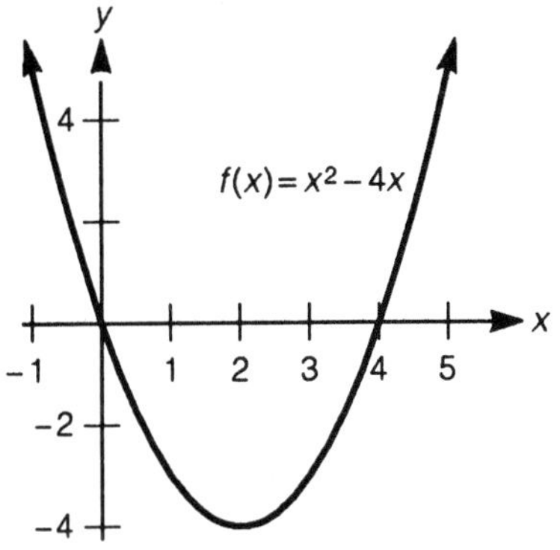

9. The vertex is $(0,1)$, the axis of symmetry is $x = 0$, the maximum value is 1, the x-intercepts are ± 1, and the y-intercept is 1. We draw the graph:

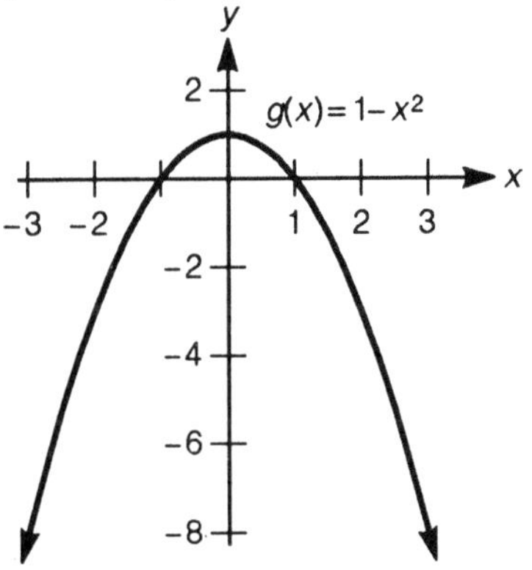

11. We first complete the square:

$$y = x^2 - 2x - 3 = \left(x^2 - 2x + 1\right) - 4 = (x-1)^2 - 4$$

The vertex is $(1,-4)$, the axis of symmetry is $x = 1$, the minimum value is -4, the x-intercepts are 3 and -1, and the y-intercept is -3. We draw the graph:

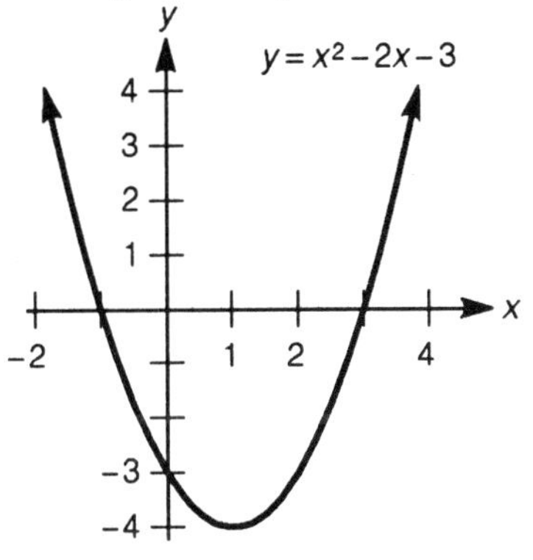

13. We first complete the square:

$$y=-x^2+6x+2=-(x^2-6x)+2=-(x^2-6x+9)+11=-(x-3)^2+11$$

The vertex is (3, 11), the axis of symmetry is $x=3$, the maximum value is 11, the x-intercepts are $3\pm\sqrt{11}$, and the y-intercept is 2. We draw the graph:

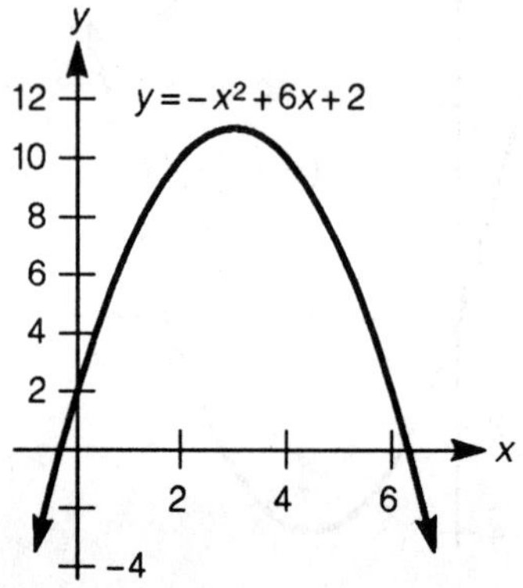

15. The vertex is (0, 0), the axis of symmetry is $t=0$, the minimum value is 0, the t-intercept is 0, and the s-intercept is 0. We draw the graph:

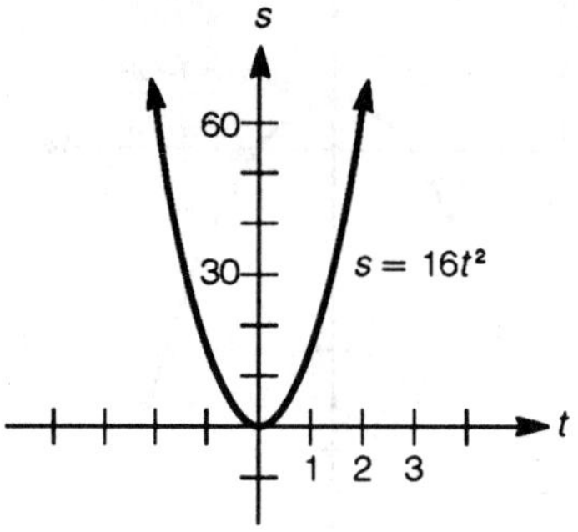

17. We first complete the square:

$$s=-9t^2+3t+2=-9\left(t^2-\tfrac{1}{3}t\right)+2=-9\left(t^2-\tfrac{1}{3}t+\tfrac{1}{36}\right)+\tfrac{9}{4}=-9\left(t-\tfrac{1}{6}\right)^2+\tfrac{9}{4}$$

The vertex is $\left(\frac{1}{6},\frac{9}{4}\right)$, the axis of symmetry is $t=\frac{1}{6}$, the maximum value is $\frac{9}{4}$, the t-intercepts are $-\frac{1}{3}$ and $\frac{2}{3}$, and the s-intercept is 2. We draw the graph:

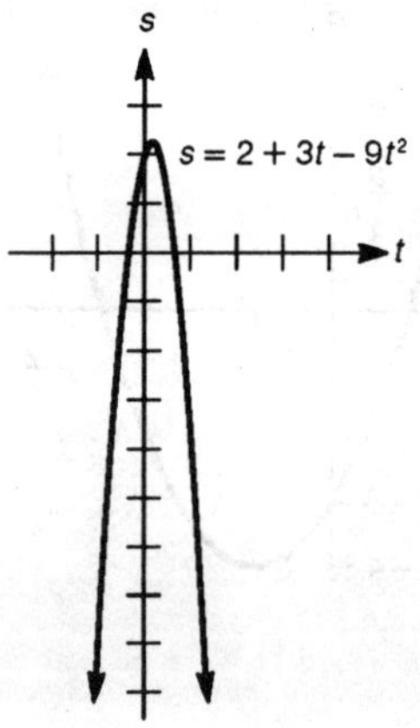

19. We complete the square:

$$y = 2x^2 - 4x + 11 = 2\left(x^2 - 2x\right) + 11 = 2\left(x^2 - 2x + 1\right) + 9 = 2(x-1)^2 + 9$$

Since the vertex is $(1, 9)$ and the parabola will be pointed up, then $x = 1$ will yield a minimum output value.

21. We complete the square:

$$g(x) = -6x^2 + 18x = -6\left(x^2 - 3x\right) = -6\left(x^2 - 3x + \tfrac{9}{4}\right) + \tfrac{27}{2} = -6\left(x - \tfrac{3}{2}\right)^2 + \tfrac{27}{2}$$

Since the vertex is $\left(\tfrac{3}{2}, \tfrac{27}{2}\right)$ and the parabola will be pointed down, then $x = \tfrac{3}{2}$ will yield a maximum output value.

23. Since the vertex is $(0, -10)$ and the parabola will be pointed up, then $x = 0$ will yield a minimum output value.

25. We complete the square:

$$y = x^2 - 8x + 3 = \left(x^2 - 8x + 16\right) - 13 = (x-4)^2 - 13$$

Since the vertex is $(4, -13)$ and the parabola will be pointed up, then the function has a minimum value of -13.

27. We complete the square:

$$\begin{aligned} y &= -2x^2 - 3x + 2 \\ &= -2\left(x^2 + \tfrac{3}{2}x\right) + 2 \\ &= -2\left(x^2 + \tfrac{3}{2}x + \tfrac{9}{16}\right) + 2 + \tfrac{9}{8} \\ &= -2\left(x + \tfrac{3}{4}\right)^2 + \tfrac{25}{8} \end{aligned}$$

Since the vertex is $\left(-\tfrac{3}{4}, \tfrac{25}{8}\right)$ and the parabola will be pointed down, then the function has a maximum value of $\tfrac{25}{8}$.

29. We do not need to complete the square, since the vertex is $(0, 1000)$ and the parabola will be pointed down. So the function has a maximum value of 1000.

31. We find the vertex of the parabola by completing the square:

$$y = x^2 - 6x + 13 = \left(x^2 - 6x + 9\right) + 4 = (x-3)^2 + 4$$

So the vertex is $(3, 4)$. We now use the distance formula with the points $(0, 0)$ and $(3, 4)$:

$$d = \sqrt{(3-0)^2 + (4-0)^2} = \sqrt{9+16} = \sqrt{25} = 5$$

So the vertex is 5 units from the origin.

33. We compute $(f \circ g)(x)$:

$$(f \circ g)(x) = 2\left(x^2 + 4x + 1\right) - 3 = 2x^2 + 8x + 2 - 3 = 2x^2 + 8x - 1$$

So $f \circ g$ is a quadratic function.

35. We compute $(g \circ h)(x)$:

$$\begin{aligned}(g \circ h)(x) &= \left(1-2x^2\right)^2 + 4\left(1-2x^2\right)+1 \\ &= 1-4x^2+4x^4+4-8x^2+1 \\ &= 4x^4-12x^2+6\end{aligned}$$

So $g \circ h$ is neither linear nor quadratic.

37. We compute $(f \circ f)(x)$:

$$(f \circ f)(x) = 2(2x-3)-3 = 4x-6-3 = 4x-9$$

So $f \circ f$ is a linear function.

39. (a) We first complete the square on $x^2 - 6x + 73$:

$$x^2-6x+73 = \left(x^2-6x+9\right)+73-9 = (x-3)^2+64$$

So $f(x) = \sqrt{(x-3)^2+64}$. This would achieve a minimum value at $\left(3,\sqrt{64}\right) = (3,8)$.

(b) Here $g(x) = \sqrt[3]{(x-3)^2+64}$, which would achieve a minimum value at

$\left(3,\sqrt[3]{64}\right) = (3,4)$.

(c) We complete the square on $x^4 - 6x^2 + 73$:

$$x^4-6x^2+73 = \left(x^4-6x^2+9\right)+73-9 = \left(x^2-3\right)^2+64$$

So $h(x) = (x^2-3)^2 + 64$, which would achieve a minimum value at $\left(\pm\sqrt{3},64\right)$.

41. (a) We complete the square on $-x^2 + 4x + 12$:

$$-x^2+4x+12 = -\left(x^2-4x\right)+12 = -\left(x^2-4x+4\right)+16 = -(x-2)^2+16$$

So $f(x) = \sqrt{-(x-2)^2+16}$, which has a maximum value at $\left(2,\sqrt{16}\right) = (2,4)$.

(b) Now $g(x) = \sqrt[3]{-(x-2)^2+16}$, which has a maximum value at $\left(2,\sqrt[3]{16}\right) = \left(2,2\sqrt[3]{2}\right)$.

(c) Here $h(x) = -\left(x^2-2\right)^2+16$, which has a maximum value at $\left(\pm\sqrt{2},16\right)$.

43. Since the vertex is $(2, 2)$, we know its equation will be $y = A(x-2)^2 + 2$. Since $(0, 0)$ must lie on this parabola, we substitute the points to find A:

$$\begin{aligned} y &= A(x-2)^2+2 \\ 0 &= A(0-2)^2+2 \\ 0 &= 4A+2 \\ -2 &= 4A \\ A &= -\tfrac{1}{2}\end{aligned}$$

So the parabola is $y = -\frac{1}{2}(x-2)^2+2$.

45. Since the vertex is $(3,-1)$, we know its equation will be $y = A(x-3)^2 - 1$. Since $(1,0)$ must lie on this parabola, we substitute the points to find A:

$$y = A(x-3)^2 - 1$$
$$0 = A(1-3)^2 - 1$$
$$0 = 4A - 1$$
$$1 = 4A$$
$$A = \tfrac{1}{4}$$

So the parabola is $y = \frac{1}{4}(x-3)^2 - 1$.

47. We complete the square:

$$y = ax^2 + bx + c$$
$$y = a\left(x^2 + \frac{b}{a}x\right) + c$$
$$y = a\left(x^2 + \frac{b}{a}x + \frac{b^2}{4a^2}\right) + c - \frac{ab^2}{4a^2}$$
$$y = a\left(x + \frac{b}{2a}\right)^2 + \frac{4ac - b^2}{4a}$$

So the vertex is $\left(-\frac{b}{2a}, \frac{4ac-b^2}{4a}\right)$.

49. We complete the square:

$$y = x^2 + 2x + c = \left(x^2 + 2x + 1\right) + c - 1 = (x+1)^2 + c - 1$$

The minimum value is $c - 1 = \sqrt{2}$, thus $c = 1 + \sqrt{2}$.

51. Since a and b will be the x-intercepts of the parabola, and all points with the same y-coordinate will be symmetric about the axis of symmetry, it follows that the midpoint $\left(\frac{a+b}{2}, 0\right)$ must lie on the axis of symmetry. Thus the x-coordinate of the vertex (which also lies on this axis) is $\frac{a+b}{2}$.

53. (a) We first complete the square to find the vertex:

$$y = p\left(x^2 + x\right) + r = p\left(x^2 + x + \frac{1}{4}\right) + r - \frac{p}{4} = p\left(x + \frac{1}{2}\right)^2 + \frac{4r-p}{4}$$

If the vertex lies on the x-axis, then $\frac{4r-p}{4} = 0$, so $4r - p = 0$, thus $p = 4r$.

(b) If $p = 4r$, then $y = 4rx^2 + 4rx + r$. We complete the square:

$$y = 4r\left(x^2 + x\right) + r = 4r\left(x^2 + x + \tfrac{1}{4}\right) + r - r = 4r\left(x + \tfrac{1}{2}\right)^2$$

But then the vertex is $\left(-\frac{1}{2}, 0\right)$, and thus it lies on the x-axis.

55. If $a + bi$ and $a - bi$ are roots of the equation, then we can write the equation in factored form as:

$$[x-(a+bi)][x-(a-bi)]=0$$
$$x^2-(a+bi)x-(a-bi)x+(a+bi)(a-bi)=0$$
$$x^2-2ax+\left(a^2+b^2\right)=0$$

We complete the square:

$$f(x)=\left(x^2-2ax\right)+\left(a^2+b^2\right)=\left(x^2-2ax+a^2\right)+\left(a^2+b^2-a^2\right)=(x-a)^2+b^2$$

So the vertex is $\left(a,b^2\right)$.

5.3 Applied Functions: More on Setting Up Equations

1. (a) We first draw the figure:

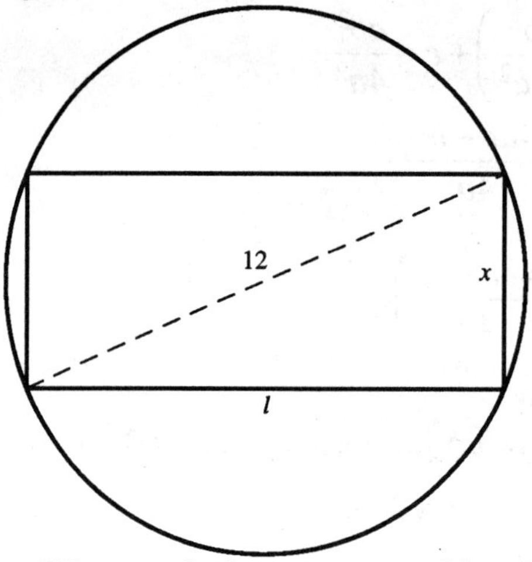

Call P the perimeter. We are asked to come up with a formula for P in terms of x. Since the diameter forms a right triangle, we use the Pythagorean theorem to get:

$$l^2+x^2=12^2$$
$$l^2=144-x^2$$
$$l=\sqrt{144-x^2}$$

Now $P = 2x + 2l = 2x+2\sqrt{144-x^2}$. Using functional notation we have $P(x)=2x+2\sqrt{144-x^2}$.

(b) Let A denote the area of the rectangle. Then $A=xl=x\sqrt{144-x^2}$. Using functional notation we have $A(x)=x\sqrt{144-x^2}$.

3. (a) We first draw the figure:

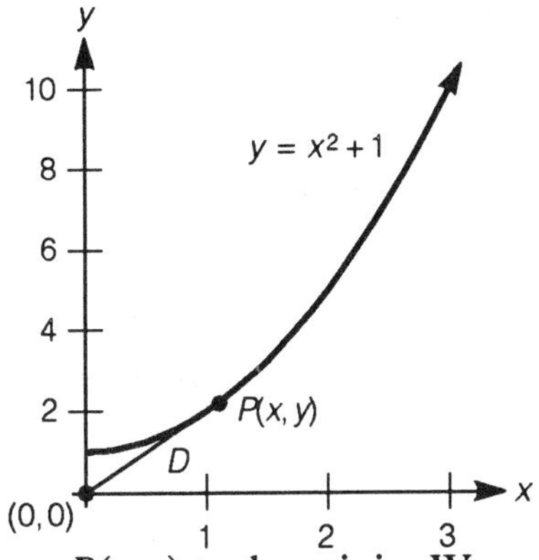

Call D the distance from $P(x, y)$ to the origin. We are asked to come up with a formula for D in terms of x. By the distance formula, we have:

$$D = \sqrt{(x-0)^2 + (y-0)^2} = \sqrt{x^2 + y^2}$$

Since $P(x, y)$ lies on the curve, then $y = x^2 + 1$. Substituting this for y in our equation for D, we have:

$$D = \sqrt{x^2 + y^2} = \sqrt{x^2 + \left(x^2+1\right)^2} = \sqrt{x^2 + x^4 + 2x^2 + 1} = \sqrt{x^4 + 3x^2 + 1}$$

Using functional notation we have $D(x) = \sqrt{x^4 + 3x^2 + 1}$.

(b) Let m denote the slope of the line segment from the origin to $P(x, y)$. Then:

$$m = \frac{y-0}{x-0} = \frac{y}{x}$$

Substituting $y = x^2 + 1$ in for y in this equation, we have:

$$m = \frac{y}{x} = \frac{x^2+1}{x}$$

Using functional notation we have $m(x) = \frac{x^2+1}{x}$.

5. (a) We first draw the figure:

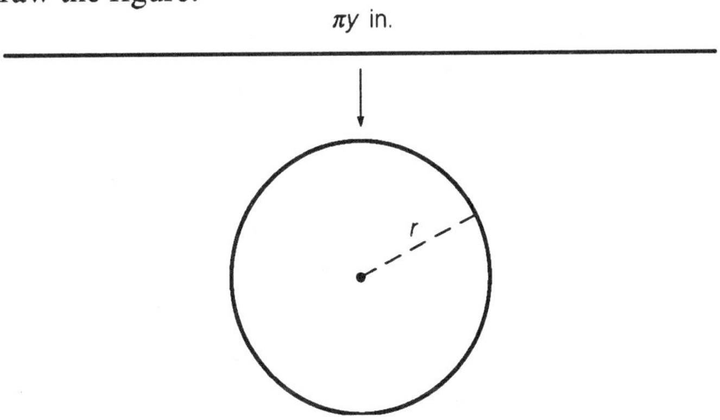

Call A the area of the circle and r its radius. We are asked to come up with a formula for A in terms of y. Since πy is the circumference C of the circle, and $C = 2\pi r$, we have:

$$2\pi r = \pi y$$
$$r = \frac{y}{2}$$

Now $A = \pi r^2 = \pi\left(\frac{y}{2}\right)^2 = \frac{\pi y^2}{4}$. Using functional notation we have $A(y) = \frac{\pi y^2}{4}$.

(b) We draw the figure:

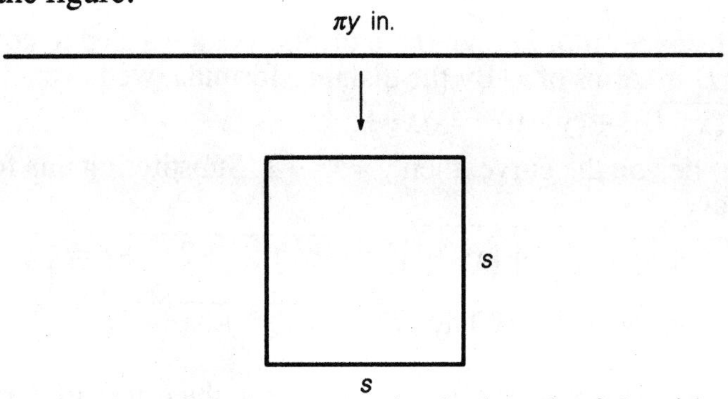

Let A denote the area of the square, and s the length of the side. Since the perimeter P is πy and $P = 4s$, we have:

$$4s = \pi y$$
$$s = \frac{\pi y}{4}$$

Now $A = s^2 = \left(\frac{\pi y}{4}\right)^2 = \frac{\pi^2 y^2}{16}$. Using functional notation we have $A(y) = \frac{\pi^2 y^2}{16}$.

7. (a) Let the two numbers be x and $16 - x$. Then the product P would be:

$$P = x(16 - x) = 16x - x^2$$

Using functional notation we have $P(x) = 16x - x^2$.

(b) Since the two numbers are x and $16 - x$, then the sum of squares S would be:

$$S = (x)^2 + (16 - x)^2 = x^2 + 256 - 32x + x^2 = 2x^2 - 32x + 256$$

Using functional notation we have $S(x) = 2x^2 - 32x + 256$.

(c) There are two ways to set this up. Since the two numbers are x and $16 - x$, then the difference of the cubes D could be:

$$D = (x)^3 - (16 - x)^3 \text{ or } D = (16 - x)^3 - x^3$$

Using functional notation we have:

$$D(x) = (x)^3 - (16 - x)^3 \text{ or } D(x) = (16 - x)^3 - x^3$$

(d) Let A denote the average of the two numbers. Since the two numbers are x and $16-x$, we have:

$$A=\frac{x+16-x}{2}=\frac{16}{2}=8$$

So $A(x)=8$. Notice that the average does not depend on what the two numbers are!

9. Let R be the revenue, x be the number of units sold, and p be the demand (price). Then:

$$R=xp=x\left(-\tfrac{1}{4}x+8\right)=-\tfrac{1}{4}x^2+8x$$

Using functional notation we have $R(x)=-\frac{1}{4}x^2+8x$.

11. (a) We complete the table:

x	1	2	3	4	5	6	7
$P(x)$	17.88	19.49	20.83	21.86	22.49	22.58	21.75

(b) The largest value for $P(x)$ is 22.58, corresponding to $x=6$.

(c) We compute $P(4\sqrt{2})$:

$$\begin{aligned}P(4\sqrt{2})&=2(4\sqrt{2})+2\sqrt{64-(4\sqrt{2})^2}\\&=8\sqrt{2}+2\sqrt{64-32}\\&=8\sqrt{2}+2\sqrt{32}\\&=8\sqrt{2}+8\sqrt{2}\\&=16\sqrt{2}\\&\approx 22.63\end{aligned}$$

This is indeed larger than any of our table values.

13. (a) We first draw the figure:

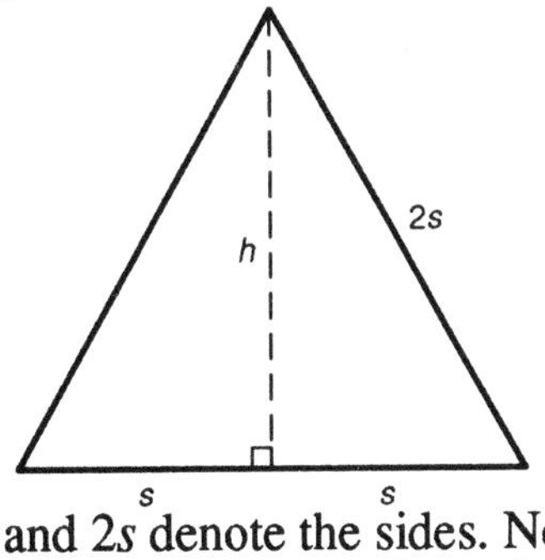

Let h denote the height and $2s$ denote the sides. Note that the height (called the altitude) bisects the base into the lengths of s and s. We are asked to find h in terms of s, so we use the Pythagorean theorem on the right triangle:

$$\begin{aligned}h^2+s^2&=(2s)^2\\h^2+s^2&=4s^2\\h^2&=3s^2\\h&=\sqrt{3s^2}\\h&=\sqrt{3}s\end{aligned}$$

Using functional notation we have $h(s)=\sqrt{3}s$.

(b) Let A denote the area of the triangle. Then:

$$A = \tfrac{1}{2}(\text{base})(\text{height}) = \tfrac{1}{2}(2s)(\sqrt{3}s) = \sqrt{3}s^2$$

Using functional notation we have $A(s) = \sqrt{3}s^2$.

(c) If each side is 8 cm, then:

$$2s = 8$$
$$s = 4$$

Using the function from (a), we have:

$$h(4) = \sqrt{3} \bullet 4 = 4\sqrt{3} \text{ cm}$$

(d) If each side is 5 in., then:

$$2s = 5$$
$$s = \frac{5}{2}$$

Using the function from (b), we have:

$$A\left(\frac{5}{2}\right) = \sqrt{3}\left(\frac{5}{2}\right)^2 = \frac{25\sqrt{3}}{4} \text{ in.}^2$$

15. Let h be the height, r be the radius, and V be the volume. We know that:

$$V = \pi r^2 h$$

We are also given that $h = 2r$, so we substitute into the formula for V:

$$V = \pi r^2(2r) = 2\pi r^3$$

Using functional notation we have $V(r) = 2\pi r^3$.

17. (a) Let h be the height, r be the radius and V be the volume. We know that $V = 12\pi$ and $V = \pi r^2 h$, so:

$$\pi r^2 h = 12\pi$$
$$h = \frac{12\pi}{\pi r^2}$$
$$h = \frac{12}{r^2}$$

Using functional notation we have $h(r) = \frac{12}{r^2}$.

(b) Let S be the total surface area. Then:

$$S = 2\pi r^2 + 2\pi rh = 2\pi r^2 + 2\pi r\left(\frac{12}{r^2}\right) = 2\pi r^2 + \frac{24\pi}{r}$$

Using functional notation we have $S(r) = 2\pi r^2 + \frac{24\pi}{r}$.

19. We solve $S = 4\pi r^2$ for r:

$$4\pi r^2 = S$$
$$r^2 = \frac{S}{4\pi}$$
$$r = \sqrt{\frac{S}{4\pi}}$$

Now since $V = \frac{4}{3}\pi r^3$, we have:

$$V = \frac{4}{3}\pi\left(\sqrt{\frac{S}{4\pi}}\right)^3 = \frac{4\pi S\sqrt{s}}{3(4\pi)\sqrt{4\pi}} = \frac{S\sqrt{S}}{3\sqrt{4\pi}} = \frac{S\sqrt{S\pi}}{6\pi}$$

Using functional notation we have $V(S) = \frac{S\sqrt{S\pi}}{6\pi}$.

21. We draw a figure:

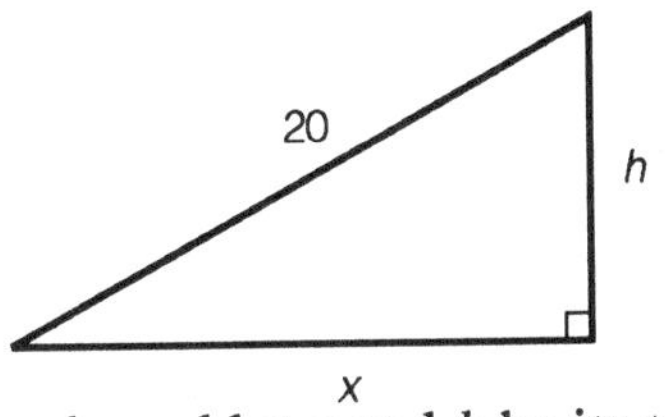

Let A be the area of the triangle and let x and h be its two legs. By the Pythagorean theorem, we have:

$$x^2 + h^2 = 20^2$$
$$h^2 = 400 - x^2$$
$$h = \sqrt{400 - x^2}$$

Therefore:

$$A = \frac{1}{2}(\text{base})(\text{height}) = \frac{1}{2}(x)(h) = \frac{1}{2}x\sqrt{400 - x^2}$$

Using functional notation we have $A(x) = \frac{1}{2}x\sqrt{400 - x^2}$.

23. We draw the figure:

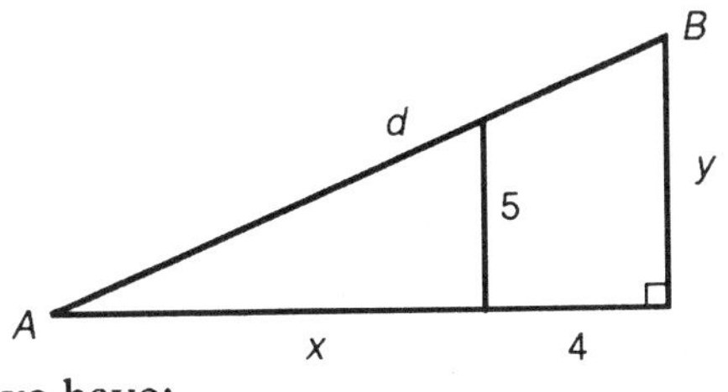

Using similar triangles, we have:

$$\frac{x}{5} = \frac{x+4}{y}$$
$$xy = 5(x+4)$$
$$y = \frac{5(x+4)}{x}$$

Therefore:

$$\begin{aligned} d^2 &= (x+4)^2 + y^2 \\ &= (x+4)^2 + \left(\frac{5(x+4)}{x}\right)^2 \\ &= (x+4)^2 + \frac{25(x+4)^2}{x^2} \\ &= \frac{x^2(x+4)^2 + 25(x+4)^2}{x^2} \\ &= \frac{(x+4)^2(x^2+25)}{x^2} \end{aligned}$$

So $d = \frac{(x+4)\sqrt{x^2+25}}{x}$. Using functional notation we have $d(x) = \frac{(x+4)\sqrt{x^2+25}}{x}$.

25. We substitute into $A(x) = 50x - x^2$:

x	5	10	20	24	24.8	24.9	25	25.1	25.2	45
$A(x)$	225	400	600	624	624.96	624.99	625	624.99	624.96	225

So $x = 25$ yields the largest area. Since $L = 50 - x = 50 - 25 = 25$, then $L = 25$ is the corresponding value.

27. (a) We substitute into $A(x) = 8x - \frac{1}{2}x^3$:

<u>Table 1</u>:

x	1	2	3	4
A	7.5	12	10.5	0

Here $x = 2$ yields the largest area.

<u>Table 2</u>:

x	1.75	2.00	2.25	2.50	2.75
A	11.3203	12.0000	12.3047	12.1875	11.6016

Here $x = 2.25$ yields the largest area.

<u>Table 3</u>:

x	2.15	2.20	2.25	2.30	2.35
A	12.2308	12.2760	12.3047	12.3165	12.3111

Here $x = 2.30$ yields the largest area.

(b) Since $x = \frac{4\sqrt{3}}{3} = 2.309$ to four significant places, then $x = 2.30$ is the closest x-value. This yields an area of 12.3168. Notice that $x = 2.30$ agrees with this to five significant digits.

29. For the square, the side is $\frac{x}{4}$. For the rectangle which has a perimeter of $3 - x$, we have:

$$\begin{aligned} 2w + 2l &= 3 - x \\ 2 \bullet \tfrac{1}{2} l + 2l &= 3 - x \\ 3l &= 3 - x \\ l &= 1 - \tfrac{1}{3} x \\ w &= \tfrac{1}{2} - \tfrac{1}{6} x \end{aligned}$$

So the combined area of the square and rectangle is:

$$A = \left(\tfrac{x}{4}\right)^2 + \left(1 - \tfrac{1}{3}x\right)\left(\tfrac{1}{2} - \tfrac{1}{6}x\right) = \tfrac{1}{16}x^2 + \tfrac{1}{2} - \tfrac{1}{3}x + \tfrac{1}{18}x^2 = \tfrac{17}{144}x^2 - \tfrac{1}{3}x + \tfrac{1}{2}$$

Using functional notation we have $A(x) = \frac{17}{144}x^2 - \frac{1}{3}x + \frac{1}{2}$.

31. (a) Since $V = \frac{1}{3}\pi r^2 h$ and $h = \sqrt{3}r$, we have:

$$V = \tfrac{1}{3}\pi r^2 \left(\sqrt{3}r\right) = \tfrac{\sqrt{3}}{3}\pi r^3$$

Using functional notation we have $V(r) = \frac{\sqrt{3}}{3}\pi r^3$.

(b) Since $S = \pi r\sqrt{r^2 + h^2}$ and $h = \sqrt{3}r$, we have:

$$S = \pi r\sqrt{r^2 + h^2} = \pi r\sqrt{r^2 + \left(\sqrt{3}r\right)^2} = \pi r\sqrt{r^2 + 3r^2} = \pi r\sqrt{4r^2} = \pi r(2r) = 2\pi r^2$$

Using functional notation we have $S(r) = 2\pi r^2$.

33. (a) Since $V = \frac{1}{3}\pi r^2 h$ and $S = \pi r\sqrt{r^2 + h^2}$, and $V = S$, we have:

$$\begin{aligned} \tfrac{1}{3}\pi r^2 h &= \pi r\sqrt{r^2 + h^2} \\ rh &= 3\sqrt{r^2 + h^2} \\ r^2h^2 &= 9\left(r^2 + h^2\right) \\ r^2h^2 &= 9r^2 + 9h^2 \\ r^2h^2 - 9r^2 &= 9h^2 \\ r^2\left(h^2 - 9\right) &= 9h^2 \\ r^2 &= \frac{9h^2}{h^2 - 9} \\ r &= \sqrt{\frac{9h^2}{h^2 - 9}} \\ r &= \frac{3h}{\sqrt{h^2 - 9}} \end{aligned}$$

Using functional notation we have $r(h) = \frac{3h}{\sqrt{h^2 - 9}}$.

(b) After squaring in (a), we had:

$$r^2h^2 = 9r^2 + 9h^2$$
$$r^2h^2 - 9h^2 = 9r^2$$
$$h^2(r^2 - 9) = 9r^2$$
$$h^2 = \frac{9r^2}{r^2 - 9}$$

Taking roots:

$$h = \sqrt{\frac{9r^2}{r^2 - 9}} = \frac{3r}{\sqrt{r^2 - 9}}$$

Using functional notation we have $h(r) = \frac{3r}{\sqrt{r^2-9}}$.

35. Let x be the length of wire used for the circle, so $14 - x$ is the length of wire used for the square. For the circle, we have $2\pi r = x$, so $r = \frac{x}{2\pi}$ and thus the area is given by:

$$\pi r^2 = \pi\left(\frac{x}{2\pi}\right)^2 = \frac{x^2}{4\pi}$$

For the square, we have $4s = 14 - x$, so $s = \frac{14-x}{4}$ and thus the area is given by:

$$\left(\frac{14-x}{4}\right)^2 = \frac{(14-x)^2}{16}$$

So the total combined area is:

$$A = \frac{x^2}{4\pi} + \frac{(14-x)^2}{16} = \frac{4x^2 + \pi(14-x)^2}{16\pi}$$

Using functional notation we have $A(x) = \frac{4x^2+\pi(14-x)^2}{16\pi}$.

37. The perimeter of each semi-circle is $\frac{1}{2}(2\pi r) = \pi r$, so the total perimeter P is given by:

$P = \pi r + \pi r + l + l = 2\pi r + 2l$, where l is the length of the rectangle

Since $P = \frac{1}{4}$, we have:

$$2\pi r + 2l = \frac{1}{4}$$
$$2l = \frac{1}{4} - 2\pi r$$
$$2l = \frac{1 - 8\pi r}{4}$$
$$l = \frac{1 - 8\pi r}{8}$$

We now find the area A. The area of each semicircle is $\frac{1}{2}\pi r^2$, and the area of the rectangle is length • width, where $w = 2r$, so:

$$\begin{aligned} A &= \tfrac{1}{2}\pi r^2 + \tfrac{1}{2}\pi r^2 + lw \\ &= \pi r^2 + \left(\frac{1-8\pi r}{8}\right)(2r) \\ &= \pi r^2 + \frac{r - 8\pi r^2}{4} \\ &= \frac{4\pi r^2 + r - 8\pi r^2}{4} \\ &= \frac{r - 4\pi r^2}{4} \end{aligned}$$

Using functional notation we have $A(r) = \frac{r(1-4\pi r)}{4}$.

39. We draw the figure:

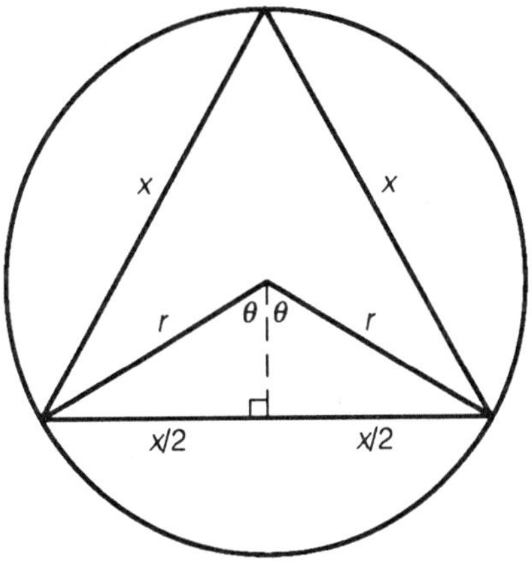

Using geometry, we see that:

$$\begin{aligned} 2\theta &= \tfrac{1}{3}(360°) \\ 2\theta &= 120° \\ \theta &= 60° \end{aligned}$$

Since $\theta = 60°$, then:

$$\begin{aligned} \frac{x}{2} &= \frac{\sqrt{3}}{2}r \\ x &= \sqrt{3}r \\ r &= \frac{x}{\sqrt{3}} \end{aligned}$$

So the area of the circle A is:

$$A = \pi r^2 = \pi\left(\frac{x}{\sqrt{3}}\right)^2 = \frac{\pi x^2}{3}$$

Using functional notation we have $A(x) = \frac{\pi}{3}x^2$.

41. Since $V = l \bullet w \bullet h$ and $l = 8 - 2w$, $w = 6 - 2x$, $h = x$, we have:

$$V = (8-2x)(6-2x)(x) = \left(48 - 28x + 4x^2\right)(x) = 4x^3 - 28x^2 + 48x$$

Using functional notation we have $V(x) = 4x^3 - 28x^2 + 48x$.

43. (a) The area of the window would be:

$$A = \tfrac{1}{2}\left(\pi r^2\right) + lw$$

It remains to find l and w in terms of r. We see that $w = 2r$, and the perimeter $P = 32$, so:

$$P = \tfrac{1}{2}(2\pi r) + 2l + w$$

Therefore:

$$\begin{aligned} \tfrac{1}{2}(2\pi r) + 2l + w &= 32 \\ \pi r + 2l + 2r &= 32 \\ 2l &= 32 - \pi r - 2r \\ l &= \frac{32 - \pi r - 2r}{2} \end{aligned}$$

We now find the area:

$$\begin{aligned} A &= \tfrac{1}{2}\left(\pi r^2\right) + lw \\ &= \tfrac{1}{2}\left(\pi r^2\right) + \left(\frac{32 - \pi r - 2r}{2}\right)(2r) \\ &= \frac{\pi r^2}{2} + 32r - \pi r^2 - 2r^2 \\ &= 32r - 2r^2 - \frac{\pi r^2}{2} \end{aligned}$$

Using functional notation we write $A(r) = 32r - 2r^2 - \frac{\pi r^2}{2}$.

(b) We have $A(r) = -\left(\frac{4+\pi}{2}\right)r^2 + 32r$, which will open downward. Since $A(0) = 0$, it does pass through the origin. We complete the square:

$$\begin{aligned} A(r) &= -\left(\frac{4+\pi}{2}\right)\left(r^2 - \frac{64}{4+\pi}r\right) \\ &= -\left(\frac{4+\pi}{2}\right)\left(r^2 - \frac{64}{4+\pi}r + \left(\frac{32}{4+\pi}\right)^2\right) + \left(\frac{4+\pi}{2}\right)\left(\frac{32}{4+\pi}\right)^2 \\ &= -\left(\frac{4+\pi}{2}\right)\left(r - \frac{32}{4+\pi}\right)^2 + \frac{512}{4+\pi} \end{aligned}$$

So the vertex is $\left(\frac{32}{4+\pi}, \frac{512}{4+\pi}\right)$.

45. (a) We use the Pythagorean theorem:

$$3^2 + y^2 = z^2$$

Taking roots:

$$z = \sqrt{y^2 + 9}$$

Now $s = \frac{y}{z} = \frac{y}{\sqrt{y^2+9}}$, so $s\sqrt{y^2+9} = y$. Squaring, we obtain:

$$s^2(y^2+9) = y^2$$
$$s^2y^2 + 9s^2 = y^2$$
$$y^2 - s^2y^2 = 9s^2$$
$$y^2(1-s^2) = 9s^2$$
$$y^2 = \frac{9s^2}{1-s^2}$$
$$y = \frac{3s}{\sqrt{1-s^2}}$$

Using functional notation we have $y(s) = \frac{3s}{\sqrt{1-s^2}}$.

(b) From part (a) we had $s(y) = \frac{y}{\sqrt{y^2+9}}$.

(c) Since $s = \frac{y}{z}$, then $z = \frac{y}{s}$. Using our result from (a), we have:

$$z = \frac{y}{s} = \frac{\frac{3s}{\sqrt{1-s^2}}}{s} = \frac{3}{\sqrt{1-s^2}}$$

Using functional notation we have $z(s) = \frac{3}{\sqrt{1-s^2}}$.

(d) Using our answer from (c), we have:

$$z = \frac{3}{\sqrt{1-s^2}}$$
$$z\sqrt{1-s^2} = 3$$

Squaring each side, we obtain:

$$z^2(1-s^2) = 9$$
$$z^2 - z^2s^2 = 9$$
$$-z^2s^2 = 9 - z^2$$
$$s^2 = \frac{z^2-9}{z^2}$$
$$s = \frac{\sqrt{z^2-9}}{z}$$

Using functional notation we have $s(z) = \frac{\sqrt{z^2-9}}{z}$.

47. (a) Using the points $(0,-1)$ and (a,a^2), we have:

$$m(a) = \frac{a^2-(-1)}{a-0} = \frac{a^2+1}{a}$$

(b) If x_0 is the x-intercept, then the area of the triangle A is:

$$A = \tfrac{1}{2}(\text{base})(\text{height}) = \tfrac{1}{2}(a - x_0)(a^2)$$

To find the x-intercept, we must find the equation of the line. We use $m = \frac{a^2+1}{a}$ (from (a) above) and $(0,-1)$ in the slope-intercept formula to get:

$$y = \frac{a^2+1}{a}x - 1$$

We find x_0 by letting $y = 0$:

$$0 = \frac{a^2+1}{a}x_0 - 1$$

$$\frac{a^2+1}{a}x_0 = 1$$

$$x_0 = \frac{a}{a^2+1}$$

Therefore:

$$\begin{aligned} A &= \frac{1}{2}(a - x_0)a^2 \\ &= \frac{1}{2}\left(a - \frac{a}{a^2+1}\right)a^2 \\ &= \frac{a^2}{2}\left(\frac{a(a^2+1) - a}{a^2+1}\right) \\ &= \frac{a^2(a^3 + a - a)}{2(a^2+1)} \\ &= \frac{a^2(a^3)}{2(a^2+1)} \\ &= \frac{a^5}{2(a^2+1)} \end{aligned}$$

49. We re-draw the figure (differently):

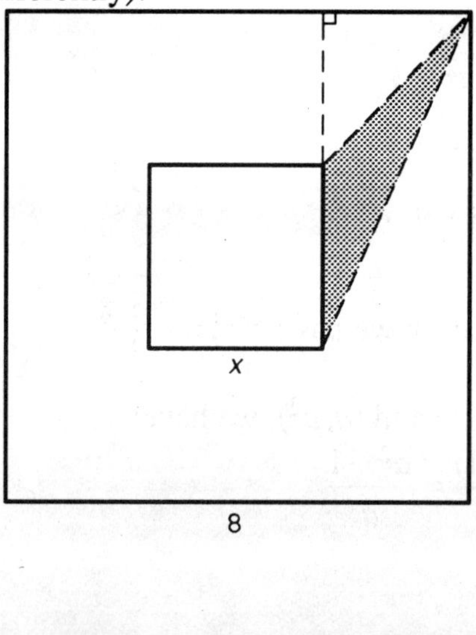

We extend the triangle to form a right triangle as pictured. We find the areas of the large and small right triangles:

$$\begin{aligned} A_{\text{large}} &= \frac{1}{2}(\text{base})(\text{height}) \\ &= \frac{1}{2}\left(x+\frac{8-x}{2}\right)\left(\frac{8-x}{2}\right) \\ &= \frac{1}{2}\left(\frac{2x+8-x}{2}\right)\left(\frac{8-x}{2}\right) \\ &= \frac{(x+8)(8-x)}{8} \\ &= \frac{64-x^2}{8} \end{aligned}$$

$$A_{\text{small}} = \frac{1}{2}(\text{base})(\text{height}) = \frac{1}{2}\left(\frac{8-x}{2}\right)\left(\frac{8-x}{2}\right) = \frac{64-16x+x^2}{8}$$

Therefore:

$$\begin{aligned} A &= A_{\text{large}} - A_{\text{small}} \\ &= \frac{64-x^2}{8} - \frac{64-16x+x^2}{8} \\ &= \frac{64-x^2-64+16x-x^2}{8} \\ &= \frac{16x-2x^2}{8} \\ &= \frac{8x-x^2}{4} \end{aligned}$$

Using functional notation we write $A(x) = \frac{8x-x^2}{4}$.

Note: A much easier approach is to realize that the altitude of the triangle need not lie on the triangle. That is:

$$\text{Area} = \frac{1}{2}(\text{base})(\text{altitude}) = \frac{1}{2}(x)\left(\frac{8-x}{2}\right) = \frac{8x-x^2}{4}$$

Both approaches are correct.

51. We draw the diagram:

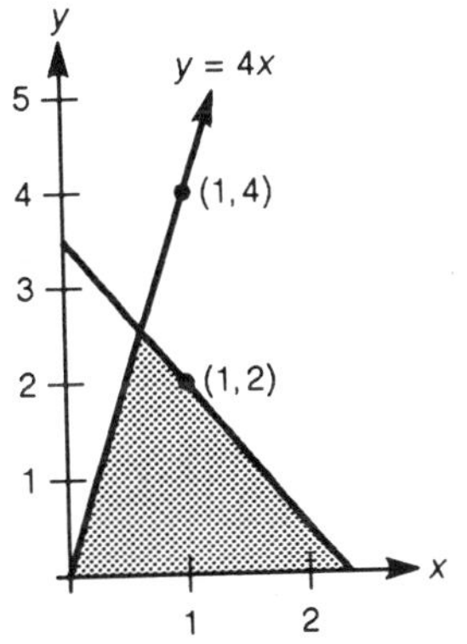

We are asked to find the area A of the shaded triangle. Since the line has slope m and passes through $(1,2)$, then by the point-slope formula, we have:

$$y-2=m(x-1)$$
$$y-2=mx-m$$
$$y=mx+(2-m)$$

Its x-intercept is where $y=0$:

$$0=mx+(2-m)$$
$$mx=m-2$$
$$x=\frac{m-2}{m}$$

This is the base of a triangle. To find its height, we must find the value of y where this line and $y=4x$ intersect. We set the two y-values equal:

$$mx+2-m=4x$$
$$mx-4x=m-2$$
$$x(m-4)=m-2$$
$$x=\frac{m-2}{m-4}$$

Since this point lies on $y=4x$, its y-coordinate is:

$$y=4x=4\left(\frac{m-2}{m-4}\right)$$

Finally, we find the area:

$$A=\frac{1}{2}(\text{base})(\text{height})$$
$$=\frac{1}{2}\left(\frac{m-2}{m}\right)(4)\left(\frac{m-2}{m-4}\right)$$
$$=2\frac{(m-2)^2}{m(m-4)}$$
$$=\frac{2\left(m^2-4m+4\right)}{m^2-4m}$$
$$=\frac{2m^2-8m+8}{m^2-4m}$$

Using functional notation we have $A(m)=\frac{2m^2-8m+8}{m^2-4m}$.

53. We draw a figure:

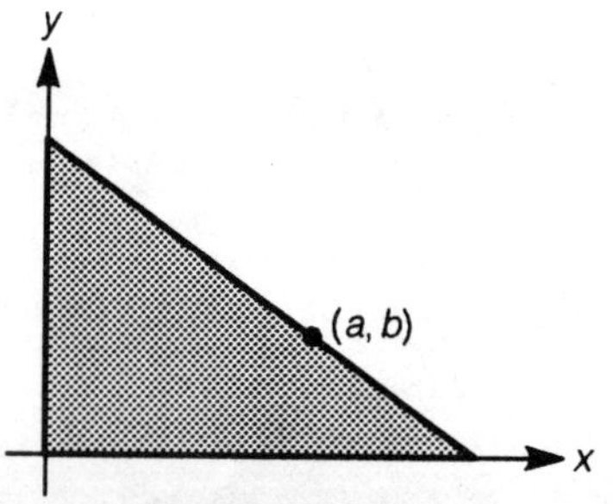

The line has the equation $y - b = m(x - a)$. Since the base and height are the x- and y-intercepts, respectively, we find each intercept:

base:
$$\begin{aligned} y &= 0 \\ -b &= m(x-a) \\ -b &= mx - ma \\ mx &= ma - b \\ x &= \frac{ma-b}{m} \end{aligned}$$

height:
$$\begin{aligned} x &= 0 \\ y - b &= m(-a) \\ y &= b - ma \end{aligned}$$

So the area of the triangle is $A = \frac{1}{2}\left(\frac{ma-b}{m}\right)(b - ma) = \frac{(ma-b)^2}{-2m}$.

5.4 Maximum and Minimum Problems

1. Call the two numbers x and y. Then $x + y = 5$, so $y = 5 - x$. So the product can be written as:

$$P = xy = x(5-x) = 5x - x^2$$

We now complete the square:

$$P = -\left(x^2 - 5x\right) = -\left(x^2 - 5x + \tfrac{25}{4}\right) + \tfrac{25}{4} = -\left(x - \tfrac{5}{2}\right)^2 + \tfrac{25}{4}$$

Since this is a parabola opening downward, it will have a maximum value of $\frac{25}{4}$.

3. Call the two numbers x and y. Then $y - x = 1$, so $y = x + 1$. The sum of their squares can be written as:

$$S = x^2 + y^2 = x^2 + (x+1)^2 = x^2 + x^2 + 2x + 1 = 2x^2 + 2x + 1$$

We now complete the square:

$$S = 2\left(x^2 + x\right) + 1 = 2\left(x^2 + x + \tfrac{1}{4}\right) + 1 - \tfrac{1}{2} = 2\left(x + \tfrac{1}{2}\right)^2 + \tfrac{1}{2}$$

Since this is a parabola opening upward, it will have a minimum value of $\frac{1}{2}$.

5. Let w and l be the width and length, respectively. Since $P = 2w + 2l$, then:

$$\begin{aligned} 2w + 2l &= 25 \\ 2l &= 25 - 2w \\ l &= \frac{25-2w}{2} \end{aligned}$$

So the area is given by:

$$A = wl = w\left(\frac{25-2w}{2}\right) = \frac{1}{2}\left(-2w^2 + 25\right)$$

We now complete the square:

$$A = -\left(w^2 - \tfrac{25}{2}w\right) = -\left(w^2 - \tfrac{25}{2}w + \tfrac{625}{16}\right) = -\left(w - \tfrac{25}{4}\right)^2 + \tfrac{625}{16}$$

This is a parabola opening downward, so it will achieve a maximum value when $w = \frac{25}{4}$.

We find l:

$$l = \frac{25 - 2\left(\frac{25}{4}\right)}{2} = \frac{25 - \frac{25}{2}}{2} = \frac{25}{4}$$

So the largest such rectangle is a square of dimensions $\frac{25}{4}$ m by $\frac{25}{4}$ m.

7. Let x and y be the lengths of the two shorter sides, so $x + y = 100$, and $y = 100 - x$. Then the area is given by:

$$A = \tfrac{1}{2}xy = \tfrac{1}{2}x(100 - x) = \tfrac{1}{2}(-x^2 + 100x)$$

We now complete the square:

$$A = -\tfrac{1}{2}\left(x^2 - 100x\right) = -\tfrac{1}{2}\left(x^2 - 100x + 2500\right) + 1250 = -\tfrac{1}{2}(x - 50)^2 + 1250$$

This is a parabola opening downward, so it will achieve a maximum value of 1250 in.^{2}.

9. Let x and y be the two numbers, so $x + y = 6$ and thus $y = 6 - x$.

(a) We have:

$$T = x^2 + y^2 = x^2 + (6 - x)^2 = x^2 + 36 - 12x + x^2 = 2x^2 - 12x + 36$$

We now complete the square:

$$T = 2\left(x^2 - 6x\right) + 36 = 2\left(x^2 - 6x + 9\right) + 36 - 18 = 2(x - 3)^2 + 18$$

This is a parabola opening upward, so it will have a minimum value of 18.

(b) We have:

$$S = x + y^2 = x + (6 - x)^2 = x + 36 - 12x + x^2 = x^2 - 11x + 36$$

We now complete the square:

$$S = \left(x^2 - 11x\right) + 36 = \left(x^2 - 11x + \tfrac{121}{4}\right) + 36 - \tfrac{121}{4} = \left(x - \tfrac{11}{2}\right)^2 + \tfrac{23}{4}$$

This is a parabola opening upward, so it will have a minimum value of $\frac{23}{4}$.

(c) We have:

$$U = x + 2y^2 = x + 2(6 - x)^2 = x + 72 - 24x + 2x^2 = 2x^2 - 23x + 72$$

We now complete the square:

$$U = 2\left(x^2 - \tfrac{23}{2}x\right) + 72 = 2\left(x^2 - \tfrac{23}{2}x + \tfrac{529}{16}\right) + 72 - \tfrac{529}{8} = 2\left(x - \tfrac{23}{4}\right)^2 + \tfrac{47}{8}$$

This is a parabola opening upward, so it will have a minimum value of $\frac{47}{8}$.

(d) We have:

$$\begin{aligned} V &= x + (2y)^2 \\ &= x + 4y^2 \\ &= x + 4(6 - x)^2 \\ &= x + 144 - 48x + 4x^2 \\ &= 4x^2 - 47x + 144 \end{aligned}$$

We now complete the square:

$$V = 4\left(x^2 - \tfrac{47}{4}x\right) + 144 = 4\left(x^2 - \tfrac{47}{4}x + \tfrac{2209}{64}\right) + 144 - \tfrac{2209}{16} = 4\left(x - \tfrac{47}{8}\right)^2 + \tfrac{95}{16}$$

This is a parabola opening upward, so it will have a minimum value of $\frac{95}{16}$.

11. (a) We compute the heights:

$$h(1) = -16(1)^2 + 32(1) = -16 + 32 = 16 \text{ ft}$$

$$h\left(\tfrac{3}{2}\right) = -16\left(\tfrac{3}{2}\right)^2 + 32\left(\tfrac{3}{2}\right) = -16\left(\tfrac{9}{4}\right) + 48 = -36 + 48 = 12 \text{ ft}$$

(b) We complete the square:

$$h = -16t^2 + 32t = -16\left(t^2 - 2t\right) = -16\left(t^2 - 2t + 1\right) + 16 = -16(t-1)^2 + 16$$

This is a parabola opening downward, so it will have a maximum height of 16 ft, attained after 1 second.

(c) We set $h = 7$:

$$\begin{aligned} 7 &= -16t^2 + 32t \\ 16t^2 - 32t + 7 &= 0 \\ (4t-7)(4t-1) &= 0 \\ t &= \tfrac{7}{4}, \tfrac{1}{4} \end{aligned}$$

So $h = 7$ when $t = \frac{7}{4}$ sec or $t = \frac{1}{4}$ sec.

13. Every point on the given curve has coordinates of the form $\left(x, \sqrt{x-2} + 1\right)$, and using the distance formula gives:

$$d = \sqrt{(4-x)^2 + \left(1 - \sqrt{x-2} - 1\right)^2} = \sqrt{16 - 8x + x^2 + x - 2} = \sqrt{x^2 - 7x + 14}$$

We complete the square:

$$\left(x^2 - 7x\right) + 14 = \left(x^2 - 7x + \tfrac{49}{4}\right) + 14 - \tfrac{49}{4} = \left(x - \tfrac{7}{2}\right)^2 + \tfrac{7}{4}$$

This is a parabola opening upward which will achieve a minimum value of $\sqrt{\frac{7}{4}} = \frac{\sqrt{7}}{2}$ at $x = \frac{7}{2}$. Then:

$$y = \sqrt{\tfrac{7}{2} - 2} + 1 = \sqrt{\tfrac{3}{2}} + 1 = \tfrac{2+\sqrt{6}}{2}$$

So the point is $\left(\frac{7}{2}, \frac{2+\sqrt{6}}{2}\right)$ and the distance is $\frac{\sqrt{7}}{2}$.

15. (a) We must find the value of x such that $x - x^2$ is as large as possible. Call $f(x) = -x^2 + x$. We complete the square:

$$f(x) = -\left(x^2 - x\right) = -\left(x^2 - x + \tfrac{1}{4}\right) + \tfrac{1}{4} = -\left(x - \tfrac{1}{2}\right)^2 + \tfrac{1}{4}$$

This is a parabola opening downward, so it will achieve a maximum value when $x = \frac{1}{2}$. So the number is $\frac{1}{2}$.

(b) We must find the value of x such that $x - 2x^2$ is as large as possible. Call $f(x) = -2x^2 + x$. We complete the square:

$$f(x) = -2\left(x^2 - \tfrac{1}{2}x\right) = -2\left(x^2 - \tfrac{1}{2}x + \tfrac{1}{16}\right) + \tfrac{1}{8} = -2\left(x - \tfrac{1}{4}\right)^2 + \tfrac{1}{8}$$

This is a parabola opening downward, so it will achieve a maximum value when $x = \frac{1}{4}$. So the number is $\frac{1}{4}$.

17. If we choose x for the depth of the pasture, then $500 - 2x$ is the length parallel to the river. The area of the pasture will then be given by:

$$A = x(500 - 2x) = -2x^2 + 500x$$

We complete the square:

$$A = -2\left(x^2 - 250x\right) = -2\left(x^2 - 250x + 125^2\right) + 2(125)^2 = -2(x-125)^2 + 31250$$

This is a parabola opening downward, so it will achieve a maximum value at $x = 125$. Then the length is $500 - 2(125) = 500 - 250 = 250$. So the dimensions are 125 ft by 250 ft.

19. We have:

$$\begin{aligned} R - C &= \left(0.4x^2 + 10x + 5\right) - \left(0.5x^2 + 2x + 101\right) \\ &= 0.4x^2 + 10x + 5 - 0.5x^2 - 2x - 101 \\ &= -0.1x^2 + 8x - 96 \end{aligned}$$

We now complete the square:

$$\begin{aligned} R - C &= -0.1\left(x^2 - 80x\right) - 96 \\ &= -0.1\left(x^2 - 80x + 1600\right) - 96 + 160 \\ &= -0.1(x-40)^2 + 64 \end{aligned}$$

This is a parabola opening downward, so it will achieve a maximum value when $x = 40$.

21. Recall that revenue R is $x \bullet p$. So:

$$R = x\left(-\tfrac{1}{4}x + 30\right) = -\tfrac{1}{4}x^2 + 30x$$

We complete the square:

$$\begin{aligned} R &= -\tfrac{1}{4}x^2 + 30x \\ &= -\tfrac{1}{4}\left(x^2 - 120x\right) \\ &= -\tfrac{1}{4}\left(x^2 - 120x + 3600\right) + 900 \\ &= -\tfrac{1}{4}(x-60)^2 + 900 \end{aligned}$$

This is a parabola opening downward, so it will achieve a maximum value at $x = 60$. The maximum revenue is \$900. The corresponding unit price p is:

$$p = -\tfrac{1}{4}(60) + 30 = -15 + 30 = \$15$$

23. (a) To use max/min methods, we need to substitute in the quantity $x^2 + y^2$ and write it strictly in terms of x or y. So take $2x + 3y = 6$ and solve for y:

$$\begin{aligned} 3y &= 6 - 2x \\ y &= \frac{6-2x}{3} \end{aligned}$$

Then substitute, and the quantity $x^2 + y^2$ becomes:

$$Q = x^2 + \left(\frac{6-2x}{3}\right)^2 = x^2 + \frac{36 - 24x + 4x^2}{9} = \tfrac{13}{9}x^2 - \tfrac{8}{3}x + 4$$

We now complete the square:

$$Q = \tfrac{13}{9}\left(x^2 - \tfrac{24}{13}x\right) + 4 = \tfrac{13}{9}\left(x^2 - \tfrac{24}{13}x + \tfrac{144}{169}\right) + 4 - \tfrac{13}{9}\left(\tfrac{144}{169}\right) = \tfrac{13}{9}\left(x - \tfrac{12}{13}\right)^2 + \tfrac{36}{13}$$

This is a parabola opening up, so it will achieve a minimum value of $\tfrac{36}{13}$.

(b) The equation of a circle with its center at the origin is $x^2 + y^2 = r^2$ where r is the radius. The line $2x + 3y = 6$ will intersect the circle in two points wherever r is sufficiently large. As we reduce r, we gradually reach a position where the circle and line are tangent and this is the minimum value of r or $\sqrt{x^2 + y^2}$. In this case, it is $\sqrt{\frac{36}{13}} = \frac{6\sqrt{13}}{13}$. This is the square root of the answer from (a).

25. (a) We substitute $y = 15 - x$:

$$Q = x^2 + y^2 = x^2 + (15 - x)^2 = x^2 + 225 - 30x + x^2 = 2x^2 - 30x + 225$$

We now complete the square:

$$\begin{aligned} Q &= 2x^2 - 30x + 225 \\ &= 2\left(x^2 - 15x\right) + 225 \\ &= 2\left(x^2 - 15x + \tfrac{225}{4}\right) + 225 - \tfrac{225}{2} \\ &= 2\left(x - \tfrac{15}{2}\right)^2 + \tfrac{225}{2} \end{aligned}$$

This is a parabola opening upward, so it will achieve a minimum value of $\frac{225}{2}$.

(b) We substitute $y = C - x$:

$$Q = x^2 + y^2 = x^2 + (C - x)^2 = x^2 + C^2 - 2Cx + x^2 = 2x^2 - 2Cx + C^2$$

We now complete the square:

$$\begin{aligned} Q &= 2x^2 - 2Cx + C^2 \\ &= 2\left(x^2 - Cx\right) + C^2 \\ &= 2\left(x^2 - Cx + \tfrac{C^2}{4}\right) + C^2 - \tfrac{C^2}{2} \\ &= 2\left(x - \tfrac{C}{2}\right)^2 + \tfrac{C^2}{2} \end{aligned}$$

This is a parabola opening upward, so it will achieve a minimum value of $\frac{C^2}{2}$. When $C = 15$, the result from (a) is verified.

27. Let the other two sides of each of the four triangles be t and $1 - t$, respectively. Then the area of the square will be a minimum when the area of these triangles is a maximum. Now write an expression for the total area of the four triangles.

$$A = 4\left(\tfrac{1}{2}\right)(t)(1 - t) = 2t - 2t^2 = -2t^2 + 2t$$

We complete the square:

$$A = -2t^2 + 2t = -2\left(t^2 - t\right) = -2\left(t^2 - t + \tfrac{1}{4}\right) + \tfrac{1}{2} = -2\left(t - \tfrac{1}{2}\right)^2 + \tfrac{1}{2}$$

This is a parabola opening downward, so it will achieve a maximum area of $\frac{1}{2}$ when $t = \frac{1}{2}$. Since the large square has area = 1, then the smaller square has area $1 - A = 1 - \frac{1}{2} = \frac{1}{2}$. We use the Pythagorean theorem to find x:

$$x^2 = t^2 + (1 - t)^2 = \tfrac{1}{4} + \tfrac{1}{4} = \tfrac{1}{2}$$

So $x = \sqrt{\frac{1}{2}} = \frac{\sqrt{2}}{2}$.

29. We draw the figure:

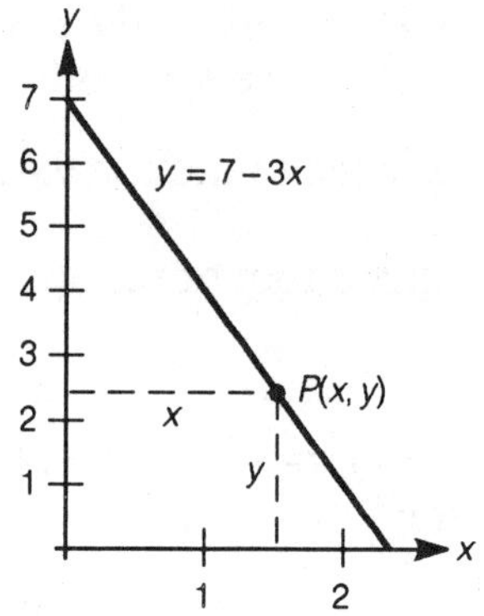

We have $A = xy = x(7-3x) = -3x^2 + 7x$. We complete the square:

$$A = -3\left(x^2 - \tfrac{7}{3}x\right) = -3\left(x^2 - \tfrac{7}{3}x + \tfrac{49}{36}\right) + \tfrac{49}{12} = -3\left(x - \tfrac{7}{6}\right)^2 + \tfrac{49}{12}$$

Since this parabola opens downward, the largest possible area is $\tfrac{49}{12}$.

31. We draw the figure:

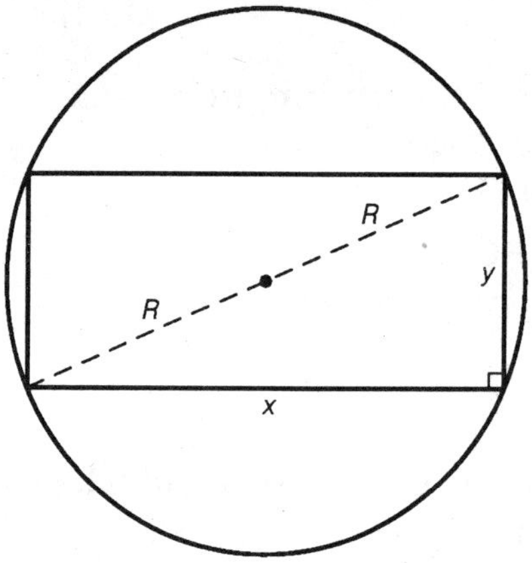

Using the Pythagorean theorem, we obtain $y = \sqrt{4R^2 - x^2}$. The area A of the rectangle is:

$$A = xy = x\sqrt{4R^2 - x^2}$$

Therefore:

$$A^2 = x^2\left(4R^2 - x^2\right) = 4R^2x^2 - x^4$$

In order to maximize the expression $4R^2x^2 - x^4$, first let $t = x^2$, so that the expression becomes $4R^2t - t^2$. We complete the square:

$$\begin{aligned} A^2 &= -t^2 + 4R^2t \\ &= -\left(t^2 - 4R^2t\right) \\ &= -\left(t^2 - 4R^2t + 4R^4\right) + 4R^4 \\ &= -\left(t - 2R^2\right)^2 + 4R^4 \end{aligned}$$

This parabola opens downward, so the maximum value of A^2 is $4R^4$. Thus the maximum area is $\sqrt{4R^4} = 2R^2$.

33. Let x = east-west dimension, and y = north-south dimension. So the cost is given by $C = 12(2x) + 8(2y) = 24x + 16y$. Since this cost is \$4800, we have $24x + 16y = 4800$, so:

$$y = \frac{4800 - 24x}{16} = \frac{600 - 3x}{2}$$

Now the area is $A = xy = x\left(\frac{600-3x}{2}\right) = \frac{-3}{2}x^2 + 300x$. So $A(x) = -\frac{3}{2}x^2 + 300x$.

This will be a parabola opening downward, so it will have a maximum value. We complete the square:

$$\begin{aligned} A(x) &= -\tfrac{3}{2}x^2 + 300x \\ &= -\tfrac{3}{2}\left(x^2 - 200x\right) \\ &= -\tfrac{3}{2}\left(x^2 - 200x + 100^2\right) + 15000 \\ &= -\tfrac{3}{2}(x-100)^2 + 15000 \end{aligned}$$

So $x = 100$ will maximize area, which is 15,000 yd^2. We find y:

$$y = \frac{600 - 3(100)}{2} = \frac{600 - 300}{2} = 150 \text{ yd}$$

So the dimensions are 100 yd by 150 yd.

35. The given function can be rewritten:

$$y = (a_1 + a_2)x^2 - 2(a_1x_1 + a_2x_2)x + \left(a_1x_1^2 + a_2x_2^2\right)$$

We complete the square:

$$\begin{aligned} y &= (a_1 + a_2)\left(x^2 - \frac{2(a_1x_1 + a_2x_2)}{a_1 + a_2}\right) + \left(a_1x_1^2 + a_2x_2^2\right) \\ &= (a_1 + a_2)\left(x - \frac{a_1x_1 + a_2x_2}{a_1 + a_2}\right)^2 + \left(a_1x_1^2 + a_2x_2^2\right) - \frac{(a_1x_1 + a_2x_2)^2}{a_1 + a_2} \end{aligned}$$

Since a_1 and a_2 are both positive, then $a_1 + a_2 > 0$ and thus this parabola opens upward.

So the minimum must occur where $x = \frac{a_1x_1 + a_2x_2}{a_1 + a_2}$.

37. (a) We have $\frac{\Delta p}{\Delta x} = \frac{10}{-5} = -2$. Also $p = 200$ when $x = 150$. We use the point-slope formula with the point $(150, 200)$:

$$\begin{aligned} p - 200 &= -2(x - 150) \\ p - 200 &= -2x + 300 \\ p &= -2x + 500 \end{aligned}$$

Using functional notation we have $p(x) = -2x + 500$.

(b) Since $R = xp$, we have $R = x(-2x + 500) = -2x^2 + 500x$. We complete the square:

$$\begin{aligned} R &= -2x^2 + 500x \\ &= -2\left(x^2 - 250x\right) \\ &= -2\left(x^2 - 250x + 15625\right) + 31250 \\ &= -2(x - 125)^2 + 31250 \end{aligned}$$

Using functional notation we have $R(x) = -2(x-125)^2 + 31250$. Since this parabola opens downward, we have a maximum revenue of \$31,250 when $x = 125$. We find $p = -2(125) + 500 = \$250$.

39. Let $x = t^2$, so $f(x) = x - x^2 = -x^2 + x$. We complete the square:

$$f(x) = -(x^2 - x) = -\left(x^2 - x + \tfrac{1}{4}\right) + \tfrac{1}{4} = -\left(x - \tfrac{1}{2}\right)^2 + \tfrac{1}{4}$$

So $f(x)$ has a maximum value when $x = \frac{1}{2}$. Then $t^2 = \frac{1}{2}$, so $t = \pm\frac{\sqrt{2}}{2}$.

41. Let $x = t^2$. Then $y = -t^4 + 6t^2 - 6 = -x^2 + 6x - 6$. We complete the square:

$$\begin{aligned} y &= -x^2 + 6x - 6 \\ &= -(x^2 - 6x) - 6 \\ &= -(x^2 - 6x + 9) - 6 + 9 \\ &= -(x-3)^2 + 3 \end{aligned}$$

So $x = 3$ will yield the largest output. Since $x = t^2$, we have:

$$\begin{aligned} t^2 &= 3 \\ t &= \pm\sqrt{3} \end{aligned}$$

So $t = \sqrt{3}$ or $t = -\sqrt{3}$ will yield the largest output.

43. (a) For the circle, we have $2\pi r = x$, so $r = \frac{x}{2\pi}$, and thus the area is:

$$A = \pi r^2 = \pi\left(\frac{x}{2\pi}\right)^2 = \frac{x^2}{4\pi}$$

For the square, we have $4s = 16 - x$, so $s = \frac{16-x}{4}$, and thus the area is:

$$A = s^2 = \left(\frac{16-x}{4}\right)^2 = \frac{256 - 32x + x^2}{16} = 16 - 2x + \tfrac{1}{16}x^2$$

So the total combined area is given by:

$$A(x) = \frac{x^2}{4\pi} + 16 - 2x + \tfrac{1}{16}x^2 = \tfrac{4+\pi}{16\pi}x^2 - 2x + 16$$

(b) This is a parabola opening upward, so it will have a minimum value at:

$$x = \frac{2}{2 \bullet \frac{4+\pi}{16\pi}} = \frac{16\pi}{4+\pi}$$

(c) This ratio is given by:

$$\frac{\frac{16\pi}{4+\pi}}{16 - \frac{16\pi}{4+\pi}} \bullet \frac{4+\pi}{4+\pi} = \frac{16\pi}{64 + 16\pi - 16\pi} = \frac{16\pi}{64} = \frac{\pi}{4}$$

45. (a) For the circle, we have $2\pi r = x$, so $r = \frac{x}{2\pi}$, and thus the area is:

$$A = \pi r^2 = \pi\left(\frac{x}{2\pi}\right)^2 = \frac{x^2}{4\pi}$$

For the square, we have $4s = L - x$, so $s = \frac{L-x}{4}$, and thus the area is:

$$A = \left(\frac{L-x}{4}\right)^2 = \frac{L^2 - 2Lx + x^2}{16} = \frac{1}{16}x^2 - \frac{L}{8}x + \frac{L^2}{16}$$

So the total combined area is given by:

$$A(x) = \left(\frac{1}{4\pi} + \frac{1}{16}\right)x^2 - \frac{L}{8}x + \frac{L^2}{16} = \frac{4+\pi}{16\pi}x^2 - \frac{L}{8}x + \frac{L^2}{16}$$

(b) This will have a minimum value when $x = -\frac{b}{2a}$, so:

$$x = \frac{\frac{L}{8}}{\frac{4+\pi}{8\pi}} \bullet \frac{8\pi}{8\pi} = \frac{\pi L}{4+\pi}$$

(c) This ratio is:

$$\frac{\frac{\pi L}{4+\pi}}{L - \frac{\pi L}{4+\pi}} \bullet \frac{4+\pi}{4+\pi} = \frac{\pi L}{4L + \pi L - \pi L} = \frac{\pi L}{4L} = \frac{\pi}{4}$$

It is interesting to note that this ratio does not depend on L, the length of the wire.

47. We re-draw the figure and label additional sides:

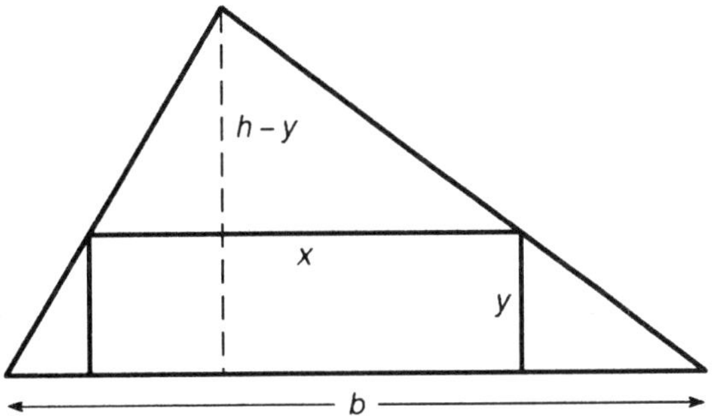

By similar triangles:

$$\frac{h-y}{h} = \frac{x}{b}$$
$$bh - by = xh$$
$$by = bh - xh$$
$$y = h - \frac{h}{b}x$$

So the area of the rectangle is given by:

$$A = xy = x\left(h - \frac{h}{b}x\right) = -\frac{h}{b}x^2 + hx$$

This is a parabola opening downward, so it will have a maximum value when:

$$x = \frac{-h}{2\left(-\frac{h}{b}\right)} = \frac{b}{2}$$

So the maximum area is given by:

$$A\left(\frac{b}{2}\right) = -\frac{h}{b} \bullet \frac{b^2}{4} + h \bullet \frac{b}{2} = -\frac{hb}{4} + \frac{hb}{2} = \frac{hb}{4}$$

Since the triangle has an area of $\frac{hb}{2}$, then the desired ratio is:

$$\frac{\frac{hb}{2}}{\frac{hb}{4}} = 2$$

49. Call h the height of the triangle (see figure), and call y the indicated value:

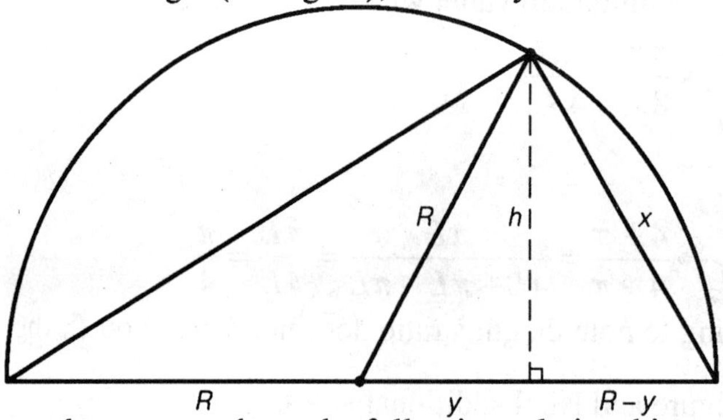

By the Pythagorean theorem, we have the following relationships:

$$y^2 + h^2 = R^2 \qquad \text{and} \qquad (R-y)^2 + h^2 = x^2$$

$$y^2 = R^2 - h^2$$

$$y = \sqrt{R^2 - h^2}$$

Substituting, we have:

$$\left(R - \sqrt{R^2 - h^2}\right)^2 + h^2 = x^2$$

$$R^2 - 2R\sqrt{R^2 - h^2} + R^2 - h^2 + h^2 = x^2$$

$$2R^2 - 2R\sqrt{R^2 - h^2} = x^2$$

We solve for h:

$$-2R\sqrt{R^2 - h^2} = x^2 - 2R^2$$

$$\sqrt{R^2 - h^2} = R - \frac{x^2}{2R}$$

$$R^2 - h^2 = R^2 - x^2 + \frac{x^4}{4R^2}$$

$$-h^2 = \frac{x^4}{4R^2} - x^2$$

$$h^2 = -\frac{x^4}{4R^2} + x^2$$

We now use the hint. Since the area of the triangle is given by $A = \frac{1}{2}(2R)(h) = Rh$, we maximize:

$$A^2 = R^2h^2 = R^2\left(\frac{-x^4}{4R^2} + x^2\right) = -\frac{1}{4}x^4 + R^2x^2$$

This will have a maximum value when:

$$x^2 = \frac{-b}{2a} = \frac{-R^2}{-\frac{1}{2}} = 2R^2$$

Therefore we have:

$$A^2 = -\tfrac{1}{4}\left(2R^2\right)^2 + R^2\left(2R^2\right) = -R^4 + 2R^4 = R^4$$

So the maximum value is $A^2 = R^4$, or $A = R^2$. Thus the shaded area is:

$$\tfrac{1}{2}\left(\pi R^2\right) - R^2 = \frac{\pi R^2}{2} - R^2 = R^2\left(\frac{\pi - 2}{2}\right)$$

5.5 Polynomial Functions

1. This graph has 4 turning points, but a polynomial function of degree 3 can have at most 2 turning points.

3. As $|x|$ gets very large, our function should be similar to $f(x) = a_3x^3$. But $f(x)$ does not have a parabolic shape like the given graph.

5. As $|x|$ gets very large with x negative, then the graph should resemble $2x^5$. But the y-values of $2x^5$ are always negative when x is negative, contrary to the given graph.

7. This graph has a corner, which cannot occur in the graph of a polynomial function.

9. This is $y = x^2$ translated 2 units to the right and 1 unit up. There are no x-intercepts and the y-intercept is 5. We draw the graph:

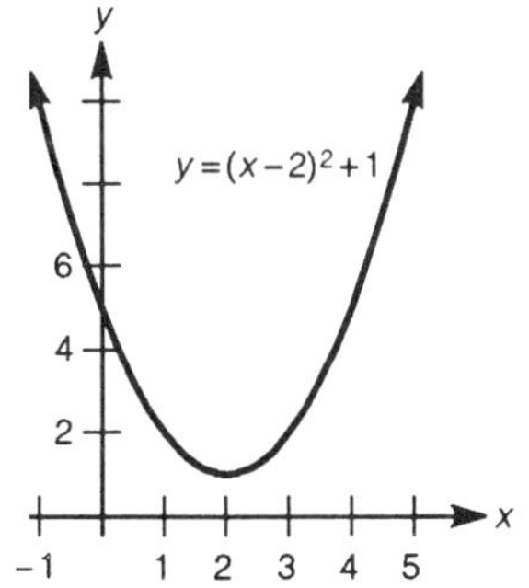

11. This is $y = x^4$ reflected across the x-axis and translated 1 unit to the right. The x-intercept is 1 and the y-intercept is -1. We draw the graph:

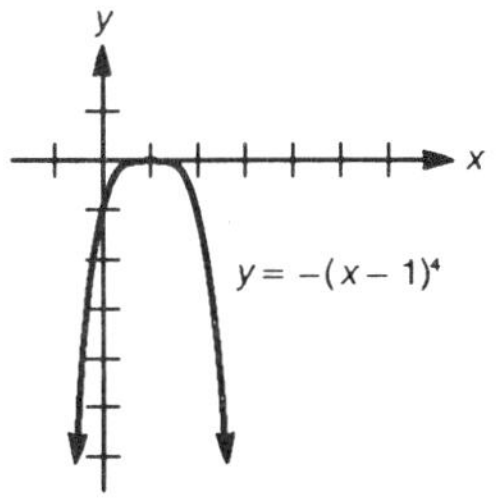

13. This is $y = x^3$ translated 4 units to the right and 2 units down. The x-intercept is $4 + \sqrt[3]{2}$ and the y-intercept is -66. We draw the graph:

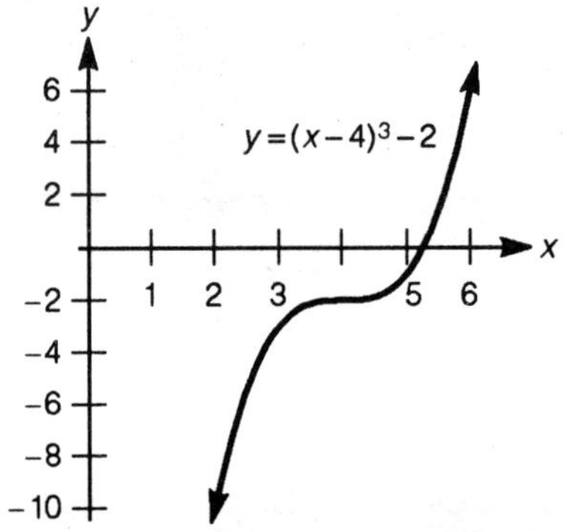

15. This is $y = 2x^4$ translated 5 units to the right and reflected across the x-axis. The x-intercept is -5 and the y-intercept is -1250. We draw the graph:

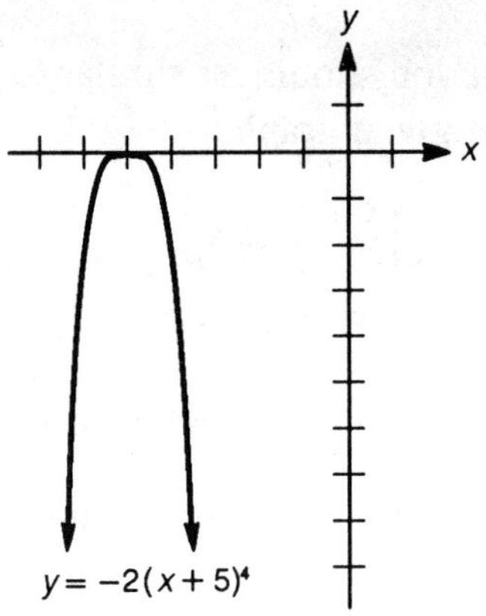

17. This is $y = \frac{1}{2}x^5$ translated 1 unit to the left. The x-intercept is -1 and the y-intercept is $\frac{1}{2}$. We draw the graph:

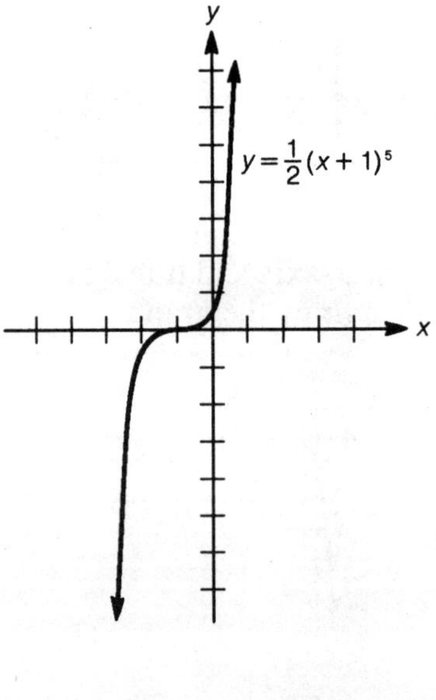

19. This is $y = x^3$ translated 1 unit to the right and 1 unit down, then reflected across the x-axis. The x- and y-intercepts are both 0. We draw the graph:

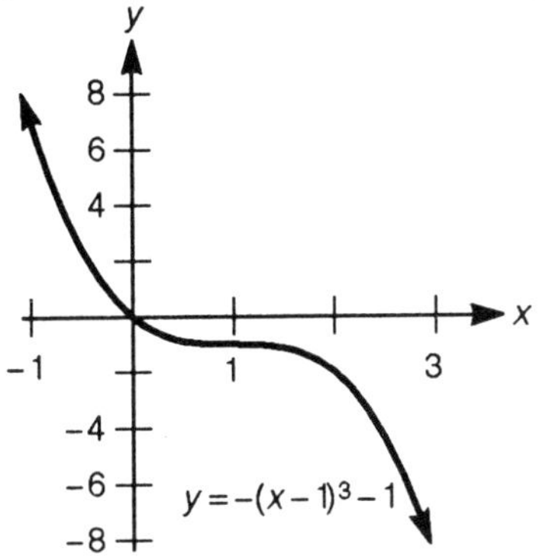

21. (a) The x-intercepts are 2, 1, and –1, and the y-intercept is 2. The signs of y are given by:

Interval	$y = (x-2)(x-1)(x+1)$
$(-\infty, -1)$	neg.
$(-1, 1)$	pos.
$(1, 2)$	neg.
$(2, \infty)$	pos.

We sketch the excluded regions:

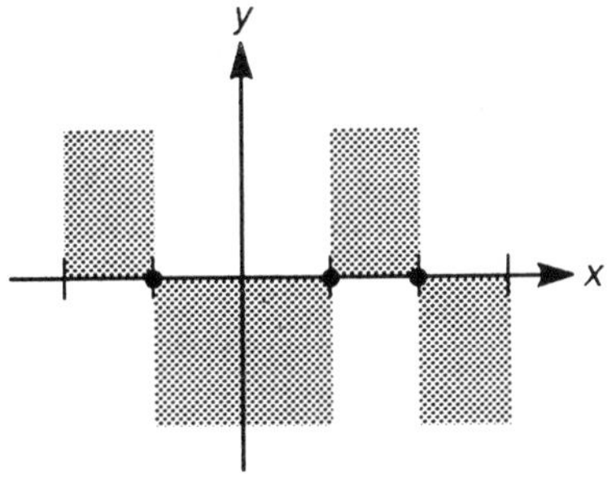

(b) We now graph the function:

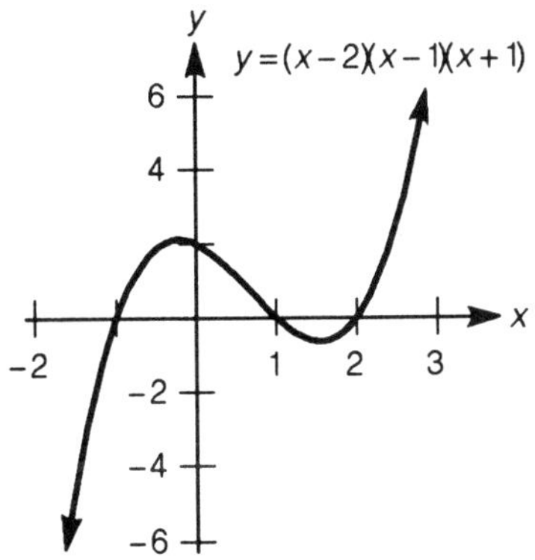

23. (a) The x-intercepts are 0, 2, and 1, and the y-intercept is 0. The signs of y are given by:

Interval	$y = 2x(x-2)(x-1)$
$(-\infty,0)$	neg.
$(0,1)$	pos.
$(1,2)$	neg.
$(2,\infty)$	pos.

We sketch the excluded regions:

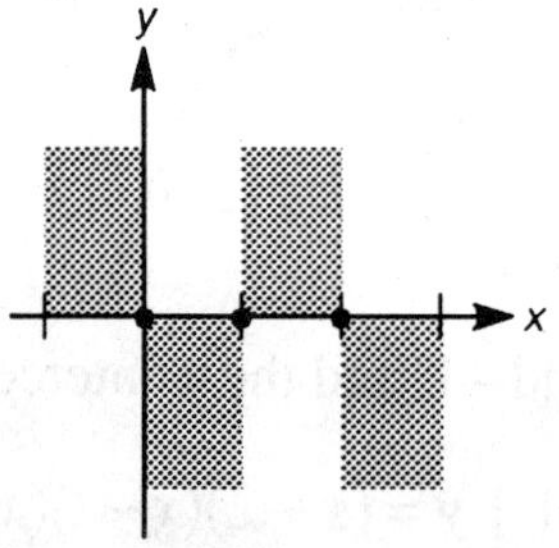

(b) We now graph the function:

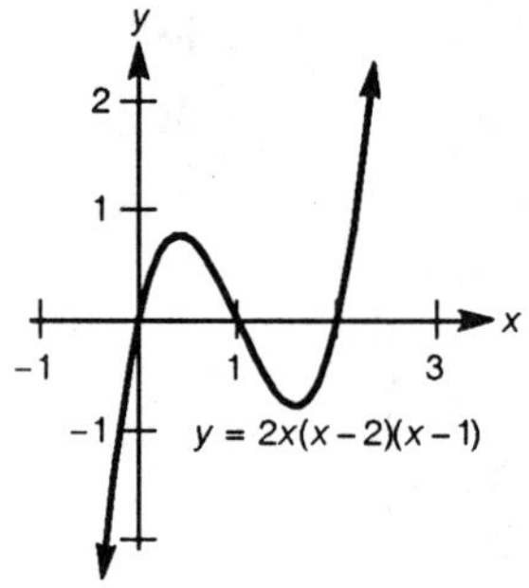

25. (a) We factor to find the x-intercepts:

$$y = x\left(x^2 - 4x - 5\right) = x(x-5)(x+1)$$

The x-intercepts are 0, 5, and –1, and the y-intercept is 0. The signs of y are given by:

Interval	$y = x(x-5)(x+1)$
$(-\infty,-1)$	neg.
$(-1,0)$	pos.
$(0,5)$	neg.
$(5,\infty)$	pos.

We sketch the excluded regions:

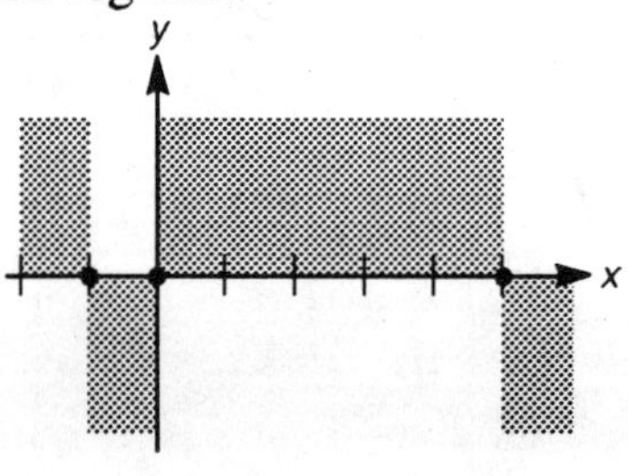

(b) We now graph the function:

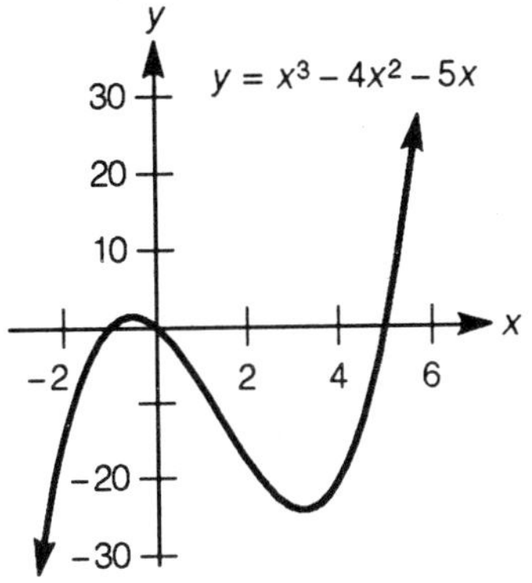

27. (a) We factor (by grouping) to find the x-intercepts:

$$y = x^2(x+3) - 4(x+3) = (x+3)(x^2-4) = (x+3)(x+2)(x-2)$$

The x-intercepts are –3, –2, and 2, and the y-intercept is –12. The signs of y are given by:

Interval	$y = (x+3)(x+2)(x-2)$
$(-\infty, -3)$	neg.
$(-3, -2)$	pos.
$(-2, 2)$	neg.
$(2, \infty)$	pos.

We sketch the excluded regions:

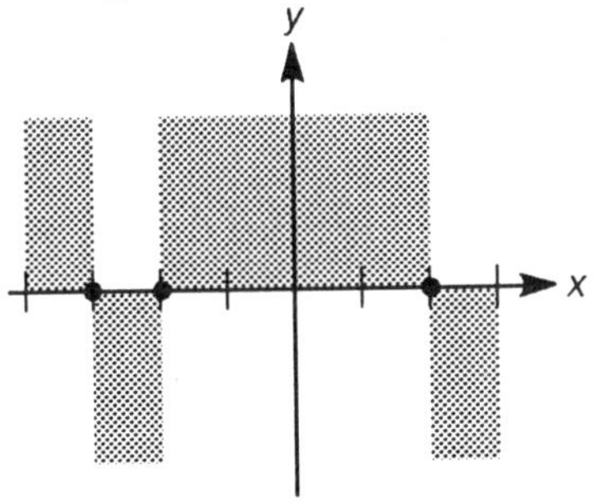

(b) We now graph the function:

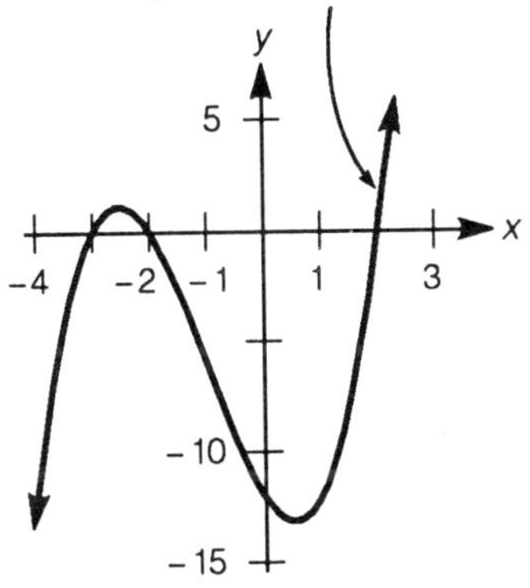

29. (a) The x-intercepts are 0 and –2, and the y-intercept is 0. The signs of y are given by:

Interval	$y = x^3(x+2)$
$(-\infty, -2)$	pos.
$(-2, 0)$	neg.
$(0, \infty)$	pos.

We sketch the excluded regions:

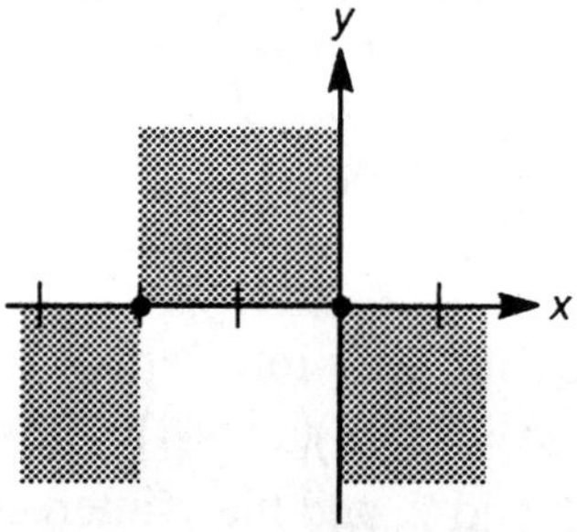

(b) Near $x = 0$, we have the approximation $y \approx x^3(0+2) = 2x^3$. So in the immediate vicinity of $x = 0$, the graph of y resembles $y = 2x^3$:

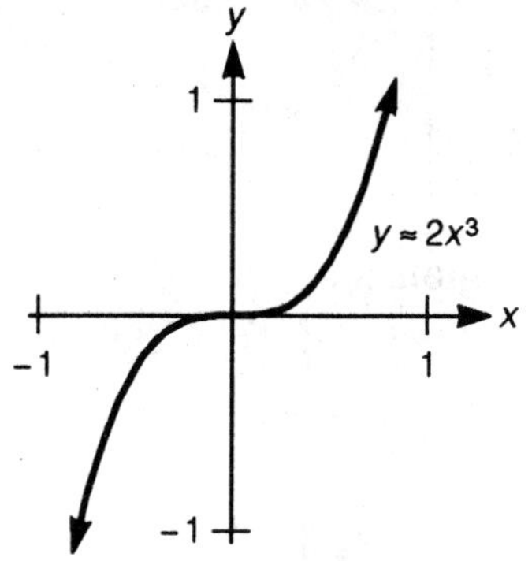

(c) We now graph the function:

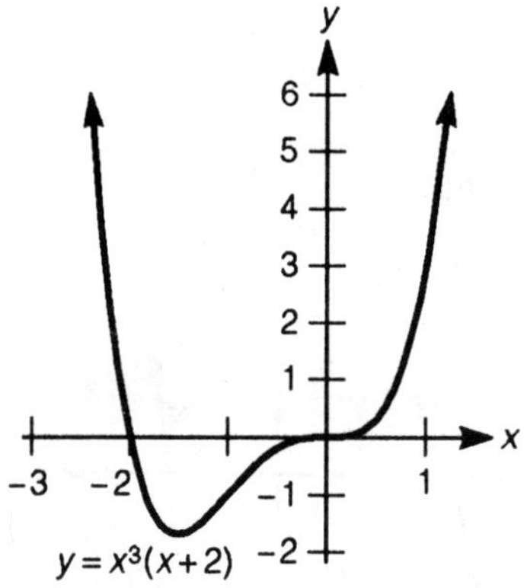

31. (a) The x-intercepts are 1 and 4, and the y-intercept is 128. The signs of y are given by:

Interval	$y = 2(x-1)(x-4)^3$
$(-\infty, 1)$	pos.
$(1, 4)$	neg.
$(4, \infty)$	pos.

We sketch the excluded regions:

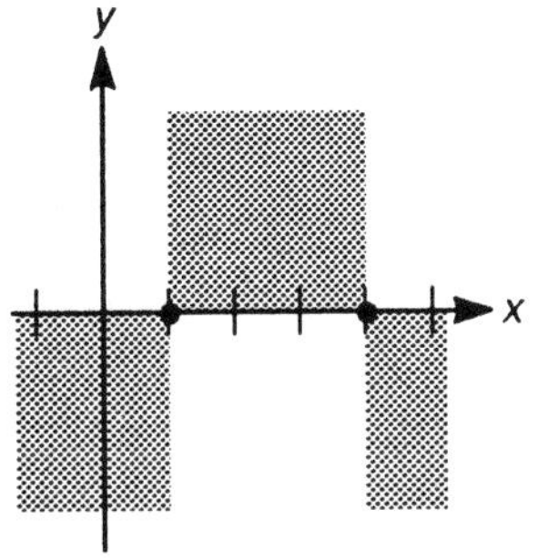

(b) Near $x = 4$, we have the approximation $y \approx 2(4-1)(x-4)^3 = 6(x-4)^3$. So in the immediate vicinity of $x = 4$, the graph of y resembles $y = 6(x-4)^3$:

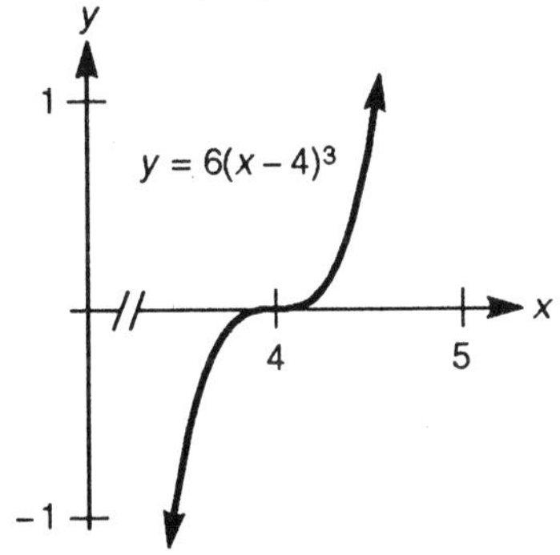

(c) We now graph the function:

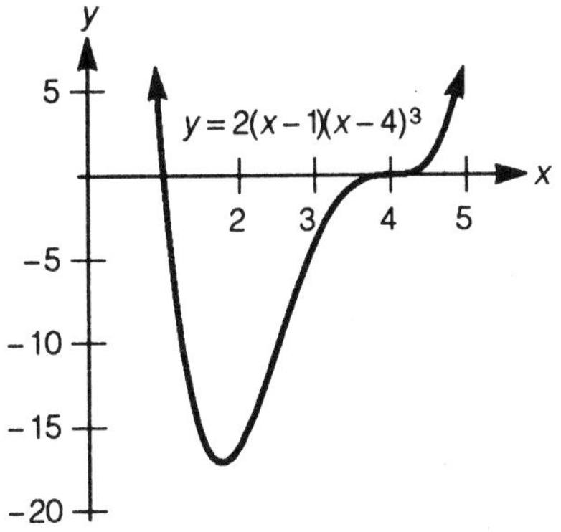

33. (a) The x-intercepts are -1, 1, and 3, and the y-intercept is 3. The signs of y are given by:

Interval	$y=(x+1)^2(x-1)(x-3)$
$(-\infty,-1)$	pos.
$(-1,1)$	pos.
$(1,3)$	neg.
$(3,\infty)$	pos.

We sketch the excluded regions:

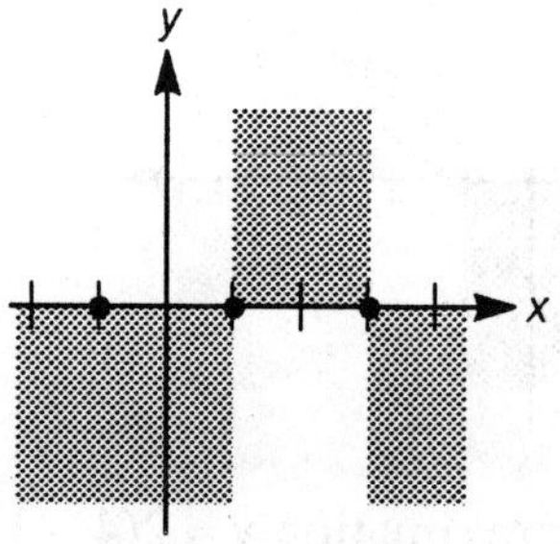

(b) Near $x=-1$, we have the approximation $y \approx (x+1)^2(-2)(-4)- = 8(x+1)^2$. So in the immediate vicinity of $x=-1$, the graph of y resembles $y = 8(x+1)^2$:

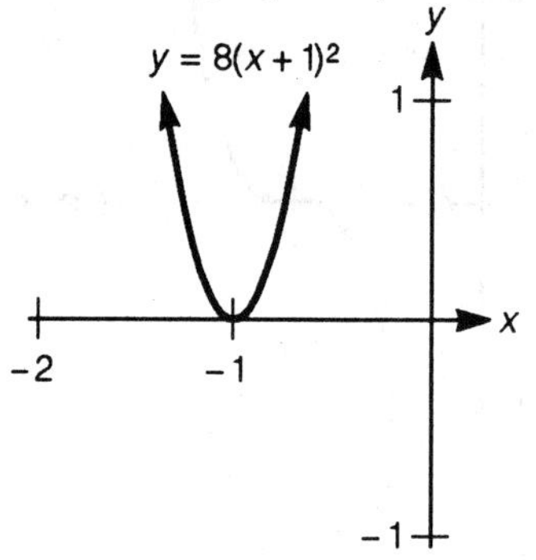

(c) We now graph the function:

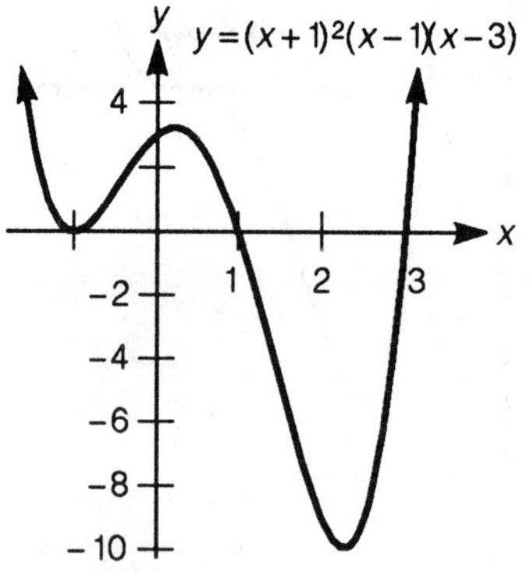

35. (a) The x-intercepts are 0, 4, and –2, and the y-intercept is 0. The signs of y are given by:

Interval	$y = -x^3(x-4)(x+2)$
$(-\infty, -2)$	pos.
$(-2, 0)$	neg.
$(0, 4)$	pos.
$(4, \infty)$	neg.

We sketch the excluded regions:

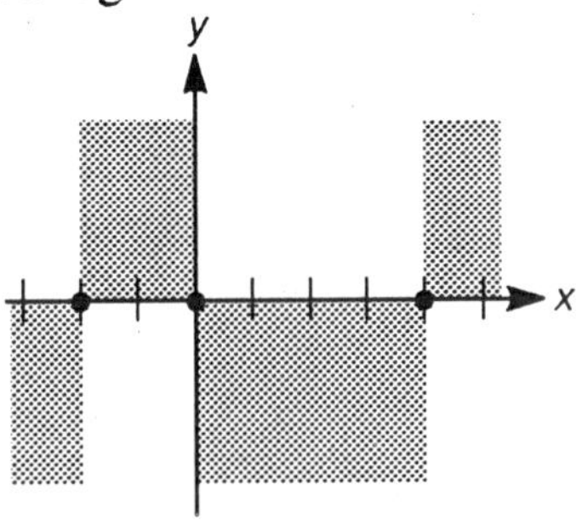

(b) Near $x = 0$, we have the approximation $y \approx -x^3(0-4)(0+2) = 8x^3$. So in the immediate vicinity of $x = 0$, the graph resembles $y = 8x^3$:

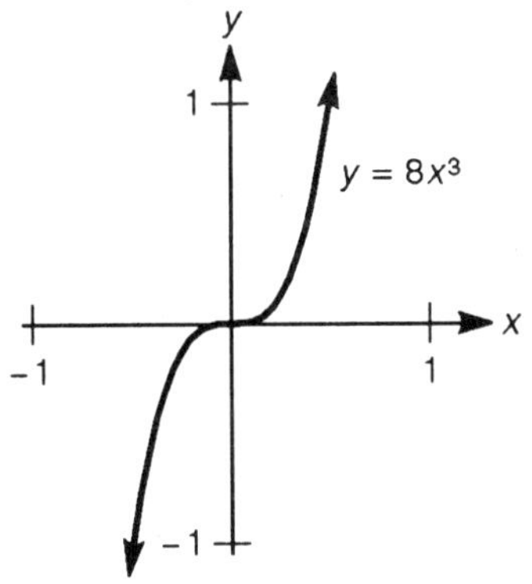

(c) We now graph the function:

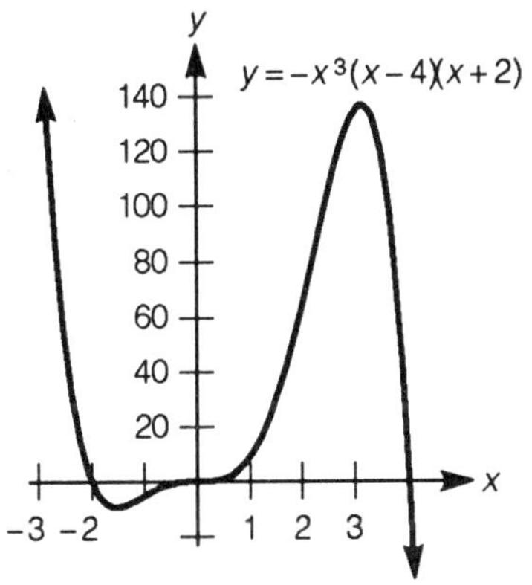

37. (a) The x-intercepts are 0, 2, and -2, and the y-intercept is 0. The signs of y are given by:

Interval	$y=-4x(x-2)^2(x+2)^3$
$(-\infty,-2)$	neg.
$(-2,0)$	pos.
$(0,2)$	neg.
$(2,\infty)$	neg.

We sketch the excluded region:

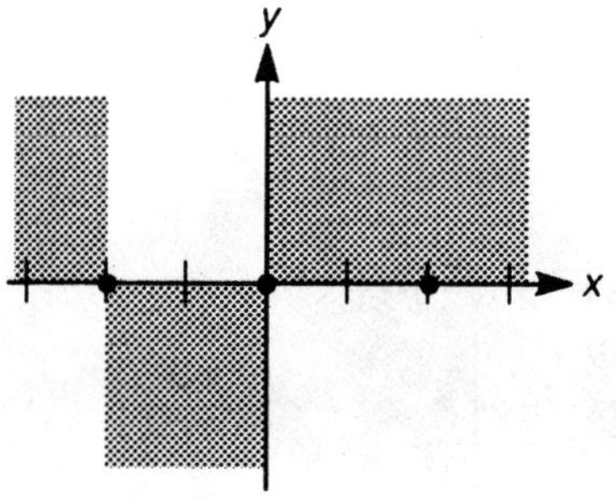

(b) Near $x=2$, we have the approximation $y\approx-4(2)(x-2)^2(2+2)^3=-512(x-2)^2$. So in the immediate vicinity of $x=2$, the graph resembles $y=-512(x-2)^2$:

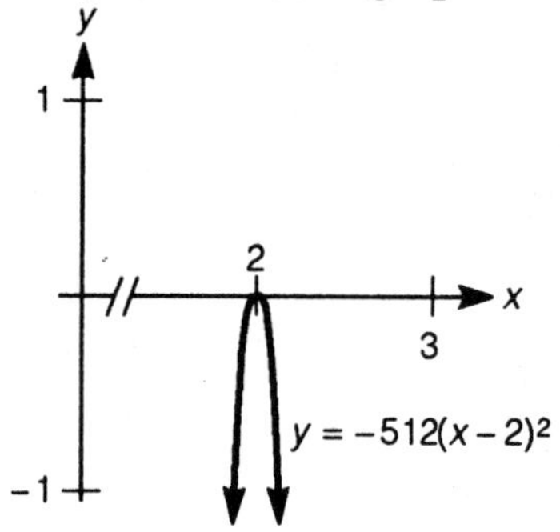

Near $x=-2$, we have the approximation $y\approx-4(-2)(-2-2)^2(x+2)^3=128(x+2)^3$. So in the immediate vicinity of $x=-2$, the graph resembles $y=128(x+2)^3$:

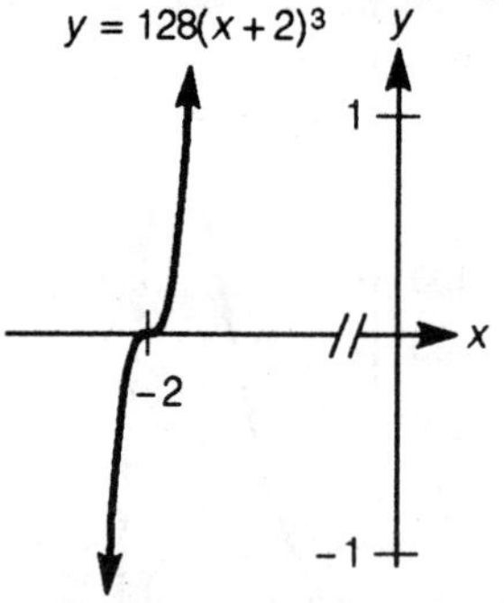

(c) We now graph the function:

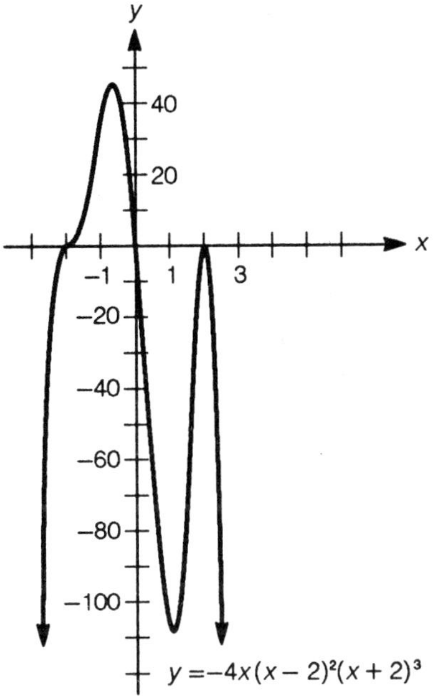

39. From left to right, they are $f(x) = x$, $g(x) = x^2$, $h(x) = x^3$, $F(x) = x^4$, $G(x) = x^5$, and $H(x) = x^6$.

41. We must find where $0 \le H(x) < 0.1$:

$$0 \le x^6 < 0.1$$
$$0 \le x < 0.68$$

When x lies in the interval $[0, 0.68)$, then $H(x)$ will lie in the interval $[0, 0.1)$.

43. We must find where:

$$g(t) - F(t) = 0.26$$
$$t^2 - t^4 = 0.26$$
$$t^4 - t^2 + 0.26 = 0$$

Using the quadratic formula:

$$t^2 = \frac{1 \pm \sqrt{1 - 4(0.26)}}{2} = \frac{1 \pm \sqrt{1 - 1.04}}{2} = \frac{1 \pm \sqrt{-0.04}}{2}$$

Since this equation has no real solutions, there is no such value of t.

45. We set the two y-coordinates equal:

$$x = \frac{1}{100}x^2$$
$$0 = \frac{1}{100}x^2 - x$$
$$0 = \frac{1}{100}\left(x^2 - 100x\right)$$
$$0 = \frac{1}{100} \bullet x(x - 100)$$

The graph intersects at the origin but also at the point $(100, 100)$.

47. (a) We first factor to obtain:

$$D(x) = x^2(1 - x^2) = x^2(1 + x)(1 - x)$$

So the x-intercepts are 0, –1, and 1. We draw the graph:

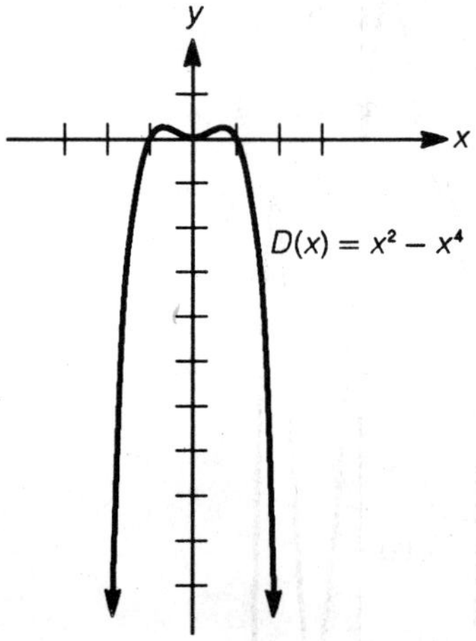

(b) We complete the square:

$$D(x) = x^2 - x^4 = -(x^4 - x^2) = -\left(x^4 - x^2 + \tfrac{1}{4}\right) + \tfrac{1}{4} = -\left(x^2 - \tfrac{1}{2}\right)^2 + \tfrac{1}{4}$$

The turning points are at $\left(\pm\frac{1}{\sqrt{2}}, \frac{1}{4}\right) = \left(\pm\frac{\sqrt{2}}{2}, \frac{1}{4}\right)$, which yield maximum values. Note that the graph also has a turning point at $(0,0)$, which is a minimum value.

(c) We graph the two functions:

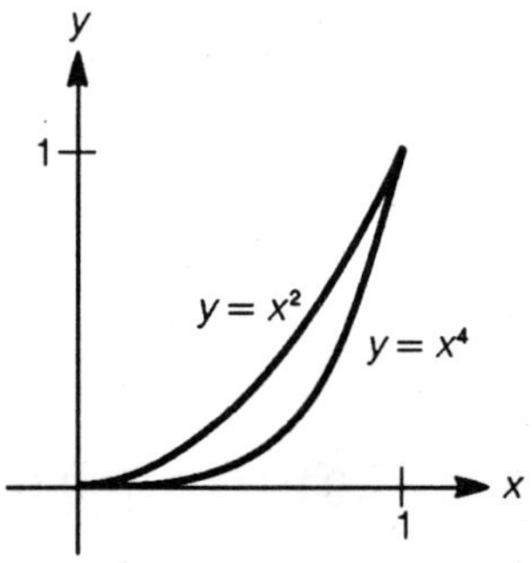

From part (b), we know that the maximum vertical distance between the two curves is $\frac{1}{4}$.

49. (a) Drawing a diagonal between two corners where the cylinder touches the circle yields the right triangle:

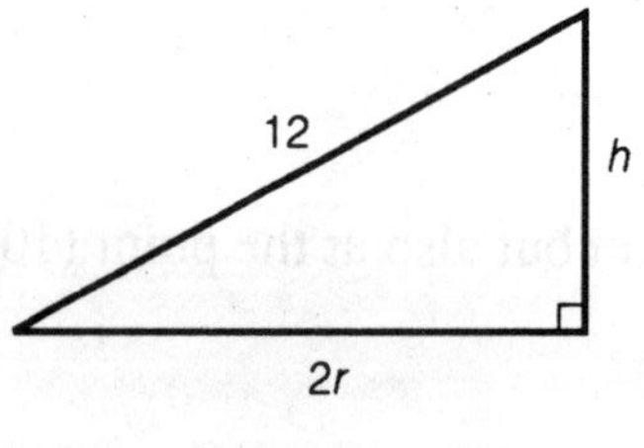

Using the Pythagorean theorem:

$$(2r)^2 + h^2 = 12^2$$
$$4r^2 + h^2 = 144$$
$$h^2 = 144 - 4r^2$$
$$h = \sqrt{144 - 4r^2}$$
$$h = 2\sqrt{36 - r^2}$$

(b) We compute the volume:

$$V = \pi r^2 h = \pi r^2\left(2\sqrt{36 - r^2}\right) = 2\pi r^2\sqrt{36 - r^2}$$

(c) We must have $36 - r^2 > 0$, so $0 < r < 6$. So the domain is $(0, 6)$.

(d) We complete the table:

r	$f(r)$
0.0	0
0.5	88
1.0	1382
1.5	6745
2.0	20213
2.5	45878
3.0	86339
3.5	140700
4.0	202129
4.5	254971
5.0	271414
5.5	207720
6.0	0

We graph $f(r) = 4\pi^2 r^4\left(36 - r^2\right)$:

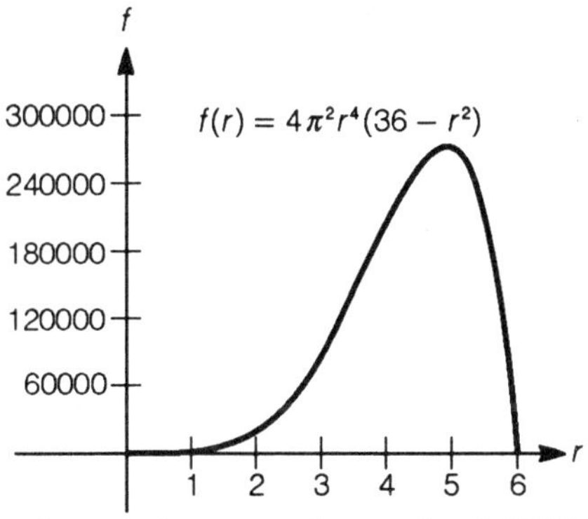

We can approximate the maximum value to be 273000. Thus the maximum possible volume is approximately $\sqrt{273000} \approx 522$ cm^3.

5.6 Graphs of Rational Functions

1. For the domain, we must exclude those values of x where:

$$\begin{aligned} 4x-12&=0\\ 4x&=12\\ x&=3 \end{aligned}$$

So the domain is all real numbers except 3, or $(-\infty,3)\cup(3,\infty)$.
For the x-intercepts, we must find where:

$$\begin{aligned} 3x+15&=0\\ 3x&=-15\\ x&=-5 \end{aligned}$$

So the x-intercept is –5. For the y-intercept, we let $x=0$ to obtain:

$$y=\frac{0+15}{0-12}=-\frac{5}{4}$$

So the y-intercept is $-\frac{5}{4}$.

3. For the domain, we must exclude those values of x where:

$$\begin{aligned} x^2-x-6&=0\\ (x-3)(x+2)&=0\\ x&=3,-2 \end{aligned}$$

So the domain is all real numbers except 3 and –2, or $(-\infty,-2)\cup(-2,3)\cup(3,\infty)$.
For the x-intercepts, we must find where:

$$\begin{aligned} x^2-8x-9&=0\\ (x-9)(x+1)&=0\\ x&=9,-1 \end{aligned}$$

So the x-intercepts are 9 and –1. For the y-intercept, we let $x=0$ to obtain:

$$y=\frac{-9}{-6}=\frac{3}{2}$$

So the y-intercept is $\frac{3}{2}$.

5. For the domain, we must exclude those values of x where:

$$\begin{aligned} x^6&=0\\ x&=0 \end{aligned}$$

So the domain is all real numbers except 0, or $(-\infty,0)\cup(0,\infty)$.
For the x-intercepts, we must find where:

$$(x^2-4)(x^3-1)=0$$

$$\begin{aligned} x^2-4&=0 &\quad\text{or}\quad&& x^3-1&=0\\ x^2&=4 &&& x^3&=1\\ x&=\pm 2 &&& x&=1 \end{aligned}$$

So the x-intercepts are –2, 2, and 1. For the y-intercept, we let $x=0$. But $x=0$ is not in the domain of the function, so there are no y-intercepts.

7. There is no x-intercept, the y-intercept is $\frac{1}{4}$, the horizontal asymptote is $y = 0$, and the vertical asymptote is $x = -4$. We draw the graph:

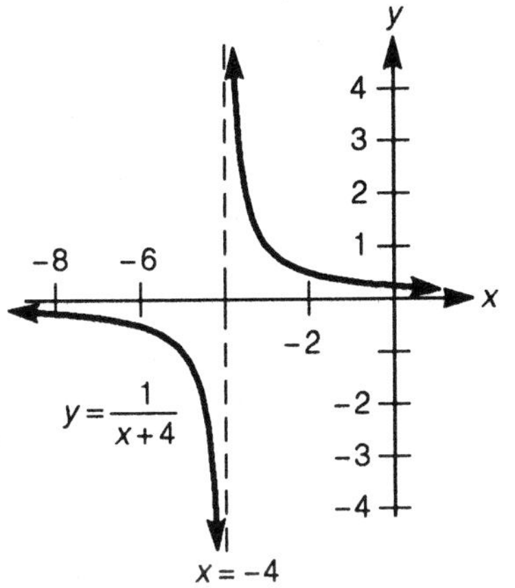

9. There is no x-intercept, the y-intercept is $\frac{3}{2}$, the horizontal asymptote is $y = 0$, and the vertical asymptote is $x = -2$. We draw the graph:

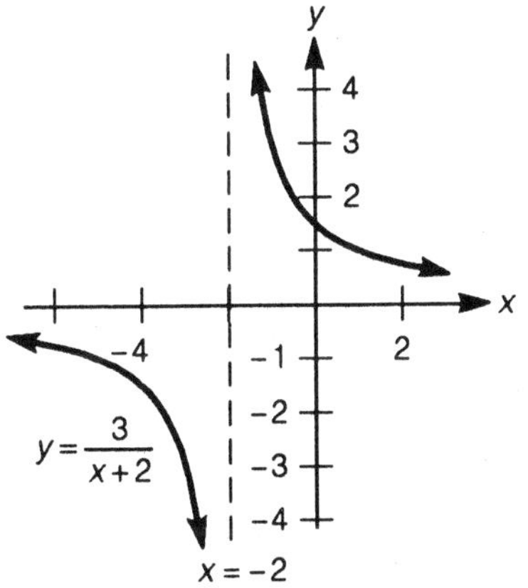

11. There is no x-intercept, the y-intercept is $\frac{2}{3}$, the horizontal asymptote is $y = 0$, and the vertical asymptote is $x = 3$. We draw the graph:

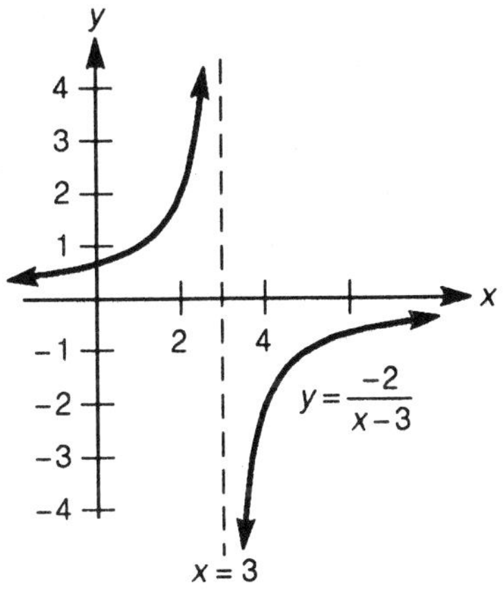

13. Using long division, we have:

$$\frac{x-3}{x-1}=1-\frac{2}{x-1}$$

The x-intercept is 3, the y-intercept is 3, the horizontal asymptote is $y = 1$, and the vertical asymptote is $x = 1$. We draw the graph:

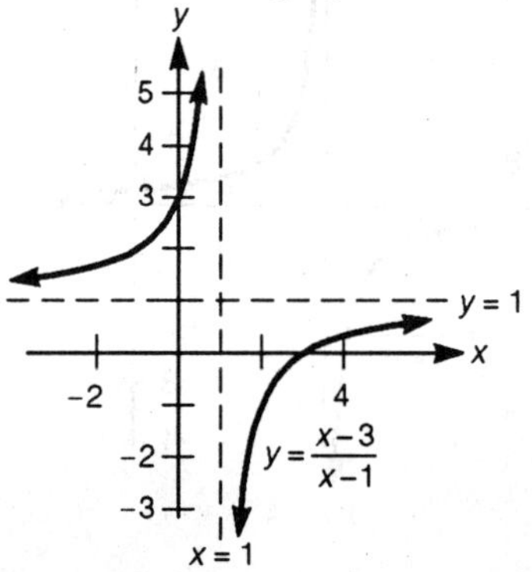

15. Using long division, we have:

$$\frac{4x-2}{2x+1}=2-\frac{4}{2x+1}$$

The x-intercept is $\frac{1}{2}$, the y-intercept is –2, the horizontal asymptote is $y = 2$, and the vertical asymptote is $x = -\frac{1}{2}$. We draw the graph:

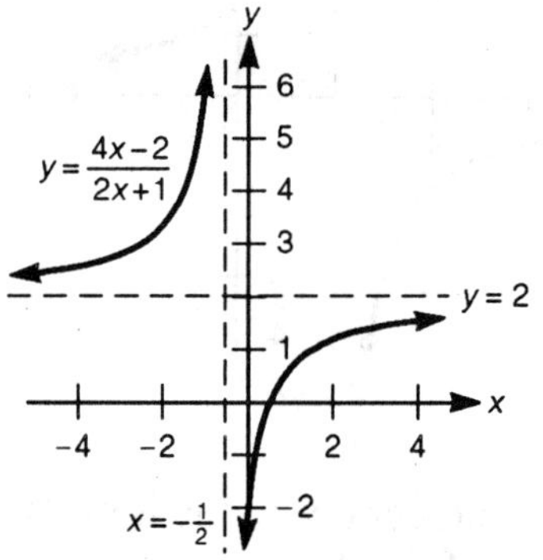

17. There is no x-intercept, the y-intercept is $\frac{1}{4}$, the horizontal asymptote is $y = 0$, and the vertical asymptote is $x = 2$. We draw the graph:

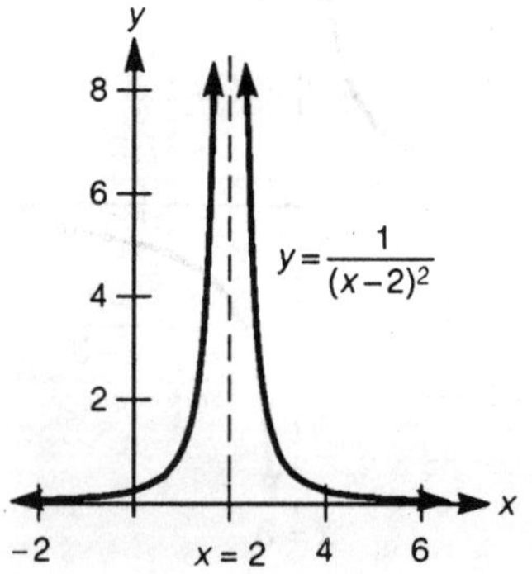

19. There is no x-intercept, the y-intercept is 3, the horizontal asymptote is $y = 0$, and the vertical asymptote is $x = -1$. We draw the graph:

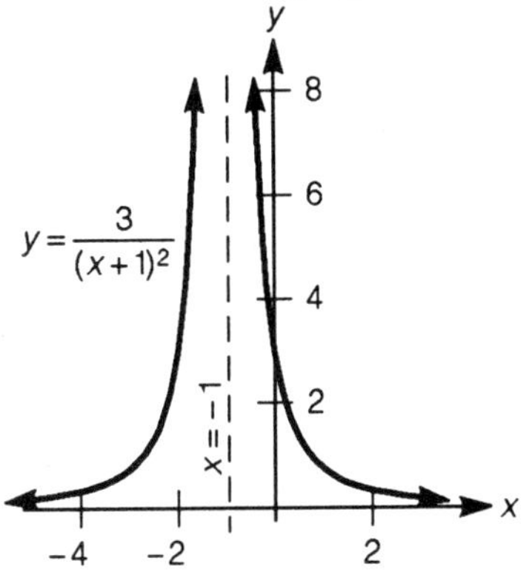

21. There is no x-intercept, the y-intercept is $\frac{1}{8}$, the horizontal asymptote is $y = 0$, and the vertical asymptote is $x = -2$. We draw the graph:

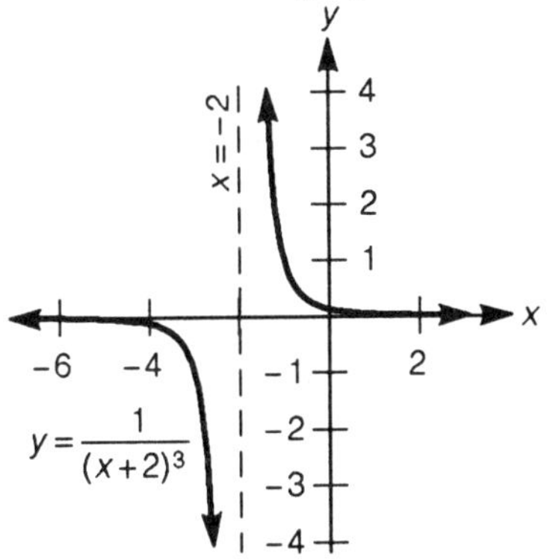

23. There is no x-intercept, the y-intercept is $-\frac{4}{125}$, the horizontal asymptote is $y = 0$, and the vertical asymptote is $x = -5$. We draw the graph:

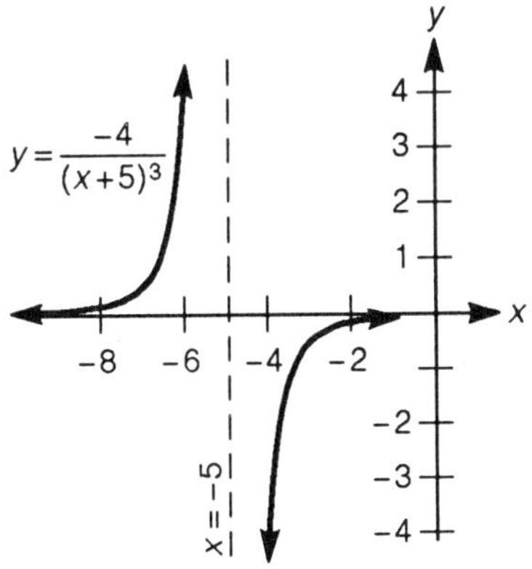

25. The x- and y-intercepts are 0, the horizontal asymptote is $y = 0$, and the vertical asymptotes are $x = -2$ and $x = 2$. We draw the graph:

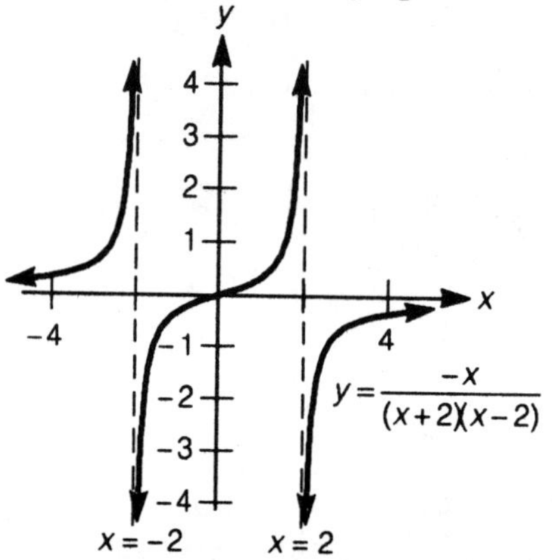

27. The x- and y-intercepts are 0, the horizontal asymptote is $y = 0$, and the vertical asymptotes are $x = -3$ and $x = 1$. We draw the graph:

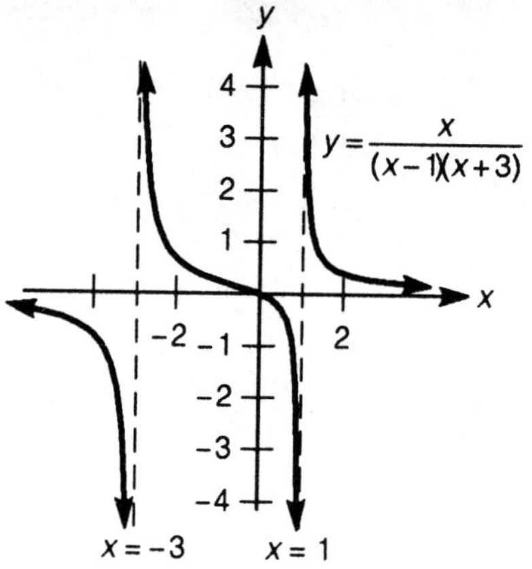

29. (a) The x-intercepts are 2 and 4, there is no y-intercept, the horizontal asymptote is $y = 1$, and the vertical asymptotes are $x = 0$ and $x = 1$. We draw the graph:

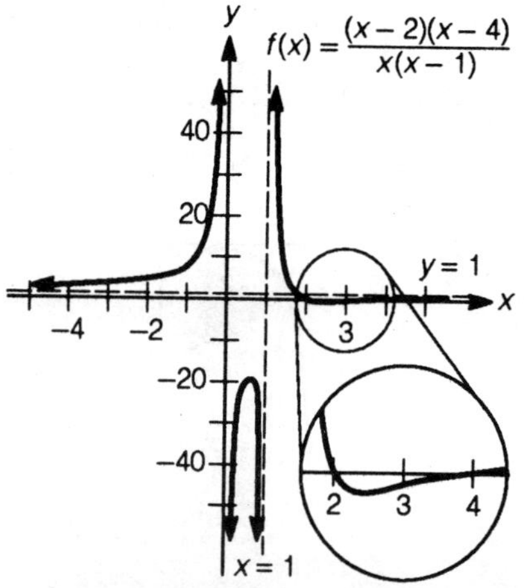

(b) The x-intercepts are 2 and 4, there is no y-intercept, the horizontal asymptote is $y = 1$, and the vertical asymptotes are $x = 0$ and $x = 3$. We draw the graph:

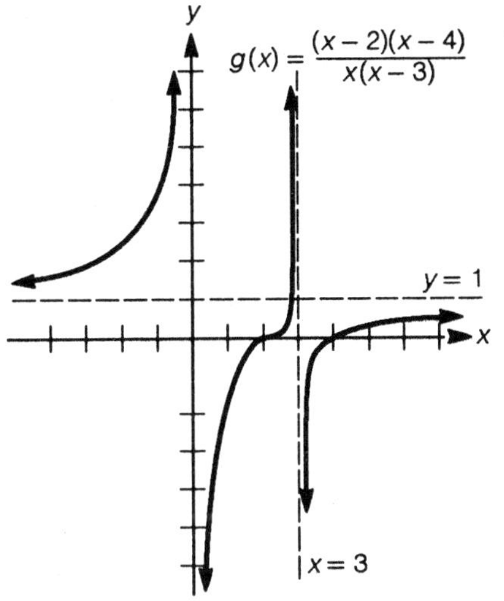

31. We find where the curve crosses the horizontal asymptote:

$$\frac{(x-4)(x+2)}{(x-1)(x-3)} = 1$$
$$(x-4)(x+2) = (x-1)(x-3)$$
$$x^2 - 2x - 8 = x^2 - 4x + 3$$
$$-2x - 8 = -4x + 3$$
$$2x = 11$$
$$x = \tfrac{11}{2}$$

The x-intercepts are -2 and 4, the y-intercept is $-\frac{8}{3}$, the horizontal asymptote is $y = 1$, and the vertical asymptotes are $x = 1$ and $x = 3$. We draw the graph:

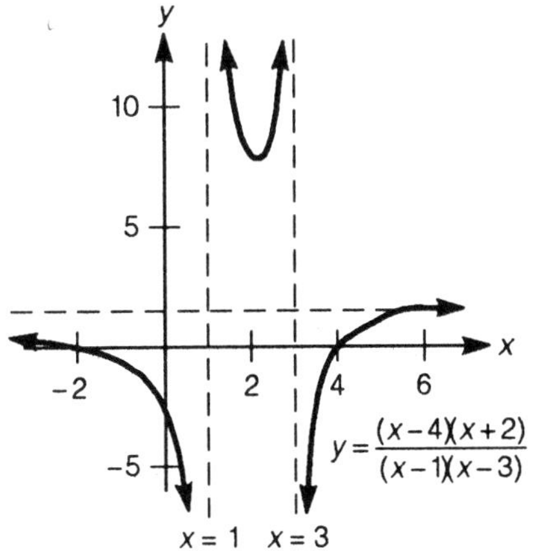

33. We find where the curve crosses the horizontal asymptote:

$$\frac{(x+1)^2}{(x-1)(x-3)}=1$$
$$(x+1)^2=(x-1)(x-3)$$
$$x^2+2x+1=x^2-4x+3$$
$$2x+1=-4x+3$$
$$6x=2$$
$$x=\tfrac{1}{3}$$

The x-intercept is -1, the y-intercept is $\frac{1}{3}$, the horizontal asymptote is $y = 1$, and the vertical asymptotes are $x = 1$ and $x = 3$. We draw the graph:

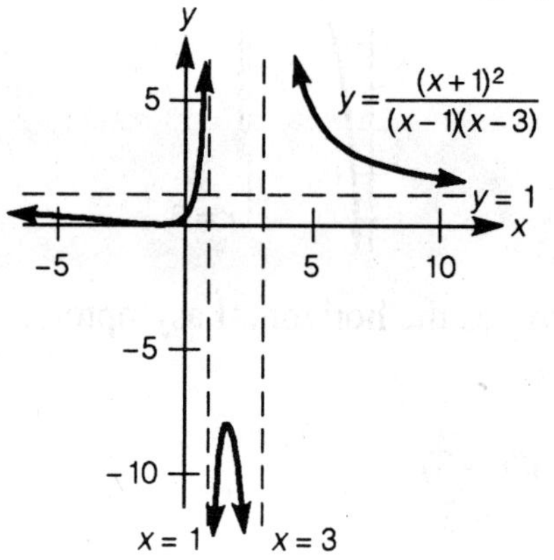

35. (a) For x near -2, we have:

$$y\approx\frac{(-5)(x+2)}{(-1)(-4)}=-\tfrac{5}{4}(x+2)$$

(b) For x near -1, we have:

$$y\approx\frac{(-4)(1)}{(x+1)(-3)}=\frac{\frac{4}{3}}{x+1}$$

(c) For x near 2, we have:

$$y\approx\frac{(-1)(4)}{(3)(x-2)}=\frac{-\frac{4}{3}}{x-2}$$

37. (a) For $x\neq-3$, we have:

$$y=\frac{x^2-9}{x+3}=\frac{(x+3)(x-3)}{x+3}=x-3$$

So this is the graph of $y = x - 3$, without the point at $(-3, -6)$:

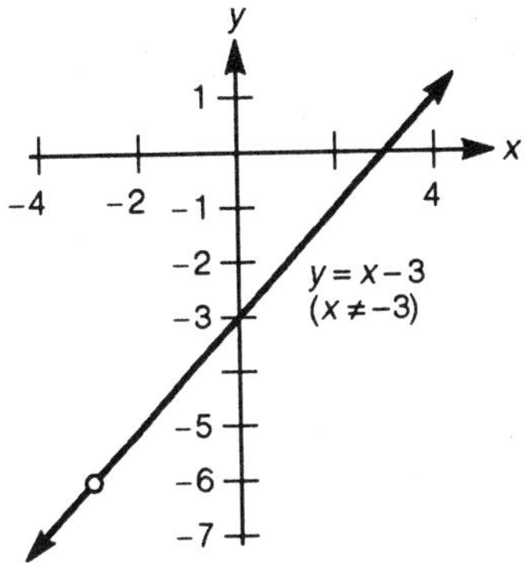

(b) For $x \neq 3$, we have:

$$y = \frac{x^2 - 5x + 6}{x^2 - 2x - 3} = \frac{(x-2)(x-3)}{(x+1)(x-3)} = \frac{x-2}{x+1}$$

So this is the graph of $y = \frac{x-2}{x+1}$, without the point at $\left(3, \frac{1}{4}\right)$.

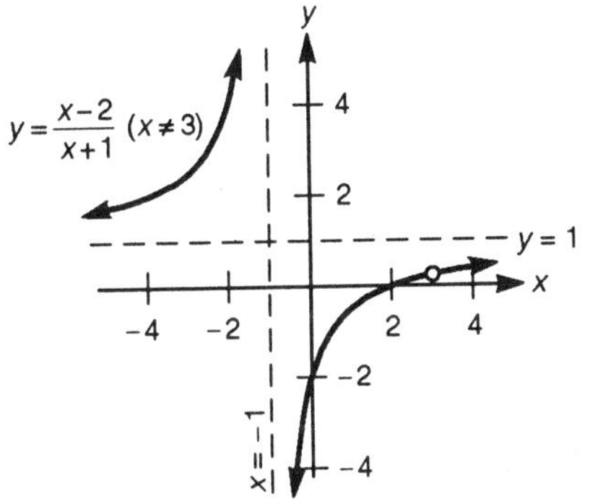

(c) For $x \neq 1, 2, 3$, we have:

$$y = \frac{(x-1)(x-2)(x-3)}{(x-1)(x-2)(x-3)(x-4)} = \frac{1}{x-4}$$

So this is the graph of $y = \frac{1}{x-4}$, without the points at $\left(1, -\frac{1}{3}\right), \left(2, -\frac{1}{2}\right), (3, -1)$.

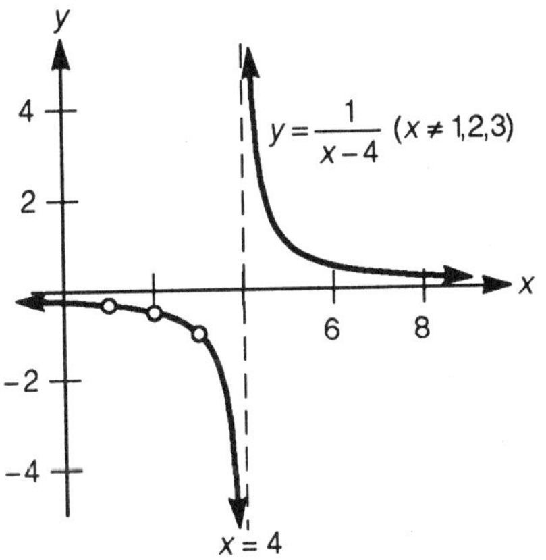

39. The horizontal asymptote is $y=0$ and the vertical asymptote is $x=3$. We find the value of k where $k=\frac{x}{(x-3)^2}$:

$$k(x-3)^2=x$$
$$kx^2-6kx+9k=x$$
$$kx^2-(6k+1)x+9k=0$$

Since this equation must have only one solution, we set the discriminant equal to zero:

$$[-(6k+1)]^2-4k(9k)=0$$
$$(6k+1)^2-36k^2=0$$
$$(6k+1+6k)(6k+1-6k)=0$$
$$(12k+1)(1)=0$$
$$k=-\tfrac{1}{12}$$

Thus $y=-\frac{1}{12}$. We find x:

$$-\frac{1}{12}=\frac{x}{(x-3)^2}$$
$$-(x-3)^2=12x$$
$$(x-3)^2=-12x$$
$$x^2-6x+9=-12x$$
$$x^2+6x+9=0$$
$$(x+3)^2=0$$
$$x=-3$$

So the low point is $\left(-3,-\frac{1}{12}\right)$. We draw the graph:

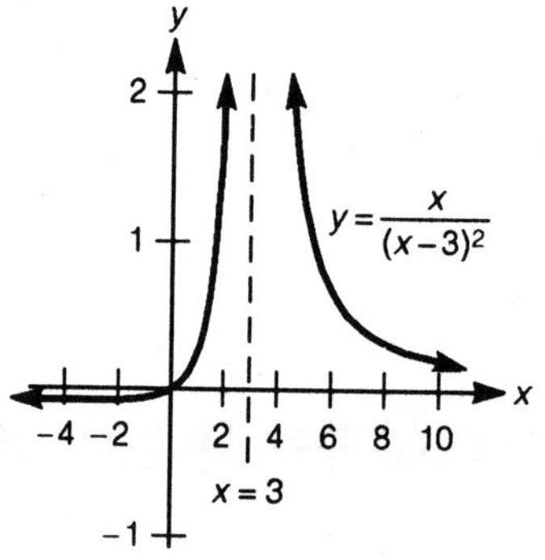

41. (a) We use long division:

$$\begin{array}{r} x+4 \\ x-3\;\overline{)\,x^2+x-6} \\ \underline{x^2-3x} \\ 4x-6 \\ \underline{4x-12} \\ 6 \end{array}$$

So $\dfrac{x^2+x-6}{x-3}=(x+4)+\dfrac{6}{x-3}$.

(b) We complete the tables:

x	$x+4$	$\frac{x^2+x-6}{x-3}$
10	14	14.8571
100	104	104.0619
1000	1004	1004.0060

x	$x+4$	$\frac{x^2+x-6}{x-3}$
−10	−6	−6.4615
−100	−96	−96.0583
−1000	−996	−996.0600

(c) The vertical asymptote is $x = 3$, the x-intercepts are −3 and 2, and the y-intercept is 2.

(d) We draw the graph:

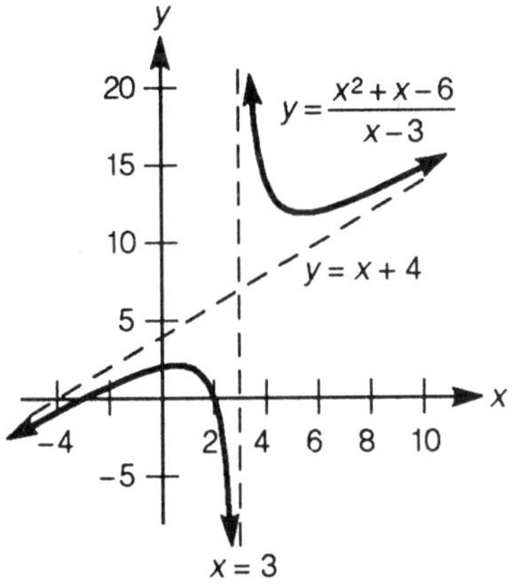

(e) We find where:

$$\frac{x^2+x-6}{x-3} = k$$
$$x^2+x-6 = kx-3k$$
$$x^2+(1-k)x+(3k-6) = 0$$

Setting the discriminant equal to 0, we have:

$$(1-k)^2 - 4(1)(3k-6) = 0$$
$$1-2k+k^2-12k+24 = 0$$
$$k^2-14k+25 = 0$$
$$(k-7)^2 = -25+49$$
$$(k-7)^2 = 24$$
$$k-7 = \pm\sqrt{24}$$
$$k = 7 \pm 2\sqrt{6}$$

So either $y=7+2\sqrt{6}$ or $y=7-2\sqrt{6}$. For each of these values, we find x:

$$\begin{aligned}\frac{x^2+x-6}{x-3}&=7+2\sqrt{6}\\ x^2+x-6&=(7+2\sqrt{6})x-21-6\sqrt{6}\\ x^2+(-6-2\sqrt{6})x+(15+6\sqrt{6})&=0\\ \left(x-(3+\sqrt{6})\right)^2&=0\\ x&=3+\sqrt{6}\end{aligned}$$

So one point is $\left(3+\sqrt{6},7+2\sqrt{6}\right)$. For $y=7-2\sqrt{6}$, we find x:

$$\begin{aligned}\frac{x^2+x-6}{x-3}&=7-2\sqrt{6}\\ x^2+x-6&=(7-2\sqrt{6})x-21+6\sqrt{6}\\ x^2+(-6+2\sqrt{6})x+(15-6\sqrt{6})&=0\\ \left(x-(3-\sqrt{6})\right)^2&=0\\ x&=3-\sqrt{6}\end{aligned}$$

So the other point is $\left(3-\sqrt{6},7-2\sqrt{6}\right)$.

43. Note that:

$$\frac{-x^2+1}{x}=-x+\frac{1}{x}$$

Thus $y=-x$ is a slant asymptote. We sketch the graph:

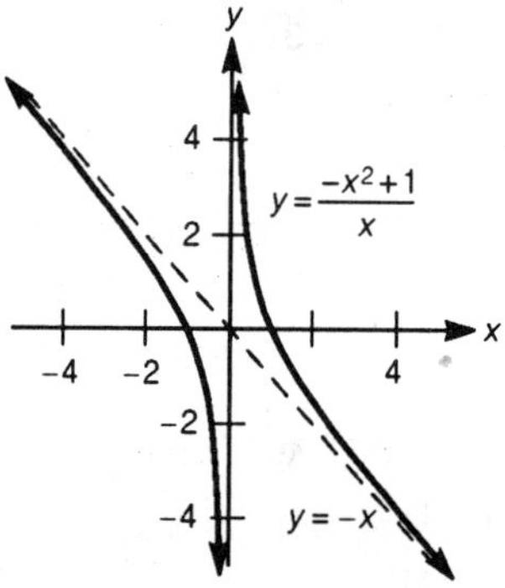

Chapter Five Review Exercises

1. We find the slope between the points $(1,-2)$ and $(-2,-11)$:

$$m=\frac{-11-(-2)}{-2-1}=\frac{-9}{-3}=3$$

Now use the point-slope formula:

$$\begin{aligned}y-(-2)&=3(x-1)\\ y+2&=3x-3\\ y&=3x-5\end{aligned}$$

So $G(x)=3x-5$, and thus $G(0)=-5$.

3. We first compute the revenue:

$$R(x) = xp = x\left(-\tfrac{1}{8}x + 100\right) = -\tfrac{1}{8}x^2 + 100x$$

Now complete the square:

$$\begin{aligned} R(x) &= -\tfrac{1}{8}\left(x^2 - 800x\right) \\ &= -\tfrac{1}{8}\left(x^2 - 800x + 160000\right) + 20000 \\ &= -\tfrac{1}{8}(x - 400)^2 + 20000 \end{aligned}$$

So the maximum possible revenue is \$20,000.

5. We graph the function:

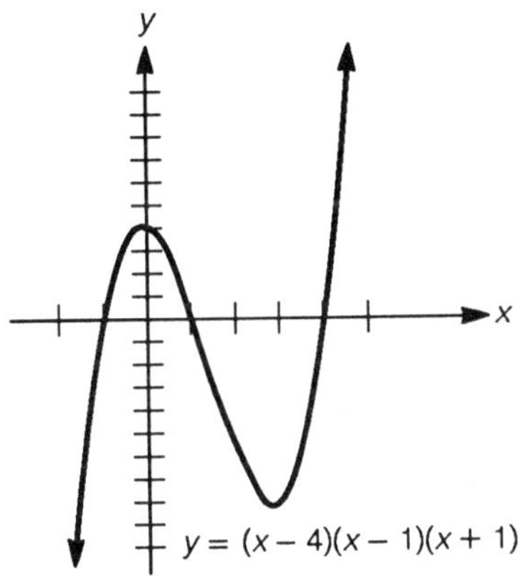

7. We have the points (0, 1000) and (5, 100). We find the slope:

$$m = \frac{100 - 1000}{5 - 0} = \frac{-900}{5} = -180$$

So $V(t) = -180t + 1000$.

9. The x-intercept is $-\frac{5}{3}$, the y-intercept is $\frac{5}{2}$, the horizontal asymptote is $y = 3$, and the vertical asymptote is $x = -2$. We graph the function:

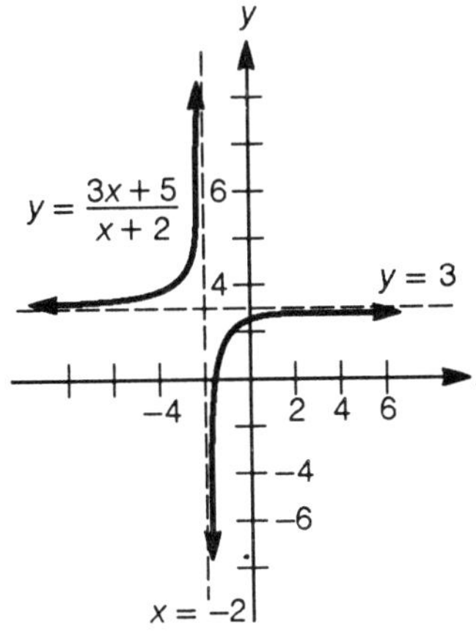

11. We graph the function:

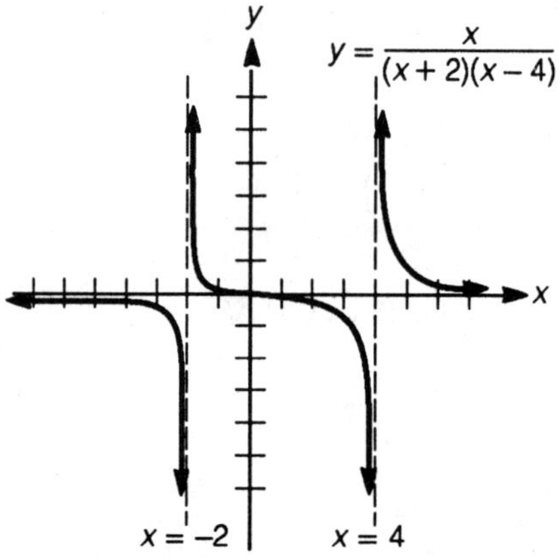

13. We draw the figure:

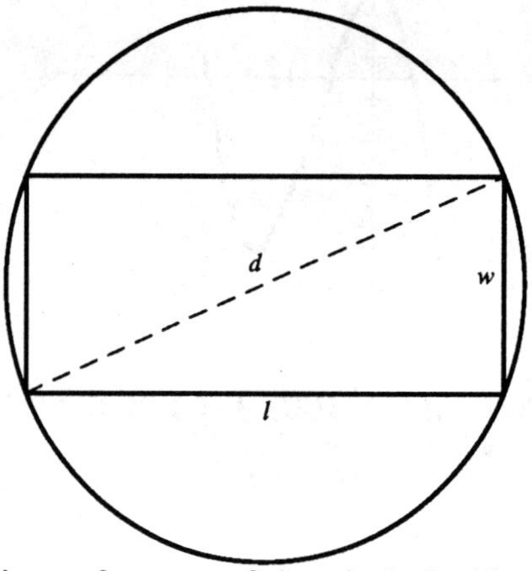

We can find r, since the circumference of the circle is 12 cm:

$$2\pi r = 12, \text{ so } r = \frac{6}{\pi} \text{ and thus } d = \frac{12}{\pi}$$

Using the Pythagorean theorem:

$$w^2 + l^2 = \left(\frac{12}{\pi}\right)^2$$

$$w^2 + l^2 = \frac{144}{\pi^2}$$

$$l^2 = \frac{144}{\pi^2} - w^2$$

$$l = \sqrt{\frac{144 - \pi^2 w^2}{\pi^2}}$$

$$l = \frac{\sqrt{144 - \pi^2 w^2}}{\pi}$$

Since the perimeter is given by $P = 2w + 2l$, we have:

$$P(w) = 2w + \frac{2\sqrt{144 - \pi^2 w^2}}{\pi}$$

15. We find the slope between the points (3, 5) and (–2, 0):

$$m = \frac{0-5}{-2-3} = \frac{-5}{-5} = 1$$

We use the point (3, 5) in the point-slope formula:

$$y-5 = 1(x-3)$$
$$y-5 = x-3$$
$$y = x+2$$

Using functional notation we have $f(x) = x + 2$.

17. We find the slope of the line $3x - 8y = 16$:

$$-8y = -3x+16$$
$$y = \tfrac{3}{8}x - 2$$

We use $m = \frac{3}{8}$ and the point (4, –1) in the point-slope formula:

$$y-(-1) = \tfrac{3}{8}(x-4)$$
$$y+1 = \tfrac{3}{8}x - \tfrac{3}{2}$$
$$y = \tfrac{3}{8}x - \tfrac{5}{2}$$

Using functional notation we have $f(x) = \frac{3}{8}x - \frac{5}{2}$.

19. If the graph of the inverse function passes through (2, 1), then (1, 2) must lie on the graph of the function. We find the slope between the points (1, 2) and (–3, 5):

$$m = \frac{5-2}{-3-1} = \frac{3}{-4} = -\frac{3}{4}$$

We use the point (1, 2) in the point-slope formula:

$$y-2 = -\tfrac{3}{4}(x-1)$$
$$y-2 = -\tfrac{3}{4}x + \tfrac{3}{4}$$
$$y = -\tfrac{3}{4}x + \tfrac{11}{4}$$

Using functional notation we have $f(x) = -\frac{3}{4}x + \frac{11}{4}$.

21. We complete the square:

$$y = x^2+2x-3 = \left(x^2+2x+1\right)-4 = (x+1)^2 - 4$$

The vertex is (–1, –4), the x-intercepts are 1 and –3, and the y-intercept is –3. We graph the function:

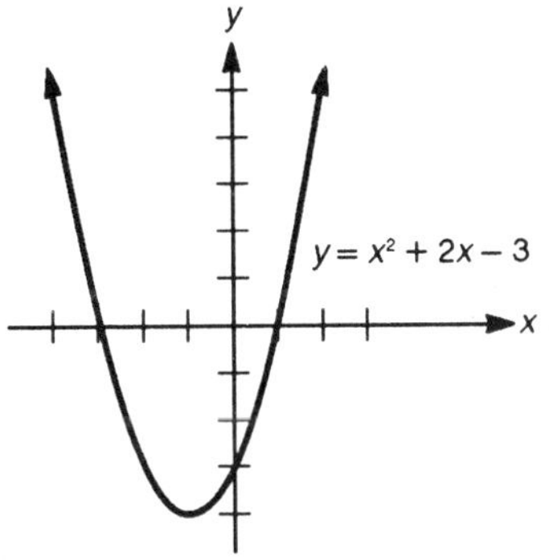

23. We complete the square:

$$y=-x^2+2\sqrt{3}x+3=-\left(x^2-2\sqrt{3}x\right)+3=-\left(x^2-2\sqrt{3}x+3\right)+6=-\left(x-\sqrt{3}\right)^2+6$$

The vertex is $\left(\sqrt{3},6\right)$, the x-intercepts are $\sqrt{3}+\sqrt{6}$ and $\sqrt{3}-\sqrt{6}$, and the y-intercept is 3. We graph the function:

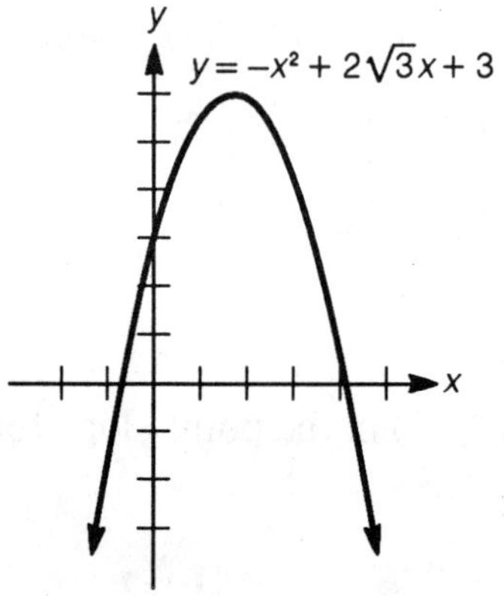

25. We complete the square:

$$y=-3x^2+12x=-3\left(x^2-4x\right)=-3\left(x^2-4x+4\right)+12=-3(x-2)^2+12$$

The vertex is $(2,12)$, the x-intercepts are 0 and 4, and the y-intercept is 0. We draw the graph:

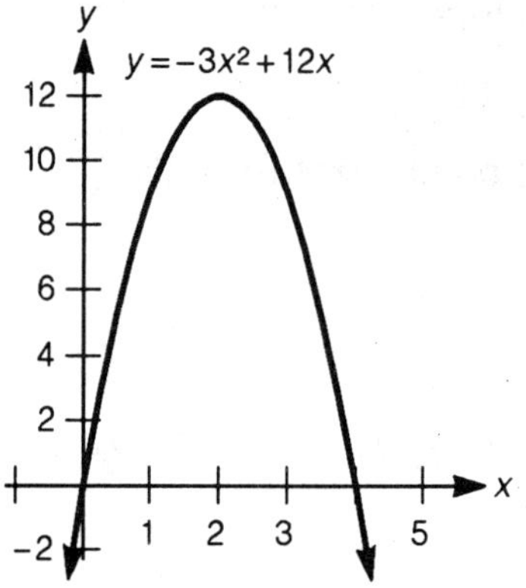

27. We find the two vertices:

$$\begin{aligned} y&=x^2-4x+6 \\ y-6&=x^2-4x \\ y-6+4&=x^2-4x+4 \\ y-2&=(x-2)^2 \end{aligned}$$

vertex: $(2,2)$

$$\begin{aligned} y&=-x^2-4x-5 \\ y+5&=-\left(x^2+4x\right) \\ y+5-4&=-\left(x^2+4x+4\right) \\ y+1&=-(x+2)^2 \end{aligned}$$

vertex: $(-2,-1)$

We now find the distance between $(2,2)$ and $(-2,-1)$:

$$d=\sqrt{(-2-2)^2+(-1-2)^2}=\sqrt{16+9}=\sqrt{25}=5$$

29. Call x and y the two numbers, so $x + y = \sqrt{3}$, and thus $y = \sqrt{3} - x$.
We find their product:

$$P = xy = x(\sqrt{3} - x) = -x^2 + \sqrt{3}x$$

Now complete the square:

$$P = -(x^2 - \sqrt{3}x) = -(x^2 - \sqrt{3}x + \tfrac{3}{4}) + \tfrac{3}{4} = -\left(x - \tfrac{\sqrt{3}}{2}\right)^2 + \tfrac{3}{4}$$

The maximum product is $\frac{3}{4}$.

31. (a) We complete the square:

$$\begin{aligned} h &= -16t^2 + v_0 t \\ &= -16\left(t^2 - \frac{v_0}{16}t\right) \\ &= -16\left(t^2 - \frac{v_0}{16}t + \frac{v_0^2}{1024}\right) + \frac{v_0^2}{64} \\ &= -16\left(t - \frac{v_0}{32}\right)^2 + \frac{v_0^2}{64} \end{aligned}$$

So the maximum height of $\frac{v_0^2}{64}$ ft. is obtained when $t = \frac{v_0}{32}$ sec.

(b) The object will strike the ground when $h(t) = 0$:

$$\begin{aligned} -16t^2 + v_0 t &= 0 \\ t(-16t + v_0) &= 0 \\ t = 0 \text{ or } t &= \frac{v_0}{16} \end{aligned}$$

So the object will strike the ground when $t = \frac{v_0}{16}$ sec.

33. (a) We find the distance between $(0, 2)$ and (x, x^2):

$$d = \sqrt{(x-0)^2 + (x^2 - 2)^2} = \sqrt{x^2 + x^4 - 4x^2 + 4} = \sqrt{x^4 - 3x^2 + 4}$$

(b) We want to minimize d. This will occur at the same x-coordinate as the minimum of d^2. We complete the square:

$$d^2 = x^4 - 3x^2 + 4 = \left(x^4 - 3x^2 + \tfrac{9}{4}\right) + 4 - \tfrac{9}{4} = \left(x^2 - \tfrac{3}{2}\right)^2 + \tfrac{7}{4}$$

So the minimum occurs where $x^2 = \frac{3}{2}$, so $x = \pm\sqrt{\frac{3}{2}} = \frac{\pm\sqrt{6}}{2}$. Since the point is in the second quadrant, we know $x < 0$, and thus the point on the parabola is $\left(\frac{-\sqrt{6}}{2}, \frac{3}{2}\right)$.

35. We find the distance between $(2,0)$ and $\left(x, \frac{4}{3}x + b\right)$:

$$\begin{aligned} d &= \sqrt{(x-2)^2 + \left(\tfrac{4}{3}x + b\right)^2} \\ &= \sqrt{x^2 - 4x + 4 + \tfrac{16}{9}x^2 + \tfrac{8b}{3}x + b^2} \\ &= \sqrt{\tfrac{25}{9}x^2 + \left(\tfrac{8b}{3} - 4\right)x + \left(4 + b^2\right)} \end{aligned}$$

We complete the square on d^2:

$$\begin{aligned} d^2 &= \tfrac{25}{9}\left(x^2 + \tfrac{9}{25}\left(\frac{8b}{3} - 4\right)x\right) + \left(4 + b^2\right) \\ &= \tfrac{25}{9}\left(x^2 + \tfrac{12}{25}(2b-3)x\right) + \left(4 + b^2\right) \\ &= \tfrac{25}{9}\left(x^2 + \tfrac{12}{25}(2b-3)x + \tfrac{36}{625}(2b-3)^2\right) + \left(4 + b^2\right) - \tfrac{4}{25}(2b-3)^2 \\ &= \tfrac{25}{9}\left(x + \tfrac{6}{25}(2b-3)\right)^2 + \frac{25\left(4 + b^2\right) - 4\left(4b^2 - 12b + 9\right)}{25} \\ &= \tfrac{25}{9}\left(x + \tfrac{6}{25}(2b-3)\right)^2 + \frac{(3b+8)^2}{25} \end{aligned}$$

Since the minimum distance is 5, then $5 = \left|\frac{3b+8}{5}\right|$. We now solve for b:

$$|3b+8| = 25$$

$$\begin{aligned} 3b + 8 &= 25 &\quad\text{or}\quad 3b + 8 &= -25 \\ 3b &= 17 & 3b &= -33 \\ b &= \tfrac{17}{3} & b &= -11 \end{aligned}$$

37. Since $x + y = \sqrt{2}$, then $y = \sqrt{2} - x$. Then:

$$s = x^2 + y^2 = x^2 + \left(\sqrt{2} - x\right)^2 = x^2 + 2 - 2\sqrt{2}x + x^2 = 2x^2 - 2\sqrt{2}x + 2$$

We complete the square:

$$s = 2\left(x^2 - \sqrt{2}x\right) + 2 = 2\left(x^2 - \sqrt{2}x + \tfrac{1}{2}\right) + 2 - 1 = 2\left(x - \tfrac{\sqrt{2}}{2}\right)^2 + 1$$

So the minimum value of s is 1.

39. Let x and h be the two legs. We have:

$$\begin{aligned} x^2 + h^2 &= 15^2 \\ h^2 &= 225 - x^2 \\ h &= \sqrt{225 - x^2} \end{aligned}$$

So $A = \frac{1}{2}(\text{base})(\text{height}) = \frac{1}{2}x\sqrt{225 - x^2}$. We find $A^2 = \frac{1}{4}x^2\left(225 - x^2\right) = -\frac{1}{4}x^4 + \frac{225}{4}x^2$.
We now complete the square on A^2:

$$A^2 = -\tfrac{1}{4}\left(x^4 - 225x^2\right) = -\tfrac{1}{4}\left[x^4 - 225x^2 + \left(\tfrac{225}{2}\right)^2\right] + \left(\tfrac{225}{4}\right)^2 = -\tfrac{1}{4}\left(x^2 - \tfrac{225}{2}\right)^2 + \left(\tfrac{225}{4}\right)^2$$

So the maximum of $A^2 = \left(\frac{225}{4}\right)^2$, thus $A = \frac{225}{4}$ cm^2.

41. We factor $f(x)$ as:

$$f(x) = x^2 - (a^2 + 2a)x + 2a^3 = (x - a^2)(x - 2a)$$

So the x-intercepts are a^2 and $2a$. Since $2a > a^2$ when $0 < a < 2$, then the distance between $(a^2, 0)$ and $(2a, 0)$ will be $D = 2a - a^2 = -a^2 + 2a$.

We now complete the square:

$$D = -(a^2 - 2a) = -(a^2 - 2a + 1) + 1 = -(a - 1)^2 + 1$$

So D is maximum when $a = 1$.

43. We compute R:

$$R = xp = x\left(160 - \tfrac{1}{5}x\right) = -\tfrac{1}{5}x^2 + 160x$$

We now complete the square:

$$R = -\tfrac{1}{5}(x^2 - 800x) = -\tfrac{1}{5}(x^2 - 800x + 160000) + 32000 = -\tfrac{1}{5}(x - 400)^2 + 32000$$

So $x = 400$ units will maximize the revenue. Then $p = 160 - \frac{1}{5}(400) = \$80$.

45. The x-intercepts are -4 and 2 and the y-intercept is -8. We draw the graph:

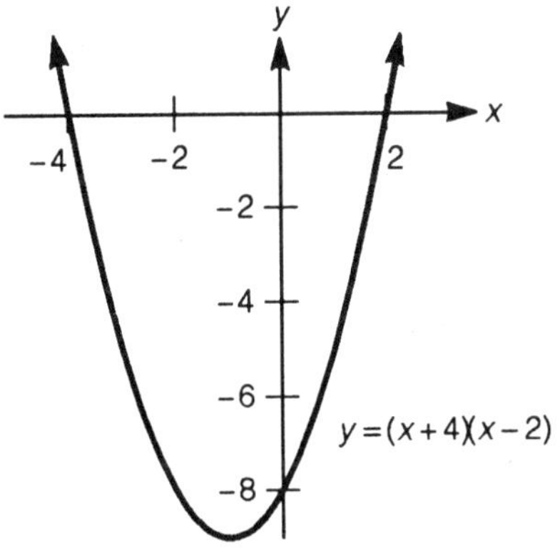

47. The x-intercept is -5 and the y-intercept is -125. We draw the graph:

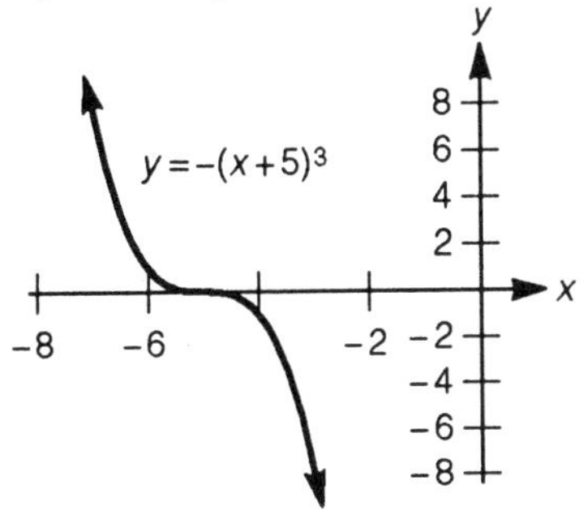

49. The x-intercepts are –1 and 0, and the y-intercept is 0. We draw the graph:

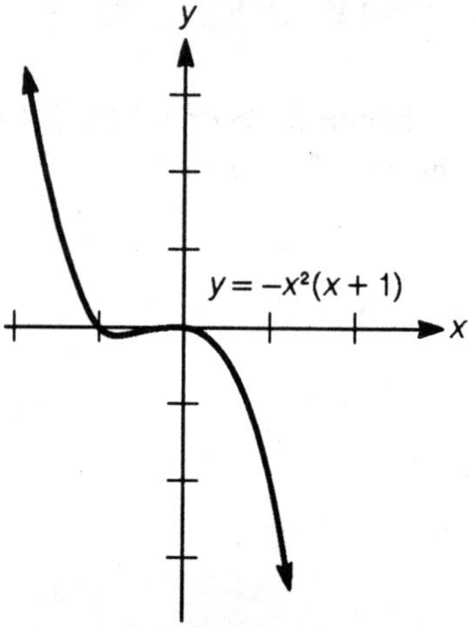

51. The x-intercepts are 0, 2, and –2, and the y-intercept is 0. We draw the graph:

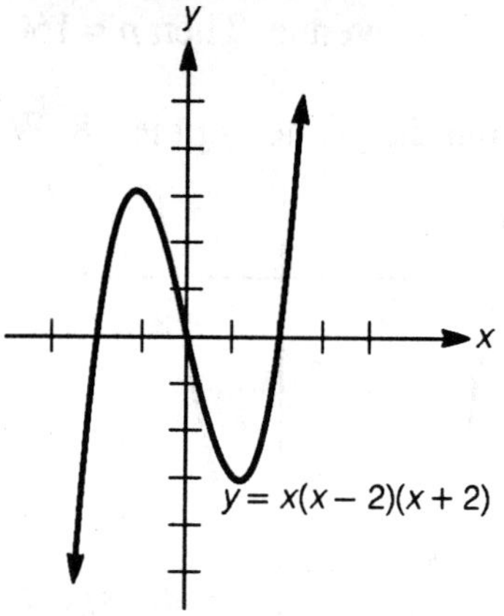

53. The x-intercept is $-\frac{1}{3}$, there is no y-intercept, the horizontal asymptote is $y = 3$, and the vertical asymptote is $x = 0$. We draw the graph:

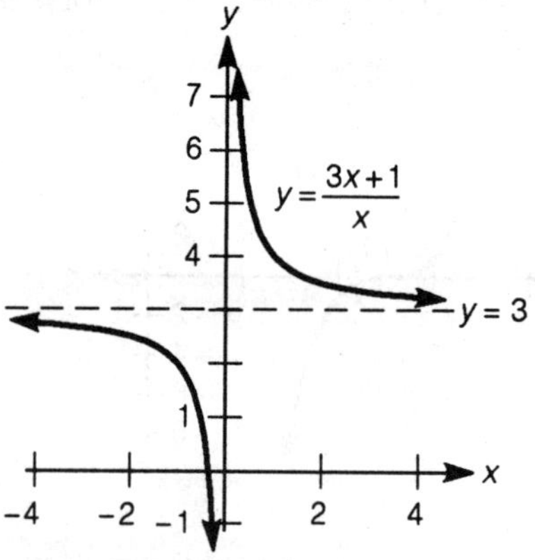

55. There is no x-intercept, the y-intercept is -1, the horizontal asymptote is $y = 0$, and the vertical asymptote is $x = 1$. We draw the graph:

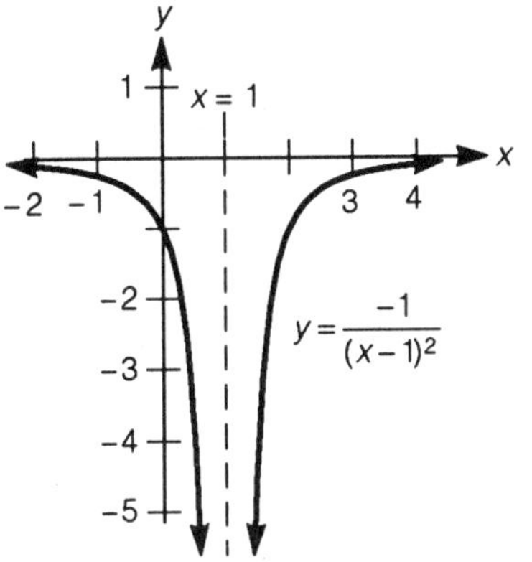

57. The x-intercept is 2, the y-intercept is $\frac{2}{3}$, the horizontal asymptote is $y = 1$, and the vertical asymptote is $x = 3$. We draw the graph:

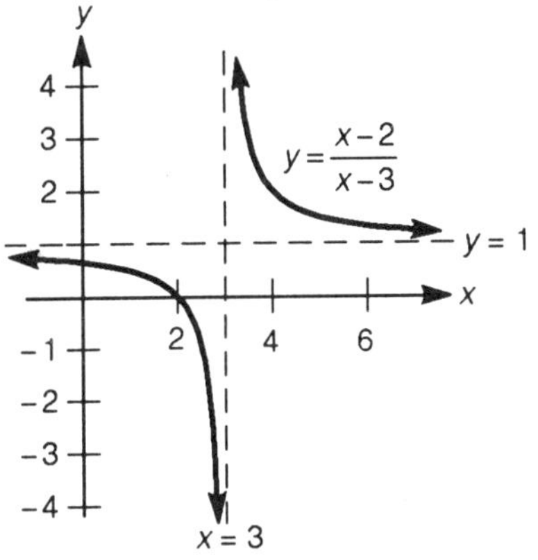

59. The x-intercept is 1, the y-intercept is $\frac{1}{4}$, the horizontal asymptote is $y = 1$, and the vertical asymptote is $x = 2$. We draw the graph:

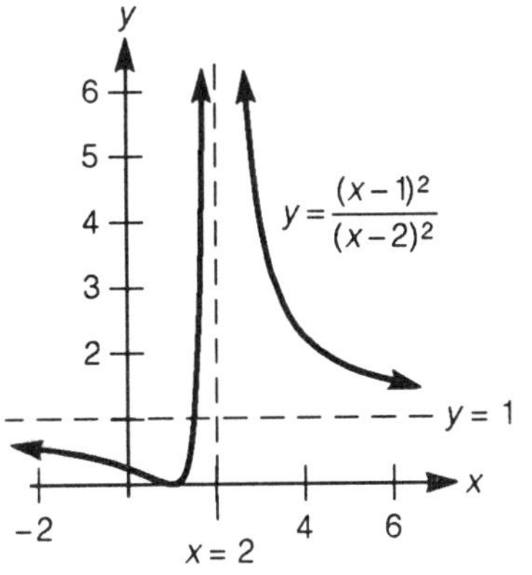

61. (a) If $b = 1$, then $f(x) = x^2 + 2x + 1 = (x + 1)^2$, so the vertex is $(-1, 0)$. This point is 1 unit from the origin.

(b) If $b = 2$, then $f(x) = x^2 + 4x + 1$. We complete the square:

$$f(x) = (x^2 + 4x + 4) + 1 - 4 = (x + 2)^2 - 3$$

So the vertex is $(-2, -3)$. By the distance formula:

$$d = \sqrt{(-2-0)^2 + (-3-0)^2} = \sqrt{4+9} = \sqrt{13}$$

(c) We complete the square:

$$f(x) = x^2 + 2bx + 1 = (x^2 + 2bx + b^2) + 1 - b^2 = (x + b)^2 + 1 - b^2$$

So the vertex is $(-b, 1 - b^2)$. By the distance formula:

$$d = \sqrt{(-b)^2 + (1 - b^2)^2} = \sqrt{b^2 + 1 - 2b^2 + b^4} = \sqrt{b^4 - b^2 + 1}$$

We now complete the square on d^2:

$$d^2 = b^4 - b^2 + 1 = \left(b^4 - b^2 + \tfrac{1}{4}\right) + 1 - \tfrac{1}{4} = \left(b^2 - \tfrac{1}{2}\right)^2 + \tfrac{3}{4}$$

So d^2 (and thus d) is minimized when $b^2 = \frac{1}{2}$, thus $b = \pm\frac{1}{\sqrt{2}} = \frac{\pm\sqrt{2}}{2}$.

63. We need to find the vertex of the parabola. Complete the square:

$$\begin{aligned} y &= x^2 - 2x + k \\ y - k &= x^2 - 2x \\ y - k + 1 &= x^2 - 2x + 1 \\ y &= (x-1)^2 + (k-1) \end{aligned}$$

Since the vertex is $(1, k - 1)$ and the parabola is opening upward, then $k - 1 = 5$, so $k = 6$.

65. We solve for x:

$$\begin{aligned} y &= \frac{(x-1)(x-3)}{x-4} \\ y(x-4) &= (x-1)(x-3) \\ yx - 4y &= x^2 - 4x + 3 \\ 0 &= x^2 - (4+y)x + (4y+3) \end{aligned}$$

Using the quadratic formula:

$$\begin{aligned} x &= \frac{4 + y \pm \sqrt{(4+y)^2 - 4(4y+3)}}{2} \\ &= \frac{4 + y \pm \sqrt{16 + 8y + y^2 - 16y - 12}}{2} \\ &= \frac{4 + y \pm \sqrt{y^2 - 8y + 4}}{2} \end{aligned}$$

So we must make sure that $y^2 - 8y + 4 \geq 0$.

We find the key numbers by using the quadratic formula:

$$y = \frac{8 \pm \sqrt{64 - 16}}{2} = \frac{8 \pm 4\sqrt{3}}{2} = 4 \pm 2\sqrt{3}$$

From a sign chart, we see that the range is $y \leq 4 - 2\sqrt{3}$ or $y \geq 4 + 2\sqrt{3}$. We write this as $(-\infty, 4 - 2\sqrt{3}] \cup [4 + 2\sqrt{3}, \infty)$.

67. We first draw a figure:

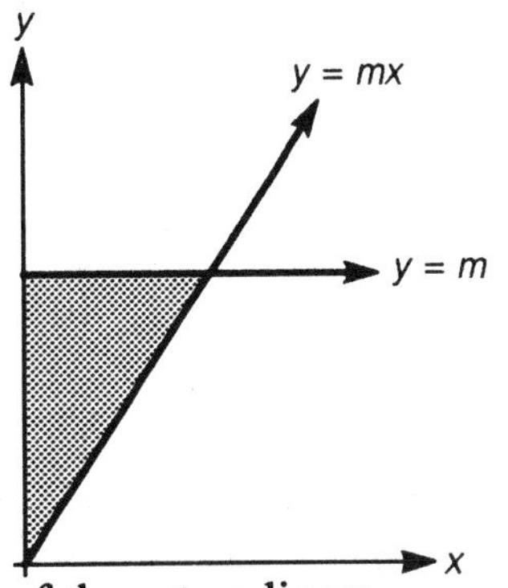

We find the intersection point of these two lines:

$$mx = m$$
$$x = 1$$
$$y = m$$

So the point is $(1, m)$. Since these are the base and height, respectively, of the triangle, we have:

$$A = \tfrac{1}{2}(1)(m) = \frac{m}{2}$$

Using functional notation we have $A(m) = \frac{m}{2}$.

69. We re-draw the figure and label essential parts:

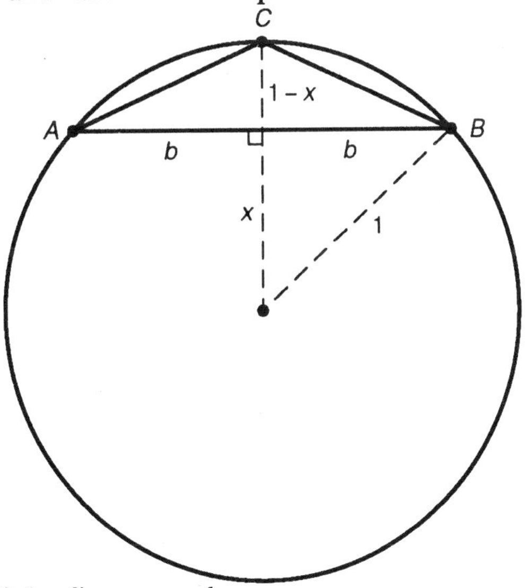

By the labeled parts of the figure, we have:

$$x^2 + b^2 = 1^2, \text{ so } b = \sqrt{1 - x^2}$$

Thus the base of the triangle is $2b = 2\sqrt{1 - x^2}$ and the height is $1 - x$, thus the area is given by:

$$A = \tfrac{1}{2} \bullet 2\sqrt{1 - x^2} \bullet (1 - x) = (1 - x)\sqrt{1 - x^2}$$

Using functional notation we have $A(x) = (1 - x)\sqrt{1 - x^2}$.

71. (a) We draw the scatter diagram (not to scale):

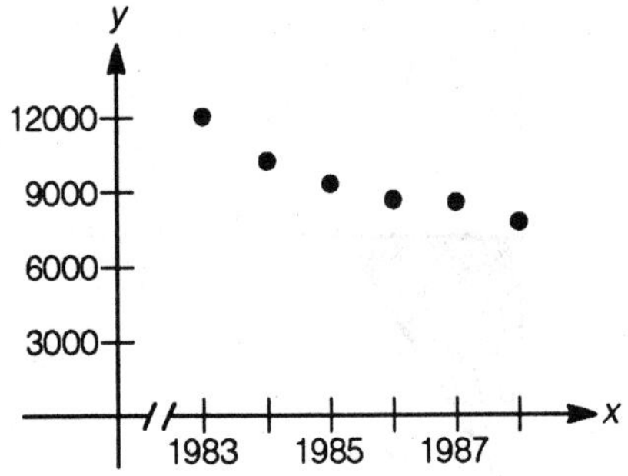

(b) We draw the best-fit line (not to scale):

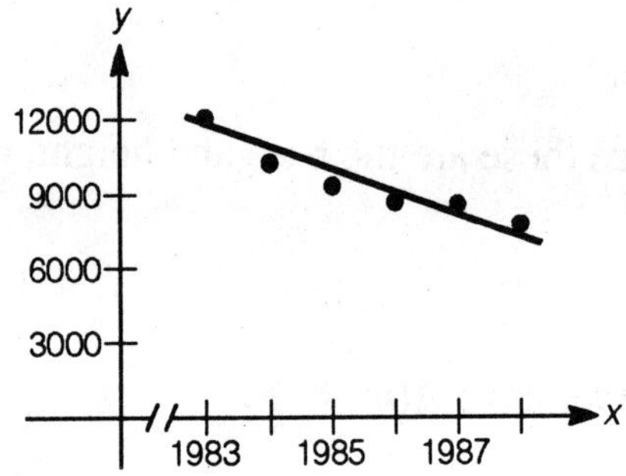

The slope appears to be approximately –770 and the y-intercept appears to be approximately 1.5 million.

(c) We let $x = 1995$:

$$f(1995) = (-771.4)(1995) + 1541090 = 2147$$

We would estimate that approximately 2100 refugees would arrive in 1995.

(d) We switch the roles of x and y, and solve the resulting equation for y:

$$x = -771.4y + 1541090$$
$$771.4y = 1541090 - x$$
$$y = \frac{1541090 - x}{771.4}$$

So $f^{-1}(x) = \frac{1541090 - x}{771.4}$.

(e) We find $f^{-1}(1000) = \frac{1541090 - 1000}{771.4} \approx 1996$. The number of refugees should dip below 1000 in the year 1996.

73. (a) The graphs are reflections of each other about the y-axis:

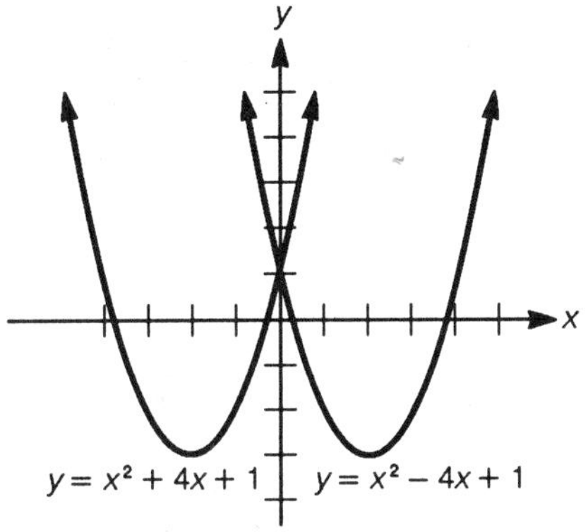

(b) We compute $f(-x)$:

$$f(-x) = a(-x)^2 + b(-x) + c = ax^2 - bx + c$$

(c) Reversing the sign of b will reflect the graph of $f(x) = ax^2 + bx + c$ across the y-axis.

Chapter Five Graphing Calculator Exercises

1. Using the indicated settings, we obtain the graph:

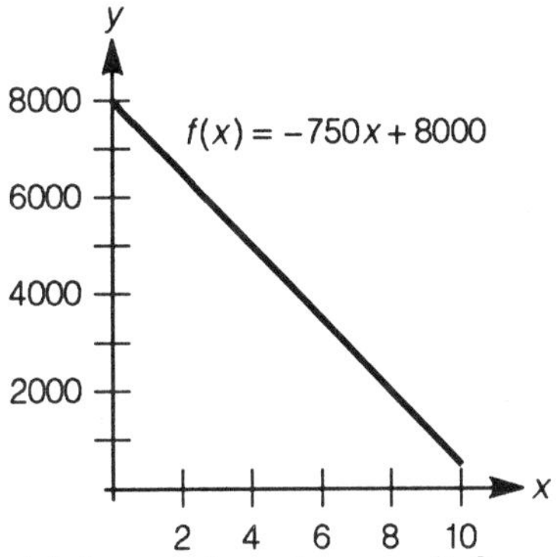

The y-coordinate is 4250, which matches the result from the text.

3. (a) We graph the two functions:

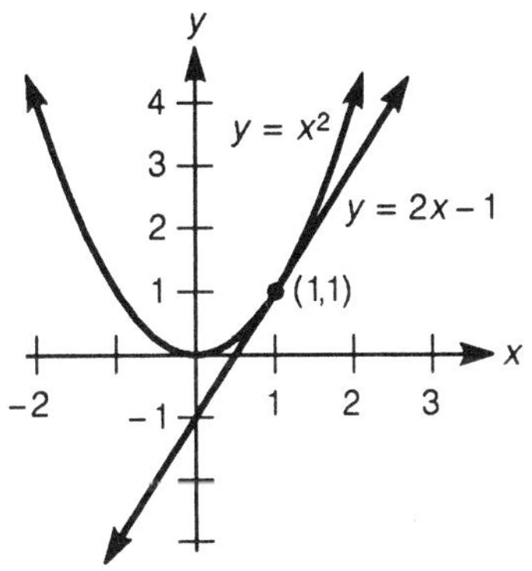

(b) We complete the tables:

x	0.9	0.99	0.999
$2x-1$	0.8	0.98	0.998
x^2	0.81	0.9801	0.998001

x	1.1	1.01	1.001
$2x-1$	1.2	1.02	1.002
x^2	1.21	1.0201	1.002001

5. (a) We graph the four functions:

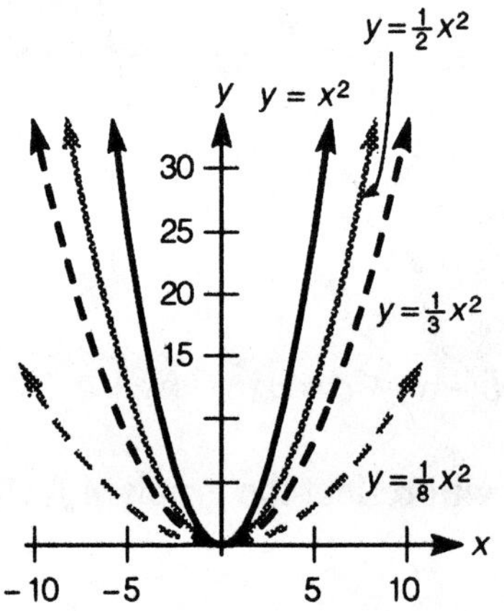

As the coefficient of x^2 decreases from 1 to $\frac{1}{8}$, the graph "flattens" and produces smaller y-coordinates.

(b) The graph should be "flatter" than the others:

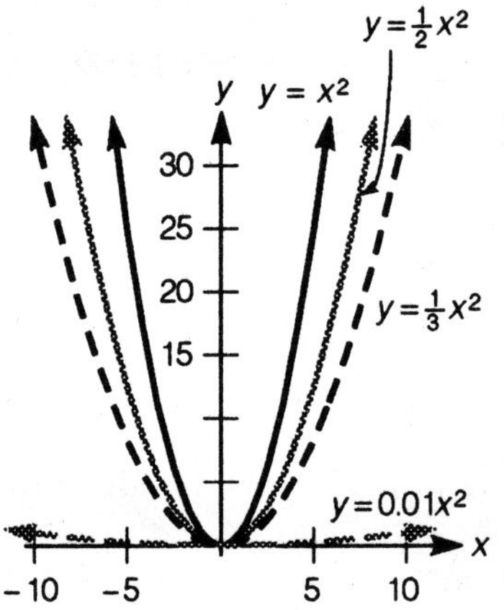

7. The equation of the line can be found using the point-slope formula:

$$y-1=m(x-2)$$
$$y=mx-2m+1$$

We now find the x- and y-intercepts. If $x=0$, then $y=-2m+1$, and if $y=0$, we have:

$$0=mx-2m+1$$
$$2m-1=mx$$
$$\frac{2m-1}{m}=x$$

So the area of the triangle is given by:

$$A(m) = \frac{1}{2} \bullet (-2m+1)\left(\frac{2m-1}{m}\right) = \frac{-4m^2+4m-1}{2m}$$

The graph of $A(m)$ appears as:

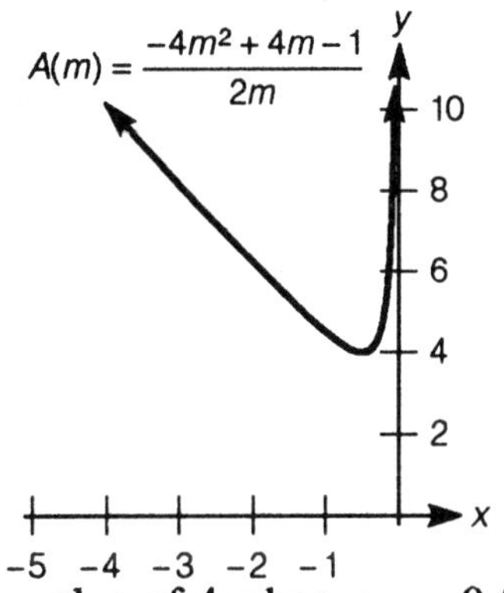

This function has a minimum value of 4 when $m \approx -0.5$.

9. Let the coordinates of P be $P(x, x^2)$. We find the slope between P and $(0,-1)$:

$$m(x) = \frac{x^2-(-1)}{x-0} = \frac{x^2+1}{x}$$

The graph of $m(x)$ appears as:

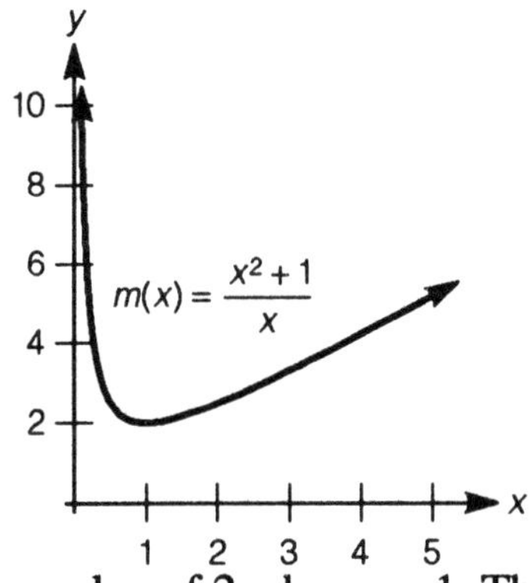

This function has a minimum value of 2 when $x \approx 1$. Thus the point P is $P(1,1)$. We draw the parabola and the line through P and $(0,-1)$:

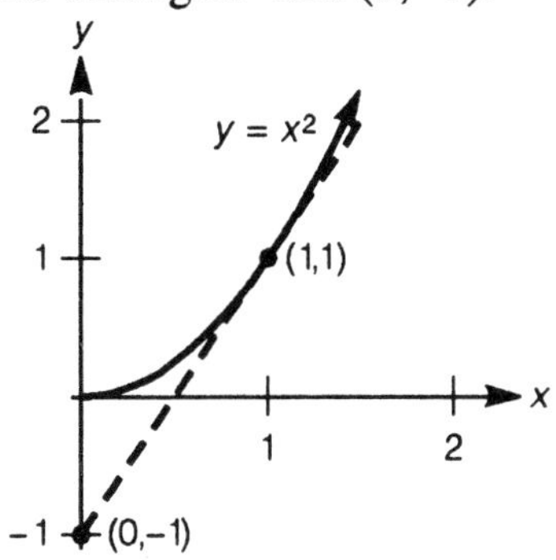

The line appears to be tangent to the parabola at $(1,1)$.

11. (a) Since the area is 6 m^2, then $xy = 6$, so $y = \frac{6}{x}$ where x is the width and y is the length. Thus the perimeter is given by:

$$P(x) = 2x + 2\left(\frac{6}{x}\right) = 2x + \frac{12}{x} = \frac{2x^2 + 12}{x}$$

(b) The graph of $P(x)$ appears as:

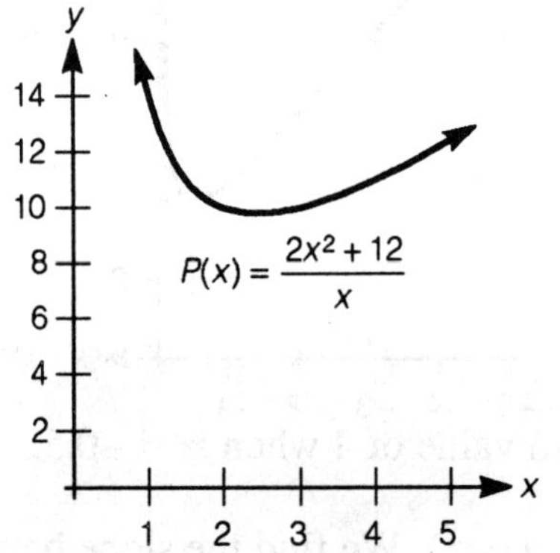

This function has a minimum value of 9.80 when $x \approx 2.45$. Substituting into $y = \frac{6}{x}$, we obtain $y \approx 2.45$. So the length and width are approximately 2.45 m.

13. (a) Since $V = \pi r^2 h$ is the formula for volume, we have:

$$\pi r^2 h = 54\pi$$
$$h = \frac{54}{r^2}$$

(b) Since the total surface area is given by $S = 2\pi r^2 + 2\pi rh$, we have:

$$S(r) = 2\pi r^2 + 2\pi r \bullet \frac{54}{r^2} = 2\pi r^2 + \frac{108\pi}{r} = \frac{2\pi r^3 + 108\pi}{r}$$

(c) The graph of $S(r)$ appears as:

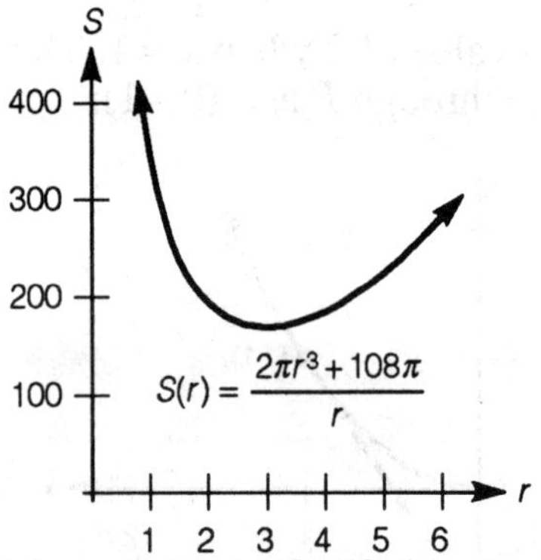

This function has a minimum value of 169.65 when $r = 3$ in.

(d) Using $h = \frac{54}{r^2}$, we obtain $h = \frac{54}{3^2} = 6$ in. The value of h is double that of r.

Chapter Five Test

1. We first find the slope between $(-2,-4)$ and $(5,1)$:

$$m=\frac{-4-1}{-2-5}=\frac{-5}{-7}=\frac{5}{7}$$

Using the point-slope formula, we have:

$$y-1=\tfrac{5}{7}(x-5)$$
$$y-1=\tfrac{5}{7}x-\tfrac{25}{7}$$
$$y=\tfrac{5}{7}x-\tfrac{18}{7}$$

So $L(x)=\frac{5}{7}x-\frac{18}{7}$, and thus $L(0)=-\frac{18}{7}$.

2. (a) We complete the square:

$$F(x)=-2x^2+4x=-2\left(x^2-2x\right)=-2\left(x^2-2x+1\right)+2=-2(x-1)^2+2$$

So the graph of $F(x)$ is a parabola pointing downward with vertex $(1,2)$, and thus it is increasing on the interval $(-\infty,1)$. The maximum value of $F(x)$ is 2.

(b) Since $9t^4\geq 0$ and $6t^2\geq 0$, then $G(t)$ will achieve a minimum value of 2 when $t=0$.

3. (a) The x-intercepts are 3 and -4, and the y-intercept is -48. The signs of $f(x)$ are given by:

Interval	$f(x)=(x-3)(x+4)^2$
$(-\infty,-4)$	neg.
$(-4,3)$	neg.
$(3,\infty)$	pos.

We sketch the excluded regions for $f(x)$:

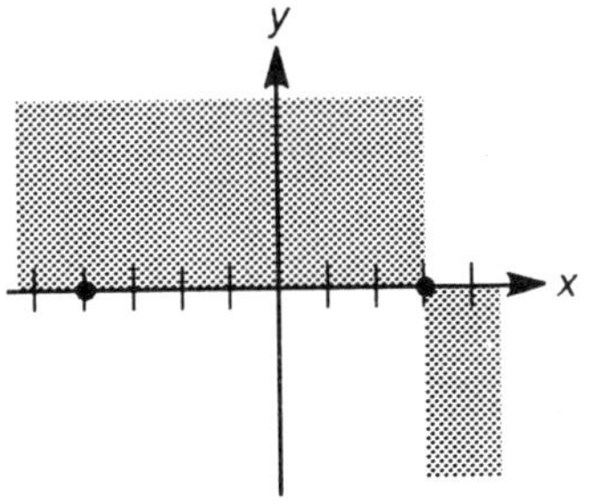

(b) For $x \approx -4$, we have $f(x) \approx (-4-3)(x+4)^2 = -7(x+4)^2$. So $f(x)$ is approximately $-7(x+4)^2$ when x is close to -4. We sketch $y = -7(x+4)^2$:

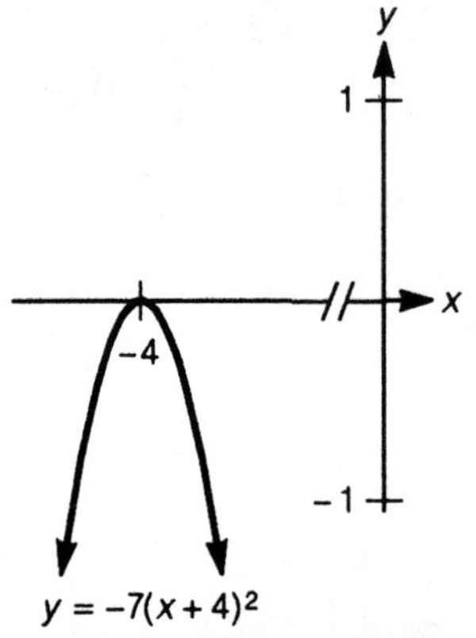

(c) We graph the function:

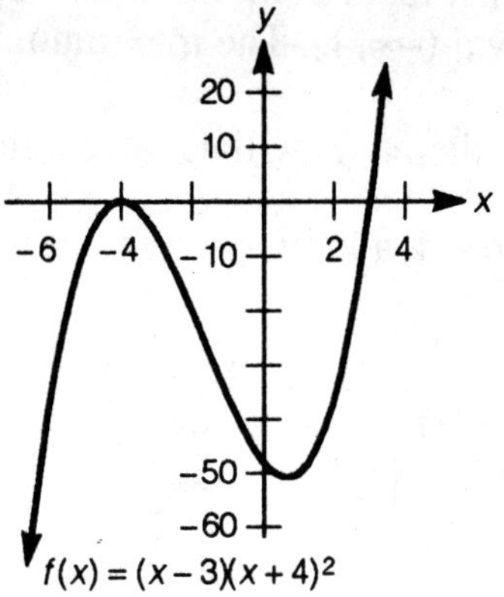

4. (a) We sketch the graph:

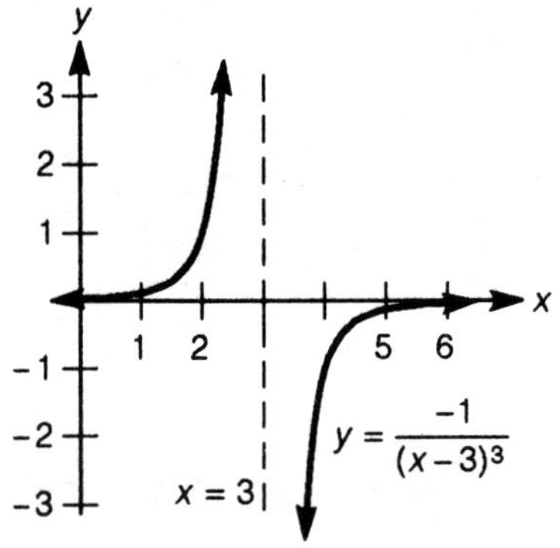

(b) We sketch the graph:

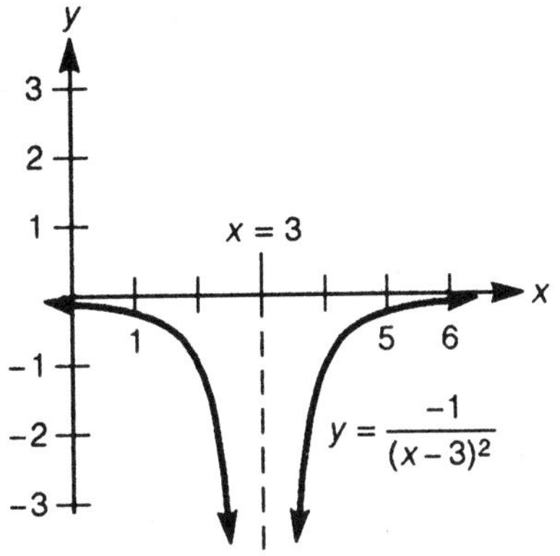

5. We first complete the square:

$$y = -x^2 + 7x + 6 = -(x^2 - 7x) + 6 = -\left(x^2 - 7x + \tfrac{49}{4}\right) + \tfrac{73}{4} = -\left(x - \tfrac{7}{2}\right)^2 + \tfrac{73}{4}$$

The turning point (vertex) is $\left(\frac{7}{2}, \frac{73}{4}\right)$, the x-intercepts are $\frac{7 \pm \sqrt{73}}{2}$, the y-intercept is 6, and the axis of symmetry is $x = \frac{7}{2}$. We draw the graph:

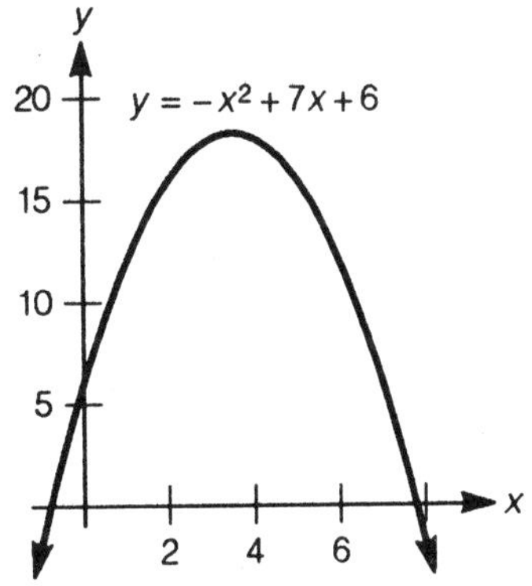

6. We set up the revenue function:

$$R = xp = x\left(-\tfrac{1}{6}x + 80\right) = -\tfrac{1}{6}x^2 + 80x$$

Now complete the square:

$$R = -\tfrac{1}{6}(x^2 - 480x) = -\tfrac{1}{6}(x^2 - 480x + 57600) + 9600 = -\tfrac{1}{6}(x - 240)^2 + 9600$$

The maximum possible revenue is \$9600, which occurs when $x = 240$.

Thus $p = -\frac{1}{6}(240) + 80 = \$40 / \text{unit}$.

7. We find the slope between the points (0, 14000) and (10, 750):

$$m = \frac{750 - 14000}{10 - 0} = \frac{-13250}{10} = -1325$$

So the value of the machine is $V(t) = -1325t + 14000$.

8. Since $y = -\frac{1}{2}(3-x)^2$ will be a reflection of $y = -\frac{1}{2}(x+3)^2$ across the y-axis, we have the graph:

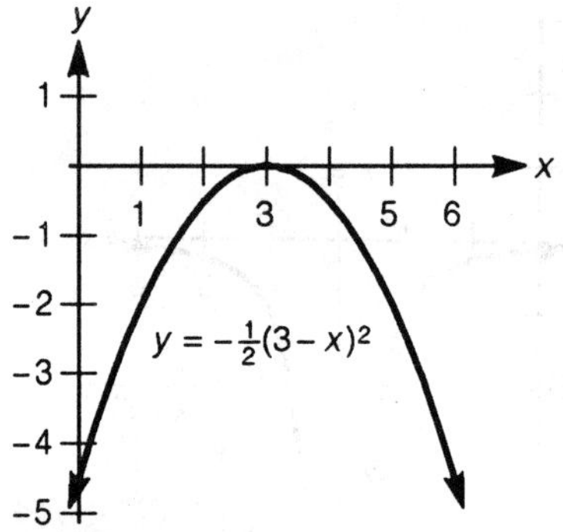

The x-intercept is 3 and the y-intercept is $-\frac{9}{2}$.

9. The x-intercept is $\frac{3}{2}$, the y-intercept is -3, the vertical asymptote is $x = -1$, and the horizontal asymptote is $y = 2$. We draw the graph:

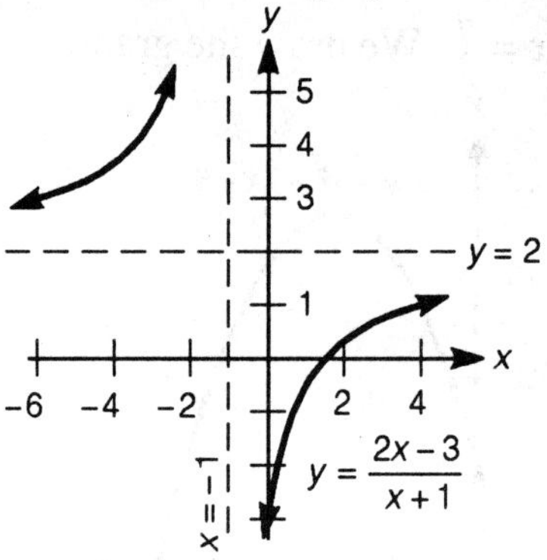

10. (a) Using the distance formula, we have:

$$\begin{aligned} L(x) &= \sqrt{(x+1)^2 + (y-3)^2} \\ &= \sqrt{(x+1)^2 + (3x-1-3)^2} \\ &= \sqrt{x^2 + 2x + 1 + 9x^2 - 24x + 16} \\ &= \sqrt{10x^2 - 22x + 17} \end{aligned}$$

(b) We complete the square inside the radical:

$$\begin{aligned} 10x^2 - 22x + 17 &= 10\left(x^2 - 2.2x\right) + 17 \\ &= 10\left(x^2 - 2.2x + 1.21\right) + 4.9 \\ &= 10(x-1.1)^2 + 4.9 \end{aligned}$$

So the length will be a minimum when L^2 is a minimum, which occurs when $x = 1.1$.

11. (a) The vertical asymptotes are $x = 3$ and $x = -3$, and the horizontal asymptote is $y = 1$.

(b) Near $x = 0$, $f(x)$ will look like:

$$y = \frac{x(0-2)}{0-9} = \frac{2}{9}x$$

Near $x = 2$, $f(x)$ will look like:

$$y = \frac{2(x-2)}{4-9} = -\frac{2}{5}(x-2)$$

We sketch the graph:

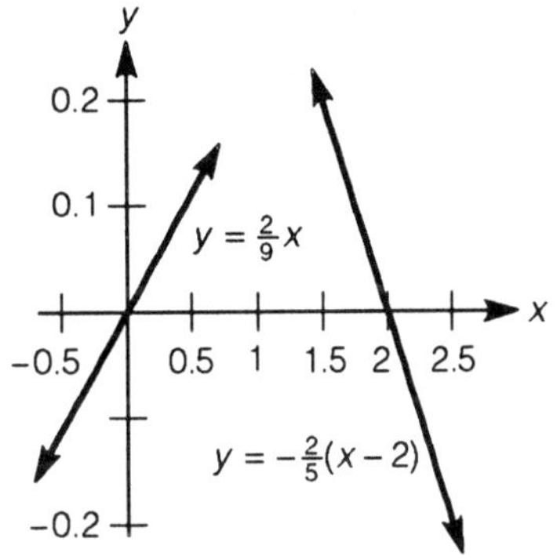

(c) Near $x = 3$, $f(x)$ will look like:

$$y = \frac{3(3-2)}{(3+3)(x-3)} = \frac{1}{2(x-3)}$$

Near $x = -3$, $f(x)$ will look like:

$$y = \frac{-3(-3-2)}{(x+3)(-3-3)} = -\frac{5}{2(x+3)}$$

We sketch the graph:

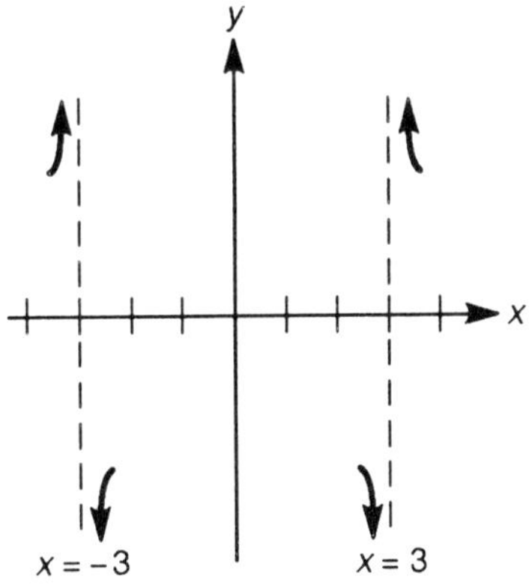

(d) We graph the function:

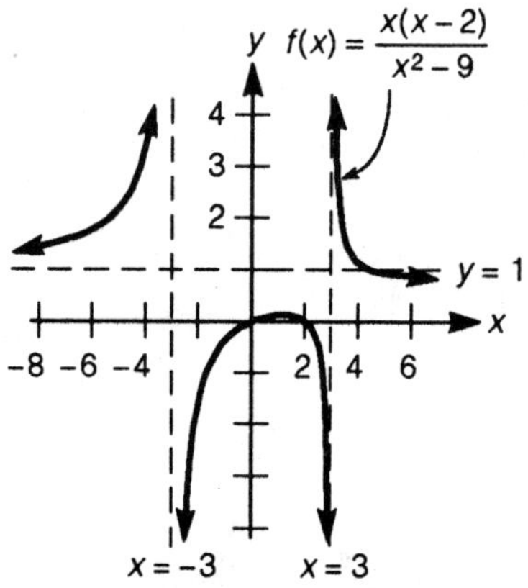

12. Let w and l represent the width and length, respectively. We draw the figure:

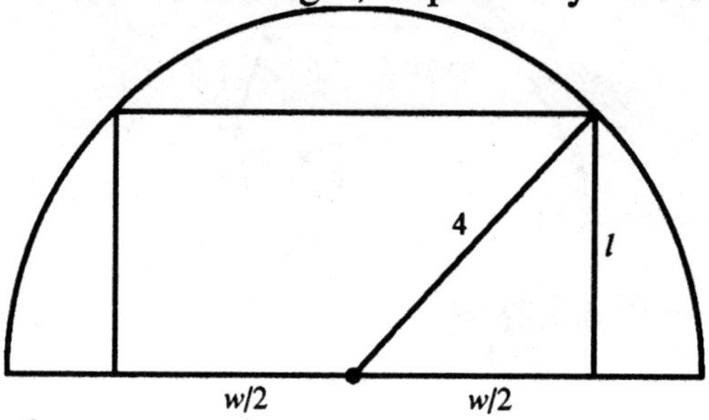

From the Pythagorean theorem, we have:

$$\left(\frac{w}{2}\right)^2 + l^2 = 4^2$$
$$\frac{w^2}{4} + l^2 = 16$$
$$l^2 = \frac{64 - w^2}{4}$$
$$l = \frac{\sqrt{64 - w^2}}{2}$$

So the area of the rectangle is given by:

$$A(w) = w \bullet l = w \bullet \frac{\sqrt{64 - w^2}}{2} = \frac{1}{2} w\sqrt{64 - w^2}$$

13. (a) As $|x|$ increases in size for positive values of x, $-x^3$ should be large negative values, which the function is not.

(b) This graph has four turning points, and a polynomial function with highest degree term $-x^3$ can have at most two turning points.

14. (a) We sketch the scatter diagram (not to scale):

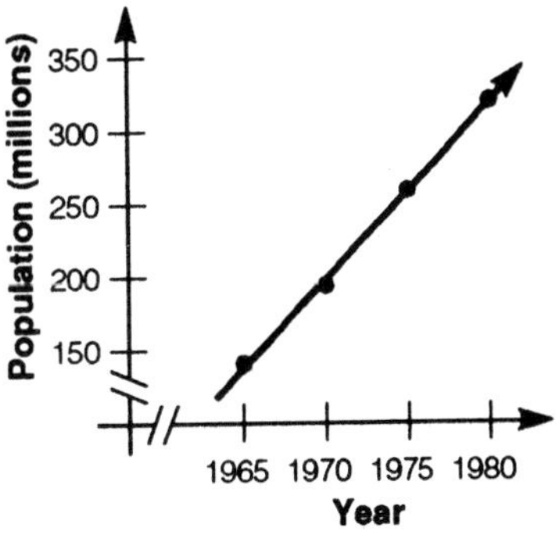

(b) Using $x = 1985$, we have:

$$y = -23757 + 12.16(1985) = 380.6$$

Our estimate is 381million. Our estimate is slightly higher than the actual amount.

Chapter Six
Exponential and Logarithmic Functions

6.1 Exponential Functions

1. (a) Since $2^{10} \approx 10^3$, then $2^{30} = (2^{10})^3 \approx (10^3)^3 = 10^9$.

(b) Since $2^{10} \approx 10^3$, then $2^{50} = (2^{10})^5 \approx (10^3)^5 = 10^{15}$.

3. (a) We solve for x:

$$\begin{aligned} 3^x &= 27 \\ 3^x &= 3^3 \\ x &= 3 \end{aligned}$$

(b) We solve for t:

$$\begin{aligned} 9^t &= 27 \\ (3^2)^t &= 3^3 \\ 3^{2t} &= 3^3 \\ 2t &= 3 \\ t &= \tfrac{3}{2} \end{aligned}$$

(c) We solve for y:

$$\begin{aligned} 3^{1-2y} &= \sqrt{3} \\ 3^{1-2y} &= 3^{1/2} \\ 1-2y &= \tfrac{1}{2} \\ -2y &= -\tfrac{1}{2} \\ y &= \tfrac{1}{4} \end{aligned}$$

(d) We solve for z:

$$3^z = 9\sqrt{3}$$
$$3^z = 3^2 \bullet 3^{1/2}$$
$$3^z = 3^{5/2}$$
$$z = \frac{5}{2}$$

5. The domain is all real numbers, or $(-\infty, \infty)$.

7. Since $2^{x-1} \neq 0$ (even if $x = 1$), then the domain is all real numbers, or $(-\infty, \infty)$.

9. We graph $y = 2^x$ and $y = 2^{-x}$:

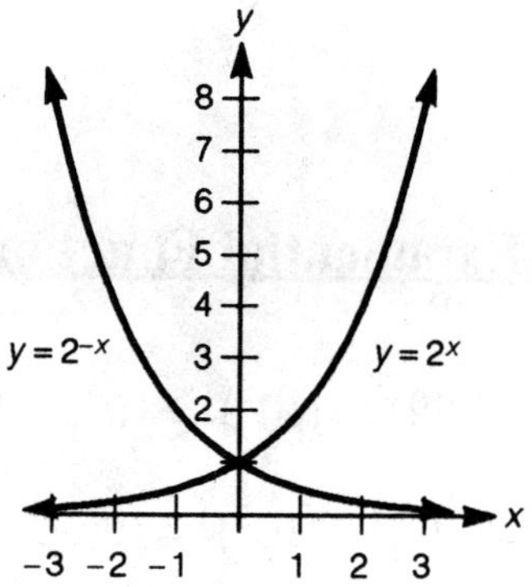

11. We graph $y = 3^x$ and $y = -3^x$ (note that $y = -3^x$ is a reflection about the x-axis):

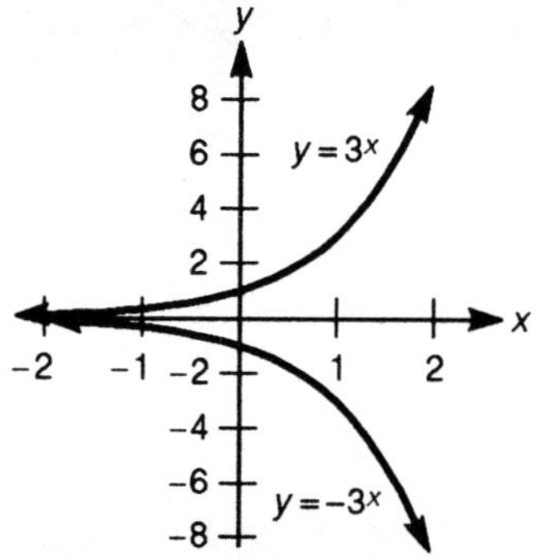

13. We graph $y = 2^x$ and $y = 3^x$:

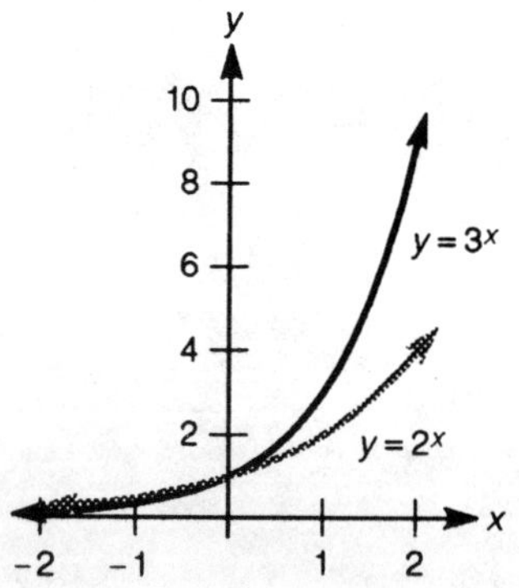

15. We graph $y=\left(\frac{1}{2}\right)^x=2^{-x}$ and $y=\left(\frac{1}{3}\right)^x=3^{-x}$:

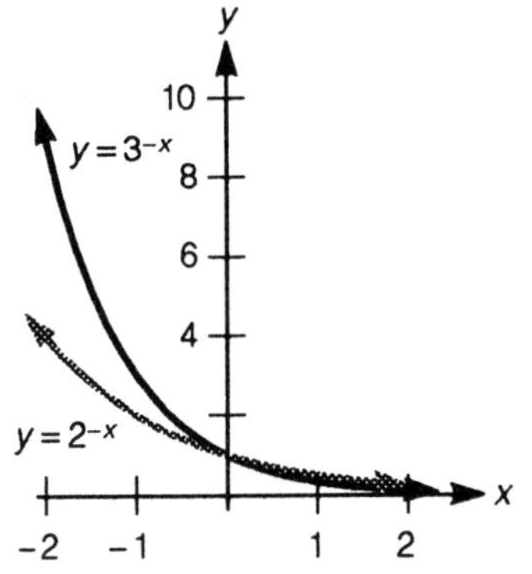

17. The domain is $(-\infty,\infty)$, the range is $(-\infty,1)$, the x- and y-intercepts are 0, and the asymptote is $y=1$. We graph the function:

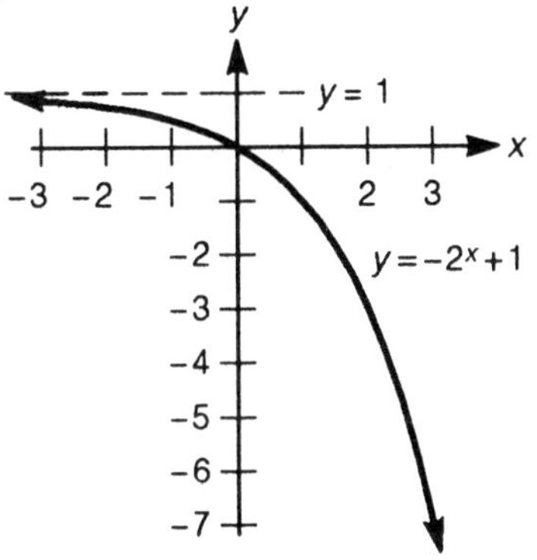

19. The domain is $(-\infty,\infty)$, the range is $(1,\infty)$, there is no x-intercept, the y-intercept is 2, and the asymptote is $y=1$. We graph the function:

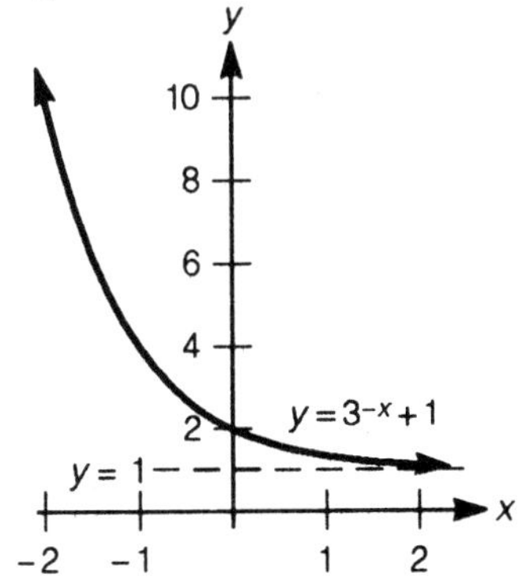

21. The domain is $(-\infty, \infty)$, the range is $(0, \infty)$, there is no x-intercept, the y-intercept is $\frac{1}{2}$, and the asymptote is $y = 0$. We graph the function:

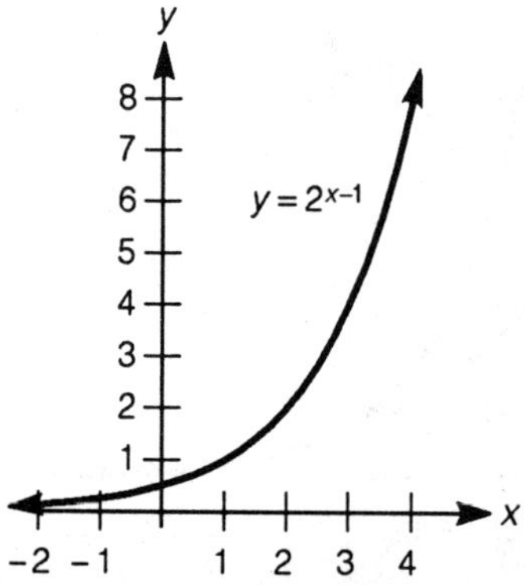

23. The domain is $(-\infty, \infty)$, the range is $(1, \infty)$, there is no x-intercept, the y-intercept is 4, and the asymptote is $y = 1$. We graph the function:

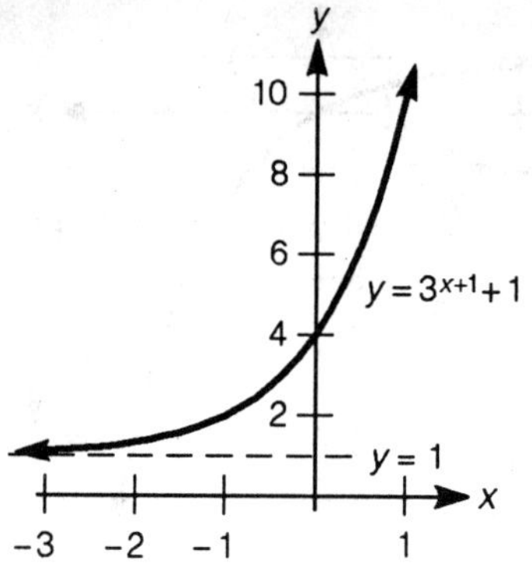

25. We solve for x:

$$
\begin{aligned}
3x(10^x) + 10^x &= 0 \\
10^x(3x+1) &= 0 \\
x &= -\tfrac{1}{3} \quad (\text{since } 10^x \neq 0)
\end{aligned}
$$

27. We solve for x:

$$
\begin{aligned}
3(3^x) - 5x(3^x) + 2x^2(3^x) &= 0 \\
3^x(3 - 5x + 2x^2) &= 0 \\
3^x(3-2x)(1-x) &= 0 \\
x &= \tfrac{3}{2}, 1 \quad (\text{since } 3^x \neq 0)
\end{aligned}
$$

29. We simplify the difference quotient:

$$\frac{f(x+h)-f(x)}{h} = \frac{2^{x+h} - 2^x}{h} = \frac{2^x 2^h - 2^x}{h} = \frac{2^x(2^h - 1)}{h} = 2^x\left(\frac{2^h - 1}{h}\right)$$

31. (a) We know the graph of $y = 2^x$ and the graph of g, the inverse of f, should contain these points with the x and y coordinates interchanged. So $g(x)$ will be the reflection of $f(x)$ across the line $y = x$:

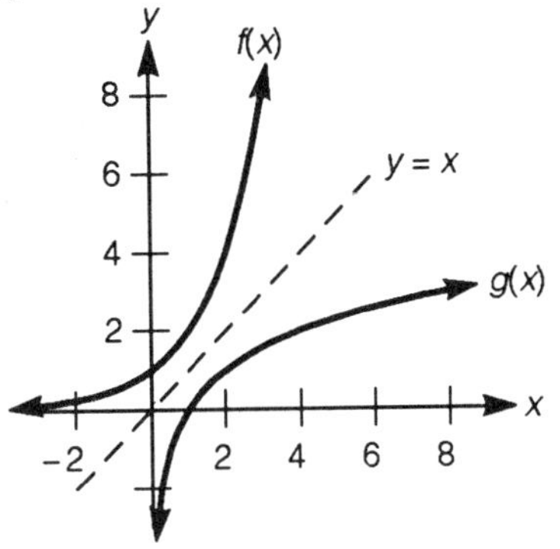

(b) The domain is $(0, \infty)$, the range is $(-\infty, \infty)$, the x-intercept is 1, there is no y-intercept, and the asymptote is $x = 0$.

33. (a) From the graph, we have $2^{1/2} \approx 1.4$.
(b) Using a calculator, we have $\sqrt{2} \approx 1.41$.

35. (a) From the graph, we have $2^{3/5} = 2^{0.6} \approx 1.5$.
(b) Using a calculator, we have $2^{3/5} \approx 1.52$.

37. (a) From the graph, we have $\sqrt{3} = 3^{1/2} \approx 1.7$.
(b) Using a calculator, we have $\sqrt{3} \approx 1.73$.

39. (a) From the graph, we have $5^{3/10} = 5^{0.3} \approx 1.6$.
(b) Using a calculator, we have $5^{3/10} \approx 1.62$.

41. From the graph, we have $x \approx 0.3$.

43. From the graph, we have $x \approx 0.7$.

45. (a) Since $10^0 = 1$, the entry in the table corresponding to $x = 1$ is 0. Since $10^1 = 10$, the entry corresponding to $x = 10$ is 1.

(b) Since $10^{0.3} \approx 2$, we have $\left(10^{0.3}\right)^2 \approx 2^2$. That is, $10^{0.6} \approx 4$. Therefore the entry in the table corresponding to $x = 4$ is 0.6. Similarly, by cubing both sides of the approximation $10^{0.3} \approx 2$, we obtain $10^{0.9} \approx 8$. Thus the entry in the table corresponding to $x = 8$ is 0.9.

(c) Using thc hint that is given, we have $5 \approx \dfrac{10}{10^{0.3}} = 10^{0.7}$. Thus, the entry in the table corresponding to $x = 5$ is 0.7.

(d) We have $7^2 \approx (5)(10) \approx (10^{0.7})(10^1) = 10^{1.7}$. Therefore, $(7^2)^{1/2} \approx (10^{1.7})^{1/2}$, or $7 \approx 10^{0.85}$. Thus the power to which 10 must be raised to yield 7 is approximately 0.85.

(e) We have $3^4 \approx (8)(10) \approx (10^{0.9})(10^1) = 10^{1.9}$. Therefore $(3^4)^{1/4} \approx (10^{1.9})^{1/4} \approx 10^{0.475}$. Thus the power to which 10 must be raised to yield 3 is approximately 0.48.

(f) We have $6 = (2)(3) \approx (10^{0.3})(10^{0.48}) = 10^{0.78}$. Thus the power to which 10 must be raised to yield 6 is approximately 0.78. We also have:

$$9 = (3)(3) \approx (10^{0.475})(10^{0.475}) = 10^{0.95}$$

Thus the power to which 10 must be raised to yield 9 is approximately 0.95. We complete the table:

x	$\log_{10} x$
1	0.00
2	0.30
3	0.48
4	0.60
5	0.70
6	0.78
7	0.85
8	0.90
9	0.95
10	1.00

6.2 The Exponential Function $y = e^x$

1. The domain is $(-\infty, \infty)$, the range is $(0, \infty)$, there is no x-intercept, the y-intercept is 1, and the asymptote is $y = 0$. We draw the graph:

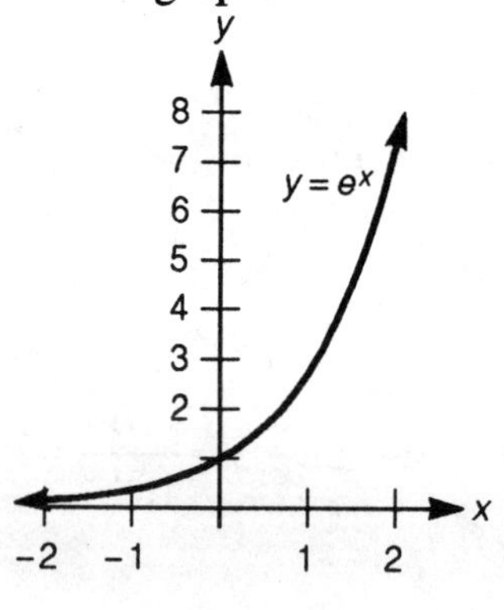

3. The domain is $(-\infty, \infty)$, the range is $(-\infty, 0)$, there is no x-intercept, the y-intercept is -1, and the asymptote is $y = 0$. We draw the graph:

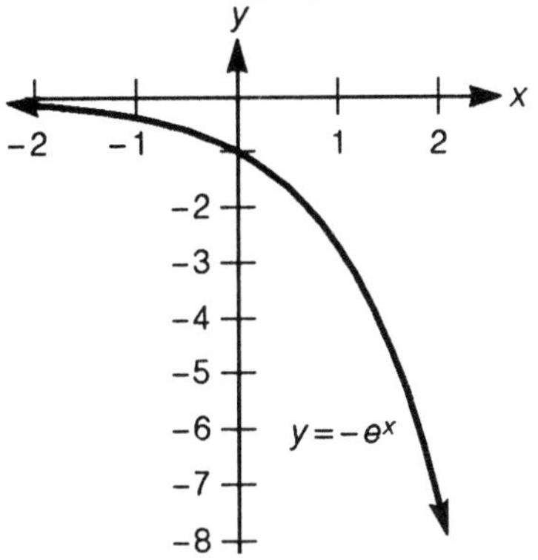

5. The domain is $(-\infty, \infty)$, the range is $(1, \infty)$, there is no x-intercept, the y-intercept is 2, and the asymptote is $y = 1$. We draw the graph:

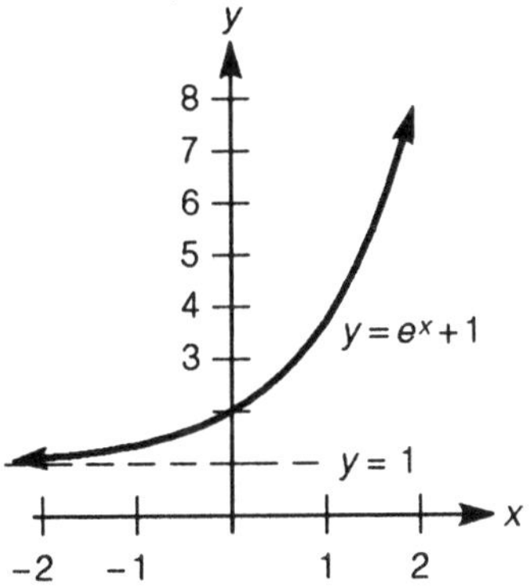

7. The domain is $(-\infty, \infty)$, the range is $(1, \infty)$, there is no x-intercept, the y-intercept is $e + 1$, and the asymptote is $y = 1$. We draw the graph:

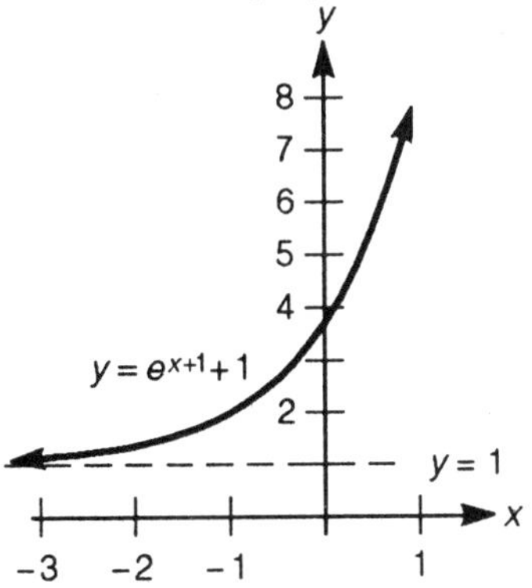

9. The domain is $(-\infty,\infty)$, the range is $(-\infty,0)$, there is no x-intercept, the y-intercept is $-\frac{1}{e^2}$, and the asymptote is $y=0$. We draw the graph:

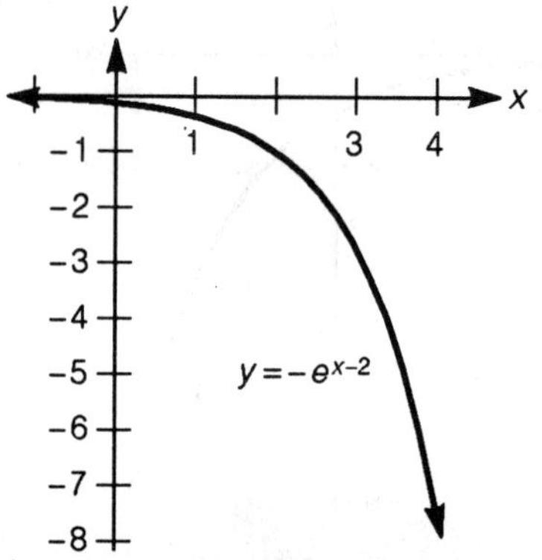

11. The domain is $(-\infty,\infty)$, the range is $(-\infty,e)$, the x-intercept is 1, the y-intercept is $e-1$, and the asymptote is $y=e$. We draw the graph:

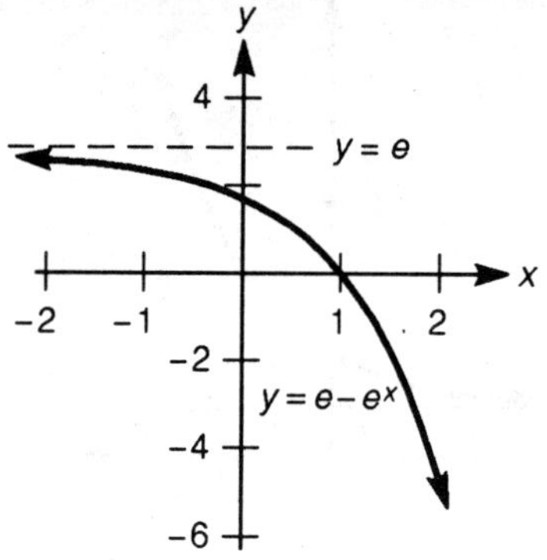

13. We graph the three functions:

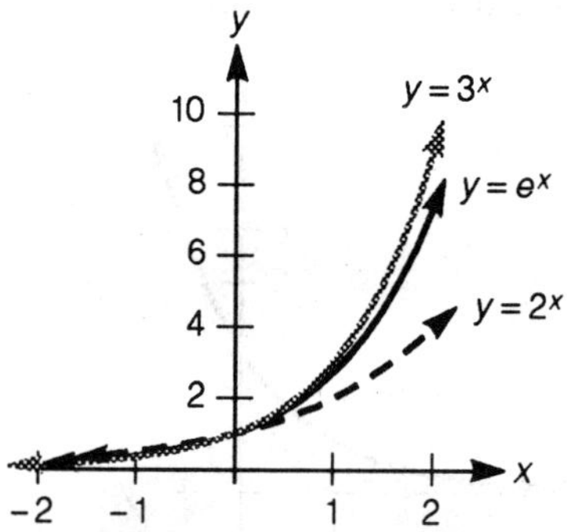

15. We solve for e^k:

$$e^{5k}=\left(e^k\right)^5, \text{ so:}$$

$$\left(e^k\right)^5=32$$

$$e^k=32^{1/5}$$

$$e^k=2$$

17. We solve for e^t:

$$e^{3t} = \left(e^t\right)^3, \text{ so:}$$
$$\left(e^t\right)^3 = 4$$
$$e^t = \sqrt[3]{4} \approx 1.587$$

19. False (since $e \approx 2.7$)

21. False (since $\sqrt{2.7} > 1$)

23. True

25. False (since $2.7^{-1} > 0$)

27. From the graph, we have $e^{0.2} \approx 1.2$. Using a calculator, we have $e^{0.2} \approx 1.22$.

29. From the graph, we have $e^{4/5} = e^{0.8} \approx 2.2$. Using a calculator, we have $e^{4/5} = e^{0.8} \approx 2.23$.

31. From the graph, we have $\sqrt{e} = e^{0.5} \approx 1.6$. Using a calculator, we have $\sqrt{e} = e^{0.5} \approx 1.65$.

33. The value of x is approximately -0.2.

35. (a) Given the growth law $N = N_0 e^{kt}$, where $N_0 = 2000$, $N = 6000$, $t = 2$, we have:

$$6000 = 2000e^{k(2)}$$
$$3 = e^{2k}$$
$$\sqrt{3} = e^k$$

(b) When $t = 10$, we have:

$$N = 2000e^{10k} = 2000\left(e^k\right)^{10} = 2000\left(\sqrt{3}\right)^{10} = 2000(243) = 486000$$

37. Given the growth law $N = N_0 e^{kt}$, where $N_0 = 3200$, $e^k = \sqrt[5]{2}$, $t = 5$, we have:

$$N = 3200e^{5k} = 3200\left(e^k\right)^5 = 3200\left(\sqrt[5]{2}\right)^5 = 3200(2) = 6400 \text{ bacteria}$$

39. (a) We complete the table:

t(seconds)	0	550	1100	1650	2200
N(grams)	8	4	2	1	0.5

(b) We complete the table:

t(years)	0	4.9×10^9	9.8×10^9	14.7×10^9	19.6×10^9
N(grams)	10	5	2.5	1.25	0.625

41. Given the decay law $N = N_0 e^{kt}$, and $N = \frac{1}{2}N_0$ when $t = 8$, we have:

$$\frac{1}{2}N_0 = N_0 e^{8k}$$
$$\frac{1}{2} = e^{8k}$$
$$\left(\frac{1}{2}\right)^{1/8} = e^k$$

So when $t = 7$ and $N_0 = 1$, we have:

$$N = 1e^{7k} = \left(e^k\right)^7 = \left(\frac{1}{2}\right)^{7/8} \approx 0.55 \text{ g}$$

43. Given the decay law $N = N_0 e^{kt}$, and $N = \frac{1}{2}N_0$ when $t = 24000$, we have:

$$\frac{1}{2}N_0 = N_0 e^{24000k}$$
$$\frac{1}{2} = e^{24000k}$$
$$\left(\frac{1}{2}\right)^{1/24000} = e^k$$

So when $t = 1000$, we have:

$$N = N_0 e^{1000k} = N_0\left(\frac{1}{2}\right)^{1/24} \approx 0.9715N_0$$

So about 97.15% will still be remaining.

45. We simplify the difference quotient:

$$\frac{E(x+h) - E(x)}{h} = \frac{e^{x+h} - e^x}{h} = \frac{e^x e^h - e^x}{h} = \frac{e^x\left(e^h - 1\right)}{h} = e^x\left(\frac{e^h - 1}{h}\right)$$

47. (a) The domain is all real numbers, or $(-\infty, \infty)$:

(b) We evaluate when $x = 0$:

$$C(0) = \frac{e^0 + e^0}{2} = \frac{1+1}{2} = 1$$

(c) We compute $C(-x)$:

$$C(-x) = \frac{e^{-x} + e^x}{2} = C(x)$$

The graph must be symmetric about the y-axis.

(d) We draw the graph:

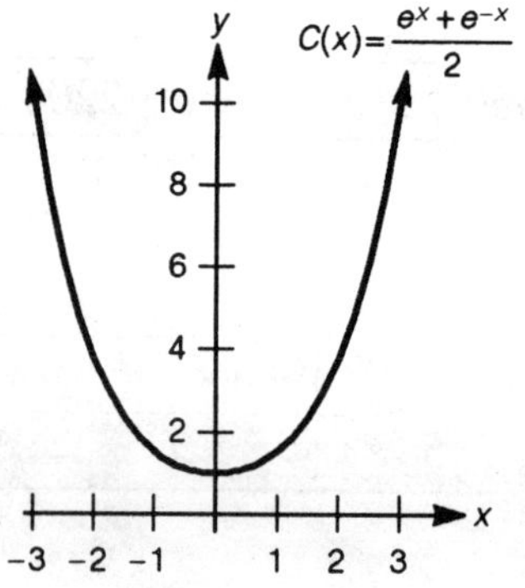

49. We complete the table:

n	$n+\frac{1}{n}$	$\left(1+\frac{1}{n}\right)^n$
1	2	2.0000
10	1.1	2.5937
100	1.01	2.7048
1000	1.001	2.7169
10000	1.0001	2.7181
100000	1.00001	2.7183

51. (a) Since $N = N_0e^{kt}$, using $N_0 = 1$ billion, $N = 2$ billion, and$t = 80$, we have:

$$N = N_0e^{kt}$$
$$2 = 1e^{k\bullet 80}$$
$$e^k = 2^{1/80}$$

So in 1990, using $t = 140$, we have:

$$N = 1\left(e^k\right)^{140} = \left(2^{1/80}\right)^{140} = 2^{7/4} \approx 3.36 \text{ billion}$$

(b) Our estimate is too low. The formula $N = N_0e^{kt}$ predicts too low a figure.

53. (a) We draw the graphs, noting that $L(x)$ will be the reflection of $f(x)$ across $y = x$:

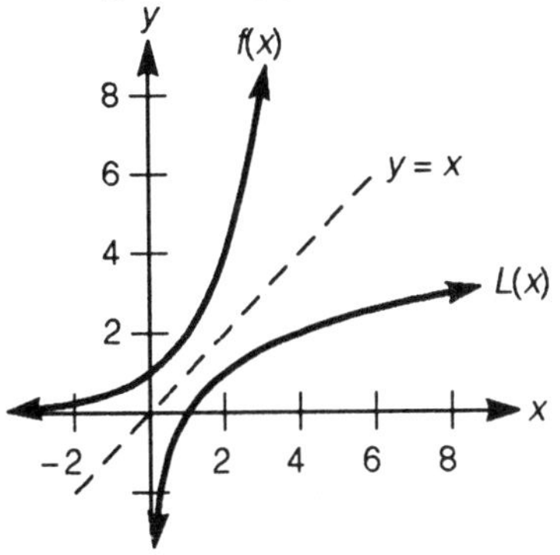

(b) The domain is $(0, \infty)$, the range is $(-\infty, \infty)$, the x-intercept is 1, and the asymptote is $x = 0$.

(c) (i) This will be a reflection of $L(x)$ across the x-axis. The x-intercept is 1 and the asymptote is $x = 0$.

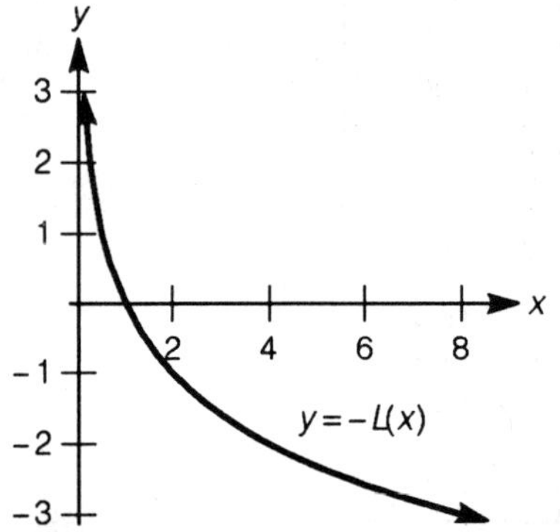

(ii) This will be a reflection of $L(x)$ across the y-axis. The x-intercept is -1 and the asymptote is $x = 0$:

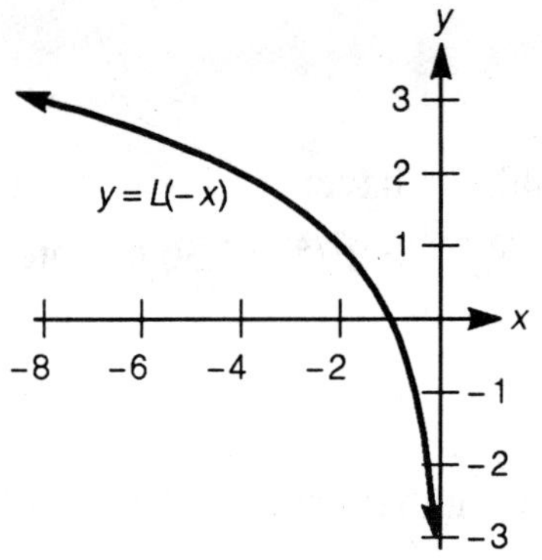

(iii) This will be a displacement of $L(x)$ to the right one unit. The x-intercept is 2 and the asymptote is $x = 1$:

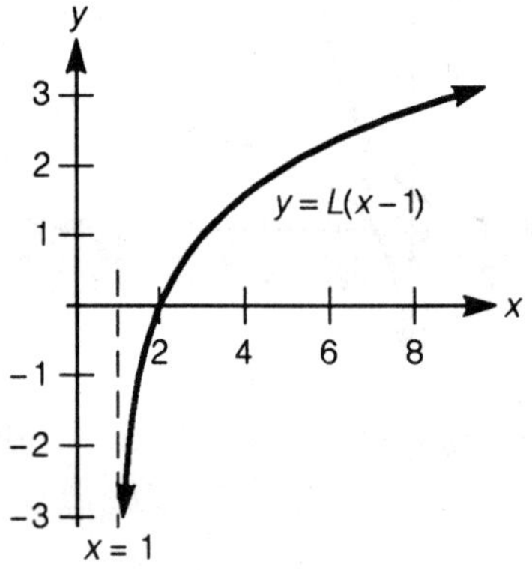

6.3 Logarithmic Functions

1. (a) No, because a horizontal line can intersect the parabola twice.
(b) Yes, because any horizontal line intersects the line only once.
(c) Yes, because any horizontal line intersects the curve only once.

3. (a) Let $y = \frac{2x-1}{3x+4}$. We switch the roles of x and y and solve the resulting equation for y:

$$\begin{aligned} x &= \frac{2y-1}{3y+4} \\ x(3y+4) &= 2y-1 \\ 3xy+4x &= 2y-1 \\ 3xy-2y &= -1-4x \\ y(3x-2) &= -1-4x \\ y &= \frac{-1-4x}{3x-2} \\ y &= \frac{4x+1}{2-3x} \end{aligned}$$

So $f^{-1}(x) = \frac{4x+1}{2-3x}$.

(b) We compute $\frac{1}{f(x)}$:

$$\frac{1}{f(x)} = \frac{3x+4}{2x-1}$$

(c) We compute $f^{-1}(0)$:

$$f^{-1}(0) = \frac{4(0)+1}{2-3(0)} = \frac{1}{2}$$

(d) We compute $\frac{1}{f(0)}$:

$$\frac{1}{f(0)} = \frac{3(0)+4}{2(0)-1} = \frac{4}{-1} = -4$$

5. Since $y = f^{-1}(x)$ would join the points $(-2, 3)$ and $(5, -1)$, then $y = f^{-1}(x-1)$ will join the points $(-1, 3)$ and $(6, -1)$.

7. (a) $\log_3 9 = 2$
(b) $\log_{10} 1000 = 3$
(c) $\log_7 343 = 3$
(d) $\log_2 \sqrt{2} = 1/2$

9. (a) $2^5 = 32$
(b) $10^0 = 1$
(c) $e^{1/2} = \sqrt{e}$
(d) $3^{-4} = 1/81$
(e) $t^v = u$

11. Since $\log_5 30$ represents the power to which 5 must be raised to get 30, it is clearly greater than 2, since $5^2 = 25$. But $\log_8 60$ is less than 2, since $8^2 = 64$. Hence $\log_5 30$ is larger.

13. (a) Since $\log_9 27$ is the power to which 9 must be raised to get 27, we can see it is between 1 and 2, since $9^1 = 9$ while $9^2 = 81$. To find it let $\log_9 27 = n$ then $9^n = 27$ in exponential form and $3^{2n} = 3^3$. So $2n = 3$ and $n = \frac{3}{2}$.

(b) If $\log_4 \frac{1}{32} = n$ then $4^n = \frac{1}{32}$, thus $2^{2n} = 2^{-5}$. So $2n = -5$, and $n = -\frac{5}{2}$. So $\log_4 \frac{1}{32} = -\frac{5}{2}$.

(c) Follow the same steps. If $\log_5 5\sqrt{5} = n$, then:

$$5^n = 5\sqrt{5}$$
$$5^n = 5^{3/2}$$
$$n = \tfrac{3}{2}$$

15. (a) Writing $\log_4 x = -2$ in expontential form, we have $x = 4^{-2} = \frac{1}{16}$.

(b) Writing $\ln x = -2$ in exponential form, we have $x = e^{-2} \approx 0.14$.

17. (a) We must have $5x > 0$, so $x > 0$. So the domain is $(0, \infty)$.

(b) We must have $3 - 4x > 0$, so $3 > 4x$ and $x < \frac{3}{4}$. So the domain is $\left(-\infty, \frac{3}{4}\right)$.

(c) We must have $x^2 > 0$, so $x \neq 0$. So the domain is all real numbers except 0, or $(-\infty, 0) \cup (0, \infty)$.

(d) We must have $x > 0$. So the domain is $(0, \infty)$.

(e) We solve the inequality:

$$x^2 - 25 > 0$$
$$(x+5)(x-5) > 0$$

The key numbers are –5 and 5. We construct the sign chart:

Inteval	Test Number	$x+5$	$x-5$	$(x+5)(x-5)$
$(-\infty, -5)$	–6	neg.	neg.	pos.
$(-5, 5)$	0	pos.	neg.	neg.
$(5, \infty)$	6	pos.	pos.	pos.

So the product is positive on the intervals $(-\infty, -5) \cup (5, \infty)$, which is the domain.

19. The coordinates for each point are:

A: $(0,1)$
B: $(1,0)$
C: $(4,\log_2 4) = (4,2)$
D: $(2,4)$

21. (a) The domain is $(0,\infty)$, the range is $(-\infty,\infty)$, the x-intercept is 1, there is no y-intercept, and the asymptote is $x = 0$:

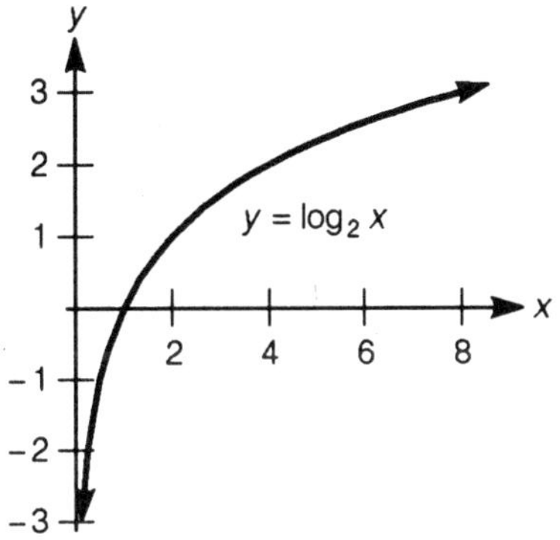

(b) The domain is $(0,\infty)$, the range is $(-\infty,\infty)$, the x-intercept is 1, there is no y-intercept, and the asymptote is $x = 0$:

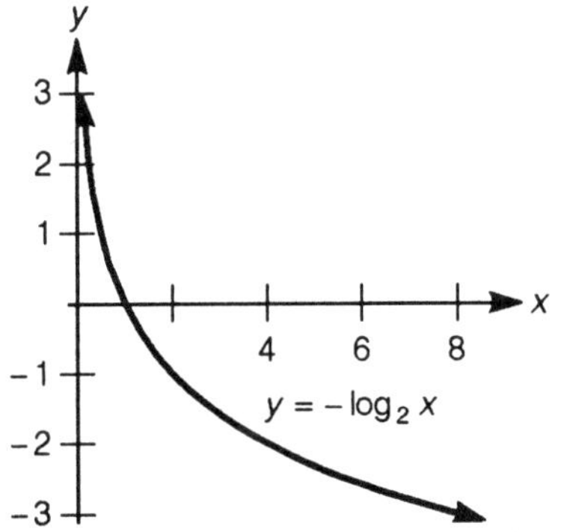

(c) The domain is $(-\infty,0)$, the range is $(-\infty,\infty)$, the x-intercept is -1, there is no y-intercept, and the asymptote is $x = 0$:

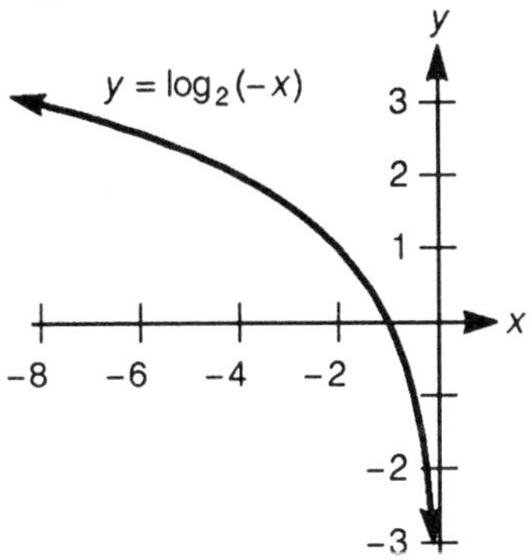

(d) The domain is $(-\infty, 0)$, the range is $(-\infty, \infty)$, the x-intercept is -1, there is no y-intercept, and the asymptote is $x = 0$:

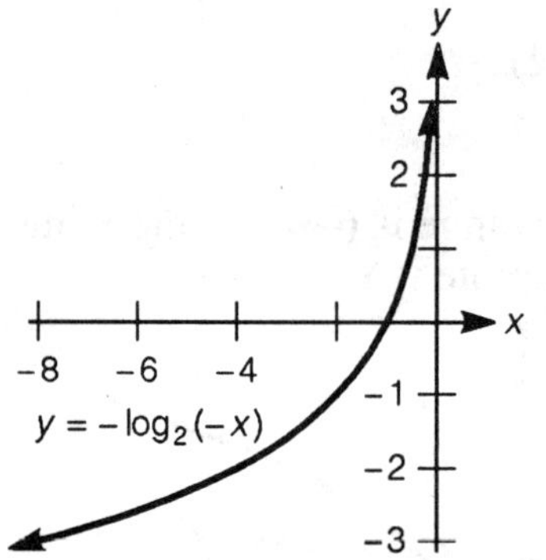

23. We translate $y = \log_3 x$ two units to the right, reflect across the x-axis, and translate 1 unit up. The domain is $(2, \infty)$, the range is $(-\infty, \infty)$, the x-intercept is 5, there is no y-intercept, and the asymptote is $x = 2$:

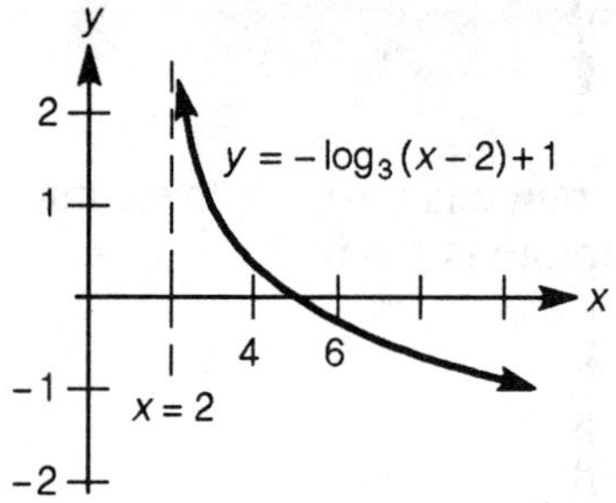

25. We translate $y = \ln x$ to the left e units. The domain is $(-e, \infty)$, the range is $(-\infty, \infty)$, the x-intercept is $-e + 1$, the y-intercept is 1, and the asymptote is $x = -e$:

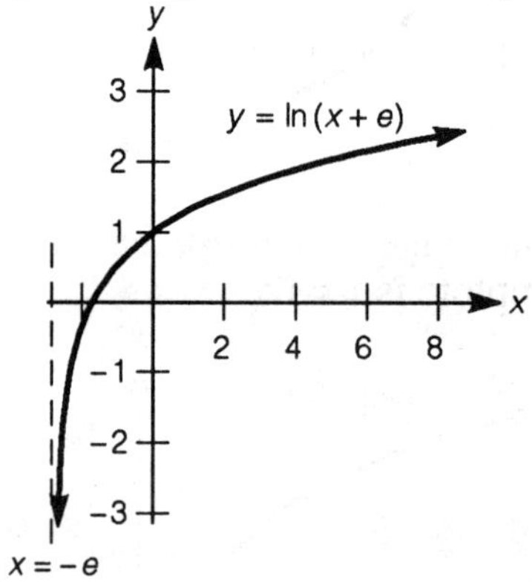

27. (a) Writing $x = \ln e^4$ in exponential form:

$$e^x = e^4$$
$$x = 4$$

(b) Writing $x = \ln\frac{1}{e}$ in exponential form:

$$\begin{aligned} e^x &= \frac{1}{e} \\ e^x &= e^{-1} \\ x &= -1 \end{aligned}$$

(c) Writing $x = \ln\sqrt{e}$ in exponential form:

$$\begin{aligned} e^x &= \sqrt{e} \\ e^x &= e^{1/2} \\ x &= \tfrac{1}{2} \end{aligned}$$

29. We solve for x:

$$\begin{aligned} 10^x &= 25 \\ x &= \log_{10} 25 \\ x &\approx 1.40 \end{aligned}$$

31. We solve for x:

$$\begin{aligned} 10^{(x^2)} &= 40 \\ x^2 &= \log_{10} 40 \\ x &= \pm\sqrt{\log_{10} 40} \\ x &= \pm\sqrt{1+\log_{10} 4} \\ x &\approx \pm 1.27 \end{aligned}$$

33. We solve for t:

$$\begin{aligned} e^{2t+3} &= 10 \\ 2t+3 &= \ln 10 \\ 2t &= -3+\ln 10 \\ t &= \frac{-3+\ln 10}{2} \\ t &\approx -0.35 \end{aligned}$$

35. We solve for t:

$$\begin{aligned} e^{1-4t} &= 12.405 \\ 1-4t &= \ln 12.405 \\ -4t &= -1+\ln 12.405 \\ t &= \frac{1-\ln 12.405}{4} \\ t &\approx -0.38 \end{aligned}$$

37. (a) We use the decay equation $N = N_0 e^{kt}$, where $N = \frac{1}{2}N_0$ when $t = 4.5\times 10^9$:

$$\frac{1}{2}N_0 = N_0 e^{(4.5\times10^9)k}$$
$$\frac{1}{2} = e^{(4.5\times10^9)k}$$
$$\ln\frac{1}{2} = (4.5\times 10^9)k$$
$$k = \frac{\ln 0.5}{4.5\times 10^9} \approx -1.54\times 10^{-10}$$

(b) Substituting $t = 1000$:

$$N = N_0 e^{1000(-1.54\times10^{-10})} = 0.999999846N_0$$

So 99.9999846% is still remaining.

39. (a) Starting with $N = N_0 e^{kt}$, we have:

$$\frac{1}{2}N_0 = N_0 e^{k\bullet 1}$$
$$e^k = \frac{1}{2}$$
$$k = \ln\frac{1}{2}$$

(b) If 90% is gone, the amount remaining will be 10%, or 0.4 g, and we have:

$$0.4 = 4e^{(\ln 0.5)t}$$
$$0.1 = e^{(\ln 0.5)t}$$
$$\ln 0.1 = \ln\left(e^{(\ln 0.5)t}\right)$$
$$\ln 0.1 = (\ln 0.5)t$$
$$t = \frac{\ln 0.1}{\ln 0.5} \approx 3.32 \text{ years}$$

41. (a) We start with:

$$3\times 10^8 = \left(2\times 10^7\right)e^{k\bullet 2}$$
$$e^{2k} = 15$$
$$2k = \ln 15$$
$$k = \frac{\ln 15}{2}$$
$$2N_0 = N_0 e^{[(\ln 15)/2]t}$$
$$2 = e^{[(\ln 15)/2]t}$$
$$\ln 2 = \frac{\ln 15}{2}\bullet t$$
$$t = \frac{2\ln 2}{\ln 15} \approx 0.51 \text{ hours}$$

(b) Since 1 billion = 10^9, we want:

$$10^9 = 2\times 10^7 e^{[(\ln 15)/2]t}$$
$$t = \frac{2\ln 50}{\ln 15} \approx 2.89 \text{ hours}$$

43. Let $y = e^{x+1}$. We switch the roles of x and y and solve the resulting equation for y:

$$x = e^{y+1}$$
$$\ln x = \ln\left(e^{y+1}\right)$$
$$\ln x = y + 1$$
$$y = -1 + \ln x$$
$$f^{-1}(x) = -1 + \ln x$$

The x-intercept is e and the asymptote is $x = 0$:

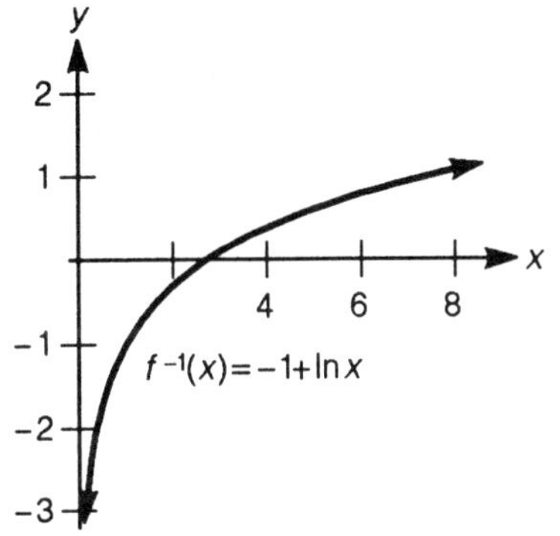

45. We graph the region:

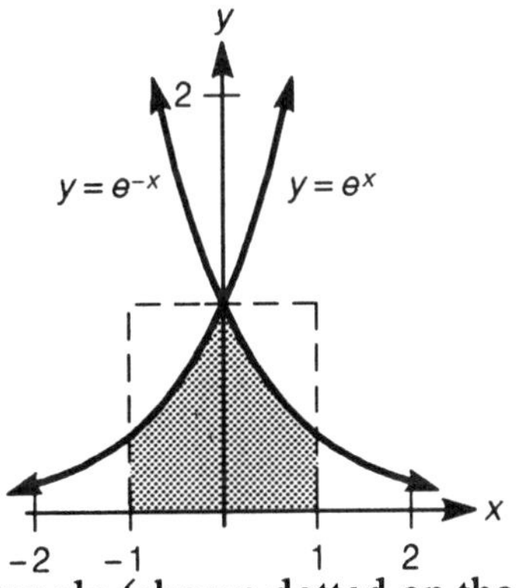

The area is less than the rectangle (shown dotted on the graph), which has an area of two square units.

47. We solve for x:

$$\left(e^x\right)^2 - 5e^x - 6 = 0$$
$$\left(e^x - 6\right)\left(e^x + 1\right) = 0$$
$$e^x = 6, -1$$

Since $e^x > 0$, then $e^x \neq -1$. Then $e^x = 6$, so $x = \ln 6$.

49. Using the approximation $2^{10} \approx 10^3$, then writing $\log_2 x = 100$ in exponential form:

$$x = 2^{100} = \left(2^{10}\right)^{10} \approx \left(10^3\right)^{10} = 10^{30}$$

51. We set $A = B$:

$$A_0 e^{k_1 t} = B_0 e^{k_2 t}$$

$$\frac{A_0}{B_0} = \frac{e^{k_2 t}}{e^{k_1 t}}$$

$$\frac{A_0}{B_0} = e^{(k_2 - k_1)t}$$

$$\ln\left(\frac{A_0}{B_0}\right) = (k_2 - k_1)t$$

$$t = \frac{\ln\left(\frac{A_0}{B_0}\right)}{k_2 - k_1}$$

This represents the amount of time it takes for two radioactive substances to decay to the same mass.

53. (a) We have $P(10) = 4$, since 2, 3, 5, 7 do not exceed 10.
We have $P(18) = 7$, since 2, 3, 5, 7, 11, 13, 17 do not exceed 18.
We have $P(19) = 8$, since 2, 3, 5, 7, 11, 13, 17, 19 do not exceed 19.

(b) We complete the table:

x	$P(x)$	$\frac{x}{\ln x}$	$\frac{P(x)}{x / \ln x}$
10^2	25	22	1.151
10^4	1229	1086	1.132
10^6	78498	72382	1.084
10^8	5761455	5428681	1.061
10^9	50847534	48254942	1.054
10^{10}	455052512	434294482	1.048

(c) We complete the table:

x	$P(x)$	$\frac{x}{\ln x - 1.08366}$	$\frac{P(x)}{x / (\ln x - 1.08366)}$
10^2	25	28	0.8804
10^4	1229	1231	0.9988
10^6	78498	78543	0.9994
10^8	5761455	5768004	0.9989
10^9	50847534	50917519	0.9986
10^{10}	455052512	455743004	0.9985

55. The easiest approach to use is to convert the equation to exponential form:

$\ln y = x$, so $y = e^x$

Thus the graph is identical to $y = e^x$:

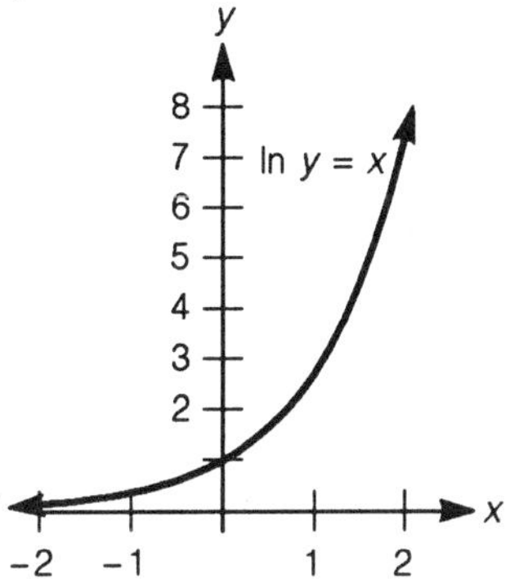

6.4 Properties of Logarithms

1. Using properties for logarithms, we have:

$\log_{10} 70 - \log_{10} 7 = \log_{10} \frac{70}{7} = \log_{10} 10 = 1$

3. Using properties for logarithms, we have:

$\log_7 \sqrt{7} = \log_7 \left(7^{1/2}\right) = \frac{1}{2}$

5. Using properties for logarithms, we have:

$\log_3 108 + \log_3 \frac{3}{4} = \log_3 \left(\frac{108 \bullet 3}{4}\right) = \log_3 81 = \log_3 3^4 = 4$

7. Using properties for logarithms, we have:

$-\frac{1}{2} + \ln \sqrt{e} = -\frac{1}{2} + \ln e^{1/2} = -\frac{1}{2} + \frac{1}{2} = 0$

9. Using properties for logarithms, we have:

$2^{\log_2 5} - 3\log_5 \sqrt[3]{5} = 5 - 3\log_5 5^{1/3} = 5 - 3\left(\frac{1}{3}\right) = 5 - 1 = 4$

11. Using properties for logarithms, we have:

$\log_{10} 30 + \log_{10} 2 = \log_{10}(30 \bullet 2) = \log_{10} 60$

13. Using properties for logarithms, we have:

$\log_5 6 + \log_5 \frac{1}{3} + \log_5 10 = \log_5\left(6 \bullet \frac{1}{3} \bullet 10\right) = \log_5 20$

15. (a) Using properties for logarithms, we have:

$$\ln 3 - 2\ln 4 + \ln 32 = \ln 3 - \ln\left(4^2\right) + \ln 32 = \ln\left(\frac{3 \bullet 32}{4^2}\right) = \ln 6$$

(b) Using properties for logarithms, we have:

$$\begin{aligned}\ln 3 - 2(\ln 4 + \ln 32) &= \ln 3 - 2[\ln(4 \bullet 32)] \\ &= \ln 3 - 2\ln 128 \\ &= \ln 3 - \ln(128^2) \\ &= \ln\left(\frac{3}{128^2}\right) \\ &= \ln\left(\frac{3}{16384}\right)\end{aligned}$$

17. Using properties for logarithms, we have:

$$\begin{aligned}\log_b 4 + 3\left[\log_b(1+x) - \tfrac{1}{2}\log_b(1-x)\right] &= \log_b 4 + 3\left(\log_b(1+x) - \log_b\sqrt{1-x}\right) \\ &= \log_b 4 + \log_b(1+x)^3 - \log_b(1-x)^{3/2} \\ &= \log_b\left(\frac{4(1+x)^3}{(1-x)^{3/2}}\right)\end{aligned}$$

19. Using properties for logarithms, we have:

$$\begin{aligned}&4\log_{10} 3 - 6\log_{10}(x^2+1) + \tfrac{1}{2}\left[\log_{10}(x+1) - 2\log_{10} 3\right] \\ &= \log_{10} 3^4 - \log_{10}(x^2+1)^6 + \tfrac{1}{2}\left[\log_{10}(x+1) - \log_{10} 3^2\right] \\ &= \log_{10} 81 - \log_{10}(x^2+1)^6 + \tfrac{1}{2}\log_{10}\left(\frac{x+1}{9}\right) \\ &= \log_{10} 81 - \log_{10}(x^2+1)^6 + \log_{10}\left(\frac{x+1}{9}\right)^{1/2} \\ &= \log_{10}\left(\frac{81\frac{\sqrt{x+1}}{3}}{(x^2+1)^6}\right) \\ &= \log_{10}\left(\frac{27\sqrt{x+1}}{(x^2+1)^6}\right)\end{aligned}$$

21. (a) Using properties for logarithms, we have:

$$\log_{10}\left(\frac{x^2}{1+x^2}\right) = \log_{10} x^2 - \log_{10}(1+x^2) = 2\log_{10} x - \log_{10}(1+x^2)$$

(b) Using properties for logarithms, we have:

$$\ln\left(\frac{x^2}{\sqrt{1+x^2}}\right) = \ln x^2 - \ln\sqrt{1+x^2} = 2\ln x - \tfrac{1}{2}\ln(1+x^2)$$

23. (a) Using properties for logarithms, we have:

$$\log_{10}\sqrt{9-x^2} = \tfrac{1}{2}\log_{10}(9-x^2)$$
$$= \tfrac{1}{2}\log_{10}[(3+x)(3-x)]$$
$$= \tfrac{1}{2}\log_{10}(3+x) + \tfrac{1}{2}\log_{10}(3-x)$$

(b) Using properties for logarithms, we have:

$$\ln\left(\frac{\sqrt{4-x^2}}{(x-1)(x+1)^{3/2}}\right) = \tfrac{1}{2}\ln(4-x^2) - \ln(x-1) - \ln(x+1)^{3/2}$$
$$= \tfrac{1}{2}\ln[(2+x)(2-x)] - \ln(x-1) - \tfrac{3}{2}\ln(x+1)$$
$$= \tfrac{1}{2}\ln(2+x) + \tfrac{1}{2}\ln(2-x) - \ln(x-1) - \tfrac{3}{2}\ln(x+1)$$

25. (a) Using properties for logarithms, we have:

$$\log_b\sqrt{\frac{x}{b}} = \frac{1}{2}\log_b\frac{x}{b} = \frac{1}{2}\log_b x - \frac{1}{2}\log_b b = \frac{1}{2}\log_b x - \frac{1}{2}$$

(b) Using properties for logarithms, we have:

$$2\ln\sqrt{(1+x^2)(1+x^4)(1+x^6)} = \ln\left[(1+x^2)(1+x^4)(1+x^6)\right]$$
$$= \ln(1+x^2) + \ln(1+x^4) + \ln(1+x^6)$$

27. (a) Using properties for logarithms, we have:

$$\log_{10}(AB^2C^3) = \log_{10}A + \log_{10}B^2 + \log_{10}C^3$$
$$= \log_{10}A + 2\log_{10}B + 3\log_{10}C$$
$$= a + 2b + 3c$$

(b) Using properties for logarithms, we have:

$$\log_{10}10\sqrt{A} = \log_{10}10 + \log_{10}\sqrt{A} = 1 + \tfrac{1}{2}\log_{10}A = 1 + \tfrac{1}{2}a$$

(c) Using properties for logarithms, we have:

$$\log_{10}\sqrt{10ABC} = \tfrac{1}{2}\log_{10}(10ABC)$$
$$= \tfrac{1}{2}\log_{10}10 + \tfrac{1}{2}\log_{10}A + \tfrac{1}{2}\log_{10}B + \tfrac{1}{2}\log_{10}C$$
$$= \tfrac{1}{2}(1) + \tfrac{1}{2}a + \tfrac{1}{2}b + \tfrac{1}{2}c$$
$$= \tfrac{1}{2}(1+a+b+c)$$

(d) Using properties for logarithms, we have:

$$\log_{10}\left(\tfrac{10A}{\sqrt{BC}}\right) = \log_{10}(10A) - \tfrac{1}{2}\log_{10}(BC)$$
$$= \log_{10}10 + \log_{10}A - \tfrac{1}{2}\log_{10}B - \tfrac{1}{2}\log_{10}C$$
$$= 1 + a - \tfrac{1}{2}b - \tfrac{1}{2}c$$

29. (a) Using properties for logarithms, we have:

$$\ln(ex) = \ln e + \ln x = 1 + t$$

(b) Using properties for logarithms, we have:

$$\ln(xy)-\ln(x^2)=\ln\frac{xy}{x^2}=\ln\frac{y}{x}=\ln y-\ln x=u-t$$

(c) Using properties for logarithms, we have:

$$\ln\sqrt{xy}+\ln\frac{x}{e}=\tfrac{1}{2}(\ln x+\ln y)+\ln x-\ln e=\tfrac{1}{2}(t+u)+t-1=\tfrac{3}{2}t+\tfrac{1}{2}u-1$$

(d) Using properties for logarithms, we have:

$$\ln\left(e^2x\sqrt{y}\right)=\ln e^2+\ln x+\ln\sqrt{y}=2\ln e+\ln x+\tfrac{1}{2}\ln y=2+t+\tfrac{1}{2}u$$

31. Taking the natural log of each side, we have:

$$\begin{aligned}\ln 5&=\ln 2+\ln e^{2x-1}\\ \ln 5&=\ln 2+(2x-1)\\ \ln 5&=\ln 2+2x-1\\ 2x&=\ln 5-\ln 2+1\\ x&=\frac{\ln 5-\ln 2+1}{2}\end{aligned}$$

33. Taking the natural log of each side, we have:

$$\begin{aligned}3e^{1+t}&=2\\ \ln 3+\ln e^{1+t}&=\ln 2\\ \ln 3+1+t&=\ln 2\\ t&=\ln 2-\ln 3-1\end{aligned}$$

35. Taking the natural log of each side, we have:

$$\begin{aligned}2^x&=9\\ \ln 2^x&=\ln 9\\ x\ln 2&=\ln 9\\ x&=\frac{\ln 9}{\ln 2}\end{aligned}$$

37. Taking the natural log of each side, we have:

$$\begin{aligned}10\bullet 2^x&=5^x\\ \ln 10+\ln 2^x&=\ln 5^x\\ \ln 10+x\ln 2&=x\ln 5\\ \ln 10&=x\ln 5-x\ln 2\\ \ln 10&=x(\ln 5-\ln 2)\\ x&=\frac{\ln 10}{\ln 5-\ln 2}\\ x&=\frac{\ln 5+\ln 2}{\ln 5-\ln 2}\end{aligned}$$

39. Solving for x, we have:

$$\begin{aligned}
\log_9(x+1) &= \tfrac{1}{2} + \log_9 x \\
\log_9(x+1) - \log_9 x &= \tfrac{1}{2} \\
\log_9\left(\frac{x+1}{x}\right) &= \tfrac{1}{2} \\
9^{1/2} &= \frac{x+1}{x} \\
3 &= \frac{x+1}{x} \\
3x &= x+1 \\
2x &= 1 \\
x &= \tfrac{1}{2}
\end{aligned}$$

41. Solving for x, we have:

$$\begin{aligned}
\log_{10}(2x+4) + \log_{10}(x-2) &= 1 \\
\log_{10}[(2x+4)(x-2)] &= 1 \\
(2x+4)(x-2) &= 10^1 \\
2x^2 - 8 &= 10 \\
2x^2 &= 18 \\
x^2 &= 9 \\
x &= \pm 3
\end{aligned}$$

Since $\log_{10}(-2)$ is undefined, $x = -3$ is an extraneous root. So the solution is $x = 3$.

43. Solving for x, we have:

$$\begin{aligned}
\log_{10}(x+3) - \log_{10}(x-2) &= 2 \\
\log_{10}\left(\frac{x+3}{x-2}\right) &= 2 \\
10^2 &= \frac{x+3}{x-2} \\
100(x-2) &= x+3 \\
100x - 200 &= x+3 \\
99x &= 203 \\
x &= \tfrac{203}{99}
\end{aligned}$$

45. Solving for x, we have:

$$\begin{aligned}
\log_b(x+1) &= 2\log_b(x-1) \\
\log_b(x+1) &= \log_b(x-1)^2 \\
x+1 &= (x-1)^2 \\
x+1 &= x^2 - 2x + 1 \\
x^2 - 3x &= 0 \\
x(x-3) &= 0 \\
x &= 0, 3
\end{aligned}$$

Since $\log_b(-1)$ is undefined, $x = 0$ is an extraneous root. So the solution is $x = 3$.

47. Solving for x, we have:

$$\log_{10}(x-6)+\log_{10}(x+3)=1$$
$$\log_{10}[(x-6)(x+3]=1$$
$$10^1=(x-6)(x+3)$$
$$x^2-3x-18=10$$
$$x^2-3x-28=0$$
$$(x-7)(x+4)=0$$
$$x=7,-4$$

Since $\log_{10}(-10)$ is undefined, $x=-4$ is an extraneous root. So the solution is $x=7$.

49. (a) Solving for x, we have:

$$\log_{10}x-y=\log_{10}(3x-1)$$
$$\log_{10}x-\log_{10}(3x-1)=y$$
$$\log_{10}\left(\frac{x}{3x-1}\right)=y$$
$$10^y=\frac{x}{3x-1}$$
$$10^y(3x-1)=x$$
$$3(10^y)x-10^y=x$$
$$3(10^y)x-x=10^y$$
$$x[3(10^y)-1]=10^y$$
$$x=\frac{10^y}{3(10^y)-1}$$

(b) Solving for x, we have:

$$\log_{10}(x-y)=\log_{10}(3x-1)$$
$$x-y=3x-1$$
$$-2x=y-1$$
$$x=\frac{y-1}{-2}\text{ or }\frac{1-y}{2}$$

51. Using the change of base formula, we have:

$$\log_2 5=\frac{\log_{10}5}{\log_{10}2}$$

53. Using the change of base formula, we have:

$$\ln 3=\log_e 3=\frac{\log_{10}3}{\log_{10}e}$$

55. Using the change of base formula, we have:

$$\log_b 2=\frac{\log_{10}2}{\log_{10}b}$$

57. Using the change of base formula, we have:

$$\log_{10} 6 = \frac{\ln 6}{\ln 10}$$

59. Using the change of base formula, we have:

$$\log_{10} e = \frac{\ln e}{\ln 10} = \frac{1}{\ln 10}$$

61. Using the change of base formula, we have:

$$\log_{10}(\log_{10} x) = \log_{10}\left(\frac{\ln x}{\ln 10}\right) = \frac{\ln\left(\frac{\ln x}{\ln 10}\right)}{\ln 10} = \frac{\ln(\ln x) - \ln(\ln 10)}{\ln 10}$$

63. (a) true
(b) true
(c) true
(d) false, since $\ln x^3 = 3\ln x$, not $\ln 3x$
(e) true
(f) false, since $\ln 2x^3 = \ln 2 + 3\ln x \neq 3\ln 2x = \ln(2x)^3$
(g) true
(h) false, since $\log_5 24$ is between 1 and 2, not 5^1 and 5^2
(i) true
(j) false, since $\log_5 24$ is close to 2 $(2 = \log_5 25)$
(k) false, since the domain is $(0, \infty)$
(l) true
(m) true

65. (a) We have:

$$\log_{10} \pi^7 \approx 3.48005$$
$$7\log_{10} \pi \approx 7(0.49715) = 3.48005$$

So $\log_{10} \pi^7 = 7\log_{10} \pi$.

(b) We have:

$$\log_b P^n = \log_{10} \pi^7 \approx 3.48005$$
$$(\log_b P)^n = (\log_{10} \pi)^7 \approx 0.00751$$

So $\log_b \pi^7 \neq (\log_b \pi)^7$.

(c) We have:

$$b^{\log_b P} = 10^{\log_{10} 1776} = 10^{3.24944} = 1776$$

So $10^{\log_{10} 1776} = 1776$.

(d) We have:

$$\ln 2 + \ln 3 + \ln 4 \approx 0.69315 + 1.09861 + 1.38629 \approx 3.17805$$
$$\ln 24 \approx 3.17805$$

So $\ln 2 + \ln 3 + \ln 4 = \ln 24$.

(e) We have:

$$\log_{10} A + \log_{10} B + \log_{10} C = \log_{10} 11 + \log_{10} 12 + \log_{10} 13$$
$$\approx 1.04139 + 1.07918 + 1.11394$$
$$= 3.23451$$
$$\log_{10}(ABC) = \log_{10}(11 \bullet 12 \bullet 13) = \log_{10} 1716 \approx 3.23451$$

So $\log_{10} A + \log_{10} B + \log_{10} C = \log_{10}(ABC)$.

(f) We compute $f[g(2345.6)]$:

$$f[g(2345.6)] = f(\ln 2345.6) \approx f(7.76030) \approx e^{7.76030} \approx 2345.6$$

(g) We compute $g[f(0.123456)]$:

$$g[f(0.123456)] = g\left(10^{0.123456}\right) = \log_{10}(1.32879) = 0.123456$$

67. Since $(-2, 324)$ and $\left(\frac{1}{2}, \frac{4}{3}\right)$ lie on the curve $N = N_0 e^{kt}$, we have:

$$324 = N_0 e^{-2k} \text{ and } \tfrac{4}{3} = N_0 e^{\frac{1}{2}k}$$

Dividing, we have:

$$\frac{324}{\frac{4}{3}} = \frac{N_0 e^{-2k}}{N_0 e^{\frac{1}{2}k}}$$
$$243 = e^{-2.5k}$$
$$\ln 243 = -2.5k$$
$$k = -\frac{\ln 243}{2.5}$$

Substituting into $324 = N_0 e^{-2k}$, we have:

$$324 = N_0 e^{(2 \ln 243)/2.5}$$
$$324 = N_0 e^{0.8 \ln 243}$$
$$324 = N_0 \left(e^{\ln 243}\right)^{0.8}$$
$$324 = N_0 \bullet 243^{0.8}$$
$$N_0 = \frac{324}{243^{0.8}} = \frac{324}{81} = 4$$

69. (a) We substitute for x:

$$3e^{1-(\ln 0.3+1)} = 3e^{-\ln 0.3} = 3e^{\ln (0.3)^{-1}} = 3(0.3)^{-1} = 3\left(\tfrac{10}{3}\right) = 10$$

(b) Simplifying x, we have:

$$\frac{\ln 0.3 + 1}{2} = \tfrac{1}{2}\ln 0.3 + \tfrac{1}{2} = \ln(0.3)^{1/2} + \tfrac{1}{2} = \tfrac{1}{2} + \ln\sqrt{0.3}$$

(c) Since $\ln e = 1$, then $\frac{1}{2} = \frac{1}{2}\ln e = \ln\sqrt{e}$. So:

$$\tfrac{1}{2} + \ln\sqrt{0.3} = \ln\sqrt{e} + \ln\sqrt{0.3} = \ln\sqrt{\frac{3e}{10}}$$

71. Working from the left-hand side, we have:

$$\begin{aligned}\log_b\left(\frac{\sqrt{3}+\sqrt{2}}{\sqrt{3}-\sqrt{2}}\right) &= \log_b\left(\frac{\sqrt{3}+\sqrt{2}}{\sqrt{3}-\sqrt{2}} \bullet \frac{\sqrt{3}+\sqrt{2}}{\sqrt{3}+\sqrt{2}}\right)\\ &= \log_b\left(\frac{(\sqrt{3}+\sqrt{2})^2}{3-2}\right)\\ &= \log_b(\sqrt{3}+\sqrt{2})^2\\ &= 2\log_b(\sqrt{3}+\sqrt{2})\end{aligned}$$

73. Using properties of logarithms, we have:

$$b^{3\log_b x} = b^{\log_b x^3} = x^3$$

75. Using properties of logarithms, we have:

$$\begin{aligned}\log_2 \sqrt[5]{4\sqrt{2}} &= \tfrac{1}{5}\log_2 4\sqrt{2}\\ &= \tfrac{1}{5}\log_2(2^2 \bullet 2^{1/2})\\ &= \tfrac{1}{5}\log_2(2^{5/2})\\ &= \tfrac{1}{5} \bullet \tfrac{5}{2}\\ &= \tfrac{1}{2}\end{aligned}$$

77. Solving for x, we have:

$$\begin{aligned}3\ln x &= \alpha + 3\ln\beta\\ 3\ln x - 3\ln\beta &= \alpha\\ 3(\ln x - \ln\beta) &= \alpha\\ 3\ln\frac{x}{\beta} &= \alpha\\ \ln\frac{x}{\beta} &= \frac{\alpha}{3}\\ \frac{x}{\beta} &= e^{\alpha/3}\\ x &= \beta e^{\alpha/3}\end{aligned}$$

79. Using the change of base formula, we have:

$$\log_b a = \frac{\log_a a}{\log_a b} = \frac{1}{\log_a b}$$

81. Using the change of base formula, we have:

$$\log_{ab} x = \frac{\log_a x}{\log_a ab} = \frac{\log_a x}{\log_a a + \log_a b} = \frac{\log_a x}{1+\log_a b}$$

Therefore:

$$\frac{\log_a x}{\log_{ab} x} = \frac{\log_a x}{\frac{\log_a x}{1+\log_a b}} = 1 + \log_a b$$

83. We work from the right-hand side:

$$\tfrac{1}{2}(\log a + \log b) = \tfrac{1}{2}\log(ab) = \log\sqrt{ab}$$

Proving the desired equality is equivalent to proving:

$$\tfrac{1}{3}(a+b) = \sqrt{ab}, \text{ since log is a one-to-one function.}$$

If a and b are both positive, we can square each side:

$$\tfrac{1}{9}\left(a^2 + 2ab + b^2\right) = ab$$

We now work with the left-hand side:

$$\begin{aligned}\tfrac{1}{9}\left(a^2 + 2ab + b^2\right) &= \tfrac{1}{9}(7ab + 2ab) \text{ by our assumption } a^2 + b^2 = 7ab\\ &= \tfrac{1}{9}(9ab)\\ &= ab\end{aligned}$$

85. Let $y = \ln\left(x + \sqrt{x^2+1}\right)$. We interchange the roles of x and y, then solve for y:

$$\begin{aligned}x &= \ln\left(y + \sqrt{y^2+1}\right)\\ e^x &= y + \sqrt{y^2+1}\\ e^x - y &= \sqrt{y^2+1}\end{aligned}$$

Squaring each side:

$$\begin{aligned}e^{2x} - 2ye^x + y^2 &= y^2 + 1\\ e^{2x} - 2ye^x &= 1\\ e^{2x} - 1 &= 2ye^x\\ \frac{e^{2x}-1}{2e^x} &= y\end{aligned}$$

So $f^{-1}(x) = \dfrac{e^{2x}-1}{2e^x}$.

87. We complete the table:

x	10^2	10^3	10^6	10^{20}	10^{50}	10^{99}
y	0.4234	0.6589	0.9654	1.3428	1.5573	1.6918

As x increases in size, so does $\ln x$, and thus $\ln(\ln x)$, and finally $\ln[\ln(\ln x)]$. The table is misleading--it only indicates that y increases in size much slower! An actual proof of the result is more complicated: Assume that y does not approach infinity, so there is some number N such that $y < N$, then $\ln[\ln(\ln x)] < N$, so $\ln(\ln x) < e^N$. But since $\ln(\ln x) < e^N$, then $\ln x < e^{(e^N)}$, and finally $x < e^{e^{(e^N)}}$. But clearly this is false, since x increases in size and thus cannot be bounded. Thus the range is the set of all real numbers.

6.5 Applications

1. For annual compounding of money we use the formula $A = P(1 + r)^t$. We find A when $P = \$800$, $r = 0.06$ and $t = 4$ yrs:

$$A = 800(1+0.06)^4 = 800(1.06)^4 \approx \$1009.98$$

3. For annual compounding of money we use the formula $A = P(1 + r)^t$. We find r when $A = \$6000$, $P = \$4000$ and $t = 5$:

$$\begin{aligned} 6000 &= 4000(1+r)^5 \\ 1.5 &= (1+r)^5 \\ (1.5)^{1/5} &= 1+r \\ r &= (1.5)^{1/5} - 1 \\ r &\approx 0.0845 \end{aligned}$$

So the interest rate is 8.45%.

5. In the first bank, we have $P = \$500$, $r = 0.05$, and $t = 4$:

$$A = 500(1+0.05)^4 = 500(1.05)^4 \approx 607.75$$

We now deposit $P = \$607.75$, $r = 0.06$, and $t = 4$:

$$A = 607.75(1+0.06)^4 = 607.75(1.06)^4 \approx 767.27$$

So the new balance will be \$767.27.

7. (a) We use $A = P(1 + r)^t$ with $P = 1000$, $r = 0.07$, and $t = 20$:

$$A = 1000(1+0.07)^{20} = 1000(1.07)^{20} \approx 3869.68$$

The new balance is \$3869.68.

(b) We use $A = P\left(1+\frac{r}{N}\right)^t$ with $P = 1000$, $r = 0.07$, $N = 4$, and $t = 20$:

$$A = 1000\left(1+\frac{0.07}{4}\right)^{4(20)} = 1000(1.0175)^{80} \approx 4006.39$$

The new balance is \$4006.39.

9. For compounding quarterly we use the formula:

$$A = P\left(1+\frac{r}{N}\right)^{Nt}$$

So here $P = \$100$, $r = 6\%$ and we have $N = 4$ compoundings per year. We find the value of t for which $A \geq \$120$:

$$\begin{aligned} 120 &\leq 100\left(1+\frac{0.06}{4}\right)^{4t} \\ 1.2 &\leq (1.015)^{4t} \\ \ln 1.2 &\leq 4t \ln 1.015 \\ 4t &\geq \frac{\ln 1.2}{\ln 1.015} \\ t &\geq \frac{\ln 1.2}{4 \ln 1.015} \\ t &\geq 3.06 \end{aligned}$$

This is slightly over 3 years, and so 13 quarters will be required.

11. We use $A = P\left(1+\frac{r}{N}\right)^{Nt}$ where $r = 0.055$, $N = 2$, $A = 6000$, and $t = 10$:

$$6000 = P\left(1+\frac{0.55}{2}\right)^{2(10)}$$
$$6000 = P(1.0275)^{20}$$
$$P = \frac{6000}{(1.0275)^{20}}$$
$$P \approx 3487.50$$

You must deposit a principal of \$3487.50.

13. We use $A = Pe^{rt}$ where $A = 5000$, $t = 10$ and $r = 0.065$:

$$5000 = Pe^{(0.065)(10)}$$
$$5000 = Pe^{0.65}$$
$$P = \frac{5000}{e^{0.65}}$$
$$P \approx 2610.23$$

A principal of \$2610.23 will grow to \$5000 in 10 yrs.

15. Since the effective rate is $r = 0.06$, we have:

$$A = P(1+0.06)^1 = 1.06P$$

The nominal rate r would yield a balance of:

$$A = Pe^{r(1)} = Pe^r$$

Setting these equal:

$$Pe^r = P(1.06)$$
$$e^r = 1.06$$
$$r = \ln 1.06$$
$$r \approx 0.0583$$

So the nominal rate is 5.83%.

17. Let's take the 6% investment first:

$$A = 10000(1+0.06)^5 = 10000(1.3382) \approx \$13382.26$$

The second choice will be:

$$A = 10000e^{0.05(5)} = 10000e^{0.25} \approx \$12840.25\text{, considerably less.}$$

19. (a) We have $T_2 \approx \frac{0.7}{r} = \frac{0.7}{0.05} = 14$ yrs.

(b) We have $T_2 = \frac{\ln 2}{r} = \frac{\ln 2}{0.05} = 13.86$ yrs.

(c) Here $d_1 = 13.86$, $d_2 = 14$, so $d = |13.86 - 14| = 0.14$.

This represents $\frac{0.14}{13.86}(100) \approx 1.01\%$ of the actual doubling time.

21. We compute:

$$A = 1000e^{(0.08)(300)} = 1000e^{24} = 1000(2.65 \times 10^{10}) = \$2.65 \times 10^{13}$$

That's \$26.5 trillion, a nice inheritance.

23. (a) We have $T_2 \approx \frac{0.7}{0.05} = 14$ yrs.

(b) We sketch the graph:

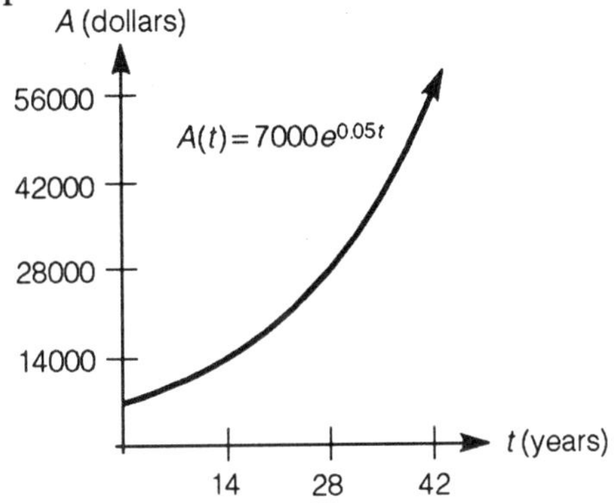

25. The growth problems are solved by assuming an exponential growth rate:

$$N = N_0 e^{kt}$$

First we find the percent distribution by dividing the total population into each of the two regions:

$$\frac{1.214}{5.321} \approx 22.8\% \text{ for more developed regions}$$

Applying the growth formula to the total world population, we have:

$$N = N_0 e^{kt} = 5.321e^{0.018(10)} = 5.321e^{0.18} \approx 6.370 \text{ billion}$$

The other calculations are similar to these two. The completed table is:

	1990 Population (billions)	% Population in 1990	Growth Rate (% per year)	Year 2000 Population (billions)	% of World Population in 2000
World	5.321	100	1.8	6.370	100
More Dev.	1.214	22.8	0.5	1.276	20.0
Less Dev.	4.107	77.2	2.1	5.067	80.0

27. (a) We complete the table:

Country	1990 Population (millions)	Relative Growth Rate (percent per year)	Year 2000 Population (millions)	Percent Increase in Population
Iraq	18.8	3.9	27.8	47.87
United Kingdom	57.4	0.2	58.6	2.09

(b) We solve the equation:

$$18.8e^{0.039t} = 57.4e^{0.002t}$$
$$\frac{e^{0.039t}}{e^{0.002t}} = \frac{57.4}{18.8}$$
$$e^{0.037t} \approx 3.05$$
$$0.037t \approx \ln 3.05$$
$$t \approx \frac{\ln 3.05}{0.037} \approx 30$$

Iraq's population will be equal to United Kingdom's in 30 years, which is the year 2020.

29. (a) Use $N = N_0e^{kt}$:

$$62947714 = 23191876e^{k(50)}$$
$$2.7142 = e^{50k}$$
$$\ln 2.7142 = 50k$$
$$k = \frac{\ln 2.7142}{50}$$
$$k \approx 0.0200$$

(b) In 1950, we have:

$$N = (62947714)e^{(0.02)(50)} \approx 170853155$$

Your answer may differ slightly due to round-off error.

(c) The actual growth over that period was slower than that predicted from (b) with a constant growth rate.

31. (a) Use $N = N_0e^{kt}$:

$$13479142 = 12588066e^{k(10)}$$
$$1.0708 = e^{10k}$$
$$\ln 1.0708 = 10k$$
$$k = \frac{\ln 1.0708}{10} \approx 0.00684$$

The growth rate is 0.684%.

(b) Use $N = N_0e^{kt}$:

$$14830192 = 13479142e^{k(10)}$$
$$1.1002 = e^{10k}$$
$$\ln 1.1002 = 10k$$
$$k = \frac{\ln 1.1002}{10} \approx 0.00955$$

The growth rate is 0.955%.

(c) Use $N = N_0e^{kt}$, $N_0 = 14830192$, $k = 0.00955$, and $t = 40$:

$$N = 14830192e^{0.00955(40)} \approx 21731258$$

(d) Our prediction is higher than the actual population.

33. (a) We draw the graph:

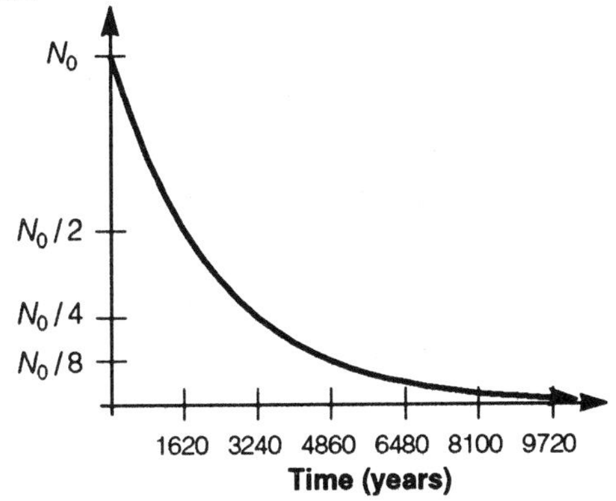

(b) We draw the graph:

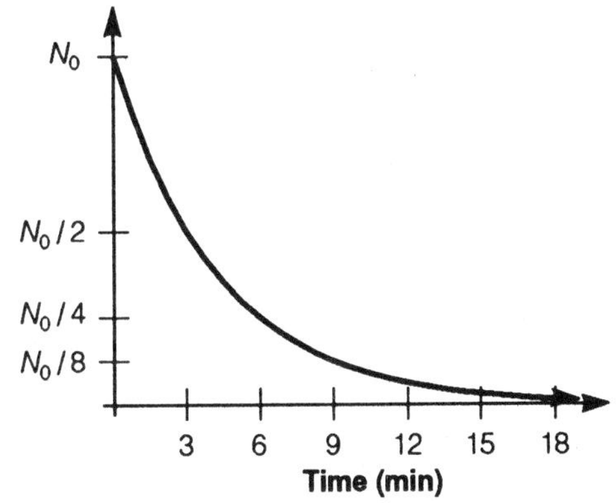

35. (a) We use $\frac{1}{2}N_0 = N_0e^{k\bullet 13}$ because after 13 years, N will be one-half of the original amount. Dividing by N_0, we obtain:

$$e^{13k} = 0.5$$
$$13k = \ln 0.5$$
$$k = \frac{\ln 0.5}{13} \approx -0.0533$$

Once you understand this, it will be convenient for you to remember that k always equals $\ln 0.5$ divided by the length of the half-life for decay problems.

(b) We have $N = N_0e^{(-0.0533)(10)}$, so $N \approx 0.587N_0$ or about 58.7% of N_0 after 10 years. If $t = 100, N = N_0e^{-5.33} \approx 0.005N_0$ so $N \approx 0.5\%$ of N_0.

37. (a) We have $k = \frac{\ln 0.5}{28} \approx -0.0248$.

(b) We solve for t:

$$\frac{N_0}{1000} = N_0 e^{kt}$$
$$0.001 = e^{-0.0248t}$$
$$\ln 0.001 = -0.0248t$$
$$t = \frac{\ln 0.001}{-0.0248} \approx 279 \text{ years}$$

(c) Since $2^{10} \approx 1000$, then after 10 half-lives, it should be reduced to $\frac{N_0}{1000}$. Since each half-life is 28 years, this is approximately 280 years. Note that this is close to our answer from (b).

39. (a) We use $N = N_0 e^{kt}$ with $N = 10 \times 10^9$, $N_0 = 3.6 \times 10^9$, and $k = 0.02$:

$$10 \times 10^9 = (3.6 \times 10^9) e^{0.02t}$$
$$\frac{10}{3.6} = e^{0.02t}$$
$$\ln \frac{10}{3.6} = 0.02t$$
$$t = \frac{\ln \frac{10}{3.6}}{0.02}$$
$$t \approx 51 \text{ years}$$

So 51 years after 1969, or 2020, the carrying capacity will be reached.

(b) We use $N = N_0 e^{kt}$ with $N = 10 \times 10^9$, $N_0 = 4.043 \times 10^9$, and $k = 0.015$:

$$10 \times 10^9 = (4.043 \times 10^9) e^{0.015t}$$
$$\frac{10}{4.043} = e^{0.015t}$$
$$\ln \frac{10}{4.043} = 0.015t$$
$$t = \frac{\ln \frac{10}{4.043}}{0.015}$$
$$t \approx 60 \text{ years}$$

So 60 years after 1975, or 2035, the carrying capacity will be reached.

(c) We use $N = N_0 e^{kt}$ with $N = 10\times10^9$, $N_0 = 4.134\times10^9$, and $k = 0.02$:

$$10\times10^9 = (4.134\times10^9)e^{0.02t}$$
$$\frac{10}{4.134} = e^{0.02t}$$
$$\ln\frac{10}{4.134} = 0.02t$$
$$t = \frac{\ln\frac{10}{4.134}}{0.02}$$
$$t \approx 44 \text{ years}$$

So 44 years after 1975, or 2019, the carrying capacity will be reached.

41. (a) We compute T:

$$T = \frac{\ln\left(\frac{Ak}{A_0}+1\right)}{k} = \frac{\ln\left(\frac{161241(0.01)}{250}+1\right)}{0.01} \approx 201 \text{ years}$$

(b) We compute T:

$$T = \frac{\ln\left(\frac{Ak}{A_0}+1\right)}{k} = \frac{\ln\left(\frac{161241(0.02)}{250}+1\right)}{0.02} \approx 132 \text{ years}$$

43. (a) Solving for E, we have:

$$\log_{10} E = 11.4 + 1.5M$$
$$E = 10^{11.4+1.5M}$$
$$E = 10^{11.4} \bullet 10^{1.5M}$$

(b) We compute the ratio:

$$\frac{E_1}{E_2} = \frac{10^{11.4} \bullet 10^{1.5M_1}}{10^{11.4} \bullet 10^{1.5M_2}} = 10^{1.5(M_1-M_2)} = 10^{1.5(8.6-7.1)} = 10^{2.25} \approx 178$$

45. We compute k:

$$k = \frac{\ln 0.5}{4.7\times10^{10}} \approx -1.4748\times10^{-11}$$

47. We compute T:

$$T = \frac{\ln\left(\frac{N_s}{N_r}+1\right)}{-k} = \frac{\ln(0.0636+1)}{-(-1.4748\times10^{-11})} \approx 4.181\times10^9 \text{ yrs}$$

The rock is approximately 4.181 billion years old.

49. We compute T:

$$T = \frac{5730\ln\frac{N}{920}}{\ln\frac{1}{2}} = \frac{5730\ln\frac{141}{920}}{\ln\frac{1}{2}} \approx 15505 \text{ yrs}$$

The two paintings are 15,505 years old.

51. We compute T:

$$T = \frac{5730 \ln \frac{N}{920}}{\ln \frac{1}{2}} = \frac{5730 \ln \frac{226}{920}}{\ln \frac{1}{2}} \approx 11605 \text{ yrs}$$

This is 11,500 years (to the nearest 500 years).

Chapter Six Review Exercises

1. If $x = \log_5 126$, then $5^x = 126$. Since $5^3 = 125$, then $x > 3$.
If $y = \log_{10} 999$, then $10^y = 999$. Since $10^3 = 1000$, then $y < 3$. So $\log_5 126$ is larger.

3. Since $N(t) = N_0 e^{kt}$ and $N_0 = 8000$, then $N(t) = 8000e^{kt}$. Since $N(4) = 10000$, we can find k:

$$\begin{aligned} N(4) &= 10000 \\ 8000e^{4k} &= 10000 \\ e^{4k} &= 1.25 \\ k &= \frac{\ln 1.25}{4} \end{aligned}$$

We now find t such that $N(t) = 12000$:

$$\begin{aligned} N(t) &= 12000 \\ 8000e^{kt} &= 12000 \\ e^{kt} &= 1.5 \\ kt &= \ln 1.5 \\ t &= \frac{\ln 1.5}{k} = \frac{\ln 1.5}{\frac{\ln 1.25}{4}} = \frac{4 \ln 1.5}{\ln 1.25} \text{ hours} \end{aligned}$$

5. We graph $f(x)$;

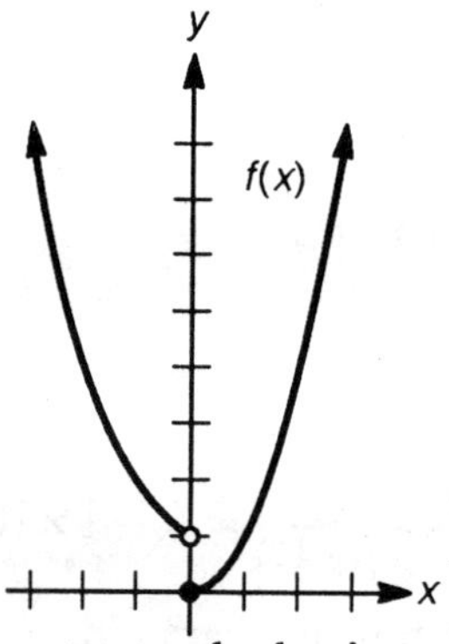

f is not one-to-one since it does not pass the horizontal line test.

7. We solve for x:

$$\begin{aligned}
\ln(x+1)-1&=\ln(x-1)\\
\ln(x+1)-\ln(x-1)&=1\\
\ln\frac{x+1}{x-1}&=1\\
\frac{x+1}{x-1}&=e\\
x+1&=ex-e\\
e+1&=ex-x\\
e+1&=x(e-1)\\
x&=\frac{e+1}{e-1}
\end{aligned}$$

9. For $y=e^x$, the domain is $(-\infty,\infty)$ and the range is $(0,\infty)$. For $y=\ln x$, the domain is $(0,\infty)$ and the range is $(-\infty,\infty)$. We graph both functions:

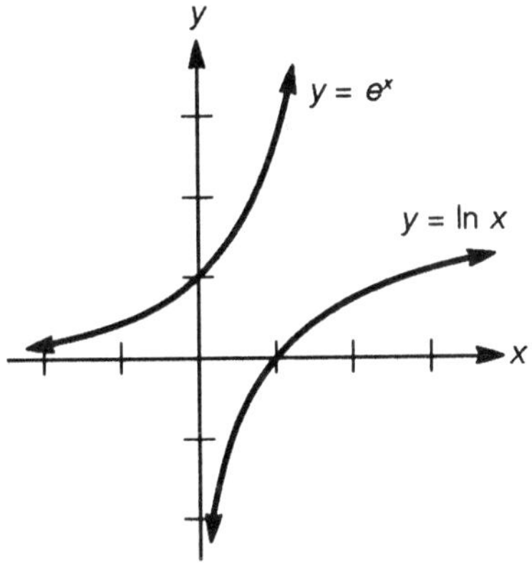

11. We simplify the log:

$$\log_9 \tfrac{1}{27}=\log_9\left(3^{-3}\right)=\log_9\left(9^{1/2}\right)^{-3}=\log_9 9^{-3/2}=-\tfrac{3}{2}$$

13. Using the decay equation $N=N_0e^{kt}$, we find k when $t=13$ and $N=\frac{1}{2}N_0$:

$$\begin{aligned}
\tfrac{1}{2}N_0&=N_0e^{13k}\\
\tfrac{1}{2}&=e^{13k}\\
\ln\tfrac{1}{2}&=13k\\
k&=\frac{\ln\frac{1}{2}}{13}\approx-0.05
\end{aligned}$$

15. We solve for x:

$$5e^{2-x} = 12$$
$$e^{2-x} = \tfrac{12}{5}$$
$$2 - x = \ln \tfrac{12}{5}$$
$$-x = -2 + \ln \tfrac{12}{5}$$
$$x = 2 - \ln \tfrac{12}{5}$$

17. We use $N = N_0 e^{kt}$ where $N_0 = 2$ and $k = 0.02$, so $N(t) = 2e^{0.02t}$. We find t when $N = 3$:

$$3 = 2e^{0.02t}$$
$$1.5 = e^{0.02t}$$
$$\ln 1.5 = 0.02t$$
$$t = \frac{\ln 1.5}{0.02} \approx 20 \text{ yrs}$$

We would expect the population to reach 3 million in the year 2000.

19. We use $A = Pe^{rt}$ where $P = \$1000$ and $r = 0.10$, so $A(t) = 1000e^{0.1t}$. The doubling time is approximately $\tfrac{70}{10} = 7$ years. We sketch the graph:

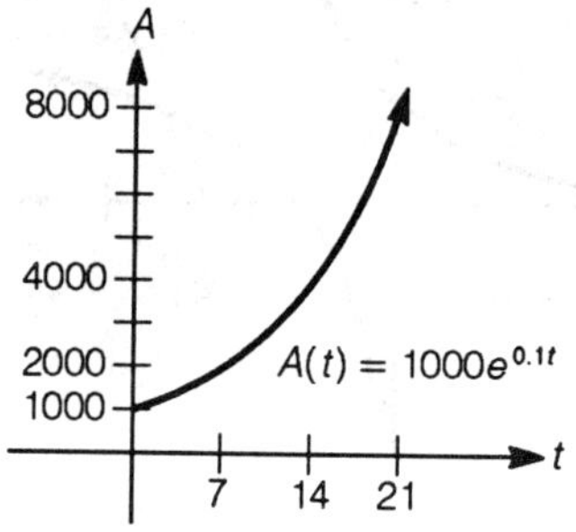

21. The horizontal asymptote is $y = 0$, there is no vertical asymptote, there is no x-intercept, and the y-intercept is 1. We draw the graph:

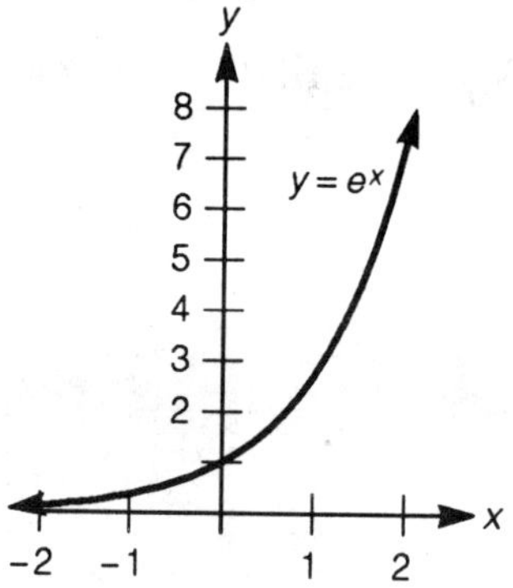

23. There is no horizontal asymptote, the vertical asymptote is $x = 0$, the x-intercept is 1, and there is no y-intercept. We draw the graph:

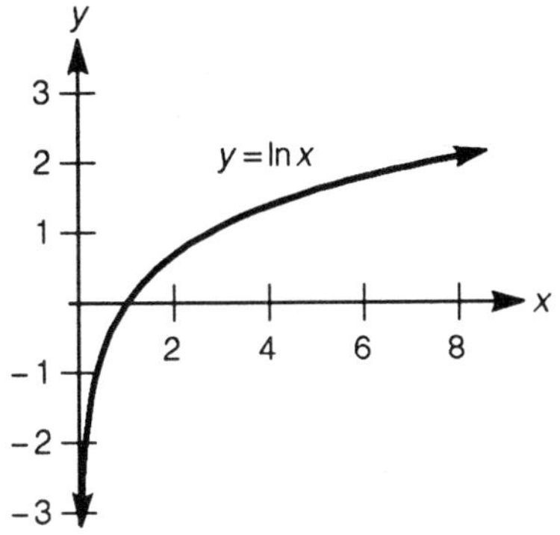

25. The horizontal asymptote is $y = 1$, there is no vertical asymptote, there is no x-intercept, and the y-intercept is 3. We draw the graph:

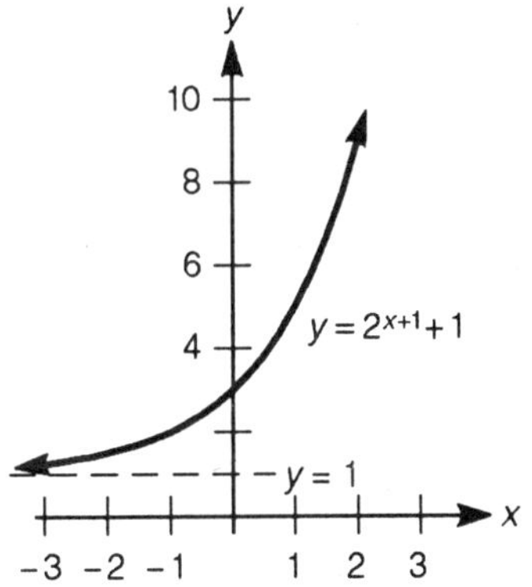

27. The horizontal asymptote is $y = 0$, there is no vertical asymptote, there is no x-intercept, and the y-intercept is 1. We draw the graph:

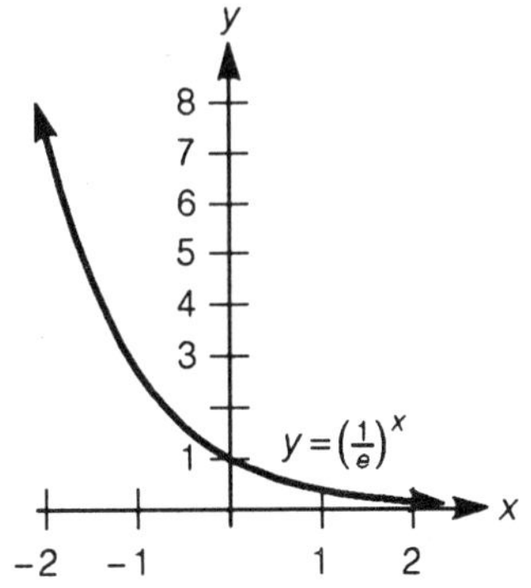

29. The horizontal asymptote is $y = 1$, there is no vertical asymptote, there is no x-intercept, and the y-intercept is $e + 1$. We draw the graph:

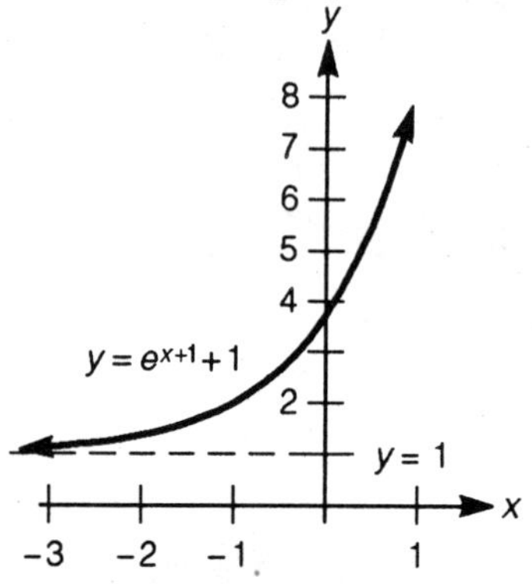

31. There are no asymptotes, and the x- and y-intercepts are both 0. We draw the graph:

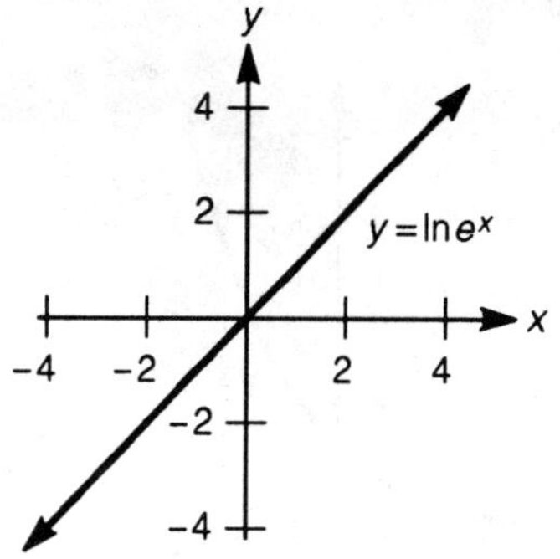

33. Using properties of logarithms, we solve:

$$\begin{aligned} \log_4 x + \log_4(x-3) &= 1 \\ \log_4[x(x-3)] &= 1 \\ x(x-3) &= 4^1 \\ x^2 - 3x &= 4 \\ x^2 - 3x - 4 &= 0 \\ (x-4)(x+1) &= 0 \\ x &= 4, -1 \end{aligned}$$

Since $\log_4(-1)$ is undefined, $x = -1$ is an extraneous root. So the solution is $x = 4$.

35. Using properties of logarithms, we solve:

$$\begin{aligned} \ln x + \ln(x+2) &= \ln 15 \\ \ln[x(x+2)] &= \ln 15 \\ x(x+2) &= 15 \\ x^2 + 2x &= 15 \\ x^2 + 2x - 15 &= 0 \\ (x+5)(x-3) &= 0 \\ x &= -5, 3 \end{aligned}$$

Since $\ln(-5)$ is undefined, $x = -5$ is an extraneous root. So the solution is $x = 3$.

37. Using properties of logarithms, we solve:

$$\begin{aligned}\log_2 x + \log_2(3x+10) - 3 &= 0\\ \log_2[x(3x+10)] &= 3\\ x(3x+10) &= 2^3\\ 3x^2+10x &= 8\\ 3x^2+10x-8 &= 0\\ (3x-2)(x+4) &= 0\\ x &= \tfrac{2}{3}, -4\end{aligned}$$

Since $\log_2(-4)$ is undefined, $x = -4$ is an extraneous root. So the solution is $x = \frac{2}{3}$.

39. We solve for x:

$$\begin{aligned}3\log_9 x &= \tfrac{1}{2}\\ \log_9 x &= \tfrac{1}{6}\\ x &= 9^{1/6}\\ x &= \sqrt[6]{9} = \sqrt[3]{3}\end{aligned}$$

41. We solve for x:

$$\begin{aligned}e^{1-5x} &= 3\sqrt{e}\\ 1-5x &= \ln 3\sqrt{e}\\ 1-5x &= \ln 3 + \tfrac{1}{2}\ln e\\ 1-5x &= \ln 3 + \tfrac{1}{2}\\ -5x &= \ln 3 - \tfrac{1}{2}\\ x &= \frac{1-2\ln 3}{10}\end{aligned}$$

43. Using properties of logarithms, we solve:

$$\begin{aligned}\log_{10} x - 2 &= \log_{10}(x-2)\\ \log_{10} x - \log_{10}(x-2) &= 2\\ \log_{10}\frac{x}{x-2} &= 2\\ \frac{x}{x-2} &= 100\\ x &= 100(x-2)\\ x &= 100x - 200\\ 200 &= 99x\\ x &= \tfrac{200}{99}\end{aligned}$$

45. Using properties of logarithms, we solve:

$$\begin{aligned}\ln(x+2) &= \ln x + \ln 2\\ \ln(x+2) &= \ln 2x\\ x+2 &= 2x\\ x &= 2\end{aligned}$$

47. Since $\ln(x^4) = 4\ln x$ is an identity, the solution is all real numbers $x > 0$, or $(0, \infty)$.

49. Using properties of logs, we have:

$$\begin{aligned} \log_{10} x &= \ln x \\ \frac{\ln x}{\ln 10} &= \ln x \\ \ln x &= (\ln 10)(\ln x) \\ \ln x - (\ln 10)(\ln x) &= 0 \\ \ln x(1 - \ln 10) &= 0 \\ \ln x &= 0, \text{ so } x = 1 \end{aligned}$$

51. Using properties of logs, we have:

$$\log_{10} \sqrt{10} = \log_{10}(10^{1/2}) = \tfrac{1}{2}$$

53. Using properties of logs, we have:

$$\ln(\sqrt[5]{e}) = \ln(e^{1/5}) = \tfrac{1}{5}$$

55. Using properties of logs, we have:

$$\begin{aligned} \log_{10} \pi - \log_{10} 10\pi &= \log_{10} \pi - (\log_{10} 10 + \log_{10} \pi) \\ &= \log_{10} \pi - \log_{10} 10 - \log_{10} \pi \\ &= -\log_{10} 10 \\ &= -1 \end{aligned}$$

57. Using properties of logs, we have:

$$10^{\log_{10} 16} = 16$$

59. We evaluate $\ln(e^4) = 4$.

61. Using properties of logs, we have:

$$\log_{12} 2 + \log_{12} 18 + \log_{12} 4 = \log_{12}(2 \bullet 18 \bullet 4) = \log_{12} 144 = \log_{12}(12^2) = 2$$

63. Using properties of logs, we have:

$$\frac{\ln 100}{\ln 10} = \frac{\ln(10^2)}{\ln 10} = \frac{2\ln 10}{\ln 10} = 2$$

65. Using properties of logs, we have:

$$\begin{aligned}\log_2 \sqrt[7]{16\sqrt[3]{2\sqrt{2}}} &= \tfrac{1}{7}\log_2 16\sqrt[3]{2\sqrt{2}}\\ &= \tfrac{1}{7}\left(\log_2 16 + \log_2 \sqrt[3]{2\sqrt{2}}\right)\\ &= \tfrac{1}{7}\left[\log_2(2^4) + \tfrac{1}{3}\log_2 2\sqrt{2}\right]\\ &= \tfrac{1}{7}\left[4 + \tfrac{1}{3}\left(\log_2 2 + \log_2 \sqrt{2}\right)\right]\\ &= \tfrac{1}{7}\left[4 + \tfrac{1}{3}\left(1 + \log_2(2^{1/2})\right)\right]\\ &= \tfrac{1}{7}\left(4 + \tfrac{1}{3}\bullet\tfrac{3}{2}\right)\\ &= \tfrac{1}{7}\left(4 + \tfrac{1}{2}\right)\\ &= \tfrac{1}{7}\bullet\tfrac{9}{2}\\ &= \tfrac{9}{14}\end{aligned}$$

67. Using properties of logs, we have:

$$\begin{aligned}\log_{10}\left(A^2B^3\sqrt{C}\right) &= \log_{10}A^2 + \log_{10}B^3 + \log_{10}\sqrt{C}\\ &= 2\log_{10}A + 3\log_{10}B + \tfrac{1}{2}\log_{10}C\\ &= 2a + 3b + \tfrac{c}{2}\end{aligned}$$

69. Using properties of logs, we have:

$$\begin{aligned}16\log_{10}\sqrt{A}\,\sqrt[4]{B} &= 16\left(\log_{10}\sqrt{A} + \log_{10}\sqrt[4]{B}\right)\\ &= 16\left(\tfrac{1}{2}\log_{10}A + \tfrac{1}{4}\log_{10}B\right)\\ &= 8\log_{10}A + 4\log_{10}B\\ &= 8a + 4b\end{aligned}$$

71. Since $\log_{10}100 = 2$ and $\log_{10}1000 = 3$, then $\log_{10}209$ lies between 2 and 3.

73. Since $\log_6 36 = 2$ and $\log_6 216 = 3$, then $\log_6 100$ lies between 2 and 3.

75. Since $\log_{10}0.010 = -2$ and $\log_{10}0.001 = -3$, then $\log_{10}0.003$ lies between –2 and –3.

77. (a) We graph $y = \ln(x + 2)$ and $y = \ln(-x) - 1$:

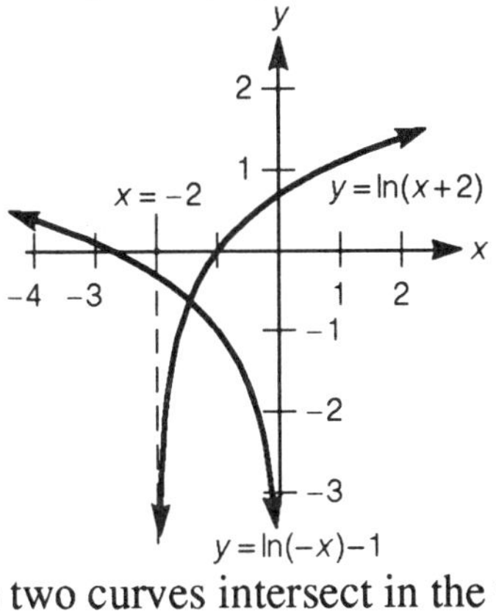

The graph shows these two curves intersect in the third quadrant.

(b) Setting the y-coordinates equal, we have:

$$\ln(x+2) = \ln(-x) - 1$$
$$\ln(x+2) - \ln(-x) = -1$$
$$\ln\frac{x+2}{-x} = -1$$
$$\frac{x+2}{-x} = e^{-1}$$
$$e(x+2) = -x$$
$$ex + 2e = -x$$
$$ex + x = -2e$$
$$x(e+1) = -2e$$
$$x = \frac{-2e}{e+1} \approx -1.46$$

79. We know $N = N_0e^{kt}$, and $N = \frac{1}{2}N_0$ when $t = T$:

$$\tfrac{1}{2}N_0 = N_0e^{kT}$$
$$\tfrac{1}{2} = e^{kT}$$
$$\ln\tfrac{1}{2} = kT$$
$$k = \frac{\ln\frac{1}{2}}{T}$$

81. Since 4 half-lives have passed, we will have $\left(\frac{1}{2}\right)^4 N_0 = \frac{1}{16}N_0$ of the substance left, or 6.25% remaining.

83. We have $k = \frac{\ln\frac{1}{2}}{d}$ and $N_0 = b$. We want to find the value of t when $N = c$:

$$N = N_0e^{kt}$$
$$c = be^{kt}$$
$$\frac{c}{b} = e^{kt}$$
$$\ln\frac{c}{b} = kt$$
$$\ln\frac{c}{b} = \frac{\ln\frac{1}{2}}{d} \bullet t$$
$$t = \frac{d\ln\frac{c}{b}}{\ln\frac{1}{2}} \text{ days}$$

85. Using properties of logs, we have:

$$\log_{10}8 + \log_{10}3 - \log_{10}12 = \log_{10}\left(\frac{8 \bullet 3}{12}\right) = \log_{10}2$$

87. Using properties of logs, we have:

$$\begin{aligned}\ln 5-3\ln 2+\ln 16&=\ln 5-\ln 2^3+\ln 16\\&=\ln 5-\ln 8+\ln 16\\&=\ln\left(\frac{5\bullet 16}{8}\right)\\&=\ln 10\end{aligned}$$

89. Using properties of logs, we have:

$$a\ln x+b\ln y=\ln x^a+\ln y^b=\ln x^a y^b$$

91. Using properties of logs, we have:

$$\ln\sqrt{(x-3)(x+4)}=\tfrac{1}{2}\ln[(x-3)(x+4)]=\tfrac{1}{2}\ln(x-3)+\tfrac{1}{2}(x+4)$$

93. Using properties of logs, we have:

$$\begin{aligned}\log_{10}\frac{x^3}{\sqrt{1+x}}&=\log_{10}x^3-\log_{10}\sqrt{1+x}\\&=\log_{10}x^3-\log_{10}(1+x)^{1/2}\\&=3\log_{10}x-\tfrac{1}{2}\log_{10}(1+x)\end{aligned}$$

95. Using properties of logs, we have:

$$\begin{aligned}\log_{10}\sqrt[3]{\frac{x}{100}}&=\log_{10}\left(\frac{x}{100}\right)^{1/3}\\&=\tfrac{1}{3}\log_{10}\frac{x}{100}\\&=\tfrac{1}{3}\log_{10}x-\tfrac{1}{3}\log_{10}100\\&=\tfrac{1}{3}\log_{10}x-\tfrac{2}{3}\end{aligned}$$

97. Using properties of logs, we have:

$$\ln\left(\frac{1+2e}{1-2e}\right)^3=3\ln\frac{1+2e}{1-2e}=3\ln(1+2e)-3\ln(1-2e)$$

99. We use $B=P(1+r)^t$, where $P=A$, $B=2A$, and $r=\frac{R}{100}$:

$$\begin{aligned}2A&=A\left(1+\tfrac{R}{100}\right)^t\\2&=\left(1+\tfrac{R}{100}\right)^t\\\ln 2&=t\ln\left(1+\tfrac{R}{100}\right)\\t&=\frac{\ln 2}{\ln\left(1+\frac{R}{100}\right)}\end{aligned}$$

101. The balance after 1 year will be $A = P\left(1+\frac{0.095}{12}\right)^{12}$. We must find the effective interest rate r where $A = P(1+r)^1$. We set these equal:

$$P(1+r) = P\left(1+\tfrac{0.095}{12}\right)^{12}$$
$$1+r = \left(1+\tfrac{0.095}{12}\right)^{12}$$
$$r = \left(1+\tfrac{0.095}{12}\right)^{12} - 1$$
$$r \approx 0.0992$$

So the effective interest rate is 9.92%.

103. We use $A = Pe^{rt}$ where $P = D$, $A = 2D$, and$r = \frac{R}{100}$:

$$2D = De^{(R/100)t}$$
$$2 = e^{(R/100)t}$$
$$\ln 2 = \frac{R}{100}(t)$$
$$t = \frac{100 \ln 2}{R}$$

105. (a) We use $A = P\left(1+\frac{r}{N}\right)^{Nt}$ where $P = \$660$, $r = 0.055$, $N = 4$, and $A = \$1000$:

$$1000 = 660\left(1+\tfrac{0.055}{4}\right)^{4t}$$
$$\tfrac{50}{33} = (1.01375)^{4t}$$
$$\ln \tfrac{50}{33} = 4t \ln 1.01375$$
$$t = \frac{\ln \frac{50}{33}}{4 \ln 1.01375} \approx 7.61 \text{ yrs}$$

So the balance will reach \$1000 after $7\frac{3}{4}$ years.

(b) We use $A = P\left(1+\frac{r}{N}\right)^{Nt}$ where $P = D, r = \frac{R}{100}, N = 4$, and $A = nD$:

$$nD = D\left(1+\tfrac{R}{400}\right)^{t}$$
$$n = \left(1+\tfrac{R}{400}\right)^{t}$$
$$\ln n = t \ln\left(1+\tfrac{R}{400}\right)$$
$$t = \frac{\ln n}{\ln\left(1+\frac{R}{400}\right)}$$

107. (a) The domain is $x > 0$, or $(0, \infty)$.

(b) We must make sure $\log_{10} x \geq 0$, so $x \geq 1$. So the domain is $x \geq 1$, or $[1, \infty)$.

109. (a) We must make sure that the expression in the log is non-zero:

$$x^2 - 2x - 15 \neq 0$$
$$(x-5)(x+3) \neq 0$$
$$x \neq 5, -3$$

So the domain is all real numbers except 5 and –3, or $(-\infty,-3)\cup(-3,5)\cup(5,\infty)$.

(b) We must make sure that the expression in the log is positive:

$$x^2 - 2x - 15 > 0$$
$$(x-5)(x+3) > 0$$

The key numbers are –3 and 5. We draw a sign chart:

Interval	Test Number	$x-5$	$x+3$	$(x-5)(x+3)$
$(-\infty,-3)$	–5	neg.	neg.	pos.
$(-3,5)$	0	neg.	pos.	neg.
$(5,\infty)$	6	pos.	pos.	pos.

So the product is positive on the intervals $(-\infty,-3)\cup(5,\infty)$. Thus the domain is $(-\infty,-3)\cup(5,\infty)$.

111. We solve for x:

$$y = \frac{e^x+1}{e^x-1}$$
$$y(e^x-1) = e^x+1$$
$$ye^x - y = e^x + 1$$
$$ye^x - e^x = y+1$$
$$e^x(y-1) = y+1$$
$$e^x = \frac{y+1}{y-1}$$
$$x = \ln\frac{y+1}{y-1}$$

We must solve the inequality $\frac{y+1}{y-1} > 0$. The key numbers are –1 and 1. We draw a sign chart:

Interval	Test Number	$y+1$	$y-1$	$\frac{y+1}{y-1}$
$(-\infty,-1)$	–4	neg.	neg.	pos.
$(-1,1)$	0	pos.	neg.	neg.
$(1,\infty)$	4	pos.	pos.	pos.

The quotient is positive on the intervals $(-\infty,-1)\cup(1,\infty)$. So the domain is $(-\infty,-1)\cup(1,\infty)$.

113. Using properties of logs, we have:

$$\ln 0.5 = \ln\left(2^{-1}\right) = -\ln 2 \approx -0.7$$

115. Using properties of logs, we have:

$$\ln\tfrac{1}{9} = \ln\left(3^{-2}\right) = -2\ln 3 \approx -2(1.1) = -2.2$$

117. Using properties of logs, we have:

$$\begin{aligned}\ln 72 &= \ln\left(3^2 \bullet 2^3\right)\\ &= \ln\left(3^2\right) + \ln\left(2^3\right)\\ &= 2\ln 3 + 3\ln 2\\ &\approx 2(1.1) + 3(0.7)\\ &= 2.2 + 2.1\\ &= 4.3\end{aligned}$$

119. Let $x = e$, then $\ln x = 1$. From the graph, we see $x \approx 2.7$.

121. Using properties of logs, we have:

$$\log_2 3 = \frac{\ln 3}{\ln 2} \approx \frac{1.1}{0.7} \approx 1.6$$

123. (a) We complete the table:

Region	1990 Population (millions)	Growth Rate (%)	2025 Population
North America	275.2	0.7	351.6
Soviet Union	291.3	0.7	372.2
Europe	499.5	0.2	535.7
Nigeria	113.3	3.1	335.3

(b) It will be 335.3 mil – 113.3 mil = 222.0 million.

(c) The net increases are given by:

North America:	351.6 mil − 275.2 mil = 76.4 mil
Soviet Union:	372.2 mil − 291.3 mil = 80.9 mil
Europe:	535.7 mil − 499.5 mil = 36.2 mil
combined:	193.5 million

(d) For Nigeria, our results support this projection.

125. (a) The corresponding x-coordinate when $y = 3.000$ is $x \approx 1.0986$.

(b) The percent error is given by:

$$\%\text{ error} = \frac{\left|1.0986 - \ln 3\right|}{\ln 3} \bullet 100 = 0.00112\%$$

(c) We estimate the quantities:

$$\ln\sqrt{3} = \tfrac{1}{2}\ln 3 \approx \tfrac{1}{2}(1.0986) \approx 0.5493$$
$$\ln 9 = \ln(3^2) = 2\ln 3 \approx 2(1.0986) \approx 2.1972$$
$$\ln\tfrac{1}{3} = \ln(3^{-1}) = -\ln 3 \approx -1.0986$$

Chapter Six Graphing Calculator Exercises

1. (a) Using the indicated settings, we draw the graphs:

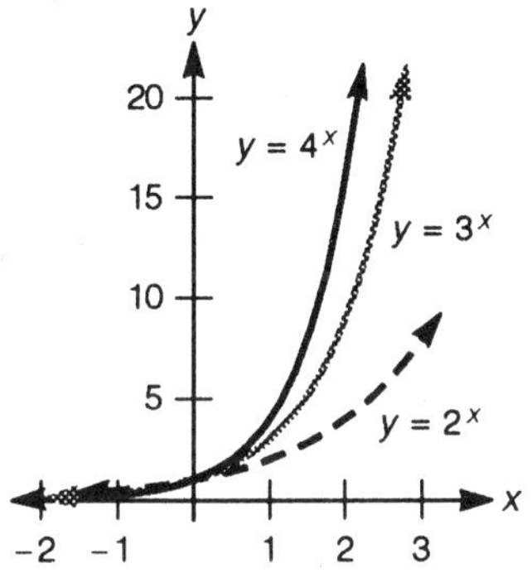

(b) It will be much "narrower". We draw the graphs:

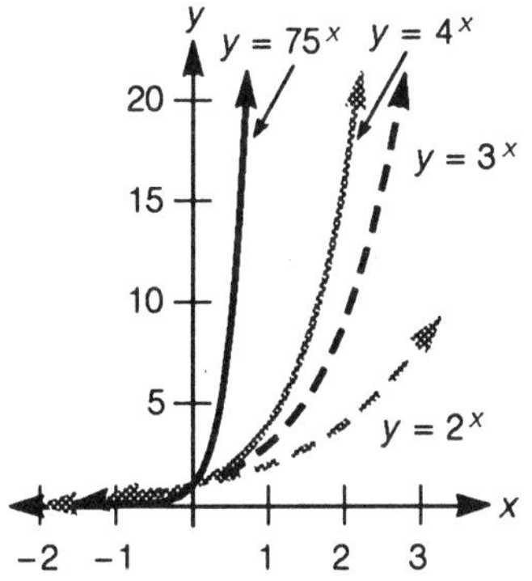

3. (a) The x-intercept is 2, the y-intercept is $\frac{8}{9}$, and the asymptote is $y = 1$.

(b) We graph the function:

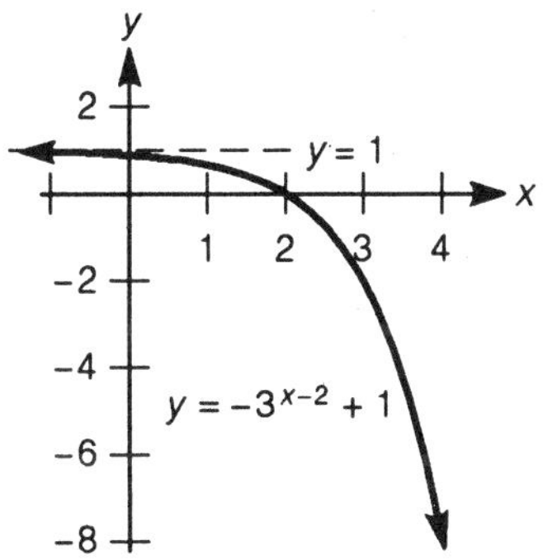

5. (a) There is no x-intercept, the y-intercept is 5, and the asymptote is $y = 4$.

(b) We graph the function:

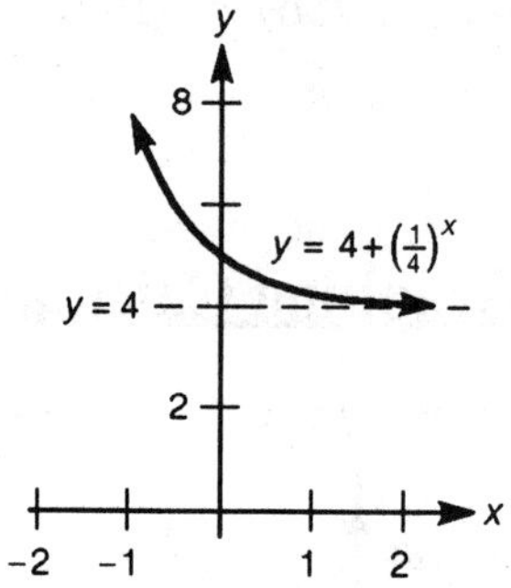

7. (a) The x-intercept is $1 - \log_{10} 2$, the y-intercept is 0.4, and the asymptote is $y = 0.5$.

(b) We graph the function:

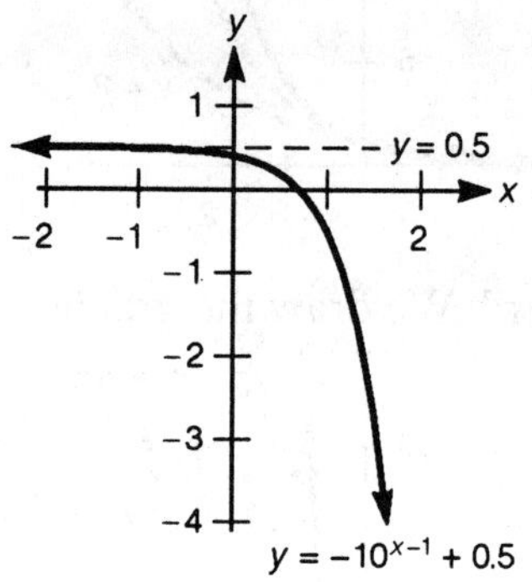

9. (a) We graph the function:

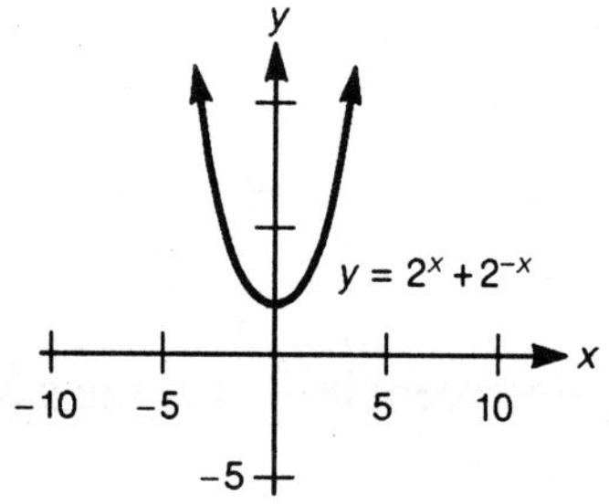

(b) Adding $y = x^3$, we draw the graph:

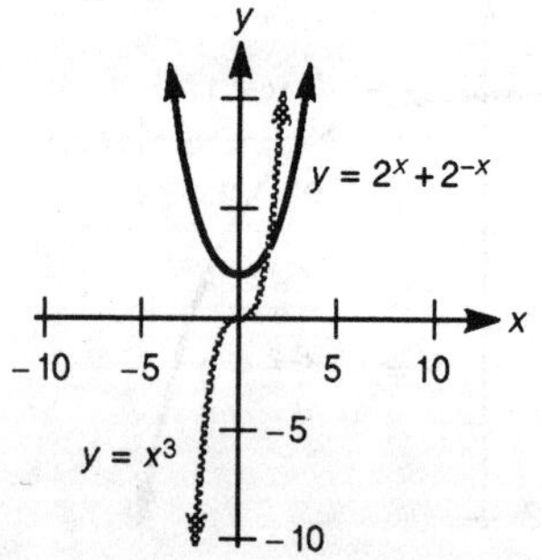

11. We sketch the graphs when $x < 0$:

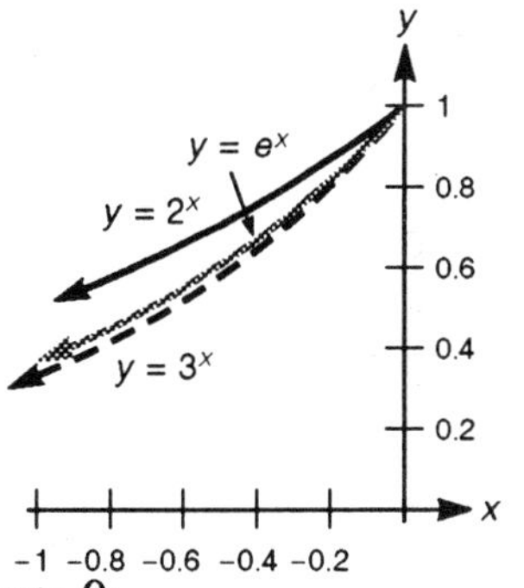

Now sketch the graphs when $x > 0$:

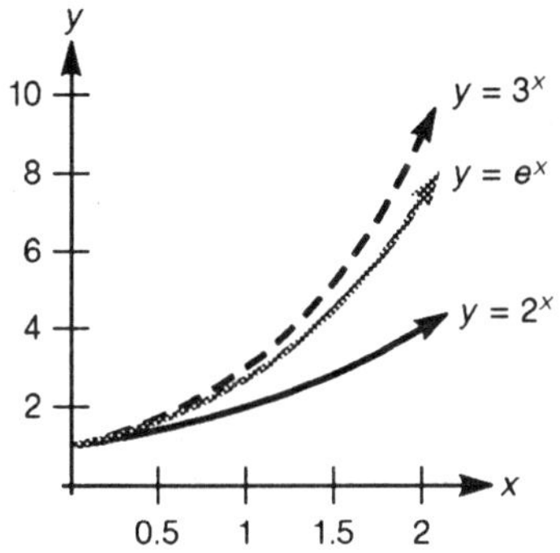

13. The graph is a reflection of $y = e^x$ across the y-axis:

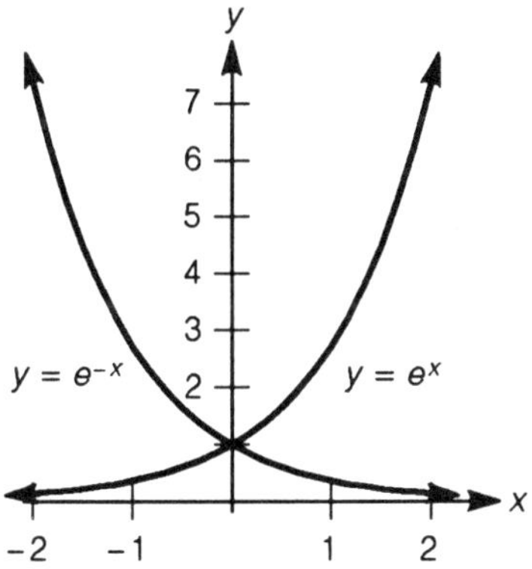

15. The graph is a reflection of $y = e^x$ across the x- and y-axes.

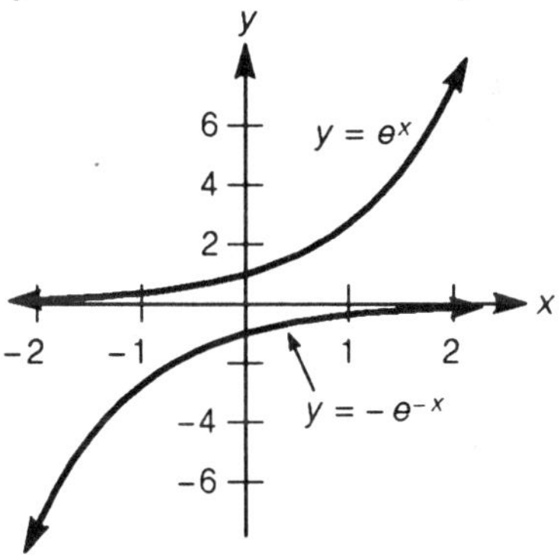

17. (a) The graph is a translation of $y = e^x$ to the right 1 unit:

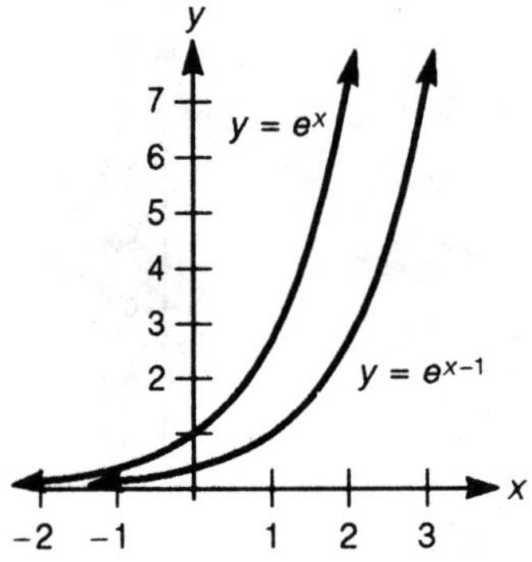

(b) The graph is a translation of $y = e^x$ to the right 1 unit, then a reflection across the y-axis:

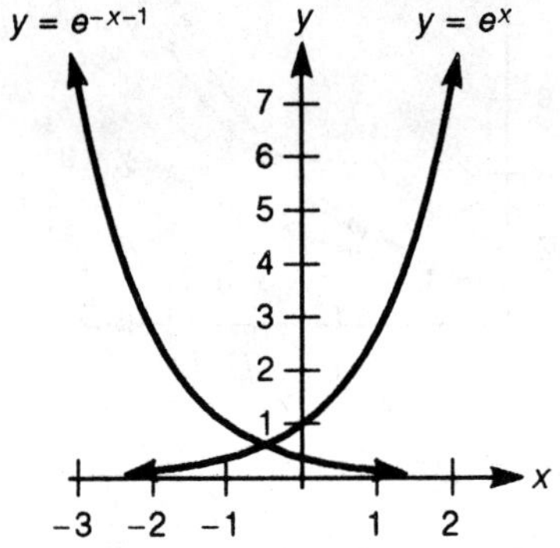

19. (a) We draw the graphs:

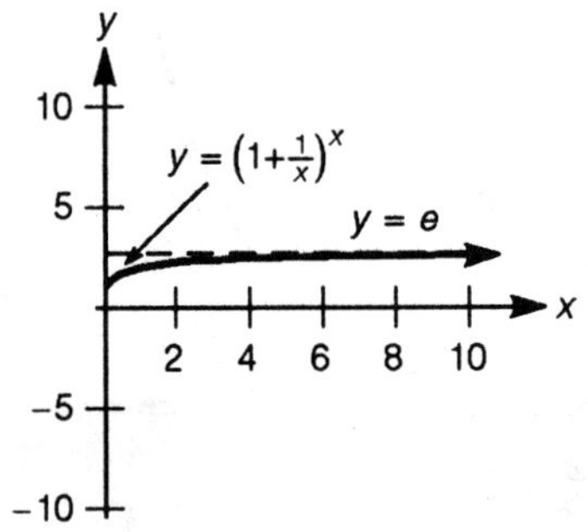

(b) We draw the graphs:

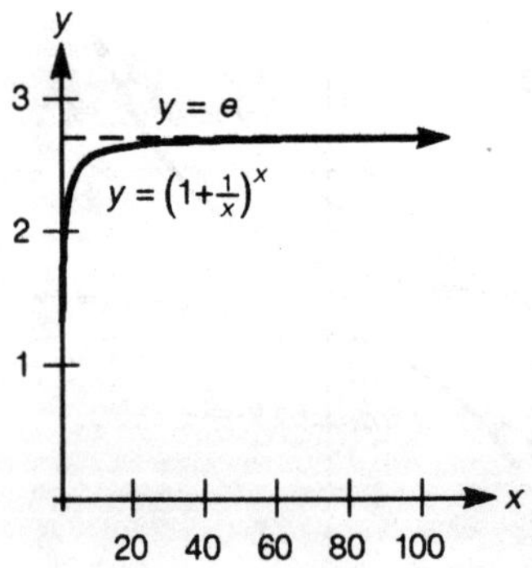

(c) We complete the table:

x	5	10	20	50
$\left(1+\frac{1}{x}\right)^x$	2.4883	2.5937	2.6533	2.6916

x	10^2	10^3	10^4	10^5	10^6
$\left(1+\frac{1}{x}\right)^x$	2.7048	2.7169	2.7181	2.7183	2.7183

Note that $e \approx 2.7183$.

21. (a) Since $y = 10^x$ and $y = \log_{10}x$ are inverse functions, we would expect their graphs to be reflections across the line $y = x$.

(b) Using Zoom-5 settings, we obtain:

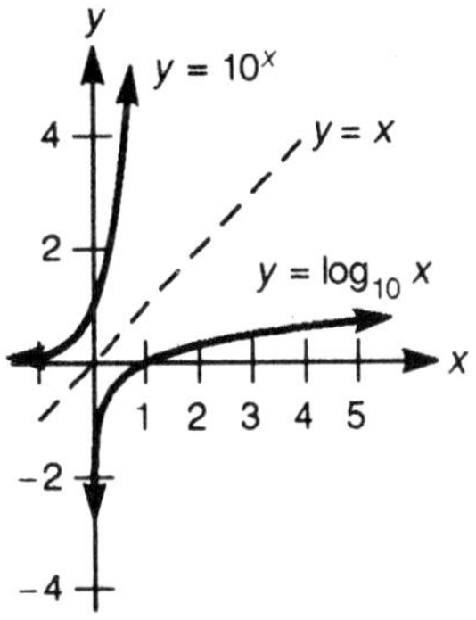

23. (a) From the graphs, we see that $\ln x < \log_{10}x$ on the interval $(0, 1)$:

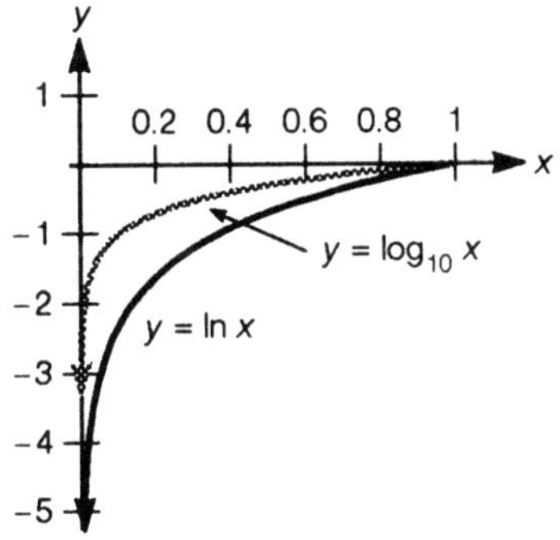

(b) From the graphs, we see that $\ln x > \log_{10}x$ on the interval $(1, \infty)$;

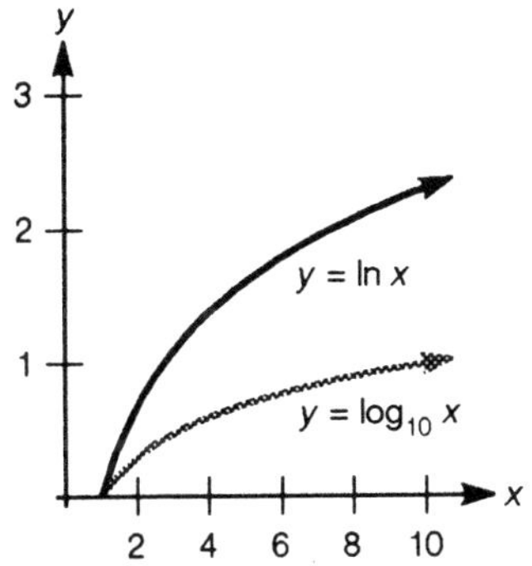

(c) From the graphs, we see that $e^x < 10^x$ on the interval $(0, \infty)$:

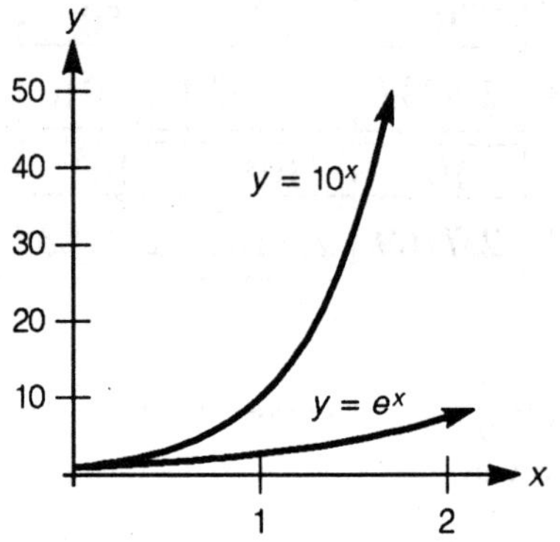

(d) From the graphs, we see that $e^x > 10^x$ on the interval $(-\infty, 0)$:

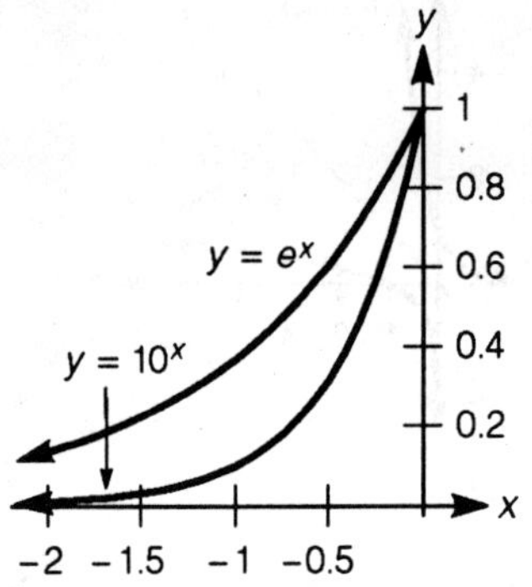

25. (c) The exponential curve values are confirmed.

(d) We draw the graph:

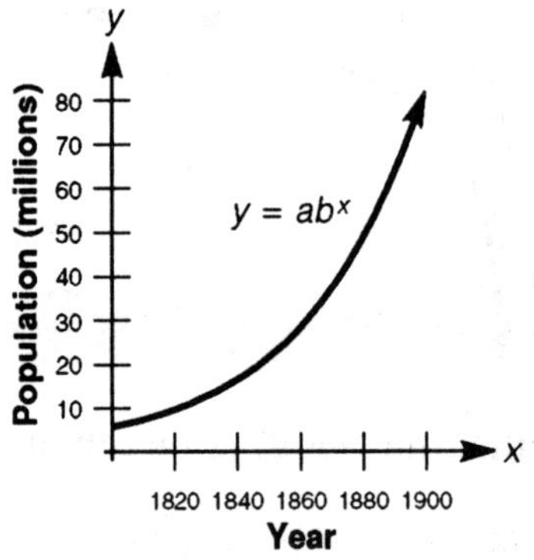

Chapter Six Test

1. The domain is $(-\infty, \infty)$, the range is $(-3, \infty)$, the x-intercept is $-\frac{\ln 3}{\ln 2}$, the y-intercept is -2, and the asymptote is $y = -3$. We draw the graph:

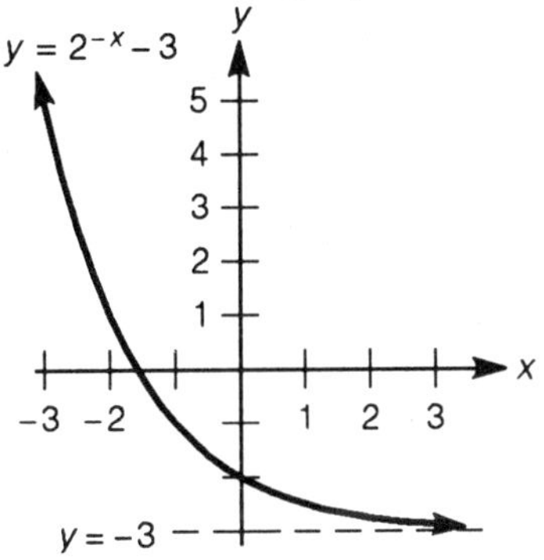

2. Using $N_0 = 6000$, $N = 6200$, and $t = 1$, we find the growth constant k:

$$6200 = 6000e^{k \cdot 1}$$
$$1.033 = e^k$$
$$k = \ln 1.033 \approx 0.03279$$

We now find the value of t when $N = 10000$:

$$10000 = 6000e^{kt}$$
$$\tfrac{5}{3} = e^{kt}$$
$$kt = \ln \tfrac{5}{3}$$
$$t = \frac{\ln \frac{5}{3}}{\ln 1.033} \approx 16 \text{ hrs}$$

3. Since $2^4 = 16$, then $\log_2 17 > 4$. Since $3^4 = 81$, then $\log_3 80 < 4$. So $\log_2 17$ is larger.

4. Using the change-of-base formula, we have:

$$\log_2 15 = \frac{\ln 15}{\ln 2}$$

5. Since $2^{10} \approx 10^3$, we have:

$$2^{40} = \left(2^{10}\right)^4 \approx \left(10^3\right)^4 = 10^{12}$$

6. Using $A = P(1 + r)^t$ for interest compounded annually, with $P = \$9500$, $A = \$12,000$ and $r = 0.06$, we have:

$$12000 = 9500(1.06)^t$$
$$1.2632 = (1.06)^t$$
$$\ln 1.2632 = t \ln 1.06$$
$$t = \frac{\ln 1.2632}{\ln 1.06} \approx 4 \text{ yrs}$$

7. (a) The identity is valid only if $\ln x$ is defined, which is the interval $(0,\infty)$.

(b) We graph the two functions:

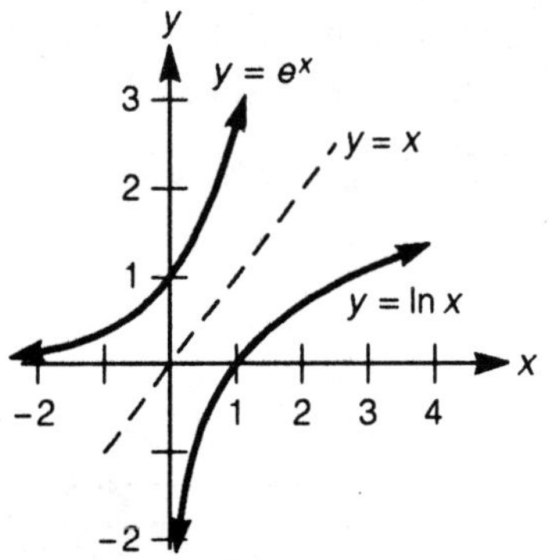

8. Using properties of logs, we have:

$$\log_{10}\frac{A^3}{\sqrt{B}} = \log_{10}A^3 - \log_{10}\sqrt{B} = 3\log_{10}A - \tfrac{1}{2}\log_{10}B = 3a - \tfrac{1}{2}b$$

9. (a) Using properties of logs, we have:

$$\log_5\tfrac{1}{\sqrt{5}} = \log_5 5^{-1/2} = -\tfrac{1}{2}$$

(b) Using properties of logs, we have:

$$\ln e^2 + \ln 1 - e^{\ln 3} = 2 + 0 - 3 = -1$$

10. Factoring, we have:

$$e^x\left(x^3 - 4x\right) = 0$$
$$xe^x\left(x^2 - 4\right) = 0$$
$$xe^x(x+2)(x-2) = 0$$
$$x = 0, -2, 2 \quad (\text{since } e^x \neq 0)$$

11. (a) Using the formula $T = \dfrac{\ln\frac{1}{2}}{k}$, we have:

$$4 = \frac{\ln\frac{1}{2}}{k}$$
$$k = \frac{\ln\frac{1}{2}}{4} \approx -0.1733$$

(b) Using $N = N_0e^{kt}$ with $N_0 = 2$, $k = -0.1733$, and $t = 10$:

$$N = 2e^{-0.1733(10)} \approx 0.35 \text{ grams}$$

12. Using properties of logs, we have:

$$2\ln x - \ln\sqrt[3]{x^2+1} = \ln x^2 - \ln(x^2+1)^{1/3} = \ln\frac{x^2}{(x^2+1)^{1/3}}$$

13. We solve for x:

$$-2e^{3x-1} = 9$$
$$e^{3x-1} = -4.5$$

But this is impossible $\left(e^{3x-1} > 0 \text{ for all } x\right)$, so there is no solution.

14. Let $y = \log_{10}(x-1)$. Switching the roles of x and y and solving the resulting equation for y yields:

$$x = \log_{10}(y-1)$$
$$10^x = y - 1$$
$$y = 10^x + 1$$

So $g^{-1}(x) = 10^x + 1$. The range of $g^{-1}(x)$ is $(1, \infty)$.

15. (a) Using the formula $T = \frac{\ln 2}{r}$, we have:

$$T = \frac{\ln 2}{0.06} \approx 12 \text{ years}$$

(b) We graph $A = 12000e^{0.06t}$:

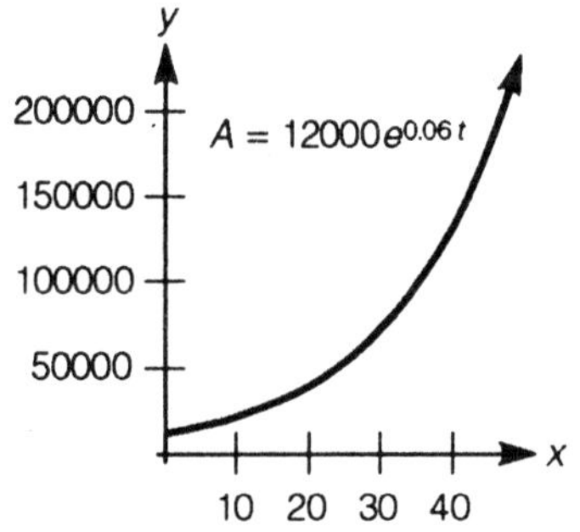

16. (a) Since $\ln x$ must be defined, this is true on the interval $(0, \infty)$. Note that $\ln(x^2) = 2\ln x$ is an identity on the interval.

(b) We solve for x:

$$(\ln x)^2 = 2\ln x$$
$$(\ln x)^2 - 2\ln x = 0$$
$$(\ln x)(\ln x - 2) = 0$$
$$\ln x = 0 \quad \text{or} \quad \ln x = 2$$
$$x = 1 \qquad\qquad x = e^2$$

17. (a) We graph $f(x)$:

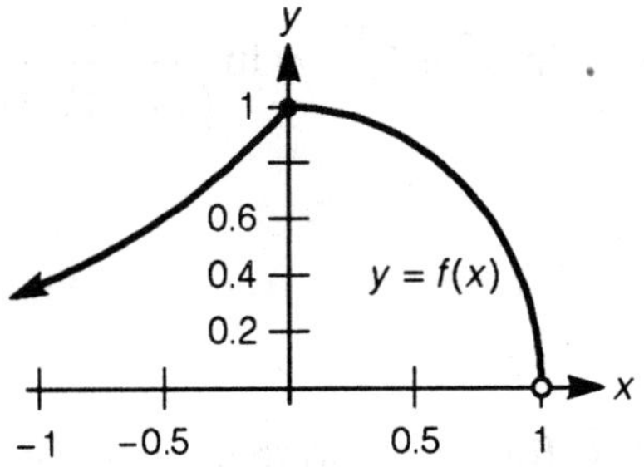

(b) By the horizontal line test, $f(x)$ is not one-to-one.

18. Using properties of logs, we solve for x:

$$\begin{aligned}
\tfrac{1}{2} - \log_{16}(x-3) &= \log_{16} x \\
\tfrac{1}{2} &= \log_{16} x + \log_{16}(x-3) \\
\tfrac{1}{2} &= \log_{16}[x(x-3)] \\
16^{1/2} &= x(x-3) \\
4 &= x^2 - 3x \\
0 &= x^2 - 3x - 4 \\
0 &= (x-4)(x+1) \\
x &= -1, 4
\end{aligned}$$

Since $\log_{10}(-1)$ is undefined, $x = -1$ is an extraneous root. So the solution is $x = 4$.

Chapter Seven
Systems of Equations

7.1 Systems of Two Linear Equations in Two Unknowns

1. (a) yes
(b) no--the xy term makes it non-linear
(c) yes
(d) yes

3. We test (5, 1) in the two equations and see if it produces true statements:

$2(5)-8(1)=2$
$10-8=2$ true
$3(5)+7(1)=22$
$15+7=22$ true

5. We test $(0,-4)$ in the two equations and see if it produces true statements:

$\frac{1}{6}(0)+\frac{1}{2}(-4)=-2$
$0-2=-2$ true
$\frac{2}{3}(0)+\frac{3}{4}(-4)=2$
$0-3=2$ false

So $(0,-4)$ is not a solution to the given system.

7. We test (3, –2) in the two equations and see if it produces true statements:

$$\tfrac{2}{7}(3)-\tfrac{1}{5}(-2)=\tfrac{44}{35}$$
$$\tfrac{6}{7}+\tfrac{2}{5}=\tfrac{44}{35}$$
$$\tfrac{30}{35}+\tfrac{14}{35}=\tfrac{44}{35} \qquad \text{true}$$
$$\tfrac{1}{3}(3)-\tfrac{5}{4}(-2)=\tfrac{7}{2}$$
$$1+\tfrac{5}{2}=\tfrac{7}{2}$$
$$\tfrac{2}{2}+\tfrac{5}{2}=\tfrac{7}{2} \qquad \text{true}$$

So (3, –2) is a solution to the given system.

9. We solve the second equation for x:

$$x+4y=-4$$
$$x=-4-4y$$

Now substitute $x=-4-4y$ for x in the first equation, and solve for y:

$$3x-2y=-19$$
$$3(-4-4y)-2y=-19$$
$$-12-12y-2y=-19$$
$$-14y-12=-19$$
$$-14y=-7$$
$$y=\tfrac{1}{2}$$

We substitute this back into the second equation:

$$x=-4-4y=-4-(4)\tfrac{1}{2}=-4-2=-6$$

So the solution is $\left(-6,\tfrac{1}{2}\right)$.

11. We solve the first equation for y:

$$4x+2y=3$$
$$2y=3-4x$$
$$y=\frac{3-4x}{2}$$

Now substitute $y=\frac{3-4x}{2}$ for y in the second equation, and solve for x:

$$10x+4y=1$$
$$10x+4\left(\frac{3-4x}{2}\right)=1$$
$$10x+2(3-4x)=1$$
$$10x+6-8x=1$$
$$2x+6=1$$
$$2x=-5, \text{ so } x=-\tfrac{5}{2}$$

We substitute this back into the first equation:

$$y=\frac{3-4x}{2}=\frac{3-4\left(-\frac{5}{2}\right)}{2}=\frac{3+10}{2}=\frac{13}{2}$$

The solution is $\left(-\tfrac{5}{2},\tfrac{13}{2}\right)$.

13. We solve the second equation for y:

$$-7x+2y=0$$
$$2y=7x$$
$$y=\frac{7x}{2}$$

Now substitute $y=\frac{7x}{2}$ for y in the first equation, and solve for x:

$$13x-8y=-3$$
$$13x-(8)\frac{7x}{2}=-3$$
$$13x-28x=-3$$
$$-15x=-3$$
$$x=\tfrac{1}{5}$$

We substitute this back into the second equation:

$$y=\frac{7x}{2}=\frac{(7)\frac{1}{5}}{2}=\frac{7}{10}$$

The solution is $\left(\frac{1}{5},\frac{7}{10}\right)$.

15. First we multiply each equation by 20 to clear fractions:

$$20\left(-\tfrac{2}{5}x+\tfrac{1}{4}y\right)=20(3)$$
$$20\left(\tfrac{1}{4}x-\tfrac{2}{5}y\right)=20(-3)$$

Therefore we have the system:

$$-8x+5y=60$$
$$5x-8y=-60$$

We solve the first equation for y:

$$-8x+5y=60$$
$$5y=8x+60$$
$$y=\frac{8x+60}{5}$$

Now substitute $y=\frac{8x+60}{5}$ for y in the second equation and solve for x:

$$5x-8y=-60$$
$$5x-8\left(\frac{8x+60}{5}\right)=-60$$

Multiply by 5 to clear fractions:

$$25x-8(8x+60)=-300$$
$$25x-64x-480=-300$$
$$-39x=180$$
$$x=-\frac{180}{39}=-\frac{60}{13}$$

We substitute this back into the first equation:

$$y=\frac{8x+60}{5}=\frac{8\left(-\frac{60}{13}\right)+60}{5}=\frac{-480+780}{65}=\frac{300}{65}=\frac{60}{13}$$

The solution is $\left(-\frac{60}{13},\frac{60}{13}\right)$.

17. We solve the first equation for x:

$$\sqrt{2}x - \sqrt{3}y = \sqrt{3}$$
$$\sqrt{2}x = \sqrt{3}y + \sqrt{3}$$
$$x = \frac{\sqrt{3}y + \sqrt{3}}{\sqrt{2}}$$

Now substitute $x = \frac{\sqrt{3}y+\sqrt{3}}{\sqrt{2}}$ for x in the second equation and solve for y:

$$\sqrt{3}x - \sqrt{8}y = \sqrt{2}$$
$$\sqrt{3}\left(\frac{\sqrt{3}y + \sqrt{3}}{\sqrt{2}}\right) - \sqrt{8}y = \sqrt{2}$$
$$\frac{3y+3}{\sqrt{2}} - \sqrt{8}y = \sqrt{2}$$

Multiply by $\sqrt{2}$ to clear the fractions:

$$3y + 3 - 4y = 2$$
$$3 - y = 2$$
$$-y = -1$$
$$y = 1$$

We substitute this back into the first equation:

$$x = \frac{\sqrt{3}y + \sqrt{3}}{\sqrt{2}} = \frac{\sqrt{3} + \sqrt{3}}{\sqrt{2}} = \frac{2\sqrt{3}}{\sqrt{2}} = \frac{2\sqrt{6}}{2} = \sqrt{6}$$

The solution is $(\sqrt{6}, 1)$.

19. We multiply the second equation by 2:

$$5x + 6y = 4$$
$$4x - 6y = -6$$

Adding, we obtain:

$$9x = -2$$
$$x = -\tfrac{2}{9}$$

We substitute $x = -\frac{2}{9}$ into the first equation:

$$5x + 6y = 4$$
$$5\left(-\tfrac{2}{9}\right) + 6y = 4$$
$$-\tfrac{10}{9} + 6y = 4$$
$$6y = \tfrac{46}{9}$$
$$y = \tfrac{46}{54} = \tfrac{23}{27}$$

The solution is $\left(-\frac{2}{9}, \frac{23}{27}\right)$.

21. We multiply the second equation by –2:

$$4x + 13y = -5$$
$$-4x + 108y = 2$$

Adding, we obtain:

$$121y = -3$$
$$y = -\frac{3}{121}$$

We substitute $y = -\frac{3}{121}$ into the first equation:

$$4x + 13y = -5$$
$$4x + 13\left(-\frac{3}{121}\right) = -5$$
$$4x - \frac{39}{121} = -5$$
$$4x = -\frac{566}{121}$$
$$x = -\frac{283}{242}$$

The solution is $\left(-\frac{283}{242}, -\frac{3}{121}\right)$.

23. We multiply the first equation by 12 and the second equation by 70 to clear fractions:

$$12\left(\frac{1}{4}x - \frac{1}{3}y\right) = 12(4)$$
$$70\left(\frac{2}{7}x - \frac{1}{7}y\right) = 70\left(\frac{1}{10}\right)$$

Therefore we have the system:

$$3x - 4y = 48$$
$$20x - 10y = 7$$

We multiply the first equation by –5 and the second equation by 2:

$$-15x + 20y = -240$$
$$40x - 20y = 14$$

Adding, we obtain:

$$25x = -226$$
$$x = -\frac{226}{25}$$

We substitute $x = -\frac{226}{25}$ into the second equation:

$$20x - 10y = 7$$
$$20\left(-\frac{226}{25}\right) - 10y = 7$$
$$-\frac{904}{5} - 10y = 7$$
$$-10y = \frac{939}{5}$$
$$y = -\frac{939}{50}$$

The solution is $\left(-\frac{226}{25}, -\frac{939}{50}\right)$.

25. We multiply the second equation by -4:

$$8x + 16y = 5$$
$$-8x - 20y = -5$$

Adding, we obtain:

$$-4y = 0, \text{ so } y = 0$$

We substitute $y = 0$ into the first equation:

$$8x+16y=5$$
$$8x=5$$
$$x=\tfrac{5}{8}$$

The solution is $\left(\tfrac{5}{8},0\right)$.

27. We divide the first equation by 5 and multiply the second equation by 10 to get:

$$25x-8y=9$$
$$x+\ y=3$$

Multiply the second equation by 8:

$$25x-8y=9$$
$$8x+8y=24$$

Adding, we obtain:

$$33x=33$$
$$x=1$$

Now substitute $x = 1$ into the second equation:

$$x+y=3$$
$$1+y=3$$
$$y=2$$

The solution is $(1,2)$.

29. The given points must satisfy the equation, so we have:

$$4=0^2=b\bullet 0+c$$
$$4=c$$

For the other point we have:

$$14=2^2+b\bullet 2+c$$
$$10=2b+c$$

This system, $c = 4$ and $2b + c = 10$, is easily solved by substitution:

$$2b+c=10$$
$$2b+4=10$$
$$2b=6$$
$$b=3$$
$$y=x^2+3x+4$$

So the equation of the parabola is $y = x^2 + 3x + 4$.

31. Again, the points satisfy the equation:

$$Ax+By=2$$
$$A(-4)+B(5)=2$$
$$A(7)+B(-9)=2$$

So we have the system:

$$-4A+5B=2$$
$$7A-9B=2$$

Multiply the first equation by 7 and the second equation by 4:

$$\begin{aligned} -28A + 35B &= 14 \\ 28A - 36B &= 8 \\ -B &= 22 \\ B &= -22 \end{aligned}$$

Substituting for B:

$$\begin{aligned} 7A - 9(-22) &= 2 \\ 7A + 198 &= 2 \\ 7A &= -196 \\ A &= -28 \end{aligned}$$

33. First we graph the region:

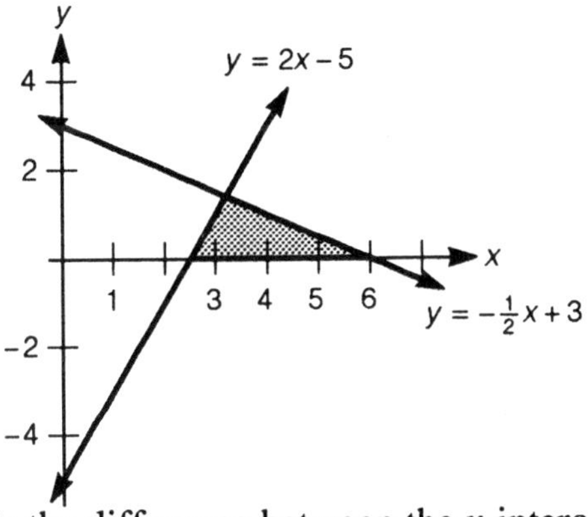

The base of the triangle is the difference between the x-intercepts, so:

$$\text{base} = 6 - \tfrac{5}{2} = \tfrac{7}{2}$$

The height is the y-coordinate of the intersection of the two lines $y = -\frac{1}{2}x + 3$ and $y = 2x - 5$, so:

$$\begin{aligned} -\tfrac{1}{2}x + 3 &= 2x - 5 \\ 8 &= \tfrac{5}{2}x \\ \tfrac{16}{5} &= x \end{aligned}$$

So $y = 2\left(\frac{16}{5}\right) - 5 = \frac{32}{5} - 5 = \frac{7}{5}$. So the area is:

$$\tfrac{1}{2}(\text{base})(\text{height}) = \tfrac{1}{2}\left(\tfrac{7}{2}\right)\left(\tfrac{7}{5}\right) = \tfrac{49}{20} \text{ square units}$$

35. Let x = the amount of the 10% solution and y = the amount of the 35% solution. Then $x + y = 200$ cc. Also we know that the amount of acid in each separate solution, 10% of x and 35% of y, must add to equal the acid in the mixture, 25% of $(x + y)$. This second equation is usually in need of simplifying:

$$\begin{aligned} 0.10x + 0.35y &= 0.25(x + y) \\ 10x + 35y &= 25(x + y) \\ 2x + 7y &= 5x + 5y \\ -3x + 2y &= 0 \end{aligned}$$

This system is now solved by either method:

$$\begin{aligned} x + y &= 200 \\ -3x + 2y &= 0 \end{aligned}$$

Multiply the first equation by 3:

$$3x+3y=600$$
$$-3x+2y=0$$

Adding, we obtain:

$$5y=600$$
$$y=120$$

So $x = 80$, and we need 80 cc of the 10% solution and 120 cc of the 35% solution.

37. Let x be the amount of \$5.20 coffee and y be the amount of \$5.80 coffee. Then $x + y = 16$ pounds. The total value of each bean is $5.20x$ and $5.80y$, respectively, and that of the mixture is $5.50(x + y)$, so we have:

$$5.20x+5.80y=5.50(x+y)$$
$$52x+58y=55(x+y)$$
$$52x+58y=55x+55y$$
$$-3x+3y=0$$
$$-x+y=0$$

We solve the system:

$$x+y=16$$
$$-x+y=0$$

Adding, we obtain:

$$2y=16$$
$$y=8$$

Since $x = 16 - y$, then $x = 16 - 8 = 8$. So we mix 8 pounds of \$5.20 coffee and 8 pounds of \$5.80 coffee.

39. Eliminating fractions, we have:

$$bx+ay=ab$$
$$ax+by=ab$$

Therefore:

$$abx+a^2y=a^2b$$
$$-abx-b^2y=-ab^2$$

Adding, we obtain:

$$(a^2-b^2)y=a^2b-ab^2$$
$$y=\frac{ab(a-b)}{a^2-b^2}$$

If we factor and reduce, we have $y=\frac{ab}{a+b}$. Now substitute to solve for x:

$$\frac{x}{a}+\frac{\frac{ab}{a+b}}{b}=1$$
$$\frac{x}{a}+\frac{a}{a+b}=1$$
$$(a+b)x+a^2=a(a+b)$$
$$(a+b)x=a^2+ab-a^2$$
$$x=\frac{ab}{a+b}$$

The solution is $\left(\frac{ab}{a+b},\frac{ab}{a+b}\right)$. Note that we cannot have $a=\pm b$ for this solution.

41. We multiply the first equation by b and the second equation by $-a$:

$$abx + a^2by = b$$
$$-abx - ab^2y = -a$$

Adding, we obtain:

$$(a^2b - ab^2)y = b - a$$
$$y = \frac{b-a}{a^2b - ab^2} = \frac{-1}{ab}$$

We substitute $y = \frac{-1}{ab}$ into the first equation:

$$ax + a^2y = 1$$
$$ax - \frac{a}{b} = 1$$
$$ax = \frac{a+b}{b}$$
$$x = \frac{a+b}{ab}$$

So the solution is $\left(\frac{a+b}{ab}, \frac{-1}{ab}\right)$. Note that we cannot have $ab = 0$, so $a \neq 0$ and $b \neq 0$.

43. Let $x = \frac{1}{s}$ and $y = \frac{1}{t}$. Then we have the system:

$$\tfrac{1}{2}x - \tfrac{1}{2}y = -10$$
$$2x + 3y = 5$$

Multiply the first equation by -4:

$$-2x + 2y = 40$$
$$2x + 3y = 5$$

Adding, we obtain:

$$5y = 45$$
$$y = 9$$

We substitute $y = 9$ into the second equation, and solve for x:

$$2x + 3y = 5$$
$$2x + 3(9) = 5$$
$$2x + 27 = 5$$
$$2x = -22$$
$$x = -11$$

Since $s = \frac{1}{x}$ and $t = \frac{1}{y}$, then $s = -\frac{1}{11}$ and $t = \frac{1}{9}$. So the solution is $\left(-\frac{1}{11}, \frac{1}{9}\right)$.

45. First we clear fractions and put into a standard form:

$$\frac{2w-1}{3} + \frac{z+2}{4} = 4$$

Multiplying by 12 we have the equation:

$$4(2w-1) + 3(z+2) = 12 \bullet 4$$
$$8w - 4 + 3z + 6 = 48$$
$$8w + 3z = 46$$

The second equation will be $w + 2z = 9$. So we have the system:

$$8w+3z=46$$
$$w+2z=9$$

Multiply the second equation by -8:

$$8w+\ \ 3z=46$$
$$-8w-16z=-72$$

Adding, we obtain:

$$-13z=-26$$
$$z=2$$
$$w=9-2(2)=5$$

So the solution is $(w, z) = (5, 2)$.

47. Using substitution, we solve the first equation for x:

$$1.03x-2.54y=5.47$$
$$1.03x=2.54y+5.47$$
$$x=2.4660y+5.3107$$

Now substitute into the second equation:

$$3.85x+4.29y=-1.84$$
$$3.85(2.466y+5.3107)+4.29y=-1.84$$
$$13.7841y=-22.2862$$
$$y\approx-1.62$$

Substituting back for x yields:

$$x=2.4660(-1.62)+5.3107\approx1.32$$

So the solution is $(1.32, -1.62)$.

49. Letting $u = \ln x$ and $v = \ln y$, we have the system:

$$2u-5v=11$$
$$u+\ v=-5$$

Using substitution, we solve the second equation for v to obtain $v = -5 - u$.
Now substitute into the first equation:

$$2u-5(-5-u)=11$$
$$7u+25=11$$
$$7u=-14$$
$$u=-2$$
$$v=-5+2=-3$$

So $\ln x = -2$ thus $x = e^{-2}$, and $\ln y = -3$ thus $y = e^{-3}$. The solution is (e^{-2}, e^{-3}), which is approximately $(0.14, 0.05)$.

51. Letting $u = e^x$ and $v = e^y$, we have the system:

$$u-3v=2$$
$$3u+\ v=16$$

Using substitution, we solve the first equation for u to obtain $u = 2 + 3v$.

Now substitute into the second equation:

$$\begin{aligned} 3(2+3v)+v&=16\\ 10v+6&=16\\ 10v&=10\\ v&=1\\ u&=5 \end{aligned}$$

So $e^x = 5$ thus $x = \ln 5$, and $e^y = 1$ thus $y = 0$. The solution is $(\ln 5, 0)$, which is approximately $(1.61, 0)$.

53. Letting $u = \sqrt{x^2 - 3x}$ and $v = \sqrt{y^2 + 6y}$, we have the system:

$$\begin{aligned} 4u - 3v &= -4\\ \tfrac{1}{2}u + \tfrac{1}{2}v &= 3 \end{aligned}$$

Multiplying the second equation by 6, we have the system:

$$\begin{aligned} 4u - 3v &= -4\\ 3u + 3v &= 18 \end{aligned}$$

Adding we obtain $7u = 14$, so $u = 2$. Substitute into the first equation:

$$\begin{aligned} 4(2) - 3v &= -4\\ 8 - 3v &= -4\\ -3v &= -12\\ v &= 4 \end{aligned}$$

So $u = 2$ and $v = 4$. Since $u = \sqrt{x^2 - 3x}$ and $v = \sqrt{y^2 + 6y}$, we have the two equations:

$$\begin{aligned} \sqrt{x^2 - 3x} &= 2 & \sqrt{y^2 + 6y} &= 4\\ x^2 - 3x &= 4 & y^2 + 6y &= 16\\ x^2 - 3x - 4 &= 0 & y^2 + 6y - 16 &= 0\\ (x-4)(x+1) &= 0 & (y+8)(y-2) &= 0\\ x &= 4, -1 & y &= -8, 2 \end{aligned}$$

So the solutions are $(4, -8)$, $(4, 2)$, $(-1, -8)$, and $(-1, 2)$.

55. If we take tu as our number, then it is important to distinguish between its value, $10t + u$, and the sum of its digits $t + u$. Here we are told that $t + u = 14$ and that $2t = u + 1$. We solve this system:

$$\begin{aligned} t + u &= 14\\ 2t - u &= 1 \end{aligned}$$

Adding, we obtain:

$$3t = 15, \text{ so } t = 5$$

Thus $u = 9$ and our original two-digit number was 59.

57. Let l and w be the length and width, respectively, of the rectangle. Since the perimeter is 34 in., then $2l + 2w = 34$, or $l + w = 17$. We are also given that $l = 2 + 2w$, so we substitute:

$$\begin{aligned} l + w &= 17\\ (2 + 2w) + w &= 17\\ 2 + 3w &= 17\\ 3w &= 15, \text{ so } w = 5 \end{aligned}$$

Thus $l = 17 - w = 17 - 5 = 12$. Thus the length is 12 in. and the width is 5 in.

59. (a) We solve the first equation for y:

$$x+y=100$$
$$y=100-x$$

Now substituting into the second equation:

$$-2x+3y=0$$
$$-2x+3(100-x)=0$$
$$-2x+300-3x=0$$
$$300-5x=0$$
$$300=5x, \text{ so } x=60$$

Then $y=100-60=40$. Thus the solution is $(60,40)$.

(b) We multiply the first equation by 2:

$$2x+2y=200$$
$$-2x+3y=0$$

Adding, we obtain:

$$5y=200, \text{ so } y=40$$

Substituting back into the first equation:

$$x+y=100$$
$$x+40=100$$
$$x=60$$

Again, the solution is $(60,40)$.

61. We solve each equation for x:

$$x=by-ab$$
$$x=cy-ac$$

Setting these equal:

$$by-ab=cy-ac$$
$$by-cy=ab-ac$$
$$y(b-c)=a(b-c)$$
$$y=a, \text{ as long as } b \neq c$$

Plugging into the second equation, we have:

$$x=cy-ac=ac-ac=0$$

So the solution is $(0,a)$.

63. Let $u=\frac{1}{x}$ and $v=\frac{1}{y}$. Then we have:

$$\tfrac{a}{b}u+\tfrac{b}{a}v=a+b$$
$$bu+av=a^2+b^2$$

Multiply the first equation by ab to clear the fractions:

$$a^2u+b^2v=a^2b+ab^2$$
$$bu+av=a^2+b^2$$

We multiply the first equation by $-b$ and the second equation by a^2:

$$-a^2bu-b^3v=-a^2b^2-ab^3$$
$$a^2bu+a^3v=a^4+a^2b^2$$

Adding, we obtain:

$$(a^3 - b^3)v = a^4 - ab^3$$
$$(a^3 - b^3)v = a(a^3 - b^3)$$
$$v = a$$

Substituting $v = a$ into the second equation, we have:

$$bu + av = a^2 + b^2$$
$$bu + a^2 = a^2 + b^2$$
$$bu = b^2$$
$$u = b$$

Since $x = \frac{1}{u}$ and $y = \frac{1}{v}$, then $x = \frac{1}{b}$ and $y = \frac{1}{a}$. So the solution is $\left(\frac{1}{b}, \frac{1}{a}\right)$.

65. Since the lines are concurrent, the following two systems must possess the same solution:

$I.\quad 7x + 5y = 4$, $x + ky = 3$ $\qquad$ $II.\quad x + ky = 3$, $5x + y = -k$

Using either method of this section, we find that the solution of system I is $\left(\frac{4k-15}{7k-5}, \frac{17}{7k-5}\right)$.

Also the solution of system II is found to be $\left(\frac{-3-k^2}{5k-1}, \frac{k+15}{5k-1}\right)$. Since the two solutions are the same, the corresponding x- and y-coordinates must be equal. Equating the y-coordinates gives us:

$$\frac{17}{7k-5} = \frac{k+15}{5k-1}$$

After clearing fractions and simplifying, this equation becomes $7k^2 + 15k - 58 = 0$. We factor to get $(7k + 29)(k - 2) = 0$ and therefore $k = -\frac{29}{7}$ or $k = 2$. These are the required values of k. If we equate the x-coordinates rather than the y-coordinates we obtain the equation $7k^3 + 15k^2 - 58k = 0$. We factor to get $k(7k + 29)(k - 2) = 0$, which has solutions $k = 0, -\frac{29}{7}$, and 2. The root $k = 0$ is extraneous, for in that case the two y-coordinates are not equal.

67. (a) We first draw the triangle:

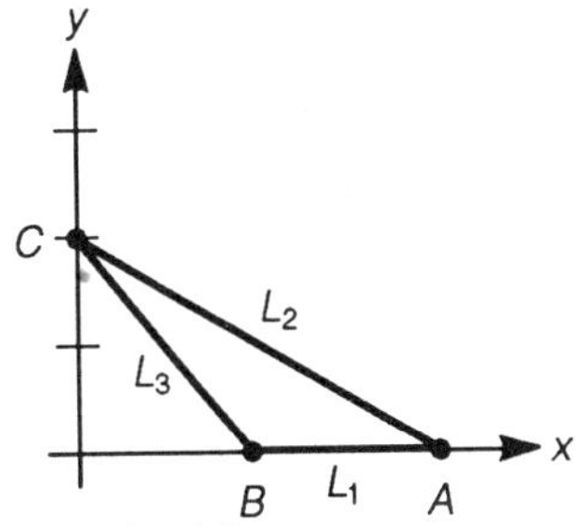

We find the equations for each side:

L_1: $\quad y = 0$

L_2: $\quad m = \dfrac{0-2}{2a-0} = -\dfrac{1}{a}$

Therefore:

$$y = -\frac{1}{a}x + 2$$
$$ay = -x + 2a$$
$$x + ay = 2a$$

L_3: $$m = \frac{0-2}{2b-0} = -\frac{1}{b}$$

Therefore:

$$y = -\frac{1}{b}x + 2$$
$$by = -x + 2b$$
$$x + by = 2b$$

(b) We first find the midpoints of each side:

L_1: $$\text{mid} = \left(\frac{2b+2a}{2}, \frac{0}{2}\right) = (a+b, 0)$$

L_2: $$\text{mid} = \left(\frac{0+2a}{2}, \frac{2+0}{2}\right) = (a, 1)$$

L_3: $$\text{mid} = \left(\frac{0+2b}{2}, \frac{2+0}{2}\right) = (b, 1)$$

Now we find the slopes from each vertex to the adjacent midpoint:

CL_1: $$m = \frac{0-2}{a+b-0} = \frac{-2}{a+b}$$

BL_2: $$m = \frac{1-0}{a-2b} = \frac{1}{a-2b}$$

AL_3: $$m = \frac{1-0}{b-2a} = \frac{1}{b-2a}$$

Finally, we use these slopes and vertices in the point-slope formula:

For CL_1, we have:

$$y - 2 = \frac{-2}{a+b}(x - 0)$$
$$y = \frac{-2}{a+b}x + 2$$
$$(a+b)y = -2x + 2(a+b)$$
$$2x + (a+b)y = 2(a+b)$$

For BL_2, we have:

$$y - 0 = \frac{1}{a-2b}(x - 2b)$$
$$y = \frac{1}{a-2b}x - \frac{2b}{a-2b}$$
$$(a-2b)y = x - 2b$$
$$-x + (a-2b)y = -2b$$
$$x + (2b-a)y = 2b$$

For AL_3, we have:

$$y-0=\frac{1}{b-2a}(x-2a)$$
$$y=\frac{1}{b-2a}x-\frac{2a}{b-2a}$$
$$(b-2a)y=x-2a$$
$$-x+(b-2a)y=-2a$$
$$x+(2a-b)y=2a$$

(c) We find the intersection of the three lines:

$$2x+\ (a+b)y=2(a+b)$$
$$x+(2b-a)y=2b$$
$$x+(2a-b)y=2a$$

We will solve the second and third equations for x:

$$x=2b-(2b-a)y$$
$$x=2a-(2a-b)y$$

Setting these equal:

$$2b-(2b-a)y=2a-(2a-b)y$$
$$(2a-b)y-(2b-a)y=2a-2b$$
$$(2a-b-2b+a)y=2a-2b$$
$$(3a-3b)y=2a-2b$$
$$y=\frac{2a-2b}{3a-3b}=\frac{2(a-b)}{3(a-b)}=\frac{2}{3}$$

We substitute $y=\frac{2}{3}$ into the second equation:

$$x=2b-\tfrac{2}{3}(2b-a)=\frac{6b-4b+2a}{3}=\frac{2b+2a}{3}=\frac{2(a+b)}{3}$$

We must still show that this point $\left(\frac{2(a+b)}{3},\frac{2}{3}\right)$ lies on the first line:

$$2x+(a+b)y=2(a+b)$$
$$4\left(\frac{a+b}{3}\right)+(a+b)\frac{2}{3}=2(a+b)$$
$$6\frac{(a+b)}{3}=2(a+b)\quad \text{true}$$

Thus the point $\left(\frac{2(a+b)}{3},\frac{2}{3}\right)$ is the intersection of all three medians.

(d) The altitudes will be perpendicular to the slopes of each side, which we found in (a). So:

For BL_2, we have:

$$m = a$$
$$y - 0 = a(x - 2b)$$
$$y = ax - 2ab$$
$$-ax + y = -2ab$$
$$ax - y = 2ab$$

For CL_2, we have:

$$m = \text{ undefined}$$
$$x = 0$$

For AL_3, we have:

$$m = b$$
$$y - 0 = b(x - 2a)$$
$$y = bx - 2ab$$
$$-bx + y = -2ab$$
$$bx - y = 2ab$$

(e) We find the intersection of the three lines:

$$ax - y = 2ab$$
$$bx - y = 2ab$$
$$x = 0$$

Since $x = 0$, substituting into the first equation yields:

$$-y = 2ab$$
$$y = -2ab$$

We must still show that this point $(0, -2ab)$ lies on the second line:

$$bx - y = 2ab$$
$$b(0) - (-2ab) = 2ab$$
$$2ab = 2ab \quad \text{true}$$

Thus, the point $(0, -2ab)$ is the intersection of all three altitudes.

(f) Recall that the perpendicular bisectors must be perpendicular and pass through the midpoints (which we found in (b)):

For L_1, we have:

$$\text{mid} = (a + b, 0)$$
$$m = \text{undefined}$$
$$x = a + b$$

For L_2, we have:

$$\text{mid} = (a, 1)$$
$$m = a$$
$$y - 1 = a(x - a)$$
$$y - 1 = ax - a^2$$
$$-ax + y = 1 - a^2$$
$$ax - y = a^2 - 1$$

For L_3, we have:

$$\text{mid} = (b,1)$$
$$m = b$$
$$y - 1 = b(x - b)$$
$$y - 1 = bx - b^2$$
$$-bx + y = 1 - b^2$$
$$bx - y = b^2 - 1$$

(g) We find the intersection of the three lines:

$$ax - y = a^2 - 1$$
$$bx - y = b^2 - 1$$
$$x = a + b$$

Substituting into the first equation yields:

$$a(a+b) - y = a^2 - 1$$
$$a^2 + ab - y = a^2 - 1$$
$$-y = -ab - 1$$
$$y = ab + 1$$

We must still show that the point $(a + b, ab + 1)$ lies on the second line:

$$bx - y = b^2 - 1$$
$$b(a+b) - (ab+1) = b^2 - 1$$
$$ab + b^2 - ab - 1 = b^2 - 1$$
$$b^2 - 1 = b^2 - 1 \quad \text{true}$$

Thus the point $(a + b, ab + 1)$ is the intersection of all three perpendicular bisectors.

(h) We find the line that passes through $\left(\frac{2(a+b)}{3}, \frac{2}{3}\right)$, $(0, -2ab)$, and $(a + b, ab + 1)$. We work with the last two points:

$$m = \frac{ab + 1 - (-2ab)}{a + b - 0} = \frac{3ab + 1}{a + b}$$

Since $(0, -2ab)$ is the y-intercept, then the line should be:

$$y = \frac{3ab + 1}{a + b}x - 2ab$$

We must still show that the point $\left(\frac{2(a+b)}{3}, \frac{2}{3}\right)$ lies on this line:

$$\tfrac{2}{3} = \left(\frac{3ab+1}{a+b}\right)\left(\frac{2(a+b)}{3}\right) - 2ab$$
$$\tfrac{2}{3} = \frac{2(3ab+1)}{3} - 2ab$$
$$\tfrac{2}{3} = \frac{6ab + 2 - 6ab}{3}$$
$$\tfrac{2}{3} = \tfrac{2}{3} \quad \text{true}$$

Thus $y = \dfrac{3ab+1}{a+b}x - 2ab$ is the Euler line.

7.2 Gaussian Elimination

1. We are given the system:

$$\begin{aligned} 2x+y+\ \ z&=-9 \\ 3y-2z&=-4 \\ 8z&=-8 \end{aligned}$$

Solve for z in the third equation:

$$\begin{aligned} 8z&=-8 \\ z&=-1 \end{aligned}$$

Substitute into the second equation:

$$\begin{aligned} 3y-2z&=-4 \\ 3y-2(-1)&=-4 \\ 3y+2&=-4 \\ 3y&=-6 \\ y&=-2 \end{aligned}$$

Substitute into the first equation:

$$\begin{aligned} 2x+y+z&=-9 \\ 2x+(-2)+(-1)&=-9 \\ 2x-3&=9 \\ 2x&=-6 \\ x&=-3 \end{aligned}$$

So the solution is $(-3,-2,-1)$.

3. We are given the system:

$$\begin{aligned} 8x+5y+3z&=1 \\ 3y+4z&=2 \\ 5z&=3 \end{aligned}$$

Solve for z in the third equation:

$$\begin{aligned} 5z&=3 \\ z&=\tfrac{3}{5} \end{aligned}$$

Substitute into the second equation:

$$\begin{aligned} 3y+4z&=2 \\ 3y+(4)\tfrac{3}{5}&=2 \\ 3y+\tfrac{12}{5}&=2 \\ 3y&=-\tfrac{2}{5} \\ y&=-\tfrac{2}{15} \end{aligned}$$

Substitute into the first equation:

$$\begin{aligned} 8x+5y+3z&=1 \\ 8x+5\left(-\tfrac{2}{15}\right)+(3)\tfrac{3}{5}&=1 \\ 8x-\tfrac{2}{3}+\tfrac{9}{5}&=1 \\ 8x-\tfrac{10}{15}+\tfrac{27}{15}&=1 \\ 8x&=-\tfrac{2}{15} \\ x&=-\tfrac{1}{60} \end{aligned}$$

So the solution is $\left(-\tfrac{1}{60},-\tfrac{2}{15},\tfrac{3}{5}\right)$.

5. We are given the system:

$$-4x+5y=0$$
$$3y+2z=1$$
$$3z=-1$$

Solve for z in the third equation:

$$3z=-1$$
$$z=-\tfrac{1}{3}$$

Substitute into the second equation:

$$3y+2z=1$$
$$3y+2\left(-\tfrac{1}{3}\right)=1$$
$$3y-\tfrac{2}{3}=1$$
$$3y=\tfrac{5}{3}$$
$$y=\tfrac{5}{9}$$

Substitute into the first equation:

$$-4x+5y=0$$
$$-4x+(5)\tfrac{5}{9}=0$$
$$-4x+\tfrac{25}{9}=0$$
$$-4x=-\tfrac{25}{9}$$
$$x=\tfrac{25}{36}$$

So the solution is $\left(\tfrac{25}{36},\tfrac{5}{9},-\tfrac{1}{3}\right)$.

7. We are given the system:

$$-x+8y+3z=0$$
$$2z=0$$

Solve for z in the second equation:

$$2z=0$$
$$z=0$$

Substitute into the first equation:

$$-x+8y+3z=0$$
$$-x+8y+3(0)=0$$
$$-x+8y=0$$
$$8y=x$$
$$y=\frac{x}{8}$$

So the solution is $\left(x,\tfrac{x}{8},0\right)$, where x is any real number.

9. We are given the system:

$$2x+3y+z+w=-6$$
$$y+3z-4w=23$$
$$6z-5w=31$$
$$-2w=10$$

Solve for w in the fourth equation:

$$-2w = 10$$
$$w = -5$$

Substitute into the third equation:

$$6z - 5w = 31$$
$$6z - 5(-5) = 31$$
$$6z + 25 = 31$$
$$6z = 6$$
$$z = 1$$

Substitute into the second equation:

$$y + 3z - 4w = 23$$
$$y + 3(1) - 4(-5) = 23$$
$$y + 23 = 23$$
$$y = 0$$

Substitute into the first equation:

$$2x + 3y + z + w = -6$$
$$2x + 3(0) + 1 + (-5) = -6$$
$$2x - 4 = -6$$
$$2x = -2$$
$$x = -1$$

So the solution is $(-1, 0, 1, -5)$.

11. We must arrange this according to echelon form. First add –2 times the first equation to the second one, and –3 times it to the third:

$$x + y + z = 12$$
$$-3y - 3z = -25$$
$$-y - 2z = -14$$

We want to eliminate the y term from the third equation so interchange the second and third equations:

$$x + y + z = 12$$
$$-y - 2z = -14$$
$$-3y - 3z = -25$$

Add –3 times the second equation to the third:

$$x + y + z = 12$$
$$-y - 2z = -14$$
$$3z = 17$$

Solve for z in the third equation:

$$3z = 17$$
$$z = \tfrac{17}{3}$$

Substitute into the second equation:

$$-y - 2z = -14$$
$$-y - (2)\tfrac{17}{3} = -14$$
$$-y - \tfrac{34}{3} = -14$$
$$-y = -\tfrac{8}{3}$$
$$y = \tfrac{8}{3}$$

Substitute into the first equation:

$$x+y+z=12$$
$$x+\tfrac{8}{3}+\tfrac{17}{3}=12$$
$$x+\tfrac{25}{3}=12$$
$$x=\tfrac{11}{3}$$

So the solution is $\left(\tfrac{11}{3},\tfrac{8}{3},\tfrac{17}{3}\right)$.

13. We must arrange the system in echelon form. Multiply the first equation by –2 and add it to the second equation:

$$-4x+6y-4z=-8$$
$$4x+2y+3z=7$$

Adding, we obtain:

$$8y-z=-1$$

Multiply the first equation by $-\tfrac{5}{2}$ and add it to the third equation:

$$-5x+\tfrac{15}{2}y-5z=-10$$
$$5x+\ 4y+2z=7$$

Adding, we obtain:

$$\tfrac{23}{2}y-3z=-3$$
$$23y-6z=-6$$

So we have the system:

$$\begin{aligned} 2x-3y+2z&=4\\ 8y-\ z&=-1\\ 23y-6z&=-6 \end{aligned}$$

Multiply the second equation by $-\tfrac{23}{8}$ and add it to the third equation:

$$-23y+\tfrac{23}{8}z=\tfrac{23}{8}$$
$$23y-\ 6z=-6$$

Adding, we obtain:

$$-\tfrac{25}{8}z=-\tfrac{25}{8}$$

So we have the system in echelon form:

$$\begin{aligned} 2x-3y+\ 2z&=4\\ 8y-\ \ z&=-1\\ -\tfrac{25}{8}z&=-\tfrac{25}{8} \end{aligned}$$

Solve for z in the third equation:

$$-\tfrac{25}{8}z=-\tfrac{25}{8}$$
$$z=1$$

Substitute into the second equation:

$$8y-z=-1$$
$$8y-1=-1$$
$$8y=0$$
$$y=0$$

Substitute into the first equation:

$$\begin{aligned} 2x-3y+2z&=4\\ 2x-3(0)+2(1)&=4\\ 2x+2&=4\\ 2x&=2\\ x&=1 \end{aligned}$$

So the solution is $(1,0,1)$.

15. We must arrange the system in echelon form. Multiply the first equation by –2 and add it to the second equation:

$$\begin{aligned} -6x-6y+4z&=-26\\ 6x+2y-5z&=13 \end{aligned}$$

Adding, we obtain:

$$-4y-z=-13$$

Multiply the first equation by $-\frac{7}{3}$ and add it to the third equation:

$$\begin{aligned} -7x-7y+\tfrac{14}{3}z&=-\tfrac{91}{3}\\ 7x+5y-\ 3z&=26 \end{aligned}$$

Adding, we obtain:

$$\begin{aligned} -2y+\tfrac{5}{3}z&=-\tfrac{13}{3}\\ -6y+5z&=-13 \end{aligned}$$

So we have the system:

$$\begin{aligned} 3x+3y-2z&=13\\ -4y-\ z&=-13\\ -6y+5z&=-13 \end{aligned}$$

Multiply the second equation by $-\frac{3}{2}$ and add it to the third equation:

$$\begin{aligned} 6y+\tfrac{3}{2}z&=\tfrac{39}{2}\\ -6y+5z&=-13 \end{aligned}$$

Adding, we obtain:

$$\tfrac{13}{2}z=\tfrac{13}{2}$$

So we have the system in echelon form:

$$\begin{aligned} 3x+3y-2z&=13\\ -4y-\ z&=-13\\ \tfrac{13}{2}z&=\tfrac{13}{2} \end{aligned}$$

Solve for z in the third equation:

$$\begin{aligned} \tfrac{13}{2}z&=\tfrac{13}{2}\\ z&=1 \end{aligned}$$

Substitute into the second equation:

$$\begin{aligned} -4y-z&=-13\\ -4y-1&=-13\\ -4y&=-12\\ y&=3 \end{aligned}$$

Substitute into the first equation:

$$\begin{aligned} 3x+3y-2z&=13 \\ 3x+3(3)-2(1)&=13 \\ 3x+7&=13 \\ 3x&=6 \\ x&=2 \end{aligned}$$

So the solution is $(2, 3, 1)$.

17. We must arrange the system in echelon form. Multiply the first equation by 2 and add it to the second equation:

$$\begin{aligned} 2x+2y+2z&=2 \\ -2x+y+z&=-2 \end{aligned}$$

Adding, we obtain:

$$\begin{aligned} 3y+3z&=0 \\ y+z&=0 \end{aligned}$$

Multiply the first equation by –3 and add it to the third equation:

$$\begin{aligned} -3x-3y-3z&=-3 \\ 3x+6y+6z&=5 \end{aligned}$$

Adding, we obtain:

$$3y+3z=2$$

So we have the system:

$$\begin{aligned} x+y+z&=1 \\ y+z&=0 \\ 3y+3z&=2 \end{aligned}$$

Multiply the second equation by –3 and add it to the third equation:

$$\begin{aligned} -3y-3z&=0 \\ 3y+3z&=2 \end{aligned}$$

Adding, we obtain:

$$0=2$$

Since this equation is false, there is no solution. This system is inconsistent.

19. We first re-arrange the system as:

$$\begin{aligned} x+3y-2z&=2 \\ 2x-y+z&=-1 \\ -5x+6y-5z&=5 \end{aligned}$$

Multiply the first equation by –2 and add it to the second equation:

$$\begin{aligned} -2x-6y+4z&=-4 \\ 2x-y+z&=-1 \end{aligned}$$

Adding, we obtain:

$$-7y+5=-5$$

Multiply the first equation by 5 and add it to the third equation:

$$\begin{aligned} 5x+15y-10z&=10 \\ -5x+6y-5z&=5 \end{aligned}$$

Adding, we obtain:

$$\begin{aligned} 21y-15z&=15 \\ 7y-5z&=5 \end{aligned}$$

So we have the system:

$$x+3y-2z=2$$
$$-7y+5z=-5$$
$$7y-5z=5$$

Adding the second and third equations yields $0=0$, so the system is dependent. Solving for x and y in terms of z yields the solution $\left(\frac{z+1}{-7}, \frac{5(z+1)}{7}, z\right)$, where z is any real number.

21. We first re-arrange the system as:

$$x+3y+2z=-1$$
$$2x-y+z=4$$
$$7x+5z=11$$

Multiply the first equation by –2 and add it to the second equation:

$$-2x-6y-4z=2$$
$$2x-y+z=4$$

Adding, we obtain:

$$-7y-3z=6$$

Multiply the first equation by –7 and add it to the third equation:

$$-7x-21y-14z=7$$
$$7x+5z=11$$

Adding, we obtain:

$$-21y-9z=18$$
$$-7y-3z=6$$

So we have the system:

$$x+3y+2z=-1$$
$$-7y-3z=6$$
$$-7y-3z=6$$

Multiply the second equation by –1 and add it to the third equation:

$$7y+3z=-6$$
$$-7y-3z=6$$

Adding, we obtain:

$$0=0$$

So we have the system:

$$x+3y+2z=-1$$
$$-7y-3z=6$$

We solve the second equation for y:

$$-7y-3z=6$$
$$-7y=3z+6$$
$$y=\frac{-3z-6}{7}$$

Substitute into the first equation:

$$\begin{aligned}x+3y+2z&=-1\\ x+3\left(\frac{-3z-6}{7}\right)+2z&=-1\\ x+\frac{-9z-18+14z}{7}&=-1\\ x+\frac{5z-18}{7}&=-1\\ x&=\frac{-7-5z+18}{7}\\ x&=\frac{-5z+11}{7}\end{aligned}$$

So the solution is $\left(\frac{11-5z}{7},\frac{-3z-6}{7},z\right)$, where z is any real number.

23. Multiply the first equation by –1 and add it to the second equation:

$$\begin{aligned}-x-\ y-z-w&=-4\\ x-2y-z-w&=3\end{aligned}$$

Adding, we obtain:

$$\begin{aligned}-3y-2z-2w&=-1\\ 3y+2z+2w&=1\end{aligned}$$

Multiply the first equation by –2 and add it to the third equation:

$$\begin{aligned}-2x-2y-2z-2w&=-8\\ 2x-\ y+\ z-\ w&=2\end{aligned}$$

Adding, we obtain:

$$\begin{aligned}-3y-z-3w&=-6\\ 3y+z+3w&=6\end{aligned}$$

Multiply the first equation by –1 and add it to the fourth equation:

$$\begin{aligned}-x-y-\ z-\ w&=-4\\ x-y+2z-2w&=-7\end{aligned}$$

Adding, we obtain:

$$\begin{aligned}-2y+z-3w&=-11\\ 2y-z+3w&=11\end{aligned}$$

So we have the system:

$$\begin{aligned}x+\ y+\ z+\ w&=4\\ 3y+2z+2w&=1\\ 3y+\ z+3w&=6\\ 2y-\ z+3w&=11\end{aligned}$$

Multiply the second equation by –1 and add it to the third equation:

$$\begin{aligned}-3y-2z-2w&=-1\\ 3y+\ z+3w&=6\end{aligned}$$

Adding, we obtain:

$$\begin{aligned}-z+w&=5\\ z-w&=-5\end{aligned}$$

Multiply the second equation by $-\frac{2}{3}$ and add it to the fourth equation:

$$-2y-\frac{4}{3}z-\frac{4}{3}w=-\frac{2}{3}$$
$$2y-z+3w=11$$

Adding, we obtain:

$$-\frac{7}{3}z+\frac{5}{3}w=\frac{31}{3}$$
$$-7z+5w=31$$

So we have the system:

$$\begin{aligned} x+y+z+w&=4\\ 3y+2z+2w&=1\\ z-w&=-5\\ -7z+5w&=31 \end{aligned}$$

Multiply the third equation by 7 and add it to the fourth equation:

$$7z-7w=-35$$
$$-7z+5w=31$$

Adding, we obtain:

$$-2w=-4$$

So we have the system:

$$\begin{aligned} x+y+z+w&=4\\ 3y+2z+2w&=1\\ z-w&=-5\\ -2w&=-4 \end{aligned}$$

We solve the fourth equation for w:

$$-2w=-4$$
$$w=2$$

Substitute into the third equation:

$$z-w=-5$$
$$z-2=-5$$
$$z=-3$$

Substitute into the second equation:

$$\begin{aligned} 3y+2z+2w&=1\\ 3y+2(-3)+2(2)&=1\\ 3y-2&=1\\ 3y&=3\\ y&=1 \end{aligned}$$

Substitute into the first equation:

$$x+y+z+w=4$$
$$x+1-3+2=4$$
$$x=4$$

So the solution is $(4,1,-3,2)$.

25. We first re-arrange the system as:

$$x+4y-3z=1$$
$$2x+3y+2z=5$$

Multiply the first equation by -2 and add it to the second equation:

$$-2x-8y+6z=-2$$
$$2x+3y+2z=5$$

Adding, we obtain:

$$-5y+8z=3$$

So we have the system:

$$\begin{aligned} x+4y-3z&=1 \\ -5y+8z&=3 \end{aligned}$$

We solve the second equation for y:

$$\begin{aligned} -5y+8z&=3 \\ -5y&=3-8z \\ y&=\frac{8z-3}{5} \end{aligned}$$

Substitute into the first equation:

$$\begin{aligned} x+4y-3z&=1 \\ x+4\left(\frac{8z-3}{5}\right)-3z&=1 \\ x+\frac{17z-12}{5}&=1 \\ x&=\frac{17-17z}{5} \end{aligned}$$

So the solution is $\left(\frac{17-17z}{5},\frac{8z-3}{5},z\right)$, where z is any real number.

27. Multiply the first equation by –3 and add it to the second equation:

$$\begin{aligned} -3x+6y+6z-6w&=30 \\ 3x+4y-\ \ z-3w&=11 \end{aligned}$$

Adding, we obtain:

$$10y+5z-9w=41$$

Multiply the first equation by 4 and add it to the third equation:

$$\begin{aligned} 4x-8y-8z+8w&=-40 \\ -4x-3y-3z+8w&=-21 \end{aligned}$$

Adding, we obtain:

$$-11y-11z+16w=-61$$

So we have the system:

$$\begin{aligned} x-2y-\ 2z+\ 2w&=-10 \\ 10y+\ 5z-\ 9w&=41 \\ -11y-11z+16w&=-61 \end{aligned}$$

Add the second and third equations:

$$\begin{aligned} 10y+\ 5z-\ 9w&=41 \\ -11y-11z+16w&=-61 \end{aligned}$$

Adding, we obtain:

$$\begin{aligned} -y-6z+7w&=-20 \\ y+6z-7w&=20 \end{aligned}$$

So we have the system:

$$\begin{aligned} x-2y-\ 2z+\ 2w&=-10 \\ y+\ 6z-\ 7w&=20 \\ -11y-11z+16w&=-61 \end{aligned}$$

Multiply the second equation by 11 and add it to the third equation:

$$11y+66z-77w=220$$
$$-11y-11z+16w=-61$$

Adding, we obtain:

$$55z-61w=159$$

So we have the system:

$$\begin{aligned} x-2y-2z+2w&=-10 \\ y+6z-7w&=20 \\ 55z-61w&=159 \end{aligned}$$

Solve the third equation for z:

$$\begin{aligned} 55z-61w&=159 \\ 55z&=61w+159 \\ z&=\frac{61w+159}{55} \end{aligned}$$

Substitute into the second equation:

$$\begin{aligned} y+6z-7w&=20 \\ y+6\left(\frac{61w+159}{55}\right)-7w&=20 \\ y+\frac{954-19w}{55}&=20 \\ y&=\frac{146+19w}{55} \end{aligned}$$

Substitute into the first equation:

$$\begin{aligned} x-2y-2z+2w&=-10 \\ x-2\left(\frac{146+19w}{55}\right)-2\left(\frac{61w+159}{55}\right)+2w&=-10 \\ x+\frac{-122-10w}{11}&=-10 \\ x&=\frac{12+10w}{11} \end{aligned}$$

So the solution is $\left(\frac{12+10w}{11},\frac{146+19w}{55},\frac{61w+159}{55},w\right)$, where w is any real number.

29. Solving the second equation for y yields $y=\frac{2z+3}{3}$. We substitute this into the first equation to get $x=-\frac{5z}{12}$. So the solution is $\left(-\frac{5z}{12},\frac{2z+3}{3},z\right)$, where z is any real number.

31. (a) Multiply by $(x-2)(x+2)$, so:

$$\begin{aligned} 1&=A(x+2)+B(x-2) \\ 1&=Ax+2A+Bx-2B \\ 1&=(A+B)x+(2A-2B) \end{aligned}$$

Since A and B are constants (cannot contain x) we must have:

$$\begin{aligned} A+B&=0 \\ 2A-2B&=1 \end{aligned}$$

Multiply the first equation by –2 and add it to the second equation:

$$-2A - 2B = 0$$
$$2A - 2B = 1$$

Adding, we obtain:

$$-4B = 1$$
$$B = -\tfrac{1}{4}$$

Substituting into the first equation:

$$A + B = 0$$
$$A - \tfrac{1}{4} = 0$$
$$A = \tfrac{1}{4}$$

So $A = \frac{1}{4}$ and $B = -\frac{1}{4}$.

(b) Multiply by $(x-2)(x+2)$, so:

$$x = A(x+2) + B(x-2)$$
$$x = Ax + 2A + Bx - 2B$$
$$x = (A+B)x + (2A - 2B)$$

Since A and B are constants we have the system:

$$A + \ \ B = 1$$
$$2A - 2B = 0$$

Multiply the first equation by –2 and add it to the second equation:

$$-2A - 2B = -2$$
$$2A - 2B = 0$$

Adding, we obtain:

$$-4B = -2$$
$$b = \tfrac{1}{2}$$

Substituting into the first equation:

$$A + B = 1$$
$$A + \tfrac{1}{2} = 1$$
$$A = \tfrac{1}{2}$$

So $A = \frac{1}{2}$ and $B = \frac{1}{2}$.

33. Multiply by $(x+4)(x-5)$, so:

$$3x = A(x-5) + B(x+4)$$
$$3x = Ax - 5A + Bx + 4B$$
$$3x = (A+B)x + (-5A + 4B)$$

Since A and B are constants we have the system:

$$A + \ \ B = 3$$
$$-5A + 4B = 0$$

Multiply the first equation by 5 and add it to the second equation:

$$5A + 5B = 15$$
$$-5A + 4B = 0$$

Adding, we obtain:

$$9B = 15$$
$$B = \tfrac{5}{3}$$

Substituting into the first equation:

$$A+B=3$$
$$A+\tfrac{5}{3}=3$$
$$A=\tfrac{4}{3}$$

So $A=\frac{4}{3}$ and $B=\frac{5}{3}$.

35. (a) Multiply by $(x-4)(x+4)$, so:

$$3=A(x+4)+B(x-4)$$
$$3=Ax+4A+Bx-4B$$
$$3=(A+B)x+(4A-4B)$$

Since A and B are constants we have:

$$A+\ B=0$$
$$4A-4B=3$$

Multiply the first equation by -4 and add it to the second equation:

$$-4A-4B=0$$
$$4A-4B=3$$

Adding, we obtain:

$$-8B=3$$
$$B=-\tfrac{3}{8}$$

Substituting into the first equation:

$$A+B=0$$
$$A-\tfrac{3}{8}=0$$
$$A=\tfrac{3}{8}$$

So $A=\frac{3}{8}$ and $B=-\frac{3}{8}$.

(b) Multiply by $(x-4)(x+4)^2$, so:

$$3=A(x+4)^2+B(x-4)(x+4)+C(x-4)$$
$$3=Ax^2+8Ax+16A+Bx^2-16B+Cx-4C$$
$$3=(A+B)x^2+(8A+C)x+(16A-16B-4C)$$

Since A, B, and C are constants we have:

$$A+\ B\ \ =0$$
$$8A\ \ +C=0$$
$$16A-16B-4C=3$$

Multiply the first equation by -8 and add it to the second equation:

$$-8A-8B=0$$
$$8A+C=0$$

Adding, we obtain:

$$-8B+C=0$$

Multiply the first equation by -16 and add it to the third equation:

$$-16A-16B=0$$
$$16A-16B-4C=3$$

Adding, we obtain:

$$-32B-4C=3$$

So we have the system:

$$\begin{aligned} A + B \quad &= 0 \\ -8B + C &= 0 \\ -32B - 4C &= 3 \end{aligned}$$

Multiply the second equation by −4 and add it to the third equation:

$$\begin{aligned} 32B - 4C &= 0 \\ -32B - 4C &= 3 \end{aligned}$$

Adding, we obtain:

$$\begin{aligned} -8C &= 3 \\ C &= -\tfrac{3}{8} \end{aligned}$$

Substitute into the second equation:

$$\begin{aligned} -8B + C &= 0 \\ -8B - \tfrac{3}{8} &= 0 \\ -8B &= \tfrac{3}{8} \\ B &= -\tfrac{3}{64} \end{aligned}$$

Substitute into the first equation:

$$\begin{aligned} A + B &= 0 \\ A - \tfrac{3}{64} &= 0 \\ A &= \tfrac{3}{64} \end{aligned}$$

So $A = \frac{3}{64}$, $B = -\frac{3}{64}$ and $C = -\frac{3}{8}$.

37. Multiply by $(x+1)(x^2 - x + 1)$, so:

$$\begin{aligned} 1 &= A(x^2 - x + 1) + (Bx + C)(x + 1) \\ 1 &= Ax^2 - Ax + A + Bx^2 + Cx + Bx + C \\ 1 &= (A + B)x^2 + (-A + B + C)x + (A + C) \end{aligned}$$

Since A, B, and C are constants we have:

$$\begin{aligned} A + B \quad &= 0 \\ -A + B + C &= 0 \\ A \quad + C &= 1 \end{aligned}$$

Add the first two equations:

$$\begin{aligned} A + B \quad &= 0 \\ -A + B + C &= 0 \end{aligned}$$

Adding, we obtain:

$$2B + C = 0$$

Multiply the first equation by −1 and add it to the third equation:

$$\begin{aligned} -A - B \quad &= 0 \\ A \quad + C &= 1 \end{aligned}$$

Adding, we obtain:

$$-B + C = 1$$

So we have the system:

$$\begin{aligned} A + B \quad &= 0 \\ -B + C &= 1 \\ 2B + C &= 0 \end{aligned}$$

Multiply the second equation by 2 and add it to the third equation:

$$-2B+2C=2$$
$$2B+\ C=0$$

Adding, we obtain:

$$3C=2$$
$$C=\tfrac{2}{3}$$

Substitute into the second equation:

$$-B+C=1$$
$$-B+\tfrac{2}{3}=1$$
$$-B=\tfrac{1}{3}$$
$$B=-\tfrac{1}{3}$$

Substitute into the first equation:

$$A+B=0$$
$$A-\tfrac{1}{3}=0$$
$$A=\tfrac{1}{3}$$

So $A=\frac{1}{3}$, $B=-\frac{1}{3}$ and $C=\frac{2}{3}$.

39. Multiply by $x(1-x)$, so:

$$4=A(1-x)+B(x)$$
$$4=(-A+B)x+A$$

Since A and B are constants we have:

$$-A+B=0$$
$$A\quad\ =4$$

Substituting yields $-4+B=0$, so $B=4$. So $A=4$ and $B=4$.

41. Multiply by $(x^2+x+1)(x^2-x+1)$, so:

$$1=(Ax+B)(x^2-x+1)+(Cx+D)(x^2+x+1)$$
$$1=(A+C)x^3+(-A+B+C+D)x^2+(A-B+C+D)x+(B+D)$$

Since A, B, C, and D are constants we have:

$$A\qquad\quad +C=0$$
$$-A+B+C+D=0$$
$$A-B+C+D=0$$
$$B\qquad +D=1$$

Add the first and second equations:

$$A\quad +C\qquad =0$$
$$-A+B+C+D=0$$

Adding, we obtain:

$$B+2C+D=0$$

Multiply the first equation by -1 and add it to the third equation:

$$-A\quad -C\qquad =0$$
$$A-B+C+D=0$$

Adding, we obtain:

$$-B+D=0$$

So we have the system:

$$\begin{aligned} A \quad + C \quad &= 0 \\ -B \quad + D &= 0 \\ B + 2C + D &= 0 \\ B \quad + D &= 1 \end{aligned}$$

Add the second and third equations:

$$\begin{aligned} -B \quad + D &= 0 \\ B + 2C + D &= 0 \end{aligned}$$

Adding, we obtain:

$$\begin{aligned} 2C + 2D &= 0 \\ C + D &= 0 \end{aligned}$$

Add the second and the fourth equations:

$$\begin{aligned} -B + D &= 0 \\ B + D &= 1 \\ 2D &= 1 \\ D &= \tfrac{1}{2} \end{aligned}$$

So we have the system:

$$\begin{aligned} A \quad + C \quad &= 0 \\ -B \quad + D &= 0 \\ C + D &= 0 \\ D &= \tfrac{1}{2} \end{aligned}$$

Substituting into the third equation yields $C = -\frac{1}{2}$. Substitution into the second equation yields $B = \frac{1}{2}$. Substituting into the first equation yields $A = \frac{1}{2}$. So $A = \frac{1}{2}$, $B = \frac{1}{2}$, $C = -\frac{1}{2}$, and $D = \frac{1}{2}$.

43. We know its form is $(x - h)^2 + (y - k)^2 = r^2$, so:

$$\begin{aligned} (-1 - h)^2 + (-5 - k)^2 &= r^2 \\ (1 - h)^2 + (2 - k)^2 &= r^2 \\ (-2 - h)^2 + k^2 &= r^2 \end{aligned}$$

Therefore:

$$\begin{aligned} 1 + 2h + h^2 + 25 + 10k + k^2 &= r^2 \\ 1 - 2h + h^2 + 4 - 4k + k^2 &= r^2 \\ 4 + 4h + h^2 + k^2 &= r^2 \end{aligned}$$

Setting the first two equations equal yields $4h + 14k = -21$ and setting the last two equations equal yields $6h + 4k = 1$. So we solve the system:

$$\begin{aligned} 4h + 14k &= -21 \\ 6h + \ \ 4k &= 1 \end{aligned}$$

Multiplying the first equation by -3 and the second equation by 2, then adding, yields $k = -\frac{65}{34}$, and substituting into the second equation yields $h = \frac{49}{34}$. We find r by the original third equation:

$$r^2 = 4 + 4h + h^2 + k^2 = 4 + \tfrac{98}{17} + \left(\tfrac{49}{34}\right)^2 + \left(-\tfrac{65}{34}\right)^2 = \tfrac{17914}{34^2} = \tfrac{8957}{578}$$

So we have $\left(x - \frac{49}{34}\right)^2 + \left(y + \frac{65}{34}\right)^2 = \frac{8957}{578}$, which simplifies to:

$$17x^2 + 17y^2 - 49x + 65y - 166 = 0$$

45. Using the hint, we let $A = e^x$, $B = e^y$, and $C = e^z$. Then we have the system:

$$A + B - 2C = 2a$$
$$A + 2B - 4C = 3a$$
$$\tfrac{1}{2}A - 3B + C = -5a$$

Multiply the first equation by –1 and add it to the second equation:

$$-A - B + 2C = -2a$$
$$A + 2B - 4C = 3a$$

Adding, we obtain:

$$B - 2C = a$$

Multiply the first equation by –1 and the third equation by 2, then add the resulting equations:

$$-A - B + 2C = -2a$$
$$A - 6B + 2C = -10a$$

Adding, we obtain:

$$-7B + 4C = -12a$$

So we have the system:

$$A + B - 2C = 2a$$
$$B - 2C = a$$
$$-7B + 4C = -12a$$

Multiply the second equation by 7 and add it to the third equation:

$$7B - 14C = 7a$$
$$-7B + 4C = -12a$$

Adding, we obtain:

$$-10C = -5a$$
$$C = \tfrac{a}{2}$$

Substituting into the second equation:

$$B - 2C = a$$
$$B - 2\left(\tfrac{a}{2}\right) = a$$
$$B - a = a$$
$$B = 2a$$

Substituting into the first equation:

$$A + B - 32C = 2a$$
$$A + 2a - a = 2a$$
$$A + a = 2a$$
$$A = a$$

So $A = a$, $B = 2a$, and $C = \tfrac{a}{2}$. Since $A = e^x$, $B = e^y$, $C = e^z$, then $x = \ln a$, $y = \ln 2a$ and $z = \ln \tfrac{a}{2}$.

47. From the figure in the text, we obtain the system of three equations:

$$r_1 + r_2 = a$$
$$r_2 + r_3 = b$$
$$r_1 + r_3 = c$$

This system can be solved by repeated substitution, or by using the following technique. Add the three equations and then divide by 2 to obtain $r_1 + r_2 + r_3 = \frac{a+b+c}{2}$. Now replace $r_2 + r_3$ in this last equation by b. The result is $r_1 = \frac{a+b+c}{2} - b$. Thus $r_1 = \frac{a-b+c}{2}$. Similarly, we find that $r_2 = \frac{a+b-c}{2}$ and $r_3 = \frac{b+c-a}{2}$.

49. (a) Multiply by $(x-a)(x-b)$, so:

$$1 = A(x-b) + B(x-a)$$
$$1 = (A+B)x + (-bA - aB)$$

Since A and B are constants we have:

$$A + \quad B = 0$$
$$-bA - aB = 1$$

Multiplying the first equation by b and adding it to the second equation yields $B = \frac{1}{b-a}$. We re-solve for A. Multiplying the first equation by a and adding it to the second equation yields $A = \frac{1}{a-b}$. So $A = \frac{1}{a-b}$ and $B = \frac{1}{b-a}$.

(b) Multiply by $(x-a)(x-b)$, so:

$$px + q = A(x-b) + B(x-a)$$
$$px + q = (A+B)x + (-bA - aB)$$

Since A and B are constants we have:

$$A + \quad B = p$$
$$-bA - aB = q$$

Multiplying the first equation by b and adding it to the second equation yields $B = \frac{bp+q}{b-a}$. We re-solve for A. Multiplying the first equation by a and adding it to the second equation yields $A = \frac{ap+q}{a-b}$. So $A = \frac{ap+q}{a-b}$ and $B = \frac{bp+q}{b-a}$.

51. We multiply by $(x-a)(x-b)(x-c)$ to get:

$$x^2 + px + q = A(x-b)(x-c) + B(x-a)(x-c) + C(x-a)(x-b)$$

Since this equation is an identity for all x, then it must be true for $x = a$, $x = b$ and $x = c$.

$x = a$: $$a^2 + pa + q = A(a-b)(a-c) + 0 + 0$$
$$A = \frac{a^2 + pa + q}{(a-b)(a-c)}$$

$x = b$: $$b^2 + pb + q = 0 + B(b-a)(b-c) + 0$$
$$B = \frac{b^2 + pb + q}{(b-a)(b-c)}$$

$x = c$: $$c^2 + pc + q = 0 + 0 + C(c-a)(c-b)$$
$$C = \frac{c^2 + pc + q}{(c-a)(c-b)}$$

So $A = \dfrac{a^2 + pa + q}{(a-b)(a-c)}$, $B = \dfrac{b^2 + pb + q}{(b-a)(b-c)}$ and $C = \dfrac{c^2 + pc + q}{(c-a)(c-b)}$.

53. Let α, β and γ be the distances opposite towns A, B and C. The walking, riding and driving rates are, respectively, $\frac{1}{a}$ mi / min, $\frac{1}{b}$ mi / min and $\frac{1}{c}$ mi / min. From the statement of the problem we can obtain the following system of three equations:

$$\alpha a+\beta b+\gamma c=60(a+c-b)$$
$$\beta a+\gamma b+\alpha c=60(b+a-c)$$
$$\gamma a+\alpha b+\beta c=60(c+b-a)$$

By adding all three equations, we obtain:

$$\alpha(a+b+c)+\beta(a+b+c)+\gamma(a+b+c)=60(a+b+c)$$

Now divide through by the quantity $a+b+c$. This yields $\alpha+\beta+\gamma=60$ miles.

7.3 Matrices

1. (a) two by three (2×3)

(b) three by two (3×2)

3. five by four (5×4)

5. The coefficient matrix is $\begin{pmatrix} 2 & 3 & 4 \\ 5 & 6 & 7 \\ 8 & 9 & 10 \end{pmatrix}$ and the augmented matrix is $\begin{pmatrix} 2 & 3 & 4 & 10 \\ 5 & 6 & 7 & 9 \\ 8 & 9 & 10 & 8 \end{pmatrix}$.

7. The coefficient matrix is $\begin{pmatrix} 1 & 0 & 1 & 1 \\ 1 & 1 & 0 & 2 \\ 0 & 1 & 1 & 1 \\ 2 & -1 & -1 & 0 \end{pmatrix}$ and the augmented matrix is

$$\begin{pmatrix} 1 & 0 & 1 & 1 & -1 \\ 1 & 1 & 0 & 2 & 0 \\ 0 & 1 & 1 & 1 & 1 \\ 2 & -1 & -1 & 0 & 2 \end{pmatrix}.$$

9. Form the augmented matrix:

$$\begin{pmatrix} 1 & -1 & 2 & 7 \\ 3 & 2 & -1 & -10 \\ -1 & 3 & 1 & -2 \end{pmatrix}$$

Adding –3 times row 1 to row 2 and adding row 1 to row 3 yields:

$$\begin{pmatrix} 1 & -1 & 2 & 7 \\ 0 & 5 & -7 & -31 \\ 0 & 2 & 3 & 5 \end{pmatrix}$$

Multiplying row 3 by –3 and adding to row 2 yields:

$$\begin{pmatrix} 1 & -1 & 2 & 7 \\ 0 & -1 & -16 & -46 \\ 0 & 2 & 3 & 5 \end{pmatrix}$$

Multiplying row 2 by 2 and adding to row 3 yields:

$$\begin{pmatrix} 1 & -1 & 2 & 7 \\ 0 & -1 & -16 & -46 \\ 0 & 0 & -29 & -87 \end{pmatrix}$$

So we have the system of equations:

$$\begin{aligned} x - y + 2z &= 7 \\ -y - 16z &= -46 \\ -29z &= -87 \end{aligned}$$

Solve equation three for z:

$$\begin{aligned} -29z &= -87 \\ z &= 3 \end{aligned}$$

Substitute into equation two:

$$\begin{aligned} -y - 48 &= -46 \\ -y &= 2 \\ y &= -2 \end{aligned}$$

Substitute into equation one:

$$\begin{aligned} x + 2 + 6 &= 7 \\ x &= -1 \end{aligned}$$

So the solution is $(-1, -2, 3)$.

11. Form the augmented matrix:

$$\begin{pmatrix} 1 & 0 & 1 & -2 \\ -3 & 2 & 0 & 17 \\ 1 & -1 & -1 & -9 \end{pmatrix}$$

Adding 3 times row 1 to row 2, and –1 times row 1 to row 3 yields:

$$\begin{pmatrix} 1 & 0 & 1 & -2 \\ 0 & 2 & 3 & 11 \\ 0 & -1 & -2 & -7 \end{pmatrix}$$

Switching row 2 and row 3:

$$\begin{pmatrix} 1 & 0 & 1 & -2 \\ 0 & -1 & -2 & -7 \\ 0 & 2 & 3 & 11 \end{pmatrix}$$

Multiplying row 2 by 2 and adding it to row 3 yields:

$$\begin{pmatrix} 1 & 0 & 1 & -2 \\ 0 & -1 & -2 & -7 \\ 0 & 0 & -1 & -3 \end{pmatrix}$$

So we have the system:

$$\begin{aligned} x \quad + z &= -2 \\ -y - 2z &= -7 \\ -z &= -3 \end{aligned}$$

Solve equation three for z:

$$\begin{aligned} -z &= -3 \\ z &= 3 \end{aligned}$$

Substitute into equation two:

$$\begin{aligned} -y - 6 &= -7 \\ -y &= -1 \\ y &= 1 \end{aligned}$$

Substitute into equation one:

$$\begin{aligned} x + 3 &= -2 \\ x &= -5 \end{aligned}$$

So the solution is $(-5, 1, 3)$.

13. Form the augmented matrix:

$$\begin{pmatrix} 1 & 1 & 1 & -4 \\ 2 & -3 & 1 & -1 \\ 4 & 2 & -3 & 33 \end{pmatrix}$$

Adding -2 times row 1 to row 2, and -4 times row 1 to row 3 yields:

$$\begin{pmatrix} 1 & 1 & 1 & -4 \\ 0 & -5 & -1 & 7 \\ 0 & -2 & -7 & 49 \end{pmatrix}$$

Adding -3 times row 3 to row 2 (to get a 1 entry) yields:

$$\begin{pmatrix} 1 & 1 & 1 & -4 \\ 0 & 1 & 20 & -140 \\ 0 & -2 & -7 & 49 \end{pmatrix}$$

Adding 2 times row 2 to row 3 yields:

$$\begin{pmatrix} 1 & 1 & 1 & -4 \\ 0 & 1 & 20 & -140 \\ 0 & 0 & 33 & -231 \end{pmatrix}$$

So we have the system of equations:

$$\begin{aligned} x + y + z &= -4 \\ y + 20z &= -140 \\ 33z &= -231 \end{aligned}$$

Solving the last equation for z:

$$\begin{aligned} 33z &= -231 \\ z &= -7 \end{aligned}$$

Substituting into the second equation:

$$\begin{aligned} y + 20z &= -140 \\ y - 140 &= -140 \\ y &= 0 \end{aligned}$$

Substituting into the first equation:

$$\begin{aligned} x + y + z &= -4 \\ x + 0 - 7 &= -4 \\ x &= 3 \end{aligned}$$

So the soltution is $(3, 0, -7)$.

15. Form the augmented matrix:

$$\begin{pmatrix} 3 & -2 & 6 & 0 \\ 1 & 3 & 20 & 15 \\ 10 & -11 & -10 & -9 \end{pmatrix}$$

Switching row 1 and row 2 yields:

$$\begin{pmatrix} 1 & 3 & 20 & 15 \\ 3 & -2 & 6 & 0 \\ 10 & -11 & -10 & -9 \end{pmatrix}$$

Adding –3 times row 1 to row 2, and –10 times row 1 to row 3 yields:

$$\begin{pmatrix} 1 & 3 & 20 & 15 \\ 0 & -11 & -54 & -45 \\ 0 & -41 & -210 & -159 \end{pmatrix}$$

Multiplying row 2 by –4 and adding it to row 3 yields:

$$\begin{pmatrix} 1 & 3 & 20 & 15 \\ 0 & -11 & -54 & -45 \\ 0 & 3 & 6 & 21 \end{pmatrix}$$

Dividing row 3 by 3 and then switching it with row 2 yields:

$$\begin{pmatrix} 1 & 3 & 20 & 15 \\ 0 & 1 & 2 & 7 \\ 0 & -11 & -54 & -45 \end{pmatrix}$$

Multiplying row 2 by 11 and adding it to row 3 yields:

$$\begin{pmatrix} 1 & 3 & 20 & 15 \\ 0 & 1 & 2 & 7 \\ 0 & 0 & -32 & 32 \end{pmatrix}$$

So we have the system of equations:

$$\begin{aligned} x+3y+20z&=15 \\ y+\ 2z&=7 \\ -32z&=32 \end{aligned}$$

Solve equation three for z:

$$\begin{aligned} -32z&=32 \\ z&=-1 \end{aligned}$$

Substitute into equation two:

$$\begin{aligned} y-2&=7 \\ y&=9 \end{aligned}$$

Substitute into equation one:

$$\begin{aligned} x+27-20&=15 \\ x+7&=15 \\ x&=8 \end{aligned}$$

So the solution is $(8,9,-1)$.

17. Form the augmented matrix:

$$\begin{pmatrix} 4 & -3 & 3 & 2 \\ 5 & 1 & -4 & 1 \\ 9 & -2 & -1 & 3 \end{pmatrix}$$

Subtracting row 1 from row 2 yields:

$$\begin{pmatrix} 4 & -3 & 3 & 2 \\ 1 & 4 & -7 & -1 \\ 9 & -2 & -1 & 3 \end{pmatrix}$$

Switching row 1 and row 2 yields:

$$\begin{pmatrix} 1 & 4 & -7 & -1 \\ 4 & -3 & 3 & 2 \\ 9 & -2 & -1 & 3 \end{pmatrix}$$

Adding -4 times row 1 to row 2, and -9 times row 1 to row 3 yields:

$$\begin{pmatrix} 1 & 4 & -7 & -1 \\ 0 & -19 & 31 & 6 \\ 0 & -38 & 62 & 12 \end{pmatrix}$$

Multiplying row 2 by -2 and adding it to row 3 yields:

$$\begin{pmatrix} 1 & 4 & -7 & -1 \\ 0 & -19 & 31 & 6 \\ 0 & 0 & 0 & 0 \end{pmatrix}$$

So we have the system of equations:

$$\begin{aligned} x+4y-7z&=-1 \\ -19y+31z&=6 \end{aligned}$$

We solve equation two for y:

$$-19y = 6 - 31z$$
$$y = \frac{31z - 6}{19}$$

Substitute into equation one:

$$x + 4\left(\frac{31z - 6}{19}\right) - 7z = -1$$
$$x + \frac{-9z - 24}{19} = -1$$
$$x = \frac{9z + 5}{19}$$

So the solution is $\left(\frac{9z+5}{19}, \frac{31z-6}{19}, z\right)$, for any real number z.

19. Form the augmented matrix:

$$\begin{pmatrix} 1 & -1 & 1 & 1 & 6 \\ 1 & 1 & -1 & 1 & 4 \\ 1 & 1 & 1 & -1 & -2 \\ -1 & 1 & 1 & 1 & 0 \end{pmatrix}$$

Adding –1 times row 1 to both row 2 and row 3, and adding row 1 to row 4 yields:

$$\begin{pmatrix} 1 & -1 & 1 & 1 & 6 \\ 0 & 2 & -2 & 0 & -2 \\ 0 & 2 & 0 & -2 & -8 \\ 0 & 0 & 2 & 2 & 6 \end{pmatrix}$$

Dividing rows 2, 3, and 4 by 2 and subtracting row 2 from row 3 yields:

$$\begin{pmatrix} 1 & -1 & 1 & 1 & 6 \\ 0 & 1 & -1 & 0 & -1 \\ 0 & 0 & 1 & -1 & -3 \\ 0 & 0 & 1 & 1 & 3 \end{pmatrix}$$

Multiplying row 3 by –1 and adding it to row 4 yields:

$$\begin{pmatrix} 1 & -1 & 1 & 1 & 6 \\ 0 & 1 & -1 & 0 & -1 \\ 0 & 0 & 1 & -1 & -3 \\ 0 & 0 & 0 & 2 & 6 \end{pmatrix}$$

So we have the system of equations:

$$x - y + z + w = 6$$
$$y - z = -1$$
$$z - w = -3$$
$$2w = 6$$

Solve equation four for w:

$$2w = 6$$
$$w = 3$$

Substitute into equation three:

$$z - 3 = -3$$
$$z = 0$$

Substitute into equation two:

$$y - 0 = -1$$
$$y = -1$$

Substitute into equation one:

$$x + 1 + 0 + 3 = 6$$
$$x + 4 = 6$$
$$x = 2$$

So the solution is $(2, -1, 0, 3)$.

21. Form the augmented matrix:

$$\begin{pmatrix} 15 & 14 & 26 & 1 \\ 18 & 17 & 32 & -1 \\ 21 & 20 & 38 & 0 \end{pmatrix}$$

We reduce coefficients by subtracting row 2 from row 3, and row 1 from row 2:

$$\begin{pmatrix} 15 & 14 & 26 & 1 \\ 3 & 3 & 6 & -2 \\ 3 & 3 & 6 & 1 \end{pmatrix}$$

Subtracting row 2 from row 3 yields:

$$\begin{pmatrix} 15 & 14 & 26 & 1 \\ 3 & 3 & 6 & -2 \\ 0 & 0 & 0 & 3 \end{pmatrix}$$

But this last row corresponds to the equation $0 = 3$, which is false. So the system has no solution.

23. Adding the corresponding terms, $A + B$ will be:

$$\begin{pmatrix} 2 & 3 \\ -1 & 4 \end{pmatrix} + \begin{pmatrix} 1 & -1 \\ 3 & 0 \end{pmatrix} = \begin{pmatrix} 3 & 2 \\ 2 & 4 \end{pmatrix}$$

25. We have:

$$2A = \begin{pmatrix} 4 & 6 \\ -2 & 8 \end{pmatrix} \text{ and } 2B = \begin{pmatrix} 2 & -2 \\ 6 & 0 \end{pmatrix}$$

Therefore:

$$2A + 2B = \begin{pmatrix} 6 & 4 \\ 4 & 8 \end{pmatrix}$$

27. The multiplication is defined, since # cols in A = # rows in B:

row 1, col 1: $2(1) + 3(3) = 2 + 9 = 11$
row 1, col 2: $2(-1) + 3(0) = -2 + 0 = -2$
row 2, col 1: $-1(1) + 4(3) = -1 + 12 = 11$
row 2, col 2: $-1(-1) + 4(0) = 1 + 0 = 1$

So $AB = \begin{pmatrix} 11 & -2 \\ 11 & 1 \end{pmatrix}$.

29. The multiplication is defined, since # cols in A = # rows in C:

row 1, col 1: $2(1) + 3(0) = 2 + 0 = 2$
row 1, col 2: $2(0) + 3(1) = 0 + 3 = 3$
row 2, col 1: $-1(1) + 4(0) = -1 + 0 = -1$
row 2, col 2: $-1(0) + 4(1) = 0 + 4 = 4$

So $AC = \begin{pmatrix} 2 & 3 \\ -1 & 4 \end{pmatrix}$.

31. This operation is not defined, since D and E do not have the same size.

33. We have:

$$2F - 3G = \begin{pmatrix} 10 & -2 \\ -8 & 0 \\ 4 & 6 \end{pmatrix} - \begin{pmatrix} 0 & 0 \\ 0 & 0 \\ 0 & 0 \end{pmatrix} = \begin{pmatrix} 10 & -2 \\ -8 & 0 \\ 4 & 6 \end{pmatrix}$$

35. The multiplication is defined, since # cols of E = # rows of D.

row 1, col 1: $2(-1) + 1(4) = -2 + 4 = 2$
row 1, col 2: $2(2) + 1(0) = 4 + 0 = 4$
row 1, col 3: $2(3) + 1(5) = 6 + 5 = 11$
row 2, col 1: $8(-1) - 1(4) = -8 - 4 = -12$
row 2, col 2: $8(2) - 1(0) = 16 - 0 = 16$
row 2, col 3: $8(3) - 1(5) = 24 - 5 = 19$
row 3, col 1: $6(-1) + 5(4) = -6 + 20 = 14$
row 3, col 2: $6(2) + 5(0) = 12 + 0 = 12$
row 3, col 3: $6(3) + 5(5) = 18 + 25 = 43$

So $ED = \begin{pmatrix} 2 & 4 & 11 \\ -12 & 16 & 19 \\ 14 & 12 & 43 \end{pmatrix}$.

37. The multiplication is defined, since # cols of F = # rows of D:

row 1, col 1: $5(-1) - 1(4) = -5 - 4 = -9$
row 1, col 2: $5(2) - 1(0) = 10 - 0 = 10$
row 1, col 3: $5(3) - 1(5) = 15 - 5 = 10$
row 2, col 1: $-4(-1) + 0(4) = 4 + 0 = 4$
row 2, col 2: $-4(2) + 0(0) = -8 + 0 = -8$
row 2, col 3: $-4(3) + 0(5) = -12 + 0 = -12$
row 3, col 1: $2(-1) + 3(4) = -2 + 12 = 10$
row 3, col 2: $2(2) + 3(0) = 4 + 0 = 4$
row 3, col 3: $2(3) + 3(5) = 6 + 15 = 21$

$$\text{So } FD = \begin{pmatrix} -9 & 10 & 10 \\ 4 & -8 & -12 \\ 10 & 4 & 21 \end{pmatrix}.$$

39. The operation is not defined, since G and A are not the same size.

41. The multiplication is defined, since # cols of G = # rows of D:

$$GD = \begin{pmatrix} 0 & 0 & 0 \\ 0 & 0 & 0 \\ 0 & 0 & 0 \end{pmatrix}$$

43. We have:

$$A + (B + C) = \begin{pmatrix} 2 & 3 \\ -1 & 4 \end{pmatrix} + \begin{pmatrix} 2 & -1 \\ 3 & 1 \end{pmatrix} = \begin{pmatrix} 4 & 2 \\ 2 & 5 \end{pmatrix}$$

45. The multiplication is not defined, since # cols of $D \neq$ # rows of C.

47. The multiplication is defined, since # cols of A = # rows of A.

row 1, col 1: $2(2) + 3(-1) = 4 - 3 = 1$
row 1, col 2: $2(3) + 3(4) = 6 + 12 = 18$
row 2, col 1: $-1(2) + 4(-1) = -2 - 4 = -6$
row 2, col 2: $-1(3) + 4(4) = -3 + 16 = 13$

$$\text{So } A^2 = \begin{pmatrix} 1 & 18 \\ -6 & 13 \end{pmatrix}.$$

49. We have:

$$AA^2 = \begin{pmatrix} 2 & 3 \\ -1 & 4 \end{pmatrix}\begin{pmatrix} 1 & 18 \\ -6 & 13 \end{pmatrix} = \begin{pmatrix} -16 & 75 \\ -25 & 34 \end{pmatrix}$$

51. (a) We have:

$$A(B+C)=\begin{pmatrix}-1 & 3 & 4\\ 3 & 2 & -3\\ 9 & 1 & 6\end{pmatrix}\begin{pmatrix}11 & 6 & 2\\ 2 & 1 & 6\\ -2 & 1 & 6\end{pmatrix}=\begin{pmatrix}-13 & 1 & 40\\ 43 & 17 & 0\\ 89 & 61 & 60\end{pmatrix}$$

(b) We have:

$$AB+AC=\begin{pmatrix}-1 & 3 & 4\\ 3 & 2 & -3\\ 9 & 1 & 6\end{pmatrix}\begin{pmatrix}7 & 0 & 1\\ 0 & 0 & 3\\ -1 & 2 & 4\end{pmatrix}+\begin{pmatrix}-1 & 3 & 4\\ 3 & 2 & -3\\ 9 & 1 & 6\end{pmatrix}\begin{pmatrix}4 & 6 & 1\\ 2 & 1 & 3\\ -1 & -1 & 2\end{pmatrix}$$

$$=\begin{pmatrix}-11 & 8 & 24\\ 24 & -6 & -3\\ 57 & 12 & 36\end{pmatrix}+\begin{pmatrix}-2 & -7 & 16\\ 19 & 23 & 3\\ 32 & 49 & 24\end{pmatrix}$$

$$=\begin{pmatrix}-13 & 1 & 40\\ 43 & 17 & 0\\ 89 & 61 & 60\end{pmatrix}$$

(c) We have:

$$(AB)C=\left[\begin{pmatrix}-1 & 3 & 4\\ 3 & 2 & -3\\ 9 & 1 & 6\end{pmatrix}\begin{pmatrix}7 & 0 & 1\\ 0 & 0 & 3\\ -1 & 2 & 4\end{pmatrix}\right]\begin{pmatrix}4 & 6 & 1\\ 2 & 1 & 3\\ -1 & -1 & 2\end{pmatrix}$$

$$=\begin{pmatrix}-11 & 8 & 24\\ 24 & -6 & -3\\ 57 & 12 & 36\end{pmatrix}\begin{pmatrix}4 & 6 & 1\\ 2 & 1 & 3\\ -1 & -1 & 2\end{pmatrix}$$

$$=\begin{pmatrix}-52 & -82 & 61\\ 87 & 141 & 0\\ 216 & 318 & 165\end{pmatrix}$$

(d) We have:

$$A(BC)=\begin{pmatrix}-1 & 3 & 4\\ 3 & 2 & -3\\ 9 & 1 & 6\end{pmatrix}\begin{pmatrix}27 & 41 & 9\\ -3 & -3 & 6\\ -4 & -8 & 13\end{pmatrix}=\begin{pmatrix}-52 & -82 & 61\\ 87 & 141 & 0\\ 216 & 318 & 165\end{pmatrix}$$

53. (a) We compute A^2 and B^2:

$$A^2=\begin{pmatrix}3 & 5\\ 7 & 9\end{pmatrix}\begin{pmatrix}3 & 5\\ 7 & 9\end{pmatrix}=\begin{pmatrix}44 & 60\\ 84 & 116\end{pmatrix}$$

$$B^2=\begin{pmatrix}2 & 4\\ 6 & 8\end{pmatrix}\begin{pmatrix}2 & 4\\ 6 & 8\end{pmatrix}=\begin{pmatrix}28 & 40\\ 60 & 88\end{pmatrix}$$

So $A^2-B^2=\begin{pmatrix}16 & 20\\ 24 & 28\end{pmatrix}$.

(b) We have $A-B=\begin{pmatrix}1 & 1\\ 1 & 1\end{pmatrix}$ and $A+B=\begin{pmatrix}5 & 9\\ 13 & 17\end{pmatrix}$, so:

$$(A-B)(A+B)=\begin{pmatrix}1 & 1\\ 1 & 1\end{pmatrix}\begin{pmatrix}5 & 9\\ 13 & 17\end{pmatrix}=\begin{pmatrix}18 & 26\\ 18 & 26\end{pmatrix}$$

(c) We compute the product:

$$(A+B)(A-B)=\begin{pmatrix}5 & 9\\ 13 & 17\end{pmatrix}\begin{pmatrix}1 & 1\\ 1 & 1\end{pmatrix}=\begin{pmatrix}14 & 14\\ 30 & 30\end{pmatrix}$$

(d) We compute AB and BA:

$$AB=\begin{pmatrix}3 & 5\\ 7 & 9\end{pmatrix}\begin{pmatrix}2 & 4\\ 6 & 8\end{pmatrix}=\begin{pmatrix}36 & 52\\ 68 & 100\end{pmatrix}$$

$$BA=\begin{pmatrix}2 & 4\\ 6 & 8\end{pmatrix}\begin{pmatrix}3 & 5\\ 7 & 9\end{pmatrix}=\begin{pmatrix}34 & 46\\ 74 & 102\end{pmatrix}$$

Therefore:

$$A^2+AB-BA-B^2=$$

$$\begin{pmatrix}44 & 60\\ 84 & 116\end{pmatrix}+\begin{pmatrix}36 & 52\\ 68 & 100\end{pmatrix}-\begin{pmatrix}34 & 46\\ 74 & 102\end{pmatrix}-\begin{pmatrix}28 & 40\\ 60 & 88\end{pmatrix}=\begin{pmatrix}18 & 26\\ 18 & 26\end{pmatrix}$$

55. (a) The product AZ is given by:

$$AZ=\begin{pmatrix}1 & 0\\ 0 & -1\end{pmatrix}\begin{pmatrix}x\\ y\end{pmatrix}=\begin{pmatrix}x\\ -y\end{pmatrix}$$

(b) The product BZ is given by:

$$BZ=\begin{pmatrix}-1 & 0\\ 0 & 1\end{pmatrix}\begin{pmatrix}x\\ y\end{pmatrix}=\begin{pmatrix}-x\\ y\end{pmatrix}$$

(c) The product AB is given by:

$$AB=\begin{pmatrix}1 & 0\\ 0 & -1\end{pmatrix}\begin{pmatrix}-1 & 0\\ 0 & 1\end{pmatrix}=\begin{pmatrix}-1 & 0\\ 0 & -1\end{pmatrix}$$

So $(AB)Z=\begin{pmatrix}-1 & 0\\ 0 & -1\end{pmatrix}\begin{pmatrix}x\\ y\end{pmatrix}=\begin{pmatrix}-x\\ -y\end{pmatrix}$. This would represent a reflection about the origin.

57. (a) We first compute the product:

$$AB=\begin{pmatrix}1 & 2\\ 3 & 4\end{pmatrix}\begin{pmatrix}3 & -1\\ 5 & 8\end{pmatrix}=\begin{pmatrix}13 & 15\\ 29 & 29\end{pmatrix}$$

Now compute the function values:

$f(A) = 1(4) - 2(3) = 4 - 6 = -2$
$f(B) = 3(8) - (-1)(5) = 24 + 5 = 29$
$f(AB) = 13(29) - 15(29) = -2(29) = -58$

So $f(A) \bullet f(B) = f(AB)$.

(b) We first compute the product:

$$AB = \begin{pmatrix} a & b \\ c & d \end{pmatrix}\begin{pmatrix} e & f \\ g & h \end{pmatrix} = \begin{pmatrix} ae+bg & af+bh \\ ce+dg & cf+dh \end{pmatrix}$$

Now compute the function values:

$f(A) = ad - bc$
$f(B) = eh - fg$
$f(AB) = (ae+bg)(cf+dh) - (af+bh)(ce+dg)$
$= (acef + bcfg + adeh + bdgh) - (acef + bceh + adfg + bdgh)$
$= bcfg + adeh - bceh - adfg$

So $f(A) \bullet f(B) = f(AB)$.

59. (a) Let $A = \begin{pmatrix} a & b \\ c & d \end{pmatrix}$ and $B = \begin{pmatrix} e & f \\ g & h \end{pmatrix}$. Then $A + B = \begin{pmatrix} a+e & b+f \\ c+g & d+h \end{pmatrix}$, so

$(A+B)^T = \begin{pmatrix} a+e & c+g \\ b+f & d+h \end{pmatrix}$. Now $A^T = \begin{pmatrix} a & c \\ b & d \end{pmatrix}$ and $B^T = \begin{pmatrix} e & g \\ f & h \end{pmatrix}$, so we have

$A^T + B^T = \begin{pmatrix} a+e & c+g \\ b+f & d+h \end{pmatrix}$. Thus $(A+B)^T = A^T + B^T$.

(b) Since $A^T = \begin{pmatrix} a & c \\ b & d \end{pmatrix}$, then $\left(A^T\right)^T = \begin{pmatrix} a & b \\ c & d \end{pmatrix}$. Thus $\left(A^T\right)^T = A$.

(c) Since $AB = \begin{pmatrix} ae+bg & af+bh \\ ce+dg & cf+dh \end{pmatrix}$, then $(AB)^T = \begin{pmatrix} ae+bg & ce+dg \\ af+bh & cf+dh \end{pmatrix}$. Also

$B^T A^T = \begin{pmatrix} e & g \\ f & h \end{pmatrix}\begin{pmatrix} a & c \\ b & d \end{pmatrix} = \begin{pmatrix} ae+bg & ce+dg \\ af+bh & cf+dh \end{pmatrix}$. So we see that $(AB)^T = B^T A^T$.

7.4 The Inverse of a Square Matrix

1. We compute AI_2 and I_2A:

$$AI_2 = \begin{pmatrix} 4 & -1 \\ -5 & 2 \end{pmatrix}\begin{pmatrix} 1 & 0 \\ 0 & 1 \end{pmatrix} = \begin{pmatrix} 4 & -1 \\ -5 & 2 \end{pmatrix} = A$$

$$I_2A = \begin{pmatrix} 1 & 0 \\ 0 & 1 \end{pmatrix}\begin{pmatrix} 4 & -1 \\ -5 & 2 \end{pmatrix} = \begin{pmatrix} 4 & -1 \\ -5 & 2 \end{pmatrix} = A$$

3. We compute CI_3 and I_3C:

$$CI_3 = \begin{pmatrix} 3 & 0 & -2 \\ 0 & 5 & 6 \\ 1 & 4 & -7 \end{pmatrix}\begin{pmatrix} 1 & 0 & 0 \\ 0 & 1 & 0 \\ 0 & 0 & 1 \end{pmatrix} = \begin{pmatrix} 3 & 0 & -2 \\ 0 & 5 & 6 \\ 1 & 4 & -7 \end{pmatrix} = C$$

$$I_3C = \begin{pmatrix} 1 & 0 & 0 \\ 0 & 1 & 0 \\ 0 & 0 & 1 \end{pmatrix}\begin{pmatrix} 3 & 0 & -2 \\ 0 & 5 & 6 \\ 1 & 4 & -7 \end{pmatrix} = \begin{pmatrix} 3 & 0 & -2 \\ 0 & 5 & 6 \\ 1 & 4 & -7 \end{pmatrix} = C$$

5. We need to find numbers a, b, c and d such that:

$$\begin{pmatrix} 7 & 9 \\ 4 & 5 \end{pmatrix}\begin{pmatrix} a & b \\ c & d \end{pmatrix} = \begin{pmatrix} 1 & 0 \\ 0 & 1 \end{pmatrix} \text{ and } \begin{pmatrix} a & b \\ c & d \end{pmatrix}\begin{pmatrix} 7 & 9 \\ 4 & 5 \end{pmatrix} = \begin{pmatrix} 1 & 0 \\ 0 & 1 \end{pmatrix}$$

We compute the left-hand product:

$$\begin{pmatrix} 7 & 9 \\ 4 & 5 \end{pmatrix}\begin{pmatrix} a & b \\ c & d \end{pmatrix} = \begin{pmatrix} 7a+9c & 7b+9d \\ 4a+5c & 4b+5d \end{pmatrix} = \begin{pmatrix} 1 & 0 \\ 0 & 1 \end{pmatrix}$$

So we have the following systems of equations:

$$\begin{aligned} 7a+9c &= 1 \\ 4a+5c &= 0 \end{aligned} \qquad \text{and} \qquad \begin{aligned} 7b+9d &= 0 \\ 4b+5d &= 1 \end{aligned}$$

Multiplying the first equation by 5 and the second equation by –9 yields:

$$\begin{aligned} 35a+45c &= 5 \\ -36a-45c &= 0 \end{aligned} \qquad \text{and} \qquad \begin{aligned} 35b+45d &= 0 \\ -36b-45d &= -9 \end{aligned}$$

Adding, we obtain:

$$\begin{aligned} -a &= 5 \\ a &= -5 \end{aligned} \qquad \qquad \begin{aligned} -b &= -9 \\ b &= 9 \end{aligned}$$

Substituting for c and d, respectively:

$$\begin{aligned} 4a+5c &= 0 \\ -20+5c &= 0 \\ 5c &= 20 \\ c &= 4 \end{aligned} \qquad \qquad \begin{aligned} 4b+5d &= 1 \\ 36+5d &= 1 \\ 5d &= -35 \\ d &= -7 \end{aligned}$$

So the inverse matrix is $A^{-1} = \begin{pmatrix} -5 & 9 \\ 4 & -7 \end{pmatrix}$.

7. We need to find numbers a, b, c, and d such that:

$$\begin{pmatrix} -3 & 1 \\ 5 & 6 \end{pmatrix}\begin{pmatrix} a & b \\ c & d \end{pmatrix} = \begin{pmatrix} 1 & 0 \\ 0 & 1 \end{pmatrix}$$

We compute the left-hand product:

$$\begin{pmatrix} -3a+c & -3b+d \\ 5a+6c & 5b+6d \end{pmatrix} = \begin{pmatrix} 1 & 0 \\ 0 & 1 \end{pmatrix}$$

So we have the following systems of equations:

$$\begin{aligned} -3a + c &= 1 \\ 5a+6c &= 0 \end{aligned} \qquad \text{and} \qquad \begin{aligned} -3b + d &= 0 \\ 5b+6d &= 1 \end{aligned}$$

Solving for c and d respectively, then substituting:

$$\begin{aligned} -3a+c &= 1 \\ c &= 3a+1 \end{aligned} \qquad \begin{aligned} -3b+d &= 0 \\ d &= 3b \end{aligned}$$

Substituting:

$$\begin{aligned} 5a+6(3a+1) &= 0 \\ 23a+6 &= 0 \\ 23a &= -6 \\ a &= -\tfrac{6}{23} \end{aligned} \qquad \begin{aligned} 5b+6(3b) &= 1 \\ 23b &= 1 \\ b &= \tfrac{1}{23} \end{aligned}$$

Now substituting to find c and d:

$$\begin{aligned} c &= 3a+1 \\ c &= -\tfrac{18}{23}+1 \\ c &= \tfrac{5}{23} \end{aligned} \qquad \begin{aligned} d &= 3b \\ d &= \tfrac{3}{23} \end{aligned}$$

So the inverse matrix is $A^{-1} = \begin{pmatrix} -\frac{6}{23} & \frac{1}{23} \\ \frac{5}{23} & \frac{3}{23} \end{pmatrix}$.

9. We need to find numbers a, b, c, and d such that:

$$\begin{pmatrix} -2 & 3 \\ -4 & 6 \end{pmatrix}\begin{pmatrix} a & b \\ c & d \end{pmatrix} = \begin{pmatrix} 1 & 0 \\ 0 & 1 \end{pmatrix}$$

We compute the left-hand product:

$$\begin{pmatrix} -2a+3c & -2b+3d \\ -4a+6c & -4b+6d \end{pmatrix} = \begin{pmatrix} 1 & 0 \\ 0 & 1 \end{pmatrix}$$

So we have the following systems of equations:

$$\begin{aligned} -2a+3c &= 1 \\ -4a+6c &= 0 \end{aligned} \qquad \text{and} \qquad \begin{aligned} -2b+3d &= 0 \\ -4b+6d &= 1 \end{aligned}$$

Multiplying the top equation by –2 yields:

$$\begin{aligned} 4a-6c &= -2 \\ -4a+6c &= 0 \end{aligned} \qquad \begin{aligned} 4b-6d &= 0 \\ -4b+6d &= 1 \end{aligned}$$

Adding, we obtain:

$$0 = -2 \qquad\qquad 0 = 1$$

Since both of these equations are false, no A^{-1} exists.

11. We need to find numbers a, b, c, and d such that:

$$\begin{pmatrix} \frac{1}{3} & \frac{1}{3} \\ -\frac{1}{9} & \frac{2}{9} \end{pmatrix}\begin{pmatrix} a & b \\ c & d \end{pmatrix}=\begin{pmatrix} 1 & 0 \\ 0 & 1 \end{pmatrix}$$

We compute the left-hand product:

$$\begin{pmatrix} \frac{1}{3}a+\frac{1}{3}c & \frac{1}{3}b+\frac{1}{3}d \\ -\frac{1}{9}a+\frac{2}{9}c & -\frac{1}{9}b+\frac{2}{9}d \end{pmatrix}=\begin{pmatrix} 1 & 0 \\ 0 & 1 \end{pmatrix}$$

So we have the following systems of equations:

$$\begin{aligned} \tfrac{1}{3}a+\tfrac{1}{3}c&=1 \\ -\tfrac{1}{9}a+\tfrac{2}{9}c&=0 \end{aligned} \qquad \text{and} \qquad \begin{aligned} \tfrac{1}{3}b+\tfrac{1}{3}d&=0 \\ -\tfrac{1}{9}b+\tfrac{2}{9}d&=1 \end{aligned}$$

Multiplying the first equation by 3 and the second by 9 yields:

$$\begin{aligned} a+\ \ c&=3 \\ -a+2c&=0 \end{aligned} \qquad\qquad \begin{aligned} b+\ \ d&=0 \\ -b+2d&=9 \end{aligned}$$

Adding, we obtain:

$$\begin{aligned} 3c&=3 \\ c&=1 \end{aligned} \qquad\qquad \begin{aligned} 3d&=9 \\ d&=3 \end{aligned}$$

Now substituting to find a and b:

$$\begin{aligned} a+c&=3 \\ a+1&=3 \\ a&=2 \end{aligned} \qquad\qquad \begin{aligned} b+d&=0 \\ b+3&=0 \\ b&=-3 \end{aligned}$$

So the inverse is $A^{-1}=\begin{pmatrix} 2 & -3 \\ 1 & 3 \end{pmatrix}$.

13. Form the matrix:

$$\begin{pmatrix} 2 & 1 & 1 & 0 \\ 3 & 2 & 0 & 1 \end{pmatrix}$$

Multiply row 1 by –1 and add to row 2:

$$\begin{pmatrix} 2 & 1 & 1 & 0 \\ 1 & 1 & -1 & 1 \end{pmatrix}$$

Switch rows 1 and 2:

$$\begin{pmatrix} 1 & 1 & -1 & 1 \\ 2 & 1 & 1 & 0 \end{pmatrix}$$

Multiply row 1 by –2 and add to row 2:

$$\begin{pmatrix} 1 & 1 & -1 & 1 \\ 0 & -1 & 3 & -2 \end{pmatrix}$$

Add row 2 to row 1:

$$\begin{pmatrix} 1 & 0 & 2 & -1 \\ 0 & -1 & 3 & -2 \end{pmatrix}$$

Multiply row 2 by −1:

$$\begin{pmatrix} 1 & 0 & 2 & -1 \\ 0 & 1 & -3 & 2 \end{pmatrix}$$

So the inverse is $\begin{pmatrix} 2 & -1 \\ -3 & 2 \end{pmatrix}$.

15. Form the matrix:

$$\begin{pmatrix} 0 & -11 & 1 & 0 \\ 1 & 6 & 0 & 1 \end{pmatrix}$$

Switch rows 1 and 2:

$$\begin{pmatrix} 1 & 6 & 0 & 1 \\ 0 & -11 & 1 & 0 \end{pmatrix}$$

Multiply row 2 by $-\frac{1}{11}$:

$$\begin{pmatrix} 1 & 6 & 0 & 1 \\ 0 & 1 & -\frac{1}{11} & 0 \end{pmatrix}$$

Multiply row 2 by −6 and add to row 1:

$$\begin{pmatrix} 1 & 0 & \frac{6}{11} & 1 \\ 0 & 1 & -\frac{1}{11} & 0 \end{pmatrix}$$

So the inverse is $\begin{pmatrix} \frac{6}{11} & 1 \\ -\frac{1}{11} & 0 \end{pmatrix}$.

17. Form the matrix:

$$\begin{pmatrix} \frac{2}{3} & -\frac{1}{4} & 1 & 0 \\ -8 & 3 & 0 & 1 \end{pmatrix}$$

Multiply row 1 by 12:

$$\begin{pmatrix} 8 & -3 & 12 & 0 \\ -8 & 3 & 0 & 1 \end{pmatrix}$$

Add row 1 to row 2:

$$\begin{pmatrix} 8 & -3 & 12 & 0 \\ 0 & 0 & 12 & 1 \end{pmatrix}$$

So the inverse does not exist.

19. Form the matrix:

$$\begin{pmatrix} -5 & 4 & -3 & 1 & 0 & 0 \\ 10 & -7 & 6 & 0 & 1 & 0 \\ 8 & -6 & 5 & 0 & 0 & 1 \end{pmatrix}$$

Multiply row 1 by 2 and add to row 2:

$$\begin{pmatrix} -5 & 4 & -3 & 1 & 0 & 0 \\ 0 & 1 & 0 & 2 & 1 & 0 \\ 8 & -6 & 5 & 0 & 0 & 1 \end{pmatrix}$$

Add row 3 to row 1 (to reduce the numbers):

$$\begin{pmatrix} 3 & -2 & 2 & 1 & 0 & 1 \\ 0 & 1 & 0 & 2 & 1 & 0 \\ 8 & -6 & 5 & 0 & 0 & 1 \end{pmatrix}$$

Multiply row 2 by 2 and add to row 1, and also multiply row 2 by 6 and add to row 3:

$$\begin{pmatrix} 3 & 0 & 2 & 5 & 2 & 1 \\ 0 & 1 & 0 & 2 & 1 & 0 \\ 8 & 0 & 5 & 12 & 6 & 1 \end{pmatrix}$$

Multiply row 1 by $-\frac{8}{3}$ and add to row 3:

$$\begin{pmatrix} 3 & 0 & 2 & 5 & 2 & 1 \\ 0 & 1 & 0 & 2 & 1 & 0 \\ 0 & 0 & -\frac{1}{3} & -\frac{4}{3} & \frac{2}{3} & -\frac{5}{3} \end{pmatrix}$$

Multiply row 3 by –3:

$$\begin{pmatrix} 3 & 0 & 2 & 5 & 2 & 1 \\ 0 & 1 & 0 & 2 & 1 & 0 \\ 0 & 0 & 1 & 4 & -2 & 5 \end{pmatrix}$$

Multiply row 3 by –2 and add to row 1:

$$\begin{pmatrix} 3 & 0 & 0 & -3 & 6 & -9 \\ 0 & 1 & 0 & 2 & 1 & 0 \\ 0 & 0 & 1 & 4 & -2 & 5 \end{pmatrix}$$

Multiply row 1 by $\frac{1}{3}$:

$$\begin{pmatrix} 1 & 0 & 0 & -1 & 2 & -3 \\ 0 & 1 & 0 & 2 & 1 & 0 \\ 0 & 0 & 1 & 4 & -2 & 5 \end{pmatrix}$$

So the inverse is $\begin{pmatrix} -1 & 2 & -3 \\ 2 & 1 & 0 \\ 4 & -2 & 5 \end{pmatrix}$.

21. Form the matrix:

$$\left(\begin{array}{ccc|ccc} 1 & 2 & -1 & 1 & 0 & 0 \\ 0 & 3 & 0 & 0 & 1 & 0 \\ -4 & 0 & 5 & 0 & 0 & 1 \end{array}\right)$$

Multiply row 1 by 4 and add to row 3:

$$\left(\begin{array}{ccc|ccc} 1 & 2 & -1 & 1 & 0 & 0 \\ 0 & 3 & 0 & 0 & 1 & 0 \\ 0 & 8 & 1 & 4 & 0 & 1 \end{array}\right)$$

Multiply row 2 by $\frac{1}{3}$:

$$\left(\begin{array}{ccc|ccc} 1 & 2 & -1 & 1 & 0 & 0 \\ 0 & 1 & 0 & 0 & \frac{1}{3} & 0 \\ 0 & 8 & 1 & 4 & 0 & 1 \end{array}\right)$$

Multiply row 2 by –2 and add to row 1, and multiply row 2 by –8 and add to row 3:

$$\left(\begin{array}{ccc|ccc} 1 & 0 & -1 & 1 & -\frac{2}{3} & 0 \\ 0 & 1 & 0 & 0 & \frac{1}{3} & 0 \\ 0 & 0 & 1 & 4 & -\frac{8}{3} & 1 \end{array}\right)$$

Add row 3 to row 1:

$$\left(\begin{array}{ccc|ccc} 1 & 0 & 0 & 5 & -\frac{10}{3} & 1 \\ 0 & 1 & 0 & 0 & \frac{1}{3} & 0 \\ 0 & 0 & 1 & 4 & -\frac{8}{3} & 1 \end{array}\right)$$

So the inverse is $\begin{pmatrix} 5 & -\frac{10}{3} & 1 \\ 0 & \frac{1}{3} & 0 \\ 4 & -\frac{8}{3} & 1 \end{pmatrix}$.

23. Form the matrix:

$$\left(\begin{array}{ccc|ccc} -7 & 5 & 3 & 1 & 0 & 0 \\ 3 & -2 & -2 & 0 & 1 & 0 \\ 3 & -2 & -1 & 0 & 0 & 1 \end{array}\right)$$

Multiply row 2 by –1 and add to row 3:

$$\left(\begin{array}{ccc|ccc} -7 & 5 & 3 & 1 & 0 & 0 \\ 3 & -2 & -2 & 0 & 1 & 0 \\ 0 & 0 & 1 & 0 & -1 & 1 \end{array}\right)$$

Multiply row 2 by 2 and add to row 1:

$$\begin{pmatrix} -1 & 1 & -1 & 1 & 2 & 0 \\ 3 & -2 & -2 & 0 & 1 & 0 \\ 0 & 0 & 1 & 0 & -1 & 1 \end{pmatrix}$$

Multiply row 1 by 3 and add to row 2:

$$\begin{pmatrix} -1 & 1 & -1 & 1 & 2 & 0 \\ 0 & 1 & -5 & 3 & 7 & 0 \\ 0 & 0 & 1 & 0 & -1 & 1 \end{pmatrix}$$

Multiply row 3 by 5 and add to row 2, and add row 3 to row 1:

$$\begin{pmatrix} -1 & 1 & 0 & 1 & 1 & 1 \\ 0 & 1 & 0 & 3 & 2 & 5 \\ 0 & 0 & 1 & 0 & -1 & 1 \end{pmatrix}$$

Multiply row 1 by –1:

$$\begin{pmatrix} 1 & -1 & 0 & -1 & -1 & -1 \\ 0 & 1 & 0 & 3 & 2 & 5 \\ 0 & 0 & 1 & 0 & -1 & 1 \end{pmatrix}$$

Add row 2 to row 1:

$$\begin{pmatrix} 1 & 0 & 0 & 2 & 1 & 4 \\ 0 & 1 & 0 & 3 & 2 & 5 \\ 0 & 0 & 1 & 0 & -1 & 1 \end{pmatrix}$$

So the inverse is $\begin{pmatrix} 2 & 1 & 4 \\ 3 & 2 & 5 \\ 0 & -1 & 1 \end{pmatrix}$.

25. Form the matrix:

$$\begin{pmatrix} 1 & 2 & 3 & 1 & 0 & 0 \\ 4 & 5 & 6 & 0 & 1 & 0 \\ 7 & 8 & 9 & 0 & 0 & 1 \end{pmatrix}$$

Multiply row 1 by –4 and add to row 2, and multiply row 1 by –7 and add to row 3:

$$\begin{pmatrix} 1 & 2 & 3 & 1 & 0 & 0 \\ 0 & -3 & -6 & -4 & 1 & 0 \\ 0 & -6 & -12 & -7 & 0 & 1 \end{pmatrix}$$

Multiply row 2 by –2 and add to row 3:

$$\begin{pmatrix} 1 & 2 & 3 & 1 & 0 & 0 \\ 0 & -3 & -6 & -4 & 1 & 0 \\ 0 & 0 & 0 & 1 & -2 & 1 \end{pmatrix}$$

So the inverse does not exist.

27. (a) Since the system can be written as $A \bullet X = B$, where $X = \begin{pmatrix} x \\ y \end{pmatrix}$ and $B = \begin{pmatrix} 5 \\ 7 \end{pmatrix}$, then

$X = A^{-1} \bullet B$. So $X = \begin{pmatrix} 11 & -8 \\ -4 & 3 \end{pmatrix}\begin{pmatrix} 5 \\ 7 \end{pmatrix} = \begin{pmatrix} -1 \\ 1 \end{pmatrix}$, thus $x = -1$ and $y = 1$.

(b) Again this system can be written as $A \bullet X = B$, where $X = \begin{pmatrix} x \\ y \end{pmatrix}$ and $B = \begin{pmatrix} -12 \\ 0 \end{pmatrix}$, then

$X = A^{-1} \bullet B$. So $X = \begin{pmatrix} 11 & -8 \\ -4 & 3 \end{pmatrix}\begin{pmatrix} -12 \\ 0 \end{pmatrix} = \begin{pmatrix} -132 \\ 48 \end{pmatrix}$, thus $x = -132$ and $y = 48$.

29. (a) Since the system can be written as $A \bullet X = B$, where $X = \begin{pmatrix} x \\ y \\ z \end{pmatrix}$ and $B = \begin{pmatrix} 28 \\ 9 \\ 22 \end{pmatrix}$, then

$X = A^{-1} \bullet B$. So $X = \begin{pmatrix} 1 & 2 & -2 \\ -1 & 3 & 0 \\ 0 & -2 & 1 \end{pmatrix}\begin{pmatrix} 28 \\ 9 \\ 22 \end{pmatrix} = \begin{pmatrix} 2 \\ -1 \\ 4 \end{pmatrix}$, thus $x = 2$, $y = -1$, $z = 4$.

(b) Again this system can be written as $A \bullet X = B$, where $X = \begin{pmatrix} x \\ y \\ z \end{pmatrix}$ and $B = \begin{pmatrix} -7 \\ -2 \\ -6 \end{pmatrix}$, then

$X = A^{-1} \bullet B$. So $X = \begin{pmatrix} 1 & 2 & -2 \\ -1 & 3 & 0 \\ 0 & -2 & 1 \end{pmatrix}\begin{pmatrix} -7 \\ -2 \\ -6 \end{pmatrix} = \begin{pmatrix} 1 \\ 1 \\ -2 \end{pmatrix}$, thus $x = 1$, $y = 1$, $z = -2$.

31. (a) We compute AA:

$$AA = \begin{pmatrix} 1 & -6 & 3 \\ 2 & -7 & 3 \\ 4 & -12 & 5 \end{pmatrix}\begin{pmatrix} 1 & -6 & 3 \\ 2 & -7 & 3 \\ 4 & -12 & 5 \end{pmatrix} = \begin{pmatrix} 1 & 0 & 0 \\ 0 & 1 & 0 \\ 0 & 0 & 1 \end{pmatrix}$$

The product is the identity matrix I_3, and thus $A^{-1} = A$.

(b) Since this can be written as $A \bullet X = B$, where $X = \begin{pmatrix} x \\ y \\ z \end{pmatrix}$ and $B = \begin{pmatrix} \frac{19}{2} \\ 11 \\ 19 \end{pmatrix}$, then

$X = A^{-1} \bullet B$. Recall from part (a) that $A^{-1} = A$, so $X = \begin{pmatrix} 1 & -6 & 3 \\ 2 & -7 & 3 \\ 4 & -12 & 5 \end{pmatrix}\begin{pmatrix} \frac{19}{2} \\ 11 \\ 19 \end{pmatrix} = \begin{pmatrix} \frac{1}{2} \\ -1 \\ 1 \end{pmatrix}$,

thus $x = \frac{1}{2}$, $y = -1$, and $z = 1$.

33. We find $\begin{pmatrix} a & b \\ c & d \end{pmatrix}$ where $\begin{pmatrix} 2 & 5 \\ 6 & 15 \end{pmatrix}\begin{pmatrix} a & b \\ c & d \end{pmatrix} = \begin{pmatrix} 1 & 0 \\ 0 & 1 \end{pmatrix}$. Carrying out the multiplication, we have:

$$\begin{pmatrix} 2a+5c & 2b+5d \\ 6a+15c & 6b+15d \end{pmatrix} = \begin{pmatrix} 1 & 0 \\ 0 & 1 \end{pmatrix}$$

So we have the following systems of equations:

$$\begin{aligned} 2a + 5c &= 1 \\ 6a + 15c &= 0 \end{aligned} \qquad \text{and} \qquad \begin{aligned} 2b + 5d &= 0 \\ 6b + 15d &= 1 \end{aligned}$$

Multiplying the first equation by –3 yields:

$$\begin{aligned} -6a - 15c &= -3 \\ 6a + 15c &= 0 \end{aligned} \qquad \begin{aligned} -6b - 15d &= 0 \\ 6b + 15d &= 1 \end{aligned}$$

Adding, we obtain:

$$0 = -3 \qquad 0 = 1$$

Neither of these systems has a solution, thus the matrix has no inverse.

35. (a) Form the matrix:

$$\begin{pmatrix} 2 & 3 & 1 & 0 \\ 4 & 5 & 0 & 1 \end{pmatrix}$$

Multiply row 1 by –2 and add to row 2:

$$\begin{pmatrix} 2 & 3 & 1 & 0 \\ 0 & -1 & -2 & 1 \end{pmatrix}$$

Multiply row 2 by 3 and add to row 1:

$$\begin{pmatrix} 2 & 0 & -5 & 3 \\ 0 & -1 & -2 & 1 \end{pmatrix}$$

Multiply row 1 by $\frac{1}{2}$ and row 2 by –1:

$$\begin{pmatrix} 1 & 0 & -\frac{5}{2} & \frac{3}{2} \\ 0 & 1 & 2 & -1 \end{pmatrix}$$

So $A^{-1} = \begin{pmatrix} -\frac{5}{2} & \frac{3}{2} \\ 2 & -1 \end{pmatrix}$. Now form the matrix:

$$\begin{pmatrix} 7 & 8 & 1 & 0 \\ 6 & 7 & 0 & 1 \end{pmatrix}$$

Subtract row 2 from row 1:

$$\begin{pmatrix} 1 & 1 & 1 & -1 \\ 6 & 7 & 0 & 1 \end{pmatrix}$$

Multiply row 1 by -6 and add to row 2:

$$\begin{pmatrix} 1 & 1 & 1 & -1 \\ 0 & 1 & -6 & 7 \end{pmatrix}$$

Subtract row 2 from row 1:

$$\begin{pmatrix} 1 & 0 & 7 & -8 \\ 0 & 1 & -6 & 7 \end{pmatrix}$$

So $B^{-1} = \begin{pmatrix} 7 & -8 \\ -6 & 7 \end{pmatrix}$.

Then $B^{-1}A^{-1} = \begin{pmatrix} 7 & -8 \\ -6 & 7 \end{pmatrix}\begin{pmatrix} -\frac{5}{2} & \frac{3}{2} \\ 2 & -1 \end{pmatrix} = \begin{pmatrix} -\frac{67}{2} & \frac{37}{2} \\ 29 & -16 \end{pmatrix}$.

(b) We first find AB:

$$AB = \begin{pmatrix} 2 & 3 \\ 4 & 5 \end{pmatrix}\begin{pmatrix} 7 & 8 \\ 6 & 7 \end{pmatrix} = \begin{pmatrix} 32 & 37 \\ 58 & 67 \end{pmatrix}$$

Now form the matrix:

$$\begin{pmatrix} 32 & 37 & 1 & 0 \\ 58 & 67 & 0 & 1 \end{pmatrix}$$

Subtract row 1 from row 2 (to reduce numbers):

$$\begin{pmatrix} 32 & 37 & 1 & 0 \\ 26 & 30 & -1 & 1 \end{pmatrix}$$

Subtract row 2 from row 1 (to reduce numbers):

$$\begin{pmatrix} 6 & 7 & 2 & -1 \\ 26 & 30 & -1 & 1 \end{pmatrix}$$

Multiply row 1 by -4 and add to row 2:

$$\begin{pmatrix} 6 & 7 & 2 & -1 \\ 2 & 2 & -9 & 5 \end{pmatrix}$$

Multiply row 2 by -3 and add to row 1:

$$\begin{pmatrix} 0 & 1 & 29 & -16 \\ 2 & 2 & -9 & 5 \end{pmatrix}$$

Multiply row 1 by –2 and add to row 2:

$$\begin{pmatrix} 0 & 1 & 29 & -16 \\ 2 & 0 & -67 & 37 \end{pmatrix}$$

Multiply row 2 by $\frac{1}{2}$, then switch rows 1 and 2:

$$\begin{pmatrix} 1 & 0 & -\frac{67}{2} & \frac{37}{2} \\ 0 & 1 & 29 & -16 \end{pmatrix}$$

So $(AB)^{-1} = \begin{pmatrix} -\frac{67}{2} & \frac{37}{2} \\ 29 & -16 \end{pmatrix}$, which is the same as $B^{-1}A^{-1}$ from part (a).

7.5 Determinants and Cramer's Rule

1. (a) We evaluate the determinant:

$$\begin{vmatrix} 2 & -17 \\ 1 & 6 \end{vmatrix} = 2(6) - (-17)(1) = 12 + 17 = 29$$

(b) We evaluate the determinant:

$$\begin{vmatrix} 1 & 6 \\ 2 & -17 \end{vmatrix} = 1(-17) - (6)(2) = -17 - 12 = -29$$

3. (a) We evaluate the determinant:

$$\begin{vmatrix} 7 & 7 \\ 500 & 700 \end{vmatrix} = 100\begin{vmatrix} 5 & 7 \\ 5 & 7 \end{vmatrix} = 100[5(7) - 7(5)] = 100(35 - 35) = 0$$

(b) We evaluate the determinant:

$$\begin{vmatrix} 5 & 500 \\ 7 & 700 \end{vmatrix} = 100\begin{vmatrix} 5 & 5 \\ 7 & 7 \end{vmatrix} = 100[5(7) - 5(7)] = 100(35 - 35) = 0$$

5. We evaluate the determinant:

$$\begin{vmatrix} \sqrt{2}-1 & \sqrt{2} \\ \sqrt{2} & \sqrt{2}+1 \end{vmatrix} = (\sqrt{2}-1)(\sqrt{2}+1) - \sqrt{2}(\sqrt{2}) = (2-1) - 2 = -1$$

7. We evaluate the minor of 3:

$$\begin{vmatrix} 5 & 1 \\ 10 & -10 \end{vmatrix} = 10\begin{vmatrix} 5 & 1 \\ 1 & -1 \end{vmatrix} = 10[5(-1) - 1(1)] = 10(-6) = -60$$

9. We evaluate the minor of –10:

$$\begin{vmatrix} -6 & 3 \\ 5 & -4 \end{vmatrix} = 3\begin{vmatrix} -2 & 1 \\ 5 & -4 \end{vmatrix} = 3[-2(-4) - 1(5)] = 3(3) = 9$$

11. (a) We compute:

$$-6\begin{vmatrix} -4 & 1 \\ 9 & -10 \end{vmatrix} + 3\begin{vmatrix} 5 & 1 \\ 10 & -10 \end{vmatrix} + 8\begin{vmatrix} 5 & -4 \\ 10 & 9 \end{vmatrix} = -6(31) + 3(-60) + 8(85) = 314$$

(b) We compute:

$$-6\begin{vmatrix} -4 & 1 \\ 9 & -10 \end{vmatrix} - 3\begin{vmatrix} 5 & 1 \\ 10 & -10 \end{vmatrix} + 8\begin{vmatrix} 5 & -4 \\ 10 & 9 \end{vmatrix} = -6(31) - 3(-60) + 8(85) = 674$$

(c) The answer in part (b) would be the determinant.

13. (a) We expand along the second row:

$$-4\begin{vmatrix} 2 & 3 \\ 8 & 9 \end{vmatrix} + 5\begin{vmatrix} 1 & 3 \\ 7 & 9 \end{vmatrix} - 6\begin{vmatrix} 1 & 2 \\ 7 & 8 \end{vmatrix} = -4(-6) + 5(-12) - 6(-6) = 0$$

(b) We expand along the third row:

$$7\begin{vmatrix} 2 & 3 \\ 5 & 6 \end{vmatrix} - 8\begin{vmatrix} 1 & 3 \\ 4 & 6 \end{vmatrix} + 9\begin{vmatrix} 1 & 2 \\ 4 & 5 \end{vmatrix} = 7(-3) - 8(-6) + 9(-3) = 0$$

(c) We expand along the first column:

$$1\begin{vmatrix} 5 & 6 \\ 8 & 9 \end{vmatrix} - 4\begin{vmatrix} 2 & 3 \\ 8 & 9 \end{vmatrix} + 7\begin{vmatrix} 2 & 3 \\ 5 & 6 \end{vmatrix} = 1(-3) - 4(-6) + 7(-3) = 0$$

(d) We expand along the third column:

$$3\begin{vmatrix} 4 & 5 \\ 7 & 8 \end{vmatrix} - 6\begin{vmatrix} 1 & 2 \\ 7 & 8 \end{vmatrix} + 9\begin{vmatrix} 1 & 2 \\ 4 & 5 \end{vmatrix} = 3(-3) - 6(-6) + 9(-3) = 0$$

15. Factoring 5 out of the first row:

$$\begin{vmatrix} 5 & 10 & 15 \\ 1 & 2 & 3 \\ -9 & 11 & 7 \end{vmatrix} = 5\begin{vmatrix} 1 & 2 & 3 \\ 1 & 2 & 3 \\ -9 & 11 & 7 \end{vmatrix} = 5\begin{vmatrix} 0 & 0 & 0 \\ 1 & 2 & 3 \\ -9 & 11 & 7 \end{vmatrix} = 0$$

17. Expanding along the third row:

$$\begin{vmatrix} 1 & 2 & -3 \\ 4 & 5 & -9 \\ 0 & 0 & 1 \end{vmatrix} = 0\begin{vmatrix} 2 & -3 \\ 5 & -9 \end{vmatrix} - 0\begin{vmatrix} 1 & -3 \\ 4 & -9 \end{vmatrix} + 1\begin{vmatrix} 1 & 2 \\ 4 & 5 \end{vmatrix} = 1(-3) = -3$$

19. Factoring 2 out of the first row and 4 out of the second column:

$$\begin{vmatrix} -6 & -8 & 18 \\ 25 & 12 & 15 \\ -9 & 4 & 13 \end{vmatrix} = 2\begin{vmatrix} -3 & -4 & 9 \\ 25 & 12 & 15 \\ -9 & 4 & 13 \end{vmatrix} = 8\begin{vmatrix} -3 & -1 & 9 \\ 25 & 3 & 15 \\ -9 & 1 & 13 \end{vmatrix}$$

Now expanding along the first row:

$$\begin{vmatrix} -3 & -1 & 9 \\ 25 & 3 & 15 \\ -9 & 1 & 13 \end{vmatrix} = -3\begin{vmatrix} 3 & 15 \\ 1 & 13 \end{vmatrix} + 1\begin{vmatrix} 25 & 15 \\ -9 & 13 \end{vmatrix} + 9\begin{vmatrix} 25 & 3 \\ -9 & 1 \end{vmatrix} = -3(24) + 1(460) + 9(52) = 856$$

Therefore:

$$8\begin{vmatrix} -3 & -1 & 9 \\ 25 & 3 & 15 \\ -9 & 1 & 13 \end{vmatrix} = 8(856) = 6848$$

21. Factoring 16 out of the first row and 10 out of the third row:

$$\begin{vmatrix} 16 & 0 & -64 \\ -8 & 15 & -12 \\ 30 & -20 & 10 \end{vmatrix} = 16\begin{vmatrix} 1 & 0 & -4 \\ -8 & 15 & -12 \\ 30 & -20 & 10 \end{vmatrix} = 160\begin{vmatrix} 1 & 0 & -4 \\ -8 & 15 & -12 \\ 3 & -2 & 1 \end{vmatrix}$$

Now expanding along the first row:

$$\begin{vmatrix} 1 & 0 & -4 \\ -8 & 15 & -12 \\ 3 & -2 & 1 \end{vmatrix} = 1\begin{vmatrix} 15 & -12 \\ -2 & 1 \end{vmatrix} - 0\begin{vmatrix} -8 & -12 \\ 3 & 1 \end{vmatrix} - 4\begin{vmatrix} -8 & 15 \\ 3 & 2 \end{vmatrix} = 1(-9) - 4(-29) = 107$$

Therefore:

$$160\begin{vmatrix} 1 & 0 & -4 \\ -8 & 15 & -12 \\ 3 & -2 & 1 \end{vmatrix} = 160(107) = 17120$$

23. We evaluate the determinant:

$$\begin{aligned}
\begin{vmatrix} 1 & x & x^2 \\ 1 & y & y^2 \\ 1 & z & z^2 \end{vmatrix} &= \begin{vmatrix} 1 & x & x^2 \\ 0 & y-x & y^2-x^2 \\ 0 & z-x & z^2-x^2 \end{vmatrix} \\
&= \begin{vmatrix} y-x & y^2-x^2 \\ z-x & z^2-x^2 \end{vmatrix} \\
&= \begin{vmatrix} y-x & (y+x)(y-x) \\ z-x & (z+x)(z-x) \end{vmatrix} \\
&= (y-x)(z-x)\begin{vmatrix} 1 & y+x \\ 1 & z+x \end{vmatrix} \\
&= (y-x)(z-x)[z+x-y-x] \\
&= (y-x)(z-x)(z-y)
\end{aligned}$$

25. Substracting row 1 from row 2 and row 1 from row 3:

$$\begin{vmatrix} 1 & 1 & 1 \\ 1 & 1+x & 1 \\ 1 & 1 & 1+y \end{vmatrix} = \begin{vmatrix} 1 & 1 & 1 \\ 0 & x & 0 \\ 0 & 0 & y \end{vmatrix} = 1\begin{vmatrix} x & 0 \\ 0 & y \end{vmatrix} = xy$$

27. Adding column 2 to column 3, we obtain:

$$\begin{vmatrix} 1 & -1 & -1 & 2 \\ 0 & 1 & 0 & 0 \\ 2 & 1 & 1 & -1 \\ -2 & 2 & 3 & 1 \end{vmatrix} = 1\begin{vmatrix} 1 & -1 & 2 \\ 2 & 1 & -1 \\ -2 & 3 & 1 \end{vmatrix}$$

Adding twice column 3 to column 1, we obtain:

$$\begin{vmatrix} 5 & -1 & 2 \\ 0 & 1 & -1 \\ 0 & 3 & 1 \end{vmatrix} = 5\begin{vmatrix} 1 & -1 \\ 3 & 1 \end{vmatrix} = 5(4) = 20$$

29. We evaluate the determinant:

$$\begin{vmatrix} 2 & 0 & 0 & 0 \\ 0 & 3 & 0 & 0 \\ 0 & 0 & 4 & 0 \\ 0 & 0 & 0 & 5 \end{vmatrix} = 2\begin{vmatrix} 3 & 0 & 0 \\ 0 & 4 & 0 \\ 0 & 0 & 5 \end{vmatrix} = 2(3)\begin{vmatrix} 4 & 0 \\ 0 & 5 \end{vmatrix} = 6(20) = 120$$

31. (a) Subtract the first column from the second and add twice the first column to the third. The result is:

$$\begin{vmatrix} -20 & 22 & -43 \\ 6 & -10 & 13 \\ -1 & 0 & 0 \end{vmatrix} = -1[22(13) - (-43)(-10)] = 144$$

(b) Multiply the second row by 2 and subtract it from the first row, and multiply the second row by 4 and subtract it from the third row. This yields:

$$\begin{vmatrix} 0 & -32 & -5 \\ 1 & 6 & 1 \\ 0 & -25 & -2 \end{vmatrix} = -1[(-32)(-2) - (-5)(-25)] = 61$$

(c) Subtract the first column from the second and add 10 times the first column to the third. This yields:

$$\begin{vmatrix} 2 & 0 & 0 \\ 1 & -5 & 16 \\ 4 & -5 & 39 \end{vmatrix} = 2[(-5)(39) - 16(-5)] = -230$$

33. We compute D:

$$D=\begin{vmatrix}3 & 4 & -1\\1 & -3 & 2\\5 & 0 & -6\end{vmatrix}=5\begin{vmatrix}4 & -1\\-3 & 2\end{vmatrix}-6\begin{vmatrix}3 & 4\\1 & -3\end{vmatrix}=5(5)-6(-13)=103$$

Now compute D_x:

$$D_x=\begin{vmatrix}5 & 4 & -1\\2 & -3 & 2\\-7 & 0 & -6\end{vmatrix}=-7\begin{vmatrix}4 & -1\\-3 & 2\end{vmatrix}-6\begin{vmatrix}5 & 4\\2 & -3\end{vmatrix}=-7(5)-6(-23)=103$$

For D_y, adding –2 times column 1 to both column 2 and column 3 yields:

$$D_y=\begin{vmatrix}3 & 5 & -1\\1 & 2 & 2\\5 & -7 & -6\end{vmatrix}=\begin{vmatrix}3 & -1 & -7\\1 & 0 & 0\\5 & -17 & -16\end{vmatrix}=-1\begin{vmatrix}-1 & -7\\-17 & -16\end{vmatrix}=-(-103)=103$$

For D_z, adding 3 times column 1 to column 2, and –2 times column 1 to column 3 yields:

$$D_z=\begin{vmatrix}3 & 4 & 5\\1 & -3 & 2\\5 & 0 & -7\end{vmatrix}=\begin{vmatrix}3 & 13 & -1\\1 & 0 & 0\\5 & 15 & -17\end{vmatrix}=-1\begin{vmatrix}13 & 1\\15 & -17\end{vmatrix}=-(-206)=206$$

So $x=\frac{D_x}{D}=1$, $y=\frac{D_y}{D}=1$, and $z=\frac{D_z}{D}=2$. So the solution is $(1,1,2)$.

35. Subtracting column 1 from both column 2 and column 3 yields:

$$D=\begin{vmatrix}3 & 2 & -1\\2 & -3 & -4\\1 & 1 & 1\end{vmatrix}=\begin{vmatrix}3 & -1 & -4\\2 & -5 & -6\\1 & 0 & 0\end{vmatrix}=1\begin{vmatrix}-1 & -4\\-5 & -6\end{vmatrix}=-14$$

Adding –5 times column 3 to column 1, and subtracting column 3 from column 2 yields:

$$D_x=\begin{vmatrix}-6 & 2 & -1\\-11 & -3 & -4\\5 & 1 & 1\end{vmatrix}=\begin{vmatrix}-1 & 3 & -1\\9 & 1 & -4\\0 & 0 & 1\end{vmatrix}=1\begin{vmatrix}-1 & 3\\9 & 1\end{vmatrix}=-28$$

Adding row 3 to row 1, and 4 times row 3 to row 2 yields:

$$D_y=\begin{vmatrix}3 & -6 & -1\\2 & -11 & -4\\1 & 5 & 1\end{vmatrix}=\begin{vmatrix}4 & -1 & 0\\6 & 9 & 0\\1 & 5 & 1\end{vmatrix}=1\begin{vmatrix}4 & -1\\6 & 9\end{vmatrix}=42$$

Adding –2 times row 3 to row 1, and 3 times row 3 to row 2 yields:

$$D_z=\begin{vmatrix}3 & 2 & -6\\2 & 3 & -11\\1 & 1 & 5\end{vmatrix}=\begin{vmatrix}1 & 0 & -16\\5 & 0 & 4\\1 & 1 & 5\end{vmatrix}=-1\begin{vmatrix}1 & -16\\5 & 4\end{vmatrix}=-(84)=-84$$

So $x=\frac{D_x}{D}=2$, $y=\frac{D_y}{D}=-3$, and $z=\frac{D_z}{D}=6$. So the solution is $(2,-3,6)$.

37. Adding –2 times row 3 to row 1, and –3 times row 3 to row 2 yields:

$$D=\begin{vmatrix}2 & 5 & 2\\ 3 & -1 & -4\\ 1 & 2 & -3\end{vmatrix}=\begin{vmatrix}0 & 1 & 8\\ 0 & -7 & 5\\ 1 & 2 & -3\end{vmatrix}=1\begin{vmatrix}1 & 8\\ -7 & 5\end{vmatrix}=61$$

Now compute:

$$D_x=\begin{vmatrix}0 & 5 & 2\\ 0 & -1 & -4\\ 0 & 2 & 3\end{vmatrix}=0$$

$$D_y=\begin{vmatrix}2 & 0 & 2\\ 3 & 0 & -4\\ 1 & 0 & -3\end{vmatrix}=0$$

$$D_z=\begin{vmatrix}2 & 5 & 0\\ 3 & -1 & 0\\ 1 & 2 & 0\end{vmatrix}=0$$

So $x=\frac{D_x}{D}=0$, $y=\frac{D_y}{D}=0$, and $z=\frac{D_z}{D}=0$. So the solution is $(0,0,0)$.

39. Factoring 6 out of column 1 and 2 out of column 2, we have:

$$D=\begin{vmatrix}12 & 0 & -11\\ 6 & 6 & -4\\ 6 & 2 & -5\end{vmatrix}=6\begin{vmatrix}2 & 0 & -11\\ 1 & 6 & -4\\ 1 & 2 & -5\end{vmatrix}=12\begin{vmatrix}2 & 0 & -11\\ 1 & 3 & -4\\ 1 & 1 & -5\end{vmatrix}$$

Adding –3 times row 3 to row 2 yields:

$$D=12\begin{vmatrix}2 & 0 & -11\\ -2 & 0 & 11\\ 1 & 1 & -5\end{vmatrix}=12(-1)\begin{vmatrix}2 & -11\\ -2 & 11\end{vmatrix}=-12(0)=0$$

So Cramer's Rule will not work. We form the augmented matrix (using equation two as row 1):

$$\begin{pmatrix}6 & 6 & -4 & 26\\ 12 & 0 & -11 & 13\\ 6 & 2 & -5 & 13\end{pmatrix}$$

Adding –2 times row 1 to row 2 and –1 times row 1 to row 3 yields:

$$\begin{pmatrix}6 & 6 & -4 & 26\\ 0 & -12 & -3 & -39\\ 0 & -4 & -1 & -13\end{pmatrix}$$

Multiply row 1 by $\frac{1}{2}$ and row 2 by $\frac{1}{3}$:

$$\begin{pmatrix}3 & 3 & -2 & 13\\ 0 & 4 & 1 & 13\\ 0 & -4 & -1 & -13\end{pmatrix}$$

Adding row 2 to row 3 yields:

$$\begin{pmatrix} 3 & 3 & -2 & 13 \\ 0 & 4 & 1 & 13 \\ 0 & 0 & 0 & 0 \end{pmatrix}$$

So we have the system:

$$\begin{aligned} 3x+3y-2z&=13 \\ 4y+z&=13 \end{aligned}$$

Solve equation 2 for z:

$$\begin{aligned} 4y+z&=13 \\ z&=13-4y \end{aligned}$$

Substitute into equation 1:

$$\begin{aligned} 3x+3y-26+8y&=13 \\ 3x&=39-11y \\ x&=13-\tfrac{11}{3}y \end{aligned}$$

So the solution is $\left(13-\frac{11}{3}y, y, 13-4y\right)$, for any real number y.

41. Adding row 1 to row 2, 3 times row 1 to row 3, and row 1 to row 4 yields:

$$D=\begin{vmatrix} 1 & 1 & 1 & 1 \\ 1 & -1 & 1 & -1 \\ 2 & -2 & -3 & -3 \\ 3 & 2 & 1 & -1 \end{vmatrix}=\begin{vmatrix} 1 & 1 & 1 & 1 \\ 2 & 0 & 2 & 0 \\ 5 & 1 & 0 & 0 \\ 4 & 3 & 2 & 0 \end{vmatrix}=-1\begin{vmatrix} 2 & 0 & 2 \\ 5 & 1 & 0 \\ 4 & 3 & 2 \end{vmatrix}$$

Subtracting column 3 from column 1 yields:

$$D=-1\begin{vmatrix} 0 & 0 & 2 \\ 5 & 1 & 0 \\ 2 & 3 & 2 \end{vmatrix}=-2\begin{vmatrix} 5 & 1 \\ 2 & 3 \end{vmatrix}=-2(13)=-26$$

Adding row 1 to row 2, 3 times row 1 to row 3, and row 1 to row 4 yields:

$$D_x=\begin{vmatrix} -7 & 1 & 1 & 1 \\ -11 & -1 & 1 & -1 \\ 26 & -2 & -3 & -3 \\ -9 & 2 & 1 & -1 \end{vmatrix}=\begin{vmatrix} -7 & 1 & 1 & 1 \\ -18 & 0 & 2 & 0 \\ 5 & 1 & 0 & 0 \\ -16 & 3 & 2 & 0 \end{vmatrix}=-1\begin{vmatrix} -18 & 0 & 2 \\ 5 & 1 & 0 \\ -16 & 3 & 2 \end{vmatrix}$$

Subtracting row 3 from row 1 yields:

$$D_x=-1\begin{vmatrix} -2 & -3 & 0 \\ 5 & 1 & 0 \\ -16 & 3 & 2 \end{vmatrix}=-2\begin{vmatrix} -2 & -3 \\ 5 & 1 \end{vmatrix}=-2(13)=-26$$

Adding row 1 to row 2, 3 times row 1 to row 3, and row 1 to row 4 yields:

$$D_y=\begin{vmatrix} 1 & -7 & 1 & 1 \\ 1 & -11 & 1 & -1 \\ 2 & 26 & -3 & -3 \\ 3 & -9 & 1 & -1 \end{vmatrix}=\begin{vmatrix} 1 & -7 & 1 & 1 \\ 2 & -18 & 2 & 0 \\ 5 & 5 & 0 & 0 \\ 4 & -16 & 2 & 0 \end{vmatrix}=-1\begin{vmatrix} 2 & -18 & 2 \\ 5 & 5 & 0 \\ 4 & -16 & 2 \end{vmatrix}$$

Subtracting row 3 from row 1 yields:

$$D_y = -1\begin{vmatrix} -2 & -2 & 0 \\ 5 & 5 & 0 \\ 4 & -16 & 2 \end{vmatrix} = -2\begin{vmatrix} -2 & -2 \\ 5 & 5 \end{vmatrix} = -2(0) = 0$$

Adding row 1 to row 2, 3 times row 1 to row 3, and row 1 to row 4 yields:

$$D_z = \begin{vmatrix} 1 & 1 & -7 & 1 \\ 1 & -1 & -11 & -1 \\ 2 & -2 & 26 & -3 \\ 3 & 2 & -9 & -1 \end{vmatrix} = \begin{vmatrix} 1 & 1 & -7 & 1 \\ 2 & 0 & -18 & 0 \\ 5 & 1 & 5 & 0 \\ 4 & 3 & -16 & 0 \end{vmatrix} = -1\begin{vmatrix} 2 & 0 & -18 \\ 5 & 1 & 5 \\ 4 & 3 & -16 \end{vmatrix}$$

Adding –3 times row 2 to row 3 yields:

$$D_z = -1\begin{vmatrix} 2 & 0 & -18 \\ 5 & 1 & 5 \\ 11 & 0 & -31 \end{vmatrix} = -1\begin{vmatrix} 2 & -18 \\ -11 & -31 \end{vmatrix} = -1(-260) = 260$$

Adding row 1 to row 2, 2 times row 1 to row 3, and –2 times row 1 to row 4 yields:

$$D_w = \begin{vmatrix} 1 & 1 & 1 & -7 \\ 1 & -1 & 1 & -11 \\ 2 & -2 & -3 & 26 \\ 3 & 2 & 1 & -9 \end{vmatrix} = \begin{vmatrix} 1 & 1 & 1 & -7 \\ 2 & 0 & 2 & -18 \\ 4 & 0 & -1 & 12 \\ 1 & 0 & -1 & 5 \end{vmatrix} = -1\begin{vmatrix} 2 & 2 & -18 \\ 4 & -1 & 12 \\ 1 & -1 & 5 \end{vmatrix} = -2\begin{vmatrix} 1 & 1 & -9 \\ 4 & -1 & 12 \\ 1 & -1 & 5 \end{vmatrix}$$

Adding row 1 to both row 2 and row 3 yields:

$$D_w = -2\begin{vmatrix} 1 & 1 & -9 \\ 5 & 0 & 3 \\ 2 & 0 & -4 \end{vmatrix} = 2\begin{vmatrix} 5 & 3 \\ 2 & -4 \end{vmatrix} = 2(-26) = -52$$

So $x = \frac{D_x}{D} = 1$, $y = \frac{D_y}{D} = 0$, $z = \frac{D_z}{D} = -10$, and $w = \frac{D_w}{D} = 2$. So the solution is $(1,0,-10,2)$.

43. Expanding the determinant, we have:

$$\begin{vmatrix} x-4 & 0 & 0 \\ 0 & x+4 & 0 \\ 0 & 0 & x+1 \end{vmatrix} = (x-4)\begin{vmatrix} x+4 & 0 \\ 0 & x+1 \end{vmatrix} = (x-4)(x+4)(x+1)$$

This will equal 0 when $x = 4$, $x = -4$, or $x = -1$.

45. Expanding along the first column, we have:

$$\begin{vmatrix} a & b & c \\ a & b & c \\ d & e & f \end{vmatrix} = a\begin{vmatrix} b & c \\ e & f \end{vmatrix} - a\begin{vmatrix} b & c \\ e & f \end{vmatrix} + d\begin{vmatrix} b & c \\ b & c \end{vmatrix}$$

$$= a(bf - ec) - a(bf - ec) + d(bc - bc)$$
$$= abf - aec - abf + aec + 0$$
$$= 0$$

47. Expanding the determinant:

$$\begin{vmatrix} a_1+A_1 & b_1 & c_1 \\ a_2+A_2 & b_2 & c_2 \\ a_3+A_3 & b_3 & c_3 \end{vmatrix} = (a_1+A_1)\begin{vmatrix} b_2 & c_2 \\ b_3 & c_3 \end{vmatrix} - (a_2+A_2)\begin{vmatrix} b_1 & c_1 \\ b_3 & c_3 \end{vmatrix} + (a_3+A_3)\begin{vmatrix} b_1 & c_1 \\ b_2 & c_2 \end{vmatrix}$$

$$= \left\{ a_1\begin{vmatrix} b_2 & c_2 \\ b_3 & c_3 \end{vmatrix} - a_2\begin{vmatrix} b_1 & c_1 \\ b_3 & c_3 \end{vmatrix} + a_3\begin{vmatrix} b_1 & c_1 \\ b_2 & c_2 \end{vmatrix} \right\}$$

$$+ \left\{ A_1\begin{vmatrix} b_2 & c_2 \\ b_3 & c_3 \end{vmatrix} - A_2\begin{vmatrix} b_1 & c_1 \\ b_3 & c_3 \end{vmatrix} + A_3\begin{vmatrix} b_1 & c_1 \\ b_2 & c_2 \end{vmatrix} \right\}$$

Now observe that the expression in the first set of braces is:

$$\begin{vmatrix} a_1 & b_1 & c_1 \\ a_2 & b_2 & c_2 \\ a_3 & b_3 & c_3 \end{vmatrix}$$

The expression in the second set of braces is:

$$\begin{vmatrix} A_1 & b_1 & c_1 \\ A_2 & b_2 & c_2 \\ A_3 & b_3 & c_3 \end{vmatrix}$$

49. Expanding the determinant on the left-hand side of the given equation along its first row, we obtain:

$$\begin{vmatrix} a_1 & b_1 & c_1 \\ a_2 & b_2 & c_2 \\ a_3 & b_3 & c_3 \end{vmatrix} = a_1\begin{vmatrix} b_2 & c_2 \\ b_3 & c_3 \end{vmatrix} - b_1\begin{vmatrix} a_2 & c_2 \\ a_3 & c_3 \end{vmatrix} + c_1\begin{vmatrix} a_2 & b_2 \\ a_3 & b_3 \end{vmatrix}$$

Next, expanding the deteerminant on the right-hand side of the given equation along its second row, we obtain:

$$\begin{vmatrix} a_2 & b_2 & c_2 \\ a_1 & b_1 & c_1 \\ a_3 & b_3 & c_3 \end{vmatrix} = -a_1\begin{vmatrix} b_2 & c_2 \\ b_3 & c_3 \end{vmatrix} + b_1\begin{vmatrix} a_2 & c_2 \\ a_3 & c_3 \end{vmatrix} - c_1\begin{vmatrix} a_2 & b_2 \\ a_3 & b_3 \end{vmatrix}$$

By inspection now, we observe that the two expressions for the determinants are negatives of one another.

51. Subtract the fourth row from each of the other three rows. After that, we have:

$$\begin{vmatrix} a & 0 & 0 & -d \\ 0 & b & 0 & -d \\ 0 & 0 & c & -d \\ 1 & 1 & 1 & 1+d \end{vmatrix} = abcd \begin{vmatrix} 1 & 0 & 0 & -1 \\ 0 & 1 & 0 & -1 \\ 0 & 0 & 1 & -1 \\ \frac{1}{a} & \frac{1}{b} & \frac{1}{c} & 1+\frac{1}{d} \end{vmatrix}$$

$$= abcd \begin{vmatrix} 1 & 0 & 0 & 0 \\ 0 & 1 & 0 & -1 \\ 0 & 0 & 1 & -1 \\ \frac{1}{a} & \frac{1}{b} & \frac{1}{c} & 1+\frac{1}{a}+\frac{1}{d} \end{vmatrix}$$

$$= abcd \begin{vmatrix} 1 & 0 & -1 \\ 0 & 1 & -1 \\ \frac{1}{b} & \frac{1}{c} & 1+\frac{1}{a}+\frac{1}{d} \end{vmatrix}$$

$$= abcd \begin{vmatrix} 1 & 0 & 0 \\ 0 & 1 & -1 \\ \frac{1}{b} & \frac{1}{c} & 1+\frac{1}{a}+\frac{1}{b}+\frac{1}{d} \end{vmatrix}$$

$$= abcd \begin{vmatrix} 1 & -1 \\ \frac{1}{c} & 1+\frac{1}{a}+\frac{1}{b}+\frac{1}{d} \end{vmatrix}$$

$$= abcd\left(1+\frac{1}{a}+\frac{1}{b}+\frac{1}{c}+\frac{1}{d}\right)$$

53. By expanding D along its first column, we obtain the equation:

$$a_1D = a_1\left[a_1(b_2c_3 - b_3c_2) - a_2(b_1c_3 - b_3c_1) + a_3(b_1c_2 - b_2c_1)\right] \quad **$$

On the other hand $\begin{vmatrix} B_2 & C_2 \\ B_3 & C_3 \end{vmatrix}$ is equal to:

$$\begin{aligned} B_2C_3 - B_3C_2 &= (a_1c_3 - a_3c_1)(a_1b_2 - a_2b_1) - (a_2c_1 - a_1c_2)(a_3b_1 - a_1b_3) \\ &= a_1{}^2b_2c_3 - a_1a_3b_2c_1 - a_1a_2b_1c_3 + a_2a_3b_1c_1 - a_1{}^2b_3c_2 \\ &\qquad + a_1a_3b_1c_2 + a_1a_2b_3c_1 - a_2a_3b_1c_1 \\ &= a_1\left[a_1b_2c_3 - a_3b_2c_1 - a_2b_1c_3 - a_1b_3c_2 + a_3b_1c_2 + a_2b_3c_1\right] \\ &= a_1\left[a_1(b_2c_3 - b_3c_2) - a_2(b_1c_3 - b_3c_1) + a_3(b_1c_2 - b_2c_1)\right] \end{aligned}$$

By inspection now, we see that this last expression agrees with the right-hand side of equation **. This proves that:

$$\begin{vmatrix} B_2 & C_2 \\ B_3 & C_3 \end{vmatrix} = a_1D$$

55. (a) Expanding along the first row:

$$\begin{vmatrix} 1 & 0 & 0 \\ x & 1 & 0 \\ x & y & 1 \end{vmatrix} = 1\begin{vmatrix} 1 & 0 \\ y & 1 \end{vmatrix} = 1(1) = 1$$

(b) Expanding along the first row:

$$\begin{vmatrix} 1 & 0 & 0 & 0 \\ x & 1 & 0 & 0 \\ x & y & 1 & 0 \\ x & y & z & 1 \end{vmatrix} = 1\begin{vmatrix} 1 & 0 & 0 \\ y & 1 & 0 \\ y & z & 1 \end{vmatrix} = 1\begin{vmatrix} 1 & 0 \\ z & 1 \end{vmatrix} = 1(1) = 1$$

57. Subtracting row 1 from each of row 2, row 3, and row 4 yields:

$$\begin{aligned} \begin{vmatrix} 1 & a & a & a \\ 1 & b & a & a \\ 1 & a & b & a \\ 1 & a & a & b \end{vmatrix} &= \begin{vmatrix} 1 & a & a & a \\ 0 & b-a & 0 & 0 \\ 0 & 0 & b-a & 0 \\ 0 & 0 & 0 & b-a \end{vmatrix} \\ &= 1\begin{vmatrix} b-a & 0 & 0 \\ 0 & b-a & 0 \\ 0 & 0 & b-a \end{vmatrix} \\ &= (b-a)\begin{vmatrix} b-a & 0 \\ 0 & b-a \end{vmatrix} \\ &= (b-a)(b-a)^2 \\ &= (b-a)^3 \end{aligned}$$

59. We form the augmented matrix:

$$\begin{pmatrix} a & b & c & k \\ a^2 & b^2 & c^2 & k^2 \\ a^3 & b^3 & c^3 & k^3 \end{pmatrix}$$

Adding $-a$ times row 1 to row 2 and $-a^2$ times row 1 to row 3:

$$\begin{pmatrix} a & b & c & k \\ 0 & b^2 - ab & c^2 - ac & k^2 - ak \\ 0 & b^3 - a^2b & c^3 - a^2c & k^3 - a^2k \end{pmatrix}$$

Factoring:

$$\begin{pmatrix} a & b & c & k \\ 0 & b(b-a) & c(c-a) & k(k-a) \\ 0 & b(b+a)(b-a) & c(c+a)(c-a) & k(k+a)(k-a) \end{pmatrix}$$

Adding $-(b+a)$ times row 2 to row 3:

$$\begin{pmatrix} a & b & c & k \\ 0 & b(b-a) & c(c-a) & k(k-a) \\ 0 & 0 & c(c-a)(c-b) & k(k-a)(k-b) \end{pmatrix}$$

So we have the system:

$$\begin{aligned} ax + by + cz &= k \\ b(b-a)y + c(c-a)z &= k(k-a) \\ c(c-a)(c-b)z &= k(k-a)(k-b) \end{aligned}$$

Solving the third equation for z:

$$\begin{aligned} c(c-a)(c-b)z &= k(k-a)(k-b) \\ z &= \frac{k(k-a)(k-b)}{c(c-a)(c-b)} \end{aligned}$$

Substitute into the second equation:

$$\begin{aligned} b(b-a)y + \frac{k(k-a)(k-b)}{c-b} &= k(k-a) \\ b(b-a)(c-b)y &= k(k-a)(c-b-k+b) \\ y &= \frac{k(k-a)(k-c)}{b(b-a)(b-c)} \end{aligned}$$

Substitute into the first equation:

$$\begin{aligned} ax + \frac{k(k-a)(k-c)}{(b-a)(b-c)} + \frac{k(k-a)(k-b)}{(c-a)(c-b)} &= k \\ ax &= \frac{k(k-b)(k-c)}{(b-a)(c-a)} \\ x &= \frac{k(k-b)(k-c)}{a(a-b)(a-c)} \end{aligned}$$

So the solution is $\left(\frac{k(k-b)(k-c)}{a(a-b)(a-c)}, \frac{k(k-a)(k-c)}{b(b-a)(b-c)}, \frac{k(k-a)(k-b)}{c(c-a)(c-b)}\right)$.

61. We evaluate the determinant by expanding along the first row:

$$\begin{vmatrix} x & y & 1 \\ -3 & -1 & 1 \\ 2 & 9 & 1 \end{vmatrix} = x(-1-9) - y(-3-2) + 1(-27+2) = -10x + 5y - 25$$

Setting this equal to 0, we have:

$$\begin{aligned} -10x + 5y - 25 &= 0 \\ 5y &= 10x + 25 \\ y &= 2x + 5 \end{aligned}$$

63. We re-draw the figure:

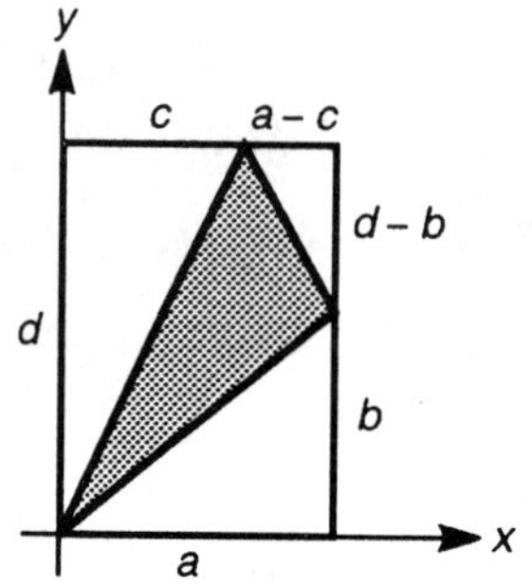

The area of the rectangle is ad, and the three triangles have areas of $\frac{1}{2}(ab)$, $\frac{1}{2}(a-c)(d-b)$, and $\frac{1}{2}(cd)$, so the area of the shaded triangle is:

$$ad - \left[\tfrac{1}{2}ab + \tfrac{1}{2}ad - \tfrac{1}{2}cd - \tfrac{1}{2}ab + \tfrac{1}{2}bc + \tfrac{1}{2}cd\right] = \tfrac{1}{2}ad - \tfrac{1}{2}bc = \tfrac{1}{2}(ad - bc) = \tfrac{1}{2}\begin{vmatrix} a & b \\ c & d \end{vmatrix}$$

65. (a) From Exercise 36 of the previous section, we found $A^{-1} = \begin{pmatrix} \frac{d}{ad-bc} & \frac{-b}{ad-bc} \\ \frac{-c}{ad-bc} & \frac{a}{ad-bc} \end{pmatrix}$. To check this, note that A times this matrix equals I_2. Since $D = ad - bc$, then

$$A^{-1} = \frac{1}{D}\begin{pmatrix} d & -b \\ -c & a \end{pmatrix}.$$

(b) We have $D = -6(9) - 7(1) = -54 - 7 = -61$, so the inverse is:

$$-\frac{1}{61}\begin{pmatrix} 9 & -7 \\ -1 & -6 \end{pmatrix} = \begin{pmatrix} -\frac{9}{61} & \frac{7}{61} \\ \frac{1}{61} & \frac{6}{61} \end{pmatrix}$$

7.6 Nonlinear Systems of Equations

1. Substituting, we obtain:

$$x^2 = 3x$$
$$x^2 - 3x = 0$$
$$x(x-3) = 0$$

From which we get $x = 0$ or $x = 3$. If $x = 0$, $y = 3 \bullet 0 = 0$ and if $x = 3$, $y = 9$. Our solutions are $(0,0)$ and $(3,9)$. Note that we have found the points of intersection of a line and a parabola.

3. Since $x^2 = 24y$, $x^2 + y^2 = 25$ becomes:

$$24y + y^2 = 25$$
$$y^2 + 24y - 25 = 0$$
$$(y+25)(y-1) = 0$$
$$y = -25 \text{ and } y = 1$$

When $y = 1$ we get $x^2 = 24$ or $x = \pm 2\sqrt{6}$, but $y = -25$ means $x^2 = -600$ which is impossible. Our two solutions are $(2\sqrt{6}, 1)$ and $(-2\sqrt{6}, 1)$.

5. Substitute into the first equation:

$$x(-x^2) = 1$$
$$-x^3 = 1$$
$$x = -1$$

Since $y = \frac{1}{x}$, then $y = -1$. So the only solution is $(-1, -1)$.

7. Multiply the first equation by -2:

$$-4x^2 - 2y^2 = -34$$
$$x^2 + 2y^2 = 22$$

Adding, we obtain:

$$-3x^2 = -12$$
$$x^2 = 4$$
$$x = \pm 2$$

When $x = 2$, we have:

$$2(4) + y^2 = 17$$
$$y^2 = 9$$
$$y = \pm 3$$

When $x = -2$, we have:

$$2(4) + y^2 = 17$$
$$y^2 = 9$$
$$y = \pm 3$$

So the solutions are $(2, 3)$, $(2, -3)$, $(-2, 3)$, and $(-2, -3)$.

9. Substitute into the first equation:

$$x^2 - 1 = 1 - x^2$$
$$2x^2 = 2$$
$$x^2 = 1$$
$$x = \pm 1$$

So $y = 1 - 1 = 0$ for each value of x. So the solutions are $(1, 0)$ and $(-1, 0)$.

11. Substitute into the first equation:

$$x(4x + 1) = 4$$
$$4x^2 + x - 4 = 0$$
$$x = \frac{-1 \pm \sqrt{1 + 64}}{8} = \frac{-1 \pm \sqrt{65}}{8}$$

When $x = \frac{-1+\sqrt{65}}{8}$, we have:

$$y = \frac{-1+\sqrt{65}}{2} + 1 = \frac{1+\sqrt{65}}{2}$$

When $x = \frac{-1-\sqrt{65}}{8}$, we have:

$$y = \frac{-1-\sqrt{65}}{2} + 1 = \frac{1-\sqrt{65}}{2}$$

So the solutions are $\left(\frac{-1+\sqrt{65}}{8}, \frac{1+\sqrt{65}}{2}\right)$ and $\left(\frac{-1-\sqrt{65}}{8}, \frac{1-\sqrt{65}}{2}\right)$.

13. Let $a = \frac{1}{x^2}$ and $b = \frac{1}{y^2}$, so:

$$a - 3b = 14$$
$$2a + b = 35$$

Multiply the first equation by –2:

$$-2a + 6b = -28$$
$$2a + b = 35$$

Adding, we obtain:

$$7b = 7$$
$$b = 1$$

So $a - 3 = 14$, and $a = 17$. Since $a = \frac{1}{x^2}$ and $b = \frac{1}{y^2}$, we have $x^2 = \frac{1}{17}$ and $y^2 = 1$.

So $x = \frac{\pm\sqrt{17}}{17}$ and $y = \pm 1$. So the solutions are $\left(\frac{\sqrt{17}}{17}, 1\right)$, $\left(\frac{\sqrt{17}}{17}, -1\right)$, $\left(\frac{-\sqrt{17}}{17}, 1\right)$, and $\left(\frac{-\sqrt{17}}{17}, -1\right)$.

15. Substitute into the second equation:

$$(x-3)^2 + \left(-\sqrt{x-1}\right)^2 = 4$$
$$x^2 - 6x + 9 + x - 1 = 4$$
$$x^2 - 5x + 4 = 0$$
$$(x-1)(x-4) = 0$$
$$x = 1 \text{ or } x = 4$$

When $x = 1$, $y = -\sqrt{1-1} = 0$ and when $x = 4$, $y = -\sqrt{4-1} = -\sqrt{3}$. So the solutions are $(1, 0)$ and $\left(4, -\sqrt{3}\right)$.

17. Since $y = 2^{2x} - 12 = \left(2^x\right)^2 - 12$, we substitute into the second equation:

$$y = y^2 - 12$$
$$0 = y^2 - y - 12$$
$$0 = (y-4)(y+3)$$
$$y = 4 \text{ or } y = -3$$

When $y = 4$, we have $2^x = 4$, or $x = 2$. But $y = -3$ will not yield a solution. So the only solution is $(2, 4)$.

19. Let $u = \log_{10}x$ and $v = \log_{10}y$, so:

$$2u^2 - v^2 = -1$$
$$4u^2 - 3v^2 = -11$$

Multiply the first equation by –2:

$$-4u^2 + 2v^2 = 2$$
$$4u^2 - 3v^2 = -11$$

Adding, we obtain:

$$-v^2 = -9$$
$$v^2 = 9$$
$$v = \pm 3$$

Substitute into the first equation:

$$2u^2 - 9 = -1$$
$$2u^2 = 8$$
$$u^2 = 4$$
$$u^2 = \pm 2$$

Since $u = \log_{10}x$, then $x = 10^{\pm 2}$. Similarly, $y = 10^{\pm 3}$. So the solutions are $(100, 1000)$, $\left(100, \frac{1}{1000}\right)$, $\left(\frac{1}{100}, 1000\right)$, and $\left(\frac{1}{100}, \frac{1}{1000}\right)$.

21. First take the logarithm of each side of the first equation:

$$\ln\left(2^x 3^y\right) = \ln 4$$
$$\ln\left(2^x\right) + \ln\left(3^y\right) = \ln 2^2$$
$$(\ln 2)x + (\ln 3)y = 2\ln 2$$

Multiply the second equation by $-\ln 2$:

$$(\ln 2)x + (\ln 3)y = 2\ln 2$$
$$(-\ln 2)x - (\ln 2)y = -5\ln 2$$

Adding, we obtain:

$$(\ln 3 - \ln 2)y = -3\ln 2$$
$$y = \frac{3\ln 2}{\ln 2 - \ln 3}$$

Substitute into the second equation:

$$x = 5 - \frac{3\ln 2}{\ln 2 - \ln 3} = \frac{2\ln 2 - 5\ln 3}{\ln 2 - \ln 3}$$

The solution is $\left(\frac{2\ln 2 - 5\ln 3}{\ln 2 - \ln 3}, \frac{3\ln 2}{\ln 2 - \ln 3}\right)$.

23. Substituting x yields:

$$y = 3x + 1 = \frac{-3 + 3\sqrt{13}}{6} + 1 = \frac{3 + 3\sqrt{13}}{6} = \frac{1 + \sqrt{13}}{2}$$
$$y = \frac{1}{x} = \frac{6}{-1 + \sqrt{13}} \bullet \frac{-1 - \sqrt{13}}{-1 - \sqrt{13}} = \frac{-6(1 + \sqrt{13})}{1 - 13} = \frac{-6(1 + \sqrt{13})}{-12} = \frac{1 + \sqrt{13}}{2}$$

They both yield the same y-value.

25. Since $ax + by = 2$, then $by = 2 - ax$. Substitute into the second equation:

$$\begin{aligned} ax(by) &= 1 \\ ax(2-ax) &= 1 \\ 2ax - a^2x^2 &= 1 \\ a^2x^2 - 2ax + 1 &= 0 \\ (ax-1)^2 &= 0 \\ ax &= 1 \\ x &= \tfrac{1}{a} \end{aligned}$$

When $ax = 1$, $by = 2 - 1 = 1$, so $y = \frac{1}{b}$. So the solution is $\left(\frac{1}{a}, \frac{1}{b}\right)$.

27. Solve the second equation for y to get $y = 23 - x$. Substitute into the first equation:

$$\begin{aligned} x^3 + (23-x)^3 &= 3473 \\ x^3 + 12167 - 1587x + 69x^2 - x^3 &= 3473 \\ 69x^2 - 1587x + 8694 &= 0 \\ x^2 - 23x + 126 &= 0 \\ (x-9)(x-14) &= 0 \\ x = 9 &\text{ or } x = 14 \end{aligned}$$

When $x = 9$, $y = 14$ and when $x = 14$, $y = 9$. So the solutions are $(9, 14)$ and $(14, 9)$.

29. First draw the rectangle:

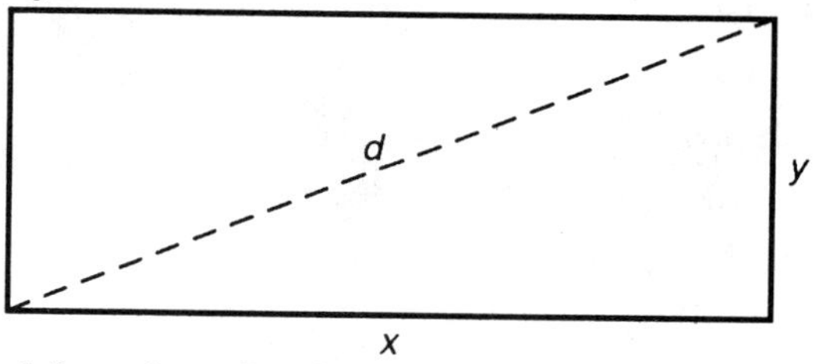

Now $x^2 + y^2 = d^2$ and $2x + 2y = 2p$. Solve the second equation for y:

$$\begin{aligned} 2x + 2y &= 2p \\ 2y &= 2p - 2x \\ y &= p - x \end{aligned}$$

Substitute into the first equation:

$$\begin{aligned} x^2 + (p-x)^2 &= d^2 \\ x^2 + p^2 - 2px + x^2 &= d^2 \\ 2x^2 - 2px + p^2 - d^2 &= 0 \end{aligned}$$

Using the quadratic formula, we have:

$$x = \frac{2p \pm \sqrt{4p^2 - 8\left(p^2 - d^2\right)}}{4} = \frac{2p \pm 2\sqrt{2d^2 - p^2}}{4} = \frac{p \pm \sqrt{2d^2 - p^2}}{2}$$

When $x = \dfrac{p + \sqrt{2d^2 - p^2}}{2}$, we have:

$$y = p - \frac{p + \sqrt{2d^2 - p^2}}{2} = \frac{p - \sqrt{2d^2 - p^2}}{2}$$

When $x = \frac{p - \sqrt{2d^2 - p^2}}{2}$, we have:

$$y = p - \frac{p - \sqrt{2d^2 - p^2}}{2} = \frac{p + \sqrt{2d^2 - p^2}}{2}$$

So the rectangle has dimenstions $\frac{p - \sqrt{2d^2 - p^2}}{2}$ by $\frac{p - \sqrt{2d^2 - p^2}}{2}$.

31. If we follow the hint we have:

$$\left(\sqrt{u+v} + \sqrt{u-v}\right)^2 = 4^2$$
$$u + v + 2\sqrt{u^2 - v^2} + u - v = 16$$

But since $u^2 - v^2 = 9$ from the first equation:

$$2u + 2\sqrt{9} = 16$$
$$2u + 6 = 16$$
$$2u = 10, \text{ so } u = 5$$

When $u = 5$, we have:

$$5^2 - v^2 = 9$$
$$v^2 = 16$$
$$v = \pm 4$$

So the solutions are $(5, 4)$ and $(5, -4)$.

33. (a) Using the point $(2, 3)$, we have $3 = N_0 e^{2k}$. Using the point $(8, 24)$, we have $24 = N_0 e^{8k}$. Dividing these two yields:

$$\frac{24}{3} = \frac{N_0 e^{8k}}{N_0 e^{2k}}$$
$$8 = e^{6k}$$
$$\ln 8 = 6k$$
$$k = \frac{\ln 8}{6}$$

Substituting into the first equation, we have:

$$3 = N_0 e^{\frac{1}{3}\ln 8}$$
$$3 = N_0 e^{\ln(8^{1/3})}$$
$$3 = N_0 \bullet 2$$
$$\tfrac{3}{2} = N_0$$

(b) Using the point $\left(\frac{1}{2},1\right)$, we have $1 = N_0e^{\frac{1}{2}k}$. Using the point (4, 10), we have $10 = N_0e^{4k}$. Dividing these two yields:

$$\frac{10}{1} = \frac{N_0e^{4k}}{N_0e^{\frac{1}{2}k}}$$
$$10 = e^{\frac{7}{2}k}$$
$$\ln 10 = \frac{7}{2}k$$
$$k = \frac{2}{7}\ln 10$$

Substituting into the first equation, we have:

$$1 = N_0e^{\frac{1}{7}\ln 10}$$
$$1 = N_0e^{\ln(10^{1/7})}$$
$$1 = N_0 \bullet 10^{1/7}$$
$$10^{-1/7} = N_0$$

35. Let $w = x + y + z$, so:

$$xw = p^2$$
$$yw = q^2$$
$$zw = r^2$$

Adding the three equations, we obtain:

$$(x+y+z)w = p^2+q^2+r^2$$
$$w^2 = p^2+q^2+r^2$$
$$w = \pm\sqrt{p^2+q^2+r^2}$$

Substitute into the first equation:

$$x = \frac{p^2}{\pm\sqrt{p^2+q^2+r^2}}$$

Similarly $y = \frac{q^2}{\pm A}$ and $z = \frac{r^2}{\pm A}$ where $A = \sqrt{p^2+q^2+r^2}$. So the solutions are $\left(\frac{p^2}{A},\frac{q^2}{A},\frac{r^2}{A}\right)$ and $\left(\frac{-p^2}{A},\frac{-q^2}{A},\frac{-r^2}{A}\right)$, where $A = \sqrt{p^2+q^2+r^2}$.

37. The area is given by $A = \frac{1}{2}bh$, where b and h are the missing legs. We have:

$180 = \frac{1}{2}bh$, so $bh = 360$ thus $b = \frac{360}{h}$

Also from the Pythagorean theorem $b^2 + h^2 = 41^2$, so:

$$\left(\frac{360}{h}\right)^2 + h^2 = 41^2$$
$$\frac{129600}{h^2} + h^2 = 1681$$
$$129600 + h^4 = 1681h^2$$
$$h^4 - 1681h^2 + 129600 = 0$$
$$(h^2 - 1600)(h^2 - 81) = 0$$
$$h^2 = 1600 \quad \text{or} \quad h^2 = 81$$
$$h = \pm 40 \quad \text{or} \quad h = \pm 9$$

Since these must be the length of the sides, we can neglect the negative values. So $h = 40$ and $h = 9$ are solutions. Since $b = \frac{360}{h}$ we obtain $b = 9$ when $h = 40$, and $b = 40$ when $h = 9$. So the legs are 9 cm and 40 cm.

39. We have $lw = 60$ and $2l + 2w = 46$, so $l + w = 23$ and $l = 23 - w$. We substitute this into the first equation:

$$(23 - w)(w) = 60$$
$$23w - w^2 = 60$$
$$w^2 - 23w + 60 = 0$$
$$(w - 20)(w - 3) = 0$$
$$w = 20 \quad \text{or} \quad w = 3$$

When $w = 20$ we have $l = 3$, and when $w = 3$ we have $l = 20$. So the rectangle must be 3 cm by 20 cm.

41. Solve $xy = 2$ to get $y = \frac{2}{x}$. Now substitute into the first equation:

$$x^2 + \frac{4}{x^2} = 5$$
$$x^4 + 4 = 5x^2$$
$$x^4 - 5x^2 + 4 = 0$$
$$(x^2 - 1)(x^2 - 4) = 0$$
$$x^2 = 1 \quad \text{or} \quad x^2 = 4$$
$$x = \pm 1 \qquad x = \pm 2$$

When $x = 1$ we have $y = 2$, when $x = -1$ we have $y = -2$, when $x = 2$ we have $y = 1$, and when $x = -2$ we have $y = -1$. So the solutions are $(1, 2)$, $(-1, -2)$, $(2, 1)$, and $(-2, -1)$.

43. Multiply the second equation by 2:

$$2xy = 6$$

Adding to the first equation, we obtain:

$$x^2 + 2xy + y^2 = 13$$
$$(x + y)^2 = 13$$
$$x + y = \pm\sqrt{13}$$

Subtracting from the first equation, we obtain:

$$x^2 - 2xy + y^2 = 1$$
$$(x-y)^2 = 1$$
$$x - y = \pm 1$$

We solve the four systems of equations:

$$\begin{aligned} x+y &= \sqrt{13} \\ x-y &= 1 \\ \hline 2x &= 1+\sqrt{13} \\ x &= \frac{1+\sqrt{13}}{2} \\ y &= \frac{-1+\sqrt{13}}{2} \end{aligned} \qquad \begin{aligned} x+y &= \sqrt{13} \\ x-y &= -1 \\ \hline 2x &= -1+\sqrt{13} \\ x &= \frac{-1+\sqrt{13}}{2} \\ y &= \frac{1+\sqrt{13}}{2} \end{aligned} \qquad \begin{aligned} x+y &= -\sqrt{13} \\ x-y &= 1 \\ \hline 2x &= 1-\sqrt{13} \\ x &= \frac{1-\sqrt{13}}{2} \\ y &= \frac{-1-\sqrt{13}}{2} \end{aligned} \qquad \begin{aligned} x+y &= -\sqrt{13} \\ x-y &= -1 \\ \hline 2x &= -1-\sqrt{13} \\ x &= \frac{-1-\sqrt{13}}{2} \\ y &= \frac{1-\sqrt{13}}{2} \end{aligned}$$

So the solutions are $\left(\frac{1+\sqrt{13}}{2}, \frac{-1+\sqrt{13}}{2}\right)$, $\left(\frac{-1+\sqrt{13}}{2}, \frac{1+\sqrt{13}}{2}\right)$, $\left(\frac{1-\sqrt{13}}{2}, \frac{-1-\sqrt{13}}{2}\right)$, and $\left(\frac{-1-\sqrt{13}}{2}, \frac{1-\sqrt{13}}{2}\right)$.

45. Solving by factoring, we have:

$$2m^2 - 7m + 6 = 0$$
$$(2m-3)(m-2) = 0$$
$$m = \tfrac{3}{2} \text{ or } m = 2$$

So we have the equations:

$$x^2\left(\tfrac{9}{2} - 4\right) = 2 \qquad \text{or} \qquad x^2(6-4) = 2$$
$$x^2 = 4 \qquad\qquad x^2 = 1$$
$$x = \pm 2 \qquad\qquad x = \pm 1$$
$$y = \pm 3 \qquad\qquad y = \pm 2$$

So the solutions are (2, 3), (–2, –3), (1, 2), and (–1, -2).

47. Using the hint, we square the first equation:

$$x^2y^2 + 2pqxy + p^2q^2 = 4p^2x^2$$
$$x^2y^2 \qquad\qquad + p^2q^2 = 2q^2y^2$$

Subtracting the second equation from the first:

$$2pqxy = 4p^2x^2 - 2q^2y^2$$
$$pqxy = 2p^2x^2 - q^2y^2$$
$$0 = 2p^2x^2 - pqxy - q^2y^2$$
$$0 = (2px + qy)(px - qy)$$

So we have the equations:

$$2px + qy = 0, \text{ so } y = -\tfrac{2p}{q}x$$
$$px - qy = 0, \text{ so } y = \tfrac{p}{q}x$$

Substituting into the first equation:

$$x\left(-\tfrac{2p}{q}x\right)+pq=2px$$
$$x\left(\tfrac{p}{q}x\right)+pq=2px$$

Multiplying each side by $\frac{q}{p}$:

$$-2x^2-2qx+q^2=0$$
$$x^2-2qx+q^2=0$$

The second equation factors to $(x-q)^2=0$, so $x=q$ and thus$y=\frac{p}{q}\bullet q=p$. Thus one solution is (q,p). This solution checks in the original (non-squared) first equation. For the first equation, we use the quadratic formula:

$$x=\frac{2q\pm\sqrt{4q^2+8q^2}}{-4}=\frac{2q\pm 2q\sqrt{3}}{-4}=\frac{-1\pm\sqrt{3}}{2}q$$
$$y=-\frac{2p}{q}\left(\frac{-1\pm\sqrt{3}}{2}\right)q=(1\pm\sqrt{3})p$$

This results in two other solutions, $\left(\frac{-1+\sqrt{3}}{2}q,(1-\sqrt{3})p\right)$ and $\left(\frac{-1-\sqrt{3}}{2}q,(1+\sqrt{3})p\right)$. Both of these solutions check in the original (non-squared) first equation.

49. Taking logs in the first equation:

$$\ln\left(x^4\right)=\ln\left(y^6\right)$$
$$4\ln x=6\ln y$$
$$2\ln x=3\ln y$$

The second equation is:

$$\ln x-\ln y=\frac{\ln x}{\ln y}$$
$$\ln x\ln y-(\ln y)^2=\ln x$$

Let $u=\ln x$ and $v=\ln y$, so we have the equations $2u=3v$ and $uv-v^2=u$. Solving the first equation for u yields $u=\frac{3v}{2}$, and substituting into the second equation yields:

$$\left(\tfrac{3v}{2}\right)v-v^2=\tfrac{3v}{2}$$
$$3v^2-2v^2=3v$$
$$v^2-3v=0$$
$$v(v-3)=0$$
$$v=0 \text{ or } v=3$$

When $v=0$, $u=0$ and when $v=3$, $u=\frac{9}{2}$. Since $v=\ln y$, $v=0$ cannot be a solution to the original second equation ($\ln y$ is the denominator). Thus $u=\ln x$ and $v=\ln y$ yields:

$$\ln x=\tfrac{9}{2} \text{ so } x=e^{9/2}$$
$$\ln y=3 \text{ so } y=e^3$$

So the only solution is $\left(e^{9/2},e^3\right)$.

7.7 Systems of Inequalities. Linear Programming

1. (a) Substitute the pair $(1,2)$:

$$4(1)-6(2)+3\geq 0$$
$$-5\geq 0$$

Then $(1,2)$ is not a solution.

(b) Substitute the pair $\left(0,\frac{1}{2}\right)$:

$$4(0)-6\left(\tfrac{1}{2}\right)+3\geq 0$$
$$0\geq 0$$

Then $\left(0,\frac{1}{2}\right)$ is a solution.

3. We graph the region:

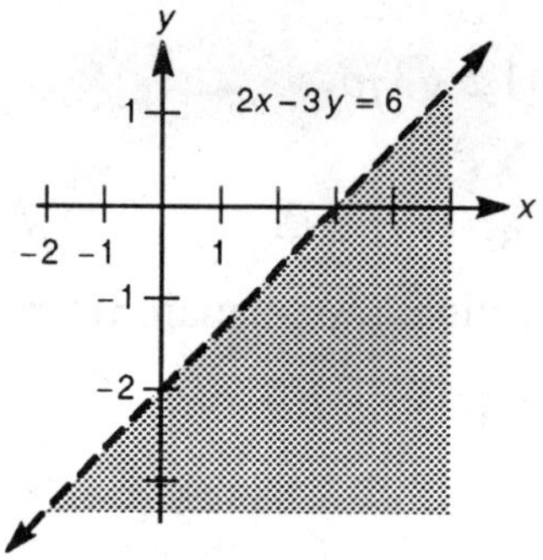

5. We graph the region:

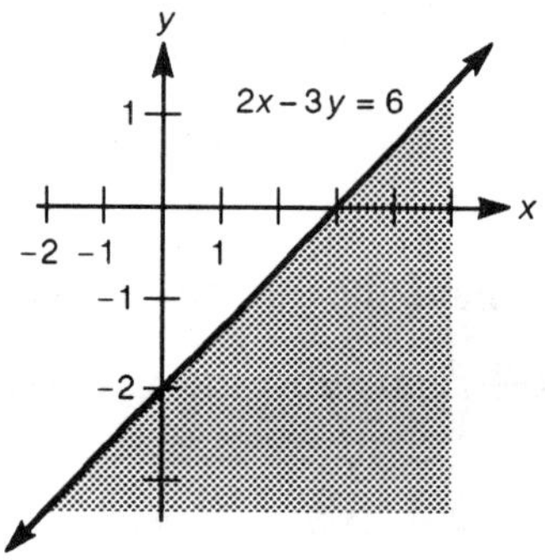

7. We graph the region:

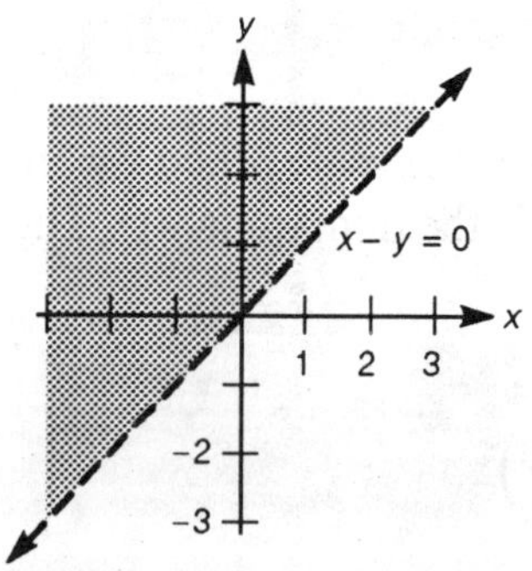

9. We graph the region:

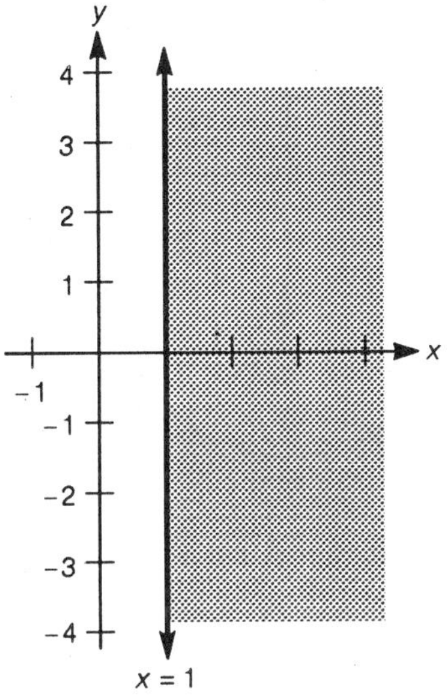

11. We graph the region:

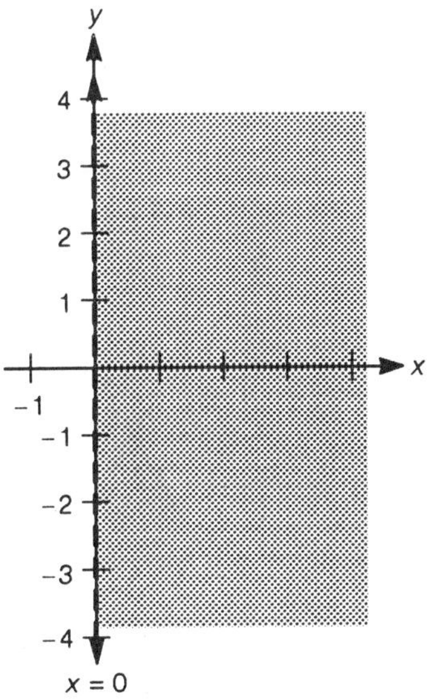

13. We graph the region:

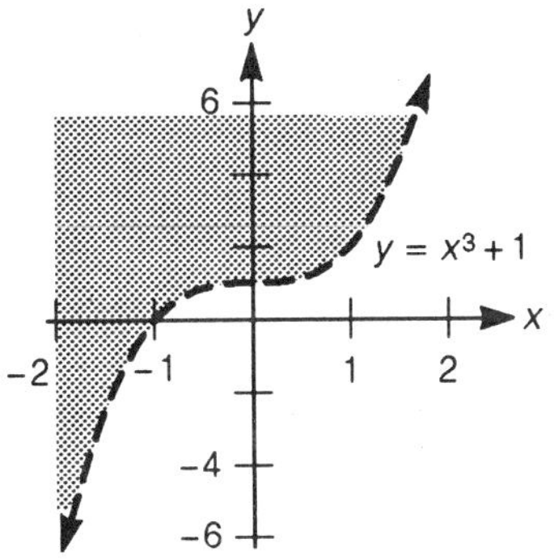

15. We graph the region:

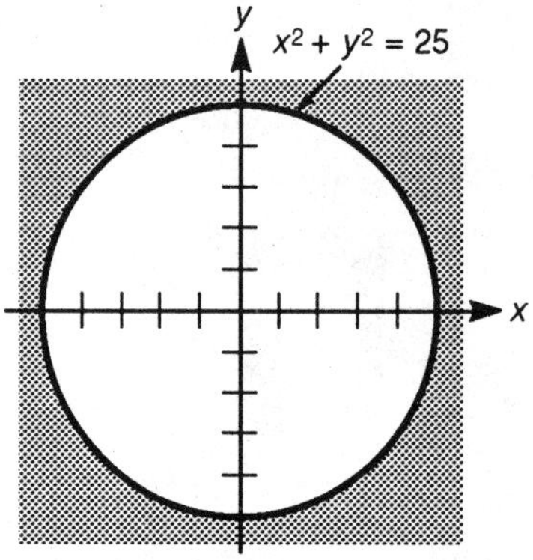

17. We graph the system of inequalities:

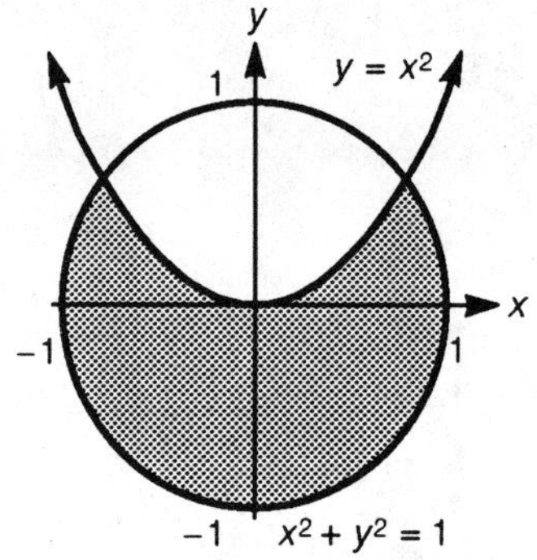

19. We graph the system of inequalities:

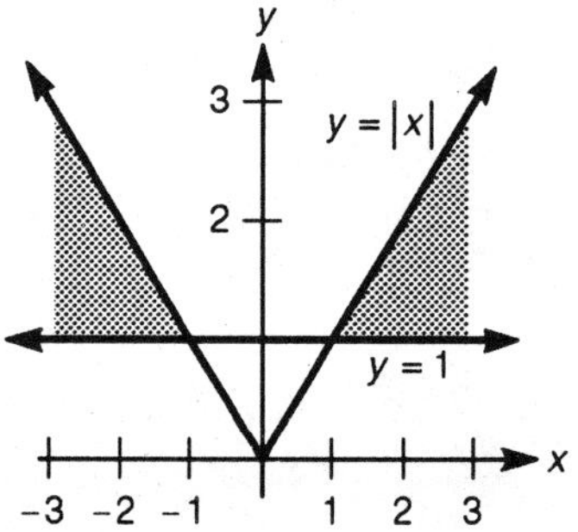

21. We graph the system of inequalities:

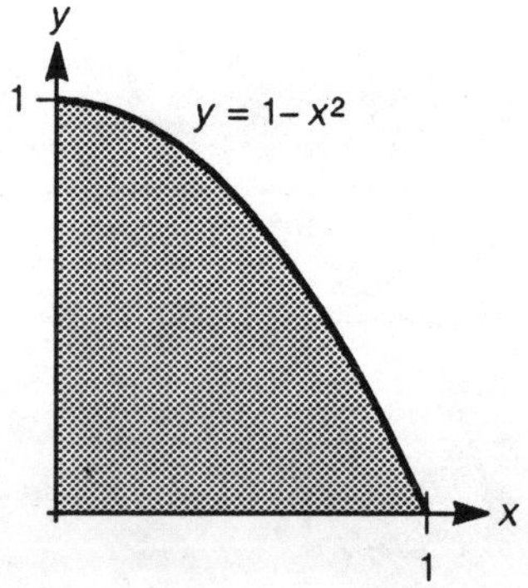

23. The region is convex and bounded. The vertices are (0,0), (7,0), (3,8), and (0,5). We graph the system of inequalities:

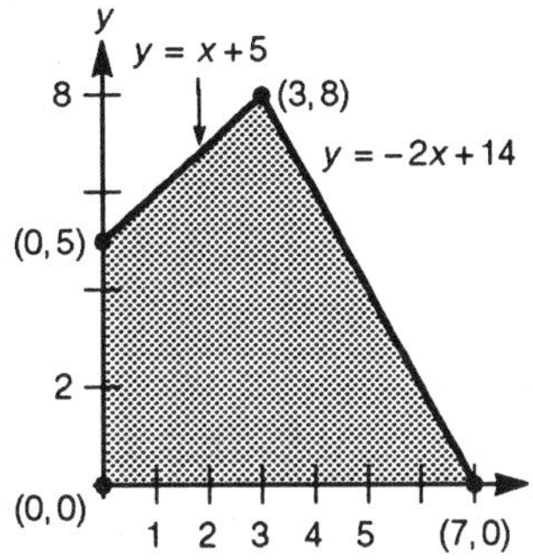

25. The region is convex and bounded. The vertices are (0,0), (0,4), (3,5), and (8,0). We graph the system of inequalities:

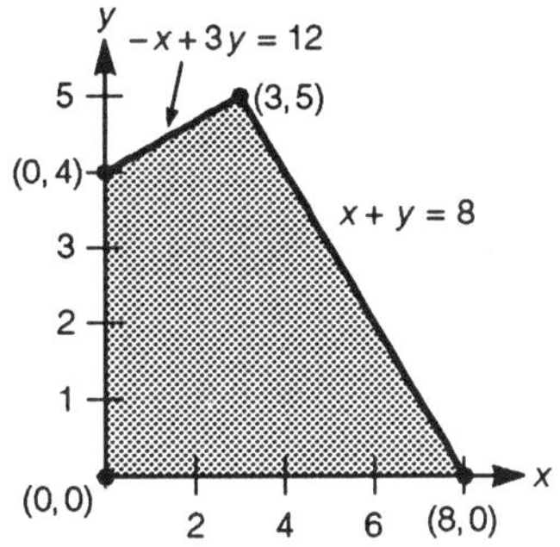

27. The region is convex but not bounded. The vertices are (2,7) and (8,5). We graph the system of inequalities:

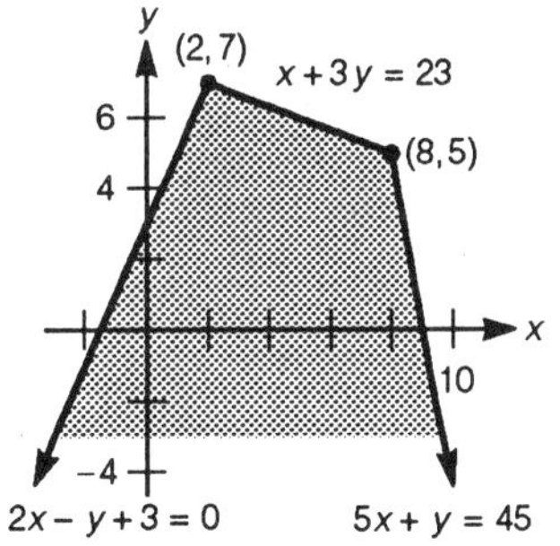

29. The region is convex but not bounded. The only vertex is (6,0). We graph the system of inequalities:

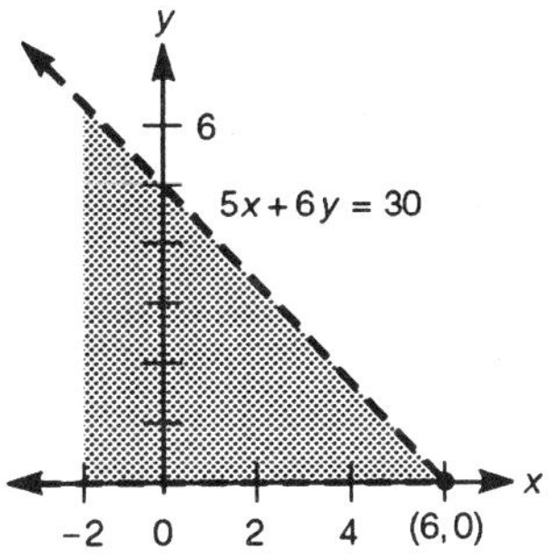

31. The region is convex and bounded. The vertices are $(0,0)$, $(0,5)$, and $(6,0)$. We graph the system of inequalities:

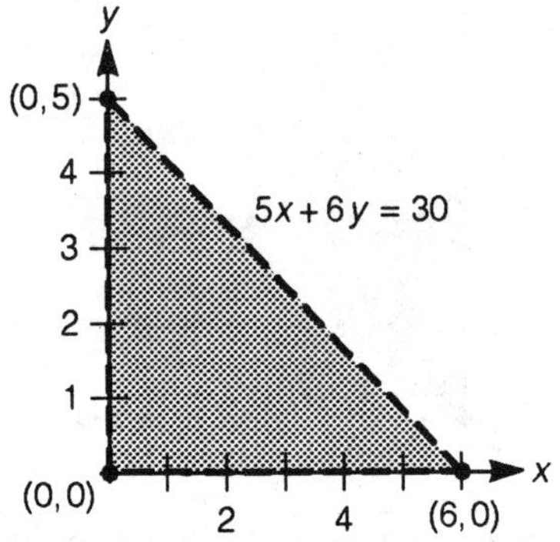

33. The region is convex and bounded. The vertices are $(5,30)$, $(10,30)$, $(20,15)$, and $(20,20)$. We graph the system of inequalities:

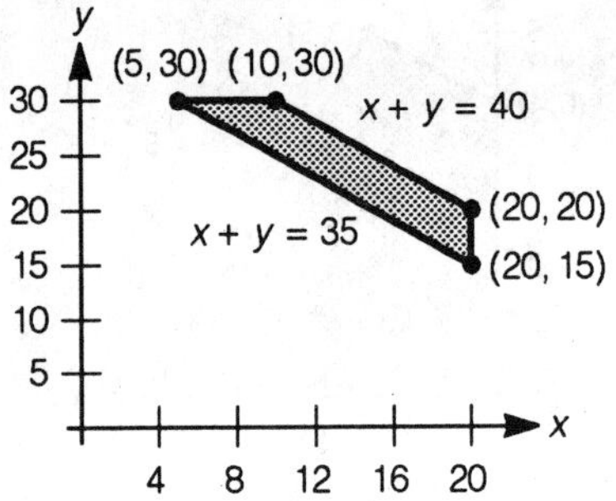

35. We first graph the conditions:

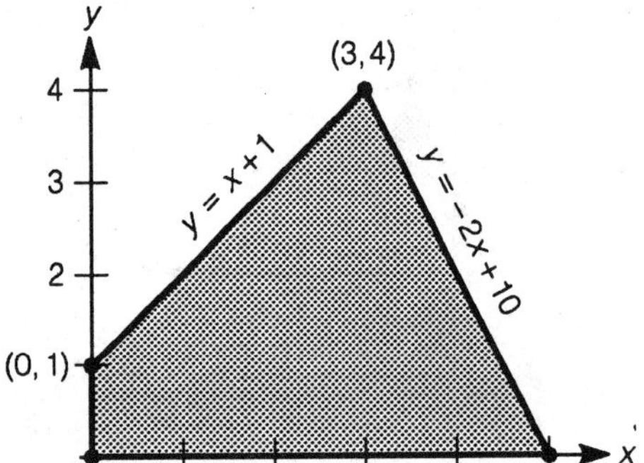

Now set up the table:

Vertices	$C = 3y - x$
(0,0)	0
(5,0)	–5
(3,4)	9
(0,1)	3

Thus we see that C has a maximum of 9 at $(3,4)$ and a minimum of –5 at $(5,0)$.

37. We first graph the conditions:

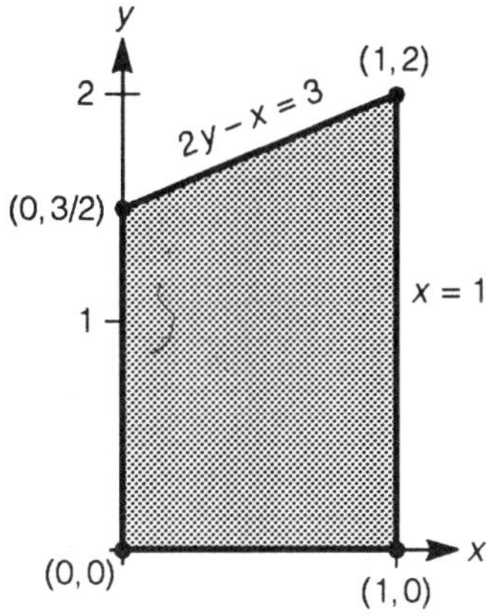

Now set up the table:

Vertices	$C = 2x + y$
(0,0)	0
$(0,\frac{3}{2})$	$\frac{3}{2}$
(1,0)	2
(1,2)	4

The maximum value is 4 at $(1,2)$ and the minimum value is 0 at $(0,0)$.

39. We first graph the conditions:

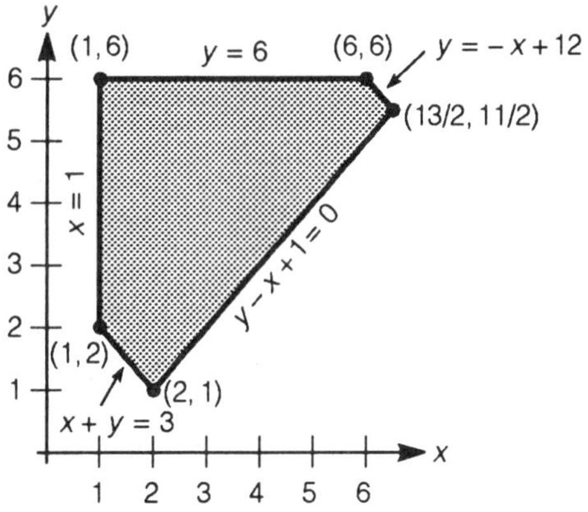

Now set up the table:

Vertices	$C = 10y + 9x - 1$
(1,2)	28
(2,1)	27
$(\frac{13}{2},\frac{11}{2})$	112.5
(6,6)	113
(1,6)	68

The maximum value is 113 at $(6,6)$ and the minimum value is 27 at $(2,1)$.

41. We first graph the conditions:

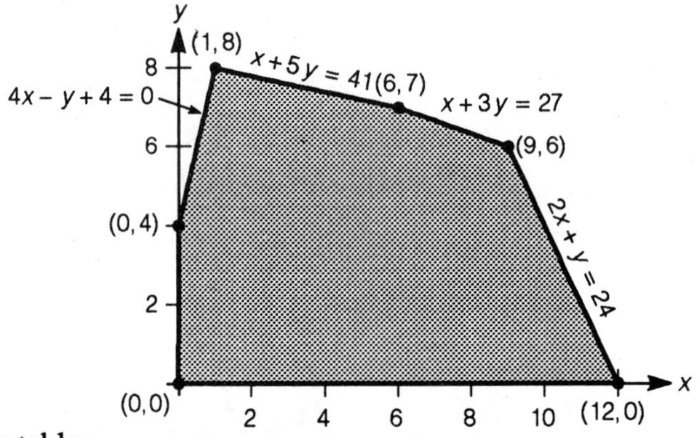

Now set up the table:

Vertices	$C=10y+3x+100$
(0,0)	100
(0,4)	140
(1,8)	183
(6,7)	188
(9,6)	187
(12,0)	136

The maximum value is 188 at $(6,7)$, and the minimum value is 100 at $(0,0)$.

43. Using the same constraints as in Exercise 41, we set up a table containing both objective functions:

Vertices	(a) $C=19x+100y$	(b) $C=21x+100y$
(0,0)	0	0
(0,4)	400	400
(1,8)	819	821
(6,7)	814	826
(9,6)	771	789
(12,0)	228	252

(a) The maximum value is 819 at $(1,8)$, and the minimum value is 0 at $(0,0)$.
(b) The maximum value is 826 at $(6,7)$, and the minimum value is 0 at $(0,0)$.

45. We first organize the data in tables:

Shipping	Cost
W_1 to D_1	180
W_1 to D_2	150
W_2 to D_1	160
W_2 to D_2	170

Supply	Demand
W_1 has 30	D_1 wants 40
W_2 has 50	D_2 wants 25

Let $x = W_1$ to D_1, so $40 - x = W_2$ to D_1. Let $y = W_1$ to D_2, so $25 - y = W_2$ to D_2. The cost function is:

$$\begin{aligned} C &= 180(x) + 150(y) + 160(40 - x) + 175(25 - y) \\ &= 180x + 150y + 6400 - 160x + 4250 - 170y \\ &= 10650 + 20x - 20y \end{aligned}$$

The constraints are:

(1) $x \geq 0$
(2) $y \geq 0$
(3) $x \leq 40$
(4) $y \leq 25$
(5) $x + y \leq 30$
(6) $(40 - x) + (25 - y) \leq 50$, or $x + y \geq 15$

We graph these constraints:

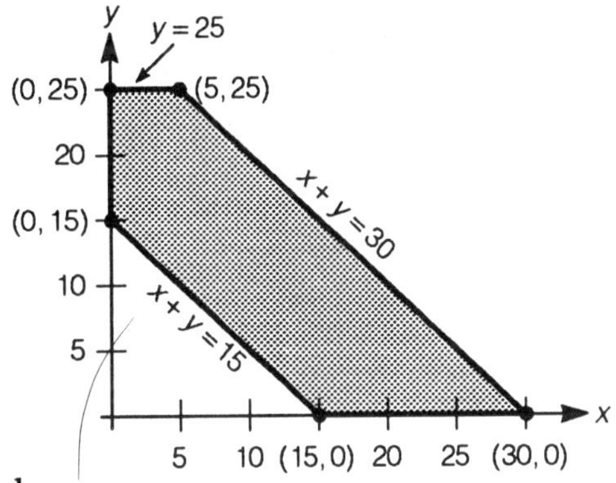

We set up a table of values:

Vertices	$C = 10650 + 20x - 20y$
(0,15)	10350
(0,25)	10150
(5,25)	10250
(30,0)	11250

So the minimum cost is \$10,150 when $x = 0$ and $y = 25$.
Shipping instructions:

W_1 to D_1: 0 cars
W_2 to D_1: 40 cars
W_1 to D_2: 25 cars
W_2 to D_2: 0 cars

47. Let x = number of A units produced and y = number of B units produced. We have $P = 0.60x + 0.80y$.
The constraints are:

(1) $x \geq 0$
(2) $y \geq 0$
(3) $4x + 5y \leq 5000$
(4) $x + 2y \leq 1500$

We graph these constraints:

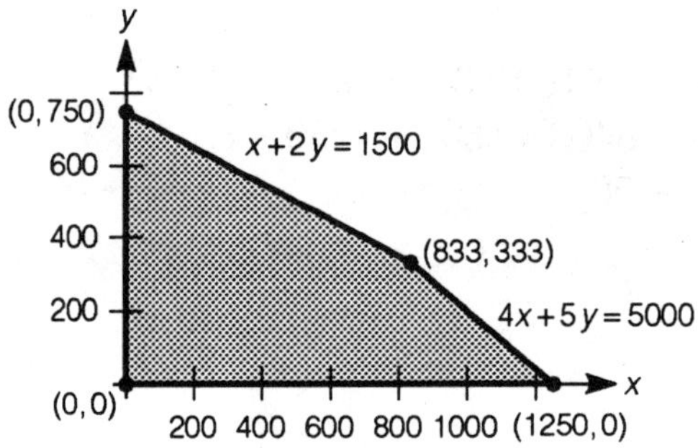

Now set up a table:

Vertices	$P = 0.60x + 0.80y$
(0,0)	0
(0,750)	600
(1250,0)	750
(833,333)	766.20

So the maximum profit is \$766.20 when 833 units of *A* and 333 units of *B* are produced.

49. Let x = acres of cherry tomatoes and y = acres of regular tomatoes. Then $P = 50x + 36y$. The constraints are:

(1) $x \geq 0$
(2) $y \geq 0$
(3) $x + y \leq 600$
(4) $3x + 2y \leq 1350$

We graph the constraints:

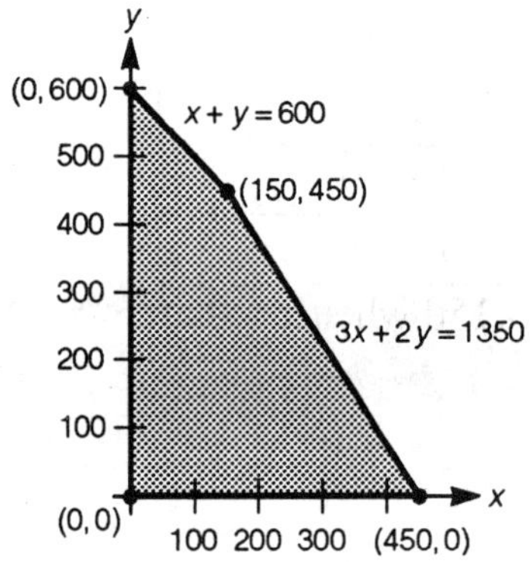

We set up a table:

Vertices	$P = 50x + 36y$
(0,0)	0
(0,600)	21600
(450,0)	22500
(150,450)	23700

So the maximum profit is \$23,700 when 150 acres of cherry tomatoes and 450 acres of regular tomatoes are planted.

51. Let x = amount to invest in A and y = amount to invest in B. Then $R = 0.06x + 0.08y$. The constraints are:

(1) $x \geq 0$
(2) $y \geq 2000$
(3) $x + y \leq 12000$
(4) $x \leq 6000$
(5) $y \leq 2x$

We graph the constraints:

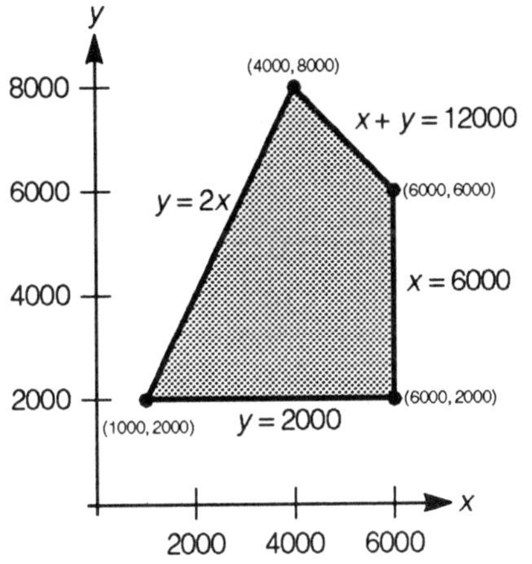

We set up a table:

Vertices	$P = 0.06x + 0.08y$
(1000,2000)	220
(4000,8000)	880
(6000,6000)	840
(6000,2000)	520

Your maximum return will be \$880 when you invest \$4000 in stock A and \$8000 in stock B.

53. We have the same constraints as from Exercise 52:

(1) $x \geq 1000$
(2) $y \geq 1000$
(3) $10000 - x - y \geq 1000, x + y \leq 9000$
(4) $x + y \geq 5000$
(5) $y \leq 5x$

The profit function is:

$$P = 0.05x + 0.04y + 0.06(10000 - x - y) = 600 - 0.01x - 0.02y$$

We graph the constraints:

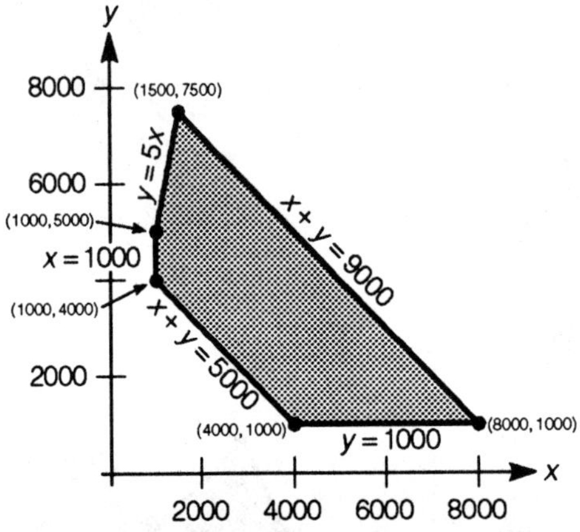

We set up a table:

Vertices	$P = 600 - 0.01x - 0.02y$
(4000,1000)	540
(8000,1000)	500
(1500,7500)	435
(1000,5000)	490
(1000,4000)	510

We invest \$4000 in *A*, \$1000 in *B* and \$5000 in *C* for a maximum return of \$540.

55. Let x = pounds of *A* and y = pounds of *B*. Then $C = 0.44x + 0.80y$. The constraints are:
(1) $x \geq 0$
(2) $y \geq 0$
(3) $3x + 5y \geq 50$
(4) $2x + 4y \geq 36$
(5) $2x + 8y \geq 40$

We graph the constraints:

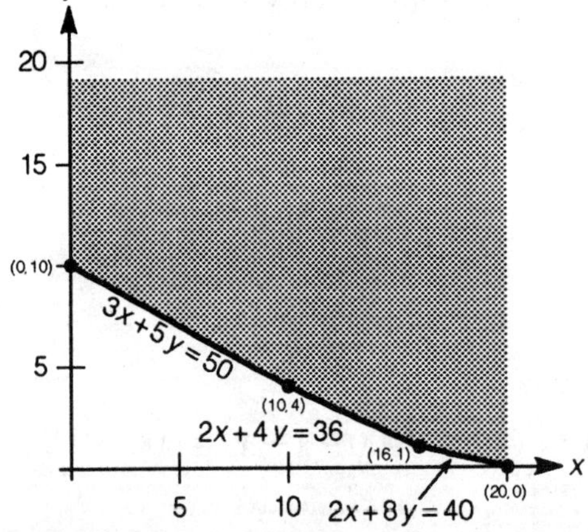

We set up a table:

Vertices	$C = 0.44x + 0.08y$
(0,10)	8
(10,4)	7.6
(16,1)	7.84
(20,0)	8.8

So 10 pounds of A and 4 pounds of B should be mixed to produce a mixture with minimum cost of \$7.60.

57. The domain results from the inequality $x + y + 2 \geq 0$, or $x + y \geq -2$. We sketch the graph:

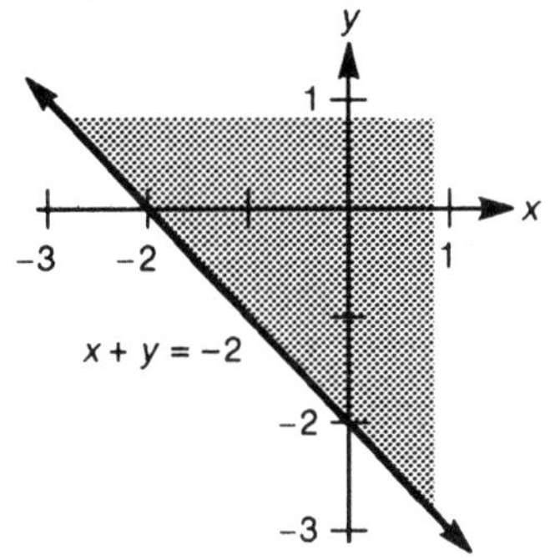

59. The domain results from the inequality $25 - x^2 - y^2 \geq 0$, or $x^2 + y^2 \leq 25$. We sketch the graph:

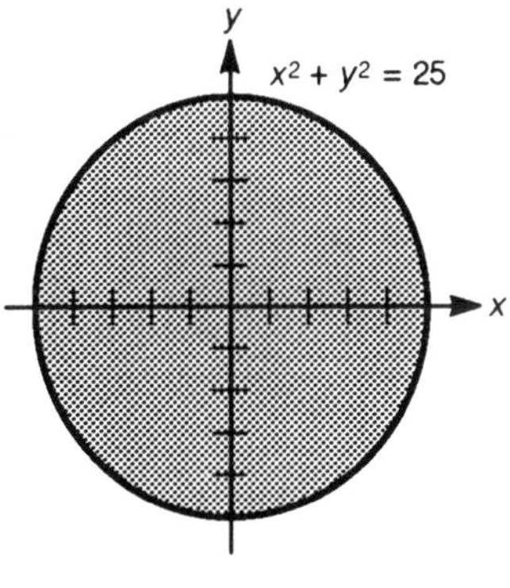

61. The domain results from the inequality $xy > 0$. This consists of two quadrants: $x > 0$ and $y > 0$ (first quadrant) or $x < 0$ and $y < 0$ (third quadrant). We sketch the graph:

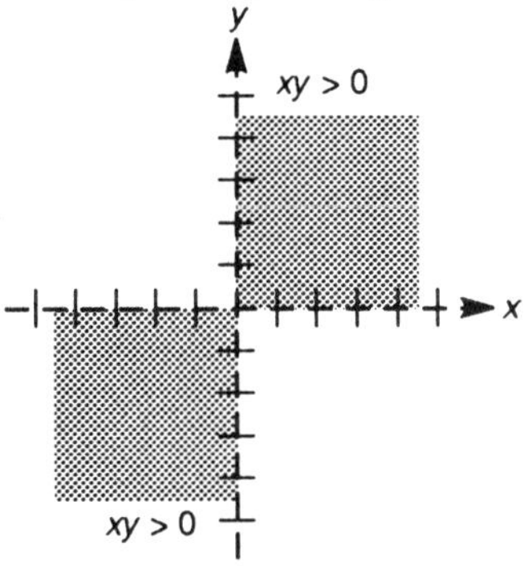

Chapter Seven Review Exercises

1. Adding the two equations yields:
$$2x = 6$$
$$x = 3$$
Substitute into equation 1:
$$3xy = -2$$
$$y = -5$$
So the solution is (3, –5).

3. Multiply the first equation by –2:
$$-4x - 2y = -4$$
$$x + 2y = 7$$
Adding, we obtain:
$$-3x = 3$$
$$x = -1$$
Substitute into equation 1:
$$-2 + y = 2$$
$$y = 4$$
So the solution is (–1, 4).

5. Multiply equation 1 by –5 and equation 2 by 2:
$$-35x - 10y = -45$$
$$8x + 10y = 126$$
Adding, we obtain:
$$-27x = 81$$
$$x = -3$$
Substitute into equation 1:
$$-21 + 2y = 9$$
$$2y = 30$$
$$y = 15$$
So the solution is (–3, 15).

7. Multiply the first equation by 2 and the second equation by 24 to clear fractions:
$$4x - y = -16$$
$$8x + 3y = -24$$
Multiply equation 1 by –2:
$$-8x + 2y = 32$$
$$8x + 3y = -24$$
Adding, we obtain:
$$5y = 8$$
$$y = \tfrac{8}{5}$$

Substitute into equation 1:

$$4x - \frac{8}{5} = -16$$
$$4x = -\frac{72}{5}$$
$$x = -\frac{18}{5}$$

So the solution is $\left(-\frac{18}{5}, \frac{8}{5}\right)$.

9. Multiply the first equation by 2:

$$6x + 10y = 2$$
$$9x - 10y = 8$$

Adding, we obtain:

$$15x = 10$$
$$x = \frac{2}{3}$$

Substitute into equation 1:

$$2 + 5y = 1$$
$$5y = -1$$
$$y = -\frac{1}{5}$$

So the solution is $\left(\frac{2}{3}, -\frac{1}{5}\right)$.

11. Multiply the first equation by 6 and the second equation by 2 to clear fractions:

$$4x + 3y = -72$$
$$x - 2y = 4$$

Multiply the second equation by -4:

$$4x + 3y = -72$$
$$-4x + 8y = -16$$

Adding, we obtain:

$$11y = -88$$
$$y = -8$$

Substitute into equation 2:

$$x + 16 = 4$$
$$x = -12$$

So the solution is $(-12, -8)$.

13. Let $a = \frac{1}{x}$ and $b = \frac{1}{y}$. So we have:

$$a + b = -1$$
$$2a + 5b = -14$$

Multiply the first equation by -2:

$$-2a - 2b = 2$$
$$2a + 5b = -14$$

Adding, we obtain:

$$3b = -12$$
$$b = -4$$

Substitute into equation 1:

$$a-4=-1$$
$$a=3$$

Since $x=\frac{1}{a}$ and $y=\frac{1}{b}$, then $x=\frac{1}{3}$ and $y=-\frac{1}{4}$. So the solution is $\left(\frac{1}{3},-\frac{1}{4}\right)$.

15. Multiply the second equation by $a-1$:

$$ax+(1-a)y=1$$
$$\left(-a^2+2a-1\right)x+(a-1)y=0$$

Adding, we obtain:

$$\left(-a^2+3a-1\right)x=1$$
$$x=\frac{-1}{a^2-3a+1}$$

Substitute into equation 2:

$$\frac{a-1}{a^2-3a+1}+y=0$$
$$y=\frac{1-a}{a^2-3a+1}$$

So the solution is $\left(\frac{-1}{a^2-3a+1},\frac{1-a}{a^2-3a+1}\right)$. We must assume that $a^2-3a+1\neq 0$, or $a\neq\frac{3\pm\sqrt{5}}{2}$.

17. Multiply the first equation by 2:

$$4x-2y=6a^2-2$$
$$x+2y=-a^2+2$$

Adding, we obtain:

$$5x=5a^2$$
$$x=a^2$$

Substitute into equation 2:

$$2y+a^2=2-a^2$$
$$2y=2-2a^2$$
$$y=1-a^2$$

So the solution is $\left(a^2,1-a^2\right)$.

19. Multiply the first equation by 3:

$$15x-3y=12a^2-18b^2$$
$$2x+3y=5a^2+b^2$$

Adding, we obtain:

$$17x=17a^2-17b^2$$
$$x=a^2-b^2$$

Substitute into equation 1:

$$5a^2-5b^2-y=4a^2-6b^2$$
$$-y=-a^2-b^2$$
$$y=a^2+b^2$$

So the solution is $\left(a^2-b^2,a^2+b^2\right)$.

21. Multiply the first equation by p and the second equation by q:

$$p^2x - pqy = pq^2$$
$$q^2x + pqy = p^2q$$

Adding, we obtain:

$$(p^2+q^2)x = pq(p+q)$$
$$x = \frac{pq(p+q)}{p^2+q^2}$$

We re-solve the system to find y. Multiply the first equation by $-q$ and the second equation by p:

$$-pqx + q^2y = -q^3$$
$$pqx + p^2y = p^3$$

Adding we obtain:

$$(p^2+q^2)y = p^3 - q^3$$
$$y = \frac{p^3-q^3}{p^2+q^2}$$

So the solution is $\left(\frac{pq(p+q)}{p^2+q^2}, \frac{p^3-q^3}{p^2+q^2}\right)$. We must assume that p and q are not both 0.

23. Let $u = \frac{1}{x}$ and $v = \frac{1}{y}$, so we have:

$$4au - 3bv = a - 7b$$
$$2a^2u - 2b^2v = 3a^2 - 5ab - 2b^2$$

Multiply the first equation by $-2b$ and the second equation by 3:

$$-8abu + 6b^2v = 14b^2 - 2ab$$
$$9a^2u - 6b^2v = 9a^2 - 15ab - 6b^2$$

Adding, we obtain:

$$(9a^2 - 8ab)u = 9a^2 - 17ab + 8b^2$$
$$u = \frac{9a^2 - 17ab + 8b^2}{9a^2 - 8ab} = \frac{(9a-8b)(a-b)}{a(9a-8b)} = \frac{a-b}{a}$$

Substitute into equation 1:

$$4a\left(\frac{a-b}{a}\right) - 3bv = a - 7b$$
$$4a - 4b - 3bv = a - 7b$$
$$-3bv = -3a - 3b$$
$$v = \frac{a+b}{b}$$

Since $x = \frac{1}{u}$ and $y = \frac{1}{v}$, we have $x = \frac{a}{a-b}$ and $y = \frac{b}{a+b}$. So the solution is $\left(\frac{a}{a-b}, \frac{b}{a+b}\right)$. We must assume that $ab \neq 0$ and $a \neq \pm b$.

25. Form the augmented matrix:

$$\begin{pmatrix} 1 & 1 & 1 & 9 \\ 1 & -1 & -1 & -5 \\ 2 & 1 & -2 & -1 \end{pmatrix}$$

Add –1 times row 1 to row 2 and –2 times row 1 to row 3:

$$\begin{pmatrix} 1 & 1 & 1 & 9 \\ 0 & -2 & -2 & -14 \\ 0 & -1 & -4 & -19 \end{pmatrix}$$

Switch row 2 and row 3, multiply each by –1:

$$\begin{pmatrix} 1 & 1 & 1 & 9 \\ 0 & 1 & 4 & 19 \\ 0 & 2 & 2 & 14 \end{pmatrix}$$

Add –2 times row 2 to row 3:

$$\begin{pmatrix} 1 & 1 & 1 & 9 \\ 0 & 1 & 4 & 19 \\ 0 & 0 & -6 & -24 \end{pmatrix}$$

So we have the system:

$$\begin{aligned} x + y + z &= 9 \\ y + 4z &= 19 \\ -6z &= -24 \end{aligned}$$

Solve equation 3 for z:

$$\begin{aligned} -6z &= -24 \\ z &= 4 \end{aligned}$$

Substitute into equation 2:

$$\begin{aligned} y + 16 &= 19 \\ y &= 3 \end{aligned}$$

Substitute into equation 1:

$$\begin{aligned} x + 3 + 4 &= 9 \\ x &= 2 \end{aligned}$$

So the solution is $(2, 3, 4)$.

27. Switching equations 1 and 3, form the augmented matrix:

$$\begin{pmatrix} 1 & 1 & 1 & -3 \\ 2 & 3 & 3 & -8 \\ 4 & -4 & 1 & 4 \end{pmatrix}$$

Add –2 times row 1 to row 2 and –4 times row 1 to row 3:

$$\begin{pmatrix} 1 & 1 & 1 & -3 \\ 0 & 1 & 1 & -2 \\ 0 & -8 & -3 & 16 \end{pmatrix}$$

Add 8 times row 2 to row 3:

$$\begin{pmatrix} 1 & 1 & 1 & -3 \\ 0 & 1 & 1 & -2 \\ 0 & 0 & 5 & 0 \end{pmatrix}$$

So we have the system:

$$\begin{aligned} x+y+z &= -3 \\ y+z &= -2 \\ 5z &= 0 \end{aligned}$$

Solve equation 3 for z:

$$\begin{aligned} 5z &= 0 \\ z &= 0 \end{aligned}$$

Substitute into equation 2:

$$\begin{aligned} y+0 &= -2 \\ y &= -2 \end{aligned}$$

Substitute into equation 1:

$$\begin{aligned} x-2+0 &= -3 \\ x &= -1 \end{aligned}$$

So the solution is $(-1,-2,0)$.

29. Using equation 3 as row 1, form the augmented matrix:

$$\begin{pmatrix} 1 & 1 & -2 & 4 \\ 1 & -2 & 1 & -2 \\ -2 & 1 & 1 & 1 \end{pmatrix}$$

Add –1 times row 1 to row 2 and 2 times row 1 to row 3:

$$\begin{pmatrix} 0 & 1 & -2 & 4 \\ 0 & -3 & 3 & -6 \\ 0 & 3 & -3 & 9 \end{pmatrix}$$

Multiply row 2 by $\frac{1}{3}$ and row 3 by $\frac{1}{3}$:

$$\begin{pmatrix} 1 & 1 & -2 & 4 \\ 0 & -1 & 1 & -2 \\ 0 & 1 & -1 & 3 \end{pmatrix}$$

Add row 2 to row 3:

$$\begin{pmatrix} 1 & 1 & -2 & 4 \\ 0 & -1 & 1 & -2 \\ 0 & 0 & 0 & 1 \end{pmatrix}$$

Since $0 = 1$ is false, there is no solution to the system.

31. Multiply the second equation by –2:

$$\begin{aligned} 4x+2y-3z &= 15 \\ -4x-2y-6z &= -6 \end{aligned}$$

Adding, we obtain:

$$-9z = 9$$
$$z = -1$$

Substitute into the original equation:

$$2x + y - 3 = 3$$
$$2x + y = 6$$
$$y = 6 - 2x$$

So the solution is $(x, 6 - 2x, -1)$, for any real number x.

33. Form the augmented matrix:

$$\begin{pmatrix} 1 & 2 & -3 & -2 \\ 2 & -1 & 1 & 1 \\ 3 & -4 & 5 & 1 \end{pmatrix}$$

Add –2 times row 1 to row 2 and –3 times row 1 to row 3:

$$\begin{pmatrix} 1 & 2 & -3 & -2 \\ 0 & -5 & 7 & 5 \\ 0 & -10 & 14 & 7 \end{pmatrix}$$

Add –2 times row 2 to row 3:

$$\begin{pmatrix} 1 & 2 & -3 & -2 \\ 0 & -5 & 7 & 5 \\ 0 & 0 & 0 & -3 \end{pmatrix}$$

Since $0 = -3$ is false, there is no solution to the system.

35. Form the augmented matrix:

$$\begin{pmatrix} 1 & 1 & 1 & a+b \\ 2 & -1 & 2 & -a+5b \\ 1 & -2 & 1 & -2a+4b \end{pmatrix}$$

Add –2 times row 1 to row 2 and –1 times row 1 to row 3:

$$\begin{pmatrix} 1 & 1 & 1 & a+b \\ 0 & -3 & 0 & -3a+3b \\ 0 & -3 & 0 & -3a+3b \end{pmatrix}$$

Add –1 times row 2 to row 3:

$$\begin{pmatrix} 1 & 1 & 1 & a+b \\ 0 & -3 & 0 & -3a+3b \\ 0 & 0 & 0 & 0 \end{pmatrix}$$

So we have the system:

$$\begin{aligned} x + y + z &= a + b \\ -3y \quad &= -3a + 3b \end{aligned}$$

Solve equation 2 for y:

$$-3y = -3a + 3b$$
$$y = a - b$$

Substitute into equation 1:

$$x+a-b+z=a+b$$
$$x=2b-z$$

So the solution is $(2b-z, a-b, z)$, for any real number z.

37. Form the augmented matrix:

$$\begin{pmatrix} 1 & 1 & 1 & 1 & 8 \\ 3 & 3 & -1 & -1 & 20 \\ 4 & -1 & -1 & 2 & 18 \\ 2 & 5 & 5 & -5 & 8 \end{pmatrix}$$

Add –3 times row 1 to row 2, –4 times row 1 to row 3, and –2 times row 1 to row 4:

$$\begin{pmatrix} 1 & 1 & 1 & 1 & 8 \\ 0 & 0 & -4 & -4 & -4 \\ 0 & -5 & -5 & -2 & -14 \\ 0 & 3 & 3 & -7 & -8 \end{pmatrix}$$

Switch row 2 and row 3, and multiply row 3 by $-\frac{1}{4}$:

$$\begin{pmatrix} 1 & 1 & 1 & 1 & 8 \\ 0 & -5 & -5 & -2 & -14 \\ 0 & 0 & 1 & 1 & 1 \\ 0 & 3 & 3 & -7 & -8 \end{pmatrix}$$

Add 2 times row 4 to row 2:

$$\begin{pmatrix} 1 & 1 & 1 & 1 & 8 \\ 0 & 1 & 1 & -16 & -30 \\ 0 & 0 & 1 & 1 & 1 \\ 0 & 3 & 3 & -7 & -8 \end{pmatrix}$$

Add –3 times row 2 to row 4:

$$\begin{pmatrix} 1 & 1 & 1 & 1 & 8 \\ 0 & 1 & 1 & -16 & -30 \\ 0 & 0 & 1 & 1 & 1 \\ 0 & 0 & 0 & 41 & 82 \end{pmatrix}$$

So we have the system:

$$\begin{aligned} x+y+z+\quad w &= 8 \\ y+z-16w &= -30 \\ z+\quad w &= 1 \\ 41w &= 82 \end{aligned}$$

Solve equation 4 for w:

$$41w=82$$
$$w=2$$

Substitute into equation 3:

$$z+2=1$$
$$z=-1$$

Substitute into equation 2:
$$y-1-32=-30$$
$$y=3$$
Substitute into equation 1:
$$x+3-1+2=8$$
$$x=4$$
So the solution is $(4,3,-1,2)$.

39. Multiply by $(x-10)(x+10)$, so:
$$1=A(x+10)+B(x-10)$$
$$1=(A+B)x+(10A-10B)$$
Since A and B are constants:
$$A+B=0$$
$$10A-10B=1$$
Multiply equation 1 by 10:
$$10A+10B=0$$
$$10A-10B=1$$
Adding, we obtain:
$$20A=1$$
$$A=\tfrac{1}{20}$$
Substitute into equation 1:
$$\tfrac{1}{20}+B=0$$
$$B=-\tfrac{1}{20}$$
So $A=\frac{1}{20}$ and $B=-\frac{1}{20}$.

41. Multiply by $(x+1)^2$, so:
$$2x=A(x+1)+B$$
$$2x=Ax+(A+B)$$
Since A and B are constants:
$$A=2$$
$$A+B=0$$
Substitute into equation 2:
$$2+B=0$$
$$B=-2$$
So $A=2$ and $B=-2$.

43. Multiply by $x(x-4)$, so:
$$5=A(x-4)+Bx$$
$$5=(A+B)x-4A$$
Since A and B are constants:
$$A+B=0$$
$$-4A=5$$
Solve equation 2 for A:
$$-4A=5$$
$$A=-\tfrac{5}{4}$$

Substitute into equation 1:

$$-\tfrac{5}{4}+B=0$$
$$B=\tfrac{5}{4}$$

So $A=-\frac{5}{4}$ and $B=\frac{5}{4}$.

45. Multiply by $(x-1)(x+3)^2$, so:

$$1=A(x+3)^2+B(x-1)(x+3)+C(x-1)$$
$$1=A\left(x^2+6x+9\right)+B\left(x^2+2x-3\right)+C(x-1)$$
$$1=(A+B)x^2+(6A+2B+C)x+(9A-3B-C)$$

Since A, B, and C are constants:

$$\begin{aligned} A+B &= 0 \\ 6A+2B+C &= 0 \\ 9A-3B-C &= 1 \end{aligned}$$

Add -6 times equation 1 to equation 2 and -9 times equation 1 to equation 3:

$$\begin{aligned} A+B &= 0 \\ -4B+C &= 0 \\ -12B-C &= 1 \end{aligned}$$

Add equation 2 and 3:

$$-16B=1$$
$$B=-\tfrac{1}{16}$$

Substitute into equation 2:

$$\tfrac{1}{4}+C=0$$
$$C=-\tfrac{1}{4}$$

Substitute into equation 1:

$$A-\tfrac{1}{16}=0$$
$$A=\tfrac{1}{16}$$

So $A=\frac{1}{16}$, $B=-\frac{1}{16}$, and $C=-\frac{1}{4}$.

47. Multiply by $(x-1)\left(x^2+x+5\right)$, so:

$$4x^2+2x+15=A\left(x^2+x+5\right)+(Bx+C)(x-1)$$
$$4x^2+2x+15=(A+B)x^2+(A-B+C)x+(5A-C)$$

Since A, B, and C are constants:

$$\begin{aligned} A+B &= 4 \\ A-B+C &= 2 \\ 5A-C &= 15 \end{aligned}$$

Add -1 times equation 1 to equation 2 and -5 times equation 1 to equation 3:

$$\begin{aligned} A+B &= 4 \\ -2B+C &= -2 \\ -5B-C &= -5 \end{aligned}$$

Add equation 2 and equation 3:

$$-7B = -7$$
$$B = 1$$

Substitute into equation 2:

$$-2 + C = -2$$
$$C = 0$$

Substitute into equation 1:

$$A + 1 = 4$$
$$A = 3$$

So $A = 3$, $B = 1$, and $C = 0$.

49. Multiply by $(x+4)(x^2-4x+16)$, so:

$$1 = A(x^2-4x+16) + (Bx+C)(x+4)$$
$$1 = (A+B)x^2 + (-4A+4B+C)x + (16A+4C)$$

Since A, B, and C are constants:

$$\begin{aligned} A + B \qquad &= 0 \\ -4A + 4B + C &= 0 \\ 16A \qquad + 4C &= 1 \end{aligned}$$

Add 4 times equation 1 to equation 2, and –16 times equation 1 to equation 3:

$$\begin{aligned} A + B \qquad &= 0 \\ 8B + C &= 0 \\ -16B + 4C &= 1 \end{aligned}$$

Add 2 times equation 2 to equation 3:

$$\begin{aligned} A + B \qquad &= 0 \\ 8B + C &= 0 \\ 6C &= 1 \end{aligned}$$

Solve equation 3 for C:

$$6C = 1$$
$$C = \tfrac{1}{6}$$

Substitute into equation 2:

$$8B + \tfrac{1}{6} = 0$$
$$8B = -\tfrac{1}{6}$$
$$B = -\tfrac{1}{48}$$

Substitute into equation 1:

$$A - \tfrac{1}{48} = 0$$
$$A = \tfrac{1}{48}$$

So $A = \frac{1}{48}$, $B = -\frac{1}{48}$, and $C = \frac{1}{6}$.

51. Multiply by $(x-a)^3$, so:

$$x = A(x-a)^2 + B(x-a) + C$$
$$x = Ax^2 + (-2aA + B)x + (a^2A - aB + C)$$

Since A, B, and C are constants:

$$\begin{aligned} A &= 0 \\ -2aA + B &= 1 \\ a^2A - aB + C &= 0 \end{aligned}$$

Substitute into equation 2:

$$\begin{aligned} 0 + B &= 1 \\ B &= 1 \end{aligned}$$

Substitute into equation 3:

$$\begin{aligned} 0 - a + C &= 0 \\ C &= a \end{aligned}$$

So $A = 0$, $B = 1$, and $C = a$.

53. Multiply by $(x - a)(x - b)$, so:

$$\begin{aligned} (a-b)(a+b-x) &= A(x-b) + B(x-a) \\ (a-b)(a+b) - (a-b)x &= (A+B)x + (-bA - aB) \end{aligned}$$

Since A and B are constants:

$$\begin{aligned} A + B &= b - a \\ -bA - aB &= a^2 - b^2 \end{aligned}$$

Multiply equation 1 by b:

$$\begin{aligned} bA + bB &= b^2 - ab \\ -bA - aB &= a^2 - b^2 \end{aligned}$$

Adding, we obtain:

$$\begin{aligned} (b-a)B &= a^2 - ab \\ B &= \frac{a(a-b)}{b-a} \\ B &= -a \end{aligned}$$

Substitute into equation 1:

$$\begin{aligned} A - a &= b - a \\ A &= b \end{aligned}$$

So $A = b$ and $B = -a$.

55. We evaluate the determinant:

$$\begin{vmatrix} 1 & 5 \\ -6 & 4 \end{vmatrix} = 1(4) - 5(-6) = 4 + 30 = 34$$

57. Adding –2 times row 2 to row 3 yields:

$$\begin{vmatrix} 4 & 0 & 3 \\ -2 & 1 & 5 \\ 0 & 2 & -1 \end{vmatrix} = \begin{vmatrix} 4 & 0 & 3 \\ -2 & 1 & 5 \\ 4 & 0 & -11 \end{vmatrix} = 1 \bullet \begin{vmatrix} 4 & 3 \\ 4 & -11 \end{vmatrix} = -56$$

59. Subtracting row 2 from row 1 yields:

$$\begin{vmatrix} 1 & 5 & 7 \\ 1 & 5 & 7 \\ 17 & 19 & 21 \end{vmatrix} = \begin{vmatrix} 0 & 0 & 0 \\ 1 & 5 & 7 \\ 17 & 19 & 21 \end{vmatrix} = 0$$

61. We evaluate the determinant:

$$\begin{vmatrix} 1 & 0 & 0 & 0 \\ 0 & 2 & 0 & 0 \\ 0 & 0 & 3 & 0 \\ 0 & 0 & 0 & 4 \end{vmatrix} = 1\begin{vmatrix} 2 & 0 & 0 \\ 0 & 3 & 0 \\ 0 & 0 & 4 \end{vmatrix} = 2\begin{vmatrix} 3 & 0 \\ 0 & 4 \end{vmatrix} = 2(12) = 24$$

63. By expanding along column 1:

$$\begin{vmatrix} a & b & c \\ b & c & a \\ c & a & b \end{vmatrix} = a\begin{vmatrix} c & a \\ a & b \end{vmatrix} - b\begin{vmatrix} b & c \\ a & b \end{vmatrix} + c\begin{vmatrix} b & c \\ c & a \end{vmatrix}$$

$$= a\left(bc - a^2\right) - b\left(b^2 - ac\right) + c\left(ab - c^2\right)$$

$$= abc - a^3 - b^3 + abc + abc - c^3$$

$$= 3abc - a^3 - b^3 - c^3$$

65. Adding b times column 2 to column 1 yields:

$$\begin{vmatrix} a^2 + x & b & c & d \\ -b & 1 & 0 & 0 \\ -c & 0 & 1 & 0 \\ -d & 0 & 0 & 1 \end{vmatrix} = \begin{vmatrix} a^2 + b^2 + x & b & c & d \\ 0 & 1 & 0 & 0 \\ -c & 0 & 1 & 0 \\ -d & 0 & 0 & 1 \end{vmatrix} = 1\begin{vmatrix} a^2 + b^2 + x & c & d \\ -c & 1 & 0 \\ -d & 0 & 1 \end{vmatrix}$$

Adding c times column 2 to column 1 yields:

$$\begin{vmatrix} a^2 + b^2 + c^2 + x & c & d \\ 0 & 1 & 0 \\ -d & 0 & 1 \end{vmatrix} = 1\begin{vmatrix} a^2 + b^2 + c^2 + x & d \\ -d & 1 \end{vmatrix}$$

$$= a^2 + b^2 + c^2 + x - \left(-d^2\right)$$

$$= a^2 + b^2 + c^2 + d^2 + x$$

67. Substituting $(x, y) = (-2, 5)$ and $(x, y) = (2, 9)$, we obtain:

$$5 = 4a - 2b - 1$$
$$9 = 4a + 2b - 1$$

Adding, we obtain:

$$14 = 8a - 2$$
$$16 = 8a$$
$$2 = a$$

Substitute into the first equation:

$$5 = 8 - 2b - 1$$
$$-2 = -2b$$
$$1 = b$$

So $a = 2$ and $b = 1$.

69. (a) We first graph the triangle:

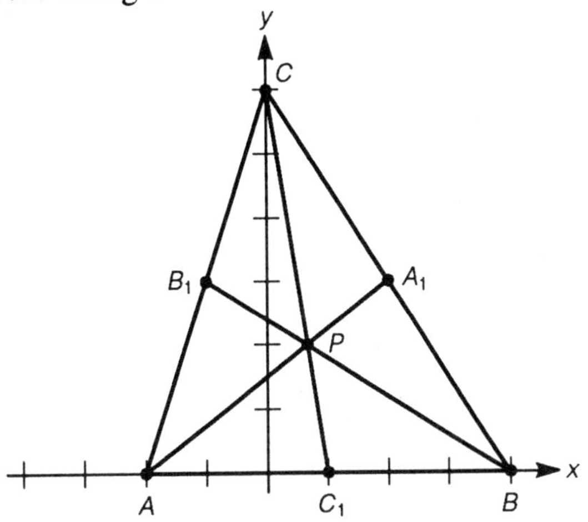

By the midpoint formula:

$$A_1 = \left(\frac{4+0}{2}, \frac{0+6}{2}\right) = (2,3)$$

$$B_1 = \left(\frac{-2+0}{2}, \frac{0+6}{2}\right) = (-1,3)$$

$$C_1 = \left(\frac{-2+4}{2}, \frac{0+0}{2}\right) = (1,0)$$

For AA_1, we have:

$$\text{slope} = \frac{3-0}{2-(-2)} = \frac{3}{4}$$

$$\text{point} = (-2,0)$$

$$y - 0 = \frac{3}{4}(x+2)$$

$$y = \frac{3}{4}x + \frac{3}{2}$$

For BB_1, we have:

$$\text{slope} = \frac{3-0}{-1-4} = -\frac{3}{5}$$

$$\text{point} = (4,0)$$

$$y - 0 = -\frac{3}{5}(x-4)$$

$$y = -\frac{3}{5}x + \frac{12}{5}$$

We solve $\frac{3}{4}x + \frac{3}{2} = -\frac{3}{5}x + \frac{12}{5}$. Multiply by 20 to clear fractions:

$$15x + 30 = -12x + 48$$

$$27x = 18$$

$$x = \tfrac{2}{3}$$

So $y = \frac{3}{4} \bullet \frac{2}{3} + \frac{3}{2} = 2$. Thus the point of intersection is $\left(\frac{2}{3}, 2\right)$.

(b) For CC_1, we have:

$$\text{slope} = \frac{6-0}{0-1} = -6$$
$$y = -6x + 6$$

We solve $-6x + 6 = -\frac{3}{5}x + \frac{12}{5}$.
Multiply by 5 to clear fractions:

$$-30x + 30 = -3x + 12$$
$$18 = 27x$$
$$\tfrac{2}{3} = x$$

So $y = -6\left(\frac{2}{3}\right) + 6 = 2$. Thus the point of intersection is $\left(\frac{2}{3}, 2\right)$.

(c) We solve $-6x + 6 = \frac{3}{4}x + \frac{3}{2}$. Multiply by 4 to clear fractions:

$$-24x + 24 = 3x + 6$$
$$18 = 27x$$
$$\tfrac{2}{3} = x$$

So $y = -6\left(\frac{2}{3}\right) + 6 = 2$. Thus the point of intersection is $\left(\frac{2}{3}, 2\right)$.

(d) We use the distance formula:

$$AP = \sqrt{\left(-2 - \tfrac{2}{3}\right)^2 + (0-2)^2} = \sqrt{\tfrac{64}{9} + \tfrac{36}{9}} = \tfrac{10}{3}$$
$$PA_1 = \sqrt{\left(\tfrac{2}{3} - 2\right)^2 + (2-3)^2} = \sqrt{\tfrac{16}{9} + \tfrac{9}{9}} = \tfrac{5}{3}$$

So $\dfrac{AP}{PA_1} = \dfrac{\frac{10}{3}}{\frac{5}{3}} = 2$. Again using the distance formula:

$$BP = \sqrt{\left(4 - \tfrac{2}{3}\right)^2 + (0-2)^2} = \sqrt{\tfrac{100}{9} + \tfrac{36}{9}} = \tfrac{2\sqrt{34}}{3}$$
$$PB_1 = \sqrt{\left(\tfrac{2}{3} + 1\right)^2 + (2-3)^2} = \sqrt{\tfrac{25}{9} + \tfrac{9}{9}} = \tfrac{\sqrt{34}}{3}$$

So $\dfrac{BP}{PB_1} = \dfrac{\frac{2\sqrt{34}}{3}}{\frac{\sqrt{34}}{3}} = 2$. Again using the distance formula:

$$CP = \sqrt{\left(0 - \tfrac{2}{3}\right)^2 + (6-2)^2} = \sqrt{\tfrac{4}{9} + \tfrac{144}{9}} = \tfrac{2\sqrt{37}}{3}$$
$$PC_1 = \sqrt{\left(\tfrac{2}{3} - 1\right)^2 + (2-0)^2} = \sqrt{\tfrac{1}{9} + \tfrac{36}{9}} = \tfrac{\sqrt{37}}{3}$$

So $\dfrac{CP}{PC_1} = \dfrac{\frac{2\sqrt{37}}{3}}{\frac{\sqrt{37}}{3}} = 2$. These ratios are all equal.

71. See the figure:

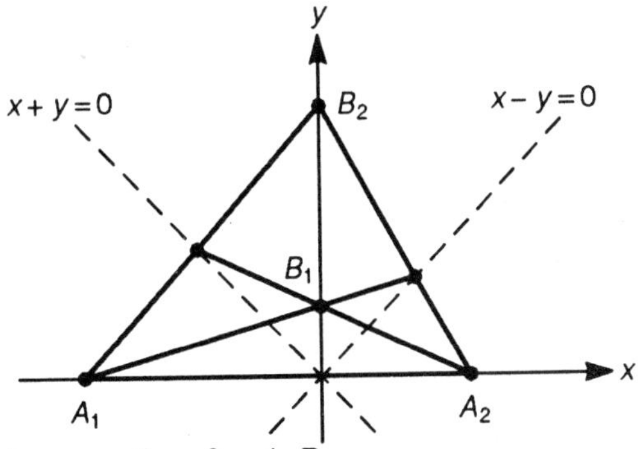

We start by finding the equation for A_2B_2:

$$m = -\frac{b_2}{a_2}$$

point: (a, a)

By the point-slope formula:

$$a_2y + b_2x = aa_2 + ab_2$$

Since $(a_2, 0)$ lies on this curve, we have $a_2b_2 = aa_2 + ab_2$. (*)

Now we find the equation for A_1B_1:

$$m = -\frac{b_1}{a_1}$$

point: (a, a)

By the point-slope formula:

$$a_1y + b_1x = aa_1 + ab_1$$

Since $(a_1, 0)$ lies on this curve, we have $a_1b_1 = aa_1 + ab_1$. (**)

Now we look at A_1B_2:

$$m = -\frac{b_2}{a_1}$$

$$b_2x + a_1y = a_1b_2$$

Also look at A_2B_1:

$$m = -\frac{b_1}{a_2}, \text{ so:}$$

$$b_1x + a_2y = a_2b_1$$

We solve the system:

$$b_2x + a_1y = a_1b_2$$
$$b_1x + a_2y = a_2b_1$$

Multiply equation 1 by $-b_1$ and equation 2 by b_2:

$$-b_1b_2x - a_1b_1y = -a_1b_1b_2$$
$$b_1b_2x + a_2b_2y = a_2b_1b_2$$

Adding, we obtain:

$$(a_2b_2 - a_1b_1)y = a_2b_1b_2 - a_1b_1b_2$$
$$y = \frac{b_1b_2(a_2 - a_1)}{a_2b_2 - a_1b_1}$$

Substituting into equation 1, we have:

$$b_2x + \frac{a_1b_1b_2(a_2 - a_1)}{a_2b_2 - a_1b_1} = a_1b_2$$
$$x = \frac{a_1a_2(b_2 - b_1)}{a_2b_2 - a_1b_1}$$

Now, we are asked to show that $x + y = 0$, so:

$$x + y = \frac{a_1a_2(b_2 - b_1) + b_1b_2(a_2 - a_1)}{a_2b_2 - a_1b_1} = \frac{a_1a_2b_2 - a_1a_2b_1 + a_2b_1b_2 - a_1b_1b_2}{a_2b_2 - a_1b_1}$$

Replacing a_1b_1 and a_2b_2 by equations (*) and (**):

$$x + y = \frac{a_1(aa_2 + ab_2) - a_2(aa_1 + ab_1) + b_1(aa_2 + ab_2) - b_2(aa_1 + ab_1)}{a_2b_2 - a_1b_1}$$
$$= \frac{a(a_1a_2 + a_1b_2 - a_1a_2 - a_2b_1 + a_2b_1 + b_1b_2 - a_1b_2 - b_1b_2)}{a_2b_2 - a_1b_1}$$
$$= 0$$

73. We find each intersection point:

$y = x - 1$	$y = x - 1$	$y = -x - 2$
$y = -x - 2$	$y = 2x + 3$	$y = 2x + 3$
$2y = -3$	$x - 1 = 2x + 3$	$-x - 2 = 2x + 3$
$y = -\frac{3}{2}$	$-4 = x$	$-5 = 3x$
		$x = -\frac{5}{3}$
$x = -\frac{1}{2}$	$y = -5$	$y = -\frac{1}{3}$
$\left(-\frac{1}{2}, -\frac{3}{2}\right)$	$(-4, -5)$	$\left(-\frac{5}{3}, -\frac{1}{3}\right)$

We now substitute into the equation $(x - h)^2 + (y - k)^2 = r^2$:

$$\left(-\tfrac{1}{2} - h\right)^2 + \left(-\tfrac{3}{2} - k\right)^2 = r^2$$
$$(-4 - h)^2 + (-5 - k)^2 = r^2$$
$$\left(-\tfrac{5}{3} - h\right)^2 + \left(-\tfrac{1}{3} - k\right)^2 = r^2$$

These equations, when multiplied out, become:

$$\tfrac{5}{2} + h + h^2 + 3k + k^2 = r^2$$
$$41 + 8h + h^2 + 10k + k^2 = r^2$$
$$\tfrac{26}{9} + \tfrac{10}{3}h + h^2 + \tfrac{2}{3}k + k^2 = r^2$$

Subtracting the first equation from the other two yields:

$$\tfrac{77}{2} + 7h + 7k = 0$$
$$\tfrac{7}{18} + \tfrac{7}{3}h - \tfrac{7}{3}k = 0$$

These simplify to:

$$6h + 6k = -33$$
$$6h - 6k = -1$$

Adding, we obtain:

$$12h = -34$$
$$h = -\tfrac{17}{6}$$

To find k, we subtract the two equations:

$$12k = -32$$
$$k = -\tfrac{8}{3}$$

Finally, we find r:

$$r^2 = (-4-h)^2 + (-5-k)^2 = \left(-\tfrac{7}{6}\right)^2 + \left(-\tfrac{7}{3}\right)^2 = \tfrac{245}{36}$$

The equation of the circle is:

$$\left(x + \tfrac{17}{6}\right)^2 + \left(y + \tfrac{8}{3}\right)^2 = \tfrac{245}{36}$$

75. $2A + 2B = \begin{pmatrix} 6 & -4 \\ 2 & 10 \end{pmatrix} + \begin{pmatrix} 4 & 2 \\ 2 & 16 \end{pmatrix} = \begin{pmatrix} 10 & -2 \\ 4 & 26 \end{pmatrix}$

77. $4B = \begin{pmatrix} 8 & 4 \\ 4 & 32 \end{pmatrix}$

79. $AB = \begin{pmatrix} 4 & -13 \\ 7 & 41 \end{pmatrix}$

81. $AB - BA = \begin{pmatrix} 7 & -13 \\ 7 & 41 \end{pmatrix} - \begin{pmatrix} 7 & 1 \\ 11 & 38 \end{pmatrix} = \begin{pmatrix} -3 & -14 \\ -4 & 3 \end{pmatrix}$

83. $C + B = \begin{pmatrix} 1 & 1 \\ 1 & 7 \end{pmatrix}$

85. $AB + AC = \begin{pmatrix} 4 & -13 \\ 7 & 41 \end{pmatrix} + \begin{pmatrix} -3 & 2 \\ -1 & -5 \end{pmatrix} = \begin{pmatrix} 1 & -11 \\ 6 & 36 \end{pmatrix}$

87. $BA + CA = \begin{pmatrix} 7 & 1 \\ 11 & 38 \end{pmatrix} + \begin{pmatrix} -3 & 2 \\ -1 & -5 \end{pmatrix} = \begin{pmatrix} 4 & 3 \\ 10 & 33 \end{pmatrix}$

89. $DE = \begin{pmatrix} -42 & 58 \\ 5 & 20 \end{pmatrix}$

91. The multiplication is undefined.

93. The multiplication is undefined.

95. $(A+B)+C = \begin{pmatrix} 5 & -1 \\ 2 & 13 \end{pmatrix} + \begin{pmatrix} -1 & 0 \\ 0 & -1 \end{pmatrix} = \begin{pmatrix} 4 & -1 \\ 2 & 12 \end{pmatrix}$

97. $(AB)C = \begin{pmatrix} 4 & -13 \\ 7 & 41 \end{pmatrix}\begin{pmatrix} -1 & 0 \\ 0 & -1 \end{pmatrix} = \begin{pmatrix} -4 & 13 \\ -7 & -41 \end{pmatrix}$

99. We compute A^2 and then A^3:

$$A^2 = \begin{pmatrix} 1 & 1 \\ 0 & 1 \end{pmatrix}\begin{pmatrix} 1 & 1 \\ 0 & 1 \end{pmatrix} = \begin{pmatrix} 1 & 2 \\ 0 & 1 \end{pmatrix}$$

$$A^3 = A \bullet A^2 = \begin{pmatrix} 1 & 1 \\ 0 & 1 \end{pmatrix}\begin{pmatrix} 1 & 2 \\ 0 & 1 \end{pmatrix} = \begin{pmatrix} 1 & 3 \\ 0 & 1 \end{pmatrix}$$

101. (a) Form the matrix:

$$\begin{pmatrix} 1 & 5 & 1 & 0 \\ 2 & 9 & 0 & 1 \end{pmatrix}$$

Multiply row 1 by –2 and add to row 2:

$$\begin{pmatrix} 1 & 5 & 1 & 0 \\ 0 & -1 & -2 & 1 \end{pmatrix}$$

Multiply row 2 by 5 and add to row 1:

$$\begin{pmatrix} 1 & 0 & -9 & 5 \\ 0 & -1 & -2 & 1 \end{pmatrix}$$

Multiply row 2 by –1:

$$\begin{pmatrix} 1 & 0 & -9 & 5 \\ 0 & 1 & 2 & -1 \end{pmatrix}$$

So the inverse is $\begin{pmatrix} -9 & 5 \\ 2 & -1 \end{pmatrix}$.

(b) Since the system is equivalent to $A \bullet X = B$, where $A = \begin{pmatrix} 1 & 5 \\ 2 & 9 \end{pmatrix}$,

$X = \begin{pmatrix} x \\ y \end{pmatrix}$, and $B = \begin{pmatrix} 3 \\ -4 \end{pmatrix}$, then:

$$X = A^{-1} \bullet B = \begin{pmatrix} -9 & 5 \\ 2 & -1 \end{pmatrix}\begin{pmatrix} 3 \\ -4 \end{pmatrix} = \begin{pmatrix} -47 \\ 10 \end{pmatrix}$$

So the solution is $(-47, 10)$.

103. (a) Form the matrix:

$$\begin{pmatrix} 1 & -2 & 3 & 1 & 0 & 0 \\ 2 & -5 & 10 & 0 & 1 & 0 \\ -1 & 2 & -2 & 0 & 0 & 1 \end{pmatrix}$$

Multiply row 1 by –2 and add to row 2, then add row 1 to row 3:

$$\begin{pmatrix} 1 & -2 & 3 & 1 & 0 & 0 \\ 0 & -1 & 4 & -2 & 1 & 0 \\ 0 & 0 & 1 & 1 & 0 & 1 \end{pmatrix}$$

Multiply row 2 by –1:

$$\begin{pmatrix} 1 & -2 & 3 & 1 & 0 & 0 \\ 0 & 1 & -4 & 2 & -1 & 0 \\ 0 & 0 & 1 & 1 & 0 & 1 \end{pmatrix}$$

Multiply row 2 by 2 and add to row 1:

$$\begin{pmatrix} 1 & 0 & -5 & 5 & -2 & 0 \\ 0 & 1 & -4 & 2 & -1 & 0 \\ 0 & 0 & 1 & 1 & 0 & 1 \end{pmatrix}$$

Multiply row 3 by 4 and add to row 2, then multiply row 3 by 5 and add to row 1:

$$\begin{pmatrix} 1 & 0 & 0 & 10 & -2 & 5 \\ 0 & 1 & 0 & 6 & -1 & 4 \\ 0 & 0 & 1 & 1 & 0 & 1 \end{pmatrix}$$

So the inverse is $\begin{pmatrix} 10 & -2 & 5 \\ 6 & -1 & 4 \\ 1 & 0 & 1 \end{pmatrix}$.

(b) Since the system is equivalent to $A \bullet X = B$, where $A = \begin{pmatrix} 1 & -2 & 3 \\ 2 & -5 & 10 \\ -1 & 2 & -2 \end{pmatrix}$,

$X = \begin{pmatrix} x \\ y \\ z \end{pmatrix}$, and $B = \begin{pmatrix} -2 \\ -3 \\ 6 \end{pmatrix}$, then:

$$X = A^{-1} \bullet B = \begin{pmatrix} 10 & -2 & 5 \\ 6 & -1 & 4 \\ 1 & 0 & 1 \end{pmatrix}\begin{pmatrix} -2 \\ -3 \\ 6 \end{pmatrix} = \begin{pmatrix} 16 \\ 15 \\ 4 \end{pmatrix}$$

So the solution is $(16, 15, 4)$.

105. Form the matrix:

$$\begin{pmatrix} 5 & 3 & 6 & -7 & 1 & 0 & 0 & 0 \\ 3 & -4 & 0 & -9 & 0 & 1 & 0 & 0 \\ 0 & 1 & -1 & -1 & 0 & 0 & 1 & 0 \\ 2 & 2 & 3 & -2 & 0 & 0 & 0 & 1 \end{pmatrix}$$

Multiply row 4 by –2 and add to row 1 (to get a 1):

$$\begin{pmatrix} 1 & -1 & 0 & -3 & 1 & 0 & 0 & -2 \\ 3 & -4 & 0 & -9 & 0 & 1 & 0 & 0 \\ 0 & 1 & -1 & -1 & 0 & 0 & 1 & 0 \\ 2 & 2 & 3 & -2 & 0 & 0 & 0 & 1 \end{pmatrix}$$

Multiply row 1 by –3 and add to row 2, and multiply row 1 by –2 and add to row 4:

$$\begin{pmatrix} 1 & -1 & 0 & -3 & 1 & 0 & 0 & -2 \\ 0 & -1 & 0 & 0 & -3 & 1 & 0 & 6 \\ 0 & 1 & -1 & -1 & 0 & 0 & 1 & 0 \\ 0 & 4 & 3 & 4 & -2 & 0 & 0 & 5 \end{pmatrix}$$

Add row 2 to row 3, then mulitply row 2 by 4 and add to row 4:

$$\begin{pmatrix} 1 & -1 & 0 & -3 & 1 & 0 & 0 & -2 \\ 0 & -1 & 0 & 0 & -3 & 1 & 0 & 6 \\ 0 & 0 & -1 & -1 & -3 & 1 & 1 & 6 \\ 0 & 0 & 3 & 4 & -14 & 4 & 0 & 29 \end{pmatrix}$$

Multiply row 3 by 3 and add to row 4:

$$\begin{pmatrix} 1 & -1 & 0 & -3 & 1 & 0 & 0 & -2 \\ 0 & -1 & 0 & 0 & -3 & 1 & 0 & 6 \\ 0 & 0 & -1 & -1 & -3 & 1 & 1 & 6 \\ 0 & 0 & 0 & 1 & -23 & 7 & 3 & 47 \end{pmatrix}$$

Add row 4 to row 3, then multiply row 4 by 3 and add to row 1:

$$\begin{pmatrix} 1 & -1 & 0 & 0 & -68 & 21 & 9 & 139 \\ 0 & -1 & 0 & 0 & -3 & 1 & 0 & 6 \\ 0 & 0 & -1 & 0 & -26 & 8 & 4 & 53 \\ 0 & 0 & 0 & 1 & -23 & 7 & 3 & 47 \end{pmatrix}$$

Multiply row 2 by –1 and row 3 by –1:

$$\begin{pmatrix} 1 & -1 & 0 & 0 & -68 & 21 & 9 & 139 \\ 0 & 1 & 0 & 0 & 3 & -1 & 0 & -6 \\ 0 & 0 & 1 & 0 & 26 & -8 & -4 & -53 \\ 0 & 0 & 0 & 1 & -23 & 7 & 3 & 47 \end{pmatrix}$$

Add row 2 to row 1:

$$\begin{pmatrix} 1 & 0 & 0 & 0 & -65 & 20 & 9 & 133 \\ 0 & 1 & 0 & 0 & 3 & -1 & 0 & -6 \\ 0 & 0 & 1 & 0 & 26 & -8 & -4 & -53 \\ 0 & 0 & 0 & 1 & -23 & 7 & 3 & 47 \end{pmatrix}$$

So the inverse is $\begin{pmatrix} -65 & 20 & 9 & 133 \\ 3 & -1 & 0 & -6 \\ 26 & -8 & -4 & -53 \\ -23 & 7 & 3 & 47 \end{pmatrix}$.

107. Adding 2 times column 2 to column 1 and column 2 to column 3 yields:

$$D = \begin{vmatrix} 2 & -1 & 1 \\ 3 & 2 & 2 \\ 1 & -5 & -3 \end{vmatrix} = \begin{vmatrix} 0 & -1 & 0 \\ 7 & 2 & 4 \\ -9 & -5 & -8 \end{vmatrix} = -(-1)\begin{vmatrix} 7 & 4 \\ -9 & -8 \end{vmatrix} = 1(-20) = -20$$

Adding 2 times row 1 to row 3 yields:

$$D_x = \begin{vmatrix} 1 & -1 & 1 \\ 0 & 2 & 2 \\ -2 & -5 & -3 \end{vmatrix} = \begin{vmatrix} 1 & -1 & 1 \\ 0 & 2 & 2 \\ 0 & -7 & -1 \end{vmatrix} = 1\begin{vmatrix} 2 & 2 \\ -7 & -1 \end{vmatrix} = 12$$

Adding 2 times row 1 to row 3 yields:

$$D_y = \begin{vmatrix} 2 & 1 & 1 \\ 3 & 0 & 2 \\ 1 & -2 & -3 \end{vmatrix} = \begin{vmatrix} 2 & 1 & 1 \\ 3 & 0 & 2 \\ 5 & 0 & -1 \end{vmatrix} = -1\begin{vmatrix} 3 & 2 \\ 5 & -1 \end{vmatrix} = -(-13) = 13$$

Adding 2 times row 1 to row 3 yields:

$$D_z = \begin{vmatrix} 2 & -1 & 1 \\ 3 & 2 & 0 \\ 1 & -5 & -2 \end{vmatrix} = \begin{vmatrix} 2 & -1 & 1 \\ 3 & 2 & 0 \\ 5 & -7 & 0 \end{vmatrix} = 1\begin{vmatrix} 3 & 2 \\ 5 & -7 \end{vmatrix} = -31$$

So $x = \frac{D_x}{D} = -\frac{12}{20} = -\frac{3}{5}$, $y = \frac{D_y}{D} = -\frac{13}{20}$, and $z = \frac{D_z}{D} = \frac{31}{20}$.

So the solution is $\left(-\frac{3}{5}, -\frac{13}{20}, \frac{31}{20}\right)$.

109. Subtracting column 2 from column 3 and column 1 from column 2 yields:

$$D = \begin{vmatrix} 1 & 2 & 3 \\ 4 & 5 & 6 \\ 7 & 8 & 9 \end{vmatrix} = \begin{vmatrix} 1 & 1 & 1 \\ 4 & 1 & 1 \\ 7 & 1 & 1 \end{vmatrix}$$

Subtracting column 3 from column 2 yields:

$$D = \begin{vmatrix} 1 & 0 & 1 \\ 4 & 0 & 1 \\ 7 & 0 & 1 \end{vmatrix} = 0$$

So Cramer's Rule will not work. Form the augmented matrix:

$$\begin{pmatrix} 1 & 2 & 3 & -1 \\ 4 & 5 & 6 & 2 \\ 7 & 8 & 9 & -3 \end{pmatrix}$$

Add -4 times row 1 to row 2 and -7 times row 1 to row 3:

$$\begin{pmatrix} 1 & 2 & 3 & -1 \\ 0 & -3 & -6 & 6 \\ 0 & -6 & -12 & 4 \end{pmatrix}$$

Add -2 times row 2 to row 3:

$$\begin{pmatrix} 1 & 2 & 3 & -1 \\ 0 & -3 & -6 & 6 \\ 0 & 0 & 0 & -8 \end{pmatrix}$$

So $0 = -8$, which is false. The system has no solution.

111. Since $D = 0$ (See Exercise 110), we form the augmented matrix:

$$\begin{pmatrix} 3 & 2 & -2 & 1 \\ 2 & 3 & -1 & -2 \\ 8 & 7 & -5 & 0 \end{pmatrix}$$

Subtract row 2 from row 1:

$$\begin{pmatrix} 1 & -1 & -1 & 3 \\ 2 & 3 & -1 & -2 \\ 8 & 7 & -5 & 0 \end{pmatrix}$$

Add –2 times row 1 to row 2 and – 8 times row 1 to row 3:

$$\begin{pmatrix} 1 & -1 & -1 & 3 \\ 0 & 5 & 1 & -8 \\ 0 & 15 & 3 & -24 \end{pmatrix}$$

Add –3 times row 2 to row 3:

$$\begin{pmatrix} 1 & -1 & -1 & 3 \\ 0 & 5 & 1 & -8 \\ 0 & 0 & 0 & 0 \end{pmatrix}$$

So we have the system:

$$x - y - z = 3$$
$$5y + z = -8$$

Solve equation 2 for z:

$$z = -8 - 5y$$

Substitute into equation 1:

$$x - y + 8 + 5y = 3$$
$$x = -5 - 4y$$

So the solution is $(-5 - 4y, y, -8 - 5y)$, for any real number y.

113. Adding 4 times column 2 to both column 1 and column 3 yields:

$$D = \begin{vmatrix} 2 & -1 & 1 & 3 \\ 1 & 2 & 0 & 2 \\ 0 & 3 & 3 & 4 \\ -4 & 1 & -4 & 0 \end{vmatrix} = \begin{vmatrix} -2 & -1 & -3 & 3 \\ 9 & 2 & 8 & 2 \\ 12 & 3 & 15 & 4 \\ 0 & 1 & 0 & 0 \end{vmatrix} = 1\begin{vmatrix} -2 & -3 & 3 \\ 9 & 8 & 2 \\ 12 & 15 & 4 \end{vmatrix}$$

Subtracting row 2 from row 3 yields:

$$D = \begin{vmatrix} -2 & -3 & 3 \\ 9 & 8 & 2 \\ 3 & 7 & 2 \end{vmatrix}$$

Adding $\frac{2}{3}$ times column 3 to column 1, and column 3 to column 2 yields:

$$D = \begin{vmatrix} 0 & 0 & 3 \\ \frac{31}{3} & 10 & 2 \\ \frac{13}{3} & 9 & 2 \end{vmatrix} = 3\begin{vmatrix} \frac{31}{3} & 10 \\ \frac{13}{3} & 9 \end{vmatrix} = \begin{vmatrix} 31 & 10 \\ 13 & 9 \end{vmatrix} = 149$$

Adding 11 times column 2 to column 1 and 4 times column 2 to column 3 yields:

$$D_x = \begin{vmatrix} 15 & -1 & 1 & 3 \\ 12 & 2 & 0 & 2 \\ 12 & 3 & 3 & 4 \\ -11 & 1 & -4 & 0 \end{vmatrix} = \begin{vmatrix} 4 & -1 & -3 & 3 \\ 34 & 2 & 8 & 2 \\ 45 & 3 & 15 & 4 \\ 0 & 1 & 0 & 0 \end{vmatrix} = 1\begin{vmatrix} 4 & -3 & 3 \\ 34 & 8 & 2 \\ 45 & 15 & 4 \end{vmatrix}$$

Subtracting row 2 from row 3 yields:

$$D_x = \begin{vmatrix} 4 & -3 & 3 \\ 34 & 8 & 2 \\ 11 & 7 & 2 \end{vmatrix}$$

Subtracting row 3 from row 2 yields:

$$D_x = \begin{vmatrix} 4 & -3 & 3 \\ 23 & 1 & 0 \\ 11 & 7 & 2 \end{vmatrix}$$

Adding –23 times column 2 to column 1:

$$D_x = \begin{vmatrix} 73 & -3 & 3 \\ 0 & 1 & 0 \\ -150 & 7 & 2 \end{vmatrix} = \begin{vmatrix} 73 & 3 \\ -150 & 2 \end{vmatrix} = 596$$

Adding –2 times row 2 to row 1 and 4 times row 2 to row 4 yields:

$$D_y = \begin{vmatrix} 2 & 15 & 1 & 3 \\ 1 & 12 & 0 & 2 \\ 0 & 12 & 3 & 4 \\ -4 & -11 & -4 & 0 \end{vmatrix} = \begin{vmatrix} 0 & -9 & 1 & -1 \\ 1 & 12 & 0 & 2 \\ 0 & 12 & 3 & 4 \\ 0 & 37 & -4 & 8 \end{vmatrix} = -1\begin{vmatrix} -9 & 1 & -1 \\ 12 & 3 & 4 \\ 37 & -4 & 8 \end{vmatrix}$$

Adding 9 times column 2 to column 1 and column 2 to column 3 yields:

$$D_y = -1\begin{vmatrix} 0 & 1 & 0 \\ 39 & 3 & 7 \\ 1 & -4 & 4 \end{vmatrix} = 1\begin{vmatrix} 39 & 7 \\ 1 & 4 \end{vmatrix} = 149$$

Adding –2 times row 2 and row 1 to 4 times row 2 to row 4 yields:

$$D_z = \begin{vmatrix} 2 & -1 & 15 & 3 \\ 1 & 2 & 12 & 2 \\ 0 & 3 & 12 & 4 \\ -4 & 1 & -11 & 0 \end{vmatrix} = \begin{vmatrix} 0 & -5 & -9 & -1 \\ 1 & 2 & 12 & 2 \\ 0 & 3 & 12 & 4 \\ 0 & 9 & 37 & 8 \end{vmatrix} = -1\begin{vmatrix} -5 & -9 & -1 \\ 3 & 12 & 4 \\ 9 & 37 & 8 \end{vmatrix}$$

Adding 4 times row 1 to row 2 and 8 times row 1 to row 3:

$$D_z = -1\begin{vmatrix} -5 & -9 & -1 \\ -17 & -24 & 0 \\ -31 & -25 & 0 \end{vmatrix} = 1\begin{vmatrix} -17 & -24 \\ -31 & -35 \end{vmatrix} = -149$$

Adding –2 times row 2 to row 1 and 4 times row 2 to row 4:

$$D_w = \begin{vmatrix} 2 & -1 & 1 & 15 \\ 1 & 2 & 0 & 12 \\ 0 & 3 & 3 & 12 \\ -4 & 1 & -4 & -11 \end{vmatrix} = \begin{vmatrix} 0 & -5 & 1 & -9 \\ 1 & 2 & 0 & 12 \\ 0 & 3 & 3 & 12 \\ 0 & 9 & -4 & 37 \end{vmatrix} = -1\begin{vmatrix} -5 & 1 & -9 \\ 3 & 3 & 12 \\ 9 & -4 & 37 \end{vmatrix}$$

Factoring 3 out of row 2 yields:

$$D_w = -3\begin{vmatrix} -5 & 1 & -9 \\ 1 & 1 & 4 \\ 9 & -4 & 37 \end{vmatrix}$$

Adding –1 times row 2 to row 1 and 4 times row 2 to row 3:

$$D_w = -3\begin{vmatrix} -6 & 0 & -13 \\ 1 & 1 & 4 \\ 13 & 0 & 53 \end{vmatrix} = -3\begin{vmatrix} -6 & -13 \\ 13 & 53 \end{vmatrix} = -3(-149) = 447$$

So $x = \frac{D_x}{D} = \frac{596}{149} = 4$, $y = \frac{D_y}{D} = \frac{149}{149} = 1$, $z = \frac{D_z}{D} = -\frac{149}{149} = -1$, and $w = \frac{D_w}{D} = \frac{447}{149} = 3$.
So the solution is $(4, 1, -1, 3)$.

115. Substitute to obtain:

$$\begin{aligned} x^2 &= 6x \\ x^2 - 6x &= 0 \\ x(x-6) &= 0 \\ x &= 0 \text{ or } x = 6 \end{aligned}$$

When $x = 0$, $y = 0$ and when $x = 6$, $y = 36$. So the solutions are $(0,0)$ and $(6,36)$.

117. Substitute to obtain:

$$\begin{aligned} x^2 - 9 &= 9 - x^2 \\ 2x^2 &= 18 \\ x^2 &= 9 \\ x &= \pm 3 \end{aligned}$$

When $x = \pm 3$, $y = 0$. So the solutions are $(3,0)$ and $(-3,0)$.

119. Adding the two equations yields:

$$\begin{aligned} 2x^2 &= 25 \\ x^2 &= \frac{25}{2} \\ x &= \frac{\pm 5\sqrt{2}}{2} \end{aligned}$$

Substituting for x yields:

$$\begin{aligned} \tfrac{25}{2} + y^2 &= 16 \\ y^2 &= \tfrac{7}{2} \\ y &= \frac{\pm\sqrt{14}}{2} \end{aligned}$$

So the solutions are $\left(\frac{5\sqrt{2}}{2}, \frac{\sqrt{14}}{2}\right)$, $\left(-\frac{5\sqrt{2}}{2}, \frac{\sqrt{14}}{2}\right)$, $\left(\frac{5\sqrt{2}}{2}, -\frac{\sqrt{14}}{2}\right)$, and $\left(-\frac{5\sqrt{2}}{2}, -\frac{\sqrt{14}}{2}\right)$.

121. Substitute to obtain:

$$x^2 + x = 1$$
$$x^2 + x - 1 = 0$$
$$x = \frac{-1 \pm \sqrt{1+4}}{2} = \frac{-1 \pm \sqrt{5}}{2}$$

Now $x = \frac{-1-\sqrt{5}}{2}$ is impossible, since $x \geq 0$. So the only solution is

$\left(\frac{-1+\sqrt{5}}{2}, \sqrt{\frac{-1+\sqrt{5}}{2}}\right)$ or $\left(\frac{-1+\sqrt{5}}{2}, \frac{\sqrt{-2+2\sqrt{5}}}{2}\right)$.

123. Substitute to obtain:

$$x^2 + (2x^2)^2 = 1$$
$$x^2 + 4x^4 = 1$$
$$4x^4 + x^2 - 1 = 0$$
$$x^2 = \frac{-1 \pm \sqrt{1+16}}{8} = \frac{-1 \pm \sqrt{17}}{8}$$

Since $x^2 \neq \frac{-1-\sqrt{17}}{8}$, we have $x^2 = \frac{-1+\sqrt{17}}{8}$, so $x = \frac{\pm\sqrt{-2+2\sqrt{17}}}{4}$. Since $y = 2x^2$, we have

$y = \frac{-1+\sqrt{17}}{4}$. So the solutions are $\left(\frac{\sqrt{-2+2\sqrt{17}}}{4}, \frac{-1+\sqrt{17}}{4}\right)$ and $\left(\frac{-\sqrt{-2+2\sqrt{17}}}{4}, \frac{-1+\sqrt{17}}{4}\right)$.

125. Multiply the second equation by –3:

$$-9x^2 + 3xy - 3y^2 = -54$$
$$x^2 + 2xy + 3y^2 = 68$$

Adding, we obtain:

$$-8x^2 + 5xy = 14$$
$$-5xy = 8x^2 + 14$$
$$y = \frac{8x^2 + 14}{5x}$$

Now substitute into the second equation:

$$3x^2 - \frac{8x^2 + 14}{5} + \frac{(8x^2 + 14)^2}{25x^2} = 18$$

Multiply by $25x^2$:

$$75x^4 - 5x^2(8x^2 + 14) + (8x^2 + 14)^2 = 450x^2$$
$$75x^4 - 40x^4 - 70x^2 + 64x^4 + 224x^2 + 196 = 450x^2$$
$$99x^4 - 296x^2 + 196 = 0$$
$$(99x^2 - 98)(x^2 - 2) = 0$$

Therefore we have the equations:

$$x^2 = \frac{98}{99} \qquad \text{or} \qquad x^2 = 2$$

$$x = \frac{\pm 7\sqrt{22}}{33} \qquad \text{or} \qquad x = \pm\sqrt{2}$$

Now $y = \dfrac{8x^2+14}{5x}$, so:

When $x = \sqrt{2}$, $y = \dfrac{16+14}{5\sqrt{2}} = 3\sqrt{2}$

When $x = -\sqrt{2}$, $y = \dfrac{16+14}{-5\sqrt{2}} = -3\sqrt{2}$

When $x = \dfrac{7\sqrt{22}}{33}$, $y = \dfrac{31\sqrt{22}}{33}$

When $x = \dfrac{-7\sqrt{22}}{33}$, $y = \dfrac{-31\sqrt{22}}{33}$

So the solutions are $(\sqrt{2}, 3\sqrt{2})$, $(-\sqrt{2}, -3\sqrt{2})$, $\left(\frac{7\sqrt{22}}{33}, \frac{31\sqrt{22}}{33}\right)$, and $\left(\frac{-7\sqrt{22}}{33}, \frac{-31\sqrt{22}}{33}\right)$.

127. Let $u = x - 3$ and $v = y + 1$, so:

$$2u^2 - v^2 = -1$$
$$-3u^2 + 2v^2 = 6$$

Multiply the first equation by 2 and add:

$$u^2 = 4$$
$$u = \pm 2$$

Substituting, we obtain:

$$8 - v^2 = -1$$
$$-v^2 = -9$$
$$v = \pm 3$$

So the solutions (u, v) are $(2, 3)$, $(2, -3)$, $(-2, 3)$, and $(-2, -3)$. Since $u = x - 3$ and $v = y + 1$, then $x = u + 3$ and $y = v - 1$. So the solutions are $(5, 2)$, $(5, -4)$, $(1, 2)$, and $(1, -4)$.

129. Call the numbers x and y. So $x + y = s$ and $\frac{x}{y} = \frac{a}{b}$. So $y = s - x$, and substituting:

$$\frac{x}{s-x} = \frac{a}{b}$$
$$bx = as - ax$$
$$(a+b)x = as$$
$$x = \frac{as}{a+b}$$

Thus $y = s - \frac{as}{a+b} = \frac{bs}{a+b}$. So the two numbers are $\frac{as}{a+b}$ and $\frac{bs}{a+b}$.

131. We have:

$$\tfrac{1}{2}x+\tfrac{1}{3}y+\tfrac{1}{4}z=62$$
$$\tfrac{1}{3}x+\tfrac{1}{4}y+\tfrac{1}{5}z=47$$
$$\tfrac{1}{4}x+\tfrac{1}{5}y+\tfrac{1}{6}z=38$$

Multiply the first equation by 12, the second equation by 60, and the third equation by 60 to clear the fractions:

$$6x+\ 4y+\ 3z=744$$
$$20x+15y+12z=2820$$
$$15x+12y+10z=2280$$

We use Cramer's Rule. Subtracting row 3 from row 2 yields:

$$D=\begin{vmatrix}6&4&3\\20&15&12\\15&12&10\end{vmatrix}=\begin{vmatrix}6&4&3\\5&3&2\\15&12&10\end{vmatrix}$$

Subtracting row 2 from row 1 and adding –3 times row 2 to row 3 yields:

$$D=\begin{vmatrix}1&1&1\\5&3&2\\0&3&4\end{vmatrix}$$

Subtracting column 1 from column 2 and column 3 yields:

$$D=\begin{vmatrix}1&0&0\\5&-2&-3\\0&3&4\end{vmatrix}=1\begin{vmatrix}-2&-3\\3&4\end{vmatrix}=1(1)=1$$

Subtracting row 3 from row 2 yields:

$$D_x=\begin{vmatrix}744&4&3\\2820&15&12\\2280&12&10\end{vmatrix}=\begin{vmatrix}744&4&3\\540&3&2\\2280&12&10\end{vmatrix}$$

Subtracting row 2 from row 1 and adding –4 times row 2 to row 3 yields:

$$D_x=\begin{vmatrix}204&1&1\\540&3&2\\120&0&2\end{vmatrix}$$

Adding –3 times row 1 to row 2 yields:

$$D_x=\begin{vmatrix}204&1&1\\-72&0&-1\\120&0&2\end{vmatrix}=-1\begin{vmatrix}-72&-1\\120&2\end{vmatrix}=-(-24)=24$$

Subtracting row 3 from row 2 yields:

$$D_y=\begin{vmatrix}6&744&3\\20&2820&12\\15&2280&10\end{vmatrix}=\begin{vmatrix}6&744&3\\5&540&2\\15&2280&10\end{vmatrix}$$

Subtracting row 2 from row 1 and adding –3 times row 2 to row 3 yields:

$$D_y = \begin{vmatrix} 1 & 204 & 1 \\ 5 & 540 & 2 \\ 0 & 660 & 4 \end{vmatrix}$$

Adding –5 times row 1 to row 2 yields:

$$D_y = \begin{vmatrix} 1 & 204 & 1 \\ 0 & -480 & -3 \\ 0 & 660 & 4 \end{vmatrix} = 1\begin{vmatrix} -480 & -3 \\ 660 & 4 \end{vmatrix} = 1(60) = 60$$

Subtracting row 3 from row 2 yields:

$$D_z = \begin{vmatrix} 6 & 4 & 744 \\ 20 & 15 & 2820 \\ 15 & 12 & 2280 \end{vmatrix} = \begin{vmatrix} 6 & 4 & 744 \\ 5 & 3 & 540 \\ 15 & 12 & 2280 \end{vmatrix}$$

Subtracting row 2 from row 1 and adding –3 times row 2 to row 3 yields:

$$D_z = \begin{vmatrix} 1 & 1 & 204 \\ 5 & 3 & 540 \\ 0 & 3 & 660 \end{vmatrix}$$

Adding –5 times row 1 to row 2 yields:

$$D_z = \begin{vmatrix} 1 & 1 & 204 \\ 0 & -2 & -480 \\ 0 & 3 & 660 \end{vmatrix} = 1\begin{vmatrix} -2 & -480 \\ 3 & 660 \end{vmatrix} = 1(120) = 120$$

So $x = \frac{D_x}{D} = 24$, $y = \frac{D_y}{D} = 60$, and $z = \frac{D_z}{D} = 120$. So the numbers are 24, 60, and 120.

133. We have:

$$xy = m$$
$$x^2 + y^2 = n$$

Add twice the first equation to the second and get:

$$x^2 + 2xy + y^2 = n + 2m$$
$$(x + y)^2 = n + 2m$$
$$x + y = \pm\sqrt{n + 2m}$$

Subtract twice the first equation from the second to get:

$$x^2 - 2xy + y^2 = n - 2m$$
$$(x - y)^2 = n - 2m$$
$$x - y = \pm\sqrt{n - 2m}$$

We solve the systems:

$$\begin{aligned} x+y&=\sqrt{n+2m} \\ x-y&=\sqrt{n-2m} \\ \hline 2x&=\sqrt{n+2m}+\sqrt{n-2m} \\ x&=\frac{\sqrt{n+2m}+\sqrt{n-2m}}{2} \\ y&=\frac{\sqrt{n+2m}-\sqrt{n-2m}}{2} \end{aligned}$$

$$\begin{aligned} x+y&=\sqrt{n+2m} \\ x-y&=-\sqrt{n-2m} \\ \hline 2x&=\sqrt{n+2m}-\sqrt{n-2m} \\ x&=\frac{\sqrt{n+2m}-\sqrt{n-2m}}{2} \\ y&=\frac{\sqrt{n+2m}+\sqrt{n-2m}}{2} \end{aligned}$$

And also the systems:

$$\begin{aligned} x+y&=-\sqrt{n+2m} \\ x-y&=\sqrt{n-2m} \\ \hline 2x&=\sqrt{n-2m}-\sqrt{n+2m} \\ x&=\frac{\sqrt{n-2m}-\sqrt{n+2m}}{2} \\ y&=\frac{-\sqrt{n-2m}-\sqrt{n+2m}}{2} \end{aligned}$$

$$\begin{aligned} x+y&=-\sqrt{n+2m} \\ x-y&=-\sqrt{n-2m} \\ \hline 2x&=-\sqrt{n+2m}-\sqrt{n-2m} \\ x&=\frac{-\sqrt{n+2m}-\sqrt{n-2m}}{2} \\ y&=\frac{\sqrt{n-2m}-\sqrt{n+2m}}{2} \end{aligned}$$

So the possible pairs of numbers are:

$$\frac{\sqrt{n+2m}+\sqrt{n-2m}}{2} \text{ and } \frac{\sqrt{n+2m}-\sqrt{n-2m}}{2},$$

$$\frac{\sqrt{n-2m}-\sqrt{n+2m}}{2} \text{ and } \frac{-\sqrt{n-2m}-\sqrt{n+2m}}{2}$$

135. The region is neither convex nor bounded. We graph the system of inequalities:

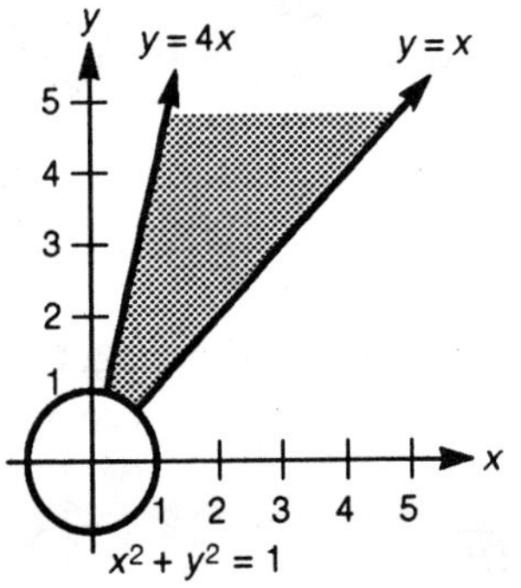

137. The region is convex and bounded. We graph the system of inequalities:

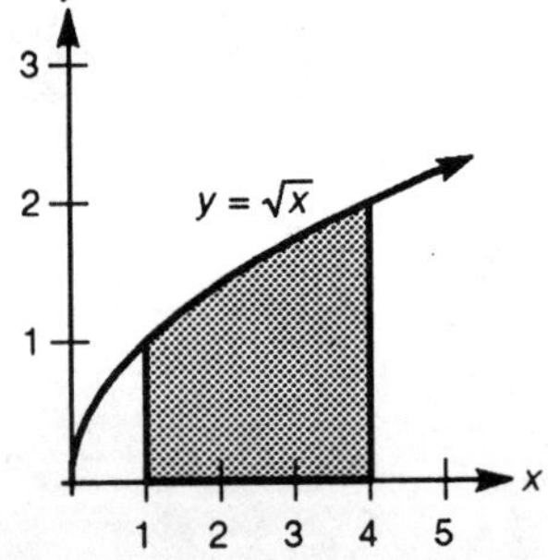

139. The region is neither convex nor bounded. We graph the system of inequalities:

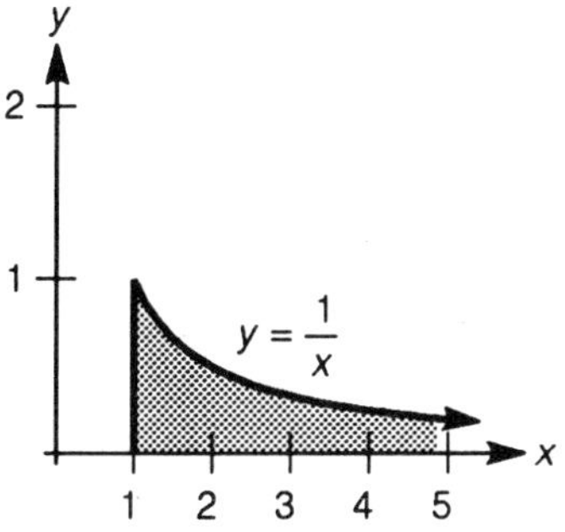

141. We graph the constraints:

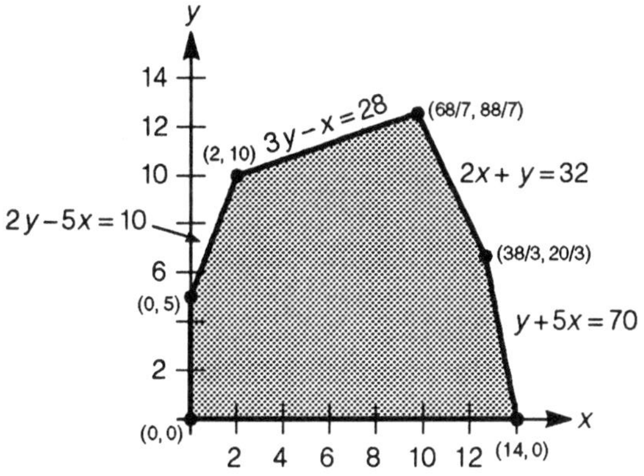

By solving systems of equations, we find the vertices to be $(2,10)$, $\left(\frac{68}{7},\frac{88}{7}\right)$, $\left(\frac{38}{3},\frac{20}{3}\right)$, $(14,0)$, $(0,0)$, and $(0,5)$. We complete the table:

Vertices	$C = 7x + 6y + 6$
(2,10)	80
$(\frac{68}{7},\frac{88}{7})$	$\frac{1046}{7} \approx 149.4$
$(\frac{38}{3},\frac{20}{3})$	$\frac{404}{3} \approx 134.7$
(14,0)	104
(0,0)	6
(0,5)	36

The maximum value is $\frac{1046}{7}$ at $\left(\frac{68}{7},\frac{88}{7}\right)$, and the minimum value is 6 at $(0,0)$.

Chapter Seven Graphing Calculator Exercises

1. (a) Solving for y, we have $y = 2 - x$ and:

$$-3y = -2x + 9$$
$$y = \tfrac{2}{3}x - 3$$

We draw the graphs:

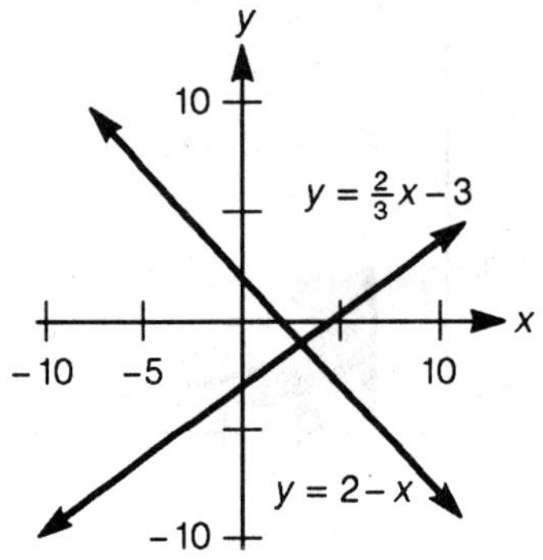

(b) The intersection point (3, –1) is confirmed.

3. (a) Solving for y, we have $y = x - 1$ and $y = 2\sqrt{x+3}$. We draw the graphs:

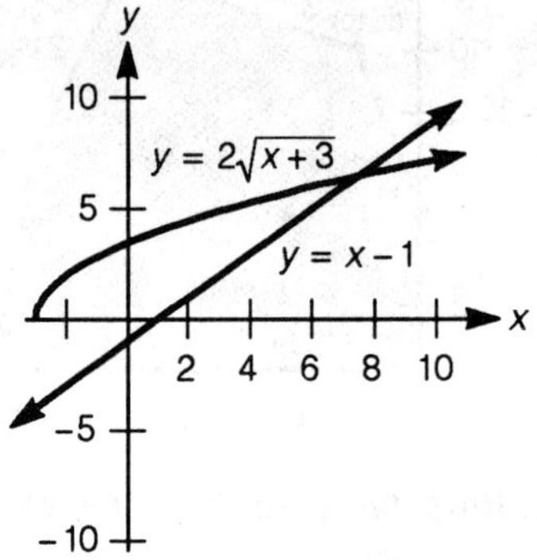

(b) Using the TRACE-ZOOM technique, the intersection point is approximately (7.47, 6.47).

5. (a) We draw the graphs:

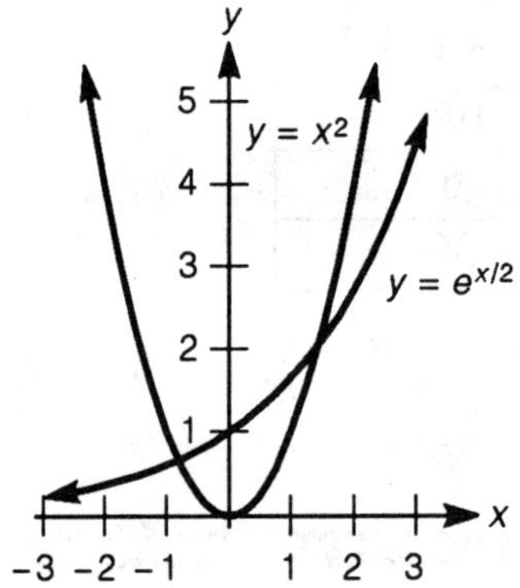

(b) Using the TRACE-ZOOM technique, the intersection points are approximately (–0.82, 0.67) and (1.43, 2.04)

7. (a) Solving for y, we have $y = 2x - x^2$ and $y = 3x^3 - x^2 - 10x$. We draw the graphs:

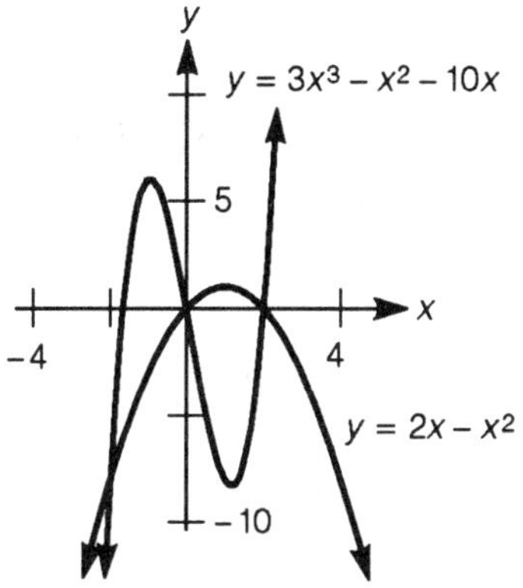

(b) Using the TRACE-ZOOM technique, the intersection points are $(-2,-8)$, $(0,0)$, and $(2,0)$. Note that these are exact coordinates.

9. (a) We draw the graphs:

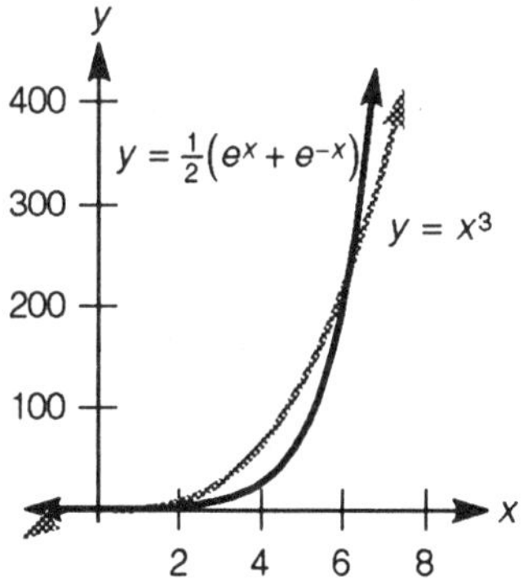

(b) Using the TRACE-ZOOM technique, the intersection points are approximately $(1.23, 1.85)$ and $(6.14, 230.95)$.

11. (a) The product is $AB = \begin{pmatrix} -12 & 17 & -37 \\ -5 & 22 & -8 \\ 7 & -2 & 26 \end{pmatrix}$.

(b) The product is $BA = \begin{pmatrix} 19 & -7 & 6 \\ 5 & 17 & 4 \\ -5 & 16 & 0 \end{pmatrix}$.

(c) The inverse is $A^{-1} = \begin{pmatrix} -3 & 2 & -4 \\ -1 & 1 & -1 \\ 8 & -5 & 10 \end{pmatrix}$, which confirms the matrix given in the text.

(d) The (rounded) inverse is $B^{-1} = \begin{pmatrix} -0.116 & 0.674 & -0.767 \\ 0.035 & 0.198 & -0.07 \\ -0.081 & -0.128 & 0.163 \end{pmatrix}$.

(e) The determinant of A is 1.

13. (a) The determinant of A is 1.
(b) The determinant of B is -452.
(c) The determinant of AB is -452.
(d) This value is $1(-452) = -452$. Note this is the same answer as (c).

15. (a) The determinant of B^{-1} is -0.0022123894.
(b) This value is -0.0022123894. Note this is the same answer as (a).

Chapter Seven Test

1. We substitute $y = x^2 + 2x + 3$ into the first equation:

$$\begin{aligned} 3x + 4y &= 12 \\ 3x + 4(x^2 + 2x + 3) &= 12 \\ 3x + 4x^2 + 8x + 12 &= 12 \\ 4x^2 + 11x &= 0 \\ x(4x + 11) &= 0 \\ x &= 0, -\tfrac{11}{4} \end{aligned}$$

When $x = 0$, $y = 3$ and when $x = -\frac{11}{4}$, $y = \frac{81}{16}$. So the solutions are $(0, 3)$ and $\left(-\frac{11}{4}, \frac{81}{16}\right)$.

2. We multiply the first equation by -3:

$$\begin{aligned} -3x + 6y &= -39 \\ 3x + 5y &= -16 \end{aligned}$$

Adding, we obtain:

$$\begin{aligned} 11y &= -55 \\ y &= -5 \end{aligned}$$

Substitute into the first equation:

$$\begin{aligned} x - 2y &= 13 \\ x - 2(-5) &= 13 \\ x + 10 &= 13 \\ x &= 3 \end{aligned}$$

So the solution is $(3, -5)$.

3. (a) Adding –3 times the first equation to the second equation, and also adding –4 times the first equation to the third equation results in the system:

$$\begin{aligned} x + 4y - z &= 0 \\ -11y + 4z &= -1 \\ -20y + 9z &= -7 \end{aligned}$$

To reduce coefficients, multiply the second equation by –2 and add the third equation to the second equation:

$$\begin{aligned} x + 4y - z &= 0 \\ 2y + z &= -5 \\ -20y + 9z &= -7 \end{aligned}$$

Multiply the second equation by 10 and add it to the third equation:

$$\begin{aligned} x + 4y - z &= 0 \\ 2y + z &= -5 \\ 19z &= -57 \end{aligned}$$

Solving the third equation for z yields $z = -3$. Substitute into the second equation:

$$\begin{aligned} 2y - 3 &= -5 \\ 2y &= -2 \\ y &= -1 \end{aligned}$$

Substitute into the first equation:

$$\begin{aligned} x + 4(-1) - (-3) &= 0 \\ x - 4 + 3 &= 0 \\ x &= 1 \end{aligned}$$

So the solution is $(1,-1,-3)$.

(b) Adding row 1 to row 2, and 5 times row 1 to row 3:

$$D = \begin{vmatrix} 1 & 4 & -1 \\ 3 & 1 & 1 \\ 4 & -4 & 5 \end{vmatrix} = \begin{vmatrix} 1 & 4 & -1 \\ 4 & 5 & 0 \\ 9 & 16 & 0 \end{vmatrix}$$

Now expand along column 3:

$$D = -1\begin{vmatrix} 4 & 5 \\ 9 & 16 \end{vmatrix} = -(64 - 45) = -19$$

Adding column 1 to columns 2 and 3:

$$D_x = \begin{vmatrix} 0 & 4 & -1 \\ -1 & 1 & 1 \\ -7 & -4 & 5 \end{vmatrix} = \begin{vmatrix} 0 & 4 & -1 \\ -1 & 0 & 0 \\ -7 & -11 & -2 \end{vmatrix}$$

Now expand along row 2:

$$D_x = -(-1)\begin{vmatrix} 4 & -1 \\ -11 & -2 \end{vmatrix} = 1(-8 - 11) = -19$$

Adding column 1 to column 3:

$$D_y = \begin{vmatrix} 1 & 0 & -1 \\ 3 & -1 & 1 \\ 4 & -7 & 5 \end{vmatrix} = \begin{vmatrix} 1 & 0 & 0 \\ 3 & -1 & 4 \\ 4 & -7 & 9 \end{vmatrix}$$

Now expand along row 1:

$$D_y = 1\begin{vmatrix} -1 & 4 \\ -7 & 9 \end{vmatrix} = -9 + 28 = 19$$

Adding -4 times column 1 to column 2:

$$D_z = \begin{vmatrix} 1 & 4 & 0 \\ 3 & 1 & -1 \\ 4 & -4 & -7 \end{vmatrix} = \begin{vmatrix} 1 & 0 & 0 \\ 3 & -11 & -1 \\ 4 & -20 & -7 \end{vmatrix}$$

Now expand along row 1:

$$D_z = 1\begin{vmatrix} -11 & -1 \\ -20 & -7 \end{vmatrix} = 77 - 20 = 57$$

So $x = \frac{D_x}{D} = \frac{-19}{-19} = 1$, $y = \frac{D_y}{D} = \frac{19}{-19} = -1$, and $z = \frac{D_z}{D} = \frac{57}{-19} = -3$. So the solution is $(1,-1,-3)$, which verifies our answer from (a).

4. (a) $2A - B = \begin{pmatrix} 2 & -6 \\ 4 & -2 \end{pmatrix} - \begin{pmatrix} 0 & 4 \\ 1 & 3 \end{pmatrix} = \begin{pmatrix} 2 & -10 \\ 3 & -5 \end{pmatrix}$

(b) $BA = \begin{pmatrix} 0 & 4 \\ 1 & 3 \end{pmatrix}\begin{pmatrix} 1 & -3 \\ 2 & -1 \end{pmatrix} = \begin{pmatrix} 8 & -4 \\ 7 & -6 \end{pmatrix}$

5. We first graph the region:

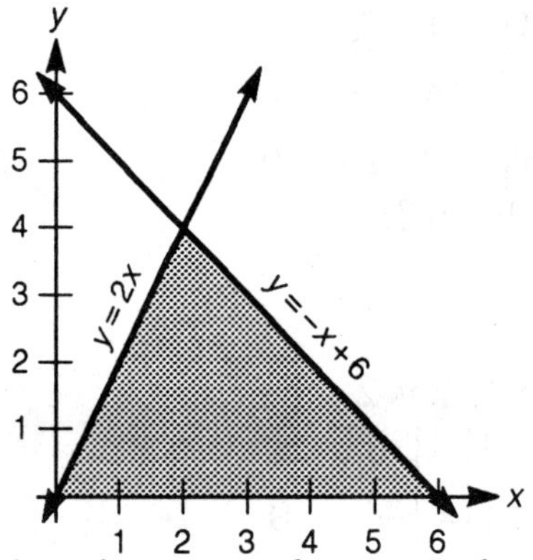

The base of the triangle is the x-intercept of $y = -x + 6$, which is 6. To find the height, we find the intersection points of the lines $y = 2x$ and $y = -x + 6$:

$$2x = -x + 6$$
$$3x = 6$$
$$x = 2$$
$$y = 4$$

So the height is 4. We find the area:

$$\text{Area} = \tfrac{1}{2}(\text{base})(\text{height}) = \tfrac{1}{2}(6)(4) = 12 \text{ sq. units}$$

6. Let $u = \frac{1}{x}$ and $v = \frac{1}{y}$, so we have the system:

$$\frac{1}{2}u + \frac{1}{3}v = 10$$
$$-5u - 4v = -4$$

Multiply the first equation by 10:

$$5u + \frac{10}{3}v = 100$$
$$-5u - 4v = -4$$

Adding, we obtain:

$$-\frac{2}{3}v = 96$$
$$v = -144$$

Substitute into the first equation to find u:

$$\frac{1}{2}u + \frac{1}{3}(-144) = 10$$
$$\frac{1}{2}u - 48 = 10$$
$$\frac{1}{2}u = 58$$
$$u = 116$$

So $\frac{1}{x} = 116$ thus $x = \frac{1}{116}$, and $\frac{1}{y} = -144$ thus $y = -\frac{1}{144}$. So the solution is $\left(\frac{1}{116}, -\frac{1}{144}\right)$.

7. The coefficient matrix is $\begin{pmatrix} 1 & 1 & -1 \\ 2 & -1 & 2 \\ 1 & -2 & 1 \end{pmatrix}$, and the augmented matrix is $\begin{pmatrix} 1 & 1 & -1 & -1 \\ 2 & -1 & 2 & 11 \\ 1 & -2 & 1 & 10 \end{pmatrix}$.

8. Using the augmented matrix formed in the preceeding exercise, add –2 times row 1 to row 2 and –1 times row 1 to row 3:

$$\begin{pmatrix} 1 & 1 & -1 & -1 \\ 0 & -3 & 4 & 13 \\ 0 & -3 & 2 & 11 \end{pmatrix}$$

Multiply row 2 by –1 and add to row 3:

$$\begin{pmatrix} 1 & 1 & -1 & -1 \\ 0 & -3 & 4 & 13 \\ 0 & 0 & -2 & -2 \end{pmatrix}$$

So we have the system of equations:

$$x + y - z = -1$$
$$-3y + 4z = 13$$
$$-2z = -2$$

Solving the third equation for z yields $z = 1$. Substitute into the second equation:

$$-3y + 4(1) = 13$$
$$-3y = 9$$
$$y = -3$$

Substitute into the first equation:

$$x - 3 - 1 = -1$$
$$x - 4 = -1$$
$$x = 3$$

So the solution is (3, –3, 1).

9. We first find the point of intersection:

$$x + y = 11$$
$$3x + 2y = 7$$

Multiply the first equation by –2:

$$-2x - 2y = -22$$
$$3x + 2y = 7$$

Adding, we obtain:

$$x = -15$$

Substitute into the first equation:

$$-15 + y = 11$$
$$y = 26$$

So the point of intersection is (–15, 26). We now find the slope of the line $2x - 4y = 7$:

$$-4y = -2x + 7$$
$$y = \frac{1}{2}x - \frac{7}{4}$$

Since this slope is $\frac{1}{2}$, the perpendicular slope must be –2. We use this slope and the point (–15, 26) in the point-slope formula:

$$y - 26 = -2(x + 15)$$
$$y - 26 = -2x - 30$$
$$y = -2x - 4$$

10. Multiply by $(x + 1)(x - 1)^2$:

$$x - 2 = A(x-1)^2 + B(x+1)(x-1) + C(x+1)$$
$$x - 2 = A(x^2 - 2x + 1) + B(x^2 - 1) + C(x+1)$$
$$x - 2 = (A+B)x^2 + (-2A+C)x + (A - B + C)$$

Equating coefficients, we have the system:

$$A + B = 0$$
$$-2A + C = 1$$
$$A - B + C = -2$$

Adding 2 times the first equation to the second equation, and –1 times the first equation to the third equation:

$$A + B = 0$$
$$2B + C = 1$$
$$-2B + C = -2$$

Adding the second equation to the third equation:

$$A + B = 0$$
$$2B + C = 1$$
$$2C = -1$$

Solving the third equation for C yields $C=-\frac{1}{2}$. Substitute into the second equation:

$$2B-\tfrac{1}{2}=1$$
$$B=\tfrac{3}{4}$$

Substitute into the first equation:

$$A+\tfrac{3}{4}=0$$
$$A=-\tfrac{3}{4}$$

So $A=-\frac{3}{4}$, $B=\frac{3}{4}$, and $C=-\frac{1}{2}$.

11. (a) The minor is $\begin{vmatrix} 2 & -1 \\ 0 & 4 \end{vmatrix}=8$.

(b) The cofactor is $-1\begin{vmatrix} 2 & -1 \\ 0 & 4 \end{vmatrix}=-1(8)=-8$.

12. Adding 2 times row 1 to row 2, and -4 times row 1 to row 3:

$$\begin{vmatrix} 4 & -5 & 0 \\ 0 & 0 & 7 \\ 0 & 40 & 14 \end{vmatrix}$$

Expanding along column 1, we have:

$$4\begin{vmatrix} 0 & 7 \\ 49 & 14 \end{vmatrix}=4(0-280)=-1120$$

13. If we multiply the second equation by 2 we have $2xy=10$. Adding the two equations, we obtain:

$$x^2+2xy+y^2=25$$
$$(x+y)^2=25$$
$$x+y=\pm 5$$

Subtracting the two equations, we obtain:

$$x^2-2xy+y^2=5$$
$$(x-y)^2=5$$
$$x-y=\pm\sqrt{5}$$

So we have the following four systems of equations:

$x+y=5$	$x+y=5$	$x+y=-5$	$x+y=-5$
$x-y=\sqrt{5}$	$x-y=-\sqrt{5}$	$x-y=\sqrt{5}$	$x-y=-\sqrt{5}$

Adding, we have the four equations:

$$2x = 5+\sqrt{5} \qquad 2x = 5-\sqrt{5} \qquad 2x = -5+\sqrt{5} \qquad 2x = -5-\sqrt{5}$$

$$x = \frac{5+\sqrt{5}}{2} \qquad x = \frac{5-\sqrt{5}}{2} \qquad x = \frac{-5+\sqrt{5}}{2} \qquad x = \frac{-5-\sqrt{5}}{2}$$

$$y = \frac{5-\sqrt{5}}{2} \qquad y = \frac{5+\sqrt{5}}{2} \qquad y = \frac{-5-\sqrt{5}}{2} \qquad y = \frac{-5+\sqrt{5}}{2}$$

So the solutions are $\left(\frac{5+\sqrt{5}}{2}, \frac{5-\sqrt{5}}{2}\right)$, $\left(\frac{5-\sqrt{5}}{2}, \frac{5+\sqrt{5}}{2}\right)$, $\left(\frac{-5+\sqrt{5}}{2}, \frac{-5-\sqrt{5}}{2}\right)$, and $\left(\frac{-5-\sqrt{5}}{2}, \frac{-5+\sqrt{5}}{2}\right)$.

14. Multiply the first equation by –2:

$$-2A - 4B - 6C = -2$$
$$2A - B - C = 2$$

Adding, we obtain:

$$-5B - 7C = 0$$
$$-5B = 7C$$
$$B = -\tfrac{7}{5}C$$

Substitute into the (original) first equation:

$$A + 2B + 3C = 1$$
$$A - \tfrac{14}{5}C + 3C = 1$$
$$A + \tfrac{1}{5}C = 1$$
$$A = 1 - \tfrac{1}{5}C$$

So the solution is $\left(1 - \frac{1}{5}C, -\frac{7}{5}C, C\right)$, where C = any real number.

15. (a) We form the matrix:

$$\begin{pmatrix} 10 & -2 & 5 & 1 & 0 & 0 \\ 6 & -1 & 4 & 0 & 1 & 0 \\ 1 & 0 & 1 & 0 & 0 & 1 \end{pmatrix}$$

Switching rows, we have:

$$\begin{pmatrix} 1 & 0 & 1 & 0 & 0 & 1 \\ 6 & -1 & 4 & 0 & 1 & 0 \\ 10 & -2 & 5 & 1 & 0 & 0 \end{pmatrix}$$

Adding –6 times row 1 to row 2, and –10 times row 1 to row 3, we have:

$$\begin{pmatrix} 1 & 0 & 1 & 0 & 0 & 1 \\ 0 & -1 & -2 & 0 & 1 & -6 \\ 0 & -2 & -5 & 1 & 0 & -10 \end{pmatrix}$$

Multiply row 2 by –1, and adding 2 times this new row 2 to row 3, we have:

$$\begin{pmatrix} 1 & 0 & 1 & 0 & 0 & 1 \\ 0 & 1 & 2 & 0 & -1 & 6 \\ 0 & 0 & -1 & 1 & -2 & 2 \end{pmatrix}$$

Adding 2 times row 3 to row 2, row 3 to row 1, then multiplying row 3 by –1, we have:

$$\begin{pmatrix} 1 & 0 & 0 & 1 & -2 & 3 \\ 0 & 1 & 0 & 2 & -5 & 10 \\ 0 & 0 & 1 & -1 & 2 & -2 \end{pmatrix}$$

So the inverse is $\begin{pmatrix} 1 & -2 & 3 \\ 2 & -5 & 10 \\ -1 & 2 & -2 \end{pmatrix}$.

(b) Calling A the coefficient matrix and $B = \begin{pmatrix} -1 \\ -2 \\ 3 \end{pmatrix}$, we know the solution will be given by $A^{-1}B$:

$$\begin{pmatrix} u \\ v \\ w \end{pmatrix} = \begin{pmatrix} 1 & -2 & 3 \\ 2 & -5 & 10 \\ -1 & 2 & -2 \end{pmatrix}\begin{pmatrix} -1 \\ -2 \\ 3 \end{pmatrix} = \begin{pmatrix} 12 \\ 38 \\ -9 \end{pmatrix}$$

So the solution is $(12, 38, -9)$.

16. We graph $5x - 6y \geq 30$:

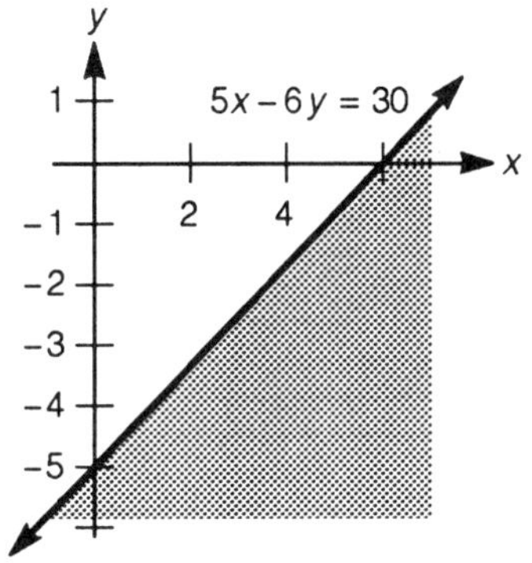

17. Substituting the two points into $y = Px^2 + Qx - 5$, we have:

$$-1 = 4P - 2Q - 5$$
$$-2 = P - Q - 5$$

So we have the system:

$$4P - 2Q = 4$$
$$P - Q = 3$$

Divide the first equation by 2, and multiply the second equation by –1:

$$2P - Q = 2$$
$$-P + Q = -3$$

Adding, we obtain $P = -1$. Substitute into the second equation:

$$-1 - Q = 3$$
$$-Q = 4$$
$$Q = -4$$

So $P = -1$ and $Q = -4$.

18. We complete the square on x:

$$x^2 - 4x + y^2 > -3$$
$$\left(x^2 - 4x + 4\right) + y^2 > -3 + 4$$
$$(x-2)^2 + y^2 > 1$$

We now graph the inequality:

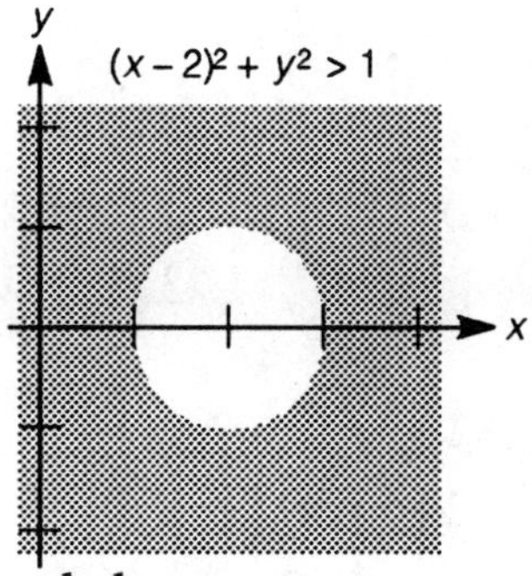

The solution set is neither bounded nor convex.

19. The vertices are $(0,0)$, $(0,7)$, $(6,10)$, $\left(\frac{261}{26}, \frac{225}{26}\right)$, and $(11,0)$. We sketch the graph:

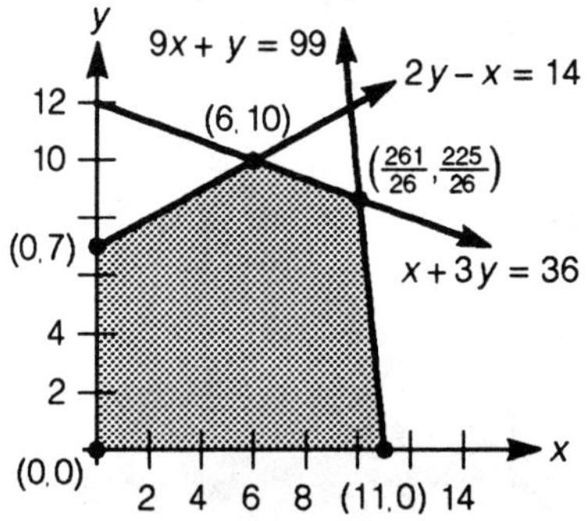

20. Let x = number of small motors produced and y = number of large motors produced. Then we have:

$$\text{Revenue} = 40x + 104y$$
$$\text{Cost} = 15x + 30y + 8(2x + 6y) = 31x + 78y$$
$$\text{Profit} = (40x + 104y) - (31x + 78y) = 9x + 26y$$

The constraints are given by:

(1) $x \geq 0$

(2) $y \geq 0$

(3) $15x + 30y \leq 1500$, or $x + 2y \leq 100$

(4) $16x + 48y \leq 2000$, or $x + 3y \leq 125$

We graph the constraints:

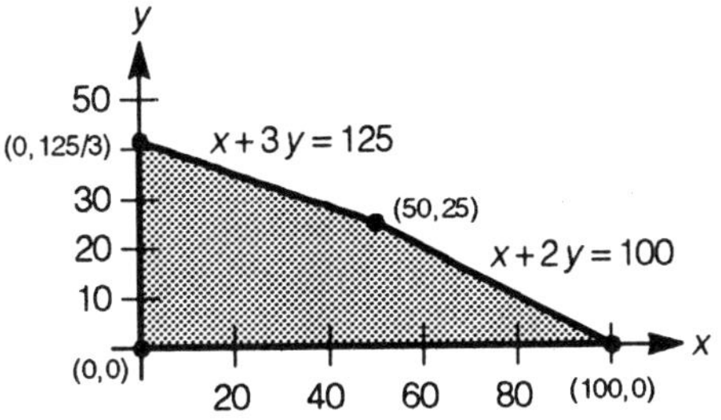

We set up the table:

Vertices	$P = 9x + 26y$
(0,0)	0
(100,0)	900
(50,25)	1100
$(0, \frac{125}{3})$	1083.33

The maximum profit is \$1100 with 50 small motors and 25 large motors being produced.

Chapter Eight
Roots of Polynomial Equations

8.1 More on Division of Polynomials

1. Using synthetic division:

5	1	–6	–2
		5	–5
	1	–1	–7

The quotient is $x - 1$ and the remainder is –7. So $x^2 - 6x - 2 = (x - 5)(x - 1) - 7$.

3. Using synthetic division:

–1	4	–1	–5
		–4	5
	4	–5	0

The quotient is $4x - 5$ and the remainder is 0. So $4x^2 - x - 5 = (x + 1)(4x - 5) + 0$.

5. Using synthetic division:

4	6	–5	2	1
		24	76	312
	6	19	78	313

The quotient is $6x^2 + 19x + 78$ and the remainder is 313.
So $6x^3 - 5x^2 + 2x + 1 = (x - 4)(6x^2 + 19x + 78) + 313$.

7. Using synthetic division:

2	1	0	0	–1
		2	4	8
	1	2	4	7

The quotient is $x^2 + 2x + 4$ and the remainder is 7. So $x^3 - 1 = (x - 2)(x^2 + 2x + 4) + 7$.

9. Using synthetic division:

−2	1	0	0	0	0	−1
		−2	4	−8	16	−32
	1	−2	4	−8	16	−33

The quotient is $x^4 - 2x^3 + 4x^2 - 8x + 16$ and the remainder is -33.
So $x^5 - 1 = (x + 2)(x^4 - 2x^3 + 4x^2 - 8x + 16) - 33$.

11. Using synthetic division:

−4	1	−6	0	0	2
		−4	40	−160	640
	1	−10	40	−160	642

The quotient is $x^3 - 10x^2 + 40x - 160$ and the remainder is 642.
So $x^4 - 6x^3 + 2 = (x + 4)(x^3 - 10x^2 + 40x - 160) + 642$.

13. Using synthetic division:

10	1	−4	−3	6
		10	60	570
	1	6	57	576

The quotient is $x^2 + 6x + 57$ and the remainder is 576.
So $x^3 - 4x^2 - 3x + 6 = (x - 10)(x^2 + 6x + 57) + 576$.

15. Using synthetic division:

−5	1	−1	0	0
		−5	30	−150
	1	−6	30	−150

The quotient is $x^2 - 6x + 30$ and the remainder is -150.
So $x^3 - x^2 = (x + 5)(x^2 - 6x + 30) - 150$.

17. Using synthetic division:

2/3	54	−27	−27	14
		36	6	−14
	54	9	−21	0

The quotient is $54x^2 + 9x - 21$ and the remainder is 0.

So $54x^3 - 27x^2 - 27x + 14 = \left(x - \frac{2}{3}\right)\left(54x^2 + 9x - 21\right) + 0$.

19. Using synthetic division:

3	1	0	3	0	12
		3	9	36	108
	1	3	12	36	120

The quotient is $x^3 + 3x^2 + 12x + 36$ and the remainder is 120.
So $x^4 + 3x^2 + 12 = (x - 3)(x^3 + 3x^2 + 12x + 36) + 120$.

21. Since $\sqrt[5]{32}=2$ is a root, we use synthetic division:

$$\begin{array}{r|rrrrrr} 2 & 1 & 0 & 0 & 0 & 0 & -32 \\ & & 2 & 4 & 8 & 16 & 32 \\ \hline & 1 & 2 & 4 & 8 & 16 & 0 \end{array}$$

So $x^5-32=(x-2)(x^4+2x^3+4x^2+8x+16)$.

23. Since $\sqrt[4]{81}=3$ is a root, we use synthetic division:

$$\begin{array}{r|rrrrr} 3 & 1 & 0 & 0 & 0 & -81 \\ & & 3 & 9 & 27 & 81 \\ \hline & 1 & 3 & 9 & 27 & 0 \end{array}$$

So $z^4-81=(z-3)(z^3+3z^2+9z+27)$.

25. Since the two quotients are the same, we can now perform synthetic division:

$$\begin{array}{r|rrr} 4/3 & 2 & -8/3 & 1/3 \\ & & 8/3 & 0 \\ \hline & 2 & 0 & 1/3 \end{array}$$

The quotient is $2x$ and the remainder is $\frac{1}{3}$. So the original problem has a quotient of $2x$ and a remainder of $3\left(\frac{1}{3}\right)=1$.

27. Adopting the hint from Exercise 25, we first divide the numerator and denominator by 2 to form the quotient:

$$\frac{3x^3+\frac{1}{2}}{x+\frac{1}{2}}$$

Using synthetic division:

$$\begin{array}{r|rrrr} -1/2 & 3 & 0 & 0 & 1/2 \\ & & -3/2 & 3/4 & -3/8 \\ \hline & 3 & -3/2 & 3/4 & 1/8 \end{array}$$

The quotient is $3x^2-\frac{3}{2}x+\frac{3}{4}$ and the remainder is $\frac{1}{8}$. So the original problem has a quotient of $3x^2-\frac{3}{2}x+\frac{3}{4}$ and a remainder of $2\left(\frac{1}{8}\right)=\frac{1}{4}$.

29. Using synthetic division:

$$\begin{array}{r|rrrr} -1 & 1 & 0 & k & 1 \\ & & -1 & 1 & -k-1 \\ \hline & 1 & -1 & k+1 & -k \end{array}$$

So $-k=-4$, and thus $k=4$.

31. Using synthetic division:

$$\begin{array}{r|rrr} p & 1 & 2p & -3q^2 \\ & & p & 3p^2 \\ \hline & 1 & 3p & 3p^2-3q^2 \end{array}$$

Since the remainder is 0, we have:

$$3p^2 - 3q^2 = 0$$
$$3p^2 = 3q^2$$
$$p^2 = q^2$$

33. Using synthetic division:

i	1	-4	1
		i	$-1-4i$
	1	$-4+i$	$-4i$

The quotient is $x + (-4 + i)$ and the remainder is $-4i$.

35. Using synthetic division:

$1+i$	1	-2	2
		$1+i$	-2
	1	$-1+i$	0

The quotient is $x + (-1 + i)$ and the remainder is 0.

37. We know that:

$$f(x) = d(x) \bullet q(x) + R(x)$$

Since $d(\sqrt{3}) = 3 - 3 = 0$, then:

$$f(\sqrt{3}) = R(\sqrt{3}) = -6\sqrt{3} + 57$$

39. Using synthetic division:

a	1	0	0	0	$-5a^4$	$4a^5$
		a	a^2	a^3	a^4	$-4a^5$
	1	a	a^2	a^3	$-4a^4$	0

The remainder is 0.

8.2 The Remainder Theorem and the Factor Theorem

1. We substitute $x = 10$:

$$12(10) - 8 = 112$$
$$120 - 8 = 112$$

Yes, it is a root.

3. We substitute $x = 1 - \sqrt{5}$:

$$(1-\sqrt{5})^2 - 2(1-\sqrt{5}) - 4 = 0$$
$$1 - 2\sqrt{5} + 5 - 2 + 2\sqrt{5} - 4 = 0$$
$$4 - 4 = 0$$

Yes, it is a root.

5. We substitute $x = \frac{1}{2}$:

$$2\left(\tfrac{1}{2}\right)^2 - 3\left(\tfrac{1}{2}\right) + 1 = 0$$
$$\tfrac{1}{2} - \tfrac{3}{2} + 1 = 0$$
$$-1 + 1 = 0$$

Yes, it is a root.

7. We compute $f\left(\frac{2}{3}\right)$:

$$f\left(\tfrac{2}{3}\right) = 3\left(\tfrac{2}{3}\right) - 2 = 2 - 2 = 0$$

So $x = \frac{2}{3}$ is a zero of $f(x)$.

9. We compute $h(-1)$:

$$h(-1) = 5(-1)^3 - (-1)^2 + 2(-1) + 8 = -5 - 1 - 2 + 8 = 0$$

So $x = -1$ is a zero of $h(x)$.

11. We compute $f(2)$:

$$f(2) = 1 + 2(2) + (2)^3 - (2)^5 = 1 + 4 + 8 - 32 = -19$$

So $t = 2$ is not a zero of $f(t)$.

13. (a) We compute $f\left(\frac{\sqrt{3}-1}{2}\right)$:

$$\begin{aligned} f\left(\frac{\sqrt{3}-1}{2}\right) &= 2\left(\frac{\sqrt{3}-1}{2}\right)^3 - 3\left(\frac{\sqrt{3}-1}{2}\right) + 1 \\ &= \frac{3\sqrt{3}-5}{2} - \frac{3\sqrt{3}-3}{2} + 1 \\ &= -1 + 1 \\ &= 0 \end{aligned}$$

So $x = \frac{\sqrt{3}-1}{2}$ is a zero of $f(x)$.

(b) We compute $f\left(\frac{\sqrt{3}+1}{2}\right)$:

$$\begin{aligned} f\left(\frac{\sqrt{3}+1}{2}\right) &= 2\left(\frac{\sqrt{3}+1}{2}\right)^3 - 3\left(\frac{\sqrt{3}+1}{2}\right) + 1 \\ &= \frac{5+3\sqrt{3}}{2} - \frac{3\sqrt{3}-3}{2} + 1 \\ &= 4 + 1 \\ &= 5 \end{aligned}$$

So $x = \frac{\sqrt{3}+1}{2}$ is not a zero of $f(x)$.

15. (a) 1, 2 (multiplicity 3), 3
(b) 1 (multiplicity 3)
(c) 5 (multiplicity 6), –1 (multiplicity 4)
(d) 0 (multiplicity 5), 1

17. Using synthetic division:

-3	4	-6	1	-5
		-12	54	-165
	4	-18	55	-170

So $f(-3) = -170$.

19. Using synthetic division:

$1/2$	6	5	-8	-10	-3
		3	4	-2	-6
	6	8	-4	-12	-9

So $f\left(\frac{1}{2}\right) = -9$.

21. Using synthetic division:

$-\sqrt{2}$	1	3	-4
		$-\sqrt{2}$	$-3\sqrt{2}+2$
	1	$3-\sqrt{2}$	$-3\sqrt{2}-2$

So $f(-\sqrt{2}) = -3\sqrt{2} - 2$.

23. Using synthetic division:

12	$1/2$	-5	-13	-10
		6	12	-12
	$1/2$	1	-1	-22

So $f(12) = -22$.

25. Using synthetic division:

-3	1	-4	-9	36
		-3	21	-36
	1	-7	12	0

So $x^3 - 4x^2 - 9x + 36 = (x+3)(x^2 - 7x + 12) = (x+3)(x-4)(x-3)$.
So the roots are ± 3, 4.

27. Using synthetic division:

1	1	1	-7	5
		1	2	-5
	1	2	-5	0

So $x^3 + x^2 - 7x + 5 = (x-1)(x^2 + 2x - 5)$. So $x = 1$ is a root. We use the quadratic formula:

$$x = \frac{-2 \pm \sqrt{4+20}}{2} = \frac{-2 \pm 2\sqrt{6}}{2} = -1 \pm \sqrt{6}$$

So the roots are $1, -1 \pm \sqrt{6}$.

29. Using synthetic division:

$$\begin{array}{r|rrrr} -2 & 3 & -5 & -16 & 12 \\ & & -6 & 22 & -12 \\ \hline & 3 & -11 & 6 & 0 \end{array}$$

So $3x^3 - 5x^2 - 16x + 12 = (x+2)(3x^2 - 11x + 6) = (x+2)(3x-2)(x-3)$.

So the roots are $-2, \frac{2}{3}$, and 3.

31. Using synthetic division:

$$\begin{array}{r|rrrr} -3/2 & 2 & 1 & -5 & -3 \\ & & -3 & 3 & 3 \\ \hline & 2 & -2 & -2 & 0 \end{array}$$

So $2x^3 + x^2 - 5x - 3 = \left(x + \frac{3}{2}\right)(2x^2 - 2x - 2) = 2\left(x + \frac{3}{2}\right)(x^2 - x - 1)$.

So $x = -\frac{3}{2}$ is a root. We use the quadratic formula:

$$x = \frac{1 \pm \sqrt{1+4}}{2} = \frac{1 \pm \sqrt{5}}{2}$$

So the roots are $-\frac{3}{2}, \frac{1 \pm \sqrt{5}}{2}$.

33. Using synthetic division:

$$\begin{array}{r|rrrrr} 5 & 1 & -15 & 75 & -125 & 0 \\ & & 5 & -50 & 125 & 0 \\ \hline & 1 & -10 & 25 & 0 & 0 \end{array}$$

Therefore:

$$\begin{aligned} x^4 - 15x^3 + 75x^2 - 125x &= (x-5)(x^3 - 10x^2 + 25x) \\ &= x(x-5)(x^2 - 10x + 25) \\ &= x(x-5)^3 \end{aligned}$$

So the roots are 0 and 5.

35. Using synthetic division:

$$\begin{array}{r|rrrrr} -4 & 1 & 2 & -23 & -24 & 144 \\ & & -4 & 8 & 60 & -144 \\ \hline & 1 & -2 & -15 & 36 & 0 \end{array}$$

Use synthetic division again:

$$\begin{array}{r|rrrr} 3 & 1 & -2 & -15 & 36 \\ & & 3 & 3 & -36 \\ \hline & 1 & 1 & -12 & 0 \end{array}$$

So $x^4 + 2x^3 - 23x^2 - 24x + 144 = (x+4)(x-3)(x^2 + x - 12) = (x+4)^2(x-3)^2$.
So the roots are -4 and 3.

37. Using synthetic division:

-9	1	7	-19	-9
		-9	18	9
	1	-2	-1	0

So $x^3 + 7x^2 - 19x - 9 = (x+9)(x^2 - 2x - 1)$. So $x = -9$ is a root. Use the quadratic formula:

$$x = \frac{2 \pm \sqrt{4+4}}{2} = \frac{2 \pm 2\sqrt{2}}{2} = 1 \pm \sqrt{2}$$

So the roots are -9 and $1 \pm \sqrt{2}$.

39. Using long division:

$$\begin{array}{r|l} & 2x^2 - 3x - 1 \\ \hline 2x^2 - 3x - 1 & 4x^4 - 12x^3 + 5x^2 + 6x + 1 \\ & \underline{4x^4 - 6x^3 - 2x^2} \\ & \quad -6x^3 + 7x^2 + 6x \\ & \quad \underline{-6x^3 + 9x^2 + 3x} \\ & \qquad -2x^2 + 3x + 1 \\ & \qquad \underline{-2x^2 + 3x + 1} \\ & \qquad\qquad 0 \end{array}$$

We use the quadratic formula to sove $2x^2 - 3x - 1 = 0$:

$$x = \frac{3 \pm \sqrt{9+8}}{4} = \frac{3 \pm \sqrt{17}}{4}$$

So the solutions are $\frac{3 \pm \sqrt{17}}{4}$ (each of multiplicity 2).

41. (a) $R(\frac{1}{2}) = f(\frac{1}{2}) = 1.125$

(b) $R(1.25) = f(1.25) = -0.046875$

(c) Since $f(1) = 0$ then $t - 1$ is a linear factor of $f(t)$.

(d) Since 1 is a solution, we use synthetic division:

1	1	0	-4	3
		1	1	-3
	1	1	-3	0

Using the quadratic formula to solve $t^2 + t - 3 = 0$:

$$t = \frac{-1 \pm \sqrt{1+12}}{2} = \frac{-1 \pm \sqrt{13}}{2}$$

So the solutions are 1, $\frac{-1 \pm \sqrt{13}}{2}$.

43. We must have $f(x) = (x - 3)(x - 5)(x + 4) = 0$. Multiplying out, we have $x^3 - 4x^2 - 17x + 60 = 0$.

45. We must have $f(x) = (x + 1)^2(x + 6) = 0$. Multiplying out, we have $x^3 + 8x^2 + 13x + 6 = 0$.

47. Since $\frac{1}{2}$ is not a root of $x^2 - 3x - 4$, we know that such an equation must have

$f(x) = (x^2 - 3x - 4)(x - \frac{1}{2})^3$, which has degree 5. Thus no such polynomial of degree 4 exists.

49. We must have $x^2 - 3x + 1$ and $x + 6$ as factors, and thus we can write $f(x) = (x^2 - 3x + 1)(x + 6)(ax + b) = 0$, for some real numbers a and b, where $a \neq 0$. Multiplying out, we have $ax^4 + (3a + b)x^3 + (-17a + 3b)x^2 + (6a - 17b)x + 6b = 0$.

51. Using synthetic division:

$$\begin{array}{r|rrrr} 1.16 & 1 & -3 & 12 & 9 \\ & & 1.16 & -2.1344 & 11.44 \\ \hline & 1 & -1.84 & 9.8656 & 20.44 \end{array}$$

So $f(1.16)$ is approximately equal to 20.44.

53. (a) Using synthetic division:

$$\begin{array}{r|rrrr} 2.41 & 1 & 0 & -5 & -2 \\ & & 2.41 & 5.8081 & 1.9475 \\ \hline & 1 & 2.41 & 0.8081 & -0.05 \end{array}$$

So $f(2.41)$ is approximately equal to -0.05.

(b) Using synthetic division:

$$\begin{array}{r|rrrr} 2.42 & 1 & 0 & -5 & -2 \\ & & 2.42 & 5.8564 & 2.0725 \\ \hline & 1 & 2.42 & 0.8564 & 0.07 \end{array}$$

So $f(2.42)$ is approximately equal to 0.07.

55. (a) Since -5 is a zero, we use synthetic division:

$$\begin{array}{r|rrrr} -5 & 1 & 1 & -18 & 10 \\ & & -5 & 20 & -10 \\ \hline & 1 & -4 & 2 & 0 \end{array}$$

So the remaining factor is $x^2 - 4x + 2 = 0$. Using the quadratic formula:

$$x = \frac{4 \pm \sqrt{16 - 4(2)}}{2(1)} = \frac{4 \pm \sqrt{8}}{2} = \frac{4 \pm 2\sqrt{2}}{2} = 2 \pm \sqrt{2}$$

So $x_2 = 2 - \sqrt{2}$ and $x_3 = 2 + \sqrt{2}$.

(b) The values are the same.

57. We evaluate the function:

$$\begin{aligned} f\left(-b-\sqrt{b^2-2c}\right) &= \tfrac{1}{2}\left(-b-\sqrt{b^2-2c}\right)^2 + b\left(-b-\sqrt{b^2-2c}\right)+c \\ &= \tfrac{1}{2}\left(b^2+2b\sqrt{b^2-2c}+b^2-2c\right)-b^2-b\sqrt{b^2-2c}+c \\ &= b^2+b\sqrt{b^2-2c}-c-b^2-b\sqrt{b^2-2c}+c \\ &= 0 \end{aligned}$$

So $x=-b-\sqrt{b^2-2c}$ is a zero of $f(x)$.

59. We evaluate the function:

$$\begin{aligned} F\left(\frac{-\sqrt{2}+\sqrt{2\sqrt{2}-2}}{2}\right) &= 2\left(\frac{-\sqrt{2}+\sqrt{2\sqrt{2}-2}}{2}\right)^4 + 4\left(\frac{-\sqrt{2}+\sqrt{2\sqrt{2}-2}}{2}\right)+1 \\ &= -1+2\sqrt{2}-2\sqrt{2\sqrt{2}-2}-2\sqrt{2}+\sqrt{2\sqrt{2}-2}+1 \\ &= 0 \end{aligned}$$

Yes, it is a zero.

61. Using synthetic division:

$$\begin{array}{r|rrrr} 1 & 1 & 1 & a & b \\ & & 1 & 2 & a+2 \\ \hline & 1 & 2 & a+2 & a+b+2 \end{array}$$

Using synthetic division again:

$$\begin{array}{r|rrrr} 1 & 1 & -1 & -a & b \\ & & 1 & 0 & -a \\ \hline & 1 & 0 & -a & -a+b \end{array}$$

If 1 is a root, then:

$$\begin{aligned} a+b+2&=0 \\ -a+b&=0 \end{aligned}$$

Adding these, we get $2b+2=0$, so $b=-1$. Substituting we get $a=-1$.
So $a=-1$ and $b=-1$.

63. Let r_1 and r_2 be the roots, so $r_2=2r_1$. Then:

$$x^2+bx+1=(x-r_1)(x-2r_1)=x^2-3r_1x+2r_1^2$$

Since r_1 is a constant, then $-3r_1=b$ and $2r_1^2=1$, so $r_1^2=\frac{1}{2}$ and thus $r_1=\frac{\pm\sqrt{2}}{2}$.

So $b=\frac{\pm 3\sqrt{2}}{2}$.

65. Let r denote the root with multiplicity 2. Using synthetic division:

$$\begin{array}{r|rrrr} r & 1 & 0 & -12 & 16 \\ & & r & r^2 & r^3-12r \\ \hline & 1 & r & r^2-12 & r^3-12r+16 \end{array}$$

Since r is a double root, then the equation $x^2 + rx + (r^2 - 12) = 0$ must also have r as a root. Substituting $x = r$, we have:

$$r^2 + r^2 + r^2 - 12 = 0$$
$$3r^2 = 12$$
$$r^2 = 4$$
$$r = \pm 2$$

But $r = -2$ does not check in the original equation, thus $r = 2$. Thus the quadratic equation becomes $x^2 + 2x - 8 = 0$, which factors to $(x + 4)(x - 2) = 0$, and thus $x = -4$ is also a solution. So the solutions are 2 (multiplicity 2) and -4.

8.3 The Fundamental Theorem of Algebra

1. (a) yes
   (b) yes
   (c) yes
   (d) no--not a polynomial equation

3. $x^2 - 2x - 3 = (x+1)(x-3) = [x-(-1)](x-3)$

5. $4x^2 + 23x - 6 = (4x-1)(x+6) = 4\left(x - \frac{1}{4}\right)[x-(-6)]$

7. $x^2 - 5 = \left(x+\sqrt{5}\right)\left(x-\sqrt{5}\right) = \left[x-\left(-\sqrt{5}\right)\right]\left(x-\sqrt{5}\right)$

9. Using the quadratic formula for $x^2 - 10x + 26 = 0$, we have:

$$x = \frac{10 \pm \sqrt{100-104}}{2} = \frac{10 \pm 2i}{2} = 5 \pm i$$

So $x^2 - 10x + 26 = [x-(5+i)][x-(5-i)]$

11. $f(x) = (x-1)^2(x+3) = x^3 + x^2 - 5x + 3$

13. $f(x) = (x-2)(x+2)(x-2i)(x+2i) = x^4 - 16$

15. $f(x) = \left(x-\sqrt{3}\right)^2\left[x-\left(-\sqrt{3}\right)\right]^2(x-4i)[x-(-4i)] = x^6 + 10x^4 - 87x^2 + 144$

17. We can write $f(x)$ as $f(x) = a(x+4)(x-9)$, for some constant a. Since $f(3) = 5$, then:

$$5 = a(7)(-6)$$
$$a = -\frac{5}{42}$$

So $f(x) = -\frac{5}{42}\left(x^2 - 5x - 36\right) = -\frac{5}{42}x^2 + \frac{25}{42}x + \frac{30}{7}$.

19. We can write $f(x)$ as $f(x) = a(x + 5)(x - 2)(x - 3)$, for some constant a. Since $f(0) = 1$, then:

$$1 = a(5)(-2)(-3)$$
$$a = \tfrac{1}{30}$$

So $f(x) = \frac{1}{30}\left(x^3 - 19x + 30\right) = \frac{1}{30}x^3 - \frac{19}{30}x + 1$.

21. We have $-b = -i - \sqrt{3}$, so $b = i + \sqrt{3}$. Also $c = -i(-\sqrt{3}) = i\sqrt{3}$.

So $x^2 + (i + \sqrt{3})x + i\sqrt{3} = 0$.

23. We have $-b = 3$, so $b = -3$. Also $c = (9)(-6) = -54$. So $x^2 - 3x - 54 = 0$.

25. We have $-b = 2$, so $b = -2$. Also $c = (1+\sqrt{5})(1-\sqrt{5}) = -4$. So $x^2 - 2x - 4 = 0$.

27. We have $-B = 2a$, so $B = -2a$. Also $C = (a+\sqrt{b})(a-\sqrt{b}) = a^2 - b$.
So $x^2 - 2ax + a^2 - b = 0$.

29. Using the hint, we have:

$$\begin{aligned} x^4 + 64 &= x^4 + 16x^2 + 64 - 16x^2 \\ &= \left(x^2 + 8\right)^2 - (4x)^2 \\ &= \left(x^2 + 8 + 4x\right)\left(x^2 + 8 - 4x\right) \\ &= \left(x^2 + 4x + 8\right)\left(x^2 - 4x + 8\right) \end{aligned}$$

Now $x^2 + 4x + 8 = 0$ when $x = \dfrac{-4 \pm \sqrt{16 - 32}}{2} = \dfrac{-4 \pm 4i}{2} = -2 \pm 2i$.

Also $x^2 - 4x + 8 = 0$ when $x = \dfrac{4 \pm \sqrt{16 - 32}}{2} = \dfrac{4 \pm 4i}{2} = 2 \pm 2i$.

Therefore:

$$\begin{aligned} x^4 + 64 &= \left(x^2 + 4x + 8\right)\left(x^2 - 4x + 8\right) \\ &= [x - (-2 + 2i)][x - (-2 - 2i)][x - (2 + 2i)][x - (2 - 2i)] \\ &= (x + 2 - 2i)(x + 2 + 2i)(x - 2 - 2i)(x - 2 + 2i) \end{aligned}$$

31. We know:

$$\begin{aligned} x^3 + bx^2 + cx + d &= (x - r_1)(x - r_2)(x - r_3) \\ &= x^3 - (r_1 + r_2 + r_3)x^2 + (r_1r_2 + r_1r_3 + r_2r_3)x - r_1r_2r_3 \end{aligned}$$

Therefore:

$$\begin{aligned} r_1 + r_2 + r_3 &= -b \\ r_1r_2 + r_1r_3 + r_2r_3 &= c \\ r_1r_2r_3 &= -d \end{aligned}$$

33. Let r_1, r_2 and r_3 be the roots and assume $r_2 = -r_1$ (their sum is therefore 0). The identities from Exercise 31 give us:

$$r_1 + r_2 + r_3 = 4 \qquad r_1 r_2 r_3 = -36$$
$$r_1 - r_1 + r_3 = 4 \qquad -r_1^2(4) = -36$$
$$r_3 = 4 \qquad r_1^2 = 9$$
$$r_1 = \pm 3$$

So $r_1 = 3$ and $r_2 = -3$ (or switch them, it doesn't matter). So the roots are 4, 3, and –3.

35. (a) Using the hint:

$$\begin{aligned}\alpha^2 + \beta^2 &= (\alpha+\beta)^2 - 2\alpha\beta \\ &= (-b)^2 - 2c \\ &= b^2 - 2c\end{aligned}$$

(b) Using our result from (a):

$$\left(\frac{1}{\alpha}\right)^2 + \left(\frac{1}{\beta}\right)^2 = \frac{\alpha^2+\beta^2}{(\alpha\beta)^2} = \frac{b^2-2c}{c^2}$$

(c) Using our result from (a):

$$\begin{aligned}\alpha^3 + \beta^3 &= (\alpha+\beta)(\alpha^2 - \alpha\beta + \beta^2) \\ &= (-b)(\alpha^2 + \beta^2 - \alpha\beta) \\ &= (-b)(b^2 - 2c - c) \\ &= -b(b^2 - 3c)\end{aligned}$$

37. (a) Suppose $A \neq 0$ or $B \neq 0$. Then $f(x) = Ax^2 + Bx + C = 0$ can have at most two distinct roots by the Linear Factor Theorem. So $A = B = 0$. But this implies $C = 0$, and thus $f(x) = 0$ for all values of x.

(b) Using the hint, let:

$$f(x) = \frac{a^2 - x^2}{(a-b)(a-c)} + \frac{b^2 - x^2}{(b-c)(b-a)} + \frac{c^2 - x^2}{(c-a)(c-b)} - 1$$

$$\begin{aligned}f(a) &= 0 + \frac{b^2 - a^2}{(b-c)(b-a)} + \frac{c^2 - a^2}{(c-a)(c-b)} - 1 \\ &= \frac{b+a}{b-c} + \frac{c+a}{c-b} - 1 \\ &= \frac{-b-a+c+a}{c-b} - 1 \\ &= \frac{c-b}{c-b} - 1 \\ &= 0\end{aligned}$$

$$
\begin{aligned}
f(b) &= \frac{a^2-b^2}{(a-b)(a-c)} + 0 + \frac{c^2-b^2}{(c-a)(c-b)} - 1 \\
&= \frac{a+b}{a-c} + \frac{c+b}{c-a} - 1 \\
&= \frac{-a-b+c+b}{c-a} - 1 \\
&= \frac{c-a}{c-a} - 1 \\
&= 0 \\
f(c) &= \frac{a^2-c^2}{(a-b)(a-c)} + \frac{b^2-c^2}{(b-c)(b-a)} + 0 - 1 \\
&= \frac{a+c}{a-b} + \frac{b+c}{b-a} - 1 \\
&= \frac{-a-c+b+c}{b-a} - 1 \\
&= \frac{b-a}{b-a} - 1 \\
&= 0
\end{aligned}
$$

Since a, b and c are distinct (for denominators to be nonzero), and all three are roots, then by part (a) $f(x) = 0$ for all values of x. Thus the equation is an identity.

8.4 Rational and Irrational Roots

1. p-factors: $\pm 1, \pm 3$
q-factors: $\pm 1, \pm 2, \pm 4$

possible rational roots $\frac{p}{q}$: $\pm 1, \pm \frac{1}{2}, \pm \frac{1}{4}, \pm 3, \pm \frac{3}{2}, \pm \frac{3}{4}$

3. p-factors: $\pm 1, \pm 3, \pm 9$

q-factors: $\pm 1, \pm 2, \pm 4, \pm 8$

possible rational roots $\frac{p}{q}$: $\pm 1, \pm \frac{1}{2}, \pm \frac{1}{4}, \pm \frac{1}{8}, \pm 3, \pm \frac{3}{2}, \pm \frac{3}{4}, \pm \frac{3}{8}, \pm 9, \pm \frac{9}{2}, \pm \frac{9}{4}, \pm \frac{9}{8}$

5. First multiply by 3 (we need integer coefficients) to get:

$$2x^3 - 3x^2 - 15x + 6 = 0$$

p-factors: $\pm 1, \pm 2, \pm 3, \pm 6$
q-factors: $\pm 1, \pm 2$

possible rational roots $\frac{p}{q}$: $\pm 1, \pm \frac{1}{2}, \pm 2, \pm 3, \pm \frac{3}{2}, \pm 6$

7. The possible rational roots are ± 1. Using synthetic division:

1	1	0	−3	1
		1	1	−2
	1	1	−2	−1

So $x = 1$ is not a root.

−1	1	0	−3	1
		−1	1	2
	1	−1	−2	3

So $x = -1$ is not a root. There are no rational roots.

9. The possible rational roots are ± 1. Using synthetic division:

1	1	1	−1	1
		1	2	1
	1	2	1	2

So $x = 1$ is not a root.

−1	1	1	−1	1
		−1	0	1
	1	0	−1	2

So $x = -1$ is not a root. There are no rational roots.

11. The possible rational roots are $\pm 1, \pm \frac{1}{2}, \pm \frac{1}{3}, \pm \frac{1}{4}, \pm \frac{1}{6}, \pm \frac{1}{12}, \pm 2, \pm \frac{2}{3}, \pm 3, \pm \frac{3}{2}, \pm \frac{3}{4}, \pm 6$. Using synthetic division:

1	12	0	−1	0	−6
		12	12	11	11
	12	12	11	11	5

So $x = 1$ is not a root. Since this row is all positive, we can exclude $x = 2$, $x = 3$, $x = \frac{3}{2}$ and $x = 6$.

−1	12	0	−1	0	−6
		−12	12	−11	11
	12	−12	11	−11	5

So $x = -1$ is not a root. Since this row alternates signs, we can exclude $x = -2$, $x = -3$, $x = -\frac{3}{2}$ and $x = -6$.

1/4	12	0	−1	0	−6
		3	3/4	−1/16	−1/64
	12	3	−1/4	−1/16	−385/64

So $x = \frac{1}{4}$ is not a root.

1/3	12	0	−1	0	−6
		4	4/3	1/9	1/27
	12	4	1/3	1/9	−161/27

Proceeding in a similar fashion, we find that none of the candidates are rational roots.

13. The possible rational roots are $\pm 1, \pm 3$. Using synthetic division:

$$\begin{array}{r|rrrr} 1 & 1 & 3 & -1 & -3 \\ & & 1 & 4 & 3 \\ \hline & 1 & 4 & 3 & 0 \end{array}$$

So $x^3 + 3x^2 - x - 3 = (x-1)(x^2+4x+3) = (x-1)(x+1)(x+3)$.
So the roots are 1, –1, and –3.

15. The possible rational roots are $\pm 1, \pm \frac{1}{2}, \pm \frac{1}{4}, \pm 5, \pm \frac{5}{2}, \pm \frac{5}{4}$. Using synthetic division:

$$\begin{array}{r|rrrr} -1/4 & 4 & 1 & -20 & -5 \\ & & -1 & 0 & 5 \\ \hline & 4 & 0 & -20 & 0 \end{array}$$

Therefore:

$$\begin{aligned} 4x^3 + x^2 - 20x - 5 &= \left(x + \tfrac{1}{4}\right)\left(4x^2 - 20\right) \\ &= 4\left(x + \tfrac{1}{4}\right)\left(x^2 - 5\right) \\ &= 4\left(x + \tfrac{1}{4}\right)\left(x + \sqrt{5}\right)\left(x - \sqrt{5}\right) \end{aligned}$$

So the roots are $-\frac{1}{4}, -\sqrt{5}$ and $\sqrt{5}$.

17. The possible rational roots are $\pm 1, \pm \frac{1}{3}, \pm \frac{1}{9}, \pm 2, \pm \frac{2}{3}, \pm \frac{2}{9}$. Using synthetic division:

$$\begin{array}{r|rrrr} 1 & 9 & 18 & 11 & 2 \\ & & 9 & 27 & 38 \\ \hline & 9 & 27 & 38 & 40 \end{array}$$

So $x = 2$ is excluded also (the row is positive).

$$\begin{array}{r|rrrr} -1 & 9 & 18 & 11 & 2 \\ & & -9 & -9 & -2 \\ \hline & 9 & 9 & 2 & 0 \end{array}$$

So $9x^3 + 18x^2 + 11x + 2 = (x+1)\left(9x^2 + 9x + 2\right) = (x+1)(3x+2)(3x+1)$.

So the roots are $-1, -\frac{2}{3}$ and $-\frac{1}{3}$.

19. The possible rational roots are $\pm 1, \pm 2, \pm 3, \pm 4, \pm 6, \pm 8, \pm 12, \pm 24$. Using synthetic division:

$$\begin{array}{r|rrrrr} 1 & 1 & 1 & -25 & -1 & 24 \\ & & 1 & 2 & -23 & -24 \\ \hline & 1 & 2 & -23 & -24 & 0 \end{array}$$

So $x = 1$ is a root. Using synthetic division again:

$$\begin{array}{r|rrrr} -1 & 1 & 2 & -23 & -24 \\ & & -1 & -1 & 24 \\ \hline & 1 & 1 & -24 & 0 \end{array}$$

So $x=-1$ is also a root. So $x^4+x^3-25x^2-x+24=(x-1)(x+1)\left(x^2+x-24\right)$. Use the quadratic formula:

$$x=\frac{-1\pm\sqrt{1+96}}{2}=\frac{-1\pm\sqrt{97}}{2}$$

So the roots are $1, -1, \frac{-1+\sqrt{97}}{2}$ and $\frac{-1-\sqrt{97}}{2}$.

21. The possible rational roots are ± 1. Using synthetic division:

1	1	−4	6	−4	1
		1	−3	3	−1
	1	−3	3	−1	0

Using synthetic division again:

1	1	−3	3	−1
		1	−2	1
	1	−2	1	0

So $x^4-4x^3+6x^2-4x+1=(x-1)^2\left(x^2-2x+1\right)=(x-1)^4$.
So the only root is 1 (with multiplicity 4).

23. First multiply by 2 to get integer coefficients:

$$2x^3-5x^2-46x+24=0$$

The possible rational roots are $\pm 1, \pm\frac{1}{2}, \pm 2, \pm 3, \pm\frac{3}{2}, \pm 4, \pm 6, \pm 8, \pm 12, \pm 24$. Using synthetic division:

1/2	2	−5	−46	24
		1	−2	−24
	2	−4	−48	0

Therefore:

$$\begin{aligned}2x^3-5x^2-46x+24&=\left(x-\tfrac{1}{2}\right)\left(2x^2-4x-48\right)\\&=2\left(x-\tfrac{1}{2}\right)\left(x^2-2x-24\right)\\&=2\left(x-\tfrac{1}{2}\right)(x-6)(x+4)\end{aligned}$$

So the roots are $\frac{1}{2}$, 6 and -4.

25. First multiply by 20 to get integer coefficients:

$$40x^4-18x^3-58x^2+27x-3=0$$

The possible rational roots are $\pm 1, \pm\frac{1}{2}, \pm\frac{1}{4}, \pm\frac{1}{5}, \pm\frac{1}{8}, \pm\frac{1}{10}, \pm\frac{1}{20}, \pm\frac{1}{40}, \pm 3, \pm\frac{3}{2}, \pm\frac{3}{4}, \pm\frac{3}{5}$, $\pm\frac{3}{8}, \pm\frac{3}{10}, \pm\frac{3}{20}, \pm\frac{3}{40}$. Using synthetic division:

1/4	40	−18	−58	27	−3
		10	−2	−15	3
	40	−8	−60	12	0

Using synthetic division again:

1/5	40	−8	−60	12
		8	0	−12
	40	0	−60	0

Therefore:

$$\begin{aligned} 40x^4 - 18x^3 - 58x^2 + 27x - 3 &= \left(x - \tfrac{1}{4}\right)\left(x - \tfrac{1}{5}\right)\left(40x^2 - 60\right) \\ &= 10\left(x - \tfrac{1}{4}\right)\left(x - \tfrac{1}{5}\right)\left(4x^2 - 6\right) \\ &= 10\left(x - \tfrac{1}{4}\right)\left(x - \tfrac{1}{5}\right)\left(2x + \sqrt{6}\right)\left(2x - \sqrt{6}\right) \end{aligned}$$

So the roots are $\frac{1}{4}, \frac{1}{5}, -\frac{\sqrt{6}}{2}$ and $\frac{\sqrt{6}}{2}$.

27. (a) The possible rational roots are $\pm 1, \pm \frac{1}{5}, \pm 2, \pm \frac{2}{5}, \pm 3, \pm \frac{3}{5}, \pm 4, \pm \frac{4}{5}, \pm 6, \pm \frac{6}{5}, \pm 12, \pm \frac{12}{5}$. Using synthetic division:

2	5	0	0	−10	−12
		10	20	40	60
	5	10	20	30	48

Since this row is all positive, $x = 2$ is an upper bound.

−1	5	0	0	−10	−12
		−5	5	−5	15
	5	−5	5	−15	3

Since this row alternates signs, $x = -1$ is a lower bound.

(b) The possible rational roots are $\pm 1, \pm \frac{1}{3}, \pm 2, \pm \frac{2}{3}, \pm 4, \pm \frac{4}{3}$. Using synthetic division:

4/3	3	−4	5	−2	−4
		4	0	20/3	56/9
	3	0	5	14/3	20/9

So $x = \frac{4}{3}$ is the upper bound.

−2/3	3	−4	5	−2	−4
		−2	4	−6	16/3
	3	−6	9	−8	4/3

So $x = -\frac{2}{3}$ is the lower bound. Since the question asked for the integral upper and lower bounds, the upper bound is 2 and the lower bound is –1.

(c) The possible rational roots are $\pm 1, \pm \frac{1}{2}, \pm 2, \pm 3, \pm \frac{3}{2}, \pm 4, \pm 6, \pm 12$. Using synthetic division:

6	2	−7	−5	28	−12
		12	30	150	1068
	2	5	25	178	1056

So $x = 6$ is an upper bound.

-2	2	-7	-5	28	-12
		-4	22	-34	12
	2	-11	17	-6	0

Actually -2 is a root, but also a lower bound.

29. Let $f(x) = x^3 + x - 1$.

$f(0) = -1$

$f(1) = 1$

So the root lies between 0 and 1.

$f(0.5) = -0.375$

$f(0.7) = 0.043$

$f(0.6) = -0.373$

So the root lies between 0.6 and 0.7.

$f(0.65) = -0.075$

$f(0.66) = -0.052$

$f(0.67) = -0.029$

$f(0.68) = -0.005$

$f(0.69) = 0.018$

So the root lies between 0.68 and 0.69.

31. Let $f(x) = x^5 - 200$.

$f(2) = -168$

$f(3) = 43$

So the root lies between 2 and 3.

$f(2.5) = -102$

$f(2.8) = -27.9$

$f(2.9) = 5.11$

So the root lies between 2.8 and 2.9.

$f(2.87) = -5.28$

$f(2.88) = -1.86$

$f(2.89) = 1.60$

So the root lies between 2.88 and 2.89.

33. Let $f(x) = x^3 - 8x^2 + 21x - 22$.

$f(4) = -2$

$f(8) = 8$

So the root lies between 4 and 5.

$f(4.5) = 1.63$

$f(4.4) = 0.704$

$f(4.3) = -0.113$

So the root lies between 4.3 and 4.4.

$f(4.33) = 0.12$

$f(4.32) = 0.04$

$f(4.31) = -0.03$

So the root lies between 4.31 and 4.32.

35. Let $f(x) = x^3 + x^2 - 2x + 1$.

$f(-2) = 1$
$f(-3) = -11$

So the root lies between –3 and –2.

$f(-2.1) = 0.35$
$f(-2.2) = -0.41$

So the root lies between –2.2 and –2.1.

$f(-2.15) = -0.02$
$f(-2.14) = 0.05$

So the root lies between –2.15 and –2.14.

37. Let $f(x) = x^3 + 2x^2 + 2x + 101$.

$f(-5) = 16$
$f(-6) = -55$

So the root lies between –6 and –5.

$f(-5.4) = -8.94$
$f(-5.3) = -2.30$
$f(-5.2) = 4.07$

So the root lies between –5.3 and –5.2.

$f(-5.26) = 0.28$
$f(-5.27) = -0.36$

So the root lies between –5.27 and –5.26.

39. (a) Since 2 is a factor of $8 \bullet 5 = 40$, and 2 is not a factor of 5, then 2 must be a factor of 8, which is true.

(b) The result guarantees only that A is a factor of B in the case where A and C have no factor in common. Here A and C have a common factor of 5, so the result does not apply.

(c) Since $x = \frac{p}{q}$ is a root of the equation, this statement must be true.

(d) Subtract a_0 and multiply by q^n to get:

$$a_n p^n + a_{n-1} q p^{n-1} + a_{n-2} q^2 p^{n-2} + \ldots + a_1 q^{n-1} p = -a_0 q^n$$

Therefore:

$$p\left(a_n p^{n-1} + a_{n-1} q p^{n-2} + a_{n-2} q^2 p^{n-3} + \ldots + a_1 q^{n-1}\right) = -a_0 q^n$$

41. We sketch the graph:

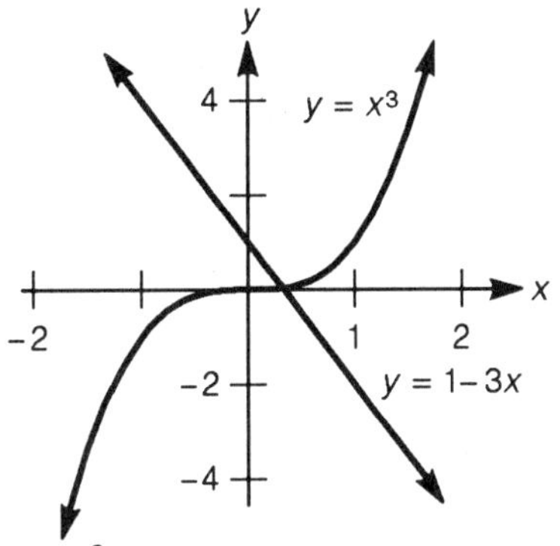

We find where $x^3 = 1 - 3x$, so $x^3 + 3x - 1 = 0$.
Let $f(x) = x^3 + 3x - 1$.

$f(0) = -1$
$f(1) = 3$

So the root lies between 0 and 1.

$f(0.3) = -0.07$
$f(0.4) = 0.264$

So the root lies between 0.3 and 0.4.

$f(0.32) = -0.007$
$f(0.33) = 0.02$

So the root lies between 0.32 and 0.33.

$f(0.325) = 0.009$

So the root lies between 0.32 and 0.325, thus the x-coordinate is $x = 0.32$, accurate to two decimal places.

43. We find where:

$$\sqrt{x} = x^2 - 1$$
$$x = x^4 - 2x^2 + 1$$
$$x^4 - 2x^2 - x + 1 = 0$$

Let $f(x) = x^4 - 2x^2 - x + 1$.

$f(1) = -1$
$f(2) = 7$

So the root lies between 1 and 2.

$f(1.4) = -0.48$
$f(1.5) = 0.06$

So the root lies between 1.4 and 1.5

$f(1.48) = -0.06$
$f(1.49) = -0.001$

So the root lies between 1.49 and 1.50.

$f(1.495) = 0.03$

So the root lies between 1.49 and 1.495, thus the x-coordinate is $x = 1.49$, accurate to two decimal places.

45. (a) Here $a_n = 9$, $a_0 = 27$ and $f(1) = 29$. Since these three numbers are odd, then the equation has no rational roots.

(b) Here $a_n = 5$, $a_0 = -25$ and $f(1) = -29$. Since these three numbers are odd, then the equation has no rational roots.

47. Since p is a prime number, the only possible rational roots are $\pm 1, \pm p$.
Let $f(x) = x^3 + x^2 + x - p$
$f(1) = 1 + 1 + 1 - p = 3 - p$
So 1 is a root only if $p = 3$.
$f(-1) = -1 + 1 - 1 - p = -1 - p$
So -1 is a root only if $p = -1$, but -1 is not prime.
$$f(p) = p^3 + p^2 + p - p = p^3 + p^2 = p^2(p+1)$$
So p is a root only if $p = 0$ or $p = -1$, neither of which is prime.
$$f(-p) = -p^3 + p^2 - p - p = -p^3 + p^2 - 2p = -p\left(p^2 - p + 2\right)$$
So $-p$ is a root only if $p = 0$ or $p = \frac{1 \pm \sqrt{1-8}}{2}$, neither of which is prime.
Thus $p = 3$ is the only prime number such that $f(x) = 0$ will have rational roots.
When $p = 3$, we have $x^3 + x^2 + x - 3 = 0$. We know $x = 1$ is a root. Now using synthetic division:

1	1	1	1	−3
		1	2	3
	1	2	3	0

Since $x^2 + 2x + 3 = 0$ has no real roots, $x = 1$ is the only root.

49. Following the suggestion, we substitute:
$x = 1$: $1 + 1 - pq = 0$, so $pq = 2$, which is impossible for prime numbers p and q.
$x = -1$: $1 - 1 - pq = 0$, so $pq = 0$, which is impossible.
$x = p$: $p^2 + p - pq = 0$
$p(p + 1 - q) = 0$
Clearly $p \neq 0$, so $p = q - 1$, which occurs when $p = 2$ and $q = 3$. Thus the equation is $x^2 + x - 6 = 0$, which factors to $(x + 3)(x - 2) = 0$, thus the solutions are -3 and 2.
$x = -p$: $p^2 - p - pq = 0$
$p(p - 1 - q) = 0$
Clearly $p \neq 0$, so $p = q + 1$, which occurs when $p = 3$ and $q = 2$. This results in identical solutions as when $x = p$.
$x = q$: $q^2 + q - pq = 0$
$q(q + 1 - p) = 0$
Clearly $q \neq 0$, so $p = q + 1$, as above.
$x = -q$: $q^2 - q - pq = 0$
$q(q - 1 - p) = 0$
Clearly $q \neq 0$, so $q = p + 1$ again, as above.
$x = pq$: $p^2q^2 + pq - pq = 0$
$p^2q^2 = 0$
Clearly this is impossible.

$x = -pq$: $p^2q^2 - pq - pq = 0$

$pq(pq - 2) = 0$

Since $pq \neq 0$, then $pq = 2$. But this is impossible for prime numbers p and q. Thus if the equation has rational roots they must be –3 and 2.

51. The possible rational roots are ±1, ±2, ±4. Using synthetic division:

$$\begin{array}{r|rrrr} 1 & 1 & -b^2 & 3b & -4 \\ & & 1 & -b^2+1 & -b^2+3b+1 \\ \hline & 1 & -b^2+1 & -b^2+3b+1 & -b^2+3b-3 \end{array}$$

Setting the remainder equal to 0:

$$-3 + 3b - b^2 = 0$$
$$b^2 - 3b + 3 = 0$$
$$b = \frac{3 \pm \sqrt{9-12}}{2}, \text{ which is not an integer}$$

We use synthetic division again:

$$\begin{array}{r|rrrr} -1 & 1 & -b^2 & 3b & -4 \\ & & -1 & b^2+1 & -b^2-3b-1 \\ \hline & 1 & -b^2-1 & b^2+3b+1 & -b^2-3b-5 \end{array}$$

Setting the remainder equal to 0:

$$-b^2 - 3b - 5 = 0$$
$$b^2 + 3b + 5 = 0$$
$$b = \frac{-3 \pm \sqrt{9-20}}{2}, \text{ which is not an integer}$$

Using synthetic division again:

$$\begin{array}{r|rrrr} 2 & 1 & -b^2 & 3b & -4 \\ & & 2 & -2b^2+4 & -4b^2+6b+8 \\ \hline & 1 & -b^2+2 & -2b^2+3b+4 & -4b^2+6b+4 \end{array}$$

Setting the remainder equal to 0:

$$4 + 6b - 4b^2 = 0$$
$$2b^2 - 3b - 2 = 0$$
$$(2b+1)(b-2) = 0$$
$$b = -\tfrac{1}{2} \text{ or } b = 2$$

Hence, if $b = 2$, the given equation has $x = 2$ as a rational root. Using synthetic division again:

$$\begin{array}{r|rrrr} -2 & 1 & -b^2 & 3b & -4 \\ & & -2 & 2b^2+4 & -4b^2-6b-8 \\ \hline & 1 & -b^2-2 & 2b^2+3b+4 & -4b^2-6b-12 \end{array}$$

Setting the remainder equal to 0:

$$-4b^2 - 6b - 12 = 0$$
$$2b^2 + 3b + 6 = 0$$
$$b = \frac{-3 \pm \sqrt{9-48}}{4}, \text{ which is not an integer}$$

Using synthetic division again:

$$\begin{array}{r|rrrr} 4 & 1 & -b^2 & 3b & -4 \\ & & 4 & -4b^2+16 & -16b^2+12b+64 \\ \hline & 1 & -b^2+4 & -4b^2+3b+16 & -16b^2+12b+60 \end{array}$$

Setting the remainder equal to 0:

$$60+12b-16b^2=0$$
$$4b^2-3b-15=0$$
$$b=\frac{3\pm\sqrt{9+240}}{8}, \text{ which is not an integer}$$

Using synthetic division again:

$$\begin{array}{r|rrrr} -4 & 1 & -b^2 & 3b & -4 \\ & & -4 & 4b^2+16 & -16b^2-12b-64 \\ \hline & 1 & -b^2-4 & 4b^2+3b+16 & -16b^2-12b-68 \end{array}$$

Setting the remainder equal to 0:

$$-16b^2-12b-68=0$$
$$4b^2+3b+17=0$$
$$b=\frac{-3\pm\sqrt{9-272}}{8}, \text{ which is not an integer}$$

So $b=2$ is the only integer value of b for which the given equation has a rational root.

8.5 Conjugate Roots and Descartes's Rule of Signs

1. The other root must be $7+2i$.

3. One other root must be $5-2i$. So $[x-(5-2i)][x-(5+2i)]$ are factors, or $x^2-10x+29$. Using long division:

$$\begin{array}{r} x-3 \\ x^2-10x+29 \overline{\smash{)}\, x^3-13x^2+59x-87} \\ \underline{x^3-10x^2+29x} \\ -3x^2+30x-87 \\ \underline{-3x^2+30x-87} \\ 0 \end{array}$$

Since $x-3$ is the other factor, then 3 is the other root. So the other two roots are $5-2i$ and 3.

5. One other root must be $-2-i$, so $[x-(-2+i)][x-(-2-i)]$ are factors, or x^2+4x+5. Using long division:

$$\begin{array}{r|l} & x^2+6x+9 \\ \hline x^2+4x+5 & x^4+10x^3+38x^2+66x+45 \\ & \underline{x^4+4x^3+5x^2} \\ & 6x^3+33x^2+66x \\ & \underline{6x^3+24x^2+30x} \\ & 9x^2+36x+45 \\ & \underline{9x^2+36x+45} \\ & 0 \end{array}$$

Since $x^2+6x+9=(x+3)^2$, the other root is -3. So the remaining roots are $-2-i$ and -3 (multiplicity 2).

7. One other root must be $6+5i$, so $[x-(6-5i)][x-(6+5i)]$ are factors, or $x^2-12x+61$. Using long division:

$$\begin{array}{r|l} & 4x+1 \\ \hline x^2-12x+61 & 4x^3-47x^2+232x+61 \\ & \underline{4x^3-48x^2+244x} \\ & x^2-12x+61 \\ & \underline{x^2-12x+61} \\ & 0 \end{array}$$

Since $4x+1$ is the other factor, then $-\frac{1}{4}$ is the other root. So the remaining roots are $6+5i$ and $-\frac{1}{4}$.

9. One other root must be $4-\sqrt{2}i$, so $\left[x-(4+\sqrt{2}i)\right]\left[x-(4-\sqrt{2}i)\right]$ are factors, or $x^2-8x+18$. Using long division:

$$\begin{array}{r|l} & 4x^2+9 \\ \hline x^2-8x+18 & 4x^4-32x^3+81x^2-72x+162 \\ & \underline{4x^4-32x^3+72x^2} \\ & 9x^2-72x+162 \\ & \underline{9x^2-72x+162} \\ & 0 \end{array}$$

Since $4x^2+9=(2x+3i)(2x-3i)$, the other roots are $-\frac{3i}{2}$ and $\frac{3i}{2}$. So the remaining roots are $4-\sqrt{2}i$, $-\frac{3i}{2}$ and $\frac{3i}{2}$.

11. One other root must be $10-2i$, so $[x-(10+2i)][x-(10-2i)]$ are factors, or $x^2-20x+104$. Using long division:

$$
\begin{array}{r|l}
 & x^2-2x-4 \\
\hline
x^2-20x+104 & x^4-22x^3+140x^2-128x-416 \\
 & \underline{x^4-20x^3+104x^2} \\
 & \quad -2x^3+36x^2-128x \\
 & \quad \underline{-2x^3+40x^2-208x} \\
 & \qquad -4x^2+80x-416 \\
 & \qquad \underline{-4x^2+80x-416} \\
 & \qquad\qquad 0
\end{array}
$$

So the other factor is x^2-2x-4. We use the quadratic formula:

$$x=\frac{2\pm\sqrt{4+16}}{2}=\frac{2\pm2\sqrt{5}}{2}=1\pm\sqrt{5}$$

So the remaining roots are $10-2i$, $1+\sqrt{5}$ and $1-\sqrt{5}$.

13. One other root must be $\frac{1-\sqrt{2}i}{3}$ so $\left(x-\frac{1+\sqrt{2}i}{3}\right)\left(x-\frac{1-\sqrt{2}i}{3}\right)$ are factors, or $x^2-\frac{2}{3}x+\frac{1}{3}$. Using long division:

$$
\begin{array}{r|l}
 & 15x-6 \\
\hline
x^2-\frac{2}{3}x+\frac{1}{3} & 15x^3-16x^2+9x-2 \\
 & \underline{15x^3-10x^2+5x} \\
 & \quad -6x^2+4x-2 \\
 & \quad \underline{-6x^2+4x-2} \\
 & \qquad 0
\end{array}
$$

So the other factor is $15x-6$, so $x=\frac{2}{5}$ is a root. The remaining roots are $\frac{1-i\sqrt{2}}{3}$ and $\frac{2}{5}$.

15. We know $3+2i$ is a root, so $[x-(3-2i)][x-(3+2i)]$ are factors, which is $x^2-6x+13$. Using long division:

$$
\begin{array}{r|l}
 & x^5+3x^4+x^3-3x^2-4x+2 \\
\hline
x^2-6x+13 & x^7-3x^6-4x^5+30x^4+27x^3-13x^2-64x+26 \\
 & \underline{x^7-6x^6+13x^5} \\
 & \quad 3x^6-17x^5+30x^4 \\
 & \quad \underline{3x^6-18x^5+39x^4} \\
 & \qquad x^5-9x^4+27x^3 \\
 & \qquad \underline{x^5-6x^4+13x^3} \\
 & \qquad\quad -3x^4+14x^3-13x^2 \\
 & \qquad\quad \underline{-3x^4+18x^3-39x^2} \\
 & \qquad\qquad -4x^3+26x^2-64x \\
 & \qquad\qquad \underline{-4x^3+24x^2-52x} \\
 & \qquad\qquad\quad 2x^2-12x+26 \\
 & \qquad\qquad\quad \underline{2x^2-12x+26} \\
 & \qquad\qquad\qquad 0
\end{array}
$$

We know $x = -1 - i$ will be a root, so $[x-(-1+i)][x-(-1-i)]$ are factors, which is $x^2 + 2x + 2$. Using long division:

$$
\begin{array}{r|l}
 & x^3 + x^2 - 3x + 1 \\
\hline
x^2 + 2x + 2 & x^5 + 3x^4 + x^3 - 3x^2 - 4x + 2 \\
 & \underline{x^5 + 2x^4 + 2x^3} \\
 & \quad x^4 - x^3 - 3x^2 \\
 & \quad \underline{x^4 + 2x^3 + 2x^2} \\
 & \qquad -3x^3 - 5x^2 - 4x \\
 & \qquad \underline{-3x^3 - 6x^2 - 6x} \\
 & \qquad\qquad x^2 + 2x + 2 \\
 & \qquad\qquad \underline{x^2 + 2x + 2} \\
 & \qquad\qquad\qquad 0
\end{array}
$$

Finally, we know $x = 1$ is a root, so we synthetically divide:

$$
\begin{array}{c|cccc}
1 & 1 & 1 & -3 & 1 \\
 & & 1 & 2 & -1 \\
\hline
 & 1 & 2 & -1 & 0
\end{array}
$$

So we are left with $x^2 + 2x - 1 = 0$. Using the quadratic formula:

$$x = \frac{-2 \pm \sqrt{4+4}}{2} = \frac{-2 \pm 2\sqrt{2}}{2} = -1 \pm \sqrt{2}$$

So the remaining roots are $3 + 2i$, $-1 - i$, $-1 + \sqrt{2}$, and $-1 - \sqrt{2}$.

17. Here $r_1 = 1 + \sqrt{6}$ and $r_2 = 1 - \sqrt{6}$, so $\left[x-\left(1+\sqrt{6}\right)\right]\left[x-\left(1-\sqrt{6}\right)\right] = 0$. Multiplied out, we obtain $x^2 - 2x - 5 = 0$.

19. Here $r_1 = \frac{2+\sqrt{10}}{3}$ and $r_2 = \frac{2-\sqrt{10}}{3}$, so $\left(x - \frac{2+\sqrt{10}}{3}\right)\left(x - \frac{2-\sqrt{10}}{3}\right) = 0$. Multiplied out, we obtain $x^2 - \frac{4}{3}x - \frac{2}{3} = 0$.

21. Let $f(x) = x^3 + 5$. Since there are no sign changes, there will be no positive roots. Now $f(-x) = -x^3 + 5$. Since there is one sign change, there is 1 negative root. So the equation has 2 complex roots and 1 negative real root.

23. Let $f(x) = 2x^5 + 3x + 4$. Since there are no sign changes, there will be no positive roots. Now $f(-x) = -2x^5 - 3x + 4$. Since there is one sign change, there is 1 negative root. So the equation has 4 complex roots and 1 negative real root.

25. Let $f(x) = 5x^4 + 2x - 7$. Since there is one sign change, there will be 1 positive root. Now $f(-x) = 5x^4 - 2x - 7$. Since there is one sign change, there will be 1 negative root. So the equation has 2 complex roots, 1 positive real root and 1 negative real root.

27. Let $f(x) = x^3 - 4x^2 - x - 1$. Since there is one sign change, there will be 1 positive root. Now $f(-x) = -x^3 - 4x^2 + x - 1$. Since there are two sign changes, there will be 0 or 2 negative roots. So the equation has either 1 positive real root and 2 negative real roots, or 1 positive real root and 2 complex roots.

29. Let $f(x) = 3x^8 + x^6 - 2x^2 - 4$. Since there is one sign change, there is 1 positive root. Now $f(-x) = 3x^8 + x^6 - 2x^2 - 4$. Since there is one sign change, there is 1 negative root. So the equation has 1 positive real root, 1 negative real root and 6 complex roots.

31. Let $f(x) = x^9 - 2$. Since there is one sign change, there is 1 positive root. Now $f(-x) = -x^9 - 2$. Since there are no sign changes, there are no negative roots. So the equation has 1 positive real root and 8 complex roots.

33. Let $f(x) = x^8 - 2$. Since there is one sign change, there is 1 positive root. Now $f(-x) = x^8 - 2$. Since there is one sign change, there is 1 negative root. So the equation has 1 positive real root, 1 negative real root and 6 complex roots.

35. Let $f(x) = x^6 + x^2 - x - 1$. Since there is one sign change, there is 1 positive root. Now $f(-x) = x^6 + x^2 + x - 1$. Since there is one sign change, there is 1 negative root. So the equation has 1 positive real root, 1 negative real root and 4 complex roots.

37. Let $f(x) = x^4 + cx^2 + dx - e$. Since there is one sign change, there is 1 positive root. Now $f(-x) = x^4 + cx^2 - dx - e$. Since there is one sign change, there is 1 negative root. So the equation has 1 positive real root, 1 negative real root and 2 complex roots.

39. The two roots $\sqrt{3} + 2i$ and $\sqrt{3} - 2i$ must be included, so:

$$f(x) = [x - (\sqrt{3} + 2i)][x - (\sqrt{3} - 2i)] = x^2 - 2\sqrt{3} + 7 = 0$$

Unfortunately, not all coefficients are rational:

$$x^2 + 7 = 2\sqrt{3}x$$

Squaring, we obtain:

$$x^4 + 14x^2 + 49 = 12x^2$$
$$x^4 + 2x^2 + 29 = 0$$

So $f(x) = x^4 + 2x^2 + 49$ is the desired polynomial.

41. (a) If $b = 0$, then $a + b\sqrt{c} = a = a - b\sqrt{c}$, so $a - b\sqrt{c}$ is also a root.

(b) We compute $d(a + b\sqrt{c})$:

$$\begin{aligned} d(a + b\sqrt{c}) &= [a + b\sqrt{c} - (a + b\sqrt{c})][a + b\sqrt{c} - (a - b\sqrt{c})] \\ &= 0(2b\sqrt{c}) \\ &= 0 \end{aligned}$$

(c) We factor $d(x) = x^2 - 2ax + a^2 - b^2c = (x - a)^2 - b^2c$.

(d) If $x = a + b\sqrt{c}$, we have:

$$\begin{aligned} f(a + b\sqrt{c}) &= d(a + b\sqrt{c})Q(a + b\sqrt{c}) + C(a + b\sqrt{c}) + D \\ 0 &= 0 \bullet Q(a + b\sqrt{c}) + Ca + D + bC\sqrt{c} \\ 0 &= Ca + D + bC\sqrt{c} \end{aligned}$$

But if C and D are rational, then $C = 0$ (otherwise $\sqrt{c}$ will not be "cancelled" out). If $C = 0$, then $D = 0$. So $C = D = 0$.

(e) We compute $f(a - b\sqrt{c})$:

$$\begin{aligned} f(a-b\sqrt{c}) &= \left[a-b\sqrt{c}-(a+b\sqrt{c})\right]\left[a-b\sqrt{c}-(a-b\sqrt{c})\right]Q(x) \\ &= (-2b\sqrt{c})(0)Q(x) \\ &= 0 \end{aligned}$$

So $x = a - b\sqrt{c}$ is also a root of $f(x) = 0$.

Chapter Eight Review Exercises

1. Using synthetic division:

−2	1	3	−1	−5	1
		−2	−2	6	−2
	1	1	−3	1	−1

So $q(x) = x^3 + x^2 - 3x + 1$ and $R(x) = -1$.

3. Using synthetic division:

3	1	0	−2	0	8
		3	9	21	63
	1	3	7	21	71

So the quotient is $x^3 + 3x^2 + 7x + 21$ and the remainder is 71.

5. Using synthetic division:

−4	2	−5	−6	−3
		−8	52	−184
	2	−13	46	−187

So the quotient is $2x^2 - 13x + 46$ and the remainder is −187.

7. Using synthetic division:

−0.2	5	−19	−4
		−1	4
	5	−20	0

So the quotient is $5x - 20$ and the remainder is zero.

9. Using synthetic division:

10	1	0	0	0	−10	4
		10	100	1000	10000	99900
	1	10	100	1000	9990	99904

So $f(10) = 99{,}904$.

11. Using synthetic division:

1/10	1	−10	1	−1
		1/10	−99/100	1/1000
	1	−99/10	1/100	−999/1000

So $f\left(\frac{1}{10}\right) = -\frac{999}{1000}$.

13. Using synthetic division:

$a-1$	1	3	3	1
		$a-1$	a^2+a-2	a^3-1
	1	$a+2$	a^2+a+1	a^3

So $f(a-1) = a^3$.

15. (a) Using synthetic division:

−0.3	1	4	−6	−8	−2
		−0.3	−1.11	2.133	1.7601
	1	3.7	−7.11	−5.867	−0.24

So $f(-0.3) \approx -0.24$.

(b) Using synthetic division:

−0.39	1	4	−6	−8	−2
		−0.39	−1.4079	2.8891	1.9933
	1	3.61	−7.4079	−5.1109	−0.007

So $f(-0.39) \approx -0.007$.

(c) Using synthetic division:

−0.394	1	4	−6	−8	−2
		−0.394	−1.420764	2.92378	2.00003
	1	3.606	−7.420764	−5.07622	0.00003

So $f(-0.394) \approx 0.00003$

17. Using synthetic division:

3	1	−4	$-a$	−6
		3	−3	$-3a-9$
	1	−1	$-a-3$	$-3a-15$

So if 3 is a root, then:

$$-3a - 15 = 0$$
$$-3a = 15$$
$$a = -5$$

19. Using synthetic division:

1	a^2	$3a$	0	2
		a^2	a^2+3a	a^2+3a
	a^2	a^2+3a	a^2+3a	a^2+3a+2

So if $x - 1$ is a factor, then:

$$a^2 + 3a + 2 = 0$$
$$(a+1)(a+2) = 0$$
$$a = -1, -2$$

So $a = -1$ or $a = -2$.

21. (a) Let $f(x + h) = 0$, and let $x = r - h$. Then $f(r - h + h) = f(r) = 0$, since r is a root of $f(x) = 0$.

(b) Let $f(-x) = 0$, and let $x = -r$. Then $f(-(-r)) = f(r) = 0$, since r is a root of $f(x) = 0$.

(c) Let $f\left(\frac{x}{k}\right) = 0$, and let $x = kr$. Then $f\left(\frac{kr}{k}\right) = f(r) = 0$, since r is a root of $f(x) = 0$.

23. p-factors: $\pm 1, \pm 2, \pm 3, \pm 6, \pm 9, \pm 18$
q-factors: ± 1
possible rational roots: $\pm 1, \pm 2, \pm 3, \pm 6, \pm 9, \pm 18$

25. p-factors: $\pm 1, \pm 2, \pm 4, \pm 8$
q-factors: $\pm 1, \pm 2$
possible rational roots: $\pm 1, \pm \frac{1}{2}, \pm 2, \pm 4, \pm 8$

27. p-factors: $\pm p, \pm 1$
q-factors: ± 1
possible rational roots: $\pm p, \pm 1$

29. The possible rational roots are $\pm 1, \pm \frac{1}{2}, \pm 2, \pm 3, \pm \frac{3}{2}, \pm 6$. Using synthetic division:

$$\begin{array}{r|rrrr} 2 & 2 & 1 & -7 & -6 \\ & & 4 & 10 & 6 \\ \hline & 2 & 5 & 3 & 0 \end{array}$$

So $2x^3 + x^2 - 7x - 6 = (x-2)(2x^2 + 5x + 3) = (x-2)(2x+3)(x+1)$.

So the roots are $2, -\frac{3}{2}$ and -1.

31. The possible rational roots are $\pm 1, \pm \frac{1}{2}, \pm 2, \pm 5, \pm \frac{5}{2}, \pm 10$. Using synthetic division:

$$\begin{array}{r|rrrr} 5/2 & 2 & -1 & -14 & 10 \\ & & 5 & 10 & -10 \\ \hline & 2 & 4 & -4 & 0 \end{array}$$

So $2x^3 - x^2 - 14x + 10 = \left(x - \frac{5}{2}\right)\left(2x^2 + 4x - 4\right) = 2\left(x - \frac{5}{2}\right)\left(x^2 + 2x - 2\right)$.

Using the quadratic formula:

$$x=\frac{-2\pm\sqrt{4+8}}{2}=\frac{-2\pm 2\sqrt{3}}{2}=-1\pm\sqrt{3}$$

So the roots are $\frac{5}{2}, -1+\sqrt{3}$ and $-1-\sqrt{3}$.

33. We first multiply by 2 to obtain $3x^3+x^2+x-2=0$. The possible rational roots are $\pm 1, \pm\frac{1}{3}, \pm 2, \pm\frac{2}{3}$. Using synthetic division:

$$\begin{array}{r|rrrr} 2/3 & 3 & 1 & 1 & -2 \\ & & 2 & 2 & 2 \\ \hline & 3 & 3 & 3 & 0 \end{array}$$

So $3x^3+x^2+x-2=\left(x-\frac{2}{3}\right)\left(3x^2+3x+3\right)=3\left(x-\frac{2}{3}\right)\left(x^2+x+1\right)$.
Using the quadratic formula:

$$x=\frac{-1\pm\sqrt{1-4}}{2}=\frac{-1\pm i\sqrt{3}}{2}$$

So the roots are $\frac{2}{3}, \frac{-1+i\sqrt{3}}{2}$, and $\frac{-1-i\sqrt{3}}{2}$.

35. The possible rational roots are $\pm 1, \pm 7, \pm 49$. Using synthetic division:

$$\begin{array}{r|rrrrrr} -1 & 1 & 1 & -14 & -14 & 49 & 49 \\ & & -1 & 0 & 14 & 0 & -49 \\ \hline & 1 & 0 & -14 & 0 & 49 & 0 \end{array}$$

Therefore:

$$\begin{aligned} x^5+x^4-14x^3-14x^2+49x+49 &= (x+1)\left(x^4-14x^2+49\right) \\ &= (x+1)\left(x^2-7\right)^2 \\ &= (x+1)\left(x+\sqrt{7}\right)^2\left(x-\sqrt{7}\right)^2 \end{aligned}$$

So the roots are $-1, -\sqrt{7}$ (multiplicity 2), and $\sqrt{7}$ (multiplicity 2).

37. Since $x^3-9x^2+24x-20=0$ has a root r with multiplicity 2, we can use synthetic division by r twice, each time the remainder must be 0:

$$\begin{array}{r|cccc} r & 1 & -9 & 24 & -20 \\ & & r & r^2-9r & r^3-9r^2+24r \\ \hline & 1 & r-9 & r^2-9r+24 & r^3-9r^2+24r-20 \end{array}$$

We continue the division:

$$\begin{array}{r|ccc} r & 1 & r-9 & r^2-9r+24 \\ & & r & 2r^2-9r \\ \hline & 1 & 2r-9 & 3r^2-18r+24 \end{array}$$

So $3r^2-18r+24=3(r^2-6r+8)=3(r-4)(r-2)=0$, thus $r=4$ or $r=2$. We check these in $r^3-9r^2+24r-20=0$:

$r=4$: $\quad (4)^3-9(4)^2+24(4)-20=64-144+96-20=-4\neq 0$

So $r=4$ cannot be a root.

$r=2$: $\quad (2)^3-9(2)^2+24(2)-20=8-36+48-20=0$

So $r = 2$ is the root with multiplicity 2. Since the synthetic division resulted in $x + (2r - 9) = 0$, then $x + (-5) = 0$ and thus $x = 5$. So the roots are 2 (multiplicity 2) and 5.
Note: Actually, an easier (and more direct) approach is to find the roots directly. The possible rational roots are $\pm1, \pm2, \pm4, \pm5, \pm10, \pm20$. Note that there are 3 sign changes, so there can be 1 or 3 positive real roots. Since $f(-x)$ has no sign changes, all 3 roots must be positive real numbers for one of them to have multiplicity of 2 (note that it cannot have a radical or be complex--why?). We use synthetic division:

2	1	−9	24	−20
		2	−14	20
	1	−7	10	0

So $x^3 - 9x^2 + 24x - 20 = (x-2)(x^2 - 7x + 10) = (x-2)^2(x-5)$.
So the roots are 2 (with multiplicity 2) and 5.

39. (a) Let $p(x)$ and $d(x)$ be the polynomials where $d(x) \neq 0$. Then there are unique polynomials $q(x)$ and $R(x)$ such that
$$p(x) = d(x) \bullet q(x) + R(x)$$
where either $R(x) = 0$ or the degree of $R(x)$ is less than the degree of $d(x)$.

(b) When a polynomial $f(x)$ is divided by $x - r$, the remainder is $f(r)$.

(c) Let $f(x)$ be a polynomial. If $f(r) = 0$, then $x - r$ is a factor of $f(x)$. Conversely, if $x - r$ is a factor of $f(x)$, then $f(r) = 0$.

(d) Every polynomial equation of the form
$$a_n x^n + a_{n-1}x^{n-1} + \ldots + a_1 x + a_0 = 0 \quad (n \geq 1,\ a_n \neq 0)$$
has at least one root among the complex numbers. (This root may be a real number.)

41. We factor:
$$6x^2 + 7x - 20 = (3x-4)(2x+5) = 6\left(x - \tfrac{4}{3}\right)\left[x - \left(-\tfrac{5}{2}\right)\right]$$

43. We factor:
$$\begin{aligned} x^4 - 4x^3 + 5x - 20 &= x^3(x-4) + 5(x-4) \\ &= (x-4)(x^3+5) \\ &= (x-4)\left(x + \sqrt[3]{5}\right)\left(x^2 - \sqrt[3]{5}x + \sqrt[3]{25}\right) \end{aligned}$$

We solve $x^2 - \sqrt[3]{5}x + \sqrt[3]{25} = 0$ using the quadratic formula, which yields:
$$x = \frac{\sqrt[3]{5} \pm i\sqrt{3\sqrt[3]{25}}}{2}$$

So $x^4 - 4x^3 + 3x - 20 = (x-4)\left(x - \left(-\sqrt[3]{5}\right)\right)\left(x - \dfrac{\sqrt[3]{5} + i\sqrt{3\sqrt[3]{25}}}{2}\right)\left(x - \dfrac{\sqrt[3]{5} - i\sqrt{3\sqrt[3]{25}}}{2}\right)$.

45. One other root is $2 + 3i$, so $[x-(2-3i)][x-(2+3i)]$ are factors, which is $x^2 - 4x + 13$. Using long division:

$$\begin{array}{r|l} & x-3 \\ \hline x^2-4x+13 & x^3-7x^2+25x-39 \\ & \underline{x^3-4x^2+13x} \\ & \quad -3x^2+12x-39 \\ & \quad \underline{-3x^2+12x-39} \\ & \qquad\qquad\qquad 0 \end{array}$$

So $x - 3$ is the other factor. So the roots are $2 - 3i$, $2 + 3i$, and 3.

47. One other root is $1 - i\sqrt{2}$, so $[x-(1+i\sqrt{2})][x-(1-i\sqrt{2})]$ are factors, which is $x^2 - 2x + 3$. Using long division:

$$\begin{array}{r|l} & x^2-4 \\ \hline x^2-2x+3 & x^4-2x^3-4x^2+14x-21 \\ & \underline{x^4-2x^3+3x^2} \\ & \qquad -7x^2+14x-21 \\ & \qquad \underline{-7x^2+14x-21} \\ & \qquad\qquad\qquad\quad 0 \end{array}$$

So $x^2 - 7 = (x+\sqrt{7})(x-\sqrt{7})$ is the other factor. So the roots are $1 + i\sqrt{2}$, $1 - i\sqrt{2}$, $\sqrt{7}$, and $-\sqrt{7}$.

49. Let $f(x) = x^3 + 8x - 7$. There is one sign change, so there is 1 positive root. Now $f(-x) = -x^3 - 8x - 7$. Since there are no sign changes, there are no negative roots. So the equation has 1 positive real root and 2 complex roots.

51. Let $f(x) = x^3 + 3x + 1$. There are no sign changes, so there are no positive roots. Now $f(-x) = -x^3 - 3x + 1$. Since there is one sign change, there is 1 negative root. So the equation has 1 negative real root and 2 complex roots.

53. Let $f(x) = x^4 - 10$. There is one sign change, so there is 1 positive root. Since $f(-x) = f(x)$, there is also 1 negative root. So the equation has 1 positive real root, 1 negative real root and 2 complex roots.

55. (a) Let $f(x) = x^3 + x^2 + x + 1$, so $f(-x) = -x^3 + x^2 - x + 1$. There are three sign changes, so there are either 1 or 3 negative roots.

(b) Using the hint, we have $(x-1)(x^3+x^2+x+1) = x^4 - 1$. Let $g(x) = x^4 - 1$. Then $g(-x) = x^4 - 1$, so this can only have 1 negative root. Since $x = 1$ is a positive root, then $x^3 + x^2 + x + 1 = 0$ can only have 1 negative root also.

(c) Factoring, we have $x^3+x^2+x+1=x^2(x+1)+(x+1)=(x+1)(x^2+1)$.
So $x=-1$ is a root. Also:

$$x^2+1=0$$
$$x^2=-1$$
$$x=\pm i$$

So the roots are -1, i, and $-i$.

57. Substituting $y=x^3$, we have $x^2+(x^3)^2=1$, so $x^6+x^2-1=0$.
Let $f(x)=x^6+x^2-1$.

$f(0)=-1$
$f(1)=1$

So there is a root between 0 and 1.

$f(0.8)=-0.09$
$f(0.9)=0.34$

So there is a root between 0.8 and 0.9.

$f(0.82)=-0.02$
$f(0.83)=0.01$

So there is a root between 0.82 and 0.83.
So the x-coordinate lies between 0.82 and 0.83.

59. (a) Let $f(x)=x^3-36x-84$. Since there is only one sign change, there is 1 positive real root.

(b) Using synthetic division:

7	1	0	−36	−84
		7	49	91
	1	7	13	7

Since the row is all positive, there are no real roots to the equation greater than 7.

(c) Here $f(6)=-84$ and $f(7)=7$, so there is a root between 6 and 7.
$f(6.8)=-14.4$ and $f(6.9)=-3.89$, so there is a root between 6.9 and 7.0.
$f(6.93)=-0.67$ and $f(6.94)=0.42$, so there is a root between 6.93 and 6.94.

61. Another root will be $4+\sqrt{5}$, so:

$$[x-(4-\sqrt{5})][x-(4+\sqrt{5})]=0$$
$$x^2-8x+11=0$$

63. Other roots will be $6+2i$ and $-\sqrt{5}$:

$$(x-\sqrt{5})[x-(-\sqrt{5})][x-(6-2i)][x-(6+2i)]=0$$
$$(x^2-5)(x^2-12x+40)=0$$
$$x^4-12x^3+35x^2+60x-200=0$$

65. We write $x - 1 = \sqrt{2} + \sqrt{3}$.
Squaring both sides, we obtain:

$$x^2 - 2x + 1 = 5 + 2\sqrt{6}$$
$$x^2 - 2x - 4 = 2\sqrt{6}$$

Squaring both sides again, we obtain:

$$x^4 - 4x^3 - 4x^2 + 16x + 16 = 24$$
$$x^4 - 4x^3 - 4x^2 + 16x - 8 = 0$$

67. Factoring, we have $y = x^3 - 2x^2 - 3x = x(x^2 - 2x - 3) = x(x-3)(x+1)$. The zeros are 0, 3, and –1.

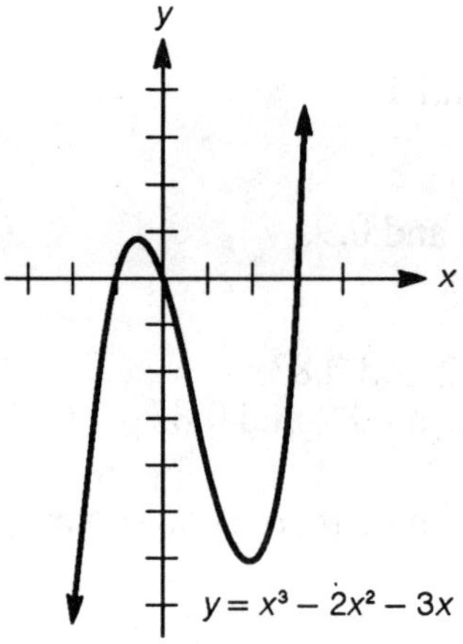

69. Factoring, we have $y = x^4 - 4x^2 = x^2(x^2 - 4) = x^2(x+2)(x-2)$. The zeros are 0, –2, and 2.

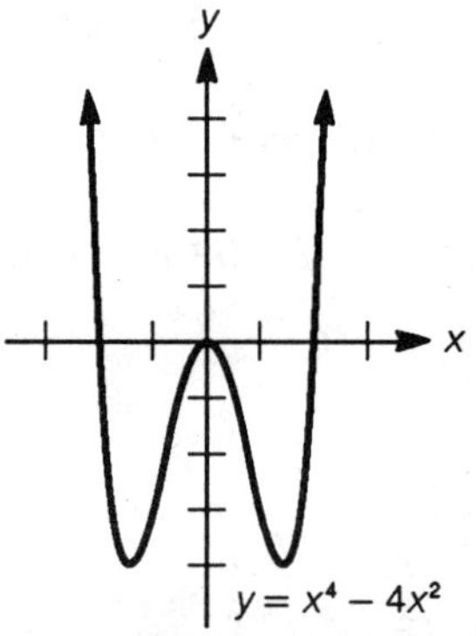

71. Using synthetic division to find rational roots:

1	1	–8	25	–38	28	–8
		1	–7	18	–20	8
	1	–7	18	–20	8	0

Using synthetic division again:

1	1	−7	18	−20	8
		1	−6	12	−8
	1	−6	12	−8	0

Using synthetic division again:

2	1	−6	12	−8
		2	−8	8
	1	−4	4	0

So the function factors as $y = (x-1)^2(x-2)(x^2-4x+4) = (x-1)^2(x-2)^3$. The zeros are 1 and 2.

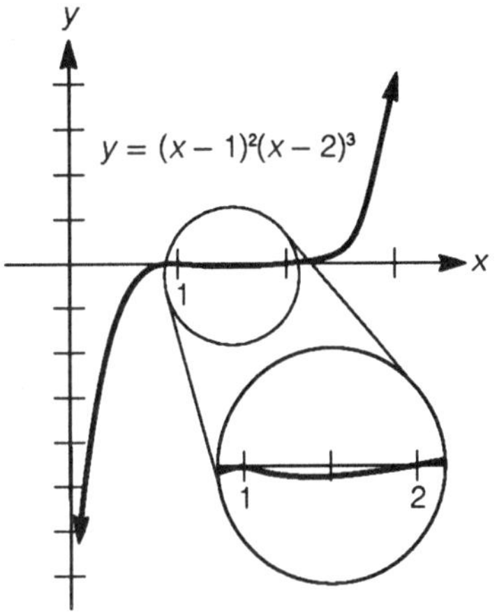

73. (a) Using synthetic division:

$\sqrt{3}$	1	0	−3	0	−16	0	48
		$\sqrt{3}$	3	0	0	$-16\sqrt{3}$	−48
	1	$\sqrt{3}$	0	0	−16	$-16\sqrt{3}$	0

Since the remainder is zero $x-\sqrt{3}$ is a factor of $f(x)$.

(b) Since $\sqrt{3}$ is a root, then $-\sqrt{3}$ must also be a root. Hence $x+\sqrt{3}$ must be a factor.

(c) Using synthetic division:

2	1	0	−3	0	−16	0	48
		2	4	2	4	−24	−48
	1	2	1	2	−12	−24	0

Since the remainder is zero $x-2$ is a factor of $f(x)$.

(d) Using synthetic division:

−2	1	0	−3	0	−16	0	48
		−2	4	−2	4	24	−48
	1	−2	1	−2	−12	24	0

Since the remainder is zero $x+2$ is a factor of $f(x)$.

Chapter Eight Graphing Calculator Exercises

1. (a) We graph the function $y = f(x)$:

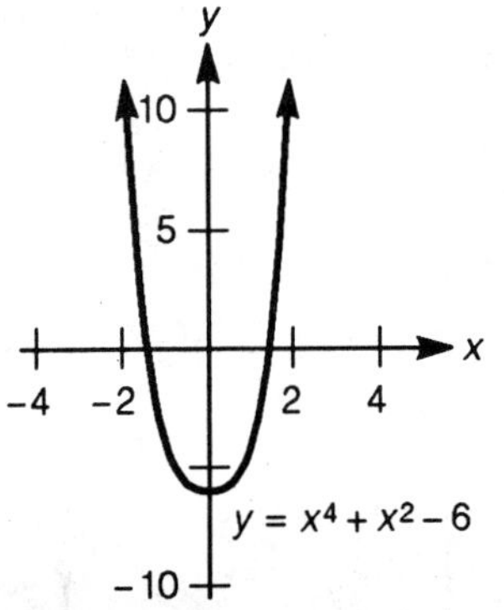

(b) Using the ZOOM-TRACE technique, the roots are ± 1.414. This agrees with the roots $\pm\sqrt{2}$ found in the text.

3. (a) We graph the function $y = f(x)$:

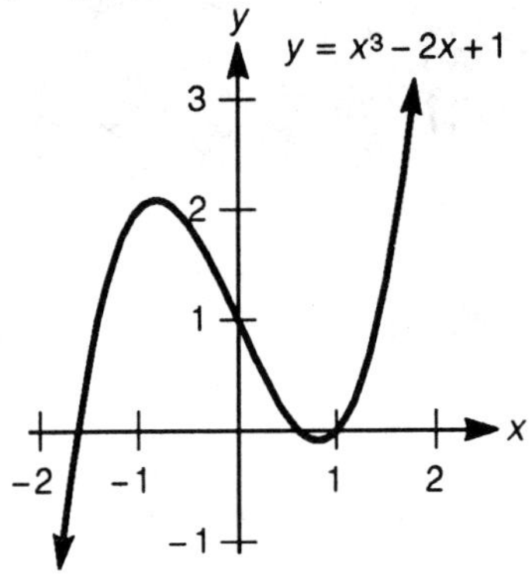

(b) Using the ZOOM-TRACE technique, the roots are -1.618, 0.618, and 1. This agrees with the roots $1, \frac{1}{2}(-1 \pm \sqrt{5})$ found in the text.

5. (a) We graph the function $y = f(x)$:

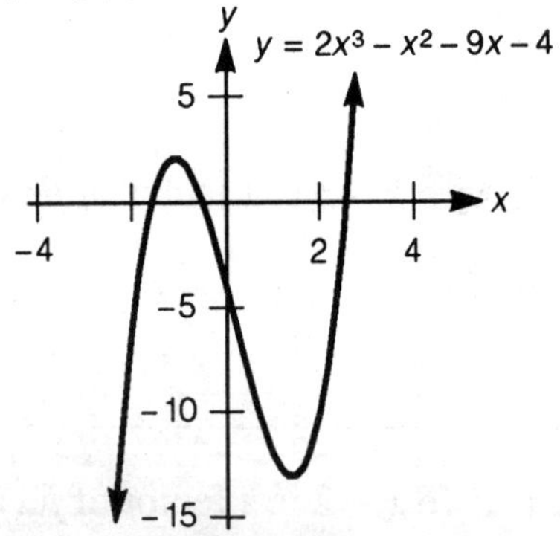

(b) Using the ZOOM-TRACE technique, the roots are –1.562, –0.5, and 2.562. This agrees with the roots $-\frac{1}{2}, \frac{1}{2}(1 \pm \sqrt{17})$ found in the text.

7. (a) We graph the function $y = f(x)$:

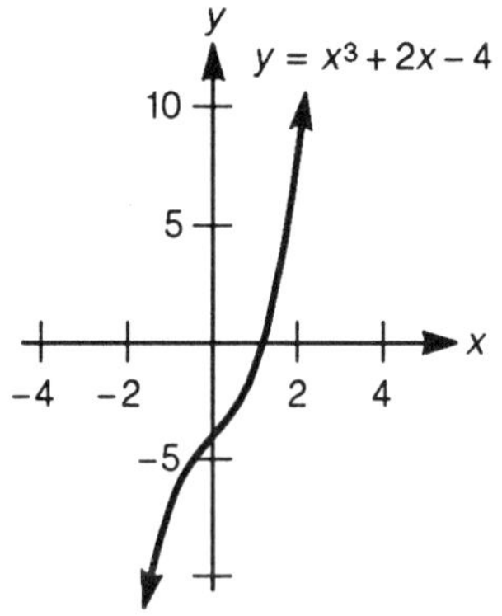

(b) Using the ZOOM-TRACE technique, the root is 1.180. This agrees with the information obtained in the text.

9. (a) We graph the function $y = f(x)$:

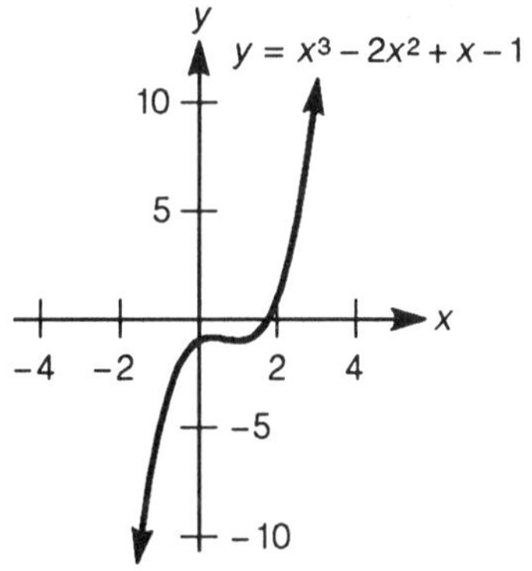

(b) Using the ZOOM-TRACE technique, the root is 1.755. This agrees with the information obtained in the text.

11. We graph the function $f(x)$:

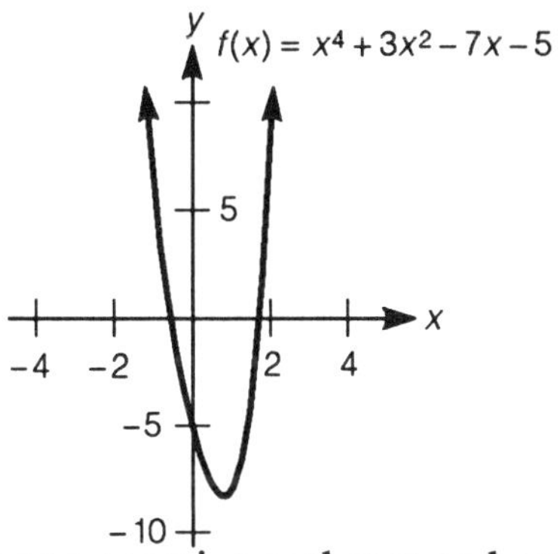

Notice that this function has one negative real root and one positive real root.

Chapter Eight Test

1. Since $f\left(\frac{1}{2}\right)$ will be the remainder after synthetic division, we synthetically divide by $\frac{1}{2}$:

1/2	6	−5	7	−2	−2
		3	−1	3	1/2
	6	−2	6	1	−3/2

So $f\left(\frac{1}{2}\right) = -\frac{3}{2}$.

2. Using synthetic division:

−3	1	1	−11	−15
		−3	6	15
	1	−2	−5	0

So $x^3 + x^2 - 11x - 15 = (x+3)(x^2 - 2x - 5)$. Using the quadratic formula, we have:

$$x = \frac{2 \pm \sqrt{4+20}}{2} = \frac{2 \pm 2\sqrt{6}}{2} = 1 \pm \sqrt{6}$$

So the roots of the equation are -3, $1 + \sqrt{6}$, and $1 - \sqrt{6}$.

3. $\pm 1, \pm\frac{1}{2}, \pm 2, \pm 3, \pm\frac{3}{2}, \pm 6$

4. Such a function would be $y = a(x-1)(x+8)$. Now $y = -24$ when $x = 0$, so $-24 = a(-1)(8)$, so $a = 3$. So the function is $y = 3(x-1)(x+8)$, or $f(x) = 3x^2 + 21x - 24$.

5. Using synthetic division:

−1	4	1	−8	3
		−4	3	5
	4	−3	−5	8

The quotient is $4x^2 - 3x - 5$ and the remainder is 8.

6. (a) Let $f(x)$ be a polynomial. If $f(r) = 0$, then $x - r$ is a factor of $f(x)$. Conversely, if $x - r$ is a factor of $f(x)$, then $f(r) = 0$.

(b) Every polynomial of the form

$$a_n x^n + a_{n-1}x^{n-1} + \ldots + a_1 x + a_0 = 0 \quad (n \geq 1,\ a_n \neq 0)$$

has at least one root among the complex numbers. (This root may be a real number.)

7. (a) Using synthetic division with 1:

1	1	−2	0	−1
		1	−1	−1
	1	−1	−1	−2

Now we use 2:

$$\begin{array}{r|rrrr} 2 & 1 & -2 & 0 & -1 \\ & & 2 & 0 & 0 \\ \hline & 1 & 0 & 0 & -1 \end{array}$$

Now use 3:

$$\begin{array}{r|rrrr} 3 & 1 & -2 & 0 & -1 \\ & & 3 & 3 & 9 \\ \hline & 1 & 1 & 3 & 8 \end{array}$$

So 3 is an upper bound for the roots.

(b) Since $f(2.2) = -0.032$ and $f(2.3) = 0.587$, the root lies between 2.2 and 2.3.

8. Two other roots are $1 - i$ and $3 + 2i$, so $x^2 - 2x + 2$ and $x^2 - 6x + 13$ are factors. Using long division:

$$\begin{array}{r|l} & x^3 - 4x^2 + x + 26 \\ \hline x^2 - 2x + 2 & x^5 - 6x^4 + 11x^3 + 16x^2 - 50x + 52 \\ & \underline{x^5 - 2x^4 + 2x^3} \\ & \quad -4x^4 + 9x^3 + 16x^2 \\ & \quad \underline{-4x^4 + 8x^3 - 8x^2} \\ & \qquad\qquad x^3 + 24x^2 - 50x \\ & \qquad\qquad \underline{x^3 - 2x^2 + 2x} \\ & \qquad\qquad\quad 26x^2 - 52x + 52 \\ & \qquad\qquad\quad \underline{26x^2 - 52x + 52} \\ & \qquad\qquad\qquad\qquad\qquad 0 \end{array}$$

Using long division again:

$$\begin{array}{r|l} & x + 2 \\ \hline x^2 - 6x + 13 & x^3 - 4x^2 + x + 26 \\ & \underline{x^3 - 6x^2 + 13x} \\ & \quad 2x^2 - 12x + 26 \\ & \quad \underline{2x^2 - 12x + 26} \\ & \qquad\qquad\qquad 0 \end{array}$$

So the roots are $1 \pm i$, $3 \pm 2i$, and -2.

9. Using long division:

$$\begin{array}{r|l} & x^2 + 2x - 1 \\ \hline x^2 + 1 & x^4 + 2x^3 + 0x^2 - x + 6 \\ & \underline{x^4 \qquad\quad + x^2} \\ & \quad 2x^3 - x^2 - x \\ & \quad \underline{2x^3 \qquad + 2x} \\ & \qquad -x^2 - 3x + 6 \\ & \qquad \underline{-x^2 \qquad - 1} \\ & \qquad\qquad -3x + 7 \end{array}$$

So $q(x) = x^2 + 2x - 1$ and $R(x) = -3x + 7$.

10. We first find the roots by the quadratic formula:

$$x = \frac{6 \pm \sqrt{36-40}}{4} = \frac{6 \pm 2i}{4} = \tfrac{3}{2} \pm \tfrac{1}{2}i$$

So $2x^2 - 6x + 5 = 2\left(x^2 - 3x + \tfrac{5}{2}\right) = 2\left[x - \left(\tfrac{3}{2} + \tfrac{1}{2}i\right)\right]\left[x - \left(\tfrac{3}{2} - \tfrac{1}{2}i\right)\right]$.

11. (a) The possible rational roots are $\pm 1, \pm 2, \pm 3, \pm 4, \pm 6, \pm 8, \pm 12$, and ± 24.

(b) Using synthetic division:

2	1	−1	0	0	24
		2	2	4	8
	1	1	2	4	32

Since this last row consists of all positive numbers, we know that 2 is an upper bound for the roots of this equation.

(c) Only $x = 1$, since $x = 2$ is an upper bound and not a root.

(d) Using synthetic division:

1	1	−1	0	0	24
		1	0	0	1
	1	0	0	0	25

Since $x = 1$ was the only possibility, there are no positive rational roots.

12. (a) The possible rational roots are $\pm 1, \pm \tfrac{1}{2}, \pm 3, \pm \tfrac{3}{2}$.
Using synthetic division:

3/2	2	−1	−1	−3
		3	3	3
	2	2	2	0

So $2x^3 - x^2 - x - 3 = \left(x - \tfrac{3}{2}\right)\left(2x^2 + 2x + 2\right) = 2\left(x - \tfrac{3}{2}\right)\left(x^2 + x + 1\right)$. Using the quadratic equation, we have:

$$x = \frac{-1 \pm \sqrt{1-4}}{2} = \frac{-1 \pm i\sqrt{3}}{2}$$

So the only rational root is $\tfrac{3}{2}$.

(b) All solutions are $\tfrac{3}{2}, \frac{-1+i\sqrt{3}}{2}$, and $\frac{-1-i\sqrt{3}}{2}$.

13. Let $f(x) = 3x^4 + x^2 - 5x - 1$. Since there is one sign change, there is 1 positive root. $f(-x) = 3x^4 + x^2 + 5x - 1$. Since there is one sign change, there is 1 negative root. So the equation has 1 positive real root, 1 negative real root, and 2 complex roots.

14. The factors are $x + 2$, $x - (1 - 3i)$, and $x - (1 + 3i)$, so:

$$(x+2)[x-(1-3i)][x-(1+3i)]=0$$
$$(x+2)(x^2-2x+10)=0$$
$$x^3+6x+20=0$$

15. Here $f(x)=(x-2)(x-3i)^3[x-(1+\sqrt{2})]^2$. Note that had a restriction of rational coefficients been added, we would also have $-3i$ (multiplicity 3) and $1-\sqrt{2}$ (multiplicity 2) as roots.

Chapter Nine
The Conic Sections

9.1 The Parabola

1. We note that $4p = 4$, so $p = 1$. So the focus is $(0, 1)$, the directrix is $y = -1$, and the focal width is 4.

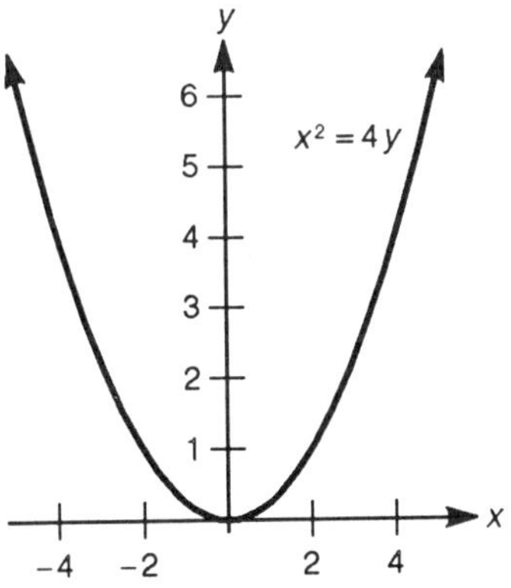

3. We note that $4p = 8$, so $p = 2$. So the focus is $(-2, 0)$, the directrix is $x = 2$, and the focal width is 8.

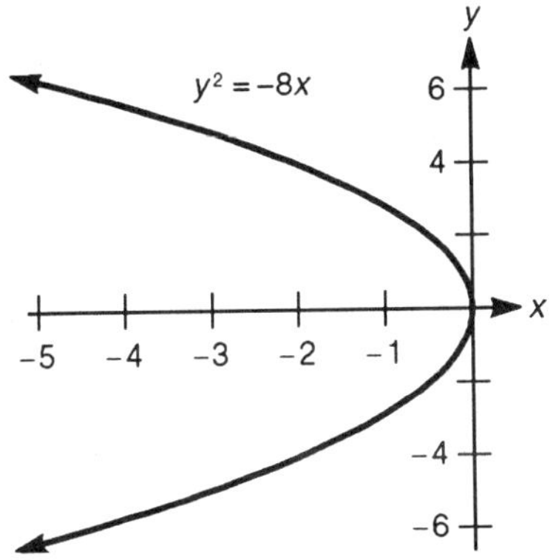

5. We note that $4p = 20$, so $p = 5$. So the focus is $(0, -5)$, the directrix is $y = 5$, and the focal width is 20.

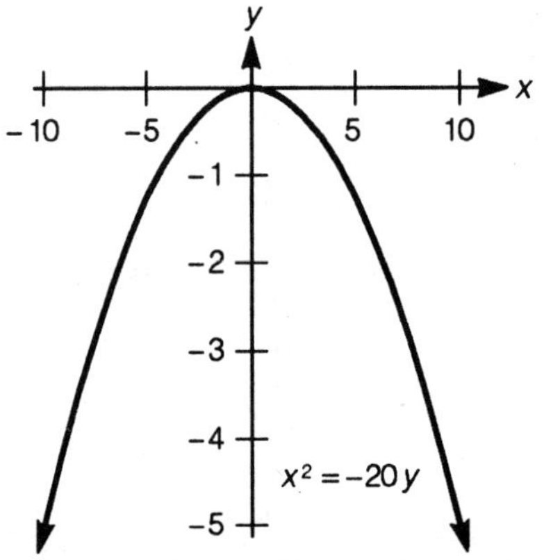

7. Since $y^2 = -28x$, then $4p = 28$ so $p = 7$. So the focus is $(-7, 0)$, the directrix is $x = 7$, and the focal width is 28.

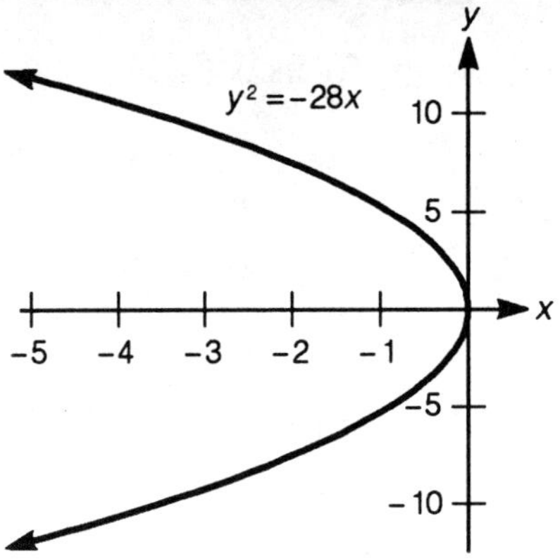

9. We note that $4p = 6$, so $p = \frac{3}{2}$. So the focus is $\left(0, \frac{3}{2}\right)$, the directrix is $y = -\frac{3}{2}$, and the focal width is 6.

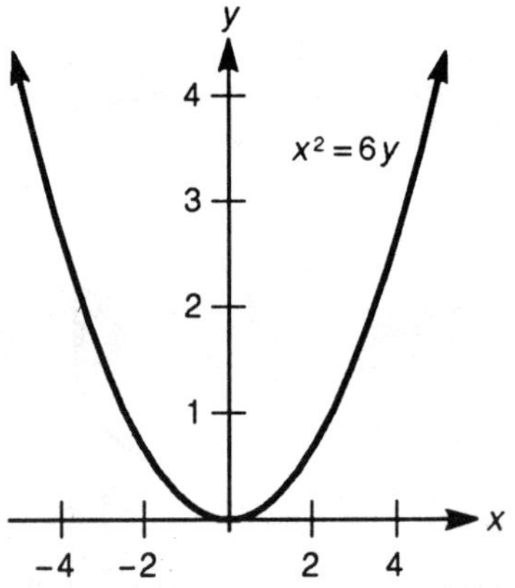

11. Since $x^2 = \frac{7}{4}y$, then $4p = \frac{7}{4}$ so $p = \frac{7}{16}$. So the focus is $\left(0, \frac{7}{16}\right)$, the directrix is $y = -\frac{7}{16}$, and the focal width is $\frac{7}{4}$.

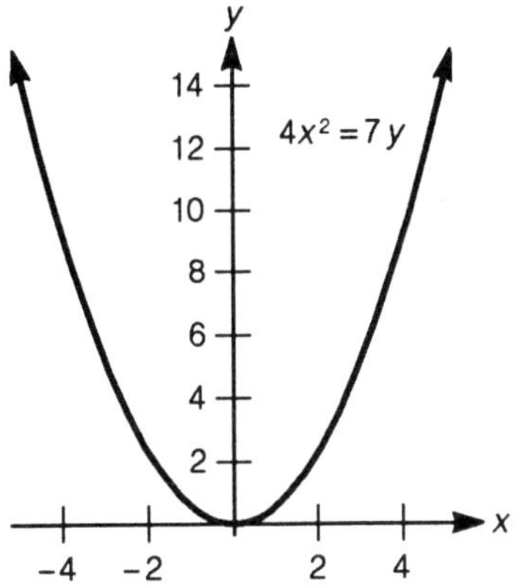

13. We first complete the square:

$$y^2 - 6y - 4x + 17 = 0$$
$$y^2 - 6y = 4x - 17$$
$$y^2 - 6y + 9 = 4x - 17 + 9$$
$$(y-3)^2 = 4x - 8$$
$$(y-3)^2 = 4(x-2)$$

We note that $4p = 4$, so $p = 1$. The vertex is $(2, 3)$, the focus is $(3, 3)$, the directrix is $x = 1$, and the focal width is 4.

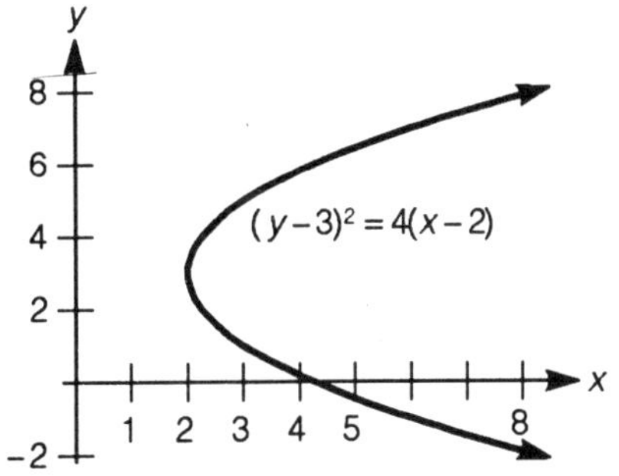

15. We first complete the square:

$$x^2 - 8x - y + 18 = 0$$
$$x^2 - 8x = y - 18$$
$$x^2 - 8x + 16 = y - 18 + 16$$
$$(x-4)^2 = y - 2$$

We note that $4p = 1$, so $p = \frac{1}{4}$. The vertex is $(4, 2)$, the focus is $\left(4, \frac{9}{4}\right)$, the directrix is $y = \frac{7}{4}$, and the focal width is 1.

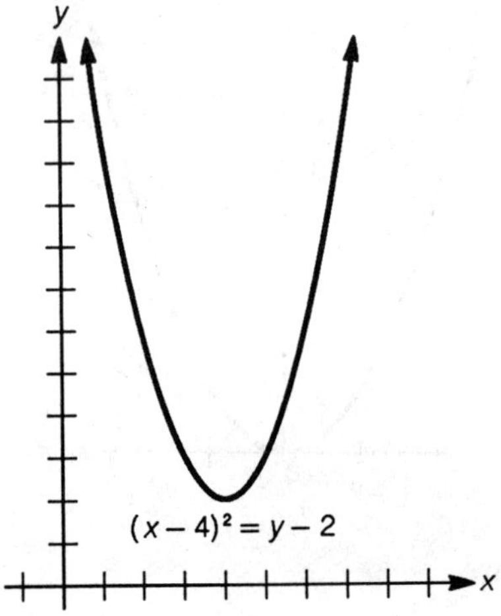

17. We first complete the square:

$$\begin{aligned} y^2 + 2y - x + 1 &= 0 \\ y^2 + 2y &= x - 1 \\ y^2 + 2y + 1 &= x - 1 + 1 \\ (y+1)^2 &= x \end{aligned}$$

We note that $4p = 1$, so $p = \frac{1}{4}$. The vertex is $(0, -1)$, the focus is $\left(\frac{1}{4}, -1\right)$, the directrix is $x = -\frac{1}{4}$, and the focal width is 1.

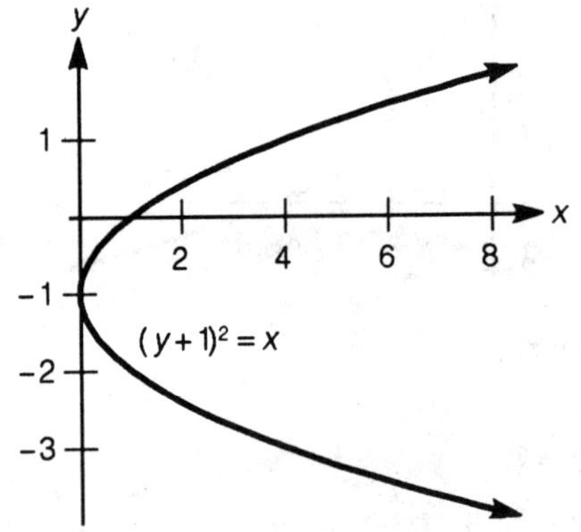

19. We first complete the square:

$$2x^2 - 12x - y + 18 = 0$$
$$2x^2 - 12x = y - 18$$
$$x^2 - 6x = \tfrac{1}{2}y - 9$$
$$x^2 - 6x + 9 = \tfrac{1}{2}y - 9 + 9$$
$$(x-3)^2 = \tfrac{1}{2}y$$

We note that $4p = \frac{1}{2}$, so $p = \frac{1}{8}$. The vertex is $(3,0)$, the focus is $\left(3, \frac{1}{8}\right)$, the directrix is $y = -\frac{1}{8}$, and the focal width is $\frac{1}{2}$.

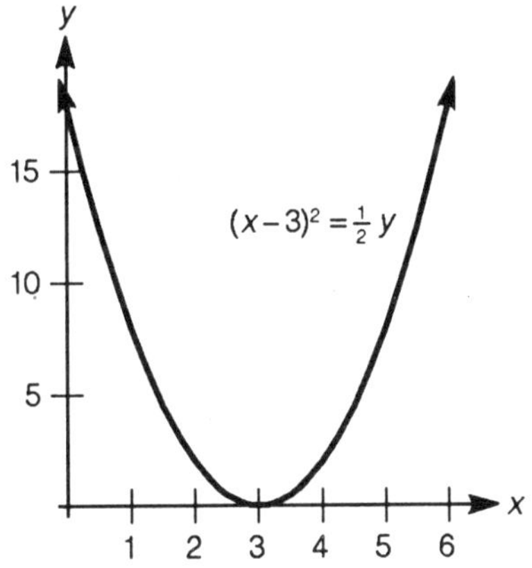

21. We first complete the square:

$$2x^2 - 16x - y + 33 = 0$$
$$2x^2 - 16x = y - 33$$
$$2(x-4)^2 = y - 1$$
$$(x-4)^2 = \tfrac{1}{2}(y-1)$$

We note that $4p = \frac{1}{2}$, so $p = \frac{1}{8}$. The vertex is $(4, 1)$, the focus is $\left(4, \frac{9}{8}\right)$, the directrix is $y = \frac{7}{8}$, and the focal width is $\frac{1}{2}$.

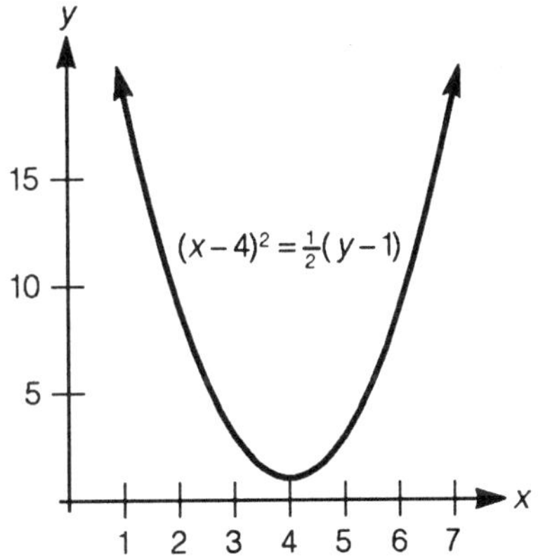

23. The line of symmetry has equation $y = 1$. The figure is graphed as:

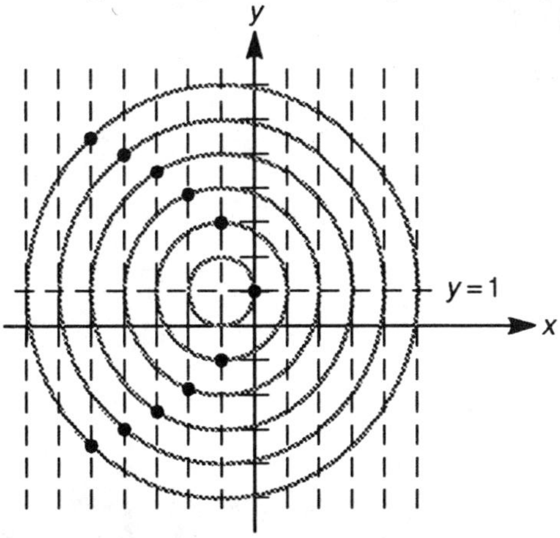

25. Given that the focus is at $(0,3)$, we see that $p = 3$ and thus $4p = 12$. Since the parabola opens up, it has the $x^2 = 4py$ form, and its equation must be $x^2 = 12y$.

27. The directrix is $x = -32$, so $(32,0)$ is the focus and the parabola opens to the right. So the parabola is of the form $y^2 = 4px$, where $p = 32$. Thus the equation is $y^2 = 128x$.

29. Since the parabola is symmetric about the x-axis, then $y^2 = \pm 4px$ is the form of the parabola. Since the x-coordinate of the focus is negative, then $y^2 = -4px$. Finally we are given $4p = 9$, so the equation is $y^2 = -9x$.

31. (a) Since $4p = 8$, then $p = 2$, and thus the focus is $(0,2)$. So the slope of the line is $\frac{8-2}{8-0} = \frac{3}{4}$, and thus the equation is $y = \frac{3}{4}x + 2$.

(b) We solve the system:

$$y = \frac{3}{4}x + 2$$
$$x^2 = 8y$$

Substituting:

$$x^2 = 8\left(\frac{3}{4}x + 2\right)$$
$$x^2 = 6x + 16$$
$$x^2 - 6x - 16 = 0$$
$$(x-8)(x+2) = 0$$
$$x = 8, -2$$
$$y = 8, \frac{1}{2}$$

Since $(8,8)$ repeats our original point, then Q must be $\left(-2, \frac{1}{2}\right)$.

(c) Use the distance formula:

$$PQ = \sqrt{(8+2)^2 + \left(8 - \tfrac{1}{2}\right)^2} = \sqrt{100 + \tfrac{225}{4}} = \sqrt{\tfrac{625}{4}} = \tfrac{25}{2}$$

(d) We know the radius is $\frac{25}{4}$, and the center of the circle will be the midpoint of the line segment PQ:

$$\text{center} = \left(\frac{8-2}{2}, \frac{8+\frac{1}{2}}{2}\right) = \left(3, \tfrac{17}{4}\right)$$

So the equation is $(x-3)^2 + \left(y - \frac{17}{4}\right)^2 = \frac{625}{16}$.

(e) The directrix is $y = -2$. The vertical distance of the center of the circle to the directrix is:

$$\left|\tfrac{17}{4} - (-2)\right| = \left|\tfrac{17}{4} + 2\right| = \tfrac{25}{4}$$

But the radius of the circle is also $\frac{25}{4}$, which implies that the directrix must be tangent to the circle.

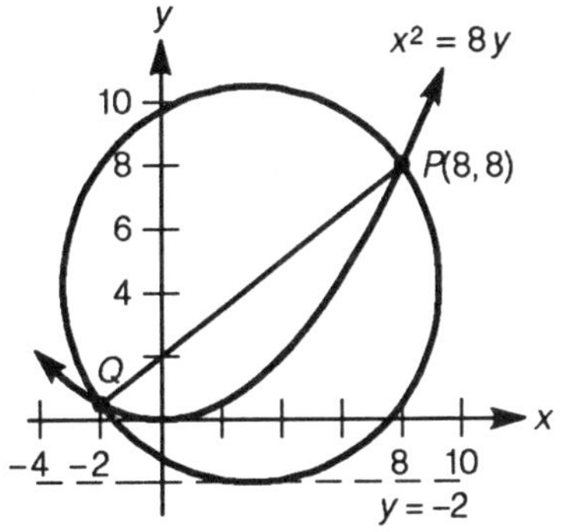

33. Call $(0,0)$ the vertex, thus the ends of the arch are at $(20,-15)$ and $(-20,-15)$. The parabola is of the form $x^2 = -4py$, so we substitute:

$$400 = -4p(-15)$$
$$-\tfrac{80}{3} = -4p$$

So the equation is $x^2 = -\frac{80}{3}y$. We wish to find the y-coordinate when the base is 20 ft, so we substitute $x = \pm 10$ into the equation:

$$100 = -\tfrac{80}{3}y$$
$$-\tfrac{15}{4} = y$$

So the height above the base is $15 - \frac{15}{4} = \frac{45}{4} = 11.25$ ft.

35. (a) We have $4p = 2$, so $p = \frac{1}{2}$, and thus the focus is $\left(0, \frac{1}{2}\right)$. To find A', we find the slope of the line containing A and the focus:

$$m = \frac{8 - \frac{1}{2}}{4 - 0} = \frac{\frac{15}{2}}{4} = \frac{15}{8}$$

So $y = \frac{15}{8}x + \frac{1}{2}$ is its equation. Now substitute into the parabola:

$$\begin{aligned} x^2 &= 2\left(\tfrac{15}{8}x + \tfrac{1}{2}\right) \\ x^2 &= \tfrac{15}{4}x + 1 \\ 4x^2 &= 15x + 4 \\ 4x^2 - 15x - 4 &= 0 \\ x &= -\tfrac{1}{4}, 4 \\ y &= \tfrac{1}{32}, 8 \end{aligned}$$

So $A' = \left(-\frac{1}{4}, \frac{1}{32}\right)$. To find B', we find the slope of the line containing B and the focus:

$$m = \frac{2 - \frac{1}{2}}{-2 - 0} = \frac{\frac{3}{2}}{-2} = -\frac{3}{4}$$

So $y = -\frac{3}{4}x + \frac{1}{2}$ is its equation. Now substitute into the parabola:

$$\begin{aligned} x^2 &= 2\left(-\tfrac{3}{4}x + \tfrac{1}{2}\right) \\ x^2 &= -\tfrac{3}{2}x + 1 \\ 2x^2 + 3x - 2 &= 0 \\ x &= \tfrac{1}{2}, -2 \\ y &= \tfrac{1}{8}, 2 \end{aligned}$$

So $B' = \left(\frac{1}{2}, \frac{1}{8}\right)$. We now find the slope through A and B':

$$m = \frac{8 - \frac{1}{8}}{4 - \frac{1}{2}} = \frac{\frac{63}{8}}{\frac{7}{2}} = \frac{9}{4}$$

Using $(4, 8)$ in the point-slope formula:

$$\begin{aligned} y - 8 &= \tfrac{9}{4}(x - 4) \\ y - 8 &= \tfrac{9}{4}x - 9 \\ y &= \tfrac{9}{4}x - 1 \end{aligned}$$

(b) We find the slope:

$$m = \frac{2 - \frac{1}{32}}{-2 + \frac{1}{4}} = \frac{\frac{63}{32}}{-\frac{7}{4}} = -\frac{9}{8}$$

Using (–2, 2) in the point-slope formula:

$$y-2=-\tfrac{9}{8}(x+2)$$
$$y-2=-\tfrac{9}{8}x-\tfrac{9}{4}$$
$$y=-\tfrac{9}{8}x-\tfrac{1}{4}$$

(c) Setting the y-coordinates equal:

$$\tfrac{9}{4}x-1=-\tfrac{9}{8}x-\tfrac{1}{4}$$
$$18x-8=-9x-2$$
$$27x=6$$
$$x=\tfrac{2}{9}$$
$$y=\tfrac{1}{2}-1=-\tfrac{1}{2}$$

So the lines intersect at the point $\left(\frac{2}{9},-\frac{1}{2}\right)$. Since the directrix is $y=-\frac{1}{2}$, then this point lies on the directrix.

37. (a) Since $4p=1$, $p=\frac{1}{4}$, thus the focus is $\left(0,\frac{1}{4}\right)$. We find the slope of the focal chord from P:

$$m=\frac{4-\frac{1}{4}}{2-0}=\frac{\frac{15}{4}}{2}=\frac{15}{8}$$

So its equation is $y=\frac{15}{8}x+\frac{1}{4}$. We now find the intersection of this line with the parabola $y=x^2$:

$$x^2=\tfrac{15}{8}x+\tfrac{1}{4}$$
$$8x^2=15x+2$$
$$8x^2-15x-2=0$$
$$(8x+1)(x-2)=0$$
$$x=-\tfrac{1}{8},2$$
$$y=\tfrac{1}{64},4$$

So the coordinates of Q are $Q\left(-\frac{1}{8},\frac{1}{64}\right)$.

(b) We find the midpoint of PQ:

$$M=\left(\frac{2-\frac{1}{8}}{2},\frac{4+\frac{1}{64}}{2}\right)=\left(\frac{15}{16},\frac{257}{128}\right)$$

(c) The coordinates of S are $\left(0, \frac{257}{128}\right)$. We can find T, since the slope of this line (which is perpendicular to PQ) is $-\frac{8}{15}$, and thus by the point-slope formula:

$$y - \tfrac{257}{128} = -\tfrac{8}{15}\left(x - \tfrac{15}{16}\right)$$
$$y - \tfrac{257}{128} = -\tfrac{8}{15}x + \tfrac{1}{2}$$
$$y = -\tfrac{8}{15}x + \tfrac{321}{128}$$

When $x = 0$, $y = \frac{321}{128}$, so the coordinates of T are $\left(0, \frac{321}{128}\right)$. Thus the length of ST is $\frac{321}{128} - \frac{257}{128} = \frac{64}{128} = \frac{1}{2}$. Since the focal width is $4p = 1$, this verifies that ST is one-half the focal width.

39. The sketch in the book will have the following coordinates:

$$O(0,0),\ B\left(x, \frac{x^2}{4p}\right),\ A\left(-x, \frac{x^2}{4p}\right)$$

We require that $AB = OB$. So:

$$2x = \sqrt{(x-0)^2 + \left(\frac{x^2}{4p} - 0\right)^2}$$
$$2x = \sqrt{x^2 + \frac{x^4}{16p^2}}$$

Squaring, we have:

$$4x^2 = x^2 + \frac{x^4}{16p^2}$$
$$0 = -3x^2 + \frac{x^4}{16p^2}$$
$$0 = -48x^2p^2 + x^4$$
$$0 = x^2\left(-48p^2 + x^2\right)$$

So $x^2 = 0$ or $x = 0$ is one root and $x^2 = 48p^2$ or $x = \pm 4\sqrt{3}p$ is the other. The distance from A to B is therefore:

$$2\left(4\sqrt{3}p\right) = 8\sqrt{3}p \text{ units}$$

To find the area we note that this is an equilateral triangle. Half of it is a 30°- 60°- 90° triangle whose height will be $\sqrt{3}$ times its base. The height is therefore $\sqrt{3} \bullet 4\sqrt{3}p = 12p$. The area is:

$$\tfrac{1}{2}\left(8\sqrt{3}p\right)(12p) = 48\sqrt{3}p^2 \text{ sq. units}$$

41. Since the focus is $(p, 0)$, then the slope of the focal chord is:

$$m = \frac{y_0 - 0}{\frac{y_0^2}{4p} - p} = \frac{4py_0}{y_0^2 - 4p^2}$$

Using the point-slope formula, the equation of this focal chord is:

$$y - y_0 = m\left(x - \frac{y_0^2}{4p}\right)$$

This line intersects the parabola when $x = \frac{y^2}{4p}$, so:

$$y - y_0 = m\left(\frac{y^2}{4p} - \frac{y_0^2}{4p}\right)$$

$$4p(y - y_0) = m(y + y_0)(y - y_0)$$

$$4p = m(y + y_0)$$

Substitute $m = \dfrac{4py_0}{y_0^2 - 4p^2}$:

$$4p = \frac{4py_0}{y_0^2 - 4p^2}(y + y_0)$$

$$\frac{y_0^2 - 4p^2}{y_0} = y + y_0$$

$$-\frac{4p^2}{y_0} = y$$

Therefore:

$$x = \frac{1}{4p}\left(-\frac{4p^2}{y_0}\right)^2 = \frac{1}{4p}\left(\frac{16p^4}{y_0^2}\right) = \frac{4p^3}{y_0^2}$$

But $y_0^2 = 4px_0$, so:

$$x = \frac{4p^3}{4px_0} = \frac{p^2}{x_0}$$

Thus the coordinates of Q are $\left(\frac{p^2}{x_0}, -\frac{4p^2}{y_0}\right)$.

43. Let P have coordinates (x_0, y_0). Then, from Exercise 41, Q has coordinates $\left(\frac{p^2}{x_0}, -\frac{4p^2}{y_0}\right)$. The following results summarize the calculations required for this problem. Note: At each step we have replaced the quantity y_0^2 by $4px_0$. This simplifies matters a great deal.

Coordinates of M: $\left(\dfrac{x_0^2 + p^2}{2x_0}, \dfrac{2p(x_0 - p)}{y_0}\right)$

Coordinates of S: $\left(\dfrac{p^2 + x_0^2}{2x_0}, 0\right)$

Slope of PQ: $\dfrac{4px_0}{y_0(x_0 - p)}$

Slope of perpendicular: $\dfrac{y_0(p - x_0)}{4px_0}$

Since $-\frac{ST}{MS}$ is the slope, m, of the line through M and T, we have:

$$ST = -(MS)m = -\frac{2p(x_0 - p)}{y_0} \bullet \frac{4px_0}{y_0(p - x_0)} = \frac{8p^2x_0}{y_0^2} = 2p$$

Thus ST is one-half of the focal width.

9.2 The Ellipse

1. Dividing by 36, the standard form is $\frac{x^2}{9} + \frac{y^2}{4} = 1$. The length of the major axis is 6 and the length of the minor axis is 4. Using $c^2 = a^2 - b^2$, we find:

$$c = \sqrt{a^2 - b^2} = \sqrt{9 - 4} = \sqrt{5}$$

So the foci are $(\pm\sqrt{5}, 0)$ and the eccentricity is $\frac{\sqrt{5}}{3}$.

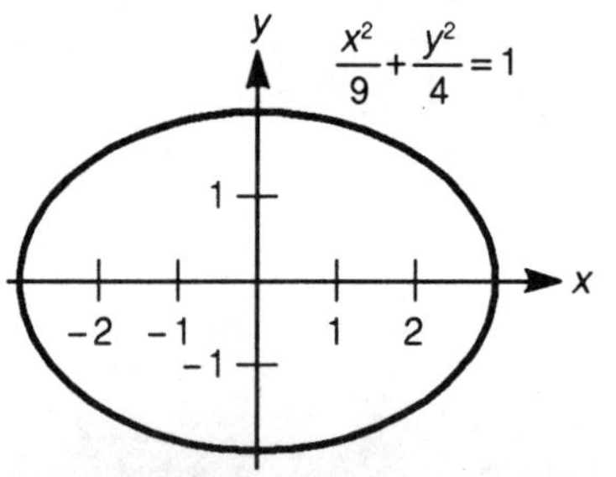

3. Dividing by 16, the standard form is $\frac{x^2}{16}+\frac{y^2}{1}=1$. The length of the major axis is 8 and the length of the minor axis is 2. Using $c^2=a^2-b^2$, we find:

$$c=\sqrt{a^2-b^2}=\sqrt{16-1}=\sqrt{15}$$

So the foci are $(\pm\sqrt{15},0)$ and the eccentricity is $\frac{\sqrt{15}}{4}$.

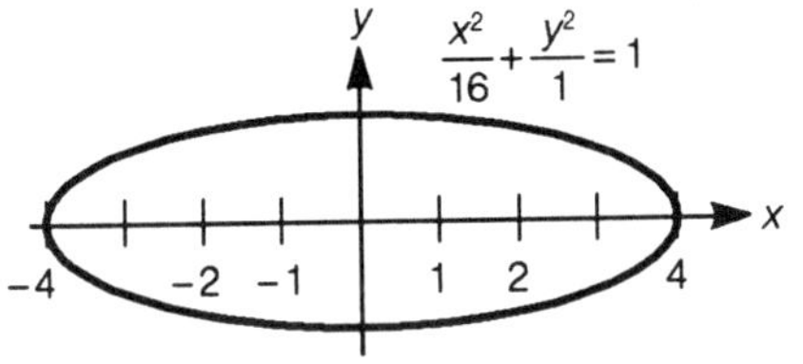

5. Dividing by 2, the standard form is $\frac{x^2}{2}+\frac{y^2}{1}=1$. The length of the major axis is $2\sqrt{2}$ and the length of the minor axis is 2. Using $c^2=a^2-b^2$, we find:

$$c=\sqrt{a^2-b^2}=\sqrt{2-1}=1$$

So the foci are $(\pm 1,0)$ and the eccentricity is $\frac{\sqrt{2}}{2}$.

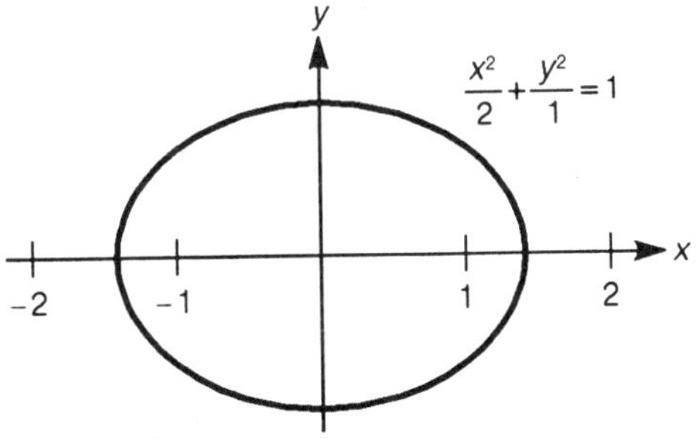

7. Dividing by 144, the standard form is $\frac{x^2}{9}+\frac{y^2}{16}=1$. The length of the major axis is 8 and the length of the minor axis is 6. Using $c^2=a^2-b^2$, we find:

$$c=\sqrt{a^2-b^2}=\sqrt{16-9}=\sqrt{7}$$

So the foci are $(0,\pm\sqrt{7})$ and the eccentricity is $\frac{\sqrt{7}}{4}$.

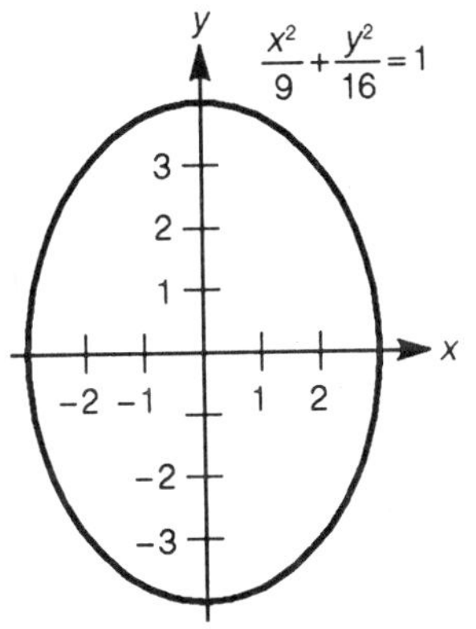

9. Dividing by 5, the standard form is $\frac{x^2}{1/3}+\frac{y^2}{5/3}=1$. The lengths of the major and minor axes are given by:

major axis: $2\sqrt{\frac{5}{3}}=\frac{2\sqrt{5}}{\sqrt{3}}=\frac{2\sqrt{15}}{3}$

minor axis: $2\sqrt{\frac{1}{3}}=\frac{2}{\sqrt{3}}=\frac{2\sqrt{3}}{3}$

Using $c^2=a^2-b^2$, we find:

$$c=\sqrt{\frac{5}{3}-\frac{1}{3}}=\sqrt{\frac{4}{3}}=\frac{2}{\sqrt{3}}=\frac{2\sqrt{3}}{3}$$

So the foci are $\left(0,\pm\frac{2\sqrt{3}}{3}\right)$. The eccentricity is given by:

$$e=\frac{\frac{2\sqrt{3}}{3}}{\frac{\sqrt{15}}{3}}=\frac{2\sqrt{3}}{\sqrt{15}}=\frac{2}{\sqrt{5}}=\frac{2\sqrt{5}}{5}$$

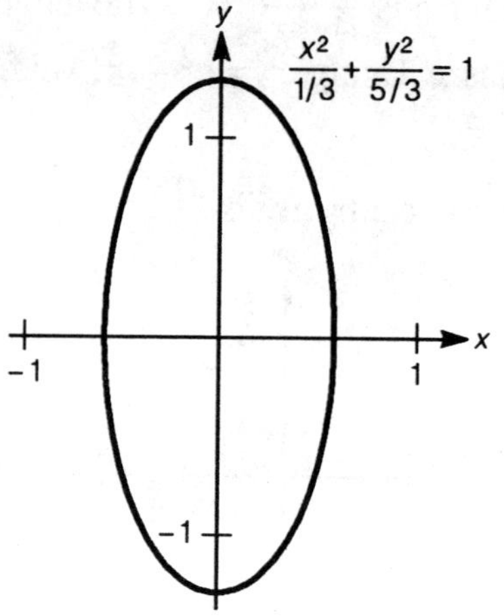

11. Dividing by 4, the standard form is $\frac{x^2}{2}+\frac{y^2}{4}=1$. The length of the major axis is 4 and the length of the minor axis is $2\sqrt{2}$. Using $c^2=a^2-b^2$, we find:

$$c=\sqrt{a^2-b^2}=\sqrt{4-2}=\sqrt{2}$$

So the foci are $(0,\pm\sqrt{2})$ and the eccentricity is $\frac{\sqrt{2}}{2}$.

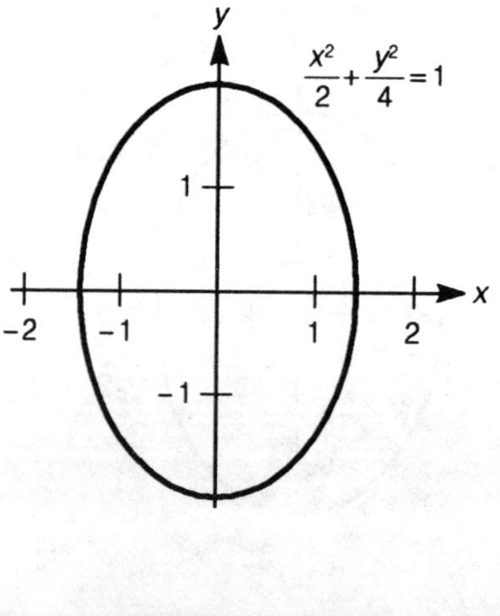

13. The equation is already in standard form with a center of $(5,-1)$. The length of the major axis is 10 and the length of the minor axis is 6. Using $c^2 = a^2 - b^2$, we find:

$$c = \sqrt{a^2 - b^2} = \sqrt{25-9} = \sqrt{16} = 4$$

So the foci are $(5+4,-1) = (9,-1)$ and $(5-4,-1) = (1,-1)$, and the eccentricity is $\frac{4}{5}$.

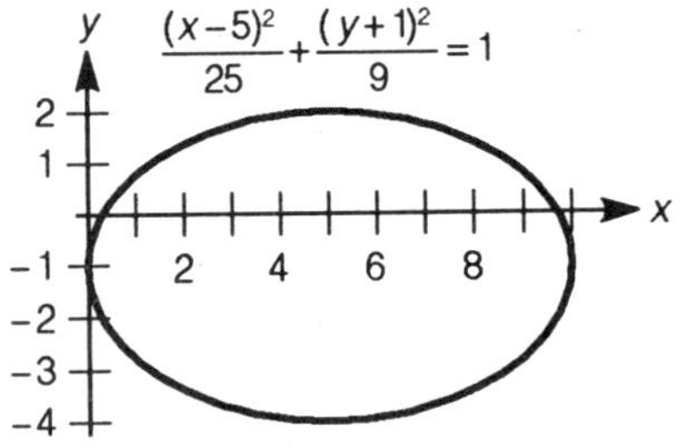

15. The equation is already in standard form with a center of $(1,2)$. The length of the major axis is 4 and the length of the minor axis is 2. Using $c^2 = a^2 - b^2$, we find:

$$c = \sqrt{a^2 - b^2} = \sqrt{4-1} = \sqrt{3}$$

So the foci are $\left(1, 2 \pm \sqrt{3}\right)$ and the eccentricity is $\frac{\sqrt{3}}{2}$.

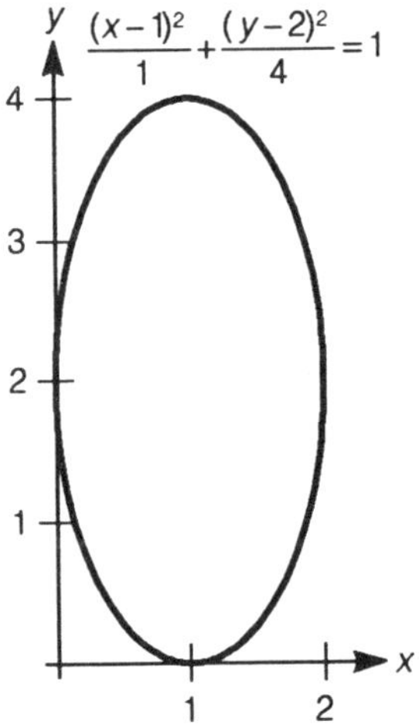

17. The equation is already in standard form with a center of $(-3,0)$. The length of the major axis is 6 and the length of the minor axis is 2. Using $c^2 = a^2 - b^2$, we find:

$$c = \sqrt{a^2 - b^2} = \sqrt{9-1} = \sqrt{8} = 2\sqrt{2}$$

So the foci are $\left(-3 \pm 2\sqrt{2}, 0\right)$ and the eccentricity is $\frac{2\sqrt{2}}{3}$.

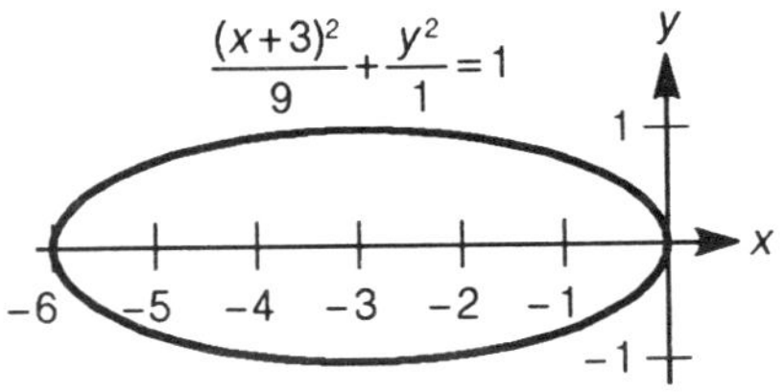

19. We complete the square to convert the equation to standard form:

$$3x^2+4y^2-6x+16y+7=0$$
$$3(x^2-2x)+4(y^2+4y)=-7$$
$$3(x^2-2x+1)+4(y^2+4y+4)=-7+3+16$$
$$3(x-1)^2+4(y+2)^2=12$$
$$\frac{(x-1)^2}{4}+\frac{(y+2)^2}{3}=1$$

The center is $(1,-2)$, the length of the major axis is 4, and the length of the minor axis is $2\sqrt{3}$. Using $c^2=a^2-b^2$, we find:

$$c=\sqrt{a^2-b^2}=\sqrt{4-3}=1$$

So the foci are $(1+1,-2)=(2,-2)$ and $(1-1,-2)=(0,-2)$, and the eccentricity is $\frac{1}{2}$.

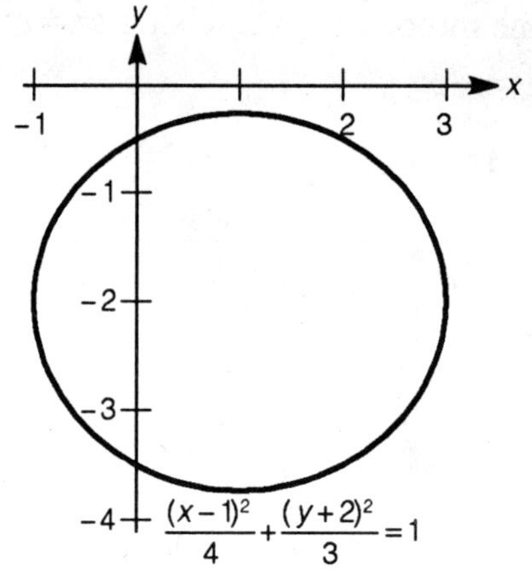

21. We complete the square to convert the equation to standard form:

$$5x^2+3y^2-40x-36y+188=0$$
$$5(x^2-8x)+3(y^2-12y)=-188$$
$$5(x^2-8x+16)+3(y^2-12y+36)=-188+80+108$$
$$5(x-4)^2+3(y-6)^2=0$$

Notice that the only solution to this is the center $(4,6)$. This is called a degenerate ellipse, or, more commonly, a point!

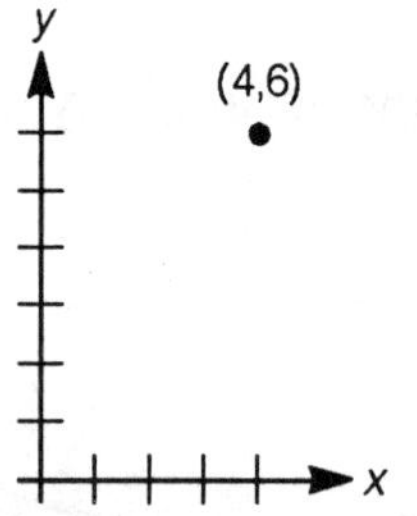

23. We complete the square to convert the equation to standard form:

$$16x^2+25y^2-64x-100y+564=0$$
$$16(x^2-4x)+25(y^2-4y)=-564$$
$$16(x^2-4x+4)+25(y^2-4y+4)=-564+64+100$$
$$16(x-2)^2+25(y-2)^2=-400$$

Notice that there is no solution to this equation, since the left-hand side is non-negative. So there is no graph.

25. We are given $c = 3$ and $a = 5$, so we have the equation in the form:

$$\frac{x^2}{5^2}+\frac{y^2}{b^2}=1$$

Since $c^2 = a^2 - b^2$, we find b:

$$9=25-b^2$$
$$16=b^2$$
$$4=b$$

So the equation is $\frac{x^2}{25}+\frac{y^2}{16}=1$, or $16x^2 + 25y^2 = 400$.

27. We are given $a = 4$, so the equation has a form of:

$$\frac{x^2}{16}+\frac{y^2}{b^2}=1$$

Now $\frac{c}{a}=\frac{1}{4}$, so $\frac{c}{4}=\frac{1}{4}$ and thus $c = 1$.
We find b:

$$c^2=a^2-b^2$$
$$1=16-b^2$$
$$b^2=15$$

So the equation is $\frac{x^2}{16}+\frac{y^2}{15}=1$, or $15x^2 + 16y^2 = 240$.

29. We have $c = 2$ and $a = 5$, so the equation has a form of:

$$\frac{x^2}{b^2}+\frac{y^2}{25}=1$$

We find b:

$$c^2=a^2-b^2$$
$$4=25-b^2$$
$$21=b^2$$

So the equation is $\frac{x^2}{21}+\frac{y^2}{25}=1$, or $25x^2+21y^2=525$.

31. We know $a = 2b$ and that the equation has a form of:

$$\frac{x^2}{a^2}+\frac{y^2}{b^2}=1$$

Using the point $(1,\sqrt{2})$ and $a = 2b$, we have:

$$\frac{(1)^2}{(2b)^2}+\frac{(\sqrt{2})^2}{b^2}=1$$

$$\frac{1}{4b^2}+\frac{2}{b^2}=1$$

Multiply by $4b^2$:

$$1+8=4b^2$$

$$\tfrac{9}{4}=b^2$$

$$\tfrac{3}{2}=b$$

Since $a = 2b$, then $a = 3$. So the equation is $\frac{x^2}{9}+\frac{y^2}{9/4}=1$, or $x^2+4y^2=9$.

33. We begin by solving for y:

$$\frac{x^2}{9}+\frac{y^2}{4}=1$$

$$\frac{y^2}{4}=1-\frac{x^2}{9}$$

$$\frac{y^2}{4}=\frac{9-x^2}{9}$$

$$y^2=\frac{36-4x^2}{9}$$

$$y=\pm\sqrt{\frac{36-4x^2}{9}}$$

$$y=\pm\frac{\sqrt{36-4x^2}}{3}$$

We use a calculator to complete the table:

x	0	0.5	1.0	1.5	2.0	2.5	3.0
y	±2	±1.97	±1.89	±1.73	±1.49	±1.11	0

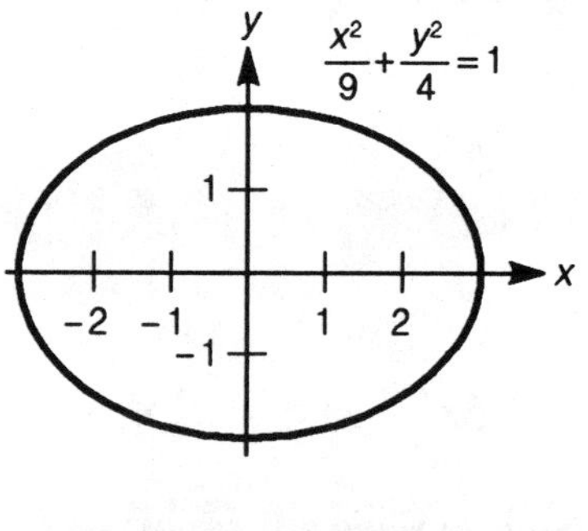

35. (a) Squaring both sides and simplifying, we obtain:

$$\sqrt{(x-c)^2+y^2}=2a-\sqrt{(x+c)^2+y^2}$$
$$(x-c)^2+y^2=4a^2-4a\sqrt{(x+c)^2+y^2}+(x+c)^2+y^2$$
$$x^2-2xc+c^2+y^2=4a^2-4a\sqrt{(x+c)^2+y^2}+x^2+2cx+c^2+y^2$$
$$-4cx-4a^2=-4a\sqrt{(x+c)^2+y^2}$$
$$a^2+cx=a\sqrt{(x+c)^2+y^2}$$

(b) Squaring each side again, we obtain:

$$\left(a^2+cx\right)^2=a^2\left((x+c)^2+y^2\right)$$
$$a^4+2a^2cx+c^2x^2=a^2\left(x^2+2cx+c^2+y^2\right)$$
$$a^4+2a^2cx+c^2x^2=a^2x^2+2a^2cx+a^2c^2+a^2y^2$$
$$a^4-a^2c^2=a^2x^2-c^2x^2+a^2y^2$$

(c) We simply factor each side of the equation to obtain:

$$a^2\left(a^2-c^2\right)=\left(a^2-c^2\right)x^2+a^2y^2$$

(d) Call $b^2=a^2-c^2$, so we can rewrite the equation from (c) as:

$$b^2x^2+a^2y^2=a^2b^2$$
$$\frac{x^2}{a^2}+\frac{y^2}{b^2}=1$$

37. We use $a=5$ and $b=3$ in each formula:

	Approximation Obtained	Percentage Error
C_1	25.531776	0.019
C_2	25.526986	0.000051
C_3	25.519489	0.029

Since C_2 has the smallest percentage error, it would be the best approximation for the circumference of the three.

9.3 The Hyperbola

1. Dividing by 4, the standard form is $\frac{x^2}{4}-\frac{y^2}{1}=1$. The vertices are $(\pm 2,0)$, the length of the transverse axis is 4, the length of the conjugate axis is 2, and the asymptotes are $y=\pm\frac{1}{2}x$. Using $c^2=a^2+b^2$, we find:

$$c=\sqrt{a^2+b^2}=\sqrt{4+1}=\sqrt{5}$$

So the foci are $(\pm\sqrt{5},0)$ and the eccentricity is $\frac{\sqrt{5}}{2}$.

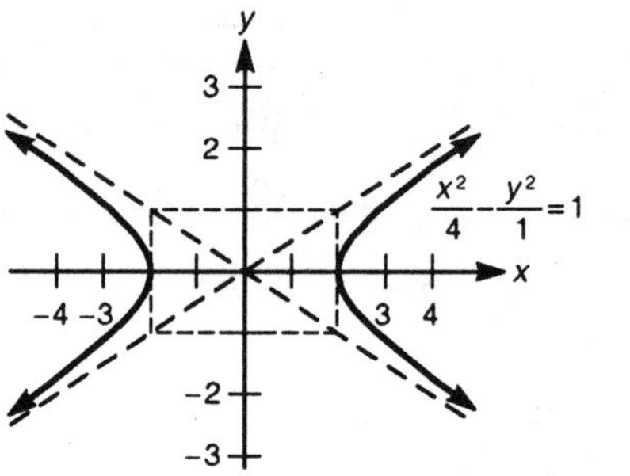

3. Dividing by 4, the standard form is $\frac{y^2}{4}-\frac{x^2}{1}=1$. The vertices are $(0,\pm 2)$, the length of the transverse axis is 4, the length of the conjugate axis is 2, and the asymptotes are $y=\pm 2x$. Using $c^2=a^2+b^2$, we find:

$$c=\sqrt{a^2+b^2}=\sqrt{4+1}=\sqrt{5}$$

So the foci are $(0,\pm\sqrt{5})$ and the eccentricity is $\frac{\sqrt{5}}{2}$.

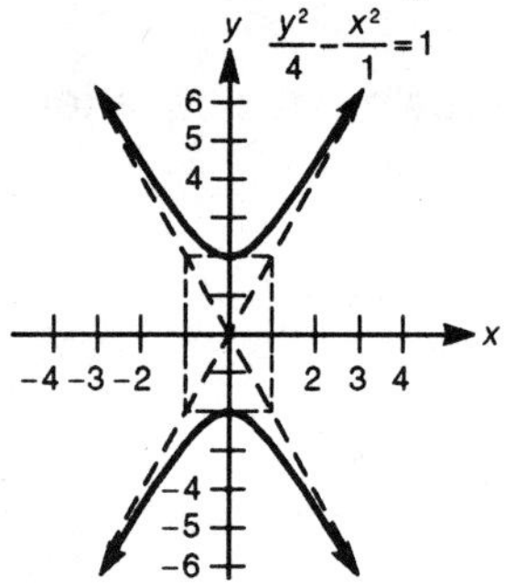

5. Dividing by 400, the standard form is $\frac{x^2}{25}-\frac{y^2}{16}=1$. The vertices are $(\pm 5,0)$, the length of the transverse axis is 10, the length of the conjugate axis is 8, and the asymptotes are $y=\pm\frac{4}{5}x$. Using $c^2=a^2+b^2$, we find:

$$c=\sqrt{a^2+b^2}=\sqrt{25+16}=\sqrt{41}$$

So the foci are $(\pm\sqrt{41},0)$ and the eccentricity is $\frac{\sqrt{41}}{5}$.

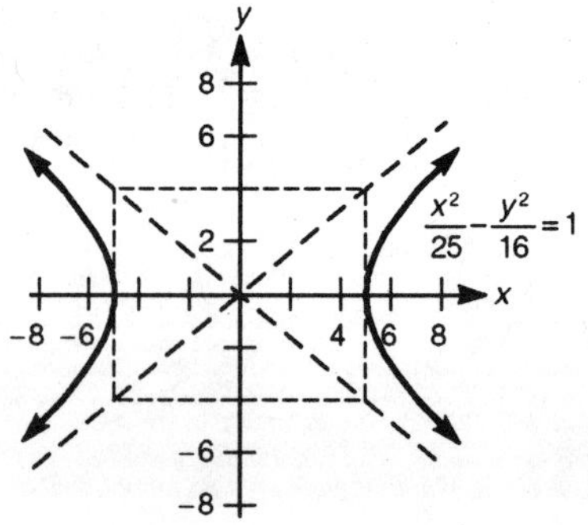

7. Rewriting the equation, the standard form is $\frac{y^2}{1/2} - \frac{x^2}{1/3} = 1$. The vertices are $\left(0, \pm\sqrt{\frac{1}{2}}\right) = \left(0, \pm\frac{\sqrt{2}}{2}\right)$, the length of the transverse axis is $\sqrt{2}$, the length of the conjugate axis is $2\sqrt{\frac{1}{3}} = \frac{2\sqrt{3}}{3}$, and the asymptotes are $y = \pm\sqrt{\frac{3}{2}}x = \pm\frac{\sqrt{6}}{2}x$. Using $c^2 = a^2 + b^2$, we find:

$$c = \sqrt{a^2 + b^2} = \sqrt{\frac{1}{2} + \frac{1}{3}} = \sqrt{\frac{5}{6}} = \frac{\sqrt{5}}{\sqrt{6}} = \frac{\sqrt{30}}{6}$$

So the foci are $\left(0, \pm\frac{\sqrt{30}}{6}\right)$ and the eccentricity is:

$$e = \frac{\frac{\sqrt{30}}{6}}{\frac{\sqrt{2}}{2}} = \frac{\sqrt{30}}{3\sqrt{2}} = \frac{\sqrt{15}}{3}$$

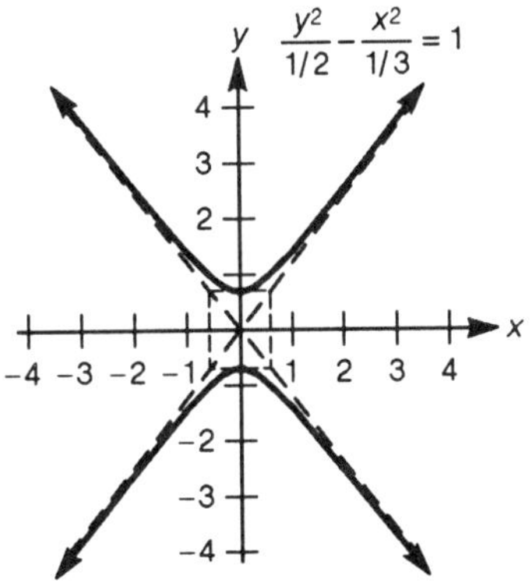

9. Dividing by 100, the standard form is $\frac{y^2}{25} - \frac{x^2}{4} = 1$. The vertices are $(0, \pm 5)$, the length of the transverse axis is 10, the length of the conjugate axis is 4, and the asymptotes are $y = \pm\frac{5}{2}x$. Using $c^2 = a^2 + b^2$, we have:

$$c = \sqrt{a^2 + b^2} = \sqrt{25 + 4} = \sqrt{29}$$

So the foci are $\left(0, \pm\sqrt{29}\right)$ and the eccentricity is $\frac{\sqrt{29}}{5}$.

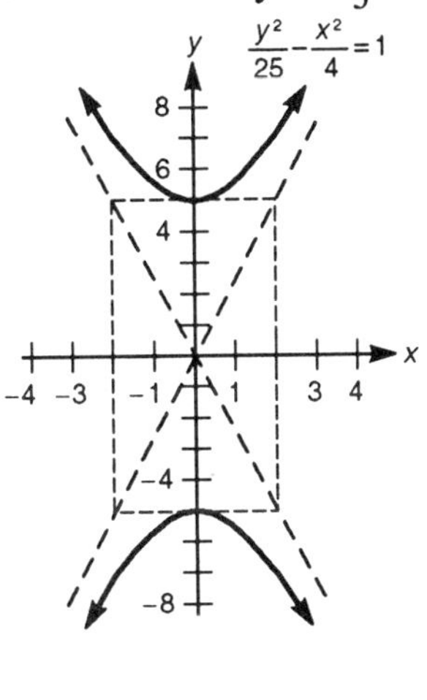

11. The equation is already in standard form with a center of $(5,-1)$. The vertices are $(5+5,-1)=(10,-1)$ and $(5-5,-1)=(0,-1)$, the length of the transverse axis is 10, and the length of the conjugate axis is 6. The asymptotes have slopes of $\pm\frac{3}{5}$, so using the point-slope formula:

$$y+1=\tfrac{3}{5}(x-5) \qquad y+1=-\tfrac{3}{5}(x-5)$$
$$y+1=\tfrac{3}{5}x-3 \qquad y+1=-\tfrac{3}{5}x+3$$
$$y=\tfrac{3}{5}x-4 \qquad y=-\tfrac{3}{5}x+2$$

Using $c^2=a^2+b^2$, we have:

$$c=\sqrt{a^2+b^2}=\sqrt{25+9}=\sqrt{34}$$

So the foci are $\left(5\pm\sqrt{34},-1\right)$ and the eccentricity is $\frac{\sqrt{34}}{5}$.

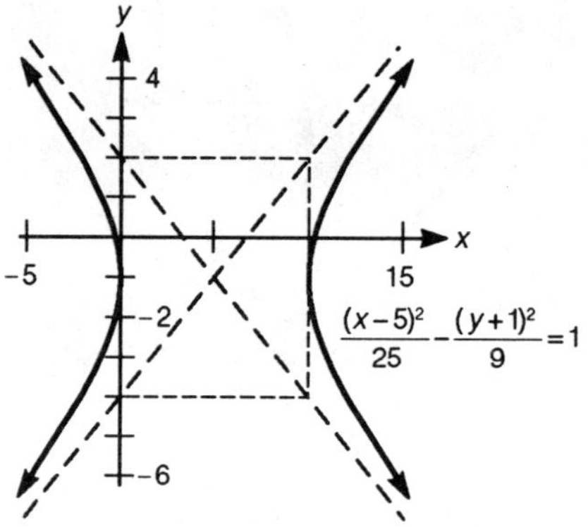

13. The equation is already in standard form with a center of $(1,2)$. The vertices are $(1,2+2)=(1,4)$ and $(1,2-2)=(1,0)$, the length of the transverse axis is 4, and the length of the conjugate axis is 2. The asymptotes have slopes of ±2, so using the point-slope formula:

$$y-2=2(x-1) \qquad y-2=-2(x-1)$$
$$y-2=2x-2 \qquad y-2=-2x+2$$
$$y=2x \qquad y=-2x+4$$

Using $c^2=a^2+b^2$, we have:

$$c=\sqrt{a^2+b^2}=\sqrt{4+1}=\sqrt{5}$$

So the foci are $\left(1,2\pm\sqrt{5}\right)$ and the eccentricity is $\frac{\sqrt{5}}{2}$.

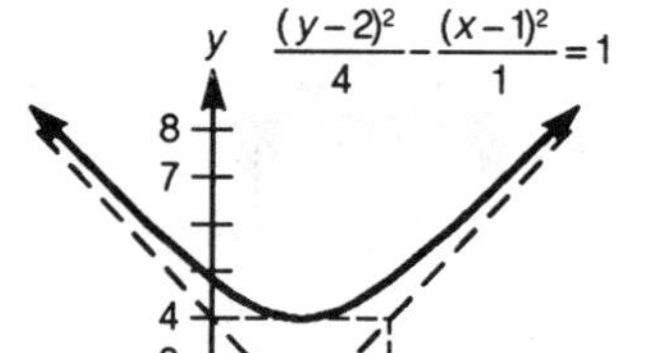

15. The equation is already in standard form with a center of $(-3,4)$. The vertices are $(-3+4,4)=(1,4)$ and $(-3-4,4)=(-7,4)$, the length of the transverse axis is 8, and the length of the conjugate axis is 8. The asymptotes have slopes of ± 1, so using the point-slope formula:

$$\begin{aligned} y-4 &= 1(x+3) \\ y-4 &= x+3 \\ y &= x+7 \end{aligned} \qquad\qquad \begin{aligned} y-4 &= -1(x+3) \\ y-4 &= -x-3 \\ y &= -x+1 \end{aligned}$$

Using $c^2 = a^2 + b^2$, we have:

$$c = \sqrt{a^2+b^2} = \sqrt{16+16} = \sqrt{32} = 4\sqrt{2}$$

So the foci are $\left(-3 \pm 4\sqrt{2}, 4\right)$ and the eccentricity is $\frac{4\sqrt{2}}{4} = \sqrt{2}$.

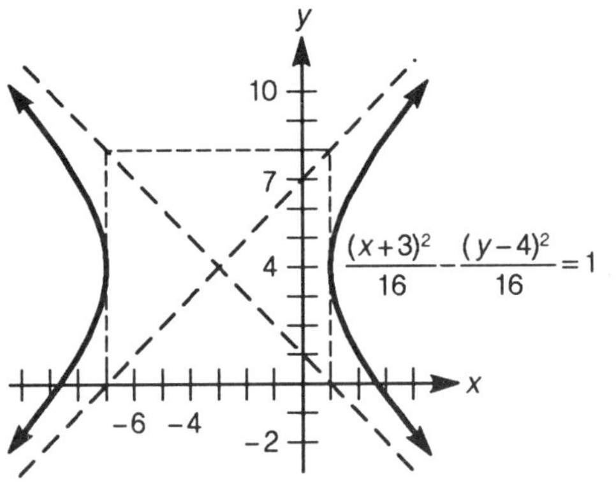

17. We complete the square to convert the equation to standard form:

$$\begin{aligned} x^2 - y^2 + 2y - 5 &= 0 \\ x^2 - \left(y^2 - 2y\right) &= 5 \\ x^2 - \left(y^2 - 2y + 1\right) &= 5 - 1 \\ x^2 - (y-1)^2 &= 4 \\ \frac{x^2}{4} - \frac{(y-1)^2}{4} &= 1 \end{aligned}$$

The center is $(0,1)$, the vertices are $(0+2,1)=(2,1)$ and $(0-2,1)=(-2,1)$, and the lengths of both the transverse and conjugate axes are 4. The asymptotes have slopes of ± 1, so using the point-slope formula:

$$\begin{aligned} y-1 &= 1(x-0) \\ y-1 &= x \\ y &= x+1 \end{aligned} \qquad\qquad \begin{aligned} y-1 &= -1(x-0) \\ y-1 &= -x \\ y &= -x+1 \end{aligned}$$

Using $c^2 = a^2 + b^2$, we have:

$$c = \sqrt{a^2+b^2} = \sqrt{4+4} = \sqrt{8} = 2\sqrt{2}$$

So the foci are $(\pm 2\sqrt{2},1)$ and the eccentricity is $\frac{2\sqrt{2}}{2} = \sqrt{2}$.

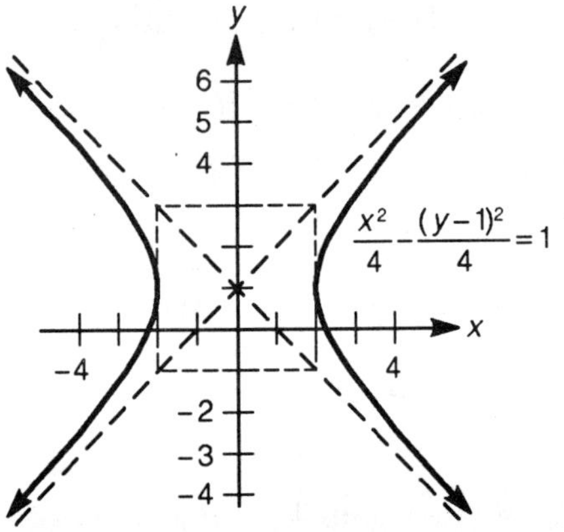

19. We complete the square to convert the equation to standard form:

$$x^2 - y^2 - 4x + 2y - 6 = 0$$
$$(x^2 - 4x) - (y^2 - 2y) = 6$$
$$(x^2 - 4x + 4) - (y^2 - 2y + 1) = 6 + 4 - 1$$
$$(x-2)^2 - (y-1)^2 = 9$$
$$\frac{(x-2)^2}{9} - \frac{(y-1)^2}{9} = 1$$

The center is (2, 1), the vertices are (2 + 3, 1) = (5, 1) and (2 – 3, 1) = (–1, 1), and the lengths of both the transverse and conjugate axes are 6. The asymptotes have slopes of ±1, so using the point-slope formula:

$$y - 1 = 1(x-2) \qquad y - 1 = -1(x-2)$$
$$y - 1 = x - 2 \qquad y - 1 = -x + 2$$
$$y = x - 1 \qquad y = -x + 3$$

Using $c^2 = a^2 + b^2$, we have:

$$c = \sqrt{a^2 + b^2} = \sqrt{9+9} = \sqrt{18} = 3\sqrt{2}$$

So the foci are $(2 \pm 3\sqrt{2},1)$ and the eccentricity is $\frac{3\sqrt{2}}{3} = \sqrt{2}$.

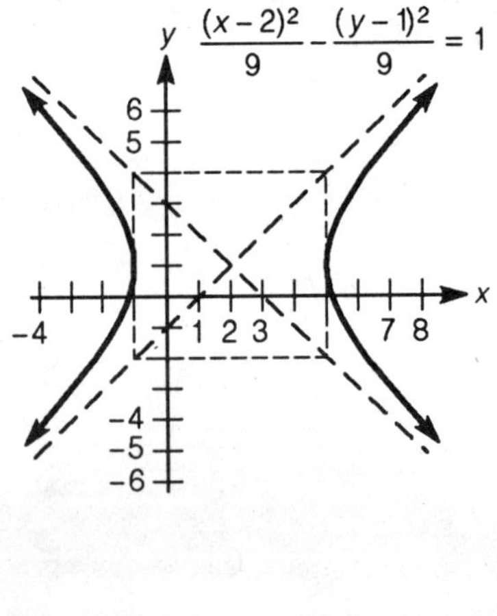

21. We complete the square to convert the equation to standard form:

$$y^2 - 25x^2 + 8y - 9 = 0$$
$$\left(y^2 + 8y\right) - 25x^2 = 9$$
$$\left(y^2 + 8y + 16\right) - 25x^2 = 9 + 16$$
$$(y+4)^2 - 25x^2 = 25$$
$$\frac{(y+4)^2}{25} - \frac{x^2}{1} = 1$$

The center is $(0,-4)$, the vertices are $(0,-4+5) = (0,1)$ and $(0,-4-5) = (0,-9)$, the length of the transverse axis is 10, and the length of the conjugate axis is 2. The asymptotes have slopes of ± 5, so using the point-slope formula:

$$y + 4 = 5(x-0) \qquad y + 4 = -5(x-0)$$
$$y + 4 = 5x \qquad y + 4 = -5x$$
$$y = 5x - 4 \qquad y = -5x - 4$$

Using $c^2 = a^2 + b^2$, we have:

$$c = \sqrt{a^2 + b^2} = \sqrt{25+1} = \sqrt{26}$$

So the foci are $\left(0, -4 \pm \sqrt{26}\right)$ and the eccentricity is $\frac{\sqrt{26}}{5}$.

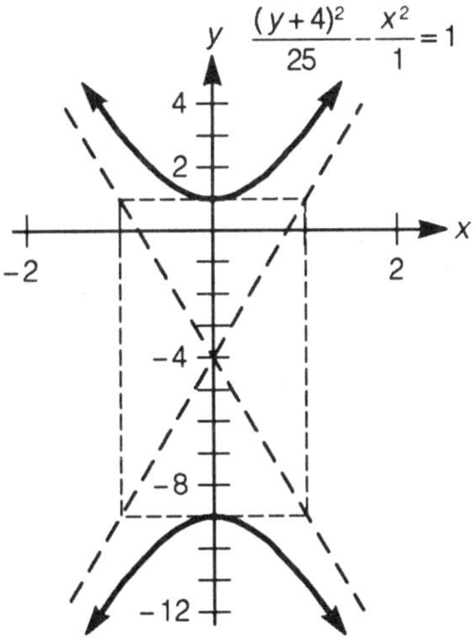

23. We complete the square to convert the equation to standard form:

$$x^2 + 7x - y^2 - y + 12 = 0$$
$$\left(x^2 + 7x\right) - \left(y^2 + y\right) = -12$$
$$\left(x^2 + 7x + \tfrac{49}{4}\right) - \left(y^2 + y + \tfrac{1}{4}\right) = -12 + \tfrac{49}{4} - \tfrac{1}{4}$$
$$\left(x + \tfrac{7}{2}\right)^2 - \left(y + \tfrac{1}{2}\right)^2 = 0$$

Notice that this is a degenerate hyperbola, and the graph consists of the "would-be" asymptotes with slopes ± 1:

$$y+\tfrac{1}{2}=1\left(x+\tfrac{7}{2}\right) \qquad y+\tfrac{1}{2}=-1\left(x+\tfrac{7}{2}\right)$$
$$y+\tfrac{1}{2}=x+\tfrac{7}{2} \qquad y+\tfrac{1}{2}=-x-\tfrac{7}{2}$$
$$y=x+3 \qquad y=-x-4$$

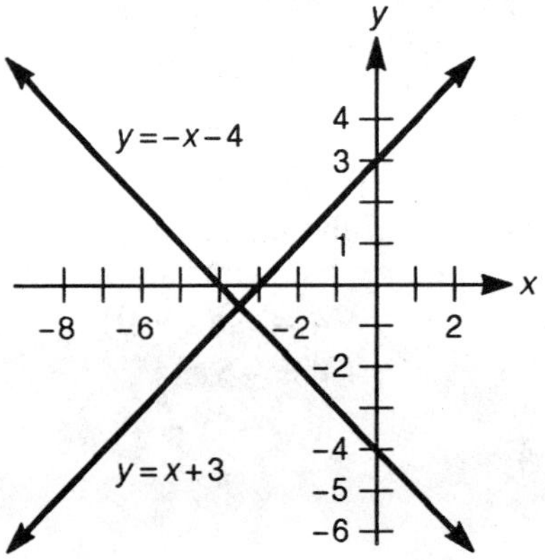

25. Since $P(x,y)$ lies on $\frac{x^2}{4}-\frac{y^2}{1}=1$, we can find y in terms of x:

$$y^2=\frac{x^2}{4}-1$$
$$y^2=\frac{x^2-4}{4}$$

Taking roots, we have $y=\frac{\sqrt{x^2-4}}{2}$, since $P(x,y)$ lies in the first quadrant. So the coordinates of P are $\left(x,\frac{\sqrt{x^2-4}}{2}\right)$. Since $Q(x,y)$ lies in the first quadrant on the asymptote, we find the equation of the asymptote:

$$y-0=\tfrac{1}{2}(x-0)$$
$$y=\tfrac{1}{2}x$$

So the coordinates of Q are $\left(x,\tfrac{1}{2}x\right)$. Since P and Q have the same x-coordinate PQ is the difference between their y-coordinates:

$$PQ=\frac{x}{2}-\frac{\sqrt{x^2-4}}{2}=\frac{x-\sqrt{x^2-4}}{2}$$

The order of subtraction is because the asymptote lies above the hyperbola in the first quadrant, and thus $\frac{x}{2}$ is larger than $\frac{\sqrt{x^2-4}}{2}$. This proves the desired result.

27. Since the foci are $(\pm 4,0)$ and the vertices are $(\pm 1,0)$, then $c = 4$, $a = 1$, and the hyperbola has the form:

$$\frac{x^2}{1} - \frac{y^2}{b^2} = 1$$

We find b:

$$c^2 = a^2 + b^2$$
$$16 = 1 + b^2$$
$$15 = b^2$$

So the equation is $\frac{x^2}{1} - \frac{y^2}{15} = 1$, or $15x^2 - y^2 = 15$.

29. The slope of the asymptotes is $\pm\frac{1}{2}$, which tells us that the ratio $\frac{b}{a} = \frac{1}{2}$ in this hyperbola. Also, since the vertices are $(\pm 2,0)$ then $a = 2$. The required ratio is therefore:

$$\frac{b}{2} = \frac{1}{2} \text{ and } b = 1$$

The equation is $\frac{x^2}{4} - \frac{y^2}{1} = 1$, or $x^2 - 4y^2 = 4$.

31. Since the asymptotes are $y = \pm\frac{\sqrt{10}}{5}x$, then $\frac{b}{a} = \frac{\sqrt{10}}{5}$, so $b = \frac{\sqrt{10}}{5}a$. Now, since the foci are $\left(\pm\sqrt{7}, 0\right)$, then $c = \sqrt{7}$ and the hyperbola has the form:

$$\frac{x^2}{a^2} - \frac{y^2}{b^2} = 1$$

Since $b = \frac{\sqrt{10}}{5}a$ and $c = \sqrt{7}$, we have:

$$c^2 = a^2 + b^2$$
$$7 = a^2 + \left(\frac{\sqrt{10}}{5}a\right)^2$$
$$7 = a^2 + \frac{2}{5}a^2$$
$$35 = 7a^2$$
$$5 = a^2$$
$$b^2 = \frac{2}{5}a^2 = \frac{2}{5}(5) = 2$$

So the equation is $\frac{x^2}{5} - \frac{y^2}{2} = 1$, or $2x^2 - 5y^2 = 10$.

33. The vertices are at $(0,\pm 7)$ so we know it is a "vertical" hyperbola. Its equation will be $\frac{y^2}{49}-\frac{x^2}{b^2}=1$, but we also know that $(1,9)$ is a point satisfying the equation. We use it to find b:

$$\frac{81}{49}-\frac{1}{b^2}=1$$
$$81b^2-49=49b^2$$
$$32b^2=49$$
$$b^2=\frac{49}{32}$$

So the equation is $\frac{y^2}{49}-\frac{x^2}{49/32}=1$, or $y^2-32x^2=49$.

35. We have $2a=6$, so $a=3$. Also $2b=2$, so $b=1$. Since the foci are on the y-axis, the hyperbola will have the form:

$$\frac{y^2}{a^2}-\frac{x^2}{b^2}=1$$

So the equation is $\frac{y^2}{9}-\frac{x^2}{1}=1$, or $y^2-9x^2=9$.

37. Writing the equation as $\frac{x^2}{16}-\frac{y^2}{16}=1$, we have $a=b=4$. So the slopes of the asymptotes are $\pm\frac{b}{a}=\pm\frac{4}{4}=\pm 1$. But these are negative reciprocals of each other, so the asymptotes are perpendicular to each other.

39. (a) Substituting $P(5,6)$ into $5y^2-4x^2=80$, we have:

$$5(6)^2-4(5)^2=5(36)-4(25)=180-100=80$$

So $P(5,6)$ lies on the hyperbola.

(b) Dividing by 80 yields $\frac{y^2}{16}-\frac{x^2}{20}=1$, so $a=4$ and $b=2\sqrt{5}$.
Using $c^2=a^2+b^2$, we have:

$$c=\sqrt{a^2+b^2}=\sqrt{16+20}=\sqrt{36}=6$$

So $c=6$ and the foci are $(0,\pm 6)$.

(c) We compute the distance:

$$F_1P=\sqrt{(5-0)^2+(6-6)^2}=5$$
$$F_2P=\sqrt{(5-0)^2+(6-(-6))^2}=\sqrt{25+144}=13$$

(d) We verify the result:

$$|F_1P-F_2P|=|5-13|=|-8|=8=2(4)=2a$$

41. (a) Since the asymptotes must have slopes of $\pm\frac{b}{a}$, then:

$$-\frac{b}{a}=\frac{-1}{\frac{b}{a}}$$
$$\frac{b^2}{a^2}=1$$
$$b^2=a^2$$
$$b=a$$

Now, since $c^2 = a^2 + b^2 = 2a^2$, then the eccentricity is:

$$\frac{c}{a}=\frac{\sqrt{2a^2}}{a}=\frac{\sqrt{2}a}{a}=\sqrt{2}$$

(b) The slopes of the asymptotes are $\pm\frac{a}{a}=\pm 1$, so the hyperbola will have perpendicular asymptotes. The eccentricity is $\frac{\sqrt{2}a}{a}=\sqrt{2}$.

43. (a) This equation is just $F_1P - F_2P = 2a$, the defining relation of a hyperbola. Since P is on the right-hand branch, $F_1P > F_2P$.

(b) Squaring each side of the equation, we obtain:

$$\sqrt{(x+c)^2+y^2}=2a+\sqrt{(x-c)^2+y^2}$$
$$(x+c)^2+y^2=4a^2+4a\sqrt{(x-c)^2+y^2}+(x-c)^2+y^2$$
$$4xc=4a^2+4a\sqrt{(x-c)^2+y^2}$$
$$xc-a^2=a\sqrt{(x-c)^2+y^2}$$
$$xc-a^2=a(F_2P)$$

(c) Dividing by a, we have:

$$\frac{xc}{a}-a=F_2P$$
$$xe-a=F_2P$$

45. For this hyperbola we find $a^2 = b^2 = k^2$, and $e = \sqrt{2}$.
Also $d^2 = x^2 + y^2 = x^2 + (x^2 - k^2) = 2x^2 - k^2$. Thus we want to show that

$F_1P \bullet F_2P = 2x^2 - k^2$. Using the formulas for F_1P and F_2P developed in Exercises 43 and 44, we have:

$$\begin{aligned} F_1P \bullet F_2P &= (xe+a)(xe-a) \\ &= x^2e^2-a^2 \\ &= x^2(2)-k^2 \\ &= 2x^2-k^2 \\ &= d^2 \end{aligned}$$

47. The coordinates of D are $\left(\frac{a}{e}, \frac{b}{e}\right)$ and those of F are $(c, 0)$. Let O denote the center $(0, 0)$. We show that $\angle ODF$ is a right angle by showing $OD^2 + DF^2 = OF^2$:

$$OF^2 = c^2$$

$$OD^2 = \left(\frac{a}{e}\right)^2 + \left(\frac{b}{e}\right)^2 = \frac{a^2 + b^2}{e^2} = \frac{c^2}{e^2} = a^2$$

$$\begin{aligned} DF^2 &= \left(\frac{a}{e} - c\right)^2 + \left(\frac{b}{e}\right)^2 \\ &= \frac{(a - ec)^2 + b^2}{e^2} \\ &= \frac{a^2 + b^2 - 2aec + e^2c^2}{e^2} \\ &= \frac{c^2 - 2c^2}{e^2} + c^2 \\ &= -\frac{c^2}{e^2} + c^2 \\ &= -a^2 + c^2 \end{aligned}$$

Thus $OD^2 + DF^2 = OF^2$, as required.

Note: An alternate approach here is to show that the slope of FD is $-\frac{a}{b}$:

$$\text{slope } = \frac{\frac{b}{e}}{\frac{a}{e} - c} = \frac{b}{a - ce} = \frac{b}{a - \frac{c^2}{a}} = \frac{ab}{a^2 - c^2} = \frac{ab}{-b^2} = -\frac{a}{b}$$

Since the asymptote has a slope of $\frac{b}{a}$, the lines are perpendicular.

9.4 The Focus-Directrix Property of Conics

1. In order to use the formulas for the focal radii, we must find a and e. Dividing by 76, the standard form is $\frac{x^2}{76} + \frac{y^2}{76/3} = 1$. So $a = \sqrt{76} = 2\sqrt{19}$. To find the eccentricity, we first find c:

$$c = \sqrt{a^2 - b^2} = \sqrt{76 - \frac{76}{3}} = \sqrt{\frac{152}{3}} = \frac{2\sqrt{38}}{\sqrt{3}} = \frac{2\sqrt{114}}{3}$$

So the eccentricity is:

$$e = \frac{c}{a} = \frac{\frac{2\sqrt{114}}{3}}{2\sqrt{19}} = \frac{\sqrt{6}}{3}$$

So the focal radii are given by:

$$F_1P = a + ex = 2\sqrt{19} + \frac{\sqrt{6}}{3}(-8) = 2\sqrt{19} - \frac{8\sqrt{6}}{3} = \frac{6\sqrt{19} - 8\sqrt{6}}{3}$$

$$F_2P = a - ex = 2\sqrt{19} - \frac{\sqrt{6}}{3}(-8) = 2\sqrt{19} + \frac{8\sqrt{6}}{3} = \frac{6\sqrt{19} + 8\sqrt{6}}{3}$$

3. The equation is already in standard form with $a = 15$ and $b = 5$. We find c:

$$c = \sqrt{a^2 - b^2} = \sqrt{225 - 25} = \sqrt{200} = 10\sqrt{2}$$

So the eccentricity is:

$$e = \frac{c}{a} = \frac{10\sqrt{2}}{15} = \frac{2\sqrt{2}}{3}$$

So the focal radii are given by:

$$F_1P = a + ex = 15 + \frac{2\sqrt{2}}{3}(9) = 15 + 6\sqrt{2}$$

$$F_2P = a - ex = 15 - \frac{2\sqrt{2}}{3}(9) = 15 - 6\sqrt{2}$$

5. (a) The ellipse is already in standard form with $a = 4$ and $b = 3$. We find c:

$$c = \sqrt{a^2 - b^2} = \sqrt{16 - 9} = \sqrt{7}$$

So the foci are $(\pm\sqrt{7}, 0)$ and the eccentricity is $\frac{\sqrt{7}}{4}$. The directricies are given by:

$$x = \pm\frac{a}{e} = \pm\frac{4}{\frac{\sqrt{7}}{4}} = \pm\frac{16}{\sqrt{7}} = \pm\frac{16\sqrt{7}}{7}$$

(b) The hyperbola is already in standard form with $a = 4$ and $b = 3$. We find c:

$$c = \sqrt{a^2 + b^2} = \sqrt{16 + 9} = \sqrt{25} = 5$$

So the foci are $(\pm 5, 0)$ and the eccentricity is $\frac{5}{4}$. The directricies are given by:

$$x = \pm\frac{a}{e} = \pm\frac{4}{\frac{5}{4}} = \pm\frac{16}{5}$$

7. (a) Dividing by 156, the standard form for the ellipse is $\frac{x^2}{13} + \frac{y^2}{12} = 1$, and so $a = \sqrt{13}$ and $b = \sqrt{12} = 2\sqrt{3}$. We find c:

$$c = \sqrt{a^2 - b^2} = \sqrt{13 - 12} = \sqrt{1} = 1$$

So the foci are $(\pm 1, 0)$ and the eccentricity is $\frac{1}{\sqrt{13}} = \frac{\sqrt{13}}{13}$. The directricies are given by:

$$x = \pm\frac{a}{e} = \pm\frac{\sqrt{13}}{\frac{\sqrt{13}}{13}} = \pm 13$$

(b) Dividing by 156, the standard form for the hyperbola is $\frac{x^2}{13} - \frac{y^2}{12} = 1$, and so $a = \sqrt{13}$ and $b = \sqrt{12} = 2\sqrt{3}$. We find c:

$$c = \sqrt{a^2 + b^2} = \sqrt{13+12} = \sqrt{25} = 5$$

So the foci are $(\pm 5, 0)$ and the eccentricity is $\frac{5}{\sqrt{13}} = \frac{5\sqrt{13}}{13}$. The directricies are given by:

$$x = \pm\frac{a}{e} = \pm\frac{\sqrt{13}}{\frac{5\sqrt{13}}{13}} = \pm\frac{13}{5}$$

9. (a) Dividing by 900, the standard form for the ellipse is $\frac{x^2}{36} + \frac{y^2}{25} = 1$, and so $a = 6$ and $b = 5$. We find c:

$$c = \sqrt{a^2 - b^2} = \sqrt{36-25} = \sqrt{11}$$

So the foci are $(\pm\sqrt{11}, 0)$ and the eccentricity is $\frac{\sqrt{11}}{6}$. The directricies are given by:

$$x = \pm\frac{a}{e} = \pm\frac{6}{\frac{\sqrt{11}}{6}} = \pm\frac{36}{\sqrt{11}} = \pm\frac{36\sqrt{11}}{11}$$

(b) Dividing by 900, the standard form for the hyperbola is $\frac{x^2}{36} - \frac{y^2}{25} = 1$, and so $a = 6$ and $b = 5$. We find c:

$$c = \sqrt{a^2 + b^2} = \sqrt{36+25} = \sqrt{61}$$

So the foci are $(\pm\sqrt{61}, 0)$ and the eccentricity is $\frac{\sqrt{61}}{6}$. The directricies are given by:

$$x = \pm\frac{a}{e} = \pm\frac{6}{\frac{\sqrt{61}}{6}} = \pm\frac{36}{\sqrt{61}} = \pm\frac{36\sqrt{61}}{61}$$

11. Since the foci are $(\pm 1, 0)$ then $c = 1$. Since the directricies are $x = \pm 4$, then $\frac{a}{e} = 4$, so $a = 4e$. But since $e = \frac{c}{a} = \frac{1}{a}$, we have:

$$a = 4 \bullet \frac{1}{a}$$
$$a^2 = 4$$

Since $a^2 - b^2 = c^2$, we can find b^2:

$$2^2 - b^2 = 1^2$$
$$-b^2 = -3$$
$$b^2 = 3$$

So the equation of the ellipse is $\frac{x^2}{4} + \frac{y^2}{3} = 1$, or $3x^2 + 4y^2 = 12$.

13. Since the foci are $(\pm 2, 0)$ then $c = 2$. Since the directricies are $x = \pm 1$, then $\frac{a}{e} = 1$, so $a = e$. But since $e = \frac{c}{a} = \frac{2}{a}$, we have:

$$a = \frac{2}{a}$$
$$a^2 = 2$$

Since $a^2 + b^2 = c^2$, we can find b^2:

$$2 + b^2 = 2^2$$
$$b^2 = 2$$

So the equation of the hyperbola is $\frac{x^2}{2} - \frac{y^2}{2} = 1$, or $x^2 - y^2 = 2$.

15. (a) By the distance formula, we have:

$$d_1 = \sqrt{(x+c)^2 + (y-0)^2} = \sqrt{(x+c)^2 + y^2}$$
$$d_2 = \sqrt{(x-c)^2 + (y-0)^2} = \sqrt{(x-c)^2 + y^2}$$

Squaring, we have:

$$d_1^2 = (x+c)^2 + y^2$$
$$d_2^2 = (x-c)^2 + y^2$$

(b) Working from the left-hand side, we have:

$$\begin{aligned} d_1^2 - d_2^2 &= (x+c)^2 - (x-c)^2 \\ &= x^2 + 2cx + c^2 - x^2 + 2cx - c^2 \\ &= 4cx \end{aligned}$$

(c) Since d_1 and d_2 represent the distances from the foci to a point on the ellipse, then $d_1 + d_2 = 2a$ by the definition of an ellipse.

(d) Factoring, we have:

$$\begin{aligned} d_1^2 - d_2^2 &= 4cx \\ (d_1 + d_2)(d_1 - d_2) &= 4cx \\ 2a(d_1 - d_2) &= 4cx \\ d_1 - d_2 &= \frac{2cx}{a} \end{aligned}$$

(e) Adding the two equations, we have:

$$2d_1 = 2a + \frac{2cx}{a}$$
$$d_1 = a + \frac{c}{a}x$$
$$d_1 = a + ex$$

(f) Substituting the result from (e), we have:

$$a + ex + d_2 = 2a$$
$$d_2 = a - ex$$

Chapter Nine Review Exercises

1. (a) We have the form $y^2 = 4px$, where $p = 4$. Thus the equation is $y^2 = 16x$.
(b) We have the form $x^2 = 4py$, where $p = 4$. Thus the equation is $x^2 = 16y$.

3. Since the parabola is symmetric about the positive y-axis, its equation must be of the form $x^2 = 4py$, where $p > 0$. Now the focal width is 12, so $4p = 12$. Thus the equation is $x^2 = 12y$.

5. We have $c = 2$ and $a = 8$, and the ellipse must have the form:

$$\frac{x^2}{8^2} + \frac{y^2}{b^2} = 1$$

Since $c^2 = a^2 - b^2$, we can find b:

$$4 = 64 - b^2$$
$$b^2 = 60$$

So the equation is $\frac{x^2}{64} + \frac{y^2}{60} = 1$, or $15x^2 + 16y^2 = 960$.

7. Since one end of the minor axis is $(-6, 0)$, then $b = 6$ and the ellipse has a form of:

$$\frac{x^2}{36} + \frac{y^2}{a^2} = 1$$

Now $\frac{c}{a} = \frac{4}{5}$, so $c = \frac{4}{5}a$. We find a:

$$c^2 = a^2 - b^2$$
$$\left(\tfrac{4}{5}a\right)^2 = a^2 - 36$$
$$\tfrac{16}{25}a^2 = a^2 - 36$$
$$36 = \tfrac{9}{25}a^2$$
$$100 = a^2$$

So the equation is $\frac{x^2}{36} + \frac{y^2}{100} = 1$, or $25x^2 + 9y^2 = 900$.

9. Since the foci are $(\pm 6, 0)$ and the vertices are $(\pm 2, 0)$, then $c = 6$, $a = 2$, and the equation has the form:

$$\frac{x^2}{4} - \frac{y^2}{b^2} = 1$$

We find b:

$$c^2 = a^2 + b^2$$
$$36 = 4 + b^2$$
$$32 = b^2$$

So the equation is $\frac{x^2}{4} - \frac{y^2}{32} = 1$, or $8x^2 - y^2 = 32$.

11. Since the foci are $(\pm 3, 0)$, then $c = 3$ and the equation has the form:

$$\frac{x^2}{a^2} - \frac{y^2}{b^2} = 1$$

Now $\frac{c}{a} = 4$, so $\frac{3}{a} = 4$, thus $a = \frac{3}{4}$. We substitute to find b:

$$c^2 = a^2 + b^2$$
$$9 = \frac{9}{16} + b^2$$
$$144 = 9 + 16b^2$$
$$135 = 16b^2$$
$$\frac{135}{16} = b^2$$

So the equation is $\dfrac{x^2}{\frac{9}{16}} - \dfrac{y^2}{\frac{135}{16}} = 1$, or $240x^2 - 16y^2 = 135$.

13. We note that $4p = 10$, so $p = \frac{5}{2}$. So the vertex is $(0,0)$, the focus is $\left(0, \frac{5}{2}\right)$, the directrix is $y = -\frac{5}{2}$, and the focal width is 10.

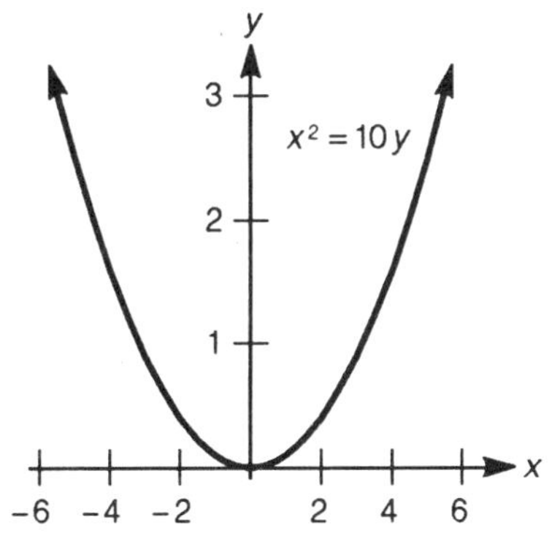

15. We note that $4p = 12$, so $p = 3$. So the vertex is $(0, 3)$, the focus is $(0, 0)$, the directrix is $y = 6$, and the focal width is 12.

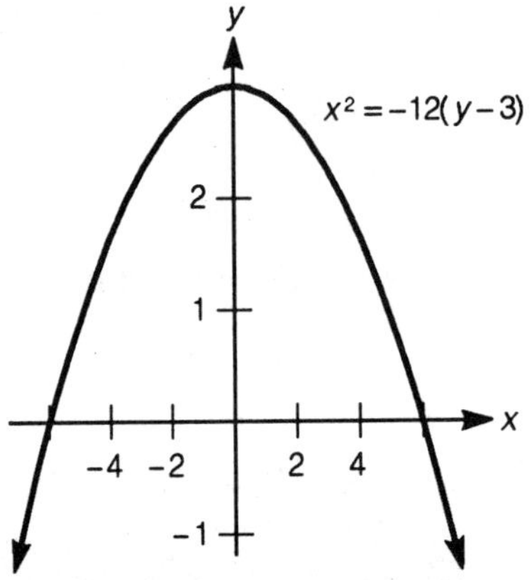

17. We note that $4p = 4$, so $p = 1$. So the vertex is $(1, 1)$, the focus is $(0, 1)$, the directrix is $x = 2$, and the focal width is 4.

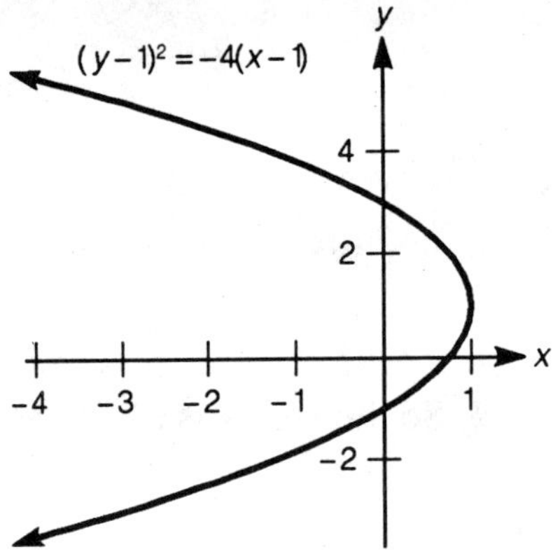

19. Dividing by 4, the standard form is $\frac{x^2}{4} + \frac{y^2}{2} = 1$. The center is $(0, 0)$, the length of the major axis is 4, and the length of the minor axis is $2\sqrt{2}$. Using $c^2 = a^2 - b^2$, we find:

$$c = \sqrt{a^2 - b^2} = \sqrt{4 - 2} = \sqrt{2}$$

So the foci are $(\pm\sqrt{2}, 0)$ and the eccentricity is $\frac{\sqrt{2}}{2}$.

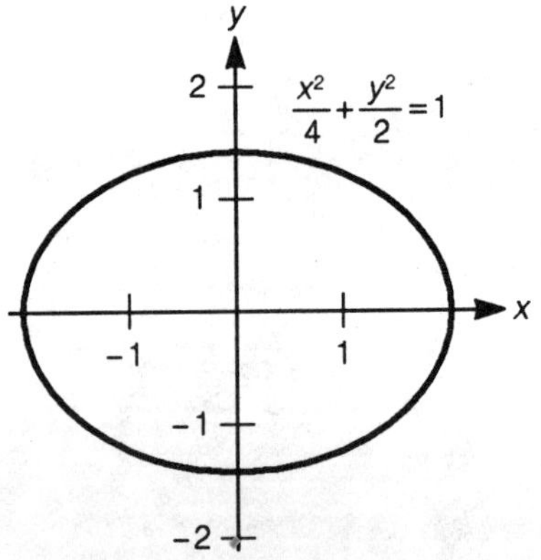

21. Dividing by 441, the standard form is $\frac{x^2}{9}+\frac{y^2}{49}=1$. The center is $(0,0)$, the length of the major axis is 14, and the length of the minor axis is 6. Using $c^2=a^2-b^2$, we find:

$$c=\sqrt{a^2-b^2}=\sqrt{49-9}=\sqrt{40}=2\sqrt{10}$$

So the foci are $\left(0,\pm 2\sqrt{10}\right)$ and the eccentricity is $\frac{2\sqrt{10}}{7}$.

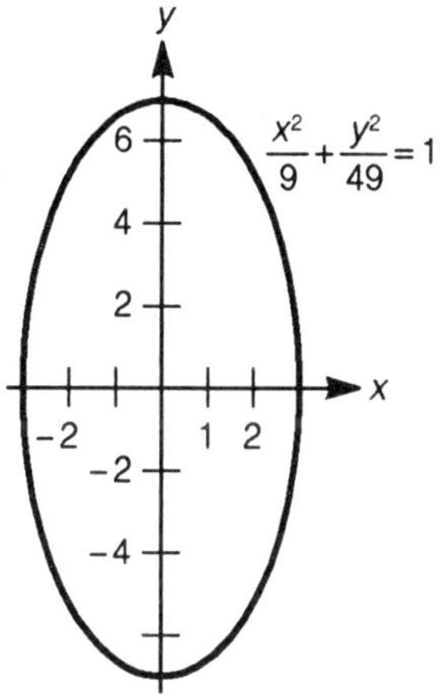

23. The equation is already in standard form where the center is $(1,-2)$, the length of the major axis is 10, and the length of the minor axis is 6. Using $c^2=a^2-b^2$, we find:

$$c=\sqrt{a^2-b^2}=\sqrt{25-9}=\sqrt{16}=4$$

So the foci are $(1+4,-2)=(5,-2)$ and $(1-4,-2)=(-3,-2)$, and the eccentricity is $\frac{4}{5}$.

25. Dividing by 4, the standard form is $\frac{x^2}{4}-\frac{y^2}{2}=1$. The center is $(0,0)$, the vertices are $(\pm 2,0)$, and the asymptotes are $y=\pm\frac{\sqrt{2}}{2}x$. Using $c^2=a^2+b^2$, we have:

$$c=\sqrt{a^2+b^2}=\sqrt{4+2}=\sqrt{6}$$

So the foci are $(\pm 6, 0)$ and the eccentricity is $\frac{\sqrt{6}}{2}$.

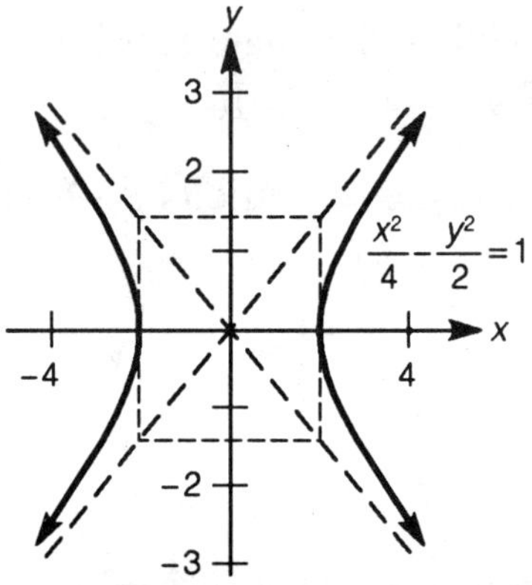

27. Dividing by 441, the standard form is $\frac{y^2}{9}-\frac{x^2}{49}=1$. The center is $(0,0)$, the vertices are $(0,\pm 3)$, and the asymptotes are $y=\pm\frac{3}{7}x$. Using $c^2=a^2+b^2$, we have:

$$c=\sqrt{a^2+b^2}=\sqrt{9+49}=\sqrt{58}$$

So the foci are $(0,\pm\sqrt{58})$ and the eccentricity is $\frac{\sqrt{58}}{3}$.

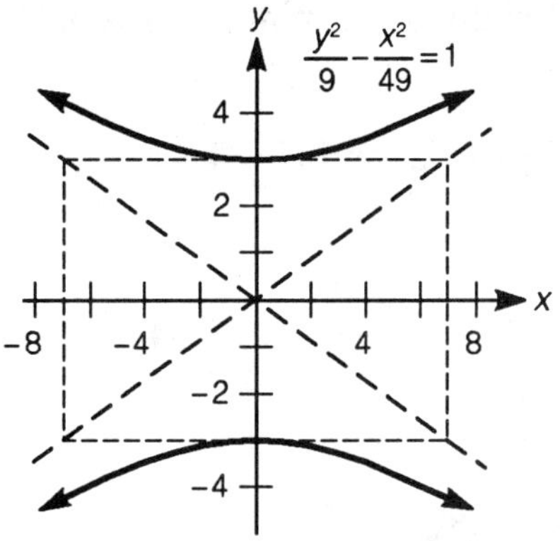

29. The equation is already in standard form where the center is $(1,-2)$ and the vertices are $(1+5,-2)=(6,-2)$ and $(1-5,-2)=(-4,-2)$. The asymptotes have slopes of $\pm\frac{3}{5}$, so using the point-slope formula:

$$y-(-2)=\tfrac{3}{5}(x-1) \qquad y-(-2)=-\tfrac{3}{5}(x-1)$$
$$y+2=\tfrac{3}{5}x-\tfrac{3}{5} \qquad y+2=-\tfrac{3}{5}x+\tfrac{3}{5}$$
$$y=\tfrac{3}{5}x-\tfrac{13}{5} \qquad y=-\tfrac{3}{5}x-\tfrac{7}{5}$$

Using $c^2=a^2+b^2$, we have:

$$c=\sqrt{a^2+b^2}=\sqrt{25+9}=\sqrt{34}$$

So the foci are $\left(1 \pm \sqrt{34}, -2\right)$ and the eccentricity is $\frac{\sqrt{34}}{5}$.

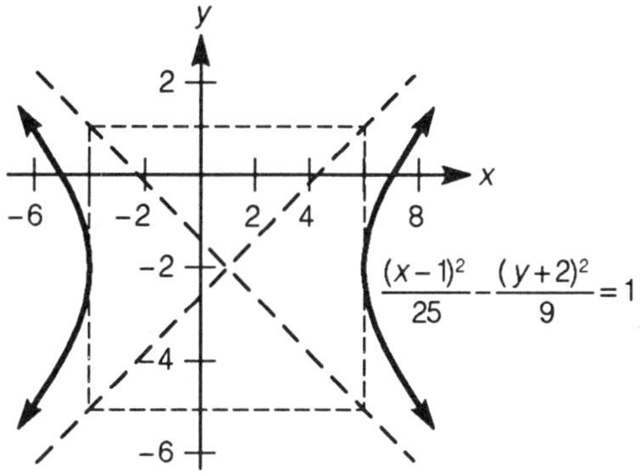

31. We complete the square to convert the equation to standard form:

$$3x^2 - 6x + 4y^2 + 16 = -7$$
$$3\left(x^2 - 2x + 1\right) + 4\left(y^2 + 4y + 4\right) = -7 + 3 + 16$$
$$3(x-1)^2 + 4(y+2)^2 = 12$$
$$\frac{(x-1)^2}{4} + \frac{(y+2)^2}{3} = 1$$

This is the equation of an ellipse. Its center is $(1, -2)$, the length of the major axis is 4, and the length of the minor axis is $2\sqrt{3}$. Using $c^2 = a^2 - b^2$, we find:

$$c = \sqrt{a^2 - b^2} = \sqrt{4-3} = 1$$

So the foci are $(1 + 1, -2) = (2, -2)$ and $(1 - 1, -2) = (0, -2)$.

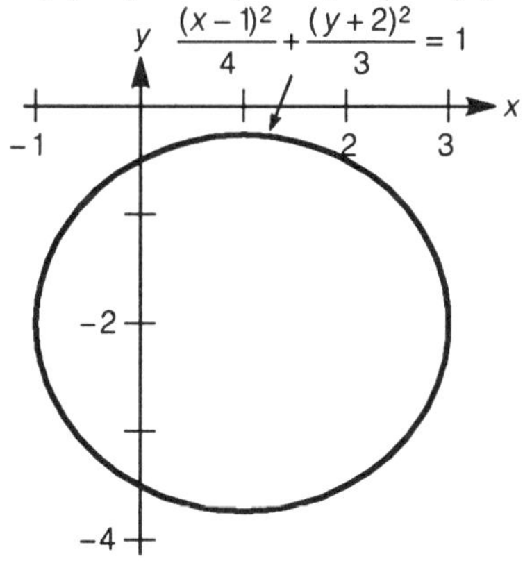

33. We complete the square to convert the equation to standard form:

$$y^2 + 2y = -4x + 15$$
$$y^2 + 2y + 1 = -4x + 15 + 1$$
$$(y+1)^2 = -4(x-4)$$

This is the equation of a parabola. Since $4p = 4$, then $p = 1$. So the vertex is $(4,-1)$, the axis of symmetry is $y = -1$, the focus is $(3,-1)$, and the directrix is $x = 5$.

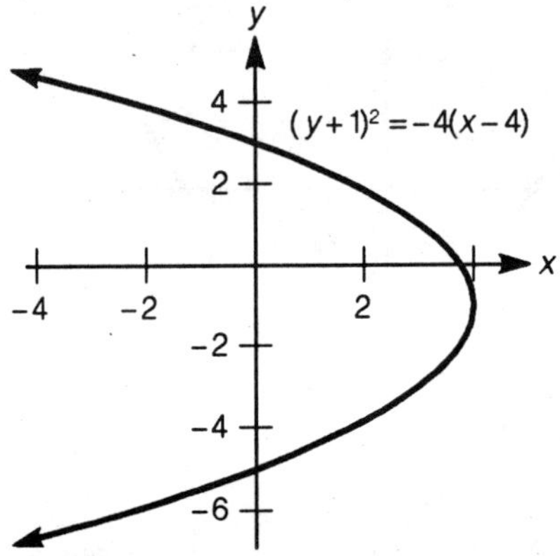

35. We complete the square to convert the equation to standard form:

$$16(x^2-2x)-9(y^2-10y)=353$$
$$16(x^2-2x+1)-9(y^2-10y+25)=353+16-225$$
$$16(x-1)^2-9(y-5)^2=144$$
$$\frac{(x-1)^2}{9}-\frac{(y-5)^2}{16}=1$$

This is the equation of a hyperbola. Its center is $(1,5)$ and its vertices are $(1+3,5)=(4,5)$ and $(1-3,5)=(-2,5)$. The asymptotes have slopes of $\pm\frac{4}{3}$, so using the point-slope formula:

$$y-5=\tfrac{4}{3}(x-1) \qquad\qquad y-5=-\tfrac{4}{3}(x-1)$$
$$y-5=\tfrac{4}{3}x-\tfrac{4}{3} \qquad\qquad y-5=-\tfrac{4}{3}x+\tfrac{4}{3}$$
$$y=\tfrac{4}{3}x+\tfrac{11}{3} \qquad\qquad y=-\tfrac{4}{3}x+\tfrac{19}{3}$$

Using $c^2 = a^2 + b^2$, we have:

$$c=\sqrt{a^2+b^2}=\sqrt{9+16}=\sqrt{25}=5$$

So the foci are $(1+5,5)=(6,5)$ and $(1-5,5)=(-4,5)$.

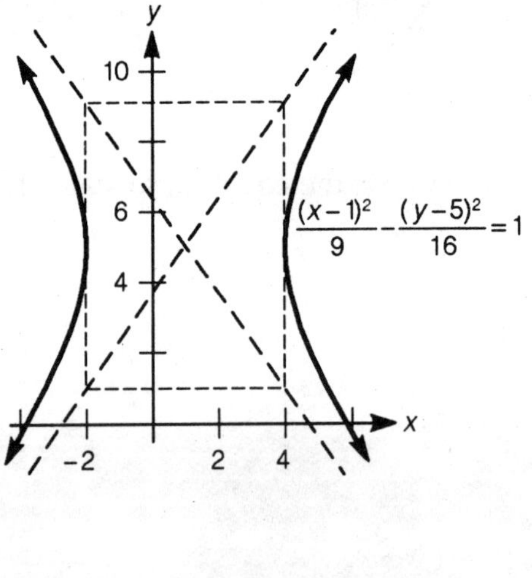

37. We complete the square to convert the equation to standard form:

$$5x^2 + 3y^2 - 40x - 36y + 188 = 0$$
$$5(x^2 - 8x) + 3(y^2 - 12y) = -188$$
$$5(x-4)^2 + 3(y-6)^2 = -188 + 80 + 108$$
$$5(x-4)^2 + 3(y-6)^2 = 0$$

This equation has only one solution, namely the point (4, 6).

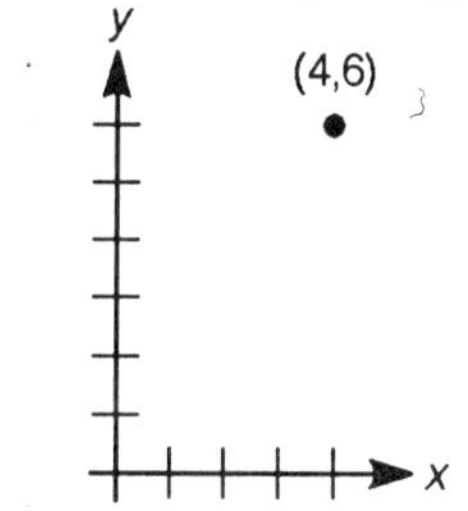

39. We complete the square to convert the equation to standard form:

$$9x^2 - 90x - 16y^2 + 32y = -209$$
$$9(x^2 - 10x) - 16(y^2 - 2y) = -209$$
$$9(x-5)^2 - 16(y-1)^2 = -209 + 225 - 16$$
$$9(x-5)^2 - 16(y-1)^2 = 0$$
$$9(x-5)^2 = 16(y-1)^2$$
$$\pm 3(x-5) = 4(y-1)$$

The graph of this equation consists of just two lines:

$3(x-5) = 4(y-1)$	$-3(x-5) = 4(y-1)$
$3x - 15 = 4y - 4$	$-3x + 15 = 4y - 4$
$3x - 4y = 11$	$3x + 4y = 19$

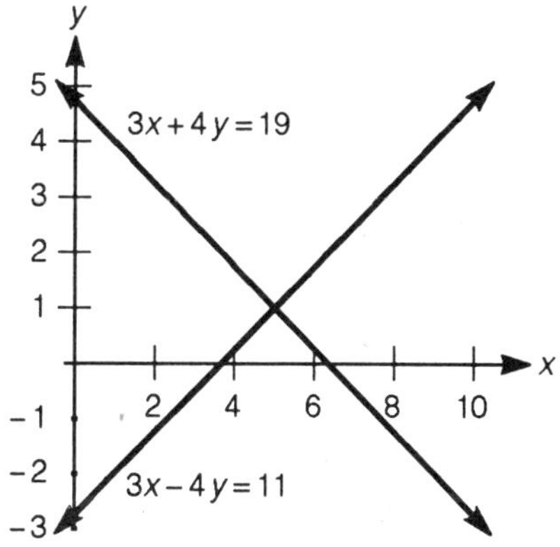

41. We complete the square to convert the equation to standard form:

$$y^2+8y-25x^2=9$$
$$(y+4)^2-25x^2=25$$
$$\frac{(y+4)^2}{25}-\frac{x^2}{1}=1$$

This is the equation of a hyperbola. Its center is $(0,-4)$ and its vertices are $(0,-4+5)=(0,1)$ and $(0,-4-5)=(0,-9)$. The asymptotes have slopes of ± 5, so using the point-slope formula:

$$y-(-4)=5(x-0) \qquad y-(-4)=-5(x-0)$$
$$y+4=5x \qquad y+4=-5x$$
$$y=5x-4 \qquad y=-5x-4$$

Using $c^2=a^2+b^2$, we have:

$$c=\sqrt{a^2+b^2}=\sqrt{25+1}=\sqrt{26}$$

So the foci are $\left(0,-4\pm\sqrt{26}\right)$.

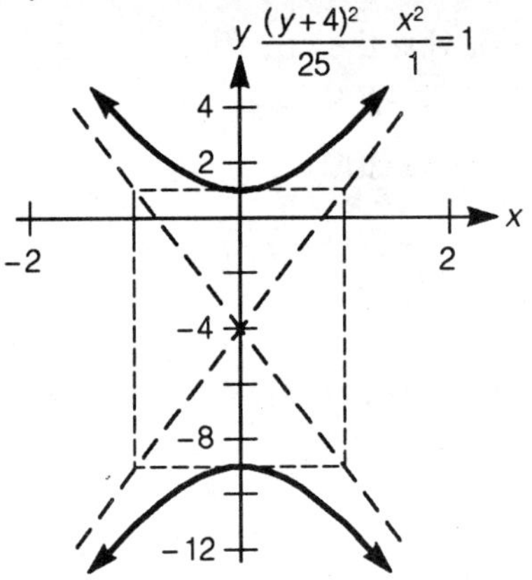

43. We complete the square to convert the equation to standard form:

$$16x^2-64x+25y^2-100y=-564$$
$$16\left(x^2-4x\right)+25\left(y^2-4y\right)=-564$$
$$16(x-2)^2+25(y-2)^2=-564+64+100$$
$$16(x-2)^2+25(y-2)^2=-400$$

Since the left-hand side of this equation is the sum of two squares, it must be non-negative and thus there are no points (x,y) which satisfy the equation. Thus there is no graph for this equation.

45. (a) Substituting $P(6,5)$ into the equation $5x^2-4y^2=80$:

$$5(6)^2-4(5)^2=5(36)-4(25)=180-100=80$$

So $P(6,5)$ lies on the hyperbola.

(b) Since P lies on the hyperbola with foci F_1, F_2, then the quantity $F_1P - F_2P = \pm 2a$ (by the definition of the hyperbola). Thus $(F_1P - F_2P)^2 = 4a^2$. It remains to find a:

$$5x^2 - 4y^2 = 80$$
$$\frac{x^2}{16} - \frac{y^2}{20} = 1$$

So $a^2 = 16$, thus $(F_1P - F_2P)^2 = 4a^2 = 4(16) = 64$.

47. Completing the square, we obtain:

$$A\left(x^2 + \frac{D}{A}x + \frac{D^2}{4A^2}\right) + C\left(y^2 + \frac{E}{C}y + \frac{E^2}{4C^2}\right) = \frac{D^2}{4A} + \frac{E^2}{4C} - F$$
$$A\left(x + \frac{D}{2A}\right)^2 + C\left(y + \frac{E}{2C}\right)^2 = \frac{CD^2 + AE^2 - 4ACF}{4AC}$$

Now if the curve represents an ellipse or a hyperbola, this last equation shows that the center is $\left(-\frac{D}{2A}, -\frac{E}{2C}\right)$.

49. For the parabola, the vertex is $(5,0)$, so we know $y^2 = -4p(x-5)$. Since the focus is $(3,0)$, then $p = 2$, so the equation is $y^2 = -8(x-5)$. For the ellipse, $a = 5$ and $c = 3$, so $b^2 = a^2 - c^2 = 25 - 9 = 16$, so $b = 4$. Thus the equation of the ellipse is $\frac{x^2}{25} + \frac{y^2}{16} = 1$.

51. (a) We first complete the square:

$$x^2 - 6x - 4(y^2 - 8y) = 59$$
$$(x^2 - 6x + 9) - 4(y^2 - 8y + 16) = 59 + 9 - 64$$
$$(x-3)^2 - 4(y-4)^2 = 4$$
$$\frac{(x-3)^2}{4} - \frac{(y-4)^2}{1} = 1$$

So $a = 2$ and $b = 1$. Therefore:

$$c = \sqrt{a^2 + b^2} = \sqrt{4+1} = \sqrt{5}$$

So the eccentricity is $e = \frac{c}{a} = \frac{\sqrt{5}}{2}$.

(b) Since the center is $(3,4)$, one focus is $(3+\sqrt{5},4)$ and the corresponding directrix is:

$$x = 3 + \frac{a}{e} = 3 + \frac{2}{\sqrt{5}/2} = 3 + \frac{4}{\sqrt{5}}$$

The other focus is $(3-\sqrt{5},4)$ and the corresponding directrix is:

$$x = 3 - \frac{a}{e} = 3 - \frac{2}{\sqrt{5}/2} = 3 - \frac{4}{\sqrt{5}}$$

(c) We sketch the graph:

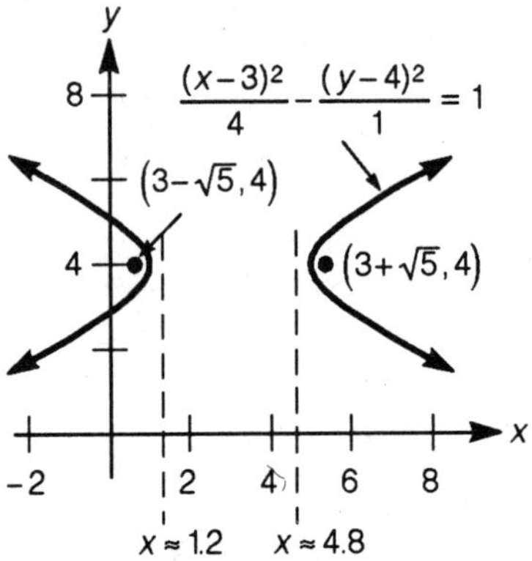

53. Since $e > 1$ and the directricies are $x = \pm 3$, this will be a hyperbola centered at the origin with equation:

$$\frac{x^2}{a^2} - \frac{y^2}{b^2} = 1$$

Now $\frac{c}{a} = \frac{3}{2}$ and $\frac{a}{3/2} = 3$, so $a = \frac{9}{2}$ and thus $c = \frac{3}{2} \bullet \frac{9}{2} = \frac{27}{4}$. Finally, we find b:

$$a^2 + b^2 = c^2$$
$$\frac{81}{4} + b^2 = \frac{729}{16}$$
$$b^2 = \frac{405}{16}$$

So the equation of the hyperbola is:

$$\frac{x^2}{81/4} - \frac{y^2}{405/16} = 1$$

Chapter Nine Graphing Calculator Exercises

1. Entering $x^2 = 8y$ as $y = \frac{1}{8}x^2$ and using the ZOOM-6 settings, we have the parabola:

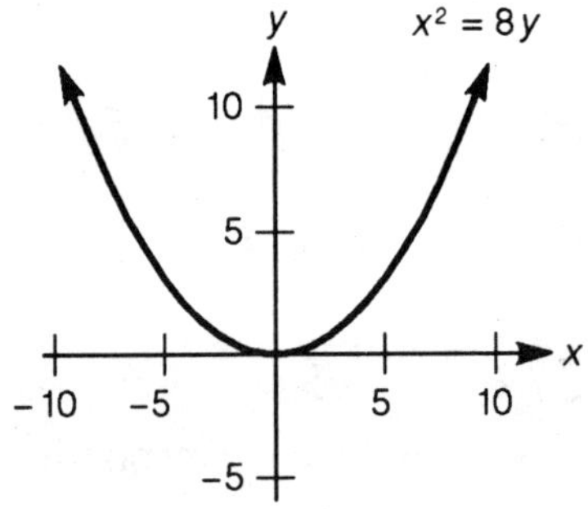

3. Entering $y^2 = 8x$ as the two functions $y = \sqrt{8x}$ and $y = -\sqrt{8x}$, and using the ZOOM-6 settings, we have the parabola:

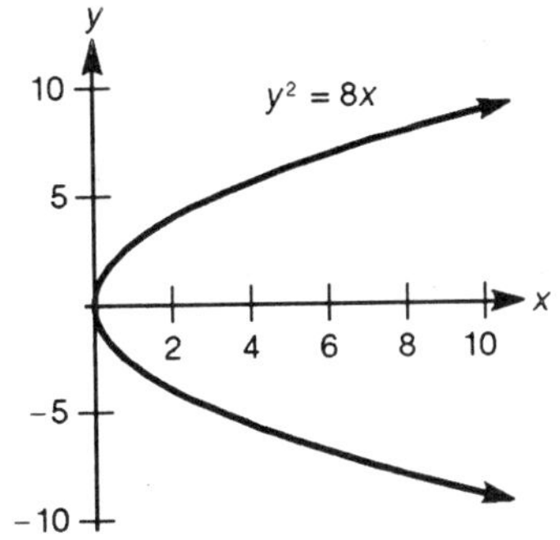

5. (a) Completing the square, we have:

$$4x + y^2 + 2y - 7 = 0$$
$$y^2 + 2y + 1 = -4x + 8$$
$$(y+1)^2 = -4(x-2)$$

(b) Entering $(y + 1)^2 = -4(x - 2)$ as the two functions $y = -1 - \sqrt{-4(x-2)}$ and $y = -1 + \sqrt{-4(x-2)}$, and using the ZOOM-6 settings, we have the parabola:

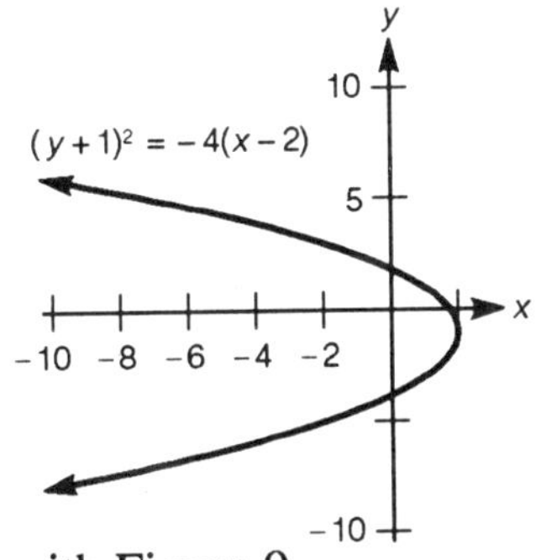

The graph is consistent with Figure 9.

7. Since the tangent line passes through the points $(4, 2)$ and $(0, -2)$, we can find the slope:

$$m = \frac{-2-2}{0-4} = \frac{-4}{-4} = 1$$

Since the y-intercept is -2, the equation of the tangent line is $y = x - 2$. Using the ZOOM-6 settings, we graph the parabola and tangent line:

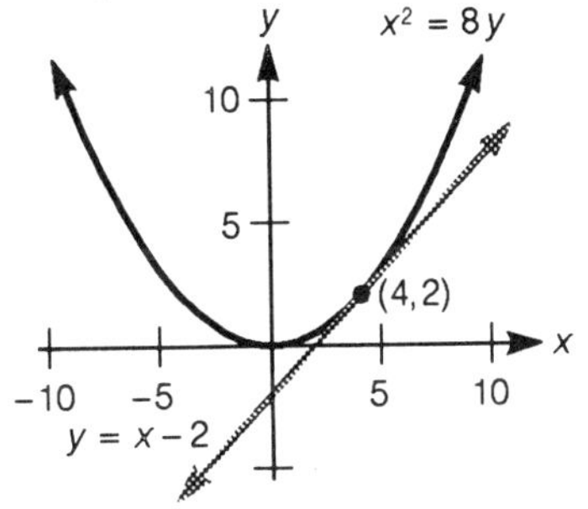

9. Since the tangent line passes through the points (–3, 9) and (0, –9), we can find the slope:

$$m = \frac{-9-9}{0-(-3)} = \frac{-18}{3} = -6$$

Since the y-intercept is –9, the equation of the tangent line is $y = -6x - 9$. Using the ZOOM-6 settings, we graph the parabola and tangent line:

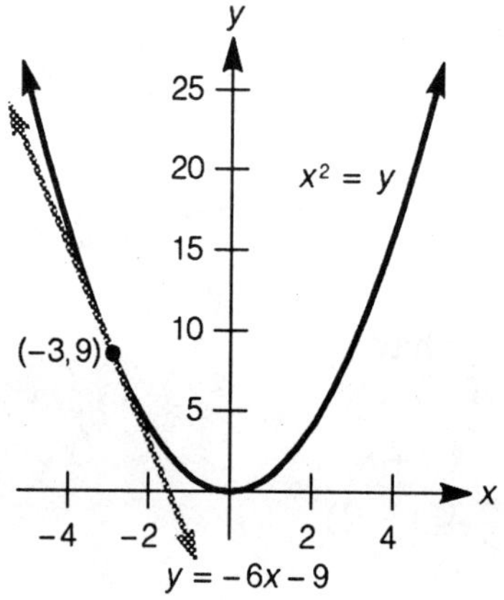

11. (a) Solving for y, we obtain:

$$\frac{x^2}{9} + \frac{y^2}{4} = 1$$
$$4x^2 + 9y^2 = 36$$
$$9y^2 = 36 - 4x^2$$
$$y^2 = \frac{36 - 4x^2}{9}$$
$$y = \pm\tfrac{1}{3}\sqrt{36 - 4x^2}$$

Graphing the two functions with the indicated settings:

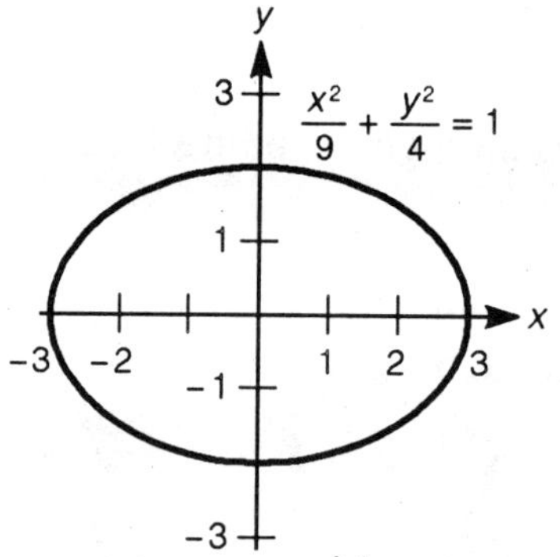

The graph appears to have symmetry with respect to the x-axis, the y-axis, and the origin.

(b) Solving for y, we obtain:

$$\frac{x^2}{1} + \frac{y^2}{16} = 1$$
$$16x^2 + y^2 = 16$$
$$y^2 = 16 - 16x^2$$
$$y = \pm 4\sqrt{1 - x^2}$$

Graphing the two functions with the indicated settings:

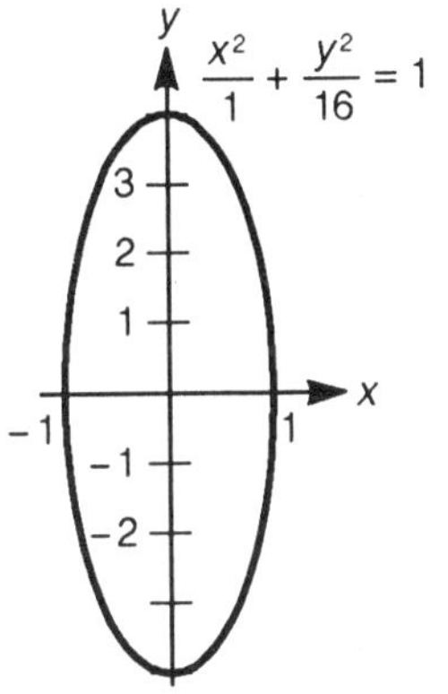

Again, the graph appears to have symmetry with respect to the x-axis, the y-axis, and the origin.

13. (a) Dividing by $16b^2$, we have:

$$\frac{x^2}{16} + \frac{y^2}{b^2} = 1$$

So the horizontal axis has length 2(4) = 8.

(b) When $b^2 = 16$, then $c^2 = 16 - 16 = 0$, so $c = 0$ and thus:

$$e = \frac{c}{a} = \frac{0}{4} = 0$$

When $b^2 = 13.44$, then $c^2 = 16 - 13.44 = 2.56$, so $c = 1.6$ and thus:

$$e = \frac{c}{a} = \frac{1.6}{4} = 0.4$$

When $b^2 = 10.24$, then $c^2 = 16 - 10.24 = 5.76$, so $c = 2.4$ and thus:

$$e = \frac{c}{a} = \frac{2.4}{4} = 0.6$$

When $b^2 = 5.76$, then $c^2 = 16 - 5.76 = 10.24$, so $c = 3.2$ and thus:

$$e = \frac{c}{a} = \frac{3.2}{4} = 0.8$$

(c) Solving for y, we have:

$$b^2x^2 + 16y^2 = 16b^2$$
$$16y^2 = 16b^2 - b^2x^2$$
$$y^2 = \frac{b^2(16 - x^2)}{16}$$
$$y = \pm 0.25\sqrt{b^2(16 - x^2)}$$

Using the indicated settings, the graphs of all four ellipses appear as:

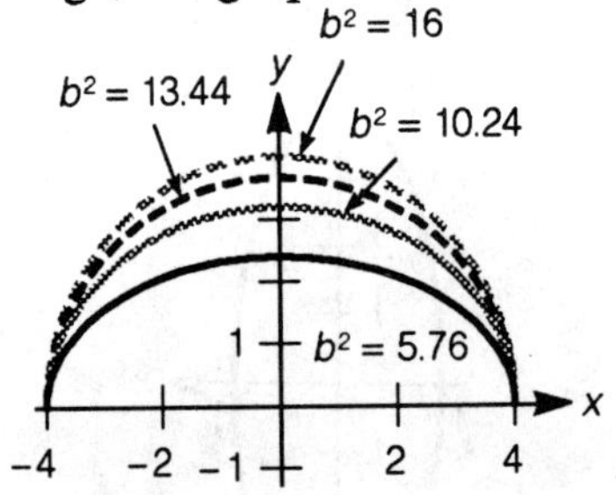

15. Dividing the ellipse equation by 12, we have:

$$\frac{x^2}{4} + \frac{y^2}{12} = 1$$

So $a^2 = 4$ and $b^2 = 12$. Using $(x_0, y_0) = (1, -3)$ in the tangent line formula yields:

$$\frac{1x}{4} + \frac{-3y}{12} = 1$$
$$3x - 3y = 12$$
$$x - y = 4$$
$$y = x - 4$$

Now graph the ellipse $\frac{x^2}{4} + \frac{y^2}{12} = 1$ and tangent line $y = x - 4$:

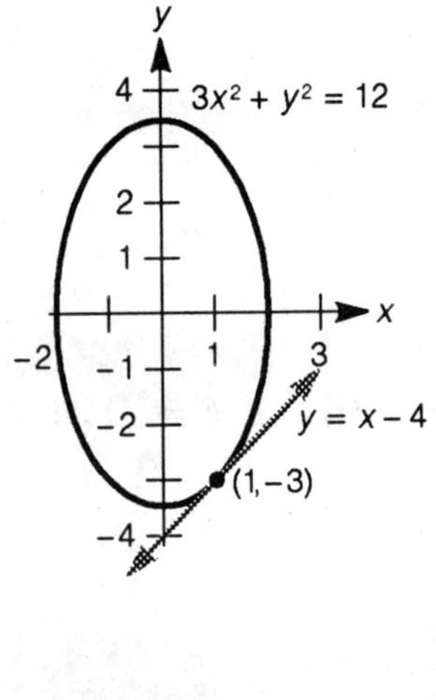

17. (a) Dividing by 12, we graph the ellipse $\frac{x^2}{12} + \frac{y^2}{4} = 1$:

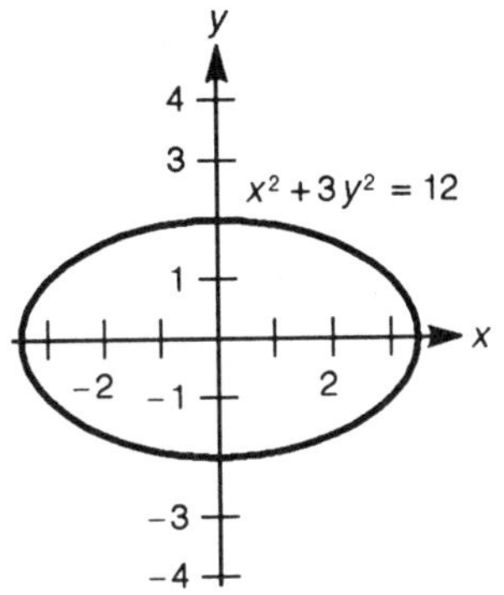

(b) Since $a^2 = 12$ and $b^2 = 4$, then $a = 2\sqrt{3}$ and $b = 2$. Therefore:

$$c^2 = a^2 - b^2 = 12 - 4 = 8, \text{ so } c = \sqrt{8} = 2\sqrt{2}$$

(c) The auxiliary circle has an equation $x^2 + y^2 = 12$. We graph the ellipse and auxiliary circle:

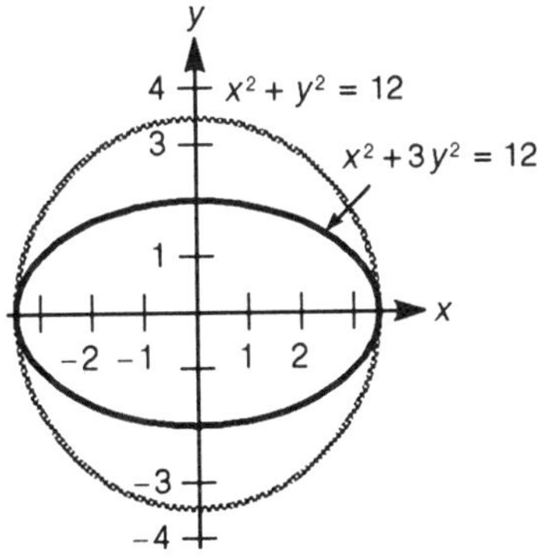

(d) We verify that the point $P(3, 1)$ lies on the ellipse:

$$(3)^2 + 3(1)^2 = 9 + 3 = 12$$

Using $(x_0, y_0) = (3, 1)$, we find the equation of the tangent line to the ellipse at P:

$$\frac{3x}{12} + \frac{1y}{4} = 1$$

$$\frac{x}{4} + \frac{y}{4} = 1$$

$$x + y = 4$$

Now we graph the upper halves of the ellipse and circle, and the line $y = -x + 4$:

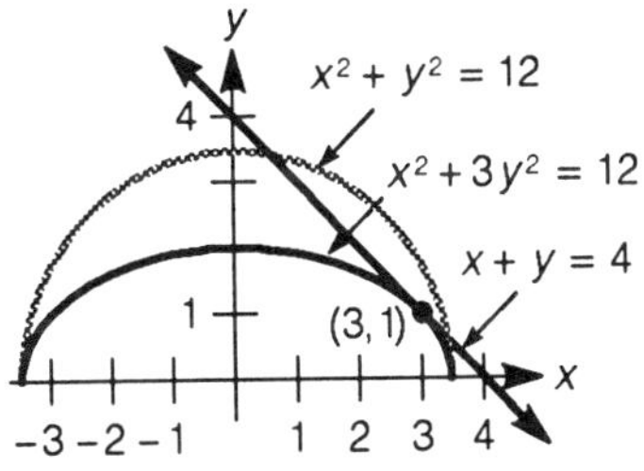

(e) The perpendicular to the tangent at $P(3,1)$ has slope $m = 1$. Using the point $(-2\sqrt{2},0)$ and the point-slope formula, we have:

$$y-0=1(x+2\sqrt{2})$$
$$y=x+2\sqrt{2}$$

Now we graph the upper halves of the ellipse and circle, as well as the lines $y=-x+4$ and $y=x+2\sqrt{2}$:

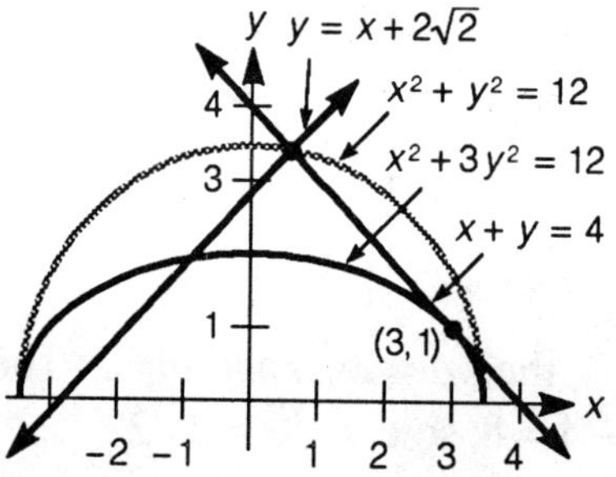

19. (a) Solving for y, we have:

$$\frac{(y-3)^2}{25}-\frac{(x-4)^2}{9}=1$$
$$9(y-3)^2-25(x-4)^2=225$$
$$9(y-3)^2=25(x-4)^2+225$$
$$(y-3)^2=\tfrac{25}{9}(x-4)^2+25$$
$$y-3=\pm 5\sqrt{1+\tfrac{1}{9}(x-4)^2}$$
$$y=3\pm 5\sqrt{1+\tfrac{1}{9}(x-4)^2}$$

Entering these two functions and making the suggested range settings, we graph the hyperbola:

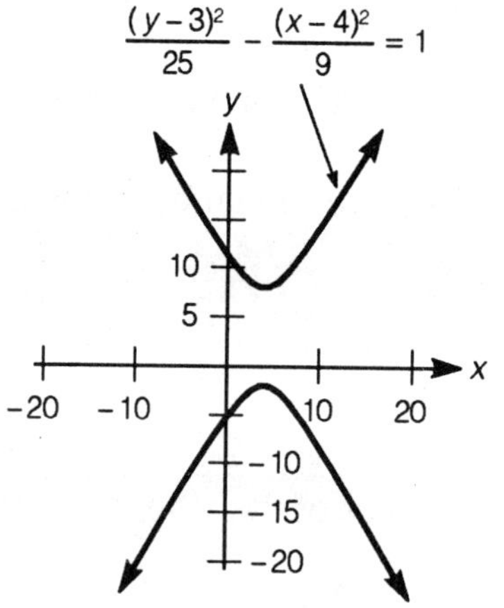

(b) Since $a^2 = 25$ and $b^2 = 9$, then $a = 5$ and $b = 3$, thus the slopes of the asymptotes are $\pm\frac{5}{3}$. Using the point $(4,3)$ in the point-slope formula, we have:

$$y-3=\pm\tfrac{5}{3}(x-4), \text{ so } y=\pm\tfrac{5}{3}(x-4)+3$$

Now we graph the hyperbola and the two asymptotes:

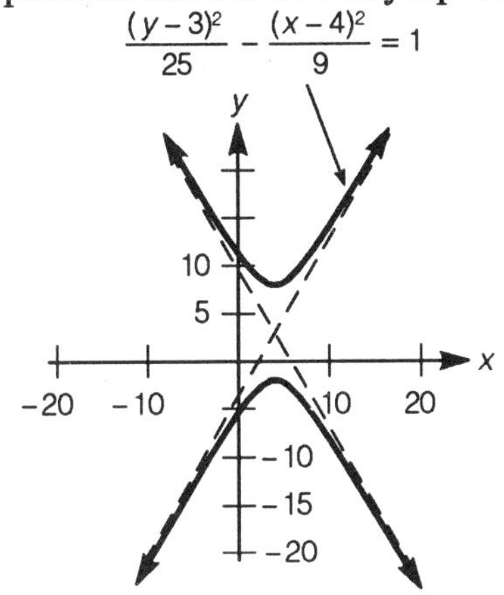

21. (a) Using the quadratic formula to solve $y^2 + xy + (x^2 - 3x - 1) = 0$, we have:

$$y = \frac{-x \pm \sqrt{x^2 - 4(1)(x^2 - 3x - 1)}}{2(1)}$$
$$= \frac{-x \pm \sqrt{x^2 - 4x^2 + 12x + 4}}{2}$$
$$= \frac{-x \pm \sqrt{-3x^2 + 12x + 4}}{2}$$

(b) Entering these two functions and making the suggested range settings, we have:

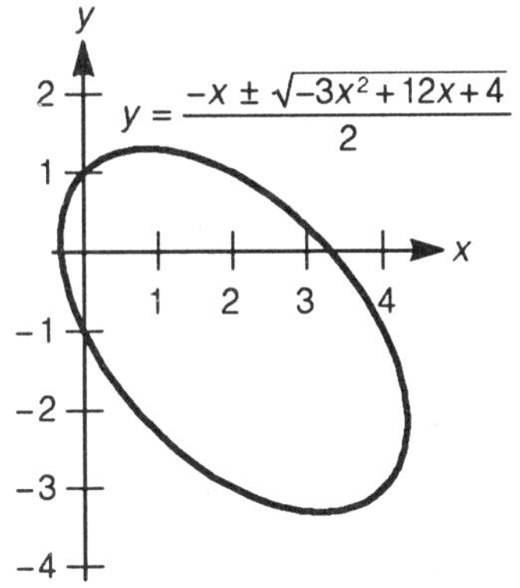

(c) The center appears to be approximately $(2, -1)$.

(d) Given $A = 1$, $B = 1$, $C = 1$, $D = -3$, $E = 0$, and $F = -1$, we compute the center using the given formulas:

$$h = \frac{2CD - BE}{B^2 - 4AC} = \frac{2(1)(-3) - (1)(0)}{(1)^2 - 4(1)(1)} = \frac{-6}{-3} = 2$$
$$k = \frac{2AE - BD}{B^2 - 4AC} = \frac{2(1)(0) - (1)(-3)}{(1)^2 - 4(1)(1)} = \frac{3}{-3} = -1$$

The center is $(2, -1)$, agreeing with our estimate from part (c).

Chapter Nine Test

1. Since $4p = 12$, then $p = 3$. So the focus is $(-3,0)$ and the directrix is $x = 3$.

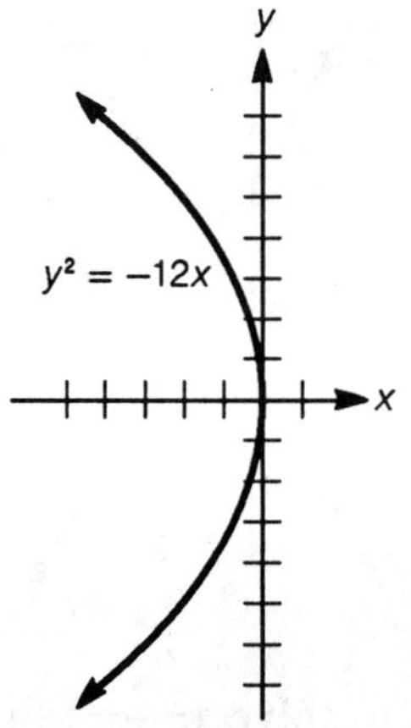

2. Dividing by 4, the standard form is $\frac{x^2}{4} - \frac{y^2}{1} = 1$. So the asymptotes are $y = \pm\frac{1}{2}x$. Using $c^2 = a^2 + b^2$, we have:

$$c = \sqrt{a^2 + b^2} = \sqrt{4+1} = \sqrt{5}$$

So the foci are $\left(\pm\sqrt{5}, 0\right)$.

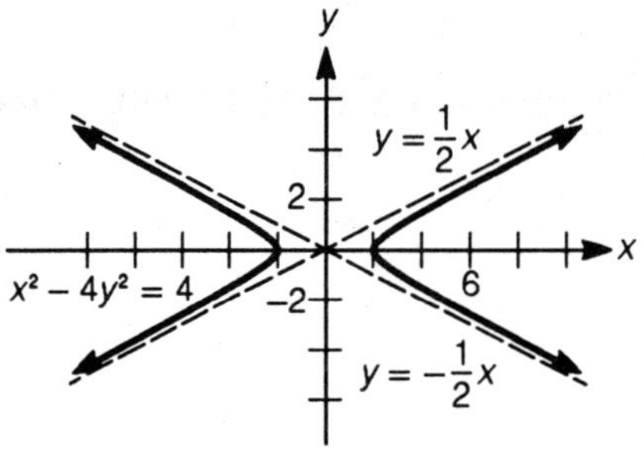

3. Since $e = \frac{c}{a}$, then $\frac{c}{a} = \frac{1}{2}$, so $a = 2c$. Since the foci are $(0, \pm 2)$, then $c = 2$ and thus $a = 4$. Now we find b^2:

$$c^2 = a^2 - b^2$$
$$4 = 16 - b^2$$
$$b^2 = 12$$

Thus the equation of the ellipse is $\frac{x^2}{12} + \frac{y^2}{16} = 1$.

4. Since the foci are $(\pm 2, 0)$, then $c = 2$. Also $\frac{b}{a} = \frac{1}{\sqrt{3}}$, so $a = b\sqrt{3}$. So:

$$c^2 = a^2 + b^2$$
$$4 = (b\sqrt{3})^2 + b^2$$
$$4 = 4b^2$$
$$1 = b^2$$
$$1 = b$$

So $a = 1\sqrt{3} = \sqrt{3}$, thus the equation is $\frac{x^2}{3} - \frac{y^2}{1} = 1$.

5. (a) Substituting $P(6,5)$ into the equation $5x^2 - 4y^2 = 80$, we have:

$$5(6)^2 - 4(5)^2 = 5(36) - 4(25) = 180 - 100 = 80$$

So $P(6,5)$ lies on the hyperbola.

(b) The quantity $(F_1P - F_2P)^2$ can be computed without determining the coordinates of F_1 and F_2. By definition we have $|F_1P - F_2P| = 2a$ for any point P on the hyperbola. Squaring both sides here yields $(F_1P - F_2P)^2 = 4a^2$. Now to compute a, we convert the equation $5x^2 - 4y^2 = 80$ to standard form. The result is $\frac{x^2}{16} - \frac{y^2}{20} = 1$. Therefore $a = 4$ and we obtain $(F_1P - F_2P)^2 = 4a^2 = 4(16) = 64$.

6. Dividing by 100 yields the standard form $\frac{x^2}{25} + \frac{y^2}{4} = 1$. The length of the major axis is 10 and the length of the minor axis is 4. Using $c^2 = a^2 - b^2$, we find:

$$c = \sqrt{a^2 - b^2} = \sqrt{25 - 4} = \sqrt{21}$$

So the foci are $(\pm\sqrt{21}, 0)$.

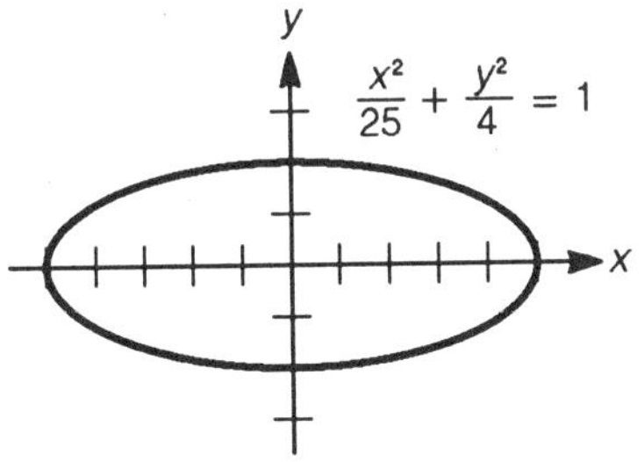

7. We complete the square to convert the equation to standard form:

$$16x^2 + y^2 - 64x + 2y + 65 = 0$$
$$16(x^2 - 4x) + (y^2 + 2y) = -65$$
$$16(x^2 - 4x + 4) + (y^2 + 2y + 1) = -65 + 64 + 1$$
$$16(x - 2)^2 + (y + 1)^2 = 0$$

Since the only solution to this equation is the point $(2,-1)$, the graph consists of a single point.

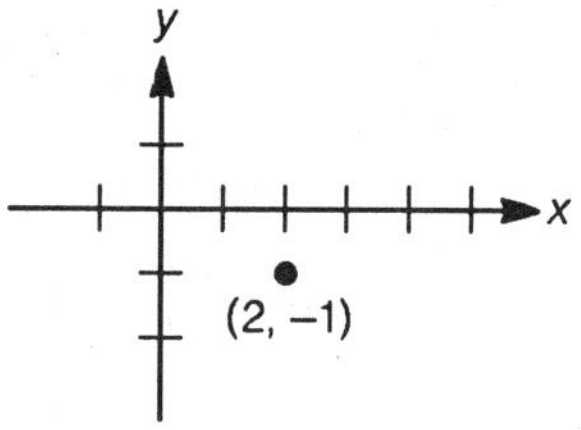

8. The equation represents a hyperbola with center $(-4,4)$. We graph the hyperbola:

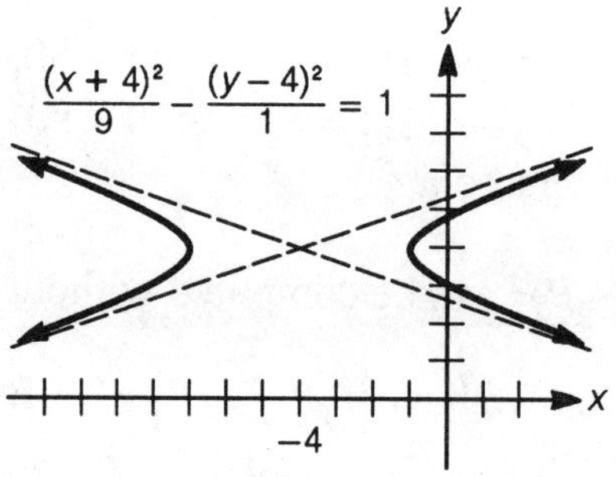

9. Since $4p = 8$, then $p = 2$. The focal width is 8 and the vertex is $(1,2)$.

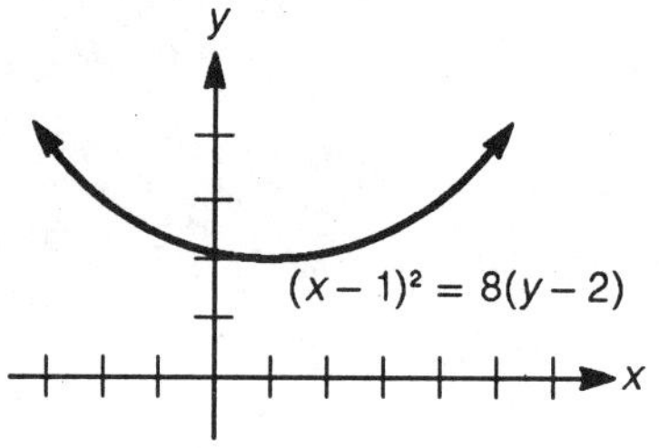

10. Dividing by 225, the ellipse has equation $\frac{x^2}{25} + \frac{y^2}{9} = 1$, so $a = 5$ and $b = 3$. We find c:

$$c = \sqrt{a^2 - b^2} = \sqrt{25 - 9} = \sqrt{16} = 4$$

Since $e = \frac{c}{a} = \frac{4}{5}$, then the directrix is:

$$x = \frac{a}{e} = \frac{5}{4/5} = \frac{25}{4}$$

11. (a) Substituting, we have:

$$\begin{aligned} 9x^2 + 25\left(\tfrac{12}{5}\right)^2 &= 225 \\ 9x^2 + 144 &= 225 \\ 9x^2 &= 81 \\ x^2 &= 9 \\ x &= 3 \qquad \text{(since } x > 0\text{)} \end{aligned}$$

(b) The distance is $\frac{25}{4} - 3 = \frac{13}{4}$.

(c) Since $c = 4$, then F_2 is $(4, 0)$. Using the distance formula:

$$F_2P = \sqrt{(4-3)^2 + \left(0 - \tfrac{12}{5}\right)^2} = \sqrt{1 + \tfrac{144}{25}} = \sqrt{\tfrac{169}{25}} = \tfrac{13}{5}$$

(d) Using the formula $F_2P = a - ex$, we have:

$$F_2P = 5 - \tfrac{4}{5} \bullet 3 = 5 - \tfrac{12}{5} = \tfrac{13}{5}$$

This checks the value from (c).

12. We compute the ratio:

$$\frac{F_2P}{PD} = \frac{\frac{13}{5}}{\frac{13}{4}} = \tfrac{13}{5} \bullet \tfrac{4}{13} = \tfrac{4}{5}$$

This confirms the value of e found in Exercise 10.

Chapter Ten
Additional Topics in Algebra

10.1 Mathematical Induction

1. Here P_n denotes the statement:

$$1+2+3+\ldots+n=\frac{n(n+1)}{2}$$

Since $1=\frac{1(1+1)}{2}=1$, then P_1 is true. It remains to show that P_k implies P_{k+1}.
Assume P_k is true:

$$1+2+\ldots+k=\frac{k(k+1)}{2}$$

Thus:

$$\begin{aligned}1+2+\ldots+k+k+1&=\frac{k(k+1)}{2}+k+1\\&=(k+1)\left(\frac{k}{2}+1\right)\\&=\frac{(k+1)(k+2)}{2}\\&=\frac{(k+1)[(k+1)+1]}{2}\end{aligned}$$

So P_{k+1} is true and the induction is complete.

3. Here P_n denotes the statement:

$$1+4+7+\ldots+(3n-2)=\frac{n(3n-1)}{2}$$

Since $1=\frac{1(3(1)-1)}{2}=1$, then P_1 is true.
Assume P_k is true:

$$1+4+\ldots+(3k-2)=\frac{k(3k-1)}{2}$$

Thus:

$$\begin{aligned}1+4+\ldots+(3k-2)+[3(k+1)-2]&=\frac{k(3k-1)}{2}+3(k+1)-2\\&=\tfrac{1}{2}[k(3k-1)+6(k+1)-4]\\&=\tfrac{1}{2}(3k^2-k+6k+2)\\&=\tfrac{1}{2}(3k^2+5k+2)\\&=\frac{(k+1)(3k+2)}{2}\\&=\frac{(k+1)[3(k+1)-1]}{2}\end{aligned}$$

So P_{k+1} is true and the induction is complete.

5. Here P_n denotes:

$$1^2+2^2+3^2+\ldots+n^2=\frac{n(n+1)(2n+1)}{6}$$

Since $1^2=1=\frac{1(1+1)[2(1)+1]}{6}=\frac{1(2)(3)}{6}=1$, then P_1 is true.
Assume P_k is true:

$$1^2+2^2+\ldots+k^2=\frac{k(k+1)(2k+1)}{6}$$

Thus:

$$\begin{aligned}1^2+2^2+\ldots+k^2+(k+1)^2&=\frac{k(k+1)(2k+1)}{6}+(k+1)^2\\&=\frac{(k+1)[k(2k+1]+6(k+1)^2}{6}\\&=\frac{(k+1)(2k^2+k+6k+6)}{6}\\&=\frac{(k+1)(2k^2+7k+6)}{6}\\&=\frac{(k+1)(k+2)(2k+3)}{6}\\&=\frac{(k+1)[(k+1)+1][2(k+1)+1]}{6}\end{aligned}$$

So P_{k+1} is true and the induction is complete.

7. Here P_n denotes:

$$1^2+3^2+5^2+\ldots+(2n-1)^2=\frac{n(2n-1)(2n+1)}{3}$$

As $1^2=1=\frac{1[2(1)-1][2(1)+1]}{3}=1$, then P_1 is true.
Assume P_k is true:

$$1^2+3^2+\ldots+(2k-1)^2=\frac{k(2k-1)(2k+1)}{3}$$

Thus:

$$\begin{aligned}1^2+3^2+\ldots+(2k-1)^2+[2(k+1)-1]^2&=\frac{k(2k-1)(2k+1)}{3}+[2(k+1)-1]^2\\&=\frac{k(2k-1)(2k+1)}{3}+\frac{3(2k+1)^2}{3}\\&=\frac{2k+1}{3}[k(2k-1)+3(2k+1]\\&=\frac{[2(k+1)-1]}{3}(2k^2+5k+3)\\&=\frac{(k+1)[2(k+1)-1](2k+3)}{3}\\&=\frac{(k+1)[2(k+1)-1][2(k+1]+1}{3}\end{aligned}$$

So P_{k+1} is true and the induction is complete.

9. Here P_n denotes:

$$3+3^2+3^3+\ldots+3^n=\tfrac{1}{2}\left(3^{n+1}-3\right)$$

Since $3=\frac{1}{2}\left(3^2-3\right)=\frac{1}{2}(6)=3$, then P_1 is true.
Assume P_k is true:

$$3+3^2+\ldots+3^k=\tfrac{1}{2}\left(3^{k+1}-3\right)$$

Thus:

$$\begin{aligned}3+3^2+\ldots+3^k+3^{k+1}&=\tfrac{1}{2}\left(3^{k+1}-3\right)+3^{k+1}\\&=\tfrac{3}{2}\left(3^{k+1}\right)-\tfrac{3}{2}\\&=\tfrac{1}{2}\left[3\left(3^{k+1}\right)-3\right]\\&=\tfrac{1}{2}\left(3^{k+2}-3\right)\\&=\tfrac{1}{2}\left[3^{(k+1)+1}-3\right]\end{aligned}$$

So P_{k+1} is true and the induction is complete.

11. Here P_n denotes:

$$1^3+2^3+3^3+\ldots+n^3=\left[\frac{n(n+1)}{2}\right]^2$$

Since $1^3=1=\left[\frac{1(2)}{2}\right]^2=1$, then P_1 is true.
Assume P_k is true:

$$1^3+2^3+\ldots+k^3=\left[\frac{k(k+1)}{2}\right]^2$$

Thus:

$$\begin{aligned}1^3+2^3+\ldots+k^3+(k+1)^3&=\left[\frac{k(k+1)}{2}\right]^2+(k+1)^3\\&=(k+1)^2\left[\frac{k^2}{4}+(k+1)\right]\\&=\frac{(k+1)^2}{4}\left[k^2+4k+4\right]\\&=\frac{(k+1)^2(k+2)^2}{4}\\&=\frac{(k+1)^2[(k+1)+1]^2}{2^2}\\&=\left(\frac{(k+1)[(k+1)+1]}{2}\right)^2\end{aligned}$$

So P_{k+1} is true and the induction is complete.

13. Here P_n denotes:

$$1^3+3^3+5^3+\ldots+(2n-1)^3=n^2\left(2n^2-1\right)$$

Since $1^3=1(1)^2\left[2(1)^2-1\right]=(1)(1)=1$, then P_1 is true.
Assume P_k is true:

$$1^3+3^3+\ldots+(2k-1)^3=k^2\left(2k^2-1\right)$$

Thus:

$$\begin{aligned}1^3+3^3+\ldots+(2k-1)^3+[2(k+1)-1]^3&=k^2\left(2k^2-1\right)+(2k+1)^3\\&=2k^4-k^2+8k^3+12k^2+6k+1\\&=2k^4+8k^3+11k^2+6k+1\\&=(k+1)^2\left(2k^2+4k+1\right)\\&=(k+1)^2\left[2\left(k^2+2k+1\right)-1\right]\\&=(k+1)^2\left[2(k+1)^2-1\right]\end{aligned}$$

So P_{k+1} is true and the induction is complete.

15. Here P_n denotes:

$$1\times 3+3\times 5+5\times 7+\ldots+(2n-1)(2n+1)=\frac{n(4n^2+6n-1)}{3}$$

Since $1\times 3=3=\frac{(1)[4(1)^2+6(1)-1]}{3}=\frac{9}{3}=3$, then P_1 is true.

Assume P_k is true:

$$1\times 3+3\times 5+\ldots+(2k-1)(2k+1)=\frac{k(4k^2+6k-1)}{3}$$

Thus:

$$1\times 3+3\times 5+\ldots+(2k-1)(2k+1)+[2(k+1)-1][2(k+1)+1]$$
$$=\frac{k(4k^2+6k-1)}{3}+(2k+1)(2k+3)$$
$$=\frac{4k^3+6k^2-k+12k^2+24k+9}{3}$$
$$=\frac{4k^3+18k^2+23k+9}{3}$$

Noting that –1 is a root:

$$=\frac{(k+1)(4k^2+14k+9)}{3}$$
$$=\frac{(k+1)\left[4(k+1)^2+6(k+1)-1\right]}{3}$$

So P_{k+1} is true and the induction is complete.

17. Here P_n denotes:

$$1+\frac{3}{2}+\frac{5}{2^2}+\ldots+\frac{2n-1}{2^{n-1}}=6-\frac{2n+3}{2^{n-1}}$$

Since $1=6-\frac{[2(1)+3]}{2^0}=6-5=1$, then P_1 is true.

Assume P_k is true:

$$1+\frac{3}{2}+\ldots+\frac{2k-1}{2^{k-1}}=6-\frac{2k+3}{2^{k-1}}$$

Thus:

$$1+\frac{3}{2}+\ldots+\frac{2k-1}{2^{k-1}}+\frac{2(k+1)-1}{2^k}=6-\frac{2k+3}{2^{k-1}}+\frac{2k+1}{2^k}$$
$$=6-\frac{4k+6-2k-1}{2^k}$$
$$=6-\frac{2k+5}{2^k}$$
$$=6-\frac{2(k+1)+3}{2^{(k+1)-1}}$$

So P_{k+1} is true and the induction is complete.

19. Let P_n denote the statement:

$n \le 2^{n-1}$

Since $2^{1-1}=2^0=1\ge 1$, then P_1 is true.

Assume P_k is true:

$k \le 2^{k-1}$

Hence:

$k+1 \le 2^{k-1}+1$

Since $1 \le 2^{k-1}$ (use induction to prove this), then:

$k+1 \le 2^{k-1}+1 \le 2^{k-1}+2^{k-1}=2\left(2^{k-1}\right)=2^k=2^{(k+1)-1}$

So P_{k+1} is true and the induction is complete.

21. For $n=2$:

$2^2+4=8<(2+1)^2$

Assuming:

$k^2+4<(k+1)^2$

Then:

$k^2+2k+4<(k+1)^2+2k$

$(k+1)^2+3<k^2+4k+1$

$(k+1)^2+4<k^2+4k+2$

$(k+1)^2+4<k^2+4k+4$

$(k+1)^2+4<(k+2)^2$

This completes the induction for $n\ge 2$.

23. For $n=7$, we have:

$(1.5)^7 \approx 17.09>2(7)$, so P_1 is true

Assume P_k is true, so:

$(1.5)^k>2k$

Then for $k \geq 7$:

$$\begin{aligned}(1.5)^{k+1} &= 1.5(1.5)^k \\ &> 1.5(2k) \\ &= 3k \\ &= 2k + k \\ &> 2k + 2 \quad (\text{since } k > 2) \\ &= 2(k+1)\end{aligned}$$

So P_{k+1} is true and the induction is complete.

25. For $n = 39$, we have:

$(1.1)^{39} \approx 41.1 > 39$, so P_{39} is true

Assume P_k is true for some $k \geq 39$, so:

$(1.1)^k > k$

Then:

$$\begin{aligned}(1.1)^{k+1} &= 1.1(1.1)^k \\ &> 1.1k \\ &= k + 0.1k \\ &> k + 1 \quad (\text{since } k > 10)\end{aligned}$$

So P_{k+1} is true and the induction is complete.

27. (a) We complete the table:

n	1	2	3	4	5
$f(n)$	$\frac{1}{2}$	$\frac{2}{3}$	$\frac{3}{4}$	$\frac{4}{5}$	$\frac{5}{6}$

(b) $\frac{6}{7}$, $f(6) = f(5) + \frac{1}{42} = \frac{5}{6} + \frac{1}{42} = \frac{36}{42} = \frac{6}{7}$

(c) $f(n) = \dfrac{1}{1\times 2} + \dfrac{1}{2\times 3} + \ldots + \dfrac{1}{n(n+1)} = \dfrac{n}{n+1}$

$f(1) = \dfrac{1}{2} = \dfrac{1}{1+1} = \dfrac{1}{2}$

Assuming $f(k)=\dfrac{k}{k+1}$, then:

$$\begin{aligned}f(k+1)&=f(k)+\frac{1}{(k+1)(k+2)}\\&=\frac{k}{k+1}+\frac{1}{(k+1)(k+2)}\\&=\frac{k(k+2)+1}{(k+1)(k+2)}\\&=\frac{k^2+2k+1}{(k+1)(k+2)}\\&=\frac{(k+1)^2}{(k+1)(k+2)}\\&=\frac{k+1}{k+2}\end{aligned}$$

This completes the induction.

29. (a) We complete the table:

n	1	2	3	4	5
$f(n)$	1	4	9	16	25

(b) 36, $f(6)=f(5)+2\sqrt{f(5)}+1=25+2\sqrt{25}+1=36$

(c) $f(n) = n^2$
$f(1) = 1 = 1^2$
Assume that for some $k \geq 1$ that $f(k) = k^2$. Then we have:

$$f(k+1)=f(k)+2\sqrt{f(k)}+1=k^2+2\sqrt{k^2}+1=k^2+2k+1=(k+1)^2$$

The induction is complete.

31. For $n=1$: 13 is prime and P_1 is true
For $n=2$: 17 is prime and P_2 is true
For $n=3$: 23 is prime and P_3 is true
For $n=4$: 31 is prime and P_4 is true
For $n=5$: 41 is prime and P_5 is true
For $n=6$: 53 is prime and P_6 is true
For $n=7$: 67 is prime and P_7 is true
For $n=8$: 83 is prime and P_8 is true
For $n=9$: 101 is prime and P_9 is true
However, for $n = 10$, we have $(10)^2 + 10 + 11 = 121 = (11)(11)$ and so P_{10} is false.

33. Let P_n denote:

$$1+r+r^2+\ldots+r^{n-1}=\frac{r^n-1}{r-1} \text{ when } r \neq 1$$

Then since $1=\dfrac{r-1}{r-1}=1$, then P_1 is true.
Assuming P_k is true, then:

$$1+r+\ldots+r^{k-1}=\frac{r^k-1}{r-1}$$

Thus:

$$1+r+\ldots+r^{k-1}+r^k=\frac{r^k-1}{r-1}+r^k=\frac{r^k-1+r^k(r-1)}{r-1}=\frac{r^{k+1}-1}{r-1}$$

So P_{k+1} is true and the induction is complete.

35. Let P_n denote:

$n^5-n=5R$ for some natural number R

Since $2^5-2=32-2=30=5(6)$, then P_2 is true.
Since:

$$(k+1)^5-(k+1)=k^5+5k^4+10k^3+10k^2+4k=k^5-k+5\left(k^4+2k^3+2k^2+k\right)$$

Assuming P_k is true results in:

$$(k+1)^5-(k+1)=5(R+k^4+2k^3+2k^2+k)$$

So P_{k+1} is true and the induction is complete.

37. For $n=0$, we have:

$2^1+3^1=5=5(1)$, so P_0 is true

Assume P_k is true, so:

$2^{2k+1}+3^{2k+1}=5A$, for some natural number A

Then:

$$\begin{aligned}2^{2k+3}+3^{2k+3}&=4\bullet 2^{2k+1}+9\bullet 3^{2k+1}\\&=4\bullet 2^{2k+1}+9\left(5A-2^{2k+1}\right)\\&=4\bullet 2^{2k+1}+45A-9\bullet 2^{2k+1}\\&=45A-5\bullet 2^{2k+1}\\&=5\left(9A-2^{2k+1}\right)\end{aligned}$$

Since 5 is a factor of this expression, P_{k+1} is true and the induction is complete.

39. For $n=0$, we have:

$2^1+(-1)^0=2+1=3=3(1)$, so P_0 is true

Assume P_k is true, so:

$2^{k+1}+(-1)^k=3A$, for some natural number A

Then:

$$\begin{aligned}2^{k+2}+(-1)^{k+1} &= 2\bullet 2^{k+1}-(-1)^k\\ &= 2\left[3A-(-1)^k\right]-(-1)^k\\ &= 6A-2(-1)^k-(-1)^k\\ &= 6A-3(-1)^k\\ &= 3\left[2A-(-1)^k\right]\end{aligned}$$

Since 3 is a factor of this expression, P_{k+1} is true and the induction is complete.

41. Let P_n denote:

$x^n - y^n = (x-y)(Q(x,y))$ where $Q(x,y)$ is a polynomial in x and y.

Then:

$x-y=(x-y)(1)$, so P_1 is true

While using the suggested identity:

$$x^{k+1}-y^{k+1}=x^k(x-y)+(x^k-y^k)y$$

So assuming P_k, we have $x^k - y^k = (x-y)Q(x,y)$:

$$x^{k+1}-y^{k+1}=\left(x^k+yQ(x,y)\right)(x-y)$$

So P_{k+1} is true and the induction is complete.

43. For $n = 1$:

$(1+p)\ge(1+p)$, so P_1 is true

When $p > -1$ then $p + 1 > 0$. Assuming $(1 + p)^k \ge 1 + kp$ then:

$$(1+p)^{k+1}\ge(1+kp)(1+p)\ge 1+p+kp+kp^2=1+(k+1)p+kp^2\ge 1+(k+1)p$$

The induction is complete.

10.2 The Binomial Theorem

1. (a) We compute:

$$(a+b)^2=(a+b)(a+b)=a^2+ab+ba+b^2=a^2+2ab+b^2$$

(b) We compute:

$$\begin{aligned}(a+b)^3 &= (a+b)(a+b)^2\\ &= (a+b)\left(a^2+2ab+b^2\right)\\ &= a^3+2a^2b+ab^2+ba^2+2ab^2+b^3\\ &= a^3+3a^2b+3ab^2+b^3\end{aligned}$$

3. Using the binomial theorem, we expand:

$$\begin{aligned}(a+b)^9 &= \binom{9}{0}a^9+\binom{9}{1}a^8b+\binom{9}{2}a^7b^2+\binom{9}{3}a^6b^3+\binom{9}{4}a^5b^4+\binom{9}{5}a^4b^5\\ &\quad+\binom{9}{6}a^3b^6+\binom{9}{7}a^2b^7+\binom{9}{8}ab^8+\binom{9}{9}b^9\\ &= a^9+9a^8b+36a^7b^2+84a^6b^3+126a^5b^4+126a^4b^5+84a^3b^6\\ &\quad+36a^2b^7+9ab^8+b^9\end{aligned}$$

5. Using the binomial theorem, we expand:

$$\begin{aligned}(2A+B)^3 &= \binom{3}{0}(2A)^3+\binom{3}{1}(2A)^2B+\binom{3}{2}(2A)B^2+\binom{3}{3}B^3\\ &= 8A^3+12A^2B+6AB^2+B^3\end{aligned}$$

7. Using the binomial theorem, we expand:

$$\begin{aligned}(1-2x)^6 &= \binom{6}{0}1^6-\binom{6}{1}1^5(2x)+\binom{6}{2}1^4(2x)^2-\binom{6}{3}1^3(2x)^3\\ &\quad+\binom{6}{4}1^2(2x)^4-\binom{6}{5}1(2x)^5+\binom{6}{6}(2x)^6\\ &= 1-12x+60x^2-160x^3+240x^4-192x^5+64x^6\end{aligned}$$

9. Using the binomial theorem, we expand:

$$\begin{aligned}&\left(\sqrt{x}+\sqrt{y}\right)^4\\ &= \binom{4}{0}\left(\sqrt{x}\right)^4+\binom{4}{1}\left(\sqrt{x}\right)^3\left(\sqrt{y}\right)+\binom{4}{2}\left(\sqrt{x}\right)^2\left(\sqrt{y}\right)^2+\binom{4}{3}\left(\sqrt{x}\right)\left(\sqrt{y}\right)^3+\binom{4}{4}\left(\sqrt{y}\right)^4\\ &= x^2+4x\sqrt{xy}+6xy+4y\sqrt{xy}+y^2\end{aligned}$$

11. Using the binomial theorem, we expand:

$$\begin{aligned}\left(x^2+y^2\right)^5 &= \binom{5}{0}\left(x^2\right)^5+\binom{5}{1}\left(x^2\right)^4y^2+\binom{5}{2}\left(x^2\right)^3\left(y^2\right)^2+\binom{5}{3}\left(x^2\right)^2\left(y^2\right)^3\\ &\quad+\binom{5}{4}\left(x^2\right)\left(y^2\right)^4+\binom{5}{5}\left(y^2\right)^5\\ &= x^{10}+5x^8y^2+10x^6y^4+10x^4y^6+5x^2y^8+y^{10}\end{aligned}$$

13. Using the binomial theorem, we expand:

$$\begin{aligned}\left(1-\frac{1}{x}\right)^6 &= \binom{6}{0}-\binom{6}{1}\frac{1}{x}+\binom{6}{2}\left(\frac{1}{x}\right)^2-\binom{6}{3}\left(\frac{1}{x}\right)^3+\binom{6}{4}\left(\frac{1}{x}\right)^4-\binom{6}{5}\left(\frac{1}{x}\right)^5+\binom{6}{6}\left(\frac{1}{x}\right)^6\\ &= 1-\frac{6}{x}+\frac{15}{x^2}-\frac{20}{x^3}+\frac{15}{x^4}-\frac{6}{x^5}+\frac{1}{x^6}\end{aligned}$$

15. Using the binomial theorem, we expand:

$$\left(\frac{x}{2}-\frac{y}{3}\right)^3=\binom{3}{0}\left(\frac{x}{2}\right)^3-\binom{3}{1}\left(\frac{x}{2}\right)^2\left(\frac{y}{3}\right)+\binom{3}{2}\left(\frac{x}{2}\right)\left(\frac{y}{3}\right)^2-\binom{3}{3}\left(\frac{y}{3}\right)^3$$
$$=\frac{x^3}{8}-\frac{x^2y}{4}+\frac{xy^2}{6}-\frac{y^3}{27}$$

17. Using the binomial theorem, we expand:

$$(ab^2+c)^7=\binom{7}{0}(ab^2)^7+\binom{7}{1}(ab^2)^6c+\binom{7}{2}(ab^2)^5c^2+\binom{7}{3}(ab^2)^4c^3+\binom{7}{4}(ab^2)^3c^4$$
$$+\binom{7}{5}(ab^2)^2c^5+\binom{7}{6}(ab^2)c^6+\binom{7}{7}c^7$$
$$=a^7b^{14}+7a^6b^{12}c+21a^5b^{10}c^2+35a^4b^8c^3+35a^3b^6c^4+21a^2b^4c^5$$
$$+7ab^2c^6+c^7$$

19. Using the binomial theorem, we expand:

$$(x+\sqrt{2})^8=\binom{8}{0}x^8+\binom{8}{1}x^7\sqrt{2}+\binom{8}{2}x^6(\sqrt{2})^2+\binom{8}{3}x^5(\sqrt{2})^3+\binom{8}{4}x^4(\sqrt{2})^4$$
$$+\binom{8}{5}x^3(\sqrt{2})^5+\binom{8}{6}x^2(\sqrt{2})^6+\binom{8}{7}x(\sqrt{2})^7+\binom{8}{8}(\sqrt{2})^8$$
$$=x^8+8\sqrt{2}x^7+56x^6+112\sqrt{2}x^5+280x^4+224\sqrt{2}x^3+224x^2$$
$$+64\sqrt{2}x+16$$

21. Using the binomial theorem,we expand:

$$(\sqrt{2}-1)^3=\binom{3}{0}(\sqrt{2})^3-\binom{3}{1}(\sqrt{2})^2+\binom{3}{2}(\sqrt{2})-\binom{3}{3}=2\sqrt{2}-6+3\sqrt{2}-1=5\sqrt{2}-7$$

23. Using the binomial theorem, we expand:

$$(\sqrt{2}+\sqrt{3})^5=\binom{5}{0}(\sqrt{2})^5+\binom{5}{1}(\sqrt{2})^4\sqrt{3}+\binom{5}{2}(\sqrt{2})^3(\sqrt{3})^2+\binom{5}{3}(\sqrt{2})^2(\sqrt{3})^3$$
$$+\binom{5}{4}(\sqrt{2})(\sqrt{3})^4+\binom{5}{5}(\sqrt{3})^5$$
$$=4\sqrt{2}+20\sqrt{3}+60\sqrt{2}+60\sqrt{3}+45\sqrt{2}+9\sqrt{3}$$
$$=89\sqrt{3}+109\sqrt{2}$$

25. Using the binomial theorem, we expand:

$$(2\sqrt[3]{2}-\sqrt[3]{4})^3=\binom{3}{0}(2\sqrt[3]{2})^3-\binom{3}{1}(2\sqrt[3]{2})^2(\sqrt[3]{4})+\binom{3}{2}(2\sqrt[3]{2})(\sqrt[3]{4})^2-\binom{3}{3}(\sqrt[3]{4})^3$$
$$=16-12\sqrt[3]{16}+6\sqrt[3]{32}-4$$
$$=12-24\sqrt[3]{2}+12\sqrt[3]{4}$$

27. Using the binomial theorem, we expand:

$$\begin{aligned}
&\left[x^2-(2x+1)\right]^5\\
&=\binom{5}{0}(x^2)^5-\binom{5}{1}(x^2)^4(2x+1)+\binom{5}{2}(x^2)^3(2x+1)^2-\binom{5}{3}(x^2)^2(2x+1)^3\\
&\qquad+\binom{5}{4}(x^2)(2x+1)^4-\binom{5}{5}(2x+1)^5\\
&=x^{10}-\left[5x^8(2x+1)\right]+\left[10x^6\left(4x^2+4x+1\right)\right]-\left[10x^4\left(8x^3+12x^2+6x+1\right)\right]\\
&\qquad+\left[5x^2\left(16x^4+32x^3+24x^2+8x+1\right)\right]\\
&\qquad-\left(32x^5+80x^4+80x^3+40x^2+10x+1\right)\\
&=x^{10}-\left(10x^9+5x^8\right)+\left(40x^8+40x^7+10x^6\right)-\left(80x^7+120x^6+60x^5+10x^4\right)\\
&\qquad+\left(80x^6+160x^5+120x^4+40x^3+5x^2\right)\\
&\qquad-\left(32x^5+80x^4+80x^3+40x^2+10x+1\right)\\
&=x^{10}-10x^9+35x^8-40x^7-30x^6+68x^5+30x^4-40x^3-35x^2-10x-1
\end{aligned}$$

29. $5!=(5)(4)(3)(2)(1)=120$

31. We compute:

$$\binom{7}{3}\binom{3}{2}=\frac{7!}{3!(4!)}\bullet\frac{3!}{2!(1!)}=\frac{7\bullet 6\bullet 5}{3\bullet 2}\bullet 3=105$$

33. (a) We compute:

$$\binom{5}{3}=\frac{5!}{3!(5-3)!}=\frac{5(4)}{2}=10$$

(b) We compute:

$$\binom{5}{4}=\frac{5!}{4!(5-4)!}=5$$

35. We simplify:

$$\frac{(n+2)!}{n!}=\frac{(n+2)(n+1)n!}{n!}=n^2+3n+2$$

37. We compute:

$$\begin{aligned}
\binom{6}{4}+\binom{6}{3}-\binom{7}{4}&=\frac{6!}{4!2!}+\frac{6!}{3!3!}-\frac{7!}{4!3!}\\
&=\frac{(6)(5)}{2}+\frac{(6)(5)(4)}{6}-\frac{(7)(6)(5)}{6}\\
&=15+20-35\\
&=0
\end{aligned}$$

39. The fifteenth term will be given by:

$$\binom{16}{14}a^2b^{14} = 120a^2b^{14}$$

41. The one-hundredth term will be given by:

$$\binom{100}{99}x^{99} = 100x^{99}$$

43. Here $n - r + 1 = 8$, so $10 - r + 1 = 8$, so $r = 3$. Hence the coefficient is:

$$\binom{10}{2} = 45$$

45. Here $r - 1 = 8$, thus $r = 9$. Hence the coefficient is:

$$\frac{1}{2}(-4)^8\binom{9}{8} = 9 \bullet \frac{1}{2}(65536) = 294912$$

47. Here $r - 1 = 6$. Hence $r = 7$, and the coefficient is:

$$\binom{8}{6} = \frac{8(7)}{2} = 28$$

49. Here $(12 - r + 1)(-1) + 2(r - 1) = 0$, so $-12 + r - 1 + 2r - 2 = 0$, then $3r = 15$, so $r = 5$ and the coefficient is:

$$(3)^4\binom{12}{4} = \frac{(12)(11)(10)(9)}{(4)(3)(2)}(3)^4 = 495(81) = 40095$$

51. Here $r - 1 = n$, so $r = n + 1$, so the coefficient is:

$$\binom{2n}{n} = \frac{(2n)!}{n!n!} = \frac{(2n)!}{(n!)^2}$$

53. (a) We complete the table:

k	0	1	2	3	4	5	6	7	8
$\binom{8}{k}$	1	8	28	56	70	56	28	8	1

(b) We have $1 + 8 + 28 + 56 + 70 + 56 + 28 + 8 + 1 = 256$, while $2^8 = 256$.

(c) Since a^L and b^L equal 1 for any L since $a = b = 1$, then from the binomial theorem:

$$2^n = (1+1)^n = \binom{n}{0} + \binom{n}{1} + \ldots + \binom{n}{n}$$

55. (a) We simplify:

$$(1+x)^n\left(1+\frac{1}{x}\right)^n=(1+x)^n\left(\frac{x+1}{x}\right)^n=(1+x)^n\bullet\frac{(1+x)^n}{x^n}=\frac{(1+x)^{2n}}{x^n}$$

(b) The nth term of expansion on the right is:

$$\frac{\binom{2n}{n}(1)^n(x)^n}{x^n}=\binom{2n}{n}$$

This verifies the result.

(c) Using the binomial theorem:

$$(1+x)^n=\binom{n}{0}+\binom{n}{1}x+\binom{n}{2}x^2+\binom{n}{3}x^3+\ldots+\binom{n}{n}x^n$$

$$\left(1+\frac{1}{x}\right)^n=\binom{n}{0}+\binom{n}{1}x^{-1}+\binom{n}{2}x^{-2}+\binom{n}{3}x^{-3}+\ldots+\binom{n}{n}x^{-n}$$

When multiplied, the terms that do not contain x come from terms with corresponding positive and negative exponents:

$$\binom{n}{1}x\bullet\binom{n}{1}x^{-1}+\binom{n}{2}x^2\bullet\binom{n}{2}x^{-2}+\ldots+\binom{n}{n}x^n\bullet\binom{n}{n}x^{-n}$$

$$=\binom{n}{1}^2+\binom{n}{2}^2+\ldots+\binom{n}{n}^2$$

This verifies the identity.

10.3 Introduction to Sequences and Series

1. Since $a_n=\frac{n}{n+1}$, we can write out the first five terms by setting n equal to the natural numbers 1 through 5:

$$a_1=\frac{1}{1+1}=\frac{1}{2},\quad a_2=\frac{2}{2+1}=\frac{2}{3},\quad a_3=\frac{3}{3+1}=\frac{3}{4},\quad a_4=\frac{4}{4+1}=\frac{4}{5},\quad a_5=\frac{5}{5+1}=\frac{5}{6}$$

So the first five terms are $\frac{1}{2},\frac{2}{3},\frac{3}{4},\frac{4}{5},\frac{5}{6}$.

3. Again, letting n be successively 1 through 5, we obtain:

$$a_1=(1-1)^2,\quad a_2=(2-1)^2,\quad a_3=(3-1)^2,\quad a_4=(4-1)^2,\quad a_5=(5-1)^2$$

So the first five terms are 0, 1, 4, 9, 16.

5. Here $b_n=\left(1+\frac{1}{n}\right)^n$, so $b_1=2, b_2=\frac{9}{4}, b_3=\frac{64}{27}, b_4=\frac{625}{256}$, and $b_5=\frac{7776}{3125}$.

7. Here $u_n=(-1)^n$, so $u_1=-1$, $u_2=1$, $u_3=-1$, $u_4=1$, and $u_5=-1$.

9. Here $a_n = \dfrac{(-1)^{n+1}}{n!}$, so $a_1 = 1$, $a_2 = -\frac{1}{2}$, $a_3 = \frac{1}{6}$, $a_4 = -\frac{1}{24}$, and $a_5 = \frac{1}{120}$.

11. Here $a_n = 3n$, so $a_1 = 3$, $a_2 = 6$, $a_3 = 9$, $a_4 = 12$, and $a_5 = 15$.

13. Here $a_1 = 1$, so $a_n = (1 + a_{n-1})^2$ for $n \ge 2$, so:

$$a_1 = 1$$
$$a_2 = (1 + a_1)^2 = 2^2 = 4$$
$$a_3 = (1 + a_2)^2 = 5^2 = 25$$
$$a_4 = (1 + a_3)^2 = 26^2 = 676$$
$$a_5 = (1 + a_4)^2 = 677^2 = 458329$$

The first five terms are 1, 4, 25, 676, 458329.

15. Here $a_1 = 2$, $a_2 = 2$, and $a_n = a_{n-1}a_{n-2}$ for $n \ge 3$, so $a_1 = 2$, $a_2 = 2$, $a_3 = 4$, $a_4 = 8$, and $a_5 = 32$.

17. Here $a_1 = 1$, $a_{n+1} = na_n$ for $n \ge 1$, so $a_1 = 1$, $a_2 = 1$, $a_3 = 2$, $a_4 = 6$, and $a_5 = 24$.

19. Here $a_1 = 0$, $a_n = 2^{a_{n-1}}$ for $n \ge 2$, so $a_1 = 0$, $a_2 = 1$, $a_3 = 2$, $a_4 = 4$, and $a_5 = 16$.

21. First we compute the required values of a_n as shown in the table:

n	1	2	3	4
a_n	−1	1	−1	1

Now graph these values:

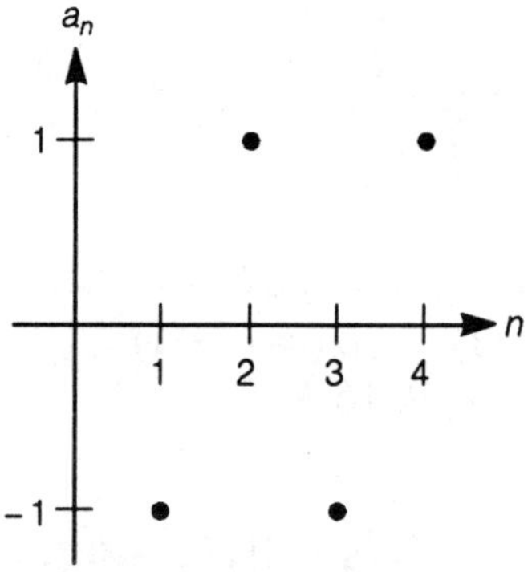

23. First we compute the required values of a_n as shown in the table:

n	1	2	3	4
a_n	1	$\frac{1}{2}$	$\frac{1}{3}$	$\frac{1}{4}$

Now graph these values:

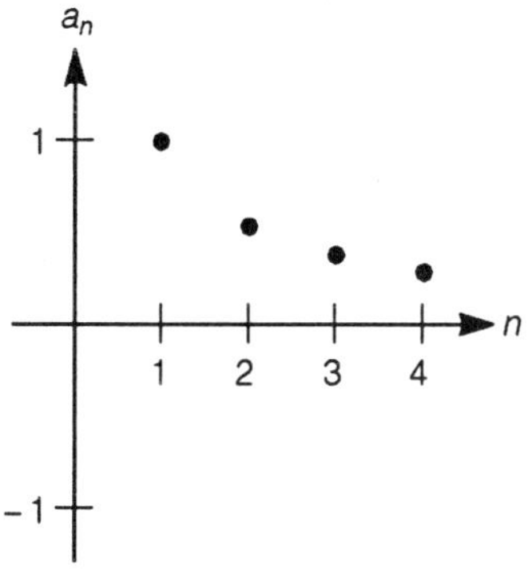

25. First we compute the required values of a_n as shown in the table:

n	1	2	3
a_n	0	$\frac{1}{3}$	$\frac{1}{2}$

Now graph these values:

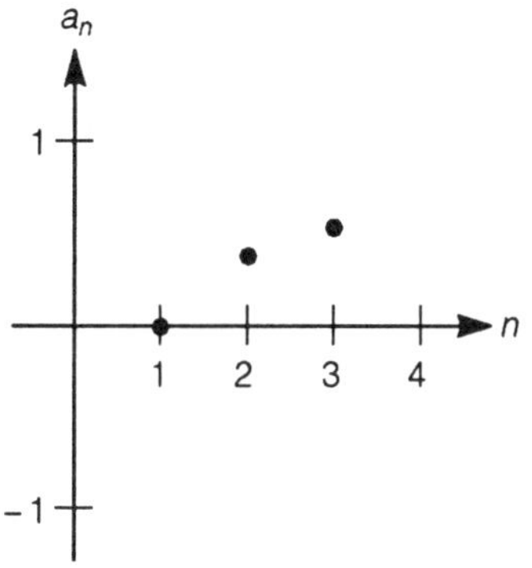

27. Since $a_n = 2^n$, the sum is $2 + 4 + 8 + 16 + 32 = 62$.

29. Since $a_n = n^2 - n$, the sum is $0 + 2 + 6 + 12 + 20 = 40$.

31. Since $a_n = \dfrac{(-1)^n}{n!}$, the sum is $-1 + \frac{1}{2} - \frac{1}{6} + \frac{1}{24} - \frac{1}{120} = -\frac{19}{30}$.

33. Here $a_1 = 1$, $a_2 = 2$, and $a_n = a^2{}_{n-1} + a^2{}_{n-2}$ for $n \geq 3$, so the sum is $1 + 2 + 5 + 29 + 866 = 903$.

35. Here $a_1 = 2$ and $a_n = (a_{n-1})^2$ for $n \geq 2$, so the sum is $2 + 4 + 16 + 256 = 278$.

37. We expand the sum:

$$\sum_{k=1}^{3}(k-1)=0+1+2=3$$

39. We expand the sum:

$$\sum_{k=4}^{5}k^2=16+25=41$$

41. We expand the sum:

$$\sum_{n=1}^{3}x^n=x+x^2+x^3$$

43. We expand the sum:

$$\sum_{n=1}^{4}\frac{1}{n}=1+\frac{1}{2}+\frac{1}{3}+\frac{1}{4}=\frac{25}{12}$$

45. We expand the sum:

$$\begin{aligned}\sum_{j=1}^{9}\log_{10}\frac{j}{j+1}&=\log_{10}\tfrac{1}{2}+\log_{10}\tfrac{2}{3}+\log_{10}\tfrac{3}{4}+\log_{10}\tfrac{4}{5}+\log_{10}\tfrac{5}{6}\\&\quad+\log_{10}\tfrac{6}{7}+\log_{10}\tfrac{7}{8}+\log_{10}\tfrac{8}{9}+\log_{10}\tfrac{9}{10}\\&=\log_{10}\left(\tfrac{1}{2}\bullet\tfrac{2}{3}\bullet\tfrac{3}{4}\bullet\tfrac{4}{5}\bullet\tfrac{5}{6}\bullet\tfrac{6}{7}\bullet\tfrac{7}{8}\bullet\tfrac{8}{9}\bullet\tfrac{9}{10}\right)\\&=\log_{10}\tfrac{1}{10}\\&=-1\end{aligned}$$

47. We expand the sum:

$$\sum_{j=1}^{6}\left(\frac{1}{j}-\frac{1}{j+1}\right)=\left(1-\tfrac{1}{2}\right)+\left(\tfrac{1}{2}-\tfrac{1}{3}\right)+\left(\tfrac{1}{3}-\tfrac{1}{4}\right)+\left(\tfrac{1}{4}-\tfrac{1}{5}\right)+\left(\tfrac{1}{5}-\tfrac{1}{6}\right)+\left(\tfrac{1}{6}-\tfrac{1}{7}\right)=1-\tfrac{1}{7}=\tfrac{6}{7}$$

49. We write the sum in sigma notation:

$$5+5^2+5^3+5^4=\sum_{j=1}^{4}5^j$$

51. We write the sum in sigma notation:

$$x+x^2+x^3+x^4+x^5+x^6=\sum_{j=1}^{6}x^j$$

53. We write the sum in sigma notation:

$$1+\tfrac{1}{2}+\tfrac{1}{3}+\ldots+\tfrac{1}{12}=\sum_{k=1}^{12}\frac{1}{k}$$

55. We write the sum in sigma notation:

$$2-2^2+2^3-2^4+2^5=\sum_{j=1}^{5}(-1)^{j+1}2^j$$

57. We write the sum in sigma notation:

$$1-2+3-4+5=\sum_{j=1}^{5}(-1)^{j+1}j$$

59. (a) Here $F_1=1, F_2=1$, and $F_{n+2}=F_n+F_{n+1}$ for $n\geq 1$.
Evaluating the first 10 terms, we obtain the table:

F_1	F_2	F_3	F_4	F_5	F_6	F_7	F_8	F_9	F_{10}
1	1	2	3	5	8	13	21	34	55

(b) We complete the chart:

n	$F_1+F_2+\ldots+F_n$	$F_{n+2}-1$
1	1	1
2	2	2
3	4	4
4	7	7
5	12	12

(c) The chart completed in (b) shows the hypothesis to be true for $n=1$. Assuming:

$$F_1+F_2+\ldots+F_k=F_{k+2}-1$$

Therefore:

$$F_1+F_2+\ldots+F_k+F_{k+1}=F_{k+2}-1+F_{k+1}=F_{k+3}-1=F_{(k+1)+2}-1$$

This completes the induction.

(d) From (a), $F_5=5\geq 5$, so the hypothesis is true for $n=5$. For some $k\geq 5$, assume:

$$F_k\geq k$$

Then:

$$F_{k+1}=F_k+F_{k-1}\geq F_k+1\geq k+1$$

This completes the induction.

(e) Since $F_1{}^2=1^2=1=F_1F_2=1(1)=1$, then the hypothesis is true for $n=1$.
Assuming:

$$F_1^2+F_2^2+\ldots+F_k^2=F_kF_{k+1}$$

Then:

$$\begin{aligned}F_1^2+F_2^2+\ldots+F_k^2+F^2{}_{k+1}&=F_kF_{k+1}+F^2{}_{k+1}\\&=F_{k+1}(F_k+F_{k+1})\\&=F_{k+1}(F_{k+2})\end{aligned}$$

This completes the induction.

(f) Since $F_2{}^2 = 1^2 = 1 = F_1F_3 + (-1)^1 = 2 - 1 = 1$, the hypothesis is true for $n = 1$.
Assume:

$$F^2{}_{k+1} = F_kF_{k+2} + (-1)^k$$

Then:

$$\begin{aligned} F_{k+1}F_{k+2} + F^2{}_{k+1} &= F_kF_{k+2} + F_{k+1}F_{k+2} + (-1)^k \\ F_{k+1}\left(F_{k+2} + F_{k+1}\right) &= F_{k+2}\left(F_k + F_{k+1}\right) + (-1)^k \\ F_{k+1}F_{k+3} &= F^2{}_{k+2} + (-1)^k \\ F^2{}_{k+2} &= F_{k+1}F_{k+3} + (-1)^{k+1} \end{aligned}$$

This completes the induction.

(g) Since $\begin{pmatrix} 1 & 1 \\ 1 & 0 \end{pmatrix}^2 = \begin{pmatrix} 2 & 1 \\ 1 & 1 \end{pmatrix} = \begin{pmatrix} F_3 & F_2 \\ F_2 & F_1 \end{pmatrix}$ then the hypothesis is true for $n = 2$. Assuming:

$$\begin{pmatrix} 1 & 1 \\ 1 & 0 \end{pmatrix}^k = \begin{pmatrix} F_{k+1} & F_k \\ F_k & F_{k-1} \end{pmatrix}$$

Then:

$$\begin{aligned} \begin{pmatrix} 1 & 1 \\ 1 & 0 \end{pmatrix}^{k+1} &= \begin{pmatrix} 1 & 1 \\ 1 & 0 \end{pmatrix}\begin{pmatrix} F_{k+1} & F_k \\ F_k & F_{k-1} \end{pmatrix} \\ &= \begin{pmatrix} F_k + F_{k+1} & F_k + F_{k-1} \\ F_{k+1} & F_k \end{pmatrix} \\ &= \begin{pmatrix} F_{k+2} & F_{k+1} \\ F_{k+1} & F_k \end{pmatrix} \end{aligned}$$

This completes the induction.

10.4 Arithmetic Sequences and Series

1. (a) To find the common difference d, we subtract any term from the succeeding term. Here that can be:

$$3 - 1 = 2$$
$$5 - 3 = 2$$
$$7 - 5 = 2$$

So 2 is the required common difference.

(b) Again we subtract terms:

$$6 - 10 = -4$$
$$2 - 6 = -4$$
$$-2 - 2 = -4$$

The common difference is -4. It is not necessary to try all three pairs, but this helps to verify that we have an arithmetic sequence.

(c) We subtract:

$$1-\tfrac{2}{3}=\tfrac{4}{3}-1=\tfrac{5}{3}-\tfrac{4}{3}=\tfrac{1}{3}$$

So $\tfrac{1}{3}$ is the common difference.

(d) We subtract:

$$1+\sqrt{2}-1=\sqrt{2} \text{ or } (1+2\sqrt{2})-(1+\sqrt{2})=\sqrt{2}$$

Here $\sqrt{2}$ is the common difference.

3. Since $a = 10$ and $d = 11$, then using $a_n = a + (n-1)d$ we find:

$$a_{12}=10+(12-1)11=131$$

5. $a_{100} = 6 + (100-1)(5) = 6 + 495 = 501$

7. $a_{1000} = -1 + (1000-1)(1) = 998$

9. Here $a_4 = -6$ and $a_{10} = 5$, thus:

$$-6=a+3d$$
$$5=a+9d$$

Subtracting the second equation from the first yields:

$$-6d=-11$$
$$d=\tfrac{11}{6}$$

Therefore:

$$a=5-9\left(\tfrac{11}{6}\right)=5-\tfrac{33}{2}=-\tfrac{23}{2}$$

11. Here $a_{60} = 105$ and $d = 5$, so:

$$105=a+59(5)$$
$$105=a+295$$
$$a=-190$$

13. Since $a_{15} = a + 14d$ and $a_7 = a + 6d$, then:

$$a_{15}-a_7=8d$$
$$-1=8d$$
$$d=-\tfrac{1}{8}$$

15. Since $a = 1$ and $d = 1$, then:

$$S_{1000}=\tfrac{1000}{2}(2+999)=500500$$

17. For $\frac{\pi}{3}+\frac{2\pi}{3}+\pi+\frac{4\pi}{3}+\ldots+\frac{13\pi}{3}$, $a=\frac{\pi}{3}$, $d=\frac{\pi}{3}$, so the sum is:

$$S_{13}=\frac{13}{2}\left(\frac{2\pi}{3}+\frac{12\pi}{3}\right)=\frac{14(13)(\pi)}{6}=\frac{91\pi}{3}$$

19. Since $d = 5$ and $S_{38} = 3534$, then:

$$\begin{aligned} 3534 &= \tfrac{38}{2}[2a + 37(5)] \\ 3534 &= 38a + 3515 \\ 19 &= 38a \\ a &= \tfrac{1}{2} \end{aligned}$$

21. Here $a = 4$, $a_{16} = -100$, therefore:

$$S_{16} = 16\left(\tfrac{4-100}{2}\right) = -768$$

So we have:

$$\begin{aligned} -768 &= \tfrac{16}{2}(2(4) + 15d) \\ -768 &= 64 + 120d \\ -832 &= 120d \\ d &= -\tfrac{104}{15} \end{aligned}$$

23. Since $a_8 = 5$ and $S_{10} = 20$, then:

$$\begin{aligned} 5 &= a + 7d \\ 20 &= 5(2a + 9d) \\ 4 &= 2a + 9d \end{aligned}$$

We solve the system:

$$\begin{aligned} a + 7d &= 5 \\ 2a + 9d &= 4 \end{aligned}$$

Multiplying the first equation by –2 and adding to the second equation:

$$\begin{aligned} -2a - 14d &= -10 \\ 2a + 9d &= 4 \\ -5d &= -6 \\ d &= \tfrac{6}{5} \end{aligned}$$

Therefore:

$$a = 5 - 7\left(\tfrac{6}{5}\right) = -\tfrac{17}{5}$$

25. $S = S_{20} = \sum_{k=1}^{20}(4k + 3)$ so $a = 7$ and $a_{20} = 83$, thus:

$$S = 20\left(\tfrac{7+83}{2}\right) = 900$$

27. Let x denote the middle term, so the three terms are $x - d, x, x + d$:

$$\begin{aligned} x - d + x + x + d &= 30 \\ 3x &= 30 \\ x &= 10 \end{aligned}$$

Therefore:

$$
\begin{aligned}
x(x-d)(x+d) &= 360\\
10\left(100-d^2\right) &= 360\\
100-d^2 &= 36\\
-d^2 &= -64\\
d &= \pm 8
\end{aligned}
$$

So the terms are 2, 10, 18 or 18, 10, 2.

29. Using equations as in Exercise 27:

$3x = 6$, so $x = 2$

$$
\begin{aligned}
(x-d)^3 + x^3 + (x+d)^3 &= 132\\
(2-d)^3 + 8 + (2+d)^3 &= 132\\
(2-d)^3 + (2+d)^3 &= 124\\
16 + 12d^2 &= 124\\
d^2 &= 9\\
d &= \pm 3
\end{aligned}
$$

So the terms are –1, 2, 5 or 5, 2, –1.

31. (a) Subtracting, we have:

$$
\begin{aligned}
a_2 - a_1 &= -1 - \frac{1}{1+\sqrt{2}} = \frac{-2-\sqrt{2}}{1+\sqrt{2}} = \frac{(1-\sqrt{2})(-2-\sqrt{2})}{-1} = -\sqrt{2}\\
a_3 - a_2 &= \frac{1}{1-\sqrt{2}} + 1 = \frac{2-\sqrt{2}}{1-\sqrt{2}} = \frac{(2-\sqrt{2})(1+\sqrt{2})}{-1} = -\sqrt{2}
\end{aligned}
$$

So $a_2 - a_1 = a_3 - a_2 = -\sqrt{2}$.

(b) Using $a = \frac{1}{1+\sqrt{2}}$ and $d = -\sqrt{2}$, we have:

$$
\begin{aligned}
S_6 &= 3\left[2\left(\frac{1}{1+\sqrt{2}}\right) + 5(-\sqrt{2})\right]\\
&= 3\left(\frac{2}{1+\sqrt{2}} - \frac{10+5\sqrt{2}}{1+\sqrt{2}}\right)\\
&= 3\left(\frac{-8-5\sqrt{2}}{1+\sqrt{2}}\right)\\
&= -\frac{24+15\sqrt{2}}{1+\sqrt{2}}\\
&= (24+15\sqrt{2})(1-\sqrt{2})\\
&= 24 - 30 - 9\sqrt{2}\\
&= -6 - 9\sqrt{2}
\end{aligned}
$$

33. Using $a = \frac{1}{1+\sqrt{b}}$, we have:

$$d = \frac{1}{2}\left(\frac{1}{1-\sqrt{b}} - \frac{1}{1+\sqrt{b}}\right) = \frac{1}{2}\left(\frac{2\sqrt{b}}{1-b}\right) = \frac{\sqrt{b}}{1-b}$$

Therefore:

$$\begin{aligned} S_n &= \frac{n}{2}\left[\frac{2}{1+\sqrt{b}} + (n-1)\frac{\sqrt{b}}{1-b}\right] \\ &= \frac{n}{2}\left[\frac{(2-2\sqrt{b}) + (n-1)\sqrt{b}}{1-b}\right] \\ &= \frac{n}{2(1-b)}\left[2 + (n-3)\sqrt{b}\right] \end{aligned}$$

35. Using the given ratio, we have:

$$\begin{aligned} \frac{n^2}{m^2} &= \frac{\frac{n}{2}[2a+(n-1)d]}{\frac{m}{2}[2a+(m-1)d]} \\ \frac{2a+(n-1)d}{2a+(m-1)d} &= \frac{n}{m} \\ m[2a+(n-1)d] &= n[2a+(m-1)d] \\ 2am + m(n-1)d &= 2an + n(m-1)d \\ 2a(m-n) &= [n(m-1) - m(n-1)]d \\ d &= \frac{2a(m-n)}{m-n} \\ d &= 2a \quad (\text{assuming } m \neq n) \end{aligned}$$

Consequently:

$$\frac{a_n}{a_m} = \frac{a+(n-1)(2a)}{a+(m-1)(2a)} = \frac{1+2n-2}{1+2m-2} = \frac{2n-1}{2m-1}$$

37. Let the three consecutive terms be $x-d$, x, and $x+d$, then using the Pythagorean theorem:

$$\begin{aligned} (x-d)^2 + x^2 &= (x+d)^2 \\ x^2 - 2xd + d^2 + x^2 &= x^2 + 2xd + d^2 \\ x^2 - 4xd &= 0 \\ x = 0 \quad &\text{or} \quad x = 4d \end{aligned}$$

Since the side of the triangle cannot have zero length, then $x = 4d$ and the three terms are $3d$, $4d$, and $5d$, which is clearly similar to a 3-4-5 right triangle.

39. We have $\frac{1}{b}-\frac{1}{a}=\frac{1}{c}-\frac{1}{b}$, from which it follows that $b=\frac{2ac}{a+c}$. Therefore:

$$\begin{aligned}\ln(a+c)+\ln(a-2b+c) &= \ln(a+c)+\ln\left(a-\frac{4ac}{a+c}+c\right)\\ &= \ln(a+c)+\ln\left(\frac{a^2+ac-4ac+ac+c^2}{a+c}\right)\\ &= \ln(a+c)+\ln\frac{(a-c)^2}{a+c}\\ &= \ln(a-c)^2\\ &= 2\ln(a-c)\end{aligned}$$

10.5 Geometric Sequences and Series

1. The first three terms have a common ratio, hence:

$$\frac{x}{9}=\frac{4}{x}, \text{ therefore } x^2=36 \text{ and } x=\pm 6$$

Since the ratio is positive the second term is 6.

3. Letting the common ratio be r, we have:

$$4,\ 4r, \text{ and } 4r^2 \text{ so } 64r^3=8000$$

Therefore, $r^3=125$, so $r=5$, and the second and third terms are 20 and 100, respectively.

5. Here $a_1=-1$ and $r=-1$, so $a_{100}=-1(-1)^{99}=1$.

7. Here $a_1=\frac{2}{3}$ and $r=\frac{2}{3}$, so $a_8=\frac{2}{3}\left(\frac{2}{3}\right)^7=\frac{256}{6561}$.

9. Here $4096=a_7=r^6$, so $r=\pm\sqrt[6]{4096}=\pm 4$.

11. We compute the sum:

$$S_{10}=\frac{7\left(1-2^{10}\right)}{1-2}=-7(1-1024)=7161$$

13. We compute the sum:

$$1+\sqrt{2}+2+\ldots+32=S_{11}=\frac{1\left[1-(\sqrt{2})^{11}\right]}{1-\sqrt{2}}=\frac{1-32\sqrt{2}}{1-\sqrt{2}}=63+31\sqrt{2}$$

15. We compute the sum:

$$\sum_{k=1}^{6}\left(\frac{3}{2}\right)^k=S_6=\frac{\frac{3}{2}\left[1-\left(\frac{3}{2}\right)^6\right]}{1-\frac{3}{2}}=-3\left(1-\frac{729}{64}\right)=\frac{1995}{64}$$

17. We compute the sum:

$$\begin{aligned}\sum_{k=2}^{6}\left(\tfrac{1}{10}\right)^k &= S_6 - a_1\\ &= \frac{\frac{1}{10}\left[1-\left(\frac{1}{10}\right)^6\right]}{1-\frac{1}{10}} - \frac{1}{10}\\ &= \tfrac{1}{9}\left(1-\frac{1}{1000000}\right) - \frac{1}{10}\\ &= \frac{1}{9}\bullet\frac{999999}{1000000} - \frac{1}{10}\\ &= \frac{111111}{1000000} - \frac{1}{10}\\ &= \frac{11111}{1000000}\end{aligned}$$

This is equivalent is 0.011111.

19. We compute the sum:

$$\frac{2}{3} - \frac{4}{9} + \frac{8}{27} - \ldots = \frac{\frac{2}{3}}{1+\frac{2}{3}} = \frac{2}{5}$$

21. We compute the sum:

$$1 + \frac{1}{1.01} + \frac{1}{(1.01)^2} + \ldots = \frac{1}{1-\frac{1}{1.01}} = 101$$

23. Writing as a geometric sum, we have:

$$0.555\ldots = \frac{5}{10} + \frac{5}{100} + \ldots = \frac{\frac{5}{10}}{1-\frac{1}{10}} = \frac{5}{9}$$

25. Writing as a geometric sum, we have:

$$0.12323\ldots = \frac{1}{10} + \frac{23}{1000} + \frac{23}{100000} + \ldots = \frac{1}{10} + \frac{\frac{23}{1000}}{1-\frac{1}{100}} = \frac{1}{10} + \frac{23}{990} = \frac{122}{990} = \frac{61}{495}$$

27. Writing as a geometric sum, we have:

$$0.432\ldots = \frac{432}{1000} + \frac{432}{1000000} + \ldots = \frac{\frac{432}{1000}}{1-\frac{1}{1000}} = \frac{432}{999} = \frac{16}{37}$$

29. For $\frac{a}{r}$, a, ar we have:

$$\begin{aligned}\frac{a}{r}(a)(ar) &= -1000\\ a^3 &= -1000\\ a &= -10\end{aligned}$$

So we have:

$$\frac{a}{r}+a+ar=15$$
$$\frac{-10}{r}-10-10r=15$$
$$2r^2+5r+2=0$$
$$(2r+1)(r+2)=0$$
$$r=-\tfrac{1}{2},-2$$

31. We compute the sum:

$$\frac{\sqrt{3}}{\sqrt{3}+1}+\frac{\sqrt{3}}{\sqrt{3}+3}+\ldots=\frac{\frac{\sqrt{3}}{\sqrt{3}+1}}{1-\frac{\sqrt{3}+1}{\sqrt{3}+3}}=\frac{\frac{\sqrt{3}}{\sqrt{3}+1}}{\frac{2}{\sqrt{3}+3}}=\frac{\sqrt{3}(\sqrt{3}+3)}{2(\sqrt{3}+1)}=\frac{3+3\sqrt{3}}{2(1+\sqrt{3})}=\frac{3}{2}$$

33. For a_1, a_2, ... a geometric sequence, with ratio r:

$$S=a_1+a_2+\ldots+a_n=\frac{a_1\left(1-r^n\right)}{1-r}$$

Also:

$$T=\frac{1}{a_1}+\frac{1}{a_2}+\ldots+\frac{1}{a_n}=\frac{\frac{1}{a_1}\left(1-\frac{1}{r^n}\right)}{1-\frac{1}{r}}$$

Therefore:

$$\frac{S}{T}=\frac{\frac{a_1\left(1-r^n\right)}{1-r}}{\frac{\frac{1}{a_1}\left(1-\frac{1}{r^n}\right)}{\frac{r-1}{r}}}=\frac{a_1\left(1-r^n\right)a_1(r-1)r^n}{(1-r)r\left(r^n-1\right)}=a_1{}^2r^{n-1}=a_1\left(a_1r^{n-1}\right)=a_1a_n$$

35. The total distance traveled by the ball consists of two infinite geometric series:

down: $6+\frac{1}{3}(6)+\frac{1}{9}(6)+\ldots$

up: $2+\frac{1}{3}(2)+\frac{1}{9}(2)+\ldots$

We now use the formula $S=\frac{a}{1-r}$ to find each sum:

down: $a=6,\ r=\dfrac{1}{3}$

$$S=\frac{6}{1-\frac{1}{3}}=\frac{6}{\frac{2}{3}}=9\text{ ft.}$$

up: $a=1,\ r=\dfrac{1}{3}$

$$S=\frac{2}{1-\frac{1}{3}}=\frac{2}{\frac{2}{3}}=3\text{ ft.}$$

So the total distance traveled is 12 ft.

10.6 Permutations and Combinations

1. There are 6 routes: $3(2) = 6$

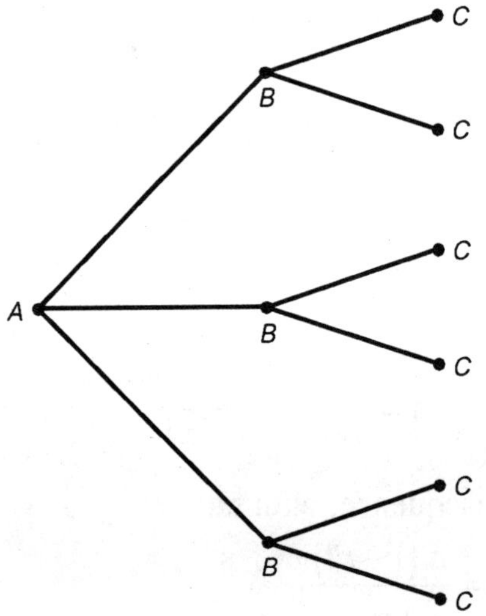

3. There are 12 routes: $3(2)(2) = 12$

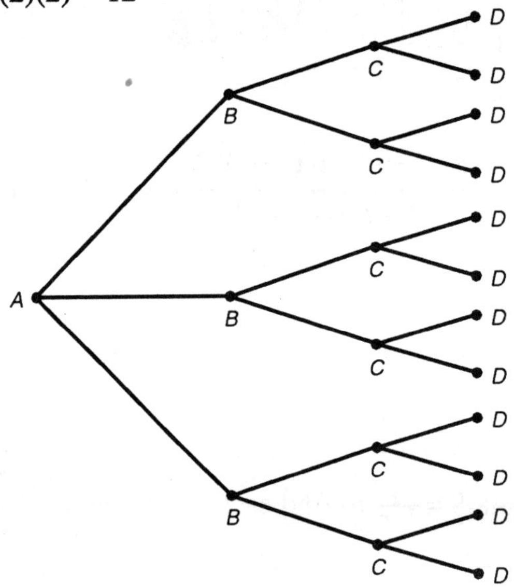

5. This is $P(4,2) = 4 \bullet 3 = 12$.

7. This is $P(5,3) = 5 \bullet 4 \bullet 3 = 60$.

9. Using the multiplication principle, the amount of numbers is:
$(3)(6)(6) = 108$

11. Using the multiplication principle, the number of license plates is:
$(10)(9)(8)(26)(25)(24) = 11{,}232{,}000$

13. This is $P(10,10) = 10! = 3{,}628{,}800$

15. Since $P(n,n) = 24$, we have $n! = 24$ and thus $n = 4$ letters.

17. This is $P(8,1) = 8$.

19. This is $P(6,3) = 6 \bullet 5 \bullet 4 = 120$.

21. We have:

$$\begin{aligned} P(n,n-1) &= n(n-1)(n-2)\ldots[n-(n-1)+1] \\ &= n(n-1)(n-2)\ldots(2) \\ &= n! \end{aligned}$$

23. Expanding the products:

$$P(100,50) = (100)(99)(98)\ldots(51)$$
$$P(101,51) = (101)(100)(99)\ldots(51) = 101 \bullet P(100,50)$$

So $P(101,51)$ is larger.

25. Following the solution of Example 9:

$$2P(5,5) = 2(5!) = 240$$

27. As in Example 9:

$$2P(3,3) = 2(3!) = 12$$

29. (a) $9(10)(10) = 900$
(b) $9(9)(8) = 648$

31. (a) $9(9)(9) = 729$
(b) $9(8)(7) = 504$
(c) $1(8)(7) = 56$
(d) $5(7)(4) = 140$

33. (a) We compute:

$$C(12,11) = \frac{12!}{11!1!} = 12$$

(b) We compute:

$$C(12,1) = \frac{12!}{1!11!} = 12$$

35. (a) We compute:

$$C(8,3) = \frac{8!}{3!5!} = 56$$

(b) We compute:

$$C(8,5) = \frac{8!}{3!5!} = 56$$

37. (a) We compute:

$$C(12,5)=\frac{12!}{5!7!}=792$$

(b) We compute:

$$C(12,7)=792$$

39. We compute:

$$C(n,n-2)=\frac{n!}{(n-2)!2!}=\frac{n(n-1)}{2}$$

41. We compute:

$$C(8,3)=\frac{8!}{3!5!}=56$$

43. (a) We compute:

$$C(10,6)=\frac{10!}{6!4!}=210$$

(b) We compute:

$$C(4,3)\bullet C(6,3)=\frac{4!}{3!1!}\bullet\frac{6!}{3!3!}=80$$

45. The number of diagonals will be $C(6,2)-6$, which is:

$$C(6,2)-6=\frac{6!}{2!4!}-6=15-6=9$$

47. We compute:

$$C(8,3)=\frac{8!}{3!5!}=56 \text{ triangles}$$

49. First notice that no two different combinations of these coins can add up to the same sum. Then we add the number of ways of combining the four coins, one at a time, two at a time, three and four at a time, thus:

$$C(4,1)+C(4,2)+C(4,3)+C(4,4)=4+6+4+1=15 \text{ sums}$$

51. (a) We choose 6 of 9 Republicans and 2 of 8 Democrats, so the number of committees is:

$$C(9,6)\bullet C(8,2)=\frac{9!}{6!3!}\bullet\frac{8!}{2!6!}=2352 \text{ committees}$$

(b) There are $C(9,6) \bullet C(8,2)$ ways to pick the committee with 6 Republicans, $C(9,7) \bullet C(8,1)$ ways to pick it with 7 Republicans, and $C(9,8) \bullet C(8,0)$ ways to pick it with 8 Republicans, so the total is:

$$\begin{aligned} &C(9,6) \bullet C(8,2) + C(9,7) \bullet C(8,1) + C(9,8) \bullet C(8,0) \\ &= \frac{9!}{6!3!} \bullet \frac{8!}{2!6!} + \frac{9!}{7!2!} \bullet \frac{8!}{1!7!} + \frac{9!}{8!1!} \bullet \frac{8!}{0!8!} \\ &= 2352 + 288 + 9 \\ &= 2649 \text{ committees} \end{aligned}$$

53. In a polygon with 10 vertices, first choose a particular vertex (there are 9 other vertices to which a line could be drawn.) If lines are drawn to the two adjacent vertices, the lines would become sides, not diagonals; thus, there are 7 diagonals from each vertex. Not wanting to count a diagonal from one vertex to another twice, we must divide by 2. Thus there are $\frac{7 \times 10}{2} = 35$ diagonals.

10.7 Introduction to Probability

1. (a) $\{2,4,6\}$
(b) $\{4,5,6\}$
(c) $\{2,4,5,6\}$
(d) $\{4,6\}$

3. (a) $\{1,3,5\}$
(b) $\{1,2,3,4,5,6\}$
(c) $\{1,2,3,4,5,6\}$
(d) $\emptyset$

5. No, since $A \cap B = \{4,6\} \neq \emptyset$.

7. Yes, since $C \cap D = \emptyset$.

9. (a) $n(A) = 3$
(b) $n(B) = 3$
(c) $n(A \cup B) = 4$

11. (a) $n(E) = 3$
(b) $n(F) = 6$
(c) $n(G) = 0$

13. (a) $P(A) = \frac{3}{6} = \frac{1}{2}$
(b) $P(B) = \frac{3}{6} = \frac{1}{2}$

15. (a) $P(E)=\frac{3}{6}=\frac{1}{2}$

(b) $P(F)=\frac{6}{6}=1$

17. (a) $P(A\cup B)=\frac{4}{6}=\frac{2}{3}$

(b) $P(A)+P(B)=\frac{3}{6}+\frac{3}{6}=1$

19. (a) $P(B\cup E)=\frac{5}{6}$

(b) $P(B)+P(E)=\frac{3}{6}+\frac{3}{6}=1$

21. $P(A)=\frac{3}{8}$

23. $P(C)=\frac{4}{8}=\frac{1}{2}$

25. $P(E)=\frac{2}{8}=\frac{1}{4}$

27. We compute:

$$P(A\cup B)=P(A)+P(B)-P(A\cap B)=\frac{3}{8}+\frac{1}{4}-0=\frac{5}{8}$$

29. We compute:

$$P(A\cup C)=P(A)+P(C)-P(A\cap C)=\frac{3}{8}+\frac{1}{2}-\frac{1}{8}=\frac{6}{8}=\frac{3}{4}$$

31. We compute:

$$P(C\cup F)=P(C)+P(F)-P(C\cap F)=\frac{1}{2}+\frac{1}{2}-0=1$$

33. We compute:

$$P(E\cup F)=P(E)+P(F)-P(E\cap F)=\frac{1}{4}+\frac{1}{2}-\frac{1}{8}=\frac{5}{8}$$

35. We compute:

$$P(B\cup C)=P(B)+P(C)-P(B\cap C)=\frac{1}{4}+\frac{1}{2}-\frac{1}{4}=\frac{1}{2}$$

37. $P(D\cap A)=0$

39. We compute:

$$P(D\cup E)=P(D)+P(E)-P(D\cap E)=\frac{1}{8}+\frac{1}{4}-\frac{1}{8}=\frac{1}{4}$$

41. We can compute this as:

$$P(1^{\text{st}}\text{ is }d)\bullet P(2^{\text{nd}}\text{ is not }d)+P(1^{\text{st}}\text{ is not }d)\bullet P(2^{\text{nd}}\text{ is }d)$$

$$=\frac{1}{6}\bullet\frac{5}{5}+\frac{5}{6}\bullet\frac{1}{5}=\frac{1}{6}+\frac{1}{6}=\frac{1}{3}$$

43. (a) The probability is:

$$\frac{12}{14} = \frac{6}{7}$$

(b) Using the multiplication principle, the probability is:

$$\frac{12}{14} \bullet \frac{11}{13} \bullet \frac{10}{12} \bullet \frac{9}{11} \bullet \frac{8}{10} = \frac{72}{182} = \frac{36}{91}$$

45. We compute:

$$P(A \cup B) = P(A) + P(B) - P(A \cap B) = \frac{1}{3} + \frac{1}{4} - 0 = \frac{7}{12}$$

47. We solve:

$$P(A \cup B) = P(A) + P(B) - P(A \cap B)$$
$$\frac{7}{9} = \frac{5}{9} + \frac{4}{9} - P(A \cap B)$$
$$\frac{7}{9} = 1 - P(A \cap B)$$
$$\frac{2}{9} = P(A \cap B)$$

49. (a) We compute:

$$\frac{6}{15} \bullet \frac{5}{15} = \frac{1}{8}$$

(b) We compute:

$$\frac{10}{16} \bullet \frac{9}{15} = \frac{3}{8}$$

(c) We compute:

$$\frac{1}{8} + \frac{3}{8} = \frac{1}{2}$$

51. We compute:

$$P\left(A\ 1^{\underline{st}} \text{ or } A\ 2^{\underline{nd}}\right) = P\left(A\ 1^{\underline{st}}\right) + P\left(A\ 2^{\underline{nd}}\right) - P\left(A \text{ both}\right) = \frac{1}{2} + \frac{3}{10} - \frac{1}{10} = \frac{7}{10}$$

53. Using the multiplication principle, we have:

$$\frac{13}{52} \bullet \frac{12}{51} \bullet \frac{11}{50} \bullet \frac{10}{49} \bullet \frac{9}{48} = \frac{33}{66640}$$

Chapter Ten Review Exercises

1. When $n = 1$:

$$5 = \frac{5}{2}(1)(1+1) = 5$$

Assuming:

$$5 + 10 + \ldots + 5k = \frac{5}{2}k(k+1)$$

Then:

$$5 + 10 + \ldots + 5k + 5(k+1) = \frac{5}{2}k(k+1) + 5(k+1) = (k+1)\left(\frac{5}{2}k + 5\right) = \frac{5}{2}(k+1)(k+2)$$

This completes the induction.

3. When $n = 1$:

$$1 \bullet 2 = 2 = \tfrac{1}{3}(1)(1+1)(1+2) = 2$$

Assuming:

$$1 \bullet 2 + 2 \bullet 3 + \ldots + k(k+1) = \tfrac{1}{3}k(k+1)(k+2)$$

Then:

$$\begin{aligned} 1 \bullet 2 + 2 \bullet 3 + \ldots + k(k+1) + (k+1)(k+2) \\ &= \tfrac{1}{3}k(k+1)(k+2) + (k+1)(k+2) \\ &= \left(\tfrac{1}{3}k + 1\right)(k+1)(k+2) \\ &= \tfrac{1}{3}(k+1)(k+2)(k+3) \end{aligned}$$

This completes the induction.

5. For $n = 1$:

$$1 = 3 + (2-3)2 = 3 - 2 = 1$$

Assuming:

$$1 + 3 \bullet 2 + 5 \bullet 2^2 + \ldots + (2k-1) \bullet 2^{k-1} = 3 + (2k-3) \bullet 2^k$$

Then:

$$\begin{aligned} 1 + 3 \bullet 2 + \ldots + (2k-1)2^{k-1} + (2k+1)2^k \\ &= 3 + (2k-3)2^k + (2k+1)2^k \\ &= 3 + (4k-2)2^k \\ &= 3 + (2k-1)2^{k+1} \\ &= 3 + (2(k+1)-3) \bullet 2^{k+1} \end{aligned}$$

The induction is complete.

7. For $n = 1$:

$$1 = \left(1^2 - 2 + 3\right)2 - 3 = 1$$

Assuming:

$$1 + 2^2 \bullet 2 + 3^2 \bullet 2^2 + 4^2 \bullet 2^3 + \ldots + k^2 \bullet 2^{k-1} = \left(k^2 - 2k + 3\right)2^k - 3$$

Then:

$$\begin{aligned} 1 + 2^2 \bullet 2 + \ldots + (k+1)^2 2^k &= \left(k^2 - 2k + 3\right)2^k - 3 + (k+1)^2 2^k \\ &= 2^k\left[k^2 - 2k + 3 + (k+1)^2\right] - 3 \\ &= 2^k\left(2k^2 + 4\right) - 3 \\ &= 2^{k+1}\left(k^2 + 2\right) - 3 \\ &= 2^{k+1}\left[(k+1)^2 - 2k + 1\right] - 3 \\ &= 2^{k+1}\left[(k+1)^2 - 2(k+1) + 3\right] - 3 \end{aligned}$$

The induction is complete.

9. For $n = 1$:

$$7^1 - 1 = 6 = 3(2)$$

Assuming:

$$7^k - 1 = 3L$$

Then:

$$7^{k+1} - 1 = 3L + 7^{k+1} - 7^k = 3L + 7^k(7-1) = 3L + 6\left(7^k\right) = 3\left(L + 2 \bullet 7^k\right)$$

The induction is complete.

11. We expand:

$$\begin{aligned}(3a+b^2)^4 &= \binom{4}{0}(3a)^4+\binom{4}{1}(3a)^3(b^2)+\binom{4}{2}(3a)^2(b^2)^2+\binom{4}{3}(3a)(b^2)^3+\binom{4}{4}(b^2)^4\\ &= 81a^4+108a^3b^2+54a^2b^4+12ab^6+b^8\end{aligned}$$

13. We expand:

$$\begin{aligned}(x+\sqrt{x})^4 &= \binom{4}{0}x^4+\binom{4}{1}x^3(\sqrt{x})+\binom{4}{2}x^2(\sqrt{x})^2+\binom{4}{3}x(\sqrt{x})^3+\binom{4}{4}(\sqrt{x})^4\\ &= x^4+4x^3\sqrt{x}+6x^3+4x^2\sqrt{x}+x^2\end{aligned}$$

15. We expand:

$$\begin{aligned}(x^2-2y^2)^5 &= \binom{5}{0}(x^2)^5-\binom{5}{1}(x^2)^4(2y^2)+\binom{5}{2}(x^2)^3(2y^2)^2-\binom{5}{3}(x^2)^2(2y^2)^3\\ &\quad+\binom{5}{4}(x^2)(2y^2)^4-\binom{5}{5}(2y^2)^5\\ &= x^{10}-10x^8y^2+40x^6y^4-80x^4y^6+80x^2y^8-32y^{10}\end{aligned}$$

17. We expand:

$$\begin{aligned}\left(1+\frac{1}{x}\right)^5 &= \binom{5}{0}+\binom{5}{1}\frac{1}{x}+\binom{5}{2}\left(\frac{1}{x}\right)^2+\binom{5}{3}\left(\frac{1}{x}\right)^3+\binom{5}{4}\left(\frac{1}{x}\right)^4+\binom{5}{5}\left(\frac{1}{x}\right)^5\\ &= 1+\frac{5}{x}+\frac{10}{x^2}+\frac{10}{x^3}+\frac{5}{x^4}+\frac{1}{x^5}\end{aligned}$$

19. We expand:

$$\begin{aligned}(a\sqrt{b}-b\sqrt{a})^4 &= a^2b^2(\sqrt{a}-\sqrt{b})^4\\ &= a^2b^2\left[(\sqrt{a})^4-4(\sqrt{a})^3\sqrt{b}+6(\sqrt{a})^2(\sqrt{b})^2-4(\sqrt{a})(\sqrt{b})^3+(\sqrt{b})^4\right]\\ &= a^2b^2(a^2-4a\sqrt{ab}+6ab-4b\sqrt{ab}+b^2)\\ &= a^4b^2-4a^3b^2\sqrt{ab}+6a^3b^3-4a^2b^3\sqrt{ab}+a^2b^4\end{aligned}$$

21. The fifth term is given by:

$$\binom{5}{5-1}(3x)(y^2)^4 = 5(3x)y^8 = 15xy^8$$

23. Here $7-r+1=5$, so $r=3$, and the coefficient of the third term is:

$$\binom{7}{2}(2)^2 = (4)\frac{7!}{5!2!} = 84$$

25. Here $r-1=6$, so $r=7$, and the coefficient of the seventh term is:

$$\binom{8}{6} = \frac{8!}{6!2!} = 28$$

27. Computing each side:

$$\binom{2}{0}^2+\binom{2}{1}^2+\binom{2}{2}^2=(1)^2+(2)^2+(1)^2=6$$

$$\binom{4}{2}=\frac{4!}{2!2!}=6$$

29. Computing each side:

$$\binom{4}{0}^2+\binom{4}{1}^2+\binom{4}{2}^2+\binom{4}{3}^2+\binom{4}{4}^2=1^2+4^2+6^2+4^2+1^2=70$$

$$\binom{8}{4}=\frac{8!}{4!4!}=70$$

31. Computing values:

$$\binom{3}{0}+\binom{3}{1}+\binom{3}{2}+\binom{3}{3}=1+3+3+1=8=2^3$$

33. Computing values, we have:

$$a_1=\frac{2}{1+1}=1,\quad a_2=\frac{4}{2+1}=\tfrac{4}{3},\quad a_3=\frac{6}{3+1}=\tfrac{3}{2},\quad a_4=\frac{8}{4+1}=\tfrac{8}{5}$$

We graph these points:

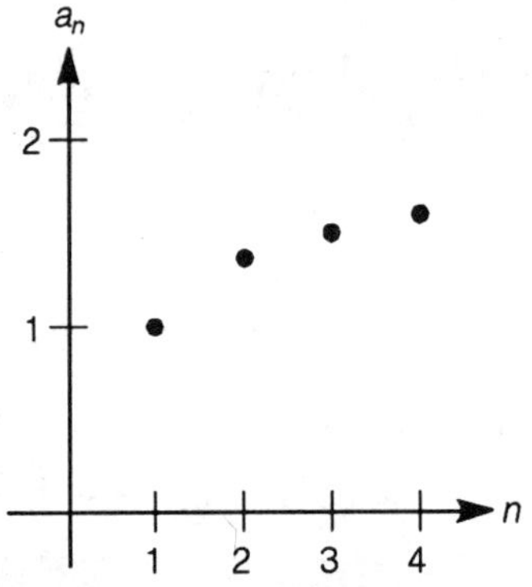

35. Computing values, we have:

$$a_1=(-1)\left(1-\tfrac{1}{2}\right)=-\tfrac{1}{2}$$
$$a_2=(1)\left(1-\tfrac{1}{3}\right)=\tfrac{2}{3}$$
$$a_3=(-1)\left(1-\tfrac{1}{4}\right)=-\tfrac{3}{4}$$
$$a_4=(1)\left(1-\tfrac{1}{5}\right)=\tfrac{4}{5}$$

We graph these points:

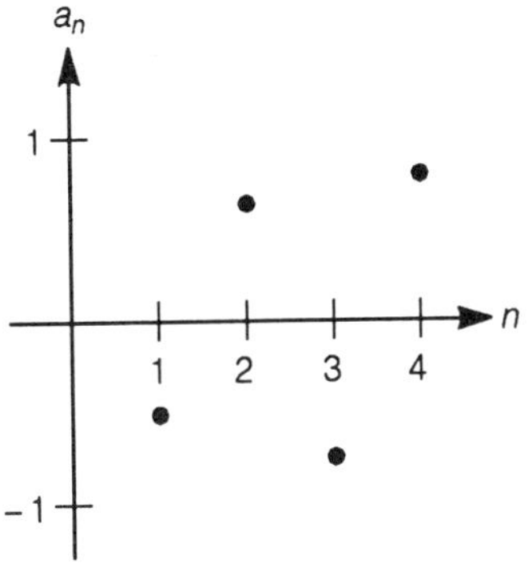

37. Computing values, we have:

$$a_0 = -3, \quad a_1 = -12, \quad a_2 = -48, \quad a_3 = -192$$

The first four terms are –3, –12, –48, and –192.

39. (a) Expanding the sum, we have:

$$\sum_{k=1}^{3} (-1)^k (2k+1) = -3 + 5 - 7 = -5$$

(b) Expanding the sum, we have:

$$\begin{aligned}\sum_{k=0}^{8}\left(\frac{1}{k+1} - \frac{1}{k+2}\right) &= \left(1 - \tfrac{1}{2}\right) + \left(\tfrac{1}{2} - \tfrac{1}{3}\right) + \left(\tfrac{1}{3} - \tfrac{1}{4}\right) + \left(\tfrac{1}{4} - \tfrac{1}{5}\right) + \left(\tfrac{1}{5} - \tfrac{1}{6}\right) \\ &\quad + \left(\tfrac{1}{6} - \tfrac{1}{7}\right) + \left(\tfrac{1}{7} - \tfrac{1}{8}\right) + \left(\tfrac{1}{8} - \tfrac{1}{9}\right) + \left(\tfrac{1}{9} - \tfrac{1}{10}\right) \\ &= 1 - \tfrac{1}{10} \\ &= \tfrac{9}{10}\end{aligned}$$

41. The sum can be written as:

$$\frac{5}{3} + \frac{5}{3^2} + \frac{5}{3^3} + \frac{5}{3^4} + \frac{5}{3^5} = \sum_{k=1}^{5} \frac{5}{3k}$$

43. Here $a_n = 1 + 4n$, so $a_{18} = 1 + 4(18) = 73$.

45. Here $a_n = \dfrac{10}{2^{n-1}}$, so $a_{12} = \dfrac{10}{2^{11}} = \dfrac{5}{1024}$.

47. The sum is given by:

$$S_{12} = (12)\left(\frac{8 + \frac{43}{2}}{2}\right) = \tfrac{59}{2}(6) = 177$$

49. By inspection $S_1 = 7$, $S_2 = 77$, ... so $S_{10} = 7{,}777{,}777{,}777$.

51. We compute:

$$a_3 = 4$$
$$a_5 = 10$$
$$\frac{a_5}{a_3} = \tfrac{10}{4} = \tfrac{5}{2} = r^2 \text{ and } r = \pm\sqrt{\tfrac{5}{2}}$$

Since r is known to be negative, $r = -\sqrt{\frac{5}{2}}$ and $a_6 = -\sqrt{\frac{5}{2}}(10) = -5\sqrt{10}$.

53. The sum is given by:

$$S = \frac{\frac{3}{5}}{1-\frac{1}{5}} = \tfrac{3}{4}$$

55. The sum is given by:

$$S = \frac{\frac{1}{9}}{1+\frac{1}{9}} = \tfrac{1}{10}$$

57. We have $0.45 = 0.45 + 0.0045 + 0.000045 + \ldots$. This is a geometric sequence with $a = 0.45$, $r = \frac{1}{100}$, so:

$$0.45 = \frac{0.45}{1-\frac{1}{100}} = \tfrac{45}{99} = \tfrac{15}{33} = \tfrac{5}{11}$$

59. The sum is given by:

$$\begin{aligned} S_n &= \frac{n}{2}\left[2(1) + (n-1)1\right] \\ &= \frac{n}{2}(2+n-1) \\ &= \frac{n}{2}(1+n) \\ &= \frac{n^2}{2} + \frac{n}{2} \\ &= n + \frac{n^2}{2} - \frac{n}{2} \\ &= n + \frac{n(n-1)}{2} \end{aligned}$$

61. The sum is given by:

$$\begin{aligned} S_n &= \frac{n}{2}[2(1)+(n-1)3] \\ &= \frac{n}{2}(2+3n-3) \\ &= \frac{n}{2}(3n-1) \\ &= \frac{3n^2}{2}-\frac{n}{2} \\ &= n+\frac{3n^2}{2}-\frac{3n}{2} \\ &= n+\frac{3n(n-1)}{2} \end{aligned}$$

63. (a) Using formula (1), we have:

$$1^2+2^2+\ldots+50^2 \approx \frac{\left(50+\frac{1}{2}\right)^{2+1}}{2+1} = \frac{\left(\frac{101}{2}\right)^3}{3} = \frac{1030301}{24} \approx 42929$$

(b) The exact sum is given by:

$$\frac{50(51)(101)}{6} = 42925$$

The percent error is:

$$100 \bullet \frac{4}{42925} \approx 0.00932\%$$

(c) Using formula (1), we have:

$$1^4+2^4+\ldots+200^4 \approx \frac{(200.5)^5}{5} \approx 6.48040 \times 10^{10}$$

(d) We compute the sum:

$$1^4+2^4+\ldots+200^4 \approx \frac{200(201)(401)(120599)}{30} \approx 6.48027 \times 10^{10}$$

The percent error is:

$$100 \bullet \frac{0.00013 \times 10^{10}}{6.48027 \times 10^{10}} = 2 \times 10^{-3}\%$$

65. There are $4 \times 4 = 16$ different numbers (assumes repetition).

67. The permutations are:

$\{abc, acb, bac, bca, cab, cba, abd, adb, bad, bda, dab, dba, acd, adc, cad, cda, dac, dca, bcd, bdc, cbd, cdb, dbc, dcb\}$

So $P(4,3) = 4(3)(2) = 24$.

69. We compute $P(12,8) = (12)(11)(10)(9)(8)(7)(6)(5) = 19{,}958{,}400$.

71. There are a total of 6! different arrangements, while 2(5!) have e, f next to one another, hence:

$$6!-2(5!)=4(5!)=4(120)=480$$

73. We compute:

$$C(9,2)\bullet C(10,2)=\frac{9!}{7!2!}\bullet\frac{10!}{8!2!}=(36)(45)=1620 \text{ subcommittees}$$

75. We compute:

$$\tfrac{9}{12}\bullet\tfrac{8}{11}=\tfrac{72}{132}=\tfrac{6}{11}$$

77. (a) Using the multiplication principle, the probability is:

$$\tfrac{5}{12}\bullet\tfrac{4}{11}=\tfrac{5}{33}$$

(b) Using the multiplication principle, the probability is:

$$\tfrac{7}{12}\bullet\tfrac{6}{11}=\tfrac{7}{22}$$

(c) Since the two events are mutually exclusive, the probability is:

$$\tfrac{5}{33}+\tfrac{7}{22}-0=\tfrac{31}{66}$$

79. The possible die combinations are $(1,2)$ and $(2,1)$, so the probability is $\frac{2}{36}=\frac{1}{18}$.

81. The possible die combinations are $(1,6)$, $(6,1)$, $(2,5)$, $(5,2)$, $(3,4)$ and $(4,3)$, so the probability is $\frac{6}{36}=\frac{1}{6}$.

83. The possible die combinations are those from Exercise 81 together with $(5,6)$ and $(6,5)$, so the probability is $\frac{8}{36}=\frac{2}{9}$.

85. The possible die combinations are $(4,6)$, $(6,4)$, $(5,5)$, $(5,6)$, $(6,5)$ and $(6,6)$, so the probability is $\frac{6}{36}=\frac{1}{6}$.

87. There are 11 die combinations resulting in a sum of either 7 or 8, so there are $36-11=25$ die combinations resulting in a sum which is not 7 or 8. So the probability is $\frac{25}{36}$.

89. (a) The probability is $\frac{7}{15}$.

(b) The probability is $\frac{8}{15}$.

(c) The probability is $\frac{10}{15}=\frac{2}{3}$.

(d) The probability is $\frac{10}{15}=\frac{2}{3}$.

91. (a) There are $9 \bullet 10 \bullet 10 = 900$ three-digit numbers (remember, the first digit cannot be 0), and there are $9 \bullet 9 \bullet 8 = 648$ three-digit numbers with no repetitions, so there are $900 - 648 = 252$ three-digit numbers containing one or more repeated digits.

(b) The probability is $\frac{252}{900} = \frac{7}{25}$.

93. Call $b = ra$ and $c = r^2a$, so $a + ar + ar^2 = 70$. Since $4a$, $5b$, $4c$ are consecutive terms in an arithmetic sequence, then their common difference is the same. So:

$$5b - 4a = 4c - 5b$$
$$10b = 4a + 4c$$
$$5b = 2a + 2c$$
$$5(ra) = 2a + 2(r^2a)$$
$$5ra = 2a + 2r^2a$$

Since a is nonzero, we divide by a:

$$5r = 2 + 2r^2$$
$$2r^2 - 5r + 2 = 0$$
$$(2r - 1)(r - 2) = 0$$
$$r = \tfrac{1}{2}, 2$$

If $r = \frac{1}{2}$, we find a:

$$a + a \bullet \tfrac{1}{2} + a \bullet \tfrac{1}{4} = 70$$
$$7a = 280$$
$$a = 40$$

So the terms are 40, 20, 10.
If $r = 2$, we find a:

$$a + 2a + 4a = 70$$
$$7a = 70$$
$$a = 10$$

So the terms are 10, 20, 40.

95. We assume:

$$\frac{1}{c+a} - \frac{1}{b+c} = \frac{1}{a+b} - \frac{1}{c+a}$$

Our goal is to prove that:

$$b^2 - a^2 = c^2 - b^2$$

Multiply the first equation by $(c + a)(b + c)(a + b)$:

$$(b+c)(a+b) - (c+a)(a+b) = (c+a)(b+c) - (b+c)(a+b)$$
$$\left(ab + ac + bc + b^2\right) - \left(ac + bc + ab + a^2\right) = \left(bc + ab + ac + c^2\right) - \left(ab + ac + bc + b^2\right)$$
$$b^2 - a^2 = c^2 - b^2$$

This proves the desired result.

97. (a) We must have $4+x=r(3+x)$ and $5+x=r^2(3+x)$, so $r=\frac{4+x}{3+x}$. Substituting into the second equality:

$$5+x=\left(\frac{4+x}{3+x}\right)^2(3+x)$$
$$5+x=\frac{(4+x)^2}{3+x}$$

Multiplying each side by $3+x$:

$$15+8x+x^2=16+8x+x^2$$
$$15=16$$

So no such value for x exists.

(b) Here $b+x=r(a+x)$ and $c+x=r^2(a+x)$, so $r=\frac{b+x}{a+x}$. Substituting:

$$c+x=\left(\frac{b+x}{a+x}\right)^2(a+x)$$
$$(c+x)(a+x)=(b+x)^2$$
$$ac+(a+c)x+x^2=b^2+2bx+x^2$$
$$(a+c-2b)x=b^2-ac$$
$$x=\frac{b^2-ac}{a+c-2b}$$

(Note that $a+c-2b=0$ for (a), explaining why no x could be found.)

99. (a) Let $S=a_1+ra_2+r^2a_3+\ldots+r^{n-1}a_n$, so:

$$rS=ra_1+r^2a_2+r^3a_3+\ldots+r^na_n$$

Then:

$$S-rS=a_1+r(a_2-a_1)+r^2(a_3-a_2)+\ldots+r^{n-1}(a_n-a_{n-1})-r^na_n$$
$$=a_1+rd+r^2d+\ldots+r^{n-1}d-r^na_n$$
$$=a_1+\frac{d(r-r^n)}{1-r}-r^na_n$$

(b) Continuing from (a), we have:

$$S(1-r)=a_1-r^na_n+\frac{d(r-r^n)}{1-r}$$
$$S=\frac{a_1-r^na_n}{1-r}+\frac{d(r-r^n)}{(1-r)^2}$$

101. (a) We start by rationalizing denominators on the left-hand side of the equation:

$$\frac{1}{\sqrt{a_1}+\sqrt{a_2}}\bullet\frac{\sqrt{a_1}-\sqrt{a_2}}{\sqrt{a_1}-\sqrt{a_2}}=\frac{\sqrt{a_1}-\sqrt{a_2}}{a_1-a_2}=\frac{\sqrt{a_1}-\sqrt{a_2}}{-d}$$
$$\frac{1}{\sqrt{a_2}+\sqrt{a_3}}\bullet\frac{\sqrt{a_2}-\sqrt{a_3}}{\sqrt{a_2}-\sqrt{a_3}}=\frac{\sqrt{a_2}-\sqrt{a_3}}{a_2-a_3}=\frac{\sqrt{a_2}-\sqrt{a_3}}{-d}$$

Adding, we have:

$$\frac{1}{\sqrt{a_1}+\sqrt{a_2}}+\frac{1}{\sqrt{a_2}+\sqrt{a_3}}=\frac{\sqrt{a_1}-\sqrt{a_3}}{-d}\bullet\frac{\sqrt{a_1}+\sqrt{a_3}}{\sqrt{a_1}+\sqrt{a_3}}$$

$$=\frac{a_1-a_3}{-d\left(\sqrt{a_1}+\sqrt{a_3}\right)}$$

$$=\frac{-2d}{-d\sqrt{a_1}+\sqrt{a_3}}$$

$$=\frac{2}{\sqrt{a_1}+\sqrt{a_3}}$$

(b) Proceeding as in part (a), we rationalize denominators on the left-hand side of the equation:

$$\frac{1}{\sqrt{a_3}+\sqrt{a_4}}\bullet\frac{\sqrt{a_3}-\sqrt{a_4}}{\sqrt{a_3}-\sqrt{a_4}}=\frac{\sqrt{a_3}-\sqrt{a_4}}{a_3-a_4}=\frac{\sqrt{a_3}-\sqrt{a_4}}{-d}$$

Adding the three fractions, we have:

$$\frac{1}{\sqrt{a_1}+\sqrt{a_2}}+\frac{1}{\sqrt{a_2}+\sqrt{a_3}}+\frac{1}{\sqrt{a_3}+\sqrt{a_4}}=\frac{\sqrt{a_1}-\sqrt{a_4}}{-d}\bullet\frac{\sqrt{a_1}+\sqrt{a_4}}{\sqrt{a_1}+\sqrt{a_4}}$$

$$=\frac{a_1-a_4}{-d\left(\sqrt{a_1}+\sqrt{a_4}\right)}$$

$$=\frac{-3d}{-d\sqrt{a_1}+\sqrt{a_4}}$$

$$=\frac{3}{\sqrt{a_1}+\sqrt{a_4}}$$

(c) We will prove this by mathematical induction. Part (a) shows the statement to be true when $n = 3$. Now assume P_k is true, so:

$$\frac{1}{\sqrt{a_1}+\sqrt{a_2}}+\ldots+\frac{1}{\sqrt{a_{k-1}}+\sqrt{a_k}}=\frac{k-1}{\sqrt{a_1}+\sqrt{a_k}}$$

We must show P_{k+1} is true. We have:

$$\frac{1}{\sqrt{a_1}+\sqrt{a_2}}+\ldots+\frac{1}{\sqrt{a_{k-1}}+\sqrt{a_k}}+\frac{1}{\sqrt{a_k}+\sqrt{a_{k+1}}}=\frac{k-1}{\sqrt{a_1}+\sqrt{a_k}}+\frac{1}{\sqrt{a_k}+\sqrt{a_{k+1}}}$$

Rationalize each denominator:

$$\frac{k-1}{\sqrt{a_1}+\sqrt{a_k}} \bullet \frac{\sqrt{a_1}-\sqrt{a_k}}{\sqrt{a_1}-\sqrt{a_k}} = \frac{(k-1)\left(\sqrt{a_1}-\sqrt{a_k}\right)}{a_1-a_k}$$

$$= \frac{(k-1)\left(\sqrt{a_1}-\sqrt{a_k}\right)}{-(k-1)d}$$

$$= \frac{\sqrt{a_1}-\sqrt{a_k}}{-d}$$

$$\frac{1}{\sqrt{a_k}+\sqrt{a_{k+1}}} \bullet \frac{\sqrt{a_k}-\sqrt{a_{k+1}}}{\sqrt{a_k}-\sqrt{a_{k+1}}} = \frac{\sqrt{a_k}-\sqrt{a_{k+1}}}{a_k-a_{k+1}} = \frac{\sqrt{a_k}-\sqrt{a_{k+1}}}{-d}$$

Adding these two fractions yields:

$$\frac{\sqrt{a_1}-\sqrt{a_{k+1}}}{-d} \bullet \frac{\sqrt{a_1}+\sqrt{a_{k+1}}}{\sqrt{a_1}+\sqrt{a_{k+1}}} = \frac{a_1-a_{k+1}}{-d\left(\sqrt{a_1}-\sqrt{a_{k+1}}\right)}$$

$$= \frac{-kd}{-d\left(\sqrt{a_1}+\sqrt{a_{k+1}}\right)}$$

$$= \frac{k}{\sqrt{a_1}+\sqrt{a_{k+1}}}$$

$$= \frac{(k+1)-1}{\sqrt{a_1}+\sqrt{a_{k+1}}}$$

So P_{k+1} is true. The identity is true for all natural numbers n.

Chapter Ten Test

1. For $n = 1$, we have:

$$1^2 = \frac{1(1+1)(2+1)}{6} = \frac{6}{6} = 1$$

Assume P_k is true, so:

$$1^2 + 2^2 + 3^2 + \ldots + k^2 = \frac{k(k+1)(2k+1)}{6}$$

Then:

$$1^2+2^2+\ldots+k^2+(k+1)^2=\frac{k(k+1)(2k+1)}{6}+(k+1)^2$$
$$=(k+1)\left[\frac{k(2k+1)}{6}+k+1\right]$$
$$=(k+1)\left(\frac{2k^2+k+6k+6}{6}\right)$$
$$=\frac{(k+1)\left(2k^2+7k+6\right)}{6}$$
$$=\frac{(k+1)(k+2)(2k+3)}{6}$$
$$=\frac{(k+1)(k+2)[2(k+1)+1]}{6}$$

So P_{k+1} is true and the induction is complete.

2. (a) Expanding the sum:

$$\sum_{k=0}^{2}(10k-1)=-1+9+19=27$$

(b) Expanding the sum:

$$\sum_{k=1}^{3}(-1)^k k^2=-1+4-9=-6$$

3. (a) $S_n=\dfrac{a\left(1-r^n\right)}{1-r}$

(b) Here $a=\frac{3}{2}$, $r=\frac{3}{2}$, and $n=10$, so:

$$S_{10}=\frac{\frac{3}{2}\left[1-\left(\frac{3}{2}\right)^{10}\right]}{1-\frac{3}{2}}=\frac{\frac{3}{2}\left(1-\frac{59049}{1024}\right)}{-\frac{1}{2}}=-3\left(-\frac{58025}{1024}\right)=\frac{174075}{1024}$$

4. (a) Here $11-r+1=3$, so $r=9$, and the coefficient is:

$$\binom{11}{8}(-2)^8=\frac{(11)(10)(9)}{(3)(2)}(256)=42240$$

(b) The fifth term is:

$$\binom{11}{4}(a)^7\left(-2b^3\right)^4=\frac{(11)(10)(9)(8)}{(4)(3)(2)}a^7\left(16b^{12}\right)=5280a^7b^{12}$$

5. Using the binomial theorem, we have:

$$(3x^2+y^3)^5=\binom{5}{0}(3x^2)^5+\binom{5}{1}(3x^2)^4(y^3)+\binom{5}{2}(3x^2)^3(y^3)^2+\binom{5}{3}(3x^2)^2(y^3)^3$$
$$+\binom{5}{4}(3x^2)(y^3)^4+\binom{5}{5}(y^3)^5$$
$$=243x^{10}+405x^8y^3+270x^6y^6+90x^4y^9+15x^2y^{12}+y^{15}$$

6. The sum is:

$$S_{12}=12\left(\frac{8+\frac{43}{2}}{2}\right)=\frac{59}{2}(6)=177$$

7. The sum is:

$$S=\frac{a}{1-r}=\frac{\frac{7}{10}}{1-\frac{1}{10}}=\frac{7}{9}$$

8. Here $a_1=1$ and $a_2=1$, so:

$$a_3=a_2^{\ 2}+a_1=1+1=2$$
$$a_4=a_3^{\ 2}+a_2=2^2+1=5$$
$$a_5=a_4^{\ 2}+a_3=5^2+2=27$$

9. Since $a_3=4$ and $a_5=10$, then $a_5=a_3r^2$, so:

$$10=4r^2$$
$$\tfrac{5}{2}=r^2$$
$$r=-\sqrt{\tfrac{5}{2}}=-\tfrac{\sqrt{10}}{2}$$

Then $a_6=ra_5=-\frac{\sqrt{10}}{2}(10)=-5\sqrt{10}$.

10. Here $a_n=-61+15(n-1)$, so:

$$a_{20}=-61+15(19)=224$$

11. Using the multiplication principle, the total number of area codes is $8\bullet 2\bullet 9=144$. Since there were 5 area codes not yet assigned, the amount of area codes in use in 1991 was 139.

12. There will be $P(14,8)=14\bullet 13\bullet 12\bullet 11\bullet 10\bullet 9\bullet 8\bullet 7=121{,}080{,}960$ possible arrangements.

13. Using the multiplication principle, the probability is:

$$\tfrac{12}{15}\bullet\tfrac{11}{14}=\tfrac{22}{35}$$

14. (a) Using the multiplication principle, the probability is:

$$\frac{3}{10} \bullet \frac{2}{9} = \frac{1}{15}$$

(b) Using the multiplication principle, the probability is:

$$\frac{7}{10} \bullet \frac{6}{9} = \frac{7}{15}$$

(c) Since the two events are mutually exclusive, the probability is:

$$\frac{1}{15} + \frac{7}{15} - 0 = \frac{8}{15}$$